P9-EEA-133

COBOL

fundamentals of structured COBOL programming

third edition

Carl Feingold
C.P.A., C.D.P.
West Los Angeles College

wcb
WM. C. BROWN COMPANY PUBLISHERS
2460 Kerper Boulevard, Dubuque, Iowa 52001

Book Team

Richard C. Crews, Publisher
Robert Stern, Editor
Ruth Richard, Production Editor
Bill Evans, Designer
James Nigg, Design Layout Assistant

Library of Congress Catalog Card Number: 77—88098

ISBN 0—697—08128—1

Seventh Printing, 1981

Printed in the United States of America

Contents

Preface

COBOL has emerged as one of the leading programming languages in use today. Support from its users, computer manufacturers, and governmental agencies, assure that the programming language will meet the needs of the user, not only today but in the future. The wide and popular acceptance of COBOL as a standard business programming language has made it essential for more COBOL programming courses to be offered at the various institutions of learning.

The relative ease of COBOL programming with its self-documenting feature makes it an ideal introductory programming course for students, whether they be programmers, business students, accountants, or others interested in learning computer programming.

The purpose of this text is to provide the general reader with a broad and comprehensive coverage of all the features of the latest American National Standard (ANSI) COBOL, which has been universally adopted by all computer manufacturers. Prior programming knowledge is not required because the introductory chapters provide sufficient background for the beginning student and serve as a review for continuing students and programmers.

The text, designed so that the user can write programs early in the course, provides an overall view of the functions and uses of all divisions of COBOL. Numerous illustrations and thorough explanations accompany each segment of the COBOL language, making the text resourceful as both a learning tool and a reference manual. Most of the illustrations are complete within themselves, freeing the user from reading through pages of narrative for an explanation of an illustration. Each segment of the COBOL language is completely explained and illustrated so that the user does not have to "wade" through numerous examples to find the basic formats, uses, functions, or rules for applying each of the COBOL elements.

Extremely comprehensive, the text provides the user with the introductory concepts of data processing through the basic components of COBOL programming and advanced COBOL programming concepts. It is a step-by-step problem-oriented approach to COBOL programming covering both basic and advanced COBOL topics. The text contains a wealth of illustrated examples that help the student progress from problem definition to solution.

This third edition represents a major revision; some chapters have been combined and new chapters have been added. All the salient features of the previous edition have been retained.

All chapters have been carefully and extensively revised with expanded explanations, numerous examples and illustrations, and new programming problems. Every COBOL entry is described and illustrated in great detail. The text is meant to be all-inclusive, with little or no need for reference to computer manufacturer's manuals.

Some of the major additions to the third edition include—

AN OVERVIEW OF COBOL PROGRAMMING—This new chapter contains brief explanations of the divisions of COBOL. The minimum number of entries necessary to write a "File-Less" COBOL program is explained. A COBOL program using ACCEPT and DISPLAY statements is illustrated. The user should be able to write the first "mini" COBOL program at this point.

WRITING COBOL PROGRAMS—This chapter combines chapters 4 and 5 of the previous edition, with content of chapter 5 preceding chapter 4. This new chapter includes all the necessary entries in the four divisions to write a simplified COBOL program using files. The user should be able to write a simple program using files at this point.

STRUCTURED PROGRAMMING—The latest technique in computer programming, structured programming is a method of organizing and coding programs that makes the programs easily understood and modified. A thorough explanation of the purpose and use of structured programming with COBOL implementation of structured programming is provided in this new chapter. The information is presented at this point becasue of its effect on Procedure Division entries. The concepts of structured programming are used throughout the text.

COBOL DIFFERENCES—ANS 1968 COBOL and ANS 1974 COBOL—This section summarizes the differences between the 1968 and 1974 American National Standard COBOL. It also summarizes the new features to be found in the ANS 1974 COBOL.

Chapters 1 and 2 provide an insight to the operation of the computer and the numerous problems encountered in the study, planning, and preparation of computer programs. In addition, it serves as an introduction to COBOL programming with explanations for the popularity of COBOL as well as its advantages and disadvantages.

Chapters 3 and 4 present an overview of COBOL programming with brief descriptions of the four divisions of COBOL and the basic elements of COBOL program writing. The reader should be able to write his first COBOL program.

The detailed explanations of the four divisions of COBOL appear in the next four chapters, with an important chapter on structured programming preceding the Procedure Division chapter.

The first eight chapters can serve as the first course in COBOL programming, with chapters 9–15 providing the advanced concepts, such as Table

Handling, Report Writer Feature, Sort Feature, Direct Access Devices, Declaratives and Linkage Sections, Additional and Optional Features, and COBOL Programming Techniques.

In the appendixes, the important features of Debugging COBOL Programs, Job-Control Language, Segmentation Feature, and the COBOL Differences, ANS 1968 and ANS 1974 COBOL, will be found.

There are exercises, questions for review, and problems following each chapter. Answers have been included for the exercises so that the reader can have immediate "feedback" to reinforce the learning process. The questions for review and problems can be assigned to fulfill the needs of the instructor. The problems that appear in the appendix may be assigned by the instructor as a term project.

The Solutions and Instructor's Resource Manual includes a brief commentary of each chapter. The commentary states the intent of the chapter, useful teaching suggestions, and a summary of the important points. The Manual also provides answers for the questions for review and the problems. In addition, there are computer printouts for the ten problems that appear in the appendix of the text.

I am indebted to the IBM Corporation for gratuitously granting permission to use the numerous illustrations, charts, photos, and diagrams that made the text more meaningful.

My special thanks to the Burroughs Corporation for permitting me to use their excellent problems and illustrations in the book.

My last special thanks is to my wife, Sylvia, who served as the chief typist, confidant, and proofreader. Without her encouragement the book would never have been written.

Another special thanks to Ms. Joan Driefus who finished typing the manuscript in a professional manner when my wife became ill.

Acknowledgments

The following information is reprinted from *COBOL Edition 1965,* published by the Conference on Data Systems Languages (CODASYL), and printed by the U. S. Government Printing Office.

"Any organization interested in reproducing the COBOL report and specifications in whole or in part, using ideas taken from this report as the basis for an instruction manual or for any other purpose is free to do so. However, all such organizations are requested to reproduce this section as part of the introduction to the document. Those using a short passage, as in a book review, are requested to mention "COBOL" in acknowledgment of the source, but need not quote this entire section.

"COBOL is an industry language and is not the property of any company or group of companies, or of any organization or group of organizations.

"No warranty, expressed or implied, is made by any contributor or by the COBOL Committee as to the accuracy and functioning of the programming system and language. Moreover, no responsibility is assumed by any contributor, or by the committee, in connection therewith.

"Procedures have been established for the maintenance of COBOL. Inquiries concerning the procedures for proposing changes should be directed to the Executive Committee of the Conference on Data Systems Languages.

"The authors and copyright holders of the copyrighted material used herein

> FLOW-MATIC (Trademark of Sperry Rand Corporation), Programming for the Univac (R) I and II, Data Automation Systems copyrighted 1958, 1959, by Sperry Rand Corporation; IBM Commercial Translator Form No. F28-8013, copyrighted 1959 by IBM; FACT, DSI 27A52602760, copyrighted 1960 by Minneapolis-Honeywell

have specifically authorized the use of this material in whole or in part, in the COBOL specifications. Such authorization extends to the reproduction and use of COBOL specifications in programming manuals of similar publications."

The following illustrations are reproduced through the courtesy of the International Business Machines Corp.

Description	Page
IBM System 370 Computer	5
IBM Card Read Punch	11
IBM Magnetic Tape Unit	12
IBM Paper Tape Reader	14
IBM Magnetic Document Sorter and Reader	15
IBM Optical Character Reader	16
IBM Visual Display Unit	17
IBM Printer	18
IBM Console	19
IBM Data Communications Terminal	19
IBM Magnetic Drum	24
IBM Magnetic Disk	25
IBM Data Cell	28
IBM 2311 Disk Storage Drive	27
IBM 2302 Disk Storage Drive	388
General Format of Control Statements	532
Job Control Procedure	532
JOB Statement	533
EXEC Statement	534
DD Statement	535
IBM American National Standard COBOL Reserved Words	581-586
Summary of File-Processing Techniques	591-594
Basic Formats	595-610
American National Standard COBOL Glossary	611-618
IBM COBOL Reference Summary	
Detail Design Level HIPO Diagram for a Two-Level Control Processing Application	491
Detailed Design Level Visual Table for Contents for the Inquiry Response Application	493
A HIPO Diagram for an Inquiry Response Application	493
Flowchart for an Inquiry Response Application	184-187
Structured Program for an Inquiry Response Application	188-189

The following illustration is reproduced from the March 1976 issue of Datamation with their permission.

Detail Design Level HIPO diagram	490

1 Basic Computer Concepts

INTRODUCTION

The dynamic introduction of the computer in the last quarter century has changed man's informational needs entirely. Inasmuch as the resources of society are limited, man has developed methods of compiling and analyzing large quantities of data with a minimal amount of human intervention. Technological advances in all fields of data processing have been dynamic and extensive, in response to man's unabated and ever-increasing thirst for information. The methods of applying data processing systems to fulfill informational needs are boundless. With each new application, data processing systems have demonstrated still newer ways in which they can be used to help man increase his productivity and so advance civilization a little farther. Data processing is not just another new industry or innovation but a giant stride forward in man's utilization of science and knowledge as a means of progress.

Computers are among the most useful tools ever invented by mankind. In this era of computers, they are used to count our votes, figure our bank accounts, help plan new buildings and bridges, guide our astronauts through space, and assist management in its everyday decisions.

Our daily lives are affected in some manner by computers. Life as we know it would not be possible without them. Patterns of consumer spending have changed. Credit cards have become a way of life. Almost any of man's needs can be satisfied with the credit card. Daily interest on bank savings would be an impossibility without the present methods of processing data. Service industries have greatly increased, providing numerous services to the consumer, making life more convenient and pleasant. To make reservations for some distant hotel or to plan a trip to another part of the world, all that is required is to reach for the telephone (fig. 1.1). Each individual is touched by the computer in some manner—some, unhappily, by the Internal Revenue Service.

One of the driving forces behind the development of the computer was the need to find faster and more efficient methods of handling paperwork. The clerical operations required to handle the ever-mounting quantities of paper to be processed had increased enormously. Paper handling alone would probably have overwhelmed all of our enormous productive capabilities if clerical mechanization had not kept pace with the technological advances in the field. Com-

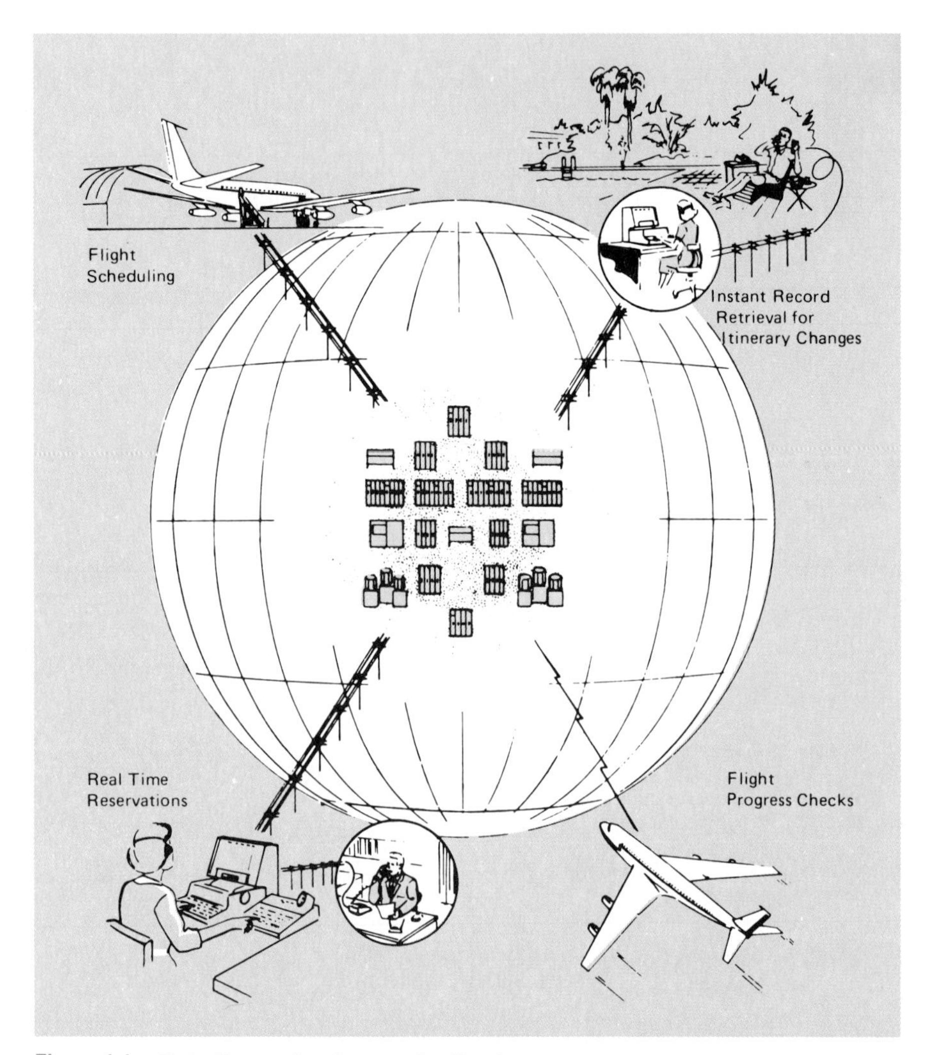

Figure 1.1 Data Processing System Applications.

puters have gone far beyond the mere takeover of jobs accomplished by paperwork and doing them faster and better. Computers are doing things for business and industry that could never have been accomplished before.

A company may test the design of a new processing plant by simulating every detail of its operation with a computer. Computer simulation frequently provides more information in a week than an expensive pilot plant could provide in a year of actual operation.

More and more management decisions now are based on information supplied by computers—information about inventories, production schedules, market forecasts, and sales analyses. Although management could always obtain this information, it was not always obtainable in time to make possible the taking of effective action. In some cases today, advanced mathematical techniques worked out on computers, are used to find answers quickly to complex marketing and production problems that a few years ago would have been solved only with extreme difficulty and a considerable expenditure of time.

Computers of the future will go even further toward freeing man from the drudgery of paperwork and repetitive manual tasks. There are many new applications now in the process of development, such as highway traffic control, automatic language translation, automatic processing of maps and pictures (fig. 1.2). It is obvious that the computer will serve us in many ways that the conventional word *paperwork* could never encompass.

All of the aforementioned activities illustrate ways in which data has become a part of our way of life, and indicate the role played by the computer in developing that function. Great opportunities lie ahead in the data processing field with the addition of the computer. Expanded markets, greater productivity, energy conservation, corporate growth, and increased governmental activity provide the data processor with new challenges day in and day out.

BASIC COMPONENTS OF DATA PROCESSING

Data processing is a planned series of actions and operations upon information, using various forms of data processing equipment to achieve a desired result. The data processing equipment came into being primarily to satisfy the need for information under increasingly complex conditions. Computer programs and physical equipment are combined into data processing systems to handle business and scientific data at high rates of speed with self-checking accuracy features. The physical data processing equipment consists of various units, such as input and output devices, storage devices, and processing devices to handle information at electronic speeds (fig. 1.3).

The computer is a major tool for implementing the solution to data processing problems. In brief, the computer accepts data, processes the data, and puts out the desired results. There are many computer systems varying in size, complexity, costs, levels of programming systems, and applications, but regardless of the nature of the information to be processed, all data processing involves these basic considerations.

1. The source data entering the system (INPUT). Input relates to the information that goes into the computer (fig. 1.4). All data—numbers, letters or symbols—must be recorded on punched cards, magnetic tape, punched paper tape, or other input media. The computer "reads" the information from these devices. Sometimes information is keyed into the computer directly through a console.
2. The planned processing steps necessary to change the data into the desired result (PROCESSING). The processing operation is carried out in a pre-established system of sequenced instructions that are followed automatically by the computer. The computer without instructions from the man who uses its capabilities is worthless. The plan of processing is of human origin and involves calculations, sorting, analysis, and other operations necessary to arrive at the desired solution. A series of instructions which accomplishes a particular job is called a *program* (fig. 1.5).

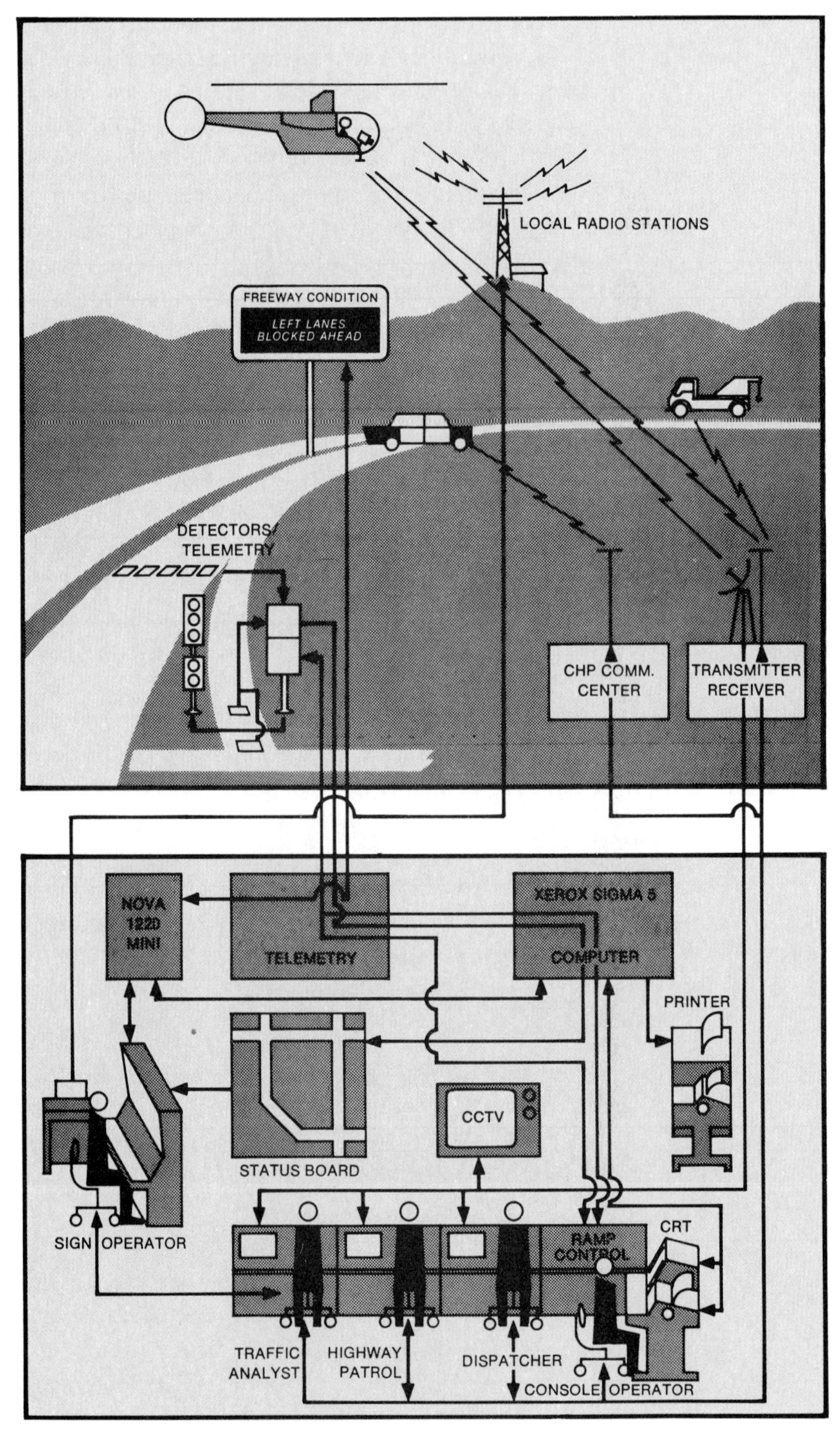

Figure 1.2 Schematic—Los Angeles Area Freeway Surveillance and Control Project.

Figure 1.3 Data Processing System.

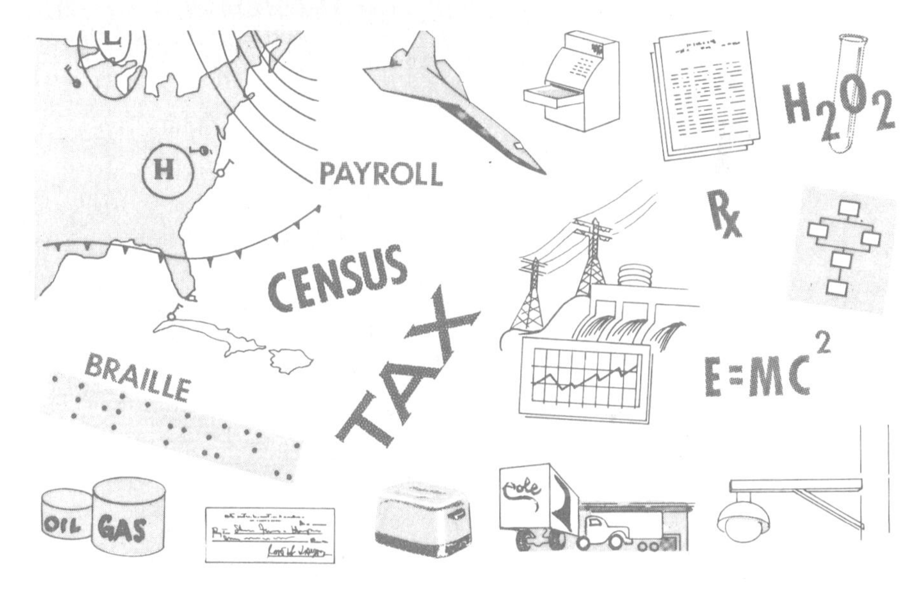

Figure 1.4 Sources of Data.

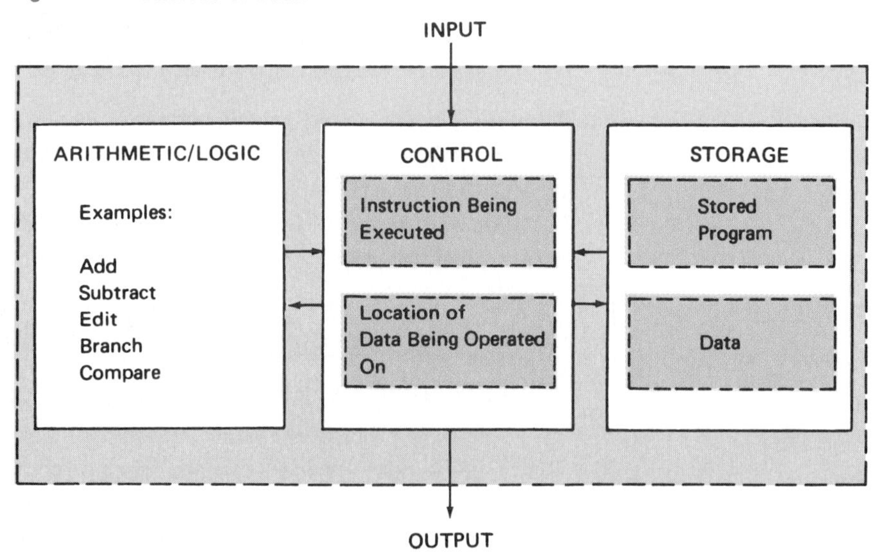

Figure 1.5 Basic Data Processing Pattern.

The Central Processing Unit (CPU) which performs the processing function is composed of two sections: Control Section, and Arithmetic and Logical Section. The Control Section manipulates the data between core storage and the Arithmetic and Logical Section according to the program. The Arithmetic and Logical Section adds, subtracts, multiplies, divides, and compares data, and directs the processing unit in making decisions when it is confronted by alternate courses in the middle of a program.

The core storage unit is the computer memory and keeps original information, intermediate results, records, reference tables, and programmed instructions. Each storage location has an "address" so the computer knows where to find the information when it is needed.

3. The finished result is the end product of the system (OUTPUT). Output is the end of the line. The computer delivers its processed information in virtually the same manner as that in which it was received; on punched cards, punched paper tape, or magnetic tape. The information is always available for further processing, when necessary, from any of these forms. Computers can also print out results—often at speeds too fast for the eye to follow (fig. 1.6).

The basic elements of a computer data processing system are its *software* and *hardware* features. *Software is a set of programs, procedures, and possibly associated documentation concerned with the operation of a data processing system.* Computers without software are merely a collection of nuts, bolts, wires, electronic devices, and inanimate hardware. The software programs and routines are used to extend the capabilities of the computer. Examples of software are compilers, library routines, subroutines, etc. The COBOL compiler is an example of software.

Hardware is defined as the physical equipment or devices of a system forming a computer and its peripheral equipment. It includes all equipment necessary for the input, processing and output functions of the system.

The major hardware elements of the computer are

1. *Input Devices*—used to enter data into the data processing system.
2. *Central Processing Unit*—accepts data for processing and makes results available to the output devices.
3. *Output Devices*—accept data from the processing unit and record it.
4. *Storage Devices*—store data before, during, and after processing.

(See figure 1.7.)

Data processing systems are divided into three types of functional units: Input/output devices, central processing unit, and storage.

INPUT/OUTPUT DEVICES

The data processing system requires, as part of its information-handling ability, input/output devices that are linked directly to the system. These devices can enter data into and record data from a data processing system, and they can read or sense coded data and make this information available to the computer.

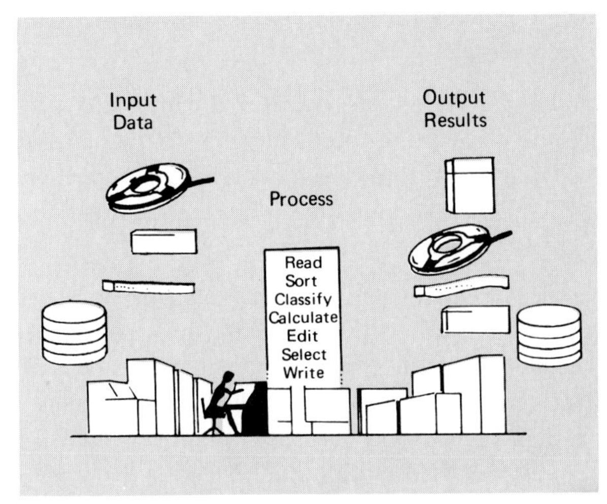

Figure 1.6 Data Processing by Computer.

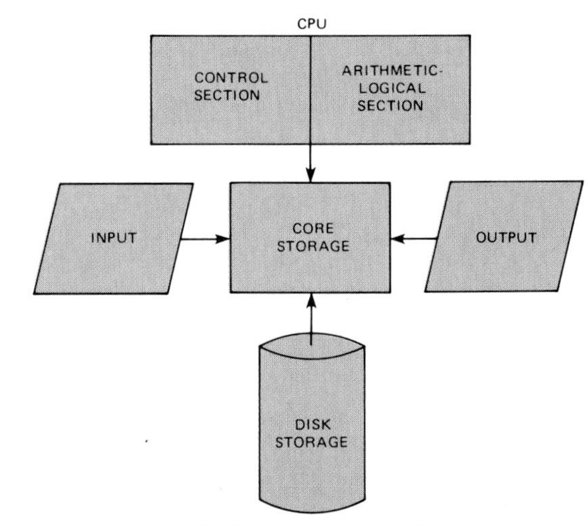

Figure 1.7 Major Hardware Elements.

The data for input may be recorded in cards, in paper tape, in magnetic tape, as characters on paper documents, or as line images created with a light pen.

Output devices record and write information from the computer into cards as punches, as holes in paper tape, or as magnetized spots on magnetic tape. These devices may also print information in the form of reports, generate signals for transmission over telephone lines, produce graphic displays on cathode tubes, and produce microfilm images.

The number and types of input/output devices will depend on the design of the particular system and the type of computer used.

An input/output device is a unit for putting or getting data out of the storage unit. The device operation is initiated by a program instruction that generates a command to a particular input or output device. A *control unit* acts as an intermediary between the command and the input/output device. The control unit decodes the command and synchronizes the device with the data processing system (fig. 1.8). The information is read by the input reader as the record moves through the input device. The data is then converted to the computer code in use and transmitted to its main storage area.

The output involves the transferring of the data from the main storage area to the particular output device. The computer code must be transcribed into the individual output medium.

The input/output devices perform their functions automatically and continue to operate as directed by the program until the entire file is processed. Program instructions select the required device, direct it to read or write, and indicate the storage locations into which the data will be entered or from which the data will be taken. Data may also be entered directly into storage by using a keyboard or switches. These input/output devices are used for manual entry of data directly into a computer without any medium for recording the data. The manual devices used are console keyboards, transmission terminals, and graphic display terminals. These terminals may be used at remote locations and the information transmitted over teleprocessing lines.

Control Unit

Because of the many types of input/output devices that can be attached to a data processing system, a unit is needed to coordinate input/output operations with the central processing unit. The control unit performs the function by acting as a traffic cop, directing information to the various input/output devices as they are read into or outputted by the system (fig. 1.9).

Channel

A channel is a separate piece of equipment devoted exclusively to managing the input/output control unit and devices assigned to it (fig. 1.10). Once the channel has been activated, it carries out its own program independent of the central processing unit. This permits the overlapping of input/output operations in computer processing. Sometimes this is performed in an interweaving pattern working with several input/output control units at one time and maintaining the proper destination for storage allocation (input) or the control unit and device (output).

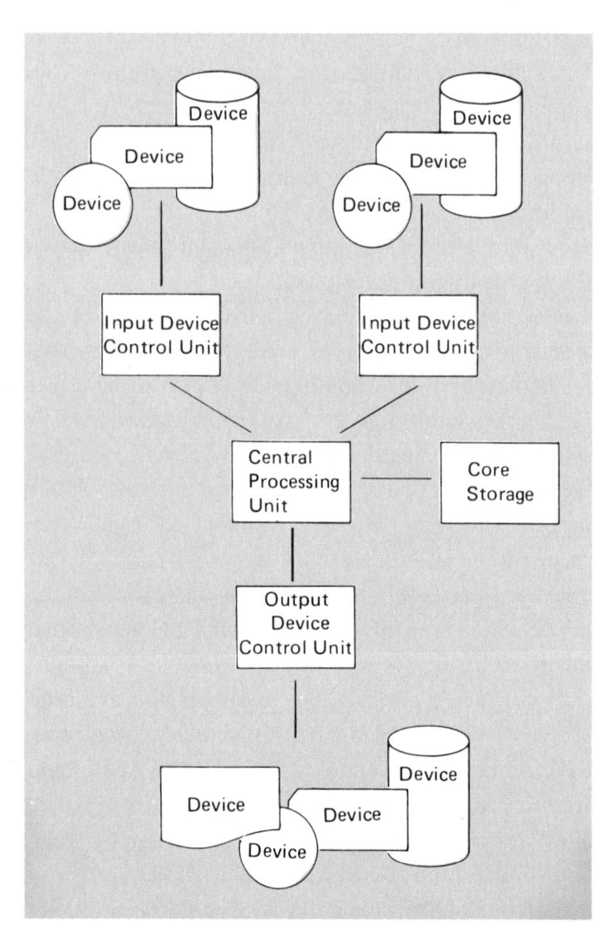

Figure 1.8 Input/Output Units in a Data Processing System.

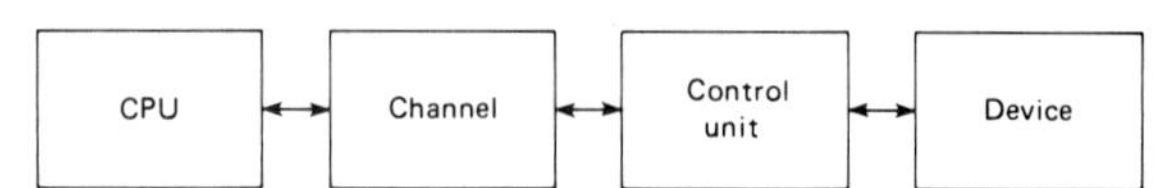

Figure 1.9 Standard Device Attachment.

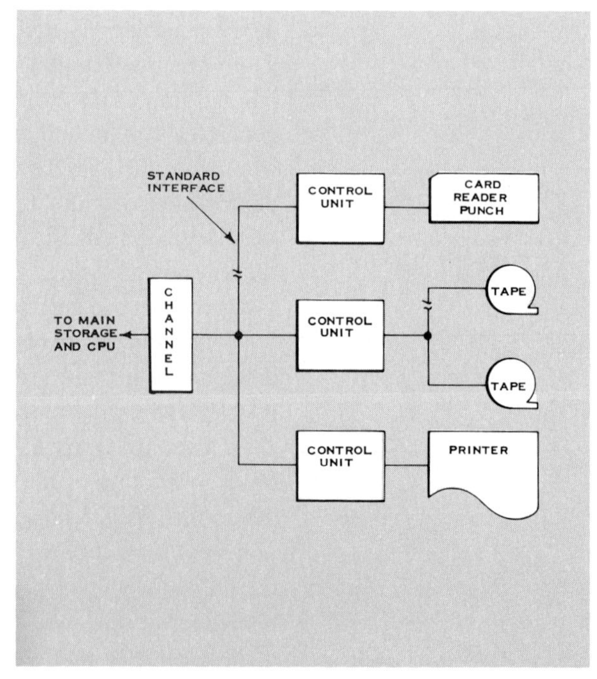

Figure 1.10 Channel Organization.

The channel thus performs the important function of permitting the simultaneous operation of processing input/output devices with the computer processing of data (fig. 1.11).

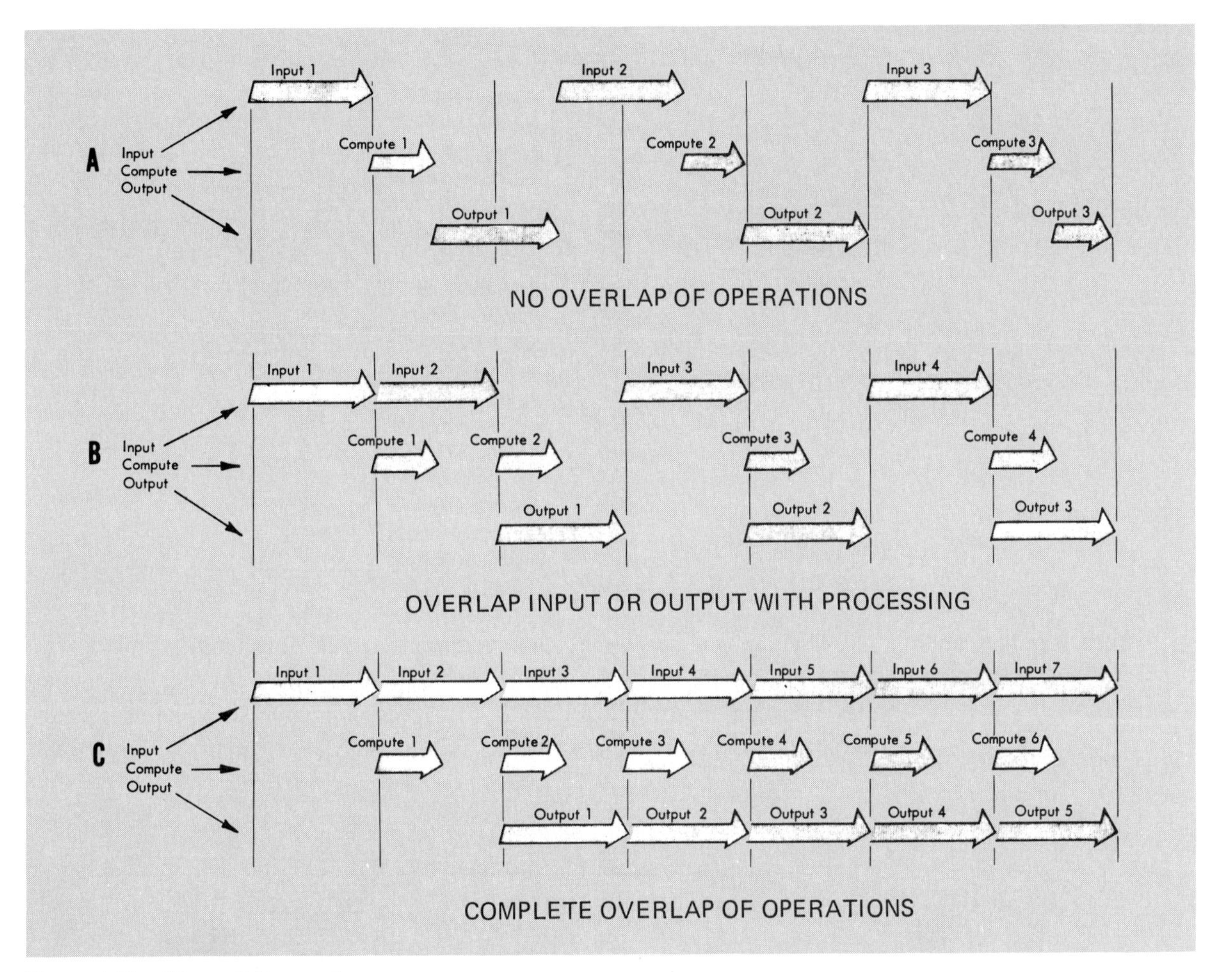

Figure 1.11 Channel Processing.

Buffer

The efficiency of any data processing system can be increased to the degree to which input, output, and internal data-handling operations can be overlapped and allowed to occur simultaneously. The usefulness of a computer is directly related to the speed at which it can complete a given procedure. The speed of input/output units should be so arranged that the central processing unit is kept busy at all times.

To synchronize the processing of input/output operations and to provide an overlap of operations, a buffering system is used. Data is first entered into an external unit known as a buffer. When the information is needed, it is transferred to main storage in a fraction of the time necessary to read the information directly from the unit. Likewise, output information is assembled in a buffer unit at high speeds until the output device is free to process it. The output device then proceeds to write the data while the central processing unit is free to continue its processing (fig. 1.12).

Large computer systems have many buffers and buffering techniques to overlap the processing operations with the many input/output devices attached to the system.

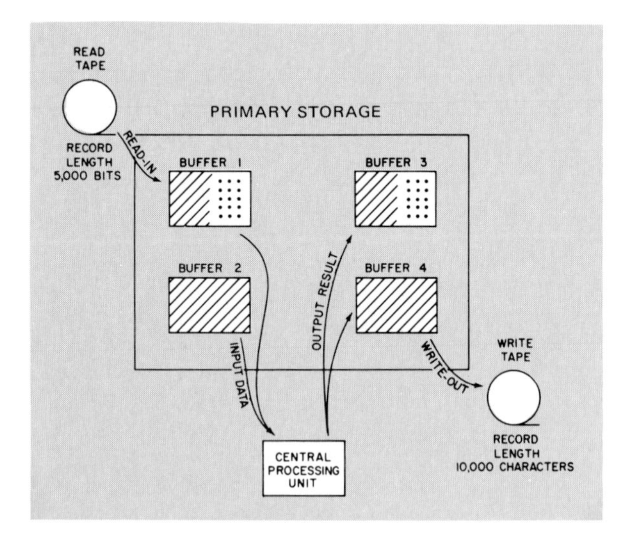

Figure 1.12
Buffer Operations.

Card Reading and Punching Devices

The punched card is the most common medium for communication with the data processing machines. Data is recorded in a pattern of small rectangular holes punched in specific locations to represent numerals, alphabetic characters, and special characters (fig. 1.13). A card reader (input) interprets the

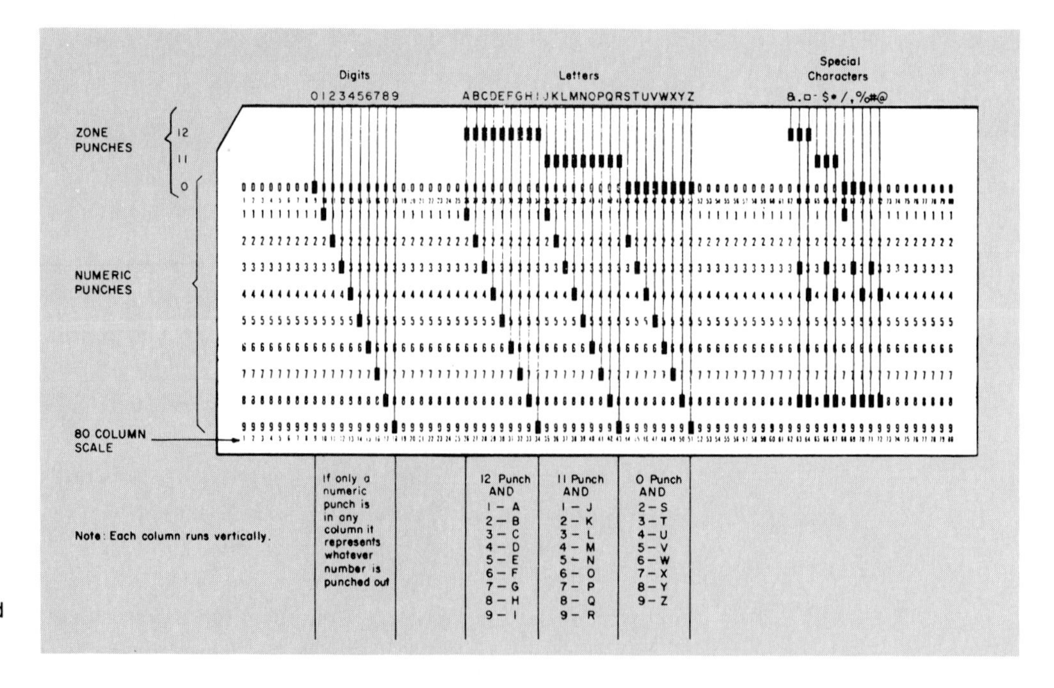

Figure 1.13
Hollerith Coded Punched Card.

pattern of holes in the card and automatically converts the data into electronic impulses, thereby entering the information into the machine. The information can also be punched into a card (output) through the use of a card punch. Thus the card can be used as original data and as a common medium of exchange of information between the machines.

Card Readers

Card-reading devices introduce punched-card records into a data processing system. The cards move past a reading unit that converts the data into a machine-processable electronic format (fig. 1.14). Two types of reading units are available; reading brushes and photoelectric cells (fig. 1.15).

In the brush-type reader, the cards are mechanically moved from the card hopper past the reading brushes that electrically sense the presence or absence of holes in each column of the punched card. There is one brush for each column of the card.

The photoelectric type of card reader acts in the same manner as the brush type; the basic difference is in the method of sensing the holes. As the punched card is read, photoelectric cells are activated by the presence of light as

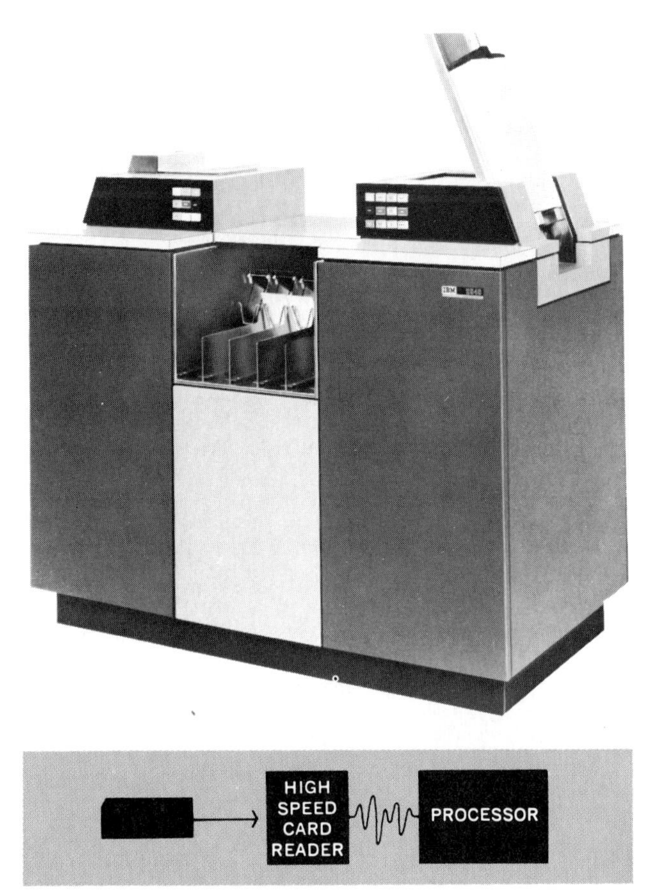

Figure 1.14 Card Read Punch.

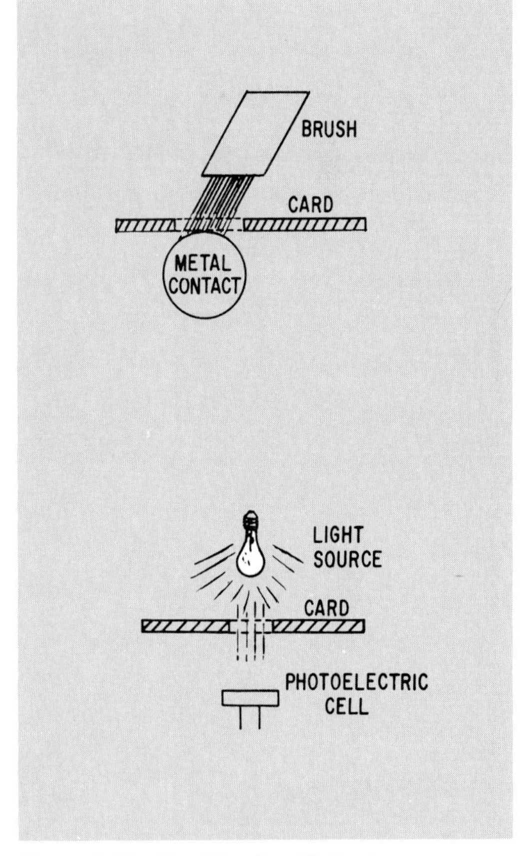

Figure 1.15 Card Reading Methods.

the light source passes over the punched card. There is one cell for each column of the card.

Card Punches

The card-punch device punches the output from the computer as a series of coded holes into a blank card.

Card-reading and card-punching devices are one of the slowest means of getting information into and out of a computer.

Magnetic Tape Devices

Magnetic tape devices, with their dual capacity of input and output, record information on tape through a read/write head by either reading the magnetized spots or by magnetizing areas in parallel tracks along the length of the tape. The writing on magnetic tape is destructive in that the new information erases the old information on the tape (figs. 1.16, 1.17).

Magnetic tape records are not restricted to any fixed record size (cards are restricted to 80 columns of data), words, or blocks. Blocks of records (which may be a single record or several records) are separated on the tape by an interblock gap, a length of blank tape averaging about .6 to .75 of an inch. This interblock gap is automatically produced at the time of the writing on the tape and provides the necessary time for starting and stopping the tape between blocks of records (fig. 1.18). Blocks of records are read into or out of buffer units.

Magnetic tape provides high-speed input and output of data to a computer.

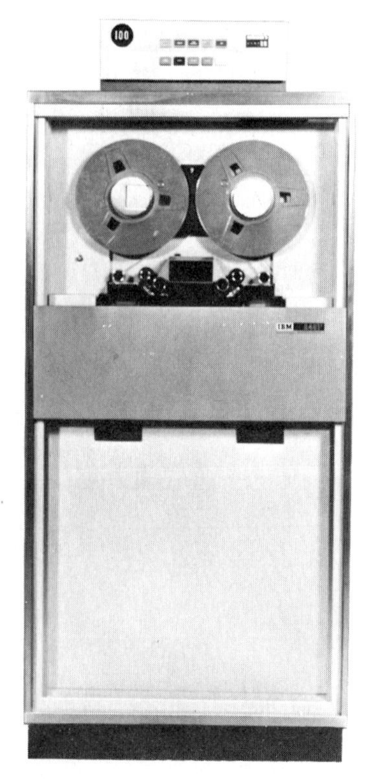

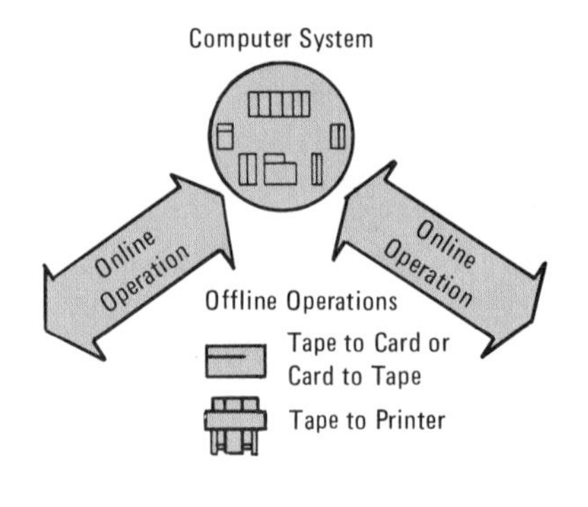

Figure 1.16 Magnetic Tape Operation.

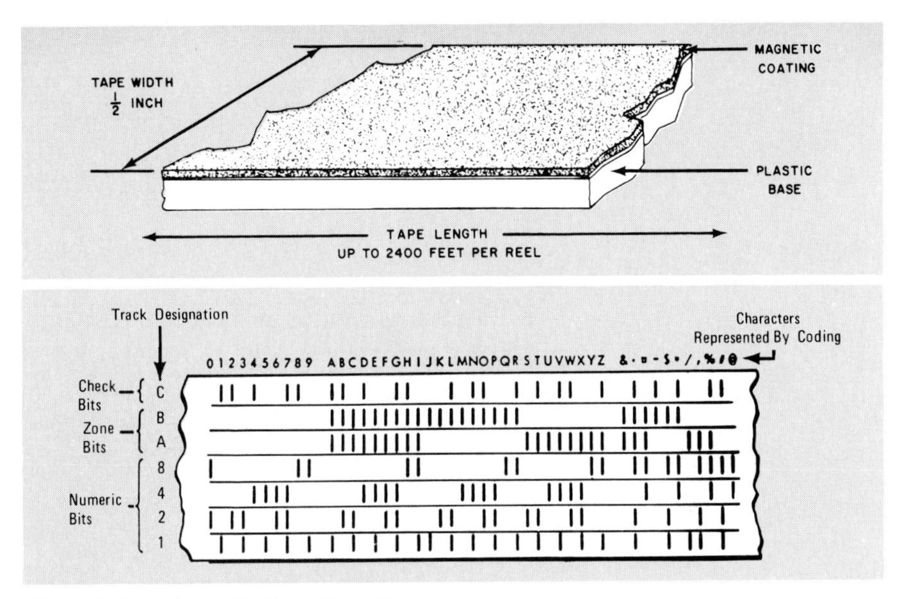

Figure 1.17 Magnetic Tape Records.

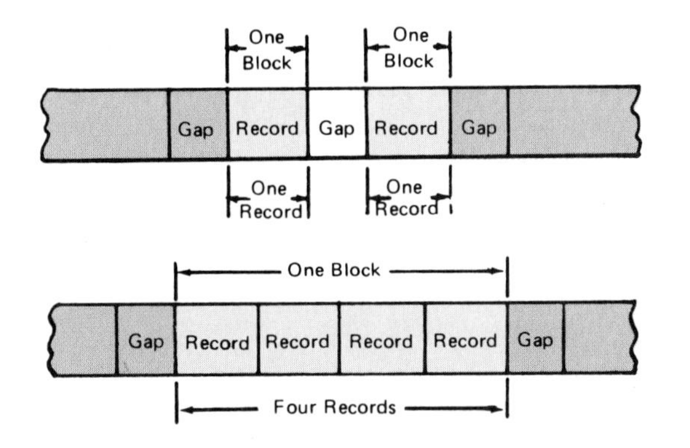

On magnetic tape, a single unit or block of information is marked by an interblock gap before and after the data. A record block may contain one record or several.

The interblock gap followed by a unique character record is used to mark the end of a file of information. The unique character, a tapemark, is generated in response to an instruction and is written on the tape following the last record of the file.

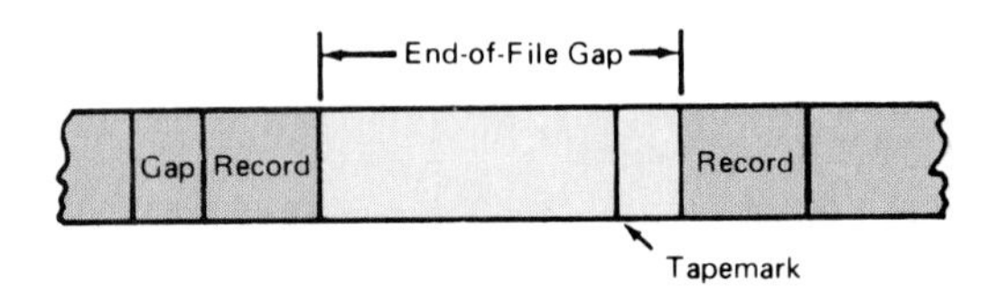

Figure 1.18 End-of-Block and End-of-File Indicators on Tape.

Paper Tape Devices

The data from main storage is converted to a tape code and is punched into a blank paper tape as the tape moves past the punch mechanism (fig. 1.19). The paper-tape reader reads the punched holes as the tape moves past the reading unit (fig. 1.20).

Paper tape may also be punched as a by-product of a cash register or some other device.

Character-Recognition Devices

Magnetic Ink-Character Readers

These machines read card and paper documents inscribed with magnetic ink characters (fig. 1.21). The special magnetic ink characters are read by a reader at high speeds and interpreted for the system.

Magnetic ink characters are used extensively in banking operations to process checks at electronic speeds (fig. 1.22).

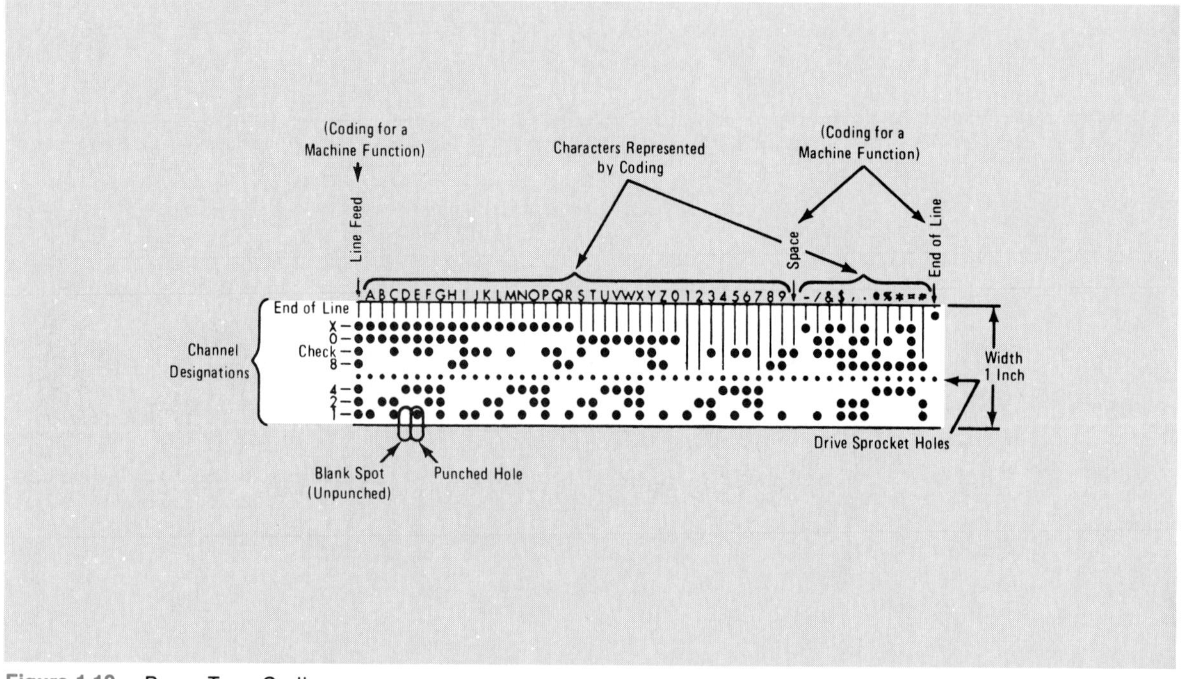

Figure 1.19 Paper Tape Coding.

Figure 1.20 Paper Tape Reader.

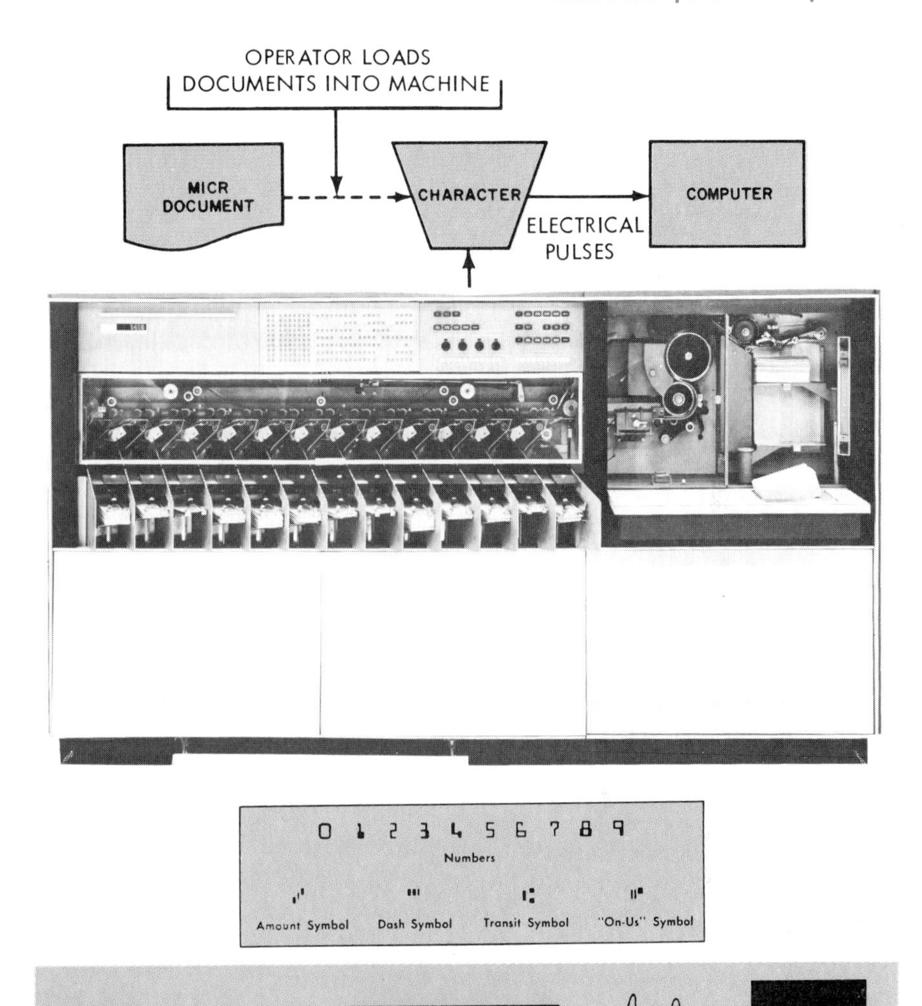

Figure 1.21
Magnetic Document Sorter and Reader.

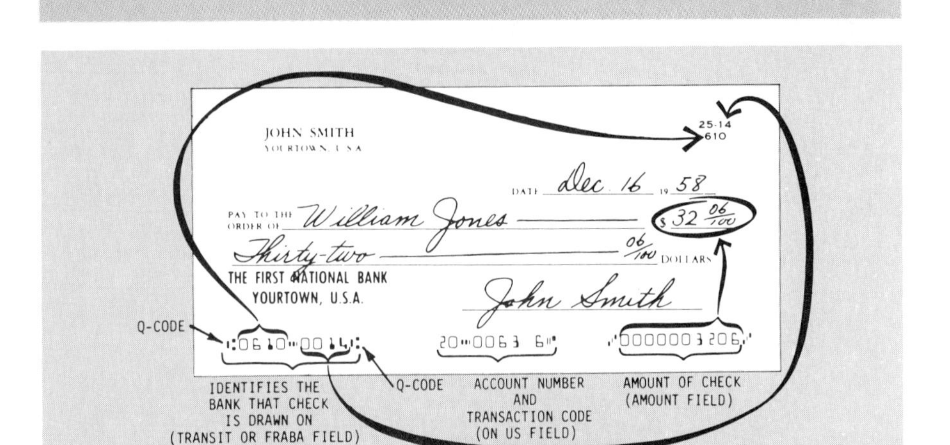

Figure 1.22
Sample Check with MICR Encoding.

Optical-Character Readers

An optical character reader can read some hand-printed or machine-printed numeric digits and certain alphabetic characters from paper and card documents. The read-and-recognition operation is automatic and takes place at electronic speeds (fig. 1.23).

Optical-character reading is used extensively with utility bills, insurance premiums, notices, and invoices.

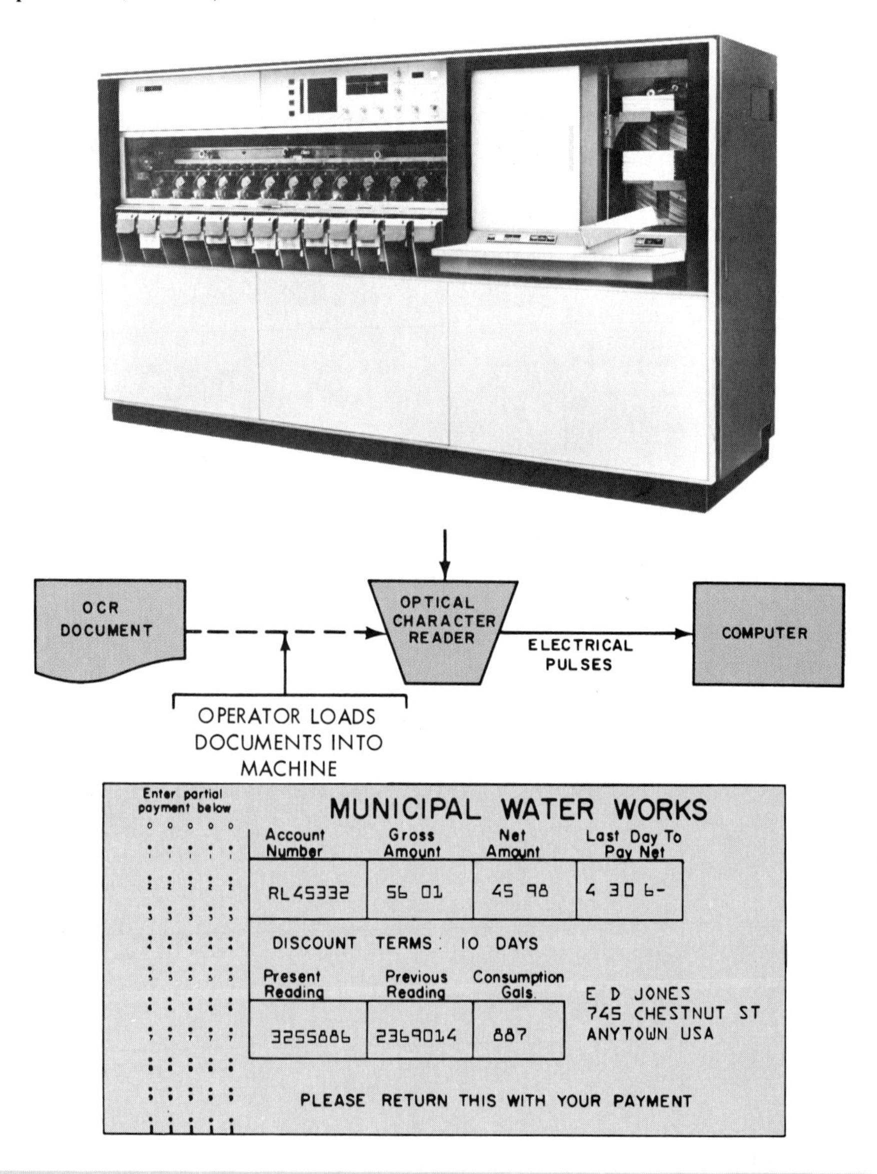

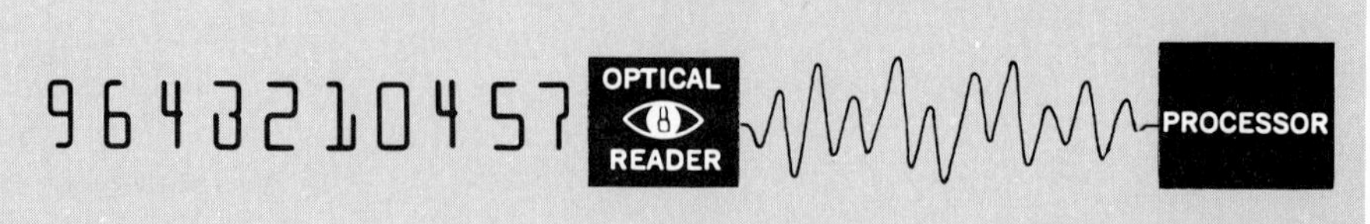

Figure 1.23
Optical Character Reader.

Visual Display Unit

A visual display unit permits the transfer of information to and from a computer. All input information to the unit is through a keyboard where the input is displayed on a video screen (fig. 1.24). Before the data is released to the computer, the operator can backspace, erase, or correct an entire input and reenter it.

Output information is written on a cleared display or added to an existing display on the tube face under operator or program control. Once information is displayed, it is available for as long as needed.

Figure 1.24
Visual Display Operation.

Printers

The printers provide the permanent visual record (hard copy) from a data processing system. As an output device, the printer receives data from the computer and prints the report (fig. 1.25). A paper transport automatically spaces the form as the report is being prepared.

There are many types of printers available, depending upon the needs of the user. Printing mechanisms can achieve speeds as high as 13,000 lines per minute.

Consoles

The console of a data processing system is used by the operator to control the system and monitor its operations. Keys, switches, audible tone signals, and display lights are some of the manual controls available to the operator for manipulation and checking of the program (fig. 1.26).

Typewriter consoles can be attached to a data processing system for communication between the operator and the system, such as operator-to-program or program-to-operator communication for program checking, and job logging (fig. 1.27).

Terminals

Terminals are used in telecommunications within a data processing system. Terminal units at the sending location accept data from cards, magnetic tape, or data entered manually into a system through a keyboard, and partially conditions this data for transmission over telephone, telegraph, radio, or microwave circuits (fig. 1.28). At the receiving station, the data is punched into cards, written on magnetic tape, printed as a report, or entered directly into a data processing system. Automatic checking features insure the validity of all transmitted data.

ACCOUNT NUMBER	BALANCE DUE	DATE OF LAST PAYMENT	
8332	$ 308.65	6/09/62	*
9818	$ .02	12/08/62	*
10003	$1,803.17	6/14/63	
10015	$ .89	10/13/62	*
11007	$1,000.56	6/01/63	
20005	$ 756.79	5/18/63	

Figure 1.25
Printer and Printed Report.

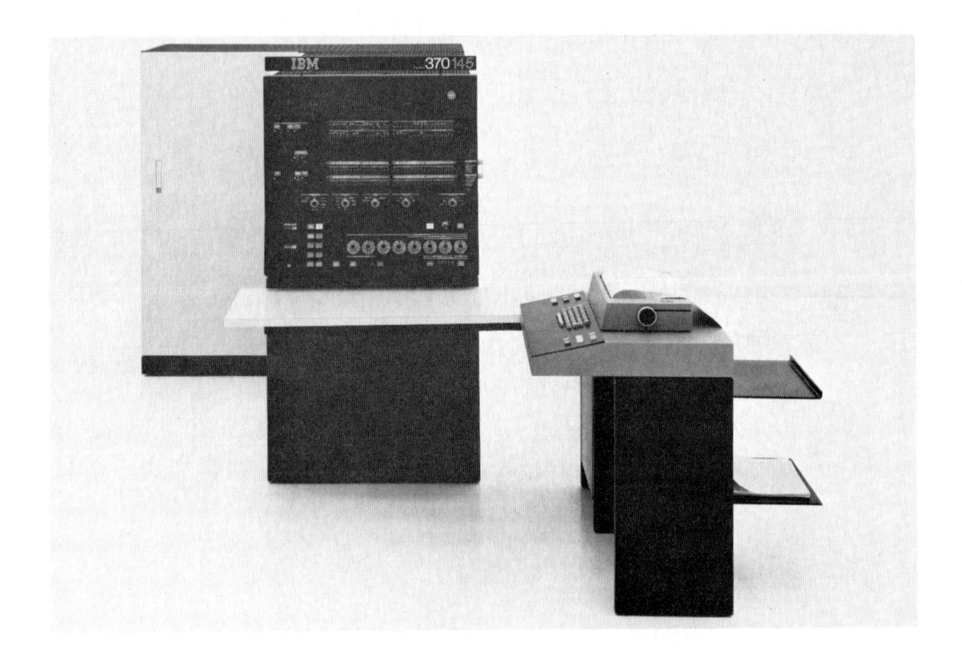

Figure 1.26
IBM 370 Console.

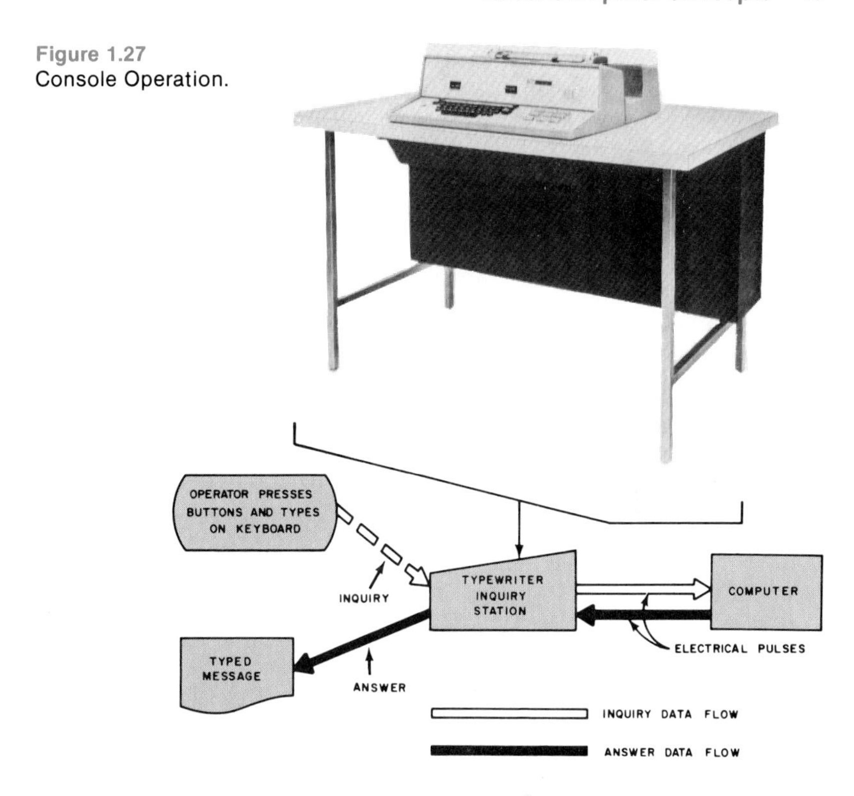

Figure 1.27
Console Operation.

Figure 1.28 IBM 3780 Data Communications Terminal.

CENTRAL PROCESSING UNIT

The Central Processing Unit is the heart of the entire data processing system. It supervises and controls the data processing components, performs the arithmetic, and makes the logical decisions (fig. 1.29).

The Central Processing Unit is divided into two sections: the Control Section, and the Arithmetic and Logical Section (fig. 1.30).

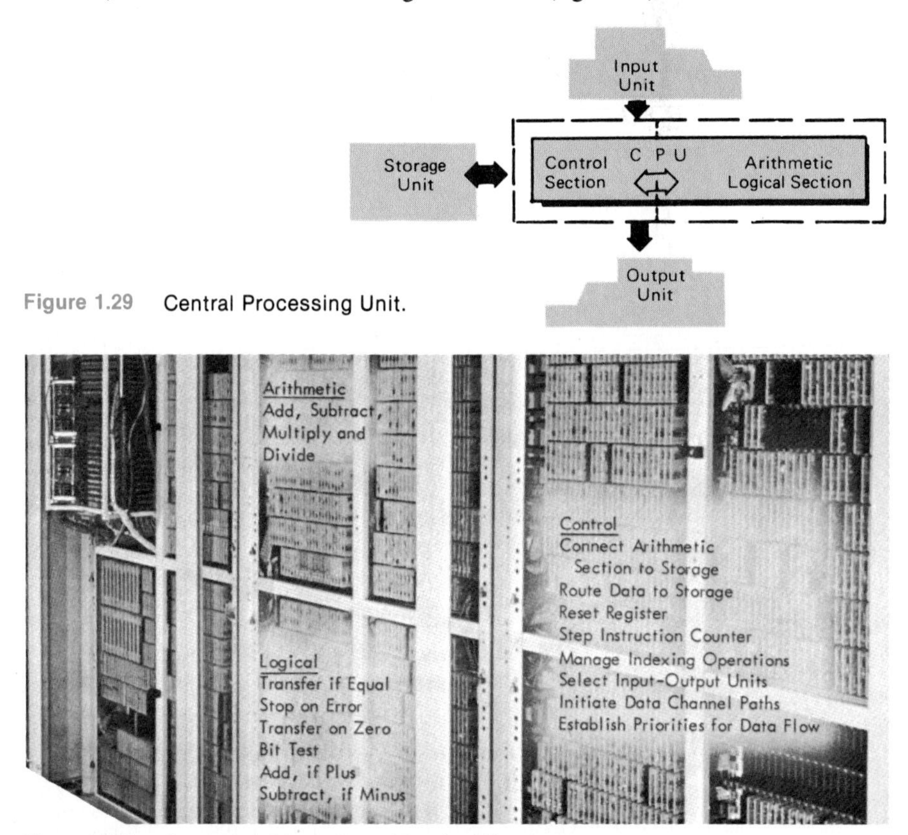

Figure 1.29 Central Processing Unit.

Figure 1.30 Control, Arithmetic and Logical Sections.

Control Section

The control section acts as a traffic manager directing and coordinating all operations called for by the instructions of the computer system. This involves control of input/output devices, entry and removal of information from storage, routing of information between storage and the Arithmetic and Logical section.

The control aspect of the processing unit comes from the individual commands contained in the program. These commands are instructions to the various devices to perform a function as specified. Each time a card is read or punched, or a line is printed on the output printer, or two amounts are added together, it is because an instruction in the processing unit caused it to happen. An instruction tells the computer what operation is to be performed (add, subtract, multiply, move, read a card) and where the data is that will be affected by this operation.

This section directs the system according to the procedure and the instructions received from its human operators and programmers. The control section automatically integrates the operation of the entire computer system.

Arithmetic and Logical Section

This section contains the circuitry to perform the necessary arithmetic and logical operations. The arithmetic portion performs operations such as addition, subtraction, multiplication, division, shifting, moving, and storing under the control of the stored program. The logical portion of the section is capable of decision-making to test various decisions encountered during the processing and to alter the sequence of the instruction execution.

STORAGE

All data entering a computer to be processed must be placed in storage first. Storage can be compared to a great electronic file cabinet, completely indexed and available for instant accessing. Information is entered into storage by an input device and is then available for internal processing. Storage is arranged so that each position has a specific location, called an *address*. For example, consider a group of numbered mail boxes in a post office. Each of these boxes is identified and located by its number. In the same manner, storage is divided into locations, each with its own assigned address. Each location holds a special character of information. In this way, the stored data can be located by the computer as it needs it.

Data may be rearranged by sorting and collating different types of information received from the various input units. Data may also be taken from storage, processed, and the result placed back in storage. The size and capacity of storage determines the amount of information that can be held within a system at one time. The larger the capacity, the more powerful and expensive the computer.

Storage may be classified as main and auxiliary storage (fig. 1.31).

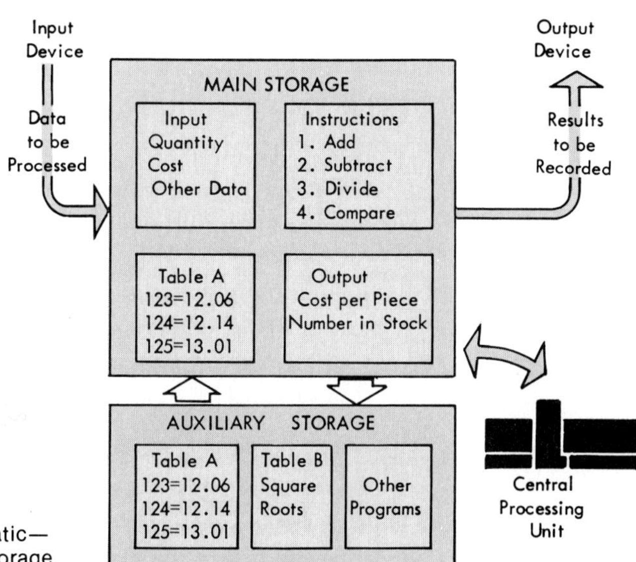

Figure 1.31 Schematic—Main and Auxiliary Storage.

Main Storage

Main storage is usually referred to as core or primary storage. Core storage consists of doughnut-shaped ferro-magnetic-coated material vertically aligned. Electric current is sent through these tiny cores (figs. 1.32, 1.33, 1.34). The direction of the current determines the polarity of the magnetic state of the core and gives each core a value of 0 or 1. All programs and data is translated into 0s and 1s and stored in the computer.

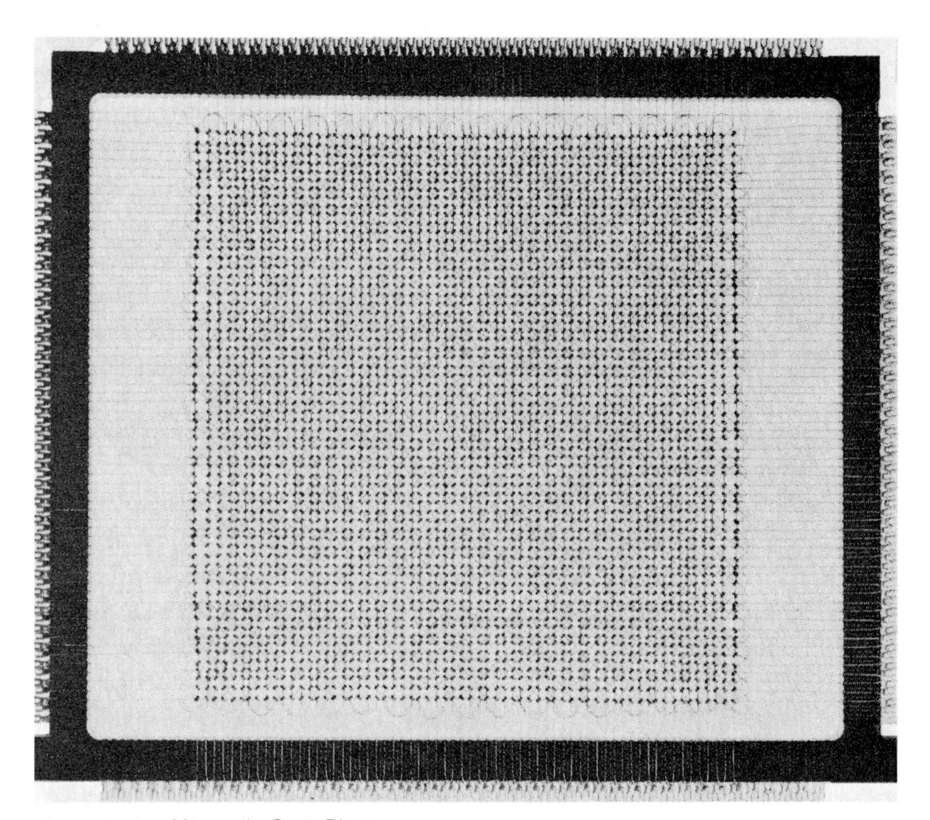

Figure 1.32 Magnetic Core Plane.

All data to be processed must pass through main storage. Main storage accepts data from the input unit, holds processed data, and can furnish data to an output unit. Since all data passes through main storage, the unit must therefore have the capacity to retain a usable amount of data and all necessary instructions for processing.

If additional storage is required, the capacity of main storage is augmented by auxiliary storage units; however, all information to and from auxiliary storage must be routed through main storage.

Auxiliary Storage

There are two types of auxiliary storage.

A. *Random-Access Units.* Drum, disk, and data cell devices can be accessed at random. That is, records can be accessed without reading from the beginning of a file to find them.

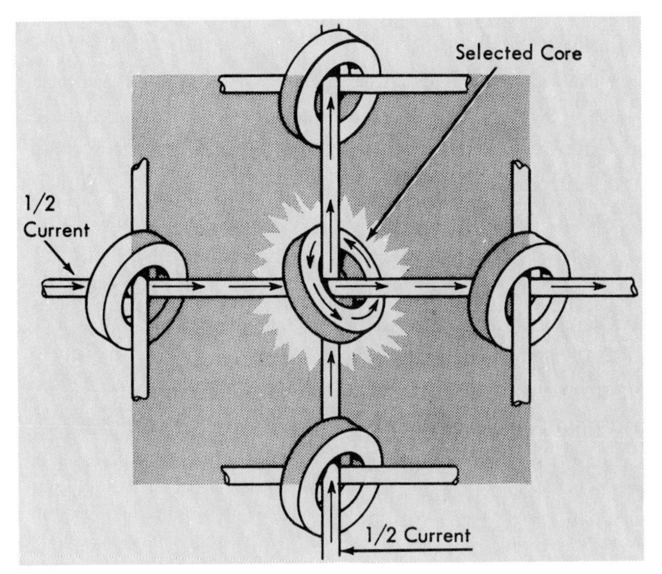

Figure 1.33 Selecting a Core.

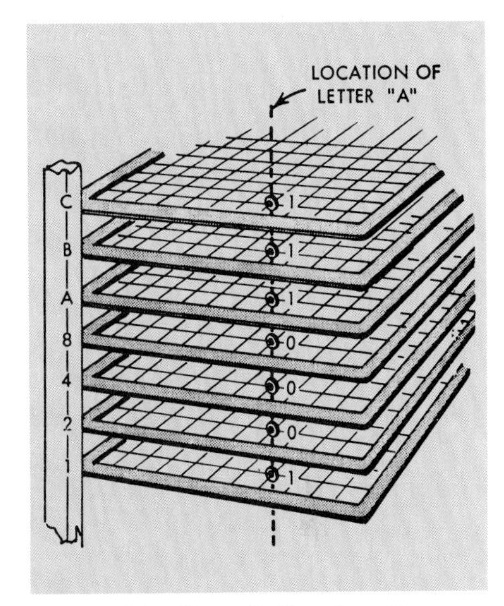

Figure 1.34 Magnetic Core Location.

B. *Sequential-Access Unit.* The magnetic tape unit is the chief type of sequential-access unit. This type of processing indicates that the tape reels must be read from the beginning of the tape to find the desired record.

(See figure 1.35.)

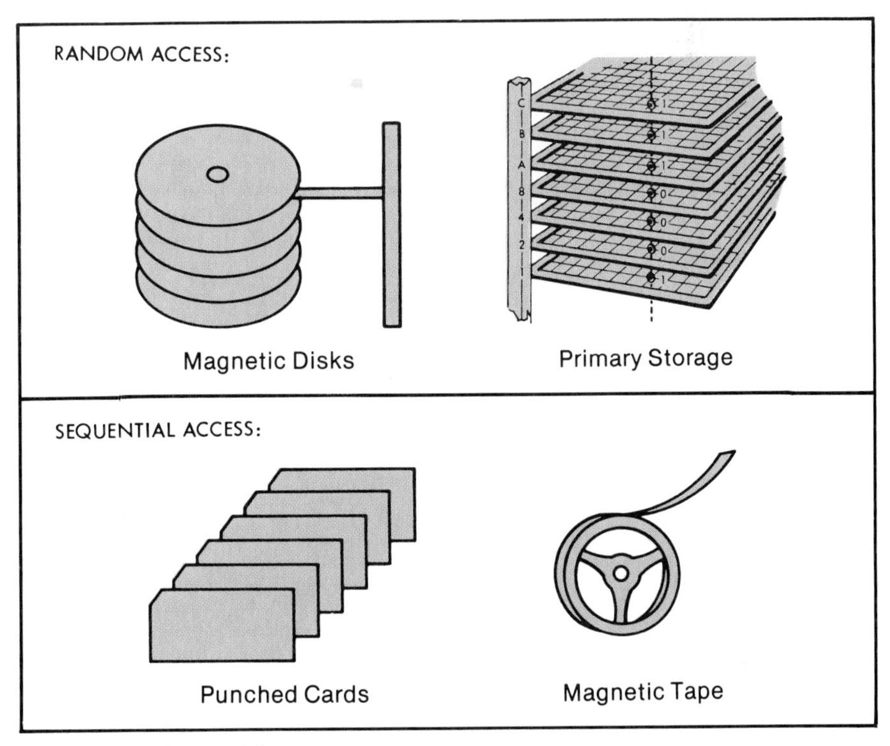

Figure 1.35 Types of Storage.

Magnetic Drum

A magnetic drum is a constant-speed rotating cylinder with an outer surface coated with a magnetic material (fig. 1.36). The chief function of the drum is to serve as a high-capacity, intermediate-access (storing results temporarily for future processing) storage device. Data is recorded as magnetized tracks around the drum. The primary uses of a magnetic drum are the storage of data that are repetitively referenced during processing (tables, rates, codes) or as a supplementary storage for core storage. Another important use today is to serve as a random-access device to provide program storage, program modification of data, and as a temporary storage for high-activity random-access operations involving limited amount of data.

The outer surface of the cylinder can be magnetized and read repeatedly as the drum rotates at a constant speed. Each time new data is read into the area, the old data is automatically erased. The data is read or written by a read/write head that is suspended at a slight distance from the drum.

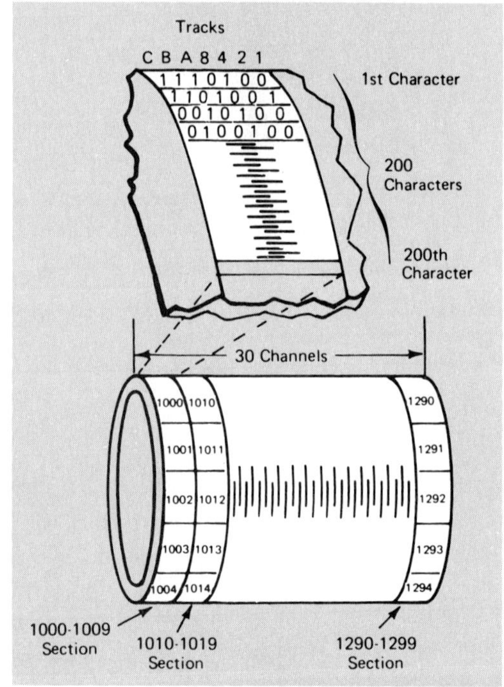

Figure 1.36 Magnetic Drum Storage.

Magnetic Disk

Disk storage is the most popular type of auxiliary storage in use today. Magnetic disk, like drum storage, provides data processing systems with the ability to read or retrieve records sequentially or randomly (direct access). The *magnetic disk* is a thin disk of metal coated on both sides with magnetic recording material (fig. 1.37).

Data is stored as magnetic spots on concentric tracks on each surface of the disk (fig. 1.38). These tracks are accessible for reading by positioning the read/write heads between the spinning disks.

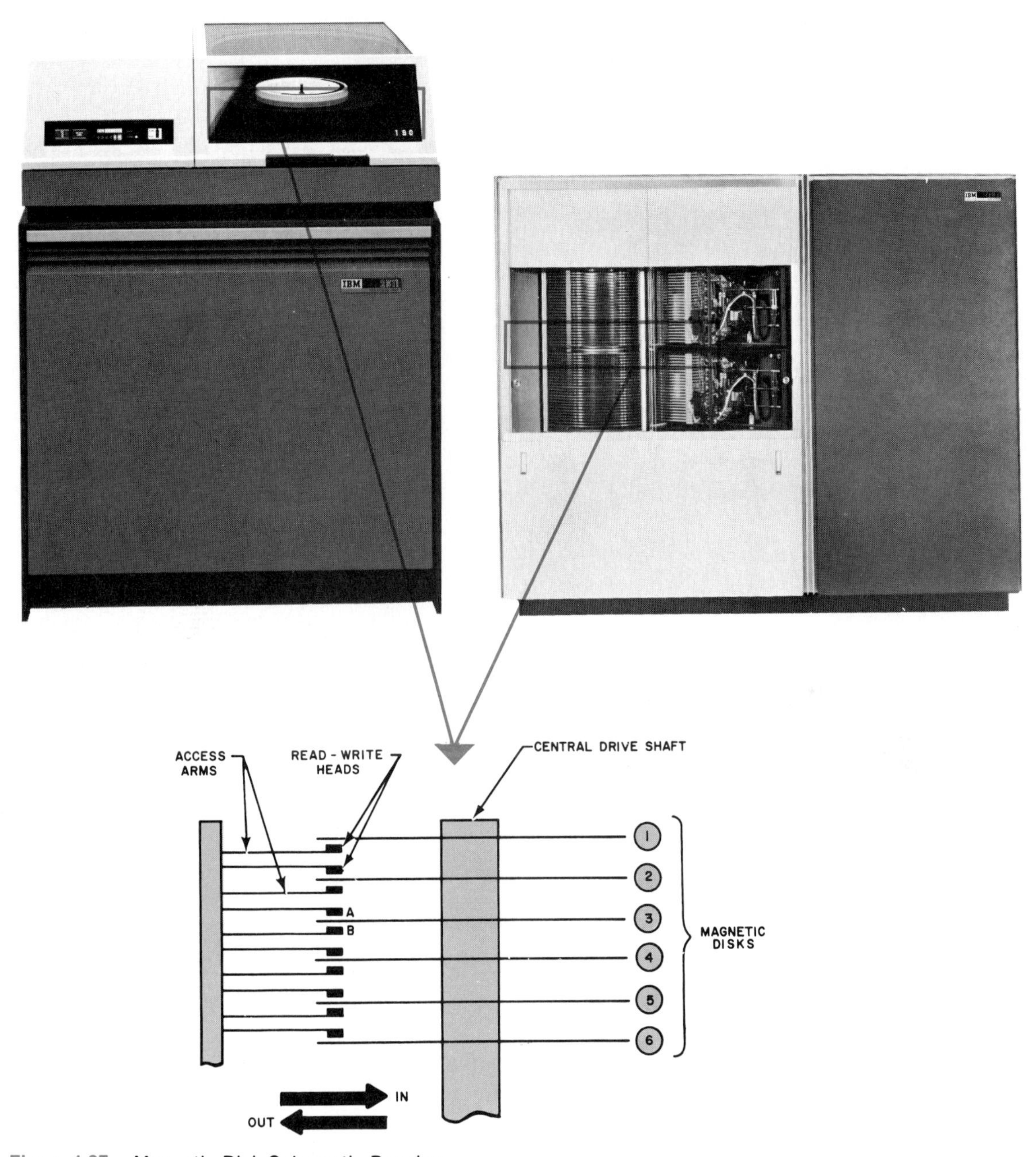

Figure 1.37 Magnetic Disk Schematic Drawing.

Disks provide the data processing systems with the ability to read and retrieve records randomly or sequentially. They permit the immediate access to specific areas of information without the need to examine each record, as in magnetic tape operations. Independent portable disks can be used with interchangeable disk packs. Six disk packs are mounted on a single unit (on some units) which can be readily removed from the disk drive and stored in a library of disk packs (fig. 1.39). Read/write heads are mounted on an access arm ar-

Figure 1.38
Magnetic Disk Recording.

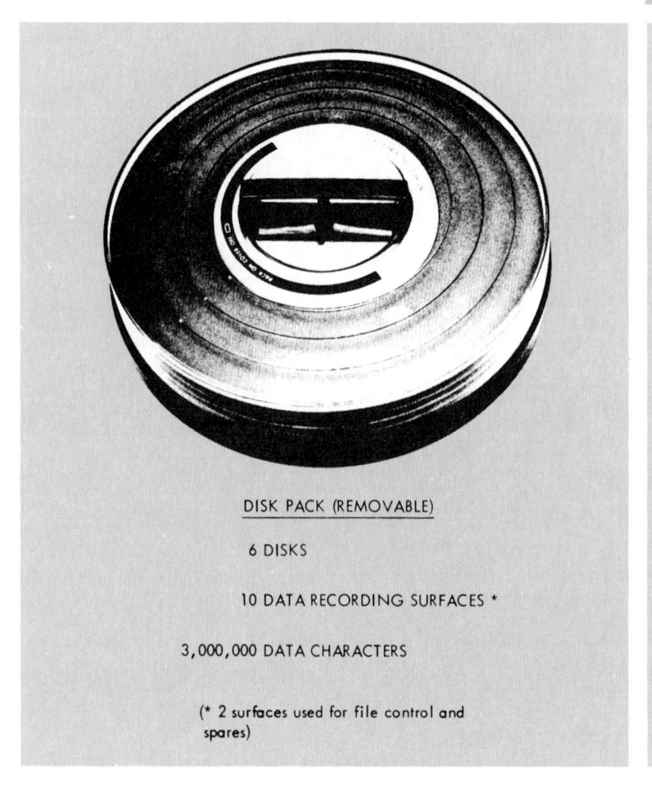

DISK PACK (REMOVABLE)
6 DISKS
10 DATA RECORDING SURFACES *
3,000,000 DATA CHARACTERS
(* 2 surfaces used for file control and spares)

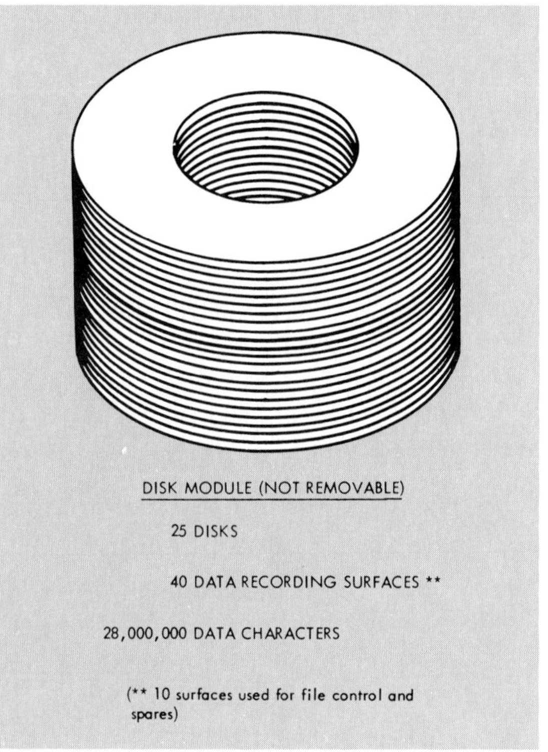

DISK MODULE (NOT REMOVABLE)
25 DISKS
40 DATA RECORDING SURFACES **
28,000,000 DATA CHARACTERS
(** 10 surfaces used for file control and spares)

Figure 1.39 Disk Packs and Modules.

ranged like teeth on a comb that moves horizontally between the disks. Two read/write heads are mounted on each arm with one head servicing the bottom surface of the top disk and the other head servicing the top surface of the lower disk. Thus it is possible to read or write on either side of the disk. Each disk pack has a capacity of over 7 million characters (fig. 1.40).

Each disk surface contains 100 tracks which are divided into 20 sectors. The capacity of each sector may be as many as 100 characters (fig. 1.41). With proper file organization, a minimum of access time is required for the retrieval of a disk record. The concept of removable disk packs means that only those disk records required for a particular application need be in use. Data records for other applications can be removed and stored.

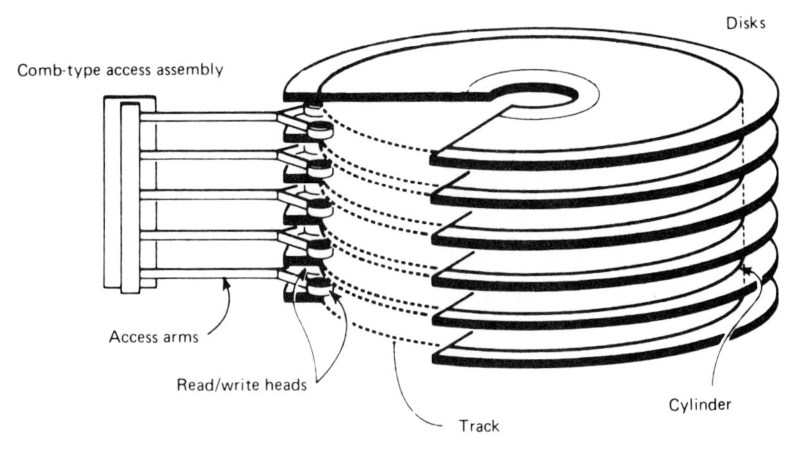

Figure 1.40 Comb-Type Access Mechanism.

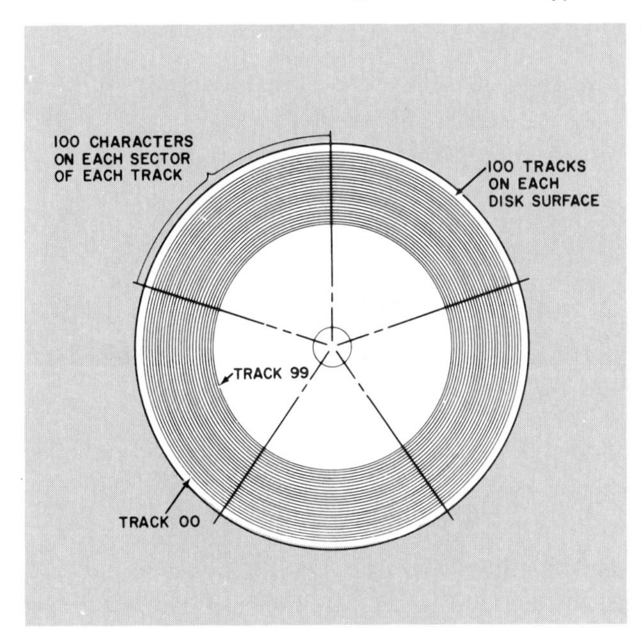

Figure 1.41 Magnetic Disk Layout.

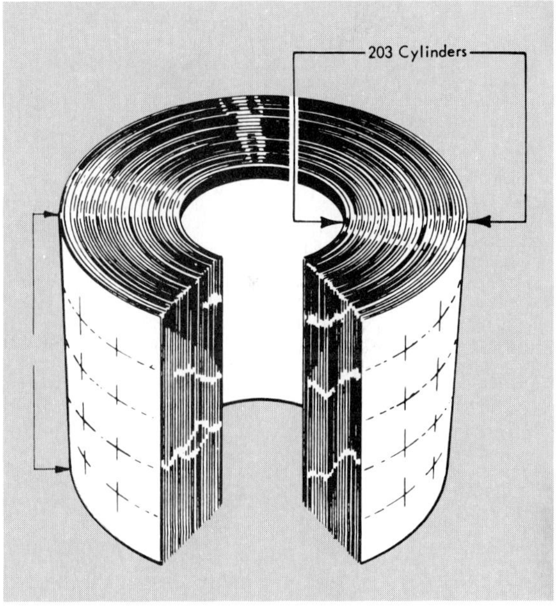

Figure 1.42 IBM 2311 Cylinders.

Each pack is divided into *cylinders* (fig. 1.42). A *cylinder* of data is that amount of information that is accessible with one position of the access mechanism. Since the movement of the access mechanism requires a significant portion of the time needed to access and transfer data, the storing of a large amount of data in a single cylinder can save time in the processing by minimizing the amount of access mechanism.

Data Cells

The data cell drive economically extends the random access storage capabilities to a volume of data beyond that of other storage devices. Each data cell device contains from one to ten data cells, each having a capacity of 40 million characters. The data cells are removable and interchangeable, permitting an open-ended capacity for libraries of data cells (fig. 1.43).

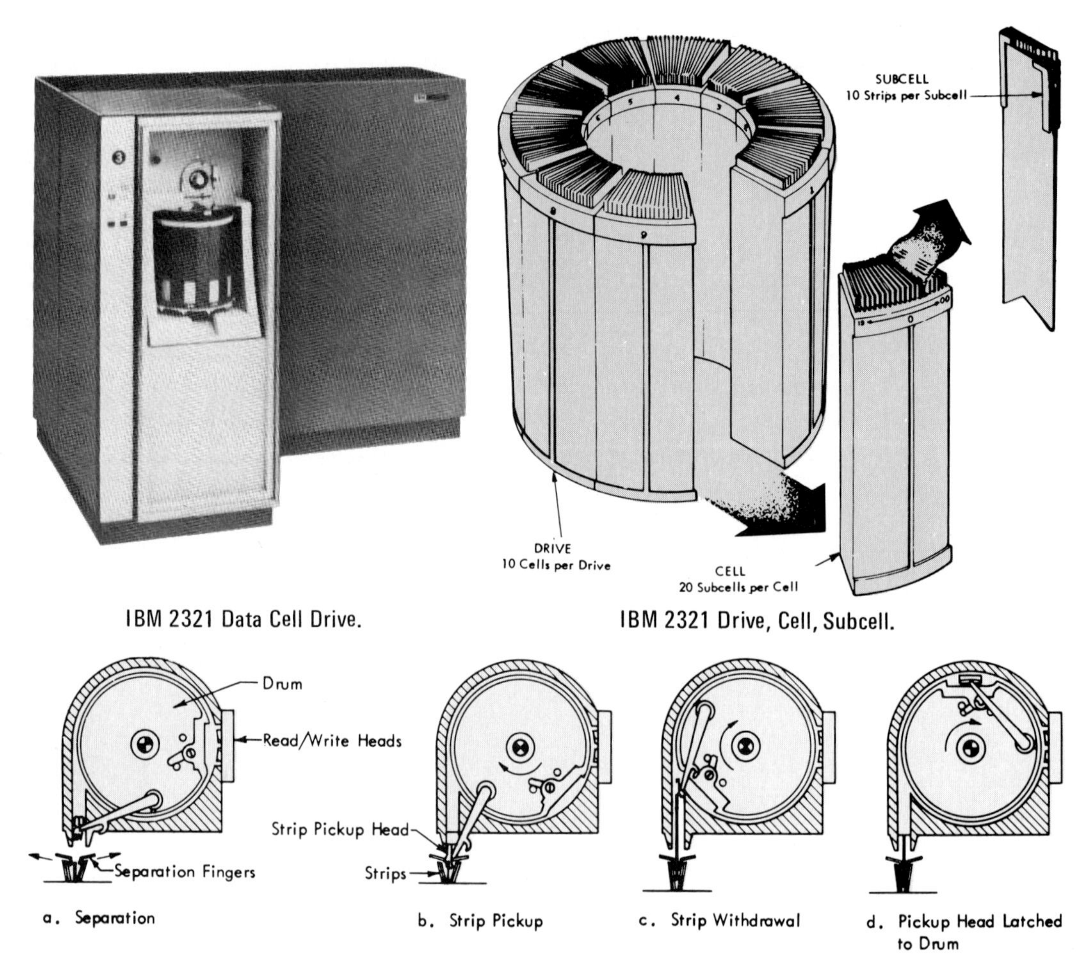

Figure 1.43 Data Cells.

The storage medium is a strip of magnetic film 2¼ inches wide by 13 inches long. Each data cell contains 200 of these strips divided into 20 subcells of 10 strips each. A rotary position aligns the selected subcell beneath the access station.

Exercises

Write your answer in the space provided (answer may be one or more words).

1. The computer was introduced in the last ________________ century.
2. ________________ industries have greatly increased making life more convenient and pleasant.
3. One of the driving forces behind the development of the computer was the need to find faster and more efficient methods of handling ________________.
4. Data processing is a planned series of actions and operations upon ________________ which uses various forms of ________________ equipment to achieve a ________________.
5. Data processing systems consists of ________________ and ________________.
6. The physical data processing equipment consists of ________________, ________________, and ________________ devices.

7. The source data entering a computer is known as ________________.
8. A series of instructions which accomplishes a particular job is called a ________________.
9. The central processing unit performs the ________________ function and is composed of two sections; ________________ and ________________ sections.
10. The ________________ unit is the computer memory.
11. The finished result is known as ________________.
12. ________________ is a set of programs, procedures and possibly associated documentation concerned with the operation of a data processing system.
13. The physical equipment or devices of a system forming a computer and its peripheral equipment is known as ________________.
14. ________________ are used to enter data into a data processing system.
15. ________________ accepts data for processing and makes results available to output devices.
16. ________________ store data, before, after, and during processing.
17. Data for input may be recorded in ________________, ________________, as ________________, or as ________________.
18. A ________________ decodes the command and synchronizes the device with the data processing system.
19. A ________________ is a separate piece of equipment devoted exclusively to managing ________________ control unit and devices assigned to it.
20. A ________________ provides the overlap operation necessary for getting information into and out of a computer.
21. ________________ and ________________ devices are the slowest means for getting information into and out of a computer.
22. Magnetic tape is not restricted to any ________________ record size and provides ________________ input and output of data to a computer.
23. ________________ devices are used to read paper documents into a computer.
24. A ________________ unit displays information on a cathode tube.
25. The permanent (hard copy) record from a data processing system is provided by a ________________.
26. The ________________ of a data processing system is used by the operator to control the system and monitor its operations.
27. ________________ are used in telecommunications within a data processing system.
28. The heart of the data processing system that supervises and controls the data processing components and performs the arithmetic and logical operations is the ________________.
29. The ________________ section directs the system according to the procedures and instructions received from its operators and programs.
30. The ________________ portion of the ________________ and ________________ section is capable of decision making to test various conditions encountered during the ________________ of the data and to ________________ the sequence of the instruction execution.
31. All data entering the computer to be processed must be placed in ________________ first.
32. Storage is arranged so that each position has a specific location, called an ________________.
33. The ________________ the capacity of storage, the more ________________ and ________________ the computer.
34. ________________ is usually referred to as core storage.
35. All data to be processed must pass through ________________.
36. ________________ storage augments main storage.
37. ________________ units can process records without the necessity of reading from the beginning of the file to find them.
38. The types of auxiliary storage are ________________, ________________, and ________________.

Answers

1. QUARTER
2. SERVICE
3. PAPERWORK
4. INFORMATION DATA PROCESSING, DESIRED RESULT
5. PROGRAMS, PHYSICAL EQUIPMENT
6. INPUT/OUTPUT, STORAGE, PROCESSING
7. INPUT
8. PROGRAM
9. PROCESSING, CONTROL, ARITHMETIC AND LOGICAL
10. CORE STORAGE
11. OUTPUT
12. SOFTWARE
13. HARDWARE
14. INPUT DEVICES
15. CENTRAL PROCESSING UNIT
16. STORAGE DEVICES
17. CARDS, MAGNETIC TAPE, CHARACTERS ON PAPER DOCUMENTS, LINE IMAGES
18. CONTROL UNIT
19. CHANNEL, INPUT/OUTPUT
20. BUFFERING SYSTEM
21. CARD READING, CARD PUNCHING
22. FIXED, HIGH-SPEED
23. CHARACTER RECOGNITION
24. VISUAL DISPLAY
25. PRINTER
26. CONSOLE
27. TERMINALS
28. CENTRAL PROCESSING UNIT
29. CONTROL
30. LOGICAL, ARITHMETIC, LOGICAL, PROCESSING, ALTER
31. STORAGE
32. ADDRESS
33. LARGER, POWERFUL, EXPENSIVE
34. MAIN STORAGE
35. MAIN STORAGE
36. AUXILIARY
37. RANDOM ACCESS
38. MAGNETIC DRUM, MAGNETIC DISK, DATA CELLS

Questions for Review

1. List the three ways our daily lives are affected by the computer.
2. What physical units constitute a data processing system?
3. What are the basic considerations in all data processing systems?
4. What is meant by "hardware"? "software"?
5. What are the major "hardware" elements of the computer?
6. What is the main function of the input/output device?
7. What main purpose does a control unit serve?
8. How does a channel perform within a data processing system?
9. How do the buffers increase the efficiency of input/output operations?
10. How is information recorded on magnetic tape?
11. What is an interblock gap and what is its purpose?
12. What are the uses of paper tape?
13. What are the two types of character recognition devices? Give examples.
14. What is a visual display device?
15. What is the main function of the printer?
16. What purpose does a console serve?
17. What is the main function of the central processing unit?
18. What are the sections of the central processing unit and what is the primary function of each section?
19. How is storage used in a data processing system?
20. What constitutes main storage? Auxiliary storage?
21. What is the difference between random access units and sequential access units?
22. Describe the following storage units and their main functions within a data processing system: magnetic core, magnetic drum, magnetic disk, data cells.

Problems

1. *Match each item with its proper description.*

_____ 1. Central Processing Unit

_____ 2. Input

_____ 3. Hardware

_____ 4. Data Processing

_____ 5. Software

_____ 6. Buffer

_____ 7. Control Unit

_____ 8. Program

_____ 9. Channel

A. Physical equipment or devices forming a computer and its peripheral equipment.

B. A series of instructions which accomplishes a particular job.

C. Decodes the command and synchronizes input/output device with the data processing system.

D. Separate piece of equipment devoted exclusively to managing the input/output control unit and devices assigned to it.

E. Source data entering the system.

F. A set of programs, procedures, and possibly documentation concerned with the operation of a data processing system.

G. Synchronizes processing of input/output operations and provides an overlap of operations.

H. A planned series of actions and operations upon information using various forms of data processing equipment to achieve a desired result.

I. The heart of the data processing system.

2. *Match each item with its proper description.*

_____ 1. Magnetic Ink Character Reader
_____ 2. Core Storage
_____ 3. Punched Cards
_____ 4. Magnetic Disk
_____ 5. Console
_____ 6. Control Section
_____ 7. Magnetic Tape
_____ 8. Magnetic Drum
_____ 9. Visual Display Unit
_____10. Printers

A. High speed input and output of data to a computer.
B. Used to control the system and monitor its operations.
C. Data is outputted on video screen.
D. Main storage.
E. Provide "hard copy" record from a data processing system.
F. One of the slowest means of getting information into and out of a computer.
G. Used extensively in banking operations to process checks.
H. Manipulates the data between core storage and the arithmetic and logical unit.
I. Most popular type of auxiliary storage in use today.
J. Used for the storage of data that is referred to repetitively such as tables, rates, codes, etc.

3. *Fill in the labels of the following units:*

4.

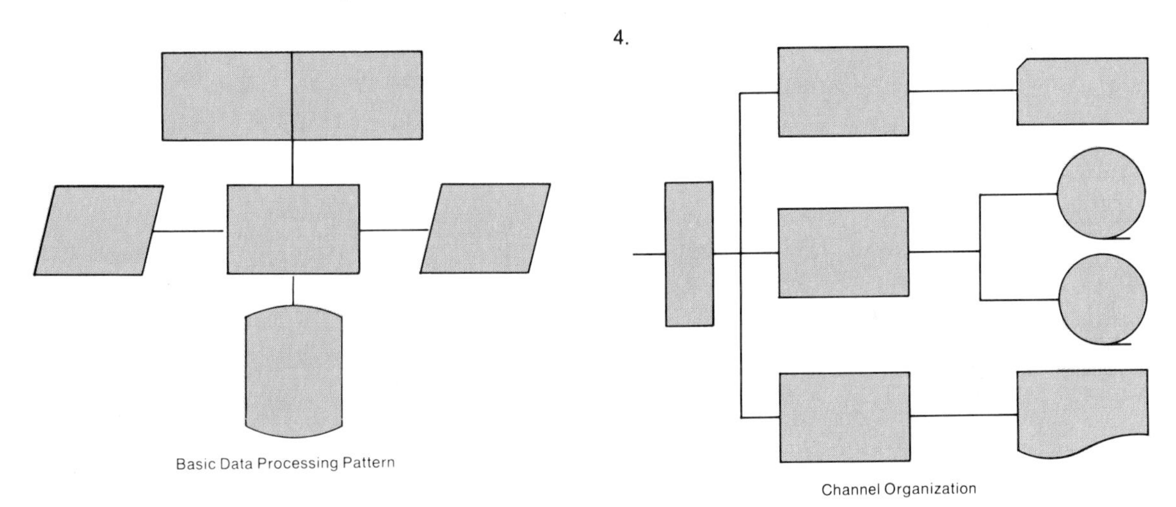

Basic Data Processing Pattern

Channel Organization

5.

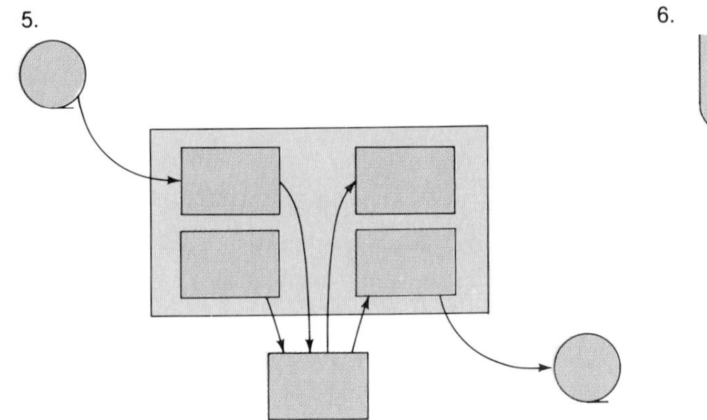

Buffer Operation

6.

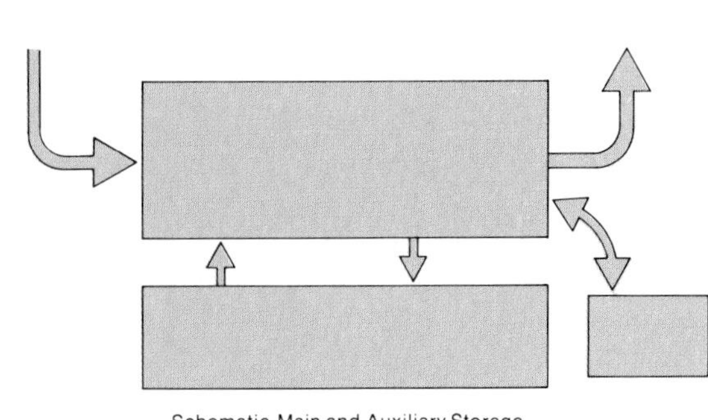

Schematic-Main and Auxiliary Storage

2 Introduction to Programming

A computer program is a set of instructions arranged in a proper sequence to cause the computer to perform a particular process.

The success of a computer program depends upon the ability of the programmer to do the following:

1. Analyze the problem.
2. Prepare a program to solve the particular problem.
3. Operate the program.

(See figures 2.1 and 2.2.)

STEP	GENERAL TERMS	PROGRAMMING TERMS
1	Defining the problem	Preparing job specifications
2	Planning the problem solution	Flowcharting (the program)
3	Describing the problem solution	Coding (the program)
4	Executing the problem solution	Program 'Translation' Program Testing Production Run
5	Documenting the problem solution	Documentation

Figure 2.1
Steps in Problem Solving.

ANALYSIS

A problem must be thoroughly analyzed before any attempt is made at a solution. This requires that boundary conditions be established so that the solution neither exceeds the objectives of management nor becomes too narrow to encompass all the necessary procedures. Output needs should be clearly stated, and the necessary input to produce the desired results should be carefully studied. If necessary, the source documents (input) should be revised so that they can be more readily converted into machine language for data processing operations. The relationship between the inputs and the outputs must be clearly shown (figs. 2.3, 2.4). Flowcharts are prepared to depict the orderly, logical steps necessary to arrive at the computer solution to the problem.

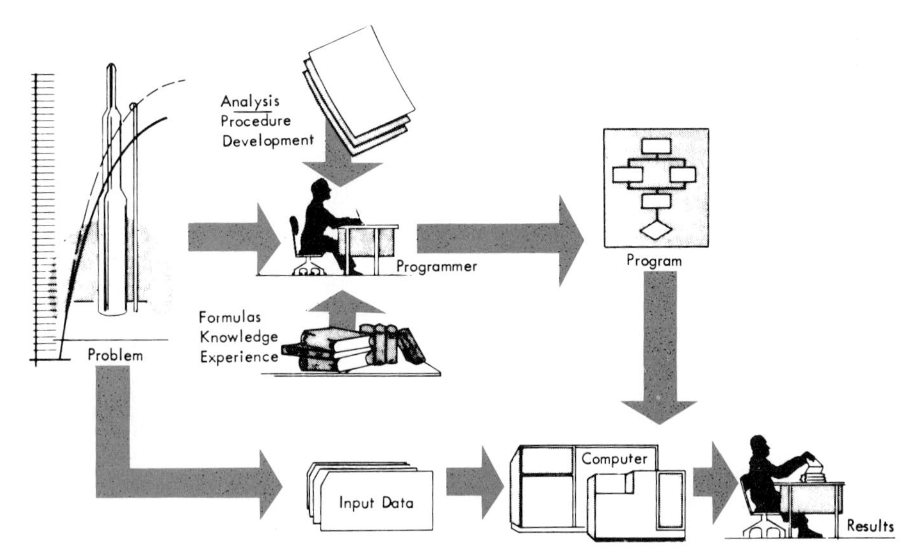

Figure 2.2 Direct Conversion of Problem to Machine Program.

PREPARATION

After both a careful analysis of the problem and flowcharting, the computer program should be written. Instructions are coded in the required computer language, using the program flowchart as a guide. The sequence of instructions will determine the computer program.

OPERATION

The next step, usually, is the placing of the program and data into the storage unit of the computer. The program must be prepared on punched cards or other media for entry into the machine through an input device. The data must also be made available to the computer through some input unit. The program must be thoroughly checked with the test data and all necessary debugging accomplished before the program is ready for operation upon actual data. (See figure 2.5.)

The following items should be checked to insure that the proper analysis and coding was made and that the computer will operate properly:

1. *Precise Statement of the Problem.* This statement must be exact, specifying what the program is to accomplish. "To compute social security tax, multiply gross pay by social security rate to arrive at the FICA tax, etc."
2. *List of inputs.* All sample copies of inputs to be used together with the size of the fields, type (alphabetic or numeric), control fields, etc., should be included.
3. *Outputs Desired.* Samples of all outputs should be included with all headings indicated. The number of copies desired, type and size of paper to be used, tape density (if used)—are some of the items of information that should be included in this section.
4. *Flowcharts.* All necessary system flowcharts, program flowcharts, and block diagrams should be included. A system flowchart represents the flow of data through all parts of a system, and a program flowchart places the em-

DEPT | SERIAL NUMBER | NAME | HOURS | NET PAY

DEPT.	SERIAL NUMBER	N A M E		HOURS WORKED	NET PAY
1	1234560	SMITH	JW	40.1	285.15
1	1892750	JONES	RA	39.6	152.16
1	8929016	MAUS	JB	62.5	182.55
					$ 619.86
2	0238648	GOLDMAN	H	31.7	100.25
2	0333367	WOLFE	DJ	9.5	26.60
					$ 126.85

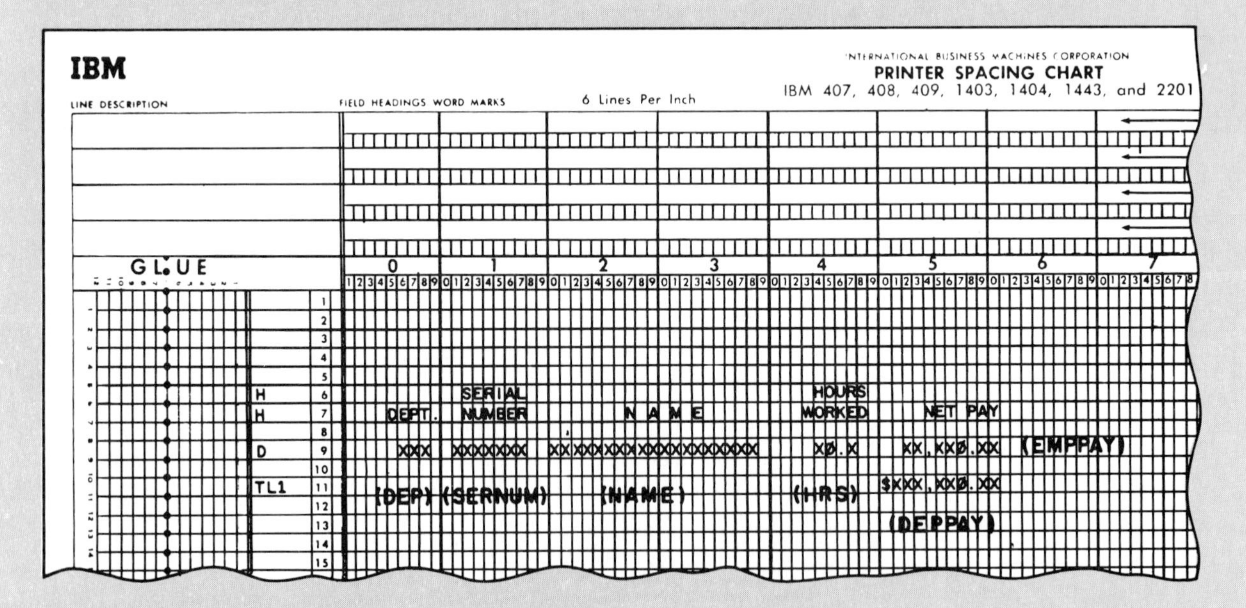

Figure 2.3 Example—Card Design and Printer Layout.

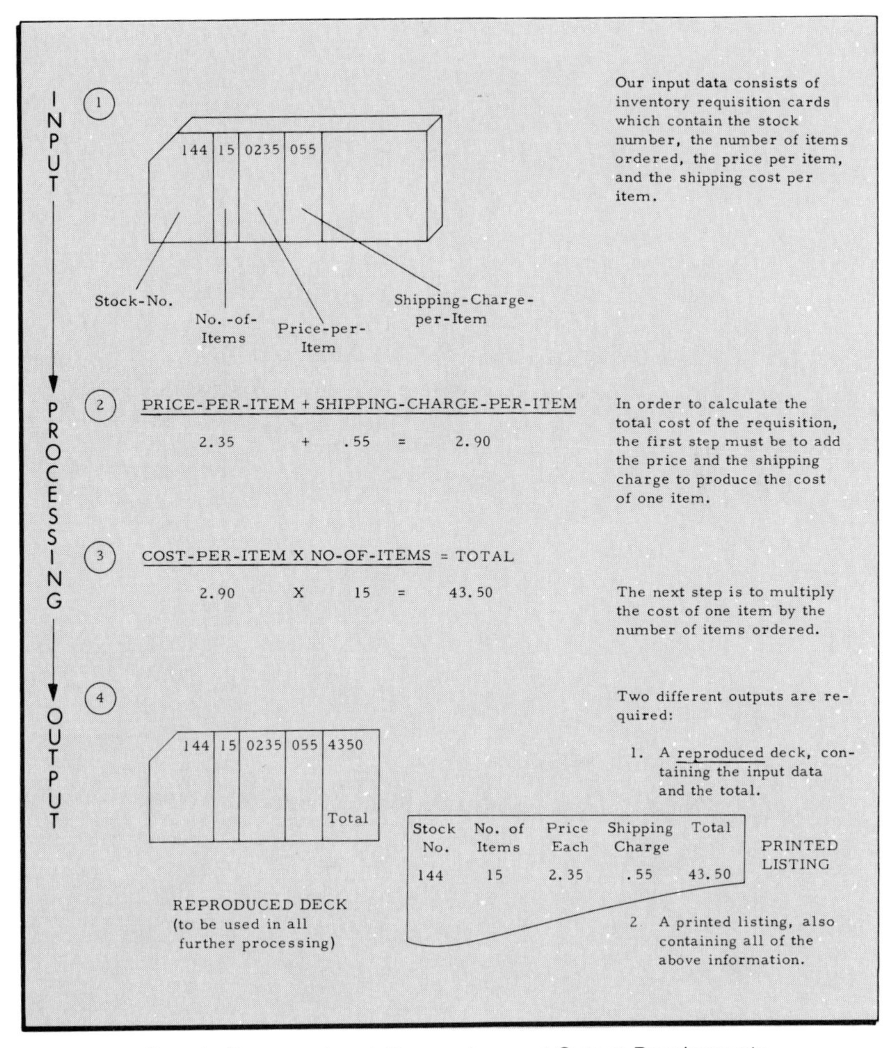
INPUT
PROCESSING
OUTPUT
1
144 15 0235 055
Stock-No.
No. -of-
Items
Price-per-
Item
Shipping-Charge-
per-Item
Our input data consists of inventory requisition cards which contain the stock number, the number of items ordered, the price per item, and the shipping cost per item.
2
PRICE-PER-ITEM + SHIPPING-CHARGE-PER-ITEM
2.35 + .55 = 2.90
In order to calculate the total cost of the requisition, the first step must be to add the price and the shipping charge to produce the cost of one item.
3
COST-PER-ITEM X NO-OF-ITEMS = TOTAL
2.90 X 15 = 43.50
The next step is to multiply the cost of one item by the number of items ordered.
4
Two different outputs are required:
144 15 0235 055 4350
Total
1. A reproduced deck, containing the input data and the total.
Stock No.
No. of Items
Price Each
Shipping Charge
Total
144 15 2.35 .55 43.50
PRINTED LISTING
REPRODUCED DECK
(to be used in all further processing)
2. A printed listing, also containing all of the above information.

Figure 2.4 Sample Program: Input, Processing, and Output Requirements.

phasis on computer decisions and processes. A block diagram is a detailed breakdown of a program flowchart.

5. *Program.* A printed copy of the computer program with all necessary comments.
6. *Test data.* Sample data to be used to test programs.
7. *Job-control cards.* All job-control cards necessary to load the program and the data into the computer. Job-control cards tell the computer what kind of job it is about to do, how to go about performing certain operations involved in doing that job, and how to recognize the end of the job. Job-control cards might vary from one program to the next as well as from one computer to another.
8. *Test results.* Output listings and/or cards used to test the accuracy of the program.

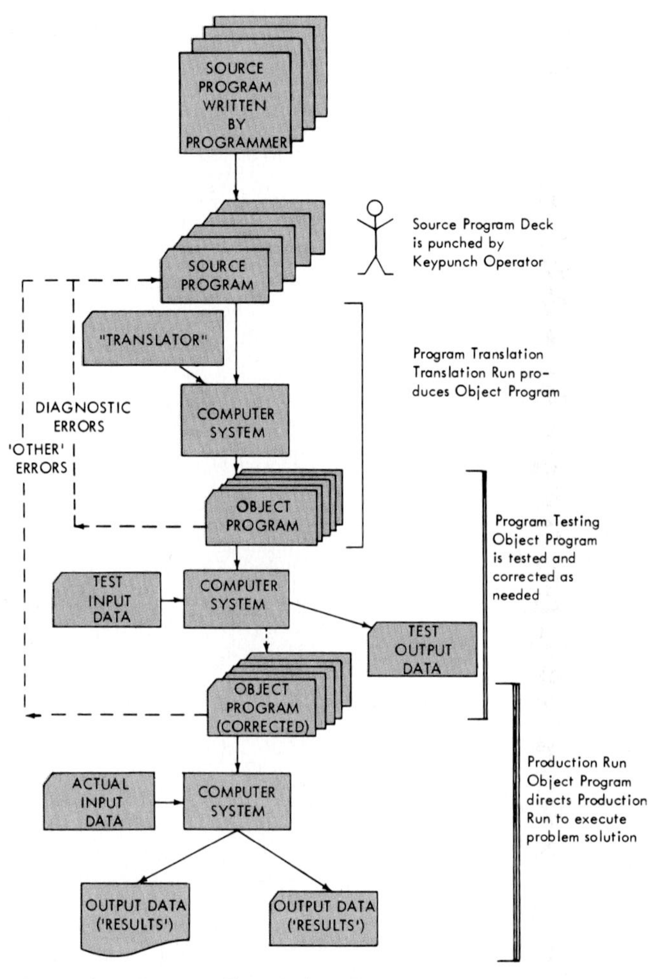

Figure 2.5 Execution: Program Translation, Testing, and Production Run.

PLANNING A PROGRAM

A computer program is the outcome of a programmer's applied knowledge of the problem and the operation of a particular computer. Problem definition, analysis, documentation, and flowcharting are just the initial steps in the preparation of a program. (See figure 2.6.)

The following must be considered even in the simplest of programs:

1. The allocation of storage locations for the storing of data, instructions, work areas, constants, etc.
2. The necessary input procedures to convert the source data into machine processable media.
3. The various reference tables and files that are essential to the program.
4. The checking of the accuracy of the data and the calculations.
5. The ability to restart the system in case of unscheduled interruptions, machine failures, or error conditions.

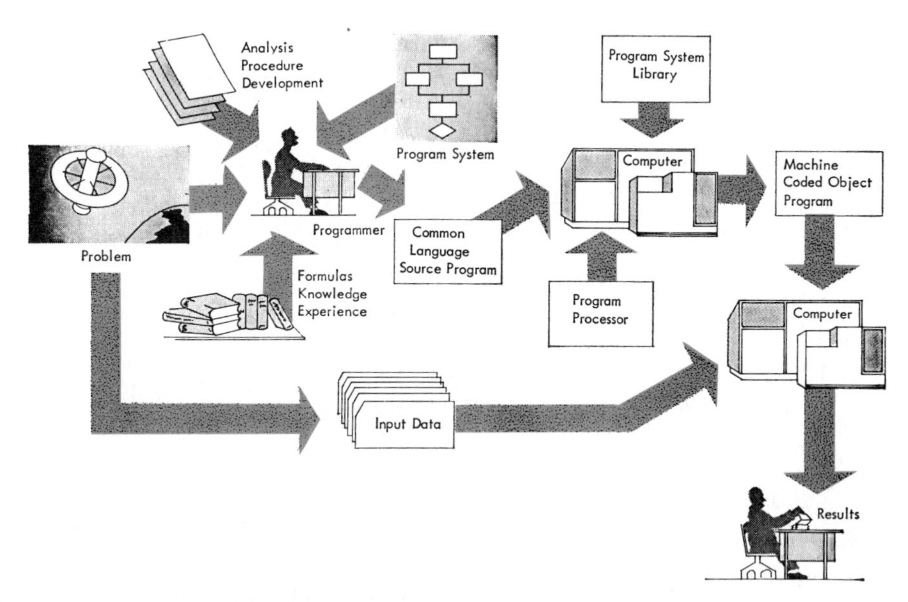

Conversion of Problem to Machine Program Using Programming Symbols

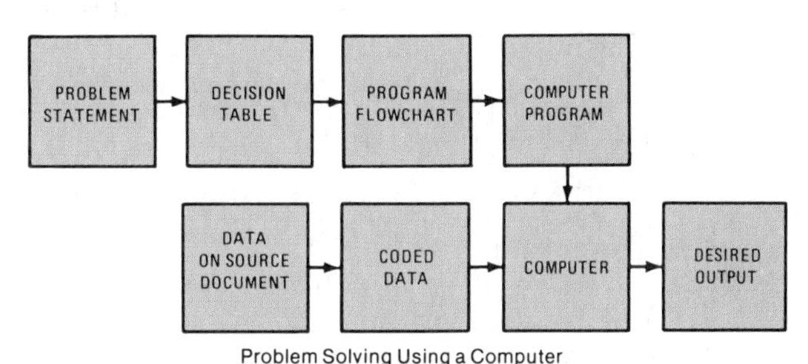

Problem Solving Using a Computer

Figure 2.6 Program Planning.

6. The necessary housekeeping procedures to clear storage areas, registers, and indicators prior to the execution of the program. Housekeeping pertains to those operations in a program or computer system which do not contribute directly to the solution of the user's program, but which are necessary in order to maintain control of the processing.
7. A thorough knowledge of the arithmetic and logical procedures to be used in the program.
8. The output formats of cards, printed reports, displayed reports, magnetic tapes, etc. (See figure 2.7.)
9. The subroutines available from other procedures to be used in the program. A subroutine is a subprogram consisting of a set of instructions that perform some subordinate function within the program. A *closed* subroutine is stored in one place and connected to the program by means of linkages at one or more points in the program. An *open* subroutine is inserted directly into a program at each point where it is to be used (fig. 2.8).

STEP 1. Preparation of Job Specifications consisting of:

A. Job Description

1. Job title

2. Summary of what is to be accomplished in run

3. System flowchart for the run

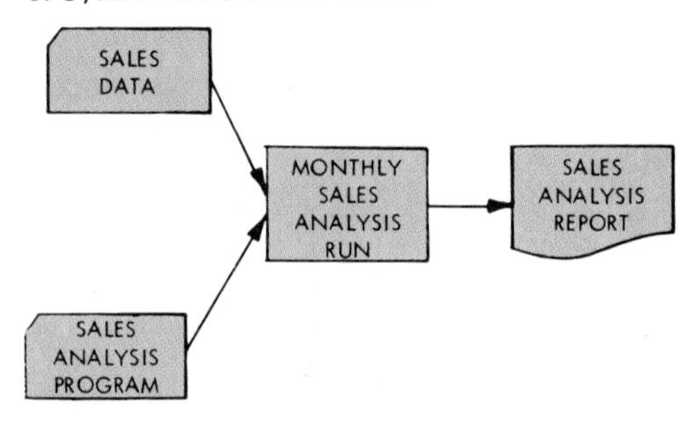

B. Input Files Description

C. Processing Requirements

D. Output Files Description

STEP 2. Flowcharting the Problem

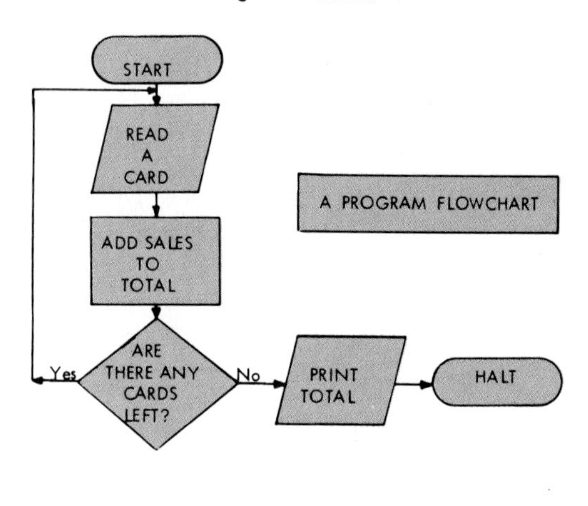

STEP 3. Coding the Program

COBOL PROGRAM SHEET

System		Punching Instructions		Sheet of
Program		Graphic	Card Form #	Identification
Programmer	Date	Punch		73 — 80

```
       PROCEDURE DIVISION.

       START-RUN.
           OPEN INPUT MASTER-FILE; OUTPUT CREDIT-LETTERS.

       READ-MASTER.
           MOVE SPACES TO OUTPUT-LINE.
           READ MASTER-FILE;
               AT END; CLOSE MASTER-FILE, CREDIT-LETTERS; STOP RUN.
           MOVE CUSTOMER-NAME TO DATA-AREA.
           WRITE OUTPUT-LINE AFTER ADVANCING 0.
           MOVE STREET-ADDRESS TO DATA-AREA.
           WRITE OUTPUT-LINE AFTER ADVANCING 1.
           MOVE CITY TO DATA-AREA.
           WRITE OUTPUT-LINE AFTER ADVANCING 1.
           GO TO READ-MASTER.
```

PROGRAMMING FORM

Figure 2.7 Documentation.

Figure 2.7 Continued

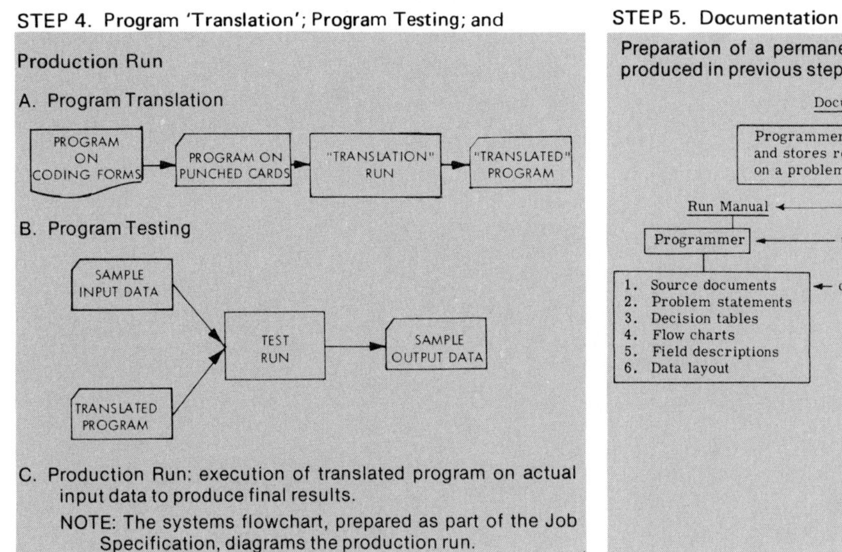

STEP 4. Program 'Translation'; Program Testing; and Production Run
A. Program Translation
PROGRAM ON CODING FORMS
PROGRAM ON PUNCHED CARDS
"TRANSLATION" RUN
"TRANSLATED" PROGRAM
B. Program Testing
SAMPLE INPUT DATA
TEST RUN
SAMPLE OUTPUT DATA
TRANSLATED PROGRAM
C. Production Run: execution of translated program on actual input data to produce final results.
NOTE: The systems flowchart, prepared as part of the Job Specification, diagrams the production run.

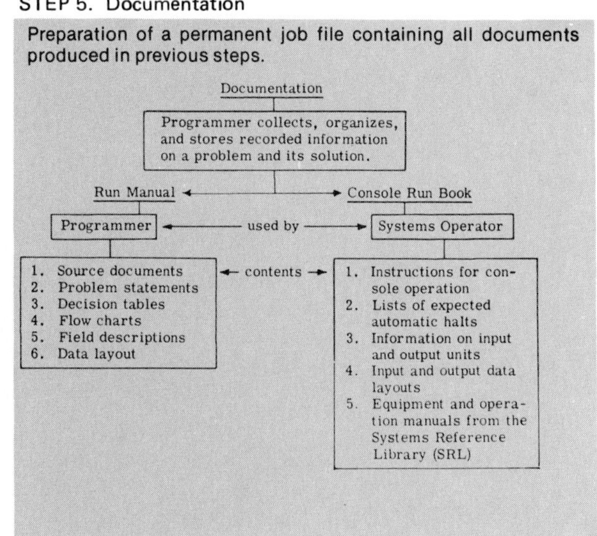

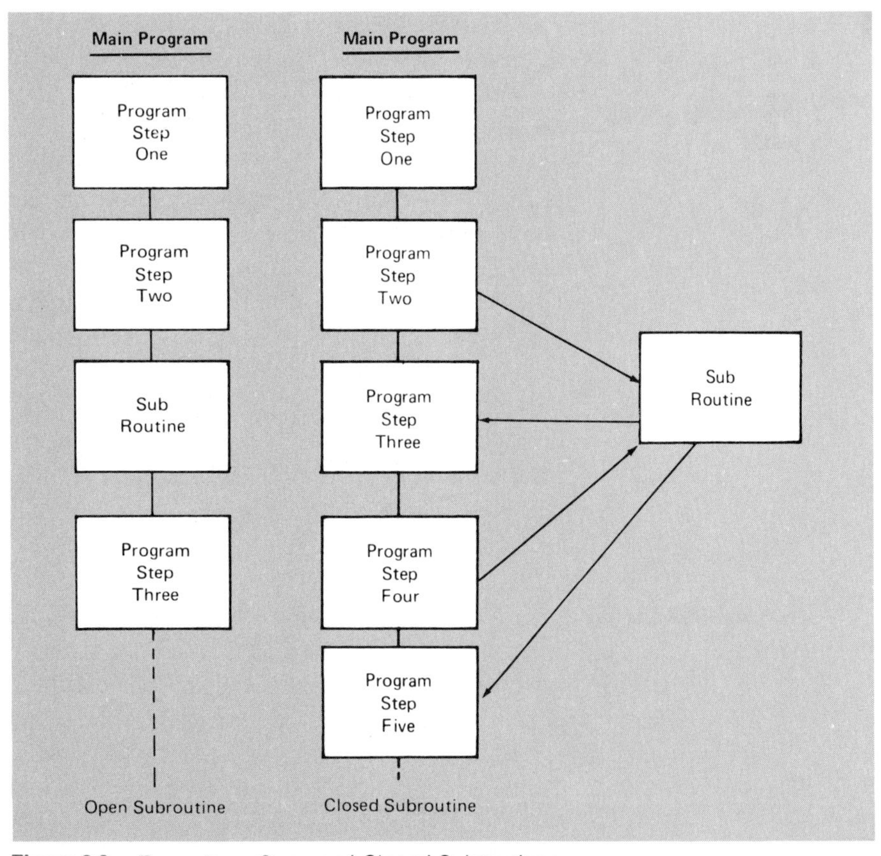

Figure 2.8 Examples—Open and Closed Subroutines.

FLOWCHARTS

The increased use of data processing has focused attention upon the need for the logical representation of data flows. Once the problem has been defined and the objectives established, the next step is the orderly presentation of procedures so that the objectives can be realized. Any successful program depends upon well-defined steps prior to the actual program. The steps involve the processes to be performed and the sequence of these processes. The processes must be stated precisely before any programming can begin (fig. 2.9).

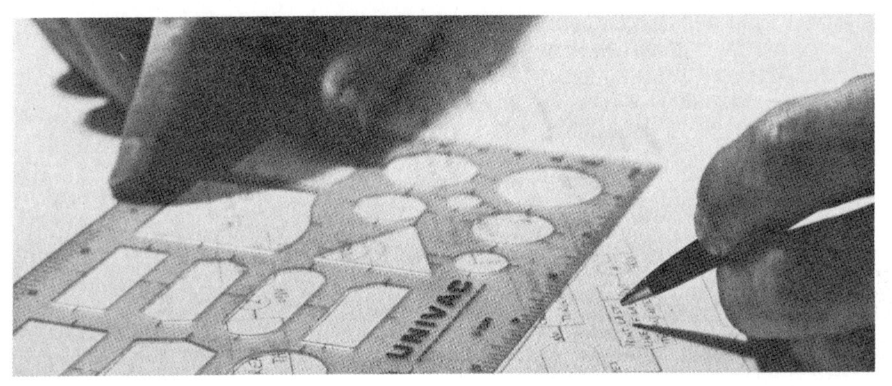

Figure 2.9 Flowcharting.

The analysis is normally accomplished by developing flowcharts. *The flowchart is a graphic representation of the flow of information through a system in which the information is converted from the source document to the final reports.* Because most data processing applications involve a large number of alternatives, decisions, exceptions, etc., it would be impractical to state these possibilities verbally. The value of a flowchart is that it can show graphically, at a glance, the organized procedures and data flows so that their apparent interrelationships are readily understood by the reader. Such relationships would be difficult to abstract from a detailed narrative text. Meaningful symbols are used in place of narrative statements (fig. 2.10). The flowchart is the "roadmap" by which the data travels through the entire system.

While flowcharts are used widely in the field of data processing, they are occasionally misinterpreted due primarily to the lack of uniformity in the meanings and use of the symbols. As a result, a uniform set of flowcharting symbols was prepared by a subcommittee of the United States of America Standards Institute. (See figure 2.11.)

System Flowcharts

There are two types of flowcharts widely used in data processing operations: the system flowchart, representing the flow of data through all parts of a system, and the program flowchart, wherein the emphasis is on the computer decisions and processes. A system flowchart is normally used to illustrate the overall objectives to data processing as well as to nondata processing personnel. The flowchart provides a picture indicating what is to be accomplished. Emphasis is on the documents and the work stations they must pass through. The flowchart

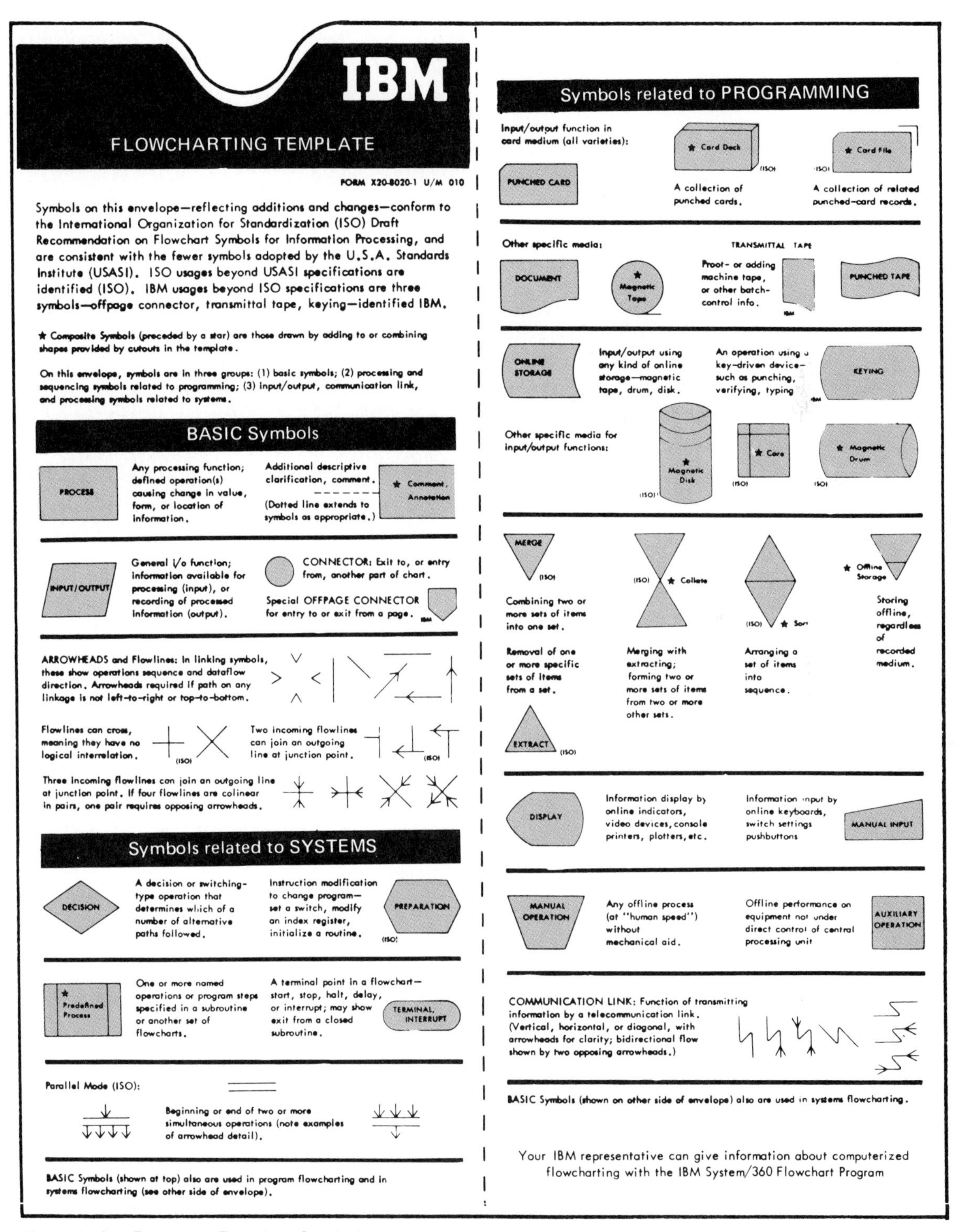

IBM

FLOWCHARTING TEMPLATE

FORM X20-8020-1 U/M 010

Symbols on this envelope—reflecting additions and changes—conform to the International Organization for Standardization (ISO) Draft Recommendation on Flowchart Symbols for Information Processing, and are consistent with the fewer symbols adopted by the U.S.A. Standards Institute (USASI). ISO usages beyond USASI specifications are identified (ISO). IBM usages beyond ISO specifications are three symbols—offpage connector, transmittal tape, keying—identified IBM.

★ Composite Symbols (preceded by a star) are those drawn by adding to or combining shapes provided by cutouts in the template.

On this envelope, symbols are in three groups: (1) basic symbols; (2) processing and sequencing symbols related to programming; (3) input/output, communication link, and processing symbols related to systems.

BASIC Symbols

PROCESS — Any processing function; defined operation(s) causing change in value, form, or location of information.

★ Comment, Annotation — Additional descriptive clarification, comment. (Dotted line extends to symbols as appropriate.)

INPUT/OUTPUT — General I/O function; information available for processing (input), or recording of processed information (output).

CONNECTOR: Exit to, or entry from, another part of chart.

Special OFFPAGE CONNECTOR for entry to or exit from a page. IBM

ARROWHEADS and Flowlines: In linking symbols, these show operations sequence and dataflow direction. Arrowheads required if path on any linkage is not left-to-right or top-to-bottom.

Flowlines can cross, meaning they have no logical interrelation. (ISO)

Two incoming flowlines can join an outgoing line at junction point. (ISO)

Three incoming flowlines can join an outgoing line at junction point. If four flowlines are colinear in pairs, one pair requires opposing arrowheads.

Symbols related to SYSTEMS

DECISION — A decision or switching-type operation that determines which of a number of alternative paths followed.

PREPARATION — Instruction modification to change program—set a switch, modify an index register, initialize a routine. (ISO)

★ Predefined Process — One or more named operations or program steps specified in a subroutine or another set of flowcharts.

TERMINAL, INTERRUPT — A terminal point in a flowchart—start, stop, halt, delay, or interrupt; may show exit from a closed subroutine.

Parallel Mode (ISO): Beginning or end of two or more simultaneous operations (note examples of arrowhead detail).

BASIC Symbols (shown at top) also are used in program flowcharting and in systems flowcharting (see other side of envelope).

Symbols related to PROGRAMMING

PUNCHED CARD — Input/output function in card medium (all varieties):

★ Card Deck (ISO) — A collection of punched cards.

★ Card File (ISO) — A collection of related punched-card records.

Other specific media:

DOCUMENT

★ Magnetic Tape

TRANSMITTAL TAPE — Proof- or adding machine tape, or other batch-control info. IBM

PUNCHED TAPE

ONLINE STORAGE — Input/output using any kind of online storage—magnetic tape, drum, disk.

KEYING — An operation using a key-driven device—such as punching, verifying, typing. IBM

Other specific media for input/output functions:

★ Magnetic Disk (ISO)

★ Core (ISO)

★ Magnetic Drum (ISO)

MERGE (ISO) — Combining two or more sets of items into one set.

★ Collate (ISO) — Merging with extracting; forming two or more sets of items from two or more other sets.

★ Sort (ISO) — Arranging a set of items into sequence.

★ Offline Storage — Storing offline, regardless of recorded medium.

EXTRACT (ISO) — Removal of one or more specific sets of items from a set.

DISPLAY — Information display by online indicators, video devices, console printers, plotters, etc.

MANUAL INPUT — Information input by online keyboards, switch settings pushbuttons

MANUAL OPERATION — Any offline process (at "human speed") without mechanical aid.

AUXILIARY OPERATION — Offline performance on equipment not under direct control of central processing unit

COMMUNICATION LINK: Function of transmitting information by a telecommunication link. (Vertical, horizontal, or diagonal, with arrowheads for clarity; bidirectional flow shown by two opposing arrowheads.)

BASIC Symbols (shown on other side of envelope) also are used in systems flowcharting.

Your IBM representative can give information about computerized flowcharting with the IBM System/360 Flowchart Program

Figure 2.10 Flowchart Template Symbols.

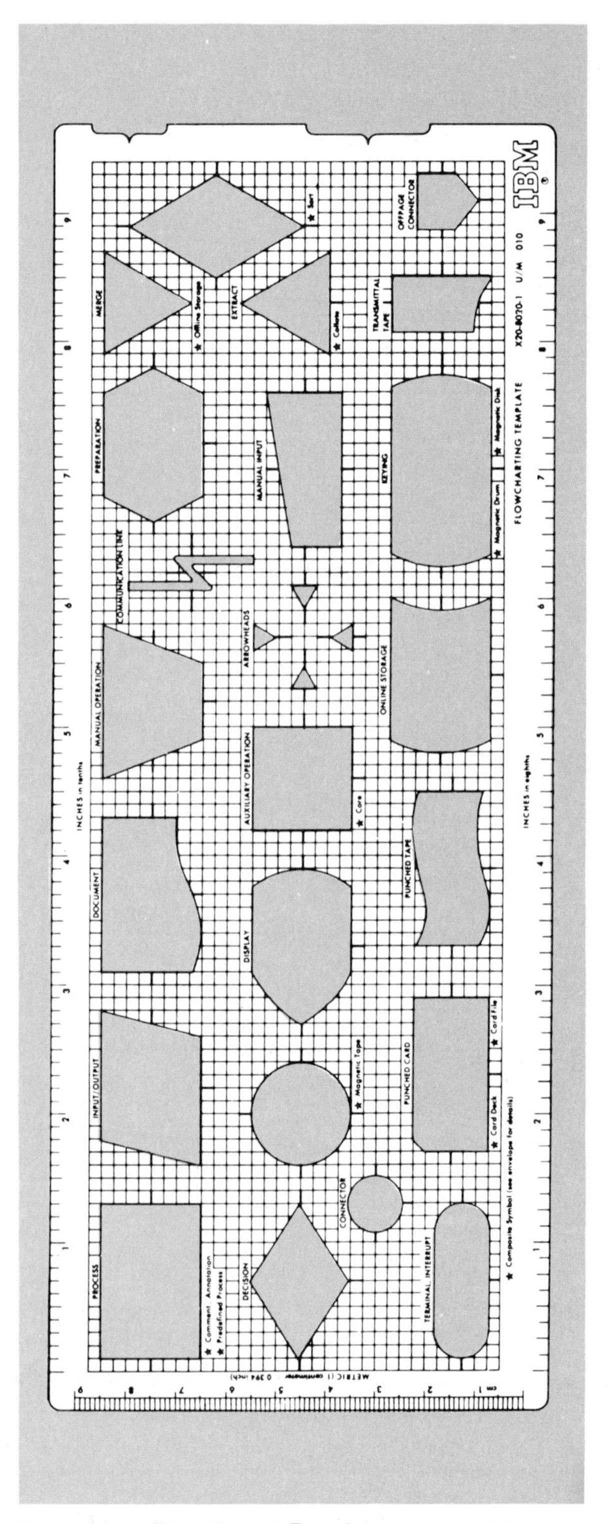

Figure 2.11 Flowcharting Template.

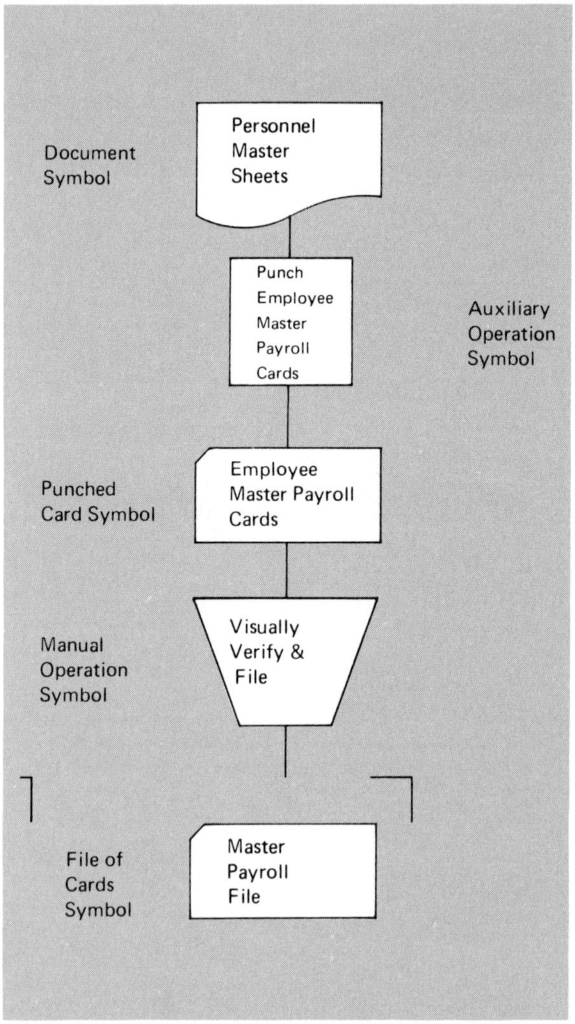

Document Symbol — Personnel Master Sheets

Auxiliary Operation Symbol — Punch Employee Master Payroll Cards

Punched Card Symbol — Employee Master Payroll Cards

Manual Operation Symbol — Visually Verify & File

File of Cards Symbol — Master Payroll File

Figure 2.12 System Flowchart Symbols—Example.

can also be applied where source media is converted to a final report or stored in files. A brief mention is made of the actual operations to be performed.

Many symbols depicting documents and operations are used throughout the system flowchart. The symbols are designed so that they are meaningful without too much further comment or text. Card symbols are used to indicate when the input or output may be a card. Document symbols are used to represent the printed reports. (See figure 2.12.)

The system flowchart is usually prepared on one sheet of paper so as to facilitate the presentation of the overall picture of the system to administrative personnel and executives. It indicates the job to be done without detailing the steps involved (fig. 2.13).

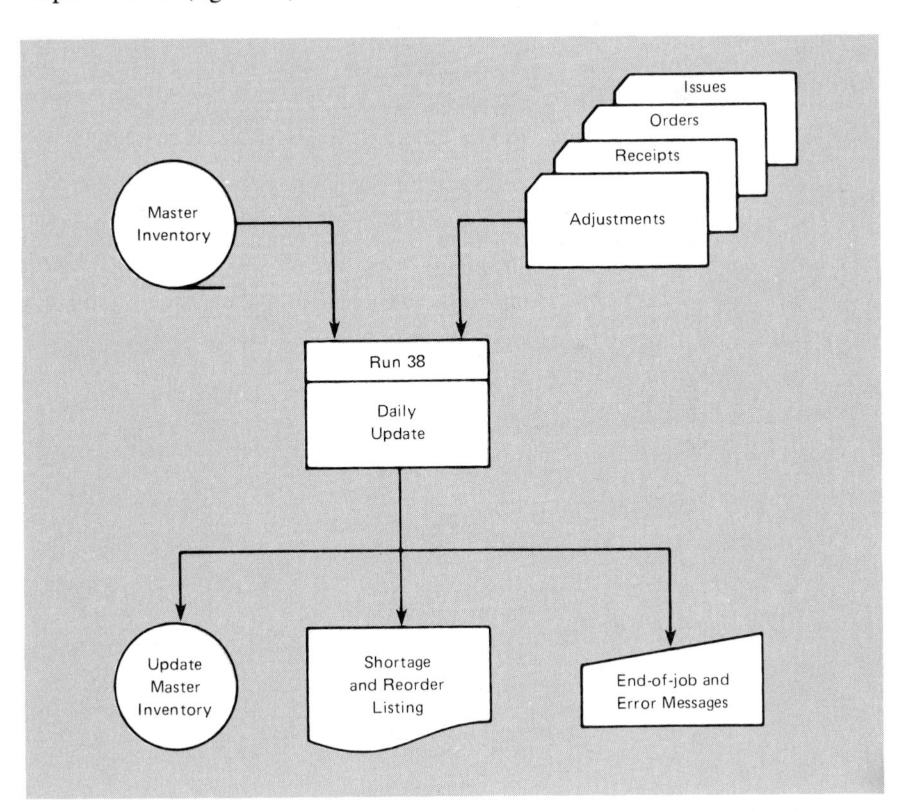

Figure 2.13 System Flowchart—Example.

Program Flowchart

A program flowchart is a graphic representation of the procedures by which data is to be processed. The chart provides a picture of the problem solution, the program logic used for coding, and the processing sequences. Specifically, it is a diagram of the operations and decisions to be made and the sequence in which they are to be performed by the machine. The major functions and sequences are shown, and if any detail is required, a *block diagram* is prepared.

The program flowchart shows the relationship of one part of the program to another. The flowchart can be used to experiment or verify the accuracy of different approaches to coding the application. Where laıge segments of the program are indicated, a single processing symbol may be used and the detail for the segment shown in a separate block diagram which would be used for the machine coding. Once the flowchart has been proven sound and the procedures developed, it may be used for coding the program. (See figures 2.14, 2.15.)

A program flowchart should provide:

1. A pictorial diagram of the problem solution to act as a map of the program.
2. A symbolic representation of the program logic used for coding, desk checking, and debugging while testing all aspects of the program.
3. Verification that all possible conditions have been considered and taken care of.
4. Documentation of the program, necessary to give an unquestionable historical reference record.
5. Aid in the development of programming and coding.

These are the important features and phases of program flowcharting:

1. It provides the programmer with a means of visualizing the entire program during its development. The sequence, the arithmetic and logical operations, the input and outputs of the system, and the relationship of one part of the program to another are all indicated.

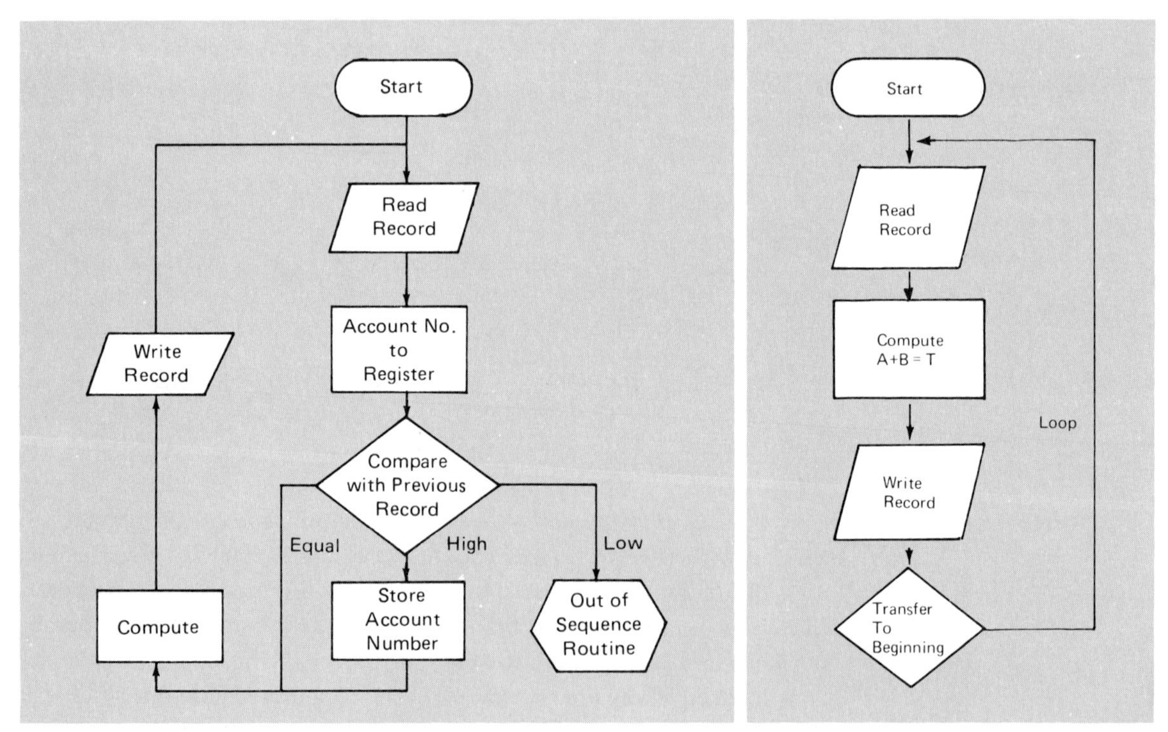

Figure 2.14 Program Flowchart—Sequence Checking.

Figure 2.15 Program Flowchart—Loop.

2. The system flowchart will provide the various inputs and outputs, the general objective of the program, and the general nature of the operation. A program flowchart will be prepared for each run and will serve as means of experimenting with the program specifically, in order to achieve the most efficient program.
3. Starting with symbols representing the major functions, the programmer must develop the overall logic by depicting blocks for input and output, identification decisions, etc.
4. After the overall logic has been developed by the programmer, he will extract the larger segments of the program and break them down into smaller, detailed block diagrams.
5. After the flowchart has been proven sound, the coding for the program will commence.
6. Upon completion of the coding, the program will be documented for further modification, which will always occur after the testing, installation, and operational stages.
7. Final documentation should involve the overall main logic, system flowcharts, program flowcharts, and the detailed block diagrams. The general system flowcharts help in the understanding of the more detailed program flowcharts.

The use of standard techniques for the preparation of flowcharts for data processing systems will greatly increase the effectiveness of the programmer's ability to convert the problem into a meaningful program. It will reduce the time necessary to program the applications and, if properly done, will provide a proper communication between the analyst and the many groups with whom he must deal.

Flowcharts are used extensively in the field and are the fundamental basis for all operations in data processing. A clear understanding of flowcharting techniques is a must for everyone who becomes involved with data processing at any level. (See figures 2.16, 2.17, 2.18.)

INTRODUCTION TO COBOL

COBOL is defined as COmmon Business Oriented Language. As such it is the result of the efforts of computer users in both industry and government to establish a language for programming business data processing applications. The committee was formed in 1959 (at the insistence of the Department of Defense) for the express purpose of producing a common business language that could be processed on the various computers without any reprogramming. The federal government, one of the largest users of data processing equipment, was being faced with the enormous task and expense of reprogramming each time a different type of computer was installed.

As data processing installations grew in size and complexity, it became apparent that a new programming tool was necessary, one in which the source language was the language of the business man. None of the existing compilers

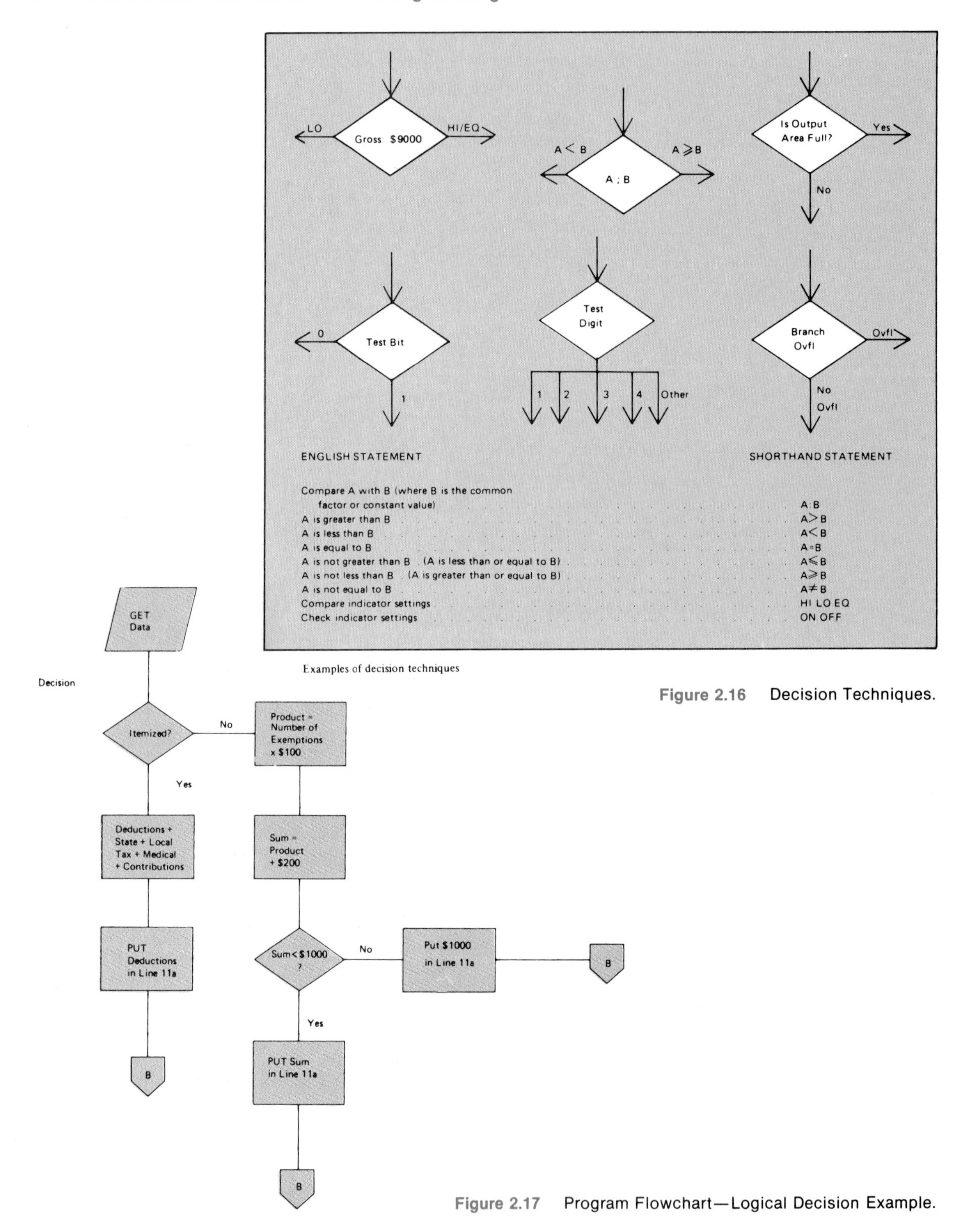

Figure 2.16 Decision Techniques.

Figure 2.17 Program Flowchart—Logical Decision Example.

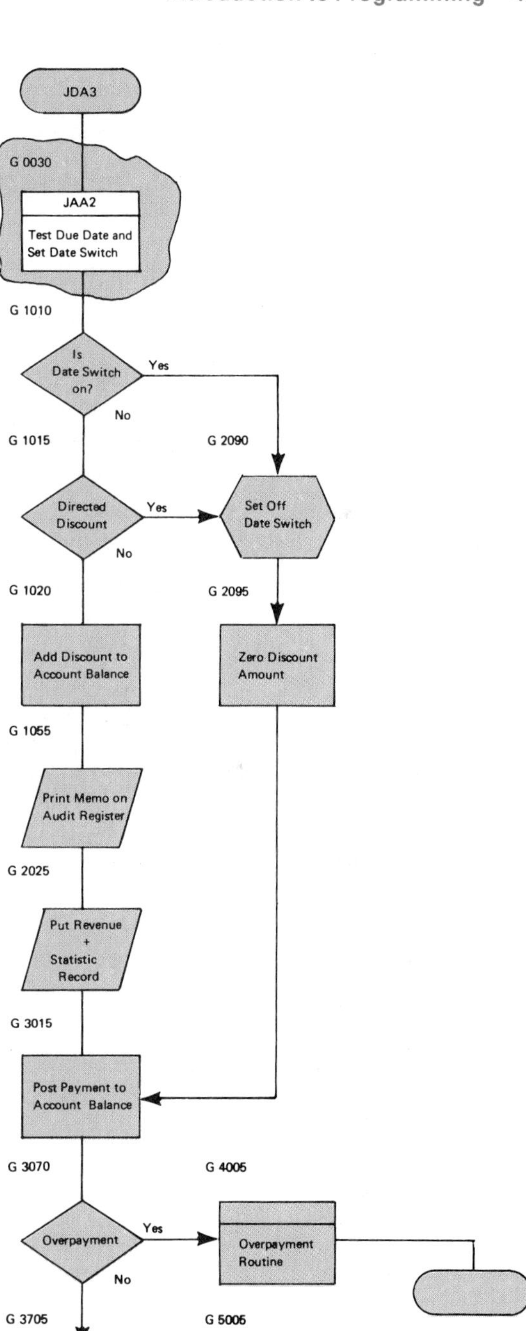

Figure 2.18 Program Flowchart—Example.

could be used since they were mathematical in nature and not geared for business applications. However, experience gained in the creation of the algebraic compilers pointed the way for the creation of the more complex data processing compilers.

The initial specifications for COBOL were presented in a report by the *CO*nference on *DA*ta *SY*stems *L*anguage (CODASYL) in April of 1960. The group consisted of computer professionals representing the United States Government, manufacturers of computer equipment, universities, and users. This group, confronted with the difficulty of program exchange among users of computer equipment, was inspired to meet the challenge presented by the situation. At the first meeting the conference agreed upon the development of a common language for the programming of commercial problems that was capable of continuous change and development. The proposed language would be problem oriented, machine independent, and would use a syntax closely resembling English-like statements, thus avoiding the use of special symbols as much as possible (fig. 2.19). The combined effort of the group was utilized to produce a business-oriented language that would permit a single expression of a program to be compiled on any computer then operative or contemplated in the future. This would reduce the reprogramming costs and provide a means for interchange of computer programs among the users.

COBOL is especially efficient in the processing of business problems. *Business data processing is characterized by the processing of many files, used repeatedly, requiring relatively few calculations and many output reports.* A payroll application is a good example of a business application owing to the fact that with a limited amount of input (time cards and personnel records, for example), many output reports are produced, such as updated personnel files, payroll checks, payroll registers, deduction reports, and various other internal and external reports (fig. 2.20). Business problems involve relatively little algebraic or logical processing; instead, they usually manipulate large files of similar records in a relatively simple manner. This means that COBOL emphasizes the description and handling of data items and input/output records.

The first special specifications for COBOL were written in 1960 and improved, refined, and standardized by subsequent meetings of the CODASYL committee. In 1970, a standard COBOL was approved by the American National Standards Institute (ANSI), an industry-wide association of computer users and manufacturers. This standard was called American National

Could a language be created for existing computers that could also be utilized on future computing systems?

Could a language be developed that would fit the rapidly changing and expanding requirements of management?

With the need to produce a large number of computer programs in a short period of time, could a language be developed that would permit existing programming staffs to be augmented with relatively inexperienced programmers?

Figure 2.19 Questions to Be Answered.

```
SUBTRACT DEDUCTIONS FROM GROSS GIVING NET-AMOUNT.

MULTIPLY UNITS BY LIST-PRICE GIVING BILLING-AMT.

IF ON-HAND IS LESS THAN MINIMUM-BALANCE GO TO
REORDER-ROUTINE.
```

Figure 2.20 Typical COBOL Statements.

Standards (ANS) COBOL. The latest 1974 ANS COBOL standard includes the following processing modules (differences between 1968 and 1974 ANS COBOL are found in the Appendix).

The *Nucleus* provides a basic language capability for the internal processing of data within the basic structure of the four divisions of a program.

The *Table Handling* module provides a capability for defining tables of contiguous data items and accessing an item relative to its position in the table. Language facility is provided for specifying how many times an item is to be repeated. Each item may be identified through use of a subscript or an index. Convenient facilities for table search are provided.

The *Sequential I-O* module provides a capability for accessing records of a file in established sequence. The sequence is established as a result of writing the records to the file. It also provides for the specification of rerun points and the sharing of memory areas among files. Each record is referred to by its physical position within a file.

The *Relative I-O* module provides a capability for accessing records of a mass storage file in either a random or sequential manner. Each record in a relative file is uniquely identified by an integer value greater than zero which specifies the record's logical ordinal position in the file.

The *Indexed I-O* module provides a capability for accessing records of a mass storage file in either a random or sequential manner. Each record in an indexed file is uniquely identified by the value of one or more keys within that record. Specifically, defined keys supplied by the programmers control successive references to the logical records.

The *Sort-Merge* module provides the capability needed to order one or more files of records, or to combine two or more identically ordered files of records according to a set of user-specified keys contained with each record. Optionally, a user may apply some special procedures to each of the individual records by input or output procedures. This special processing may be applied before and/or after the records are ordered by the SORT or after the records have been combined by the MERGE.

The *Report Writer* module provides the facility for producing reports by specifying the physical appearance of a report rather than by requiring specification of the detailed procedures necessary to produce that report.

A hierarchy of levels is used in defining the logical organization of a report. Each report is divided into report groups which are, in turn, divided into sequences of items. Such a hierarchial structure permits explicit reference to a report group with implicit reference to other levels in the hierarchy. The amount of Procedure Division coding is thus minimized.

The *Segmentation* module provides a capability for specifying object program overlay requirements. This allows a large problem program to be split into segments that can be overlaid in the Procedure Division at object time. These can be designated as permanent or overlayable core storage. This would assure the more efficient use of core storage at object time.

The *Library* module provides a capability for specifying text that is to be copied from a library. This supports the retrieval and updating of prewritten source program entries from the users library through the use of a COPY verb, for inclusion in a COBOL program at compile time. The effect of the compilation of the library text is the same as if the program were written at the same time as the source program, thereby eliminating the necessity for recopying an existing program definition.

The *Debug* module provides a means by which the user can describe his debugging algorithm, including the conditions under which data items or procedures are to be monitored during the execution of the object program.

The decisions as to what to monitor and what information to display on the output device are explicitly in the domain of the user. The COBOL debug facility simply provides a convenient access to pertinent information.

The *Inter-Program Communication* module provides the facility by which a program can communicate with one or more programs. This communication is provided by (a) the ability to transfer control from one program to another within a run unit, and (b) the ability for both programs to have access to the same data items.

The *Communication* module provides the capability for accessing, processing, and creating messages or portions thereof, and also provides the capability of communicating through a Message Control System with local and remote communication devices.

COBOL has emerged as the leading processing language in the business world and is enjoying a wide and popular acceptance in the data processing market.

POPULARITY

COBOL is a high-level computer language that is problem-oriented and relatively machine independent. COBOL was designed with the programmer in mind in that it frees him from the many machine-oriented instructions of other languages and allows him to concentrate on the logical aspect of a program. The program is written in an English-like syntax that looks and reads like ordinary business English. Organization of the language is simple in comparison to machine-oriented language—much easier to teach to new programmers, thus reducing training time (fig. 2.21).

The following are some of the reasons advanced for the popularity of COBOL today.

1. COBOL has been continuously standardized by repeated meetings of the CODASYL committee to improve the language and to guarantee its responsiveness to the data processing needs of the community.
2. COBOL has been designed to meet the needs of users today and in the future at decreasing costs to all concerned.
3. COBOL is the only language translator supported by the users, including the federal government.
4. COBOL users will be skeptical of any new equipment without COBOL

COMPATIBILITY—COBOL makes it possible for the first time to use the same program on different computers with a minimum of change. Reprogramming can be reduced to making minor modifications in the COBOL source program, and re-compiling for the new computer.

STANDARDIZATION—The standardization of a computer programming language overcomes the communication barrier which exists among programming language systems which are oriented to a single computer or a single family of computers.

COMMUNICATION—Easier communication between decision-making management, the systems analyst, the programmer, the coding technician, and the operator is established.

AUTOMATIC UNIFORM DOCUMENTATION—Easily understandable English documentation, provided automatically by the compiler, facilitates program analysis and thus simplifies any future modifications in the program.

COMPLETELY DEBUGGED PROGRAMS—Programs produced by the COBOL compiler are free from clerical errors.

CORRECTIONS AT ENGLISH LEVEL—Corrections and modifications in program logic may be made at the English level.

EASE OF TRAINING—New programming personnel can be trained to write productive programs with COBOL in substantially less time than it takes to train them in machine coding.

FASTER AND MORE ACCURATE PROGRAMMING—The English language notation expressed by the user and the computer-acceptable language produced by the COBOL computer ensure greater programming accuracy and a reduction in programming time.

REDUCTION IN PROGRAMMING COSTS—The ability to program a problem faster reduces the cost of programming. Also, reprogramming costs are greatly reduced since a program run on one system may be easily modified to run on another without being entirely recoded.

Figure 2.21
Benefits Derived from the Use of COBOL.

capabilities; therefore, it is incumbent upon the computer manufacturers to participate wholeheartedly in all technological progress in COBOL.

5. COBOL is the major data processing language available today. It is included in more software packages of computer manufacturers than any other language.
6. COBOL has proven that it is machine independent in that it can be processed through various computer configurations with the minimum of program change.
7. Although COBOL was primarily designed for commercial users, it has evolved as a highly sophisticated language in other areas of data processing.
8. COBOL is not plagued by computer obsolesence since it is constantly being revised to accommodate the newer computers.
9. COBOL has a self-documentary feature in that the English language statements are easily understood by managers and nonprogramming personnel.

After a decade of dedicated effort by a small group of data processing professionals, COBOL has emerged as the leading language in the data processing community. The continued voluntary efforts at standardization and technological improvements in the COBOL language will guarantee its responsiveness to the needs of information by management, and its ability to survive the ever-changing data processing field.

AS A COMMON LANGUAGE

A great deal of controversy arises when considering whether COBOL is truly a common language—that is, a programming language that can be compiled on any configuration of any computer. COBOL programs are written for computers of a certain minimum storage capacity. Certain small computers are thus excluded from the use of COBOL.

Can a programmer write a more efficient program in COBOL if he is familiar with the hardware? The answer is, of course, yes. He can take advantage of many programming approaches offered by the different computer manufacturers to reduce programming time and the number of storage positions required.

COBOL is not completely common as yet, but it is rapidly approaching this objective. It offers more commonality than any other processor presently in use. It is hoped that the continuing meetings of the CODASYL committee will make COBOL a more useful tool in the future.

The efficiency of COBOL has steadily increased, to the point whereby a COBOL program is more efficient than that of a new programmer who codes a program on a one-for-one basis in some assembly language for a particular machine. However, the COBOL program is not as efficient as an object program produced by an experienced programmer in the symbolic language of the individual computer.

ADVANTAGES

1. The principal advantage of the COBOL system is its advancement of communication. The ability to use English-like statements solves language difficulties that have often existed between the experienced programmer and decision-making management.
2. The program is written in the English language, thus removing the programmer from the individual machine or symbolic language instructions required in the program. Although a knowledge of the individual instructions (symbolic and machine) are not required in COBOL programming, that knowledge is very useful in the writing of an efficient program if the programmer possesses some knowledge of the hardware and coding of the particular computer.
3. Pretested modules of input and output are included in the COBOL processor which relieve the programmer of the tedious task of writing input and output specifications and testing them.
4. The programmer is writing in a language that is familiar to him, which reduces the documentation required since the chance for clerical error is diminished. Generally, the quality and the quantity of documentation provided by the COBOL compiler is far superior to that of other language processors. The printed output resulting from the compilation provides a highly desirable simplification of man-to-man and man-to-machine communication problems—a welcome improvement.
5. While COBOL is not completely machine-independent, a program written for one type of machine can be easily converted for use on another with minimum modification. The standardization of a COBOL program provides this benefit.

6. Because of the separate divisions in COBOL, a large program can be broken down into various segments, and each programmer may write one division. The format definition can be made available to all programmers engaged in the problem.
7. Nonprogrammers and managers can read the COBOL program in English, which provides them with the opportunity of judging the logic of the program.
8. During the compilation phase, the COBOL language processor generates a list of diagnostics. A diagnostic is a statement provided by the compiler that indicates all errors (except the logic errors) in a source program. Because diagnostics effect the measurement of compiling efficiency, these as well as compiling speed become important conditions for measuring the superior attributes of COBOL. This advantage derived from the attribute of COBOL can materially reduce the "debugging" time.

DISADVANTAGES

Most of the disadvantages result from the failure of personnel to fully understand the language and its use—as, for example:

1. The expectation that a single COBOL program will provide a permanent solution without ever reprogramming.
2. Assuming that the programmer need be taught only the COBOL language without any knowledge of the hardware or the operation of the computer.
3. COBOL will not generate a sophisticated program similar to one written in the actual language of the particular computer.
4. COBOL processors will operate only with a computer having a certain storage capacity. The newer COBOL compilers have drastically reduced the storage requirements. With the introduction of ever-larger storage units, this problem has been greatly reduced.

OBJECTIVES

1. To provide standardized elements in entry format that can be used on all computers regardless of make or model—a single common language that can be used by all.
2. To provide a source program that is easy to understand because it is written in the English language. Nonprogrammers can understand the logic of the program as well as the programmers can.
3. To provide a language that is oriented primarily toward commercial applications. Thus the opportunity is provided for business people to participate in the programming.

Although COBOL is oriented toward the problem rather than toward the particular machine, there are major differences in computers that have to be allowed for and adjusted to within the framework of the common language. These adjustments are usually minor, and the programmer with a COBOL knowledge can learn these on the job.

Because it uses English-language descriptions of application requirements, COBOL is especially designed for those who can best define their data processing needs. With a minimum of training and only a basic familiarity with the computing system as prerequisites, accountants, systems and procedures analysts, and many other members of operating management can use COBOL and the computer effectively.

Exercises

Write your answer in the space provided (answer may be one or more words).

1. The success of a computer program depends upon the ability of the programmer to ______________, ______________, ______________, and ______________.
2. Boundary conditions should be established so that the solution ______________ nor be ______________.
3. The relationship between ______________ and ______________ must be clearly shown.
4. The ______________ will show the orderly ______________ steps to arrive at the ______________ solution to the problem.
5. ______________ are coded in the particular ______________ language using the ______________ as a guide.
6. The program must be prepared on ______________ or ______________ before entry into the machine.
7. The ______________ must be thoroughly checked with the ______________ and all necessary ______________ accomplished before the program is ready for operation upon ______________.
8. Samples of all ______________ desired should be included with all ______________ indicated.
9. ______________ cards are necessary to load the ______________ and the ______________ into the computer.
10. A computer program is the outcome of a ______________ and the ______________.
11. ______________, ______________, ______________, and ______________ are the initial steps in the preparation of a program.
12. Housekeeping pertains to those operations that are used to clear ______________, ______________, and ______________, prior to the ______________ of the program.
13. A ______________ subroutine is stored in one location and connected to the program by means of linkage at one or more points in the program while an ______________ subroutine is inserted directly into the program where needed.
14. The steps involved in flowcharting are ______________ and ______________.
15. The flowchart is a ______________ representation of the flow of ______________ through a ______________ in which the ______________ is converted from the ______________ to the ______________.
16. The flowchart is the ______________ by which the ______________ travels through the entire ______________.
17. A ______________ flowchart represents the flow of data through all parts of a ______________ while in a ______________ flowchart the emphasis is on the ______________ and the ______________.
18. A system flowchart indicates the job to be done without ______________.
19. The system flowchart is usually prepared to present the ______________ of the system to ______________ and ______________.
20. The program flowchart is a ______________ representation of the ______________ by which data is to be processed.
21. The program flowchart is a diagram of the ______________ and the ______________ to be made and the ______________ in which they are to be performed by the machine.
22. A ______________ is a detailed breakdown of a program flowchart diagram.
23. The program flowchart can be used to ______________ and ______________ the accuracy of different ______________ to coding the application.
24. Once a program flowchart has been proven sound, it may be used to ______________ the program.
25. ______________ is necessary to give an unquestionable historical reference record to the program.
26. A program flowchart should provide ______________ that all possible conditions have been considered and taken care of.

27. The flowchart will provide a proper ______________ between the ______________ and the many groups he must deal with.
28. The use of flowcharts will greatly increase the ______________ of the ______________ ability to convert the ______________ into a meaningful solution.
29. COBOL is defined as ______________.
30. The initial specifications for COBOL was presented in a report by the ______________ committee in ______________.
31. COBOL is a ______________ oriented and machine ______________ programming language.
32. COBOL is especially efficient in the processing of ______________ problems.
33. Business data processing is characterized by the ______________, used repeatedly, ______________ and ______________.
34. The Standard COBOL adopted in 1970 is called ______________.
35. The ______________ provides the basic language capability for the internal processing of data within the basic structure of the four divisions of COBOL.
36. The Table Handling module provides a capability for defining ______________ of ______________ data items and accessing an item ______________ to its position in the ______________.
37. The ______________ module provides a capability to access records of a file in established sequence.
38. The Relative I-O module provides a capability to access records of a mass storage file in either a ______________ or ______________ manner.
39. COBOL is written in ______________ syntax.
40. COBOL is included in more ______________ packages of computer manufacturers than any other programming language.
41. COBOL is the only language translator supported by the ______________.
42. COBOL statements are easily understood by ______________ and ______________ personnel.
43. A programmer can write a more efficient program in COBOL if he is familiar with the ______________ of the particular computer.
44. Pretested modules of ______________ and ______________ are included in the COBOL processor.
45. A COBOL program written for a particular type of computer can be easily converted for use on another with ______________.
46. A ______________ is a statement provided by the compiler indicating all ______________ in a source program excluding the ______________ errors.
47. COBOL will not generate a ______________ program similar to one written in the ______________ language of the particular computer.
48. COBOL ______________ will operate only with computers having certain ______________ capacity.
49. ______________ can understand the logic of COBOL programs as well as the programmers themselves.
50. COBOL provides the opportunity for ______________ people to participate in the programming.

Answers

1. ANALYZE THE PROBLEM, PREPARE A PROGRAM, SOLVE PROBLEM, OPERATE THE PROGRAM
2. DOES NOT EXCEED OBJECTIVES OF MANAGEMENT, TOO NARROW TO ENCOMPASS ALL NECESSARY PROCEDURES
3. INPUT, OUTPUTS
4. FLOWCHARTS, LOGICAL, COMPUTER
5. INSTRUCTIONS, COMPUTER, PROGRAM FLOWCHART
6. PUNCHED CARDS, OTHER MEDIA
7. PROGRAM, TEST DATA, DEBUGGING, ACTUAL DATA
8. OUTPUTS, HEADINGS
9. JOB CONTROL, PROGRAM, DATA
10. PROGRAMMERS APPLIED KNOWLEDGE OF THE PROBLEM, OPERATION OF THE PARTICULAR COMPUTER
11. PROBLEM DEFINITION, ANALYSIS, DOCUMENTATION, FLOWCHARTING
12. STORAGE AREAS, REGISTERS, INDICATORS, EXECUTION
13. CLOSED, OPEN
14. THE PROCESSES TO BE PERFORMED, THE SEQUENCE OF THESE PROCESSES
15. GRAPHIC, INFORMATION, SYSTEM, DATA, SOURCE DOCUMENT, FINAL RESULT
16. ROADMAP, DATA, SYSTEM

17. SYSTEM, SYSTEM, PROGRAM, COMPUTER DECISIONS, PROCESSES
18. DETAILING THE STEPS INVOLVED
19. OVERALL PICTURE, ADMINISTRATIVE PERSONNEL, EXECUTIVES
20. GRAPHIC, PROCEDURES
21. OPERATIONS, DECISIONS, SEQUENCE
22. BLOCK DIAGRAM
23. EXPERIMENT, VERIFY, APPROACHES
24. CODE
25. DOCUMENTATION
26. VERIFICATION
27. COMMUNICATION, ANALYST
28. EFFECTIVENESS, PROGRAMMER'S, PROBLEM
29. COMMON BUSINESS ORIENTED LANGUAGE
30. CODASYL, APRIL 1960
31. PROBLEM, INDEPENDENT
32. BUSINESS
33. PROCESSING OF MANY FILES, REQUIRING FEW CALCULATIONS, MANY OUTPUT REPORTS
34. AMERICAN NATIONAL STANDARD COBOL
35. NUCLEUS
36. TABLES, CONTIGUOUS, RELATIVE, TABLE
37. SEQUENTIAL I-O
38. RANDOM, SEQUENTIAL
39. ENGLISHLIKE
40. SOFTWARE
41. USERS
42. MANAGERS, NONPROGRAMMING
43. HARDWARE
44. INPUT, OUTPUT
45. MINIMUM MODIFICATION
46. DIAGNOSTIC, ERRORS, LOGIC
47. SOPHISTICATED, ACTUAL
48. COMPILERS, STORAGE
49. NONPROGRAMMERS
50. BUSINESS

Questions for Review

1. Describe the steps involved in programming.
2. List the steps that are necessary to insure that the proper analysis and coding was made and that the computer will operate properly.
3. What are the important considerations in planning a program?
4. What is a flowchart and what is its importance?
5. Explain the two types of flowcharts widely used in data processing, and their purpose.
6. List the important points in program flowcharting.
7. Why was the federal government interested in developing a common business language?
8. When did the first CODASYL committee meet and what were their major objectives?
9. List and briefly describe the following processing modules of the American National Standard COBOL; Nucleus, Table Handling, Sequential I-O, Relative I-O, Indexed I-O, Sort-Merge, Report Writer, Segmentation, Library, Debug, Inter-Program Communication, and Communication.
10. List the main reasons for the popularity of COBOL.
11. What is a common language? Is COBOL a common language?
12. Why can a programmer write a more efficient COBOL program if he is familiar with the hardware of the particular computer?
13. Why has COBOL emerged as the leading programming language in the data processing community?
14. What are the main advantages of COBOL as a programming language?
15. Why isn't COBOL as efficient as a program written in a symbolic language tailored for a particular computer?

Problems

1. *Match each item with its proper description.*

______ 1. System Flowchart	A. Characterized by the processing of many files, used repeatedly, requiring relatively few calculations and many output reports.
______ 2. Nucleus	B. Load program and data into the computer.
______ 3. Open Subroutine	C. Detailed breakdown of program flowchart.
______ 4. Flowchart	D. Basic language capability for the internal processing of data within basic structures of the four divisions of COBOL.
______ 5. Test Results	E. Used to illustrate the overall objectives to personnel.
______ 6. Program Flowchart	F. Capability to access records of mass storage files either in a random or sequential manner.
______ 7. Relative I-O	G. Inserted into program where needed.
______ 8. Job Control Cards	H. Graphic representation of the procedures by which data is to be processed.
______ 9. Business Data Processing	I. Graphic flow of information through a system.
______10. Block Diagram	J. Output listing and/or cards used to test the accuracy of the program.

2. *Identify the following flowcharting symbols.*

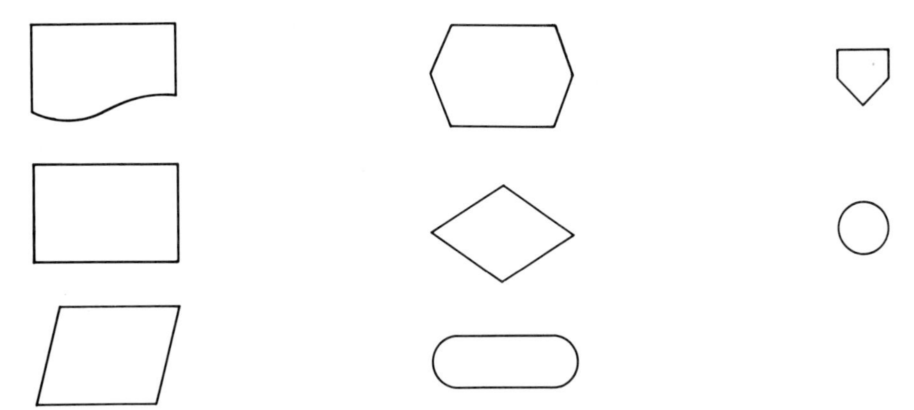

3. *In registering for classes, prepare a flowchart of the procedures and decisions necessary to enroll in the correct courses.*
4. *Given a file of records of students containing the following information:*

   Student ID Number
   Names
   Sex
   Age
   Class code: Freshman, sophomore, junior, senior.
   Grade point average.

   a. Prepare a program flowchart which will list the freshman female students between the ages of 18-20.
   b. Prepare a flowchart that will list all 21 year old junior students with a grade point average of (B) 3.0 or better.

5. *Prepare a system flowchart showing the following:*

Inputs	Magnetic tape	— Inventory file.
	Punched cards	— Transaction cards.
Outputs	Magnetic tape	— Updated Inventory file.
	Printer report	— Transaction register.

6. *In an Inventory file containing quantity, class of stock, stock number and amount, prepare a program flowchart that will*

   a. Print all items of the file.
   b. Accumulate all the amounts in class stock 34.
   c. Count the number of items in the file.

7. *Prepare a program flowchart to calculate FICA tax in a payroll procedure. Some of the information in the payroll record include the following:*

   a. Social Security Number.
   b. Employee Name.
   c. Accumulated Earnings—previous week.
   d. Current Weekly Earnings.

   Required:

   a. Read all records.
   b. Check to see if accumulated earnings exceed $9,000 this week or last week.
   c. Calculate FICA tax for
      (1) Employees who have not exceeded FICA limit this week or last week.
      (2) Employees who have reached FICA limit this week.
      (3) Employees who have reached FICA limit last week.

3 An Overview of COBOL Programming

The COmmon Business Oriented Language (COBOL) is a near-English programming language designed primarily for programming business applications on computers. It is described as near-English because its free form enables a programmer to write in such a way that the final result can be read easily and the general flow of logic understood by persons not as closely allied with the details of the problems as the programmer himself.

COBOL is similar to the English language in the use of words, sentences, and paragraphs. The programmer can use English words and conventional arithmetic symbols to direct and control the operations of a computer.

```
ADD QUANTITY TO ON-HAND.
MULTIPLY GROSS-EARN BY SS-RATE GIVING SS-TAX.
IF Y-T-D-EARN IS LESS THAN SS-LIMIT, GO TO SS-PROC.
```

Each of the above sentences is understandable by the computer, but they must first be translated into the particular machine language of the computer before the program can be executed (fig. 3.1). During the compilation stage, a special system program known as a compiler is first entered into the computer.

The COBOL system consists of two basic elements: the *source program,* which is a set of rules or instructions that carry out the logic of the particular data processing application; and the *compiler,* the intermediate routine that converts the English-like statements of COBOL into computer-acceptable instructions. Since the COBOL language is directed primarily at those unfamiliar with machine coding, terms common to business applications rather than to computing systems are used in the language.

THE COMPILER

Obviously, neither the computer nor the method of operating it is an end in itself. Rather, the purpose of a data processing system is to achieve, in the most efficient and economical way possible, solutions to the various applications that occur in the normal functioning of any business, educational, or governmental installation. With the advent of larger, more complex data processing systems, the burden of the programmer could conceivably increase to the point where problem solution becomes subordinate to the intricate methods of computer operation and direction. To preclude this possibility, innovations are constantly being made in an area that has come to be known as *software.*

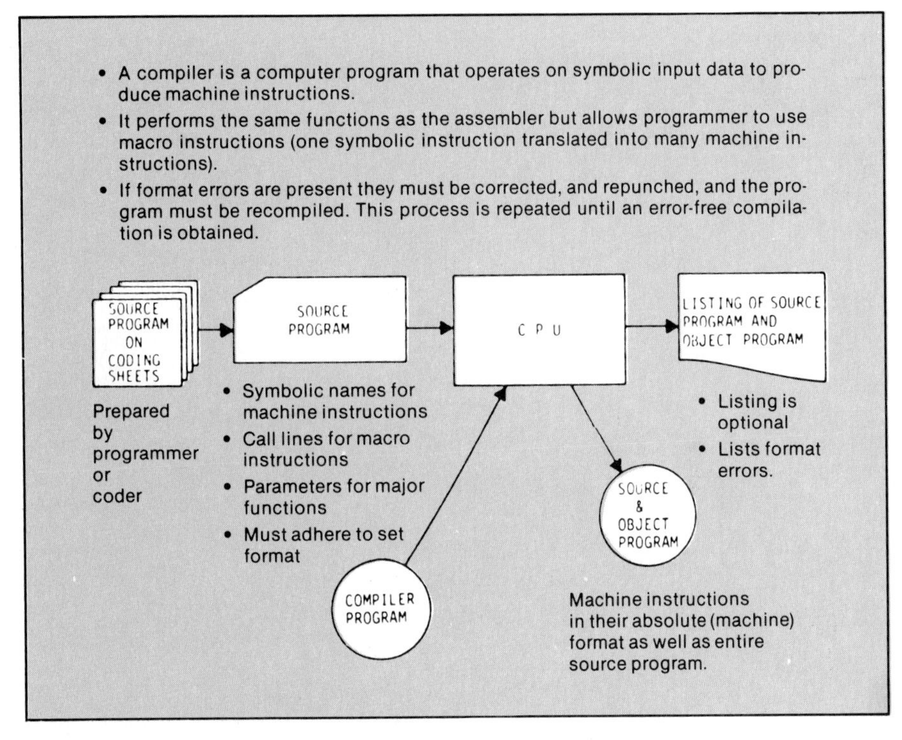

• A compiler is a computer program that operates on symbolic input data to produce machine instructions.
• It performs the same functions as the assembler but allows programmer to use macro instructions (one symbolic instruction translated into many machine instructions).
• If format errors are present they must be corrected, and repunched, and the program must be recompiled. This process is repeated until an error-free compilation is obtained.
SOURCE PROGRAM ON CODING SHEETS
SOURCE PROGRAM
C P U
LISTING OF SOURCE PROGRAM AND OBJECT PROGRAM
Prepared by programmer or coder
• Symbolic names for machine instructions
• Call lines for macro instructions
• Parameters for major functions
• Must adhere to set format
• Listing is optional
• Lists format errors.
SOURCE & OBJECT PROGRAM
COMPILER PROGRAM
Machine instructions in their absolute (machine) format as well as entire source program.

Figure 3.1 Compilation Process.

Precoded software programs, which are in large measure often considered extensions of hardware capabilities, free the programmer from exacting machine considerations and allow him to devote more time to the logic of the problem. These routines, which may vary from simple input-output and diagnostic routines to the more sophisticated routines that effect mass conversions of data being fed to the computer, are provided as special software packages with each data processing system. Certainly, the software packages effect a great saving in coding time and in problem preparation by performing many jobs normally undertaken by the programmer. Furthermore, because many sources of programming errors are removed, costly machine time is conserved.

The compiler, the most intricate of the software routines, is a master routine which takes a program, one written in some elementary form of problem statement (e.g., COBOL English-like statements), and translates it to instructions acceptable to the computer. The translated program is then fed back to the computer to be processed. The initial program that the compiler translates is called the *source program*. The machine-coded program produced from the translation is referred to as the *object program*.

The source program (COBOL symbolic program) is read into the computer and translated into a usable set of machine instructions (fig. 3.2). Thus the combination of COBOL reserved words and symbols are transformed into a machine language program (object program). This object program will be used to process the data at execution time to provide the desired outputs. The

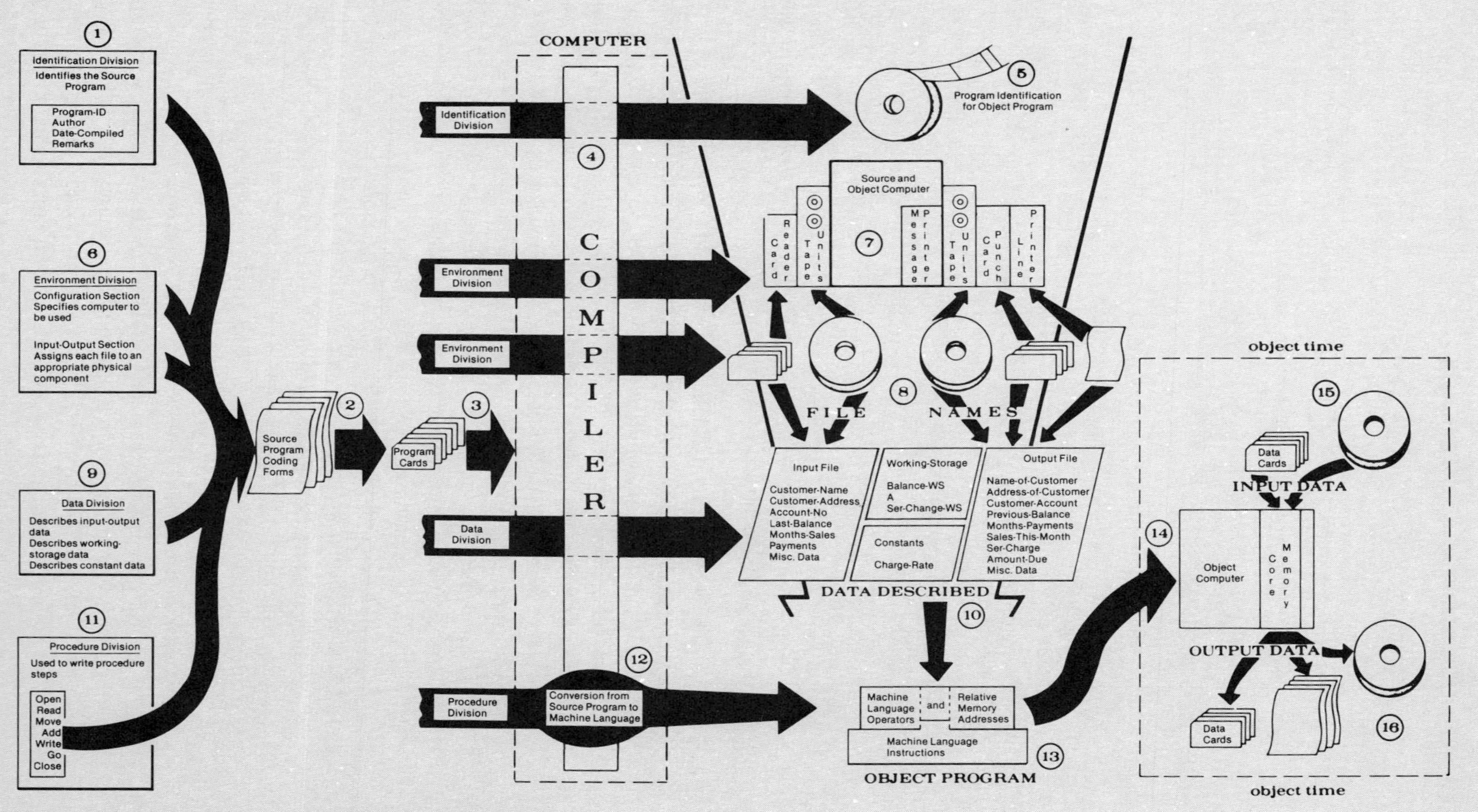

Figure 3.2 Compilation of a COBOL Source Program.

machine language program produced may be used at once or may be stored on an external medium where it may be called in when needed. This object program may be used repeatedly to process data without any further compiling.

The source program is used to specify the method of obtaining a solution to a particular data processing problem. There are four elements that must be provided in the source program, namely

1. The identification of the program,
2. The description of the equipment being used to process the data,
3. The description of the data to be processed, and
4. The set of procedures which determines how the data is to be processed.

The COBOL system has a separate division within the source program for each of these elements. The names of the four divisions are, Identification, Environment, Data, and Procedure Divisions.

DIVISIONS OF COBOL

Each of the four divisions must be placed in its proper sequence, begin with a division header, and abide by all the format rules for the particular division (fig. 3.3). The divisions listed in sequence with their main functions are:

Identification Division	Identifies the program to the computer.
Environment Division	Describes the computer to be used and the hardware features to be used in the program.
Data Division	Defines the characteristics of the data to be used, including the files, record layouts, and storage areas.
Procedure Division	Consists of a series of statements directing the processing of the data according to the program logic as expressed in the detailed program flowchart.

Identification Division

The Identification Division contains the information necessary to identify the program that is written and compiled. A unique data-name is assigned to the source program.

The Identification Division must appear first at compile time. The intended use of the division is to supply information to the reader. Usually it contains information as to when the program was written, by whom, and any security information relative to the program. The REMARKS paragraph usually contains the purpose of the program, a brief description of the processing to be performed, and the outputs produced. This information will serve as documentation for the program.

Environment Division

Although COBOL is to a large degree machine-independent, there are some aspects of programming that depend upon the particular computer to be used and the associated input and output devices. The Environment Division is the one division that is machine-dependent, since it contains the necessary infor-

IDENTIFICATION DIVISION. Provides all of the necessary documentation for the program such as:

the program name and number,
the programmer's name,
the system or application to which the program belongs,
the security restrictions on the use of the program,
a brief description of the processing performed and the output produced,
the dates on which the program was written and compiled.

ENVIRONMENT DIVISION.

CONFIGURATION SECTION.
SOURCE-COMPUTER. What computer will be used for compilation?
OBJECT-COMPUTER. What computer will be used for running the compiled object program?
SPECIAL-NAMES. What names have you assigned to the sense (alteration) switches and the channels of the paper tape loop on the printer?
INPUT-OUTPUT SECTION.
FILE-CONTROL. What name and hardware device have you assigned to each file used by the object program?

DATA DIVISION.

FILE SECTION. For each file named in the FILE-CONTROL paragraph above:
the file name,
the record name,
the layout of the record - the name, location, size, and format of each field.
WORKING-STORAGE SECTION. The size, format, and content of every counter, storage area, or constant value used by the program.

PROCEDURE DIVISION. The individual processing steps, written as COBOL-language statements. This division is divided into programmer-created paragraphs, each containing all of the procedure statements which constitute one particular routine.

Figure 3.3 The Four Divisions of a COBOL Program.

mation about the equipment that will be used to compile and execute the source program. To transfer a COBOL program from one computer to another, the Environment Division would have to be modified or even replaced to make the source program compatible with the new computer.

The Environment Division describes the hardware features of the source as well as the object computer. Each data file to be used in the program must be assigned to an input or output device. If special input or output techniques are to be used in the program, they have to be specified in this division. Any special-names assigned to hardware devices must be stipulated here.

Data Division

The Data Division describes the formats and the detailed characteristics of the input and output data to be processed by the object program. The programmer attaches unique names to the files, the records within the files, and the items within the records. All files that are named in the Environment Division must be described therein.

In addition to the file and record descriptions of data, work areas and constants to be used in the program must be described in the Working-Storage Section of the division.

The Environment Division describes the computer upon which the source program will be compiled and the computer that will be used to execute the object program, the Data Division describes the characteristics of the data, and the Procedure Division will describe the logical steps necessary to process the data.

Procedure Division

The Procedure Division specifies the actions expected of the object program to process the data to achieve the desired outputs. The division indicates the sequential order of the processing steps and also any alternate paths of actions where necessitated by decisions encountered during the processing.

This division is usually written from the program flowchart. The names of the data described in the Data Division are used to write sentences, employing program verbs to direct the computer to some action. The main types of action that may be specified are input and output, arithmetic, data transmission, and sequence control. All sentences are imperative even though they may be preceded by IF, since they direct the computer to perform some action. (See figure 3.4.)

Examples of Program Verbs

Input and output	OPEN, READ, WRITE, CLOSE, ACCEPT, DISPLAY
Arithmetic	ADD, SUBTRACT, MULTIPLY, DIVIDE, COMPUTE
Data Transmission	MOVE, EXAMINE
Sequence Control	GO TO, PERFORM, ALTER, STOP

These program verbs and other verbs are explained later in the text with the formats for each.

PROBLEM STATEMENT

A program is to be written to process data on an IBM 370 Model 155 with a I core size. The program is also to be compiled on this computer.] A

The I/O devices to be used are a 2540 card reader and a 2400 tape drive.] B

Two input files are to be used. The first file, an old master file on cards, contains customer number, name, address, balance, and the maximum balance ever carried for the customer.] C

The second input file is a transaction tape file. It contains only the customer number and the month's total purchase.] D

For each record in the old master file there is a record in the transaction file.

A new master file is to be created on tape (the output file). This file is to contain customer number, name, address, present balance, and maximum balance.] E

The new present balance is to be computed by adding the total from the transaction file to the balance in the old master file.] F

If this new balance is greater than the old maximum balance, the maximum balance is to be changed accordingly.] G

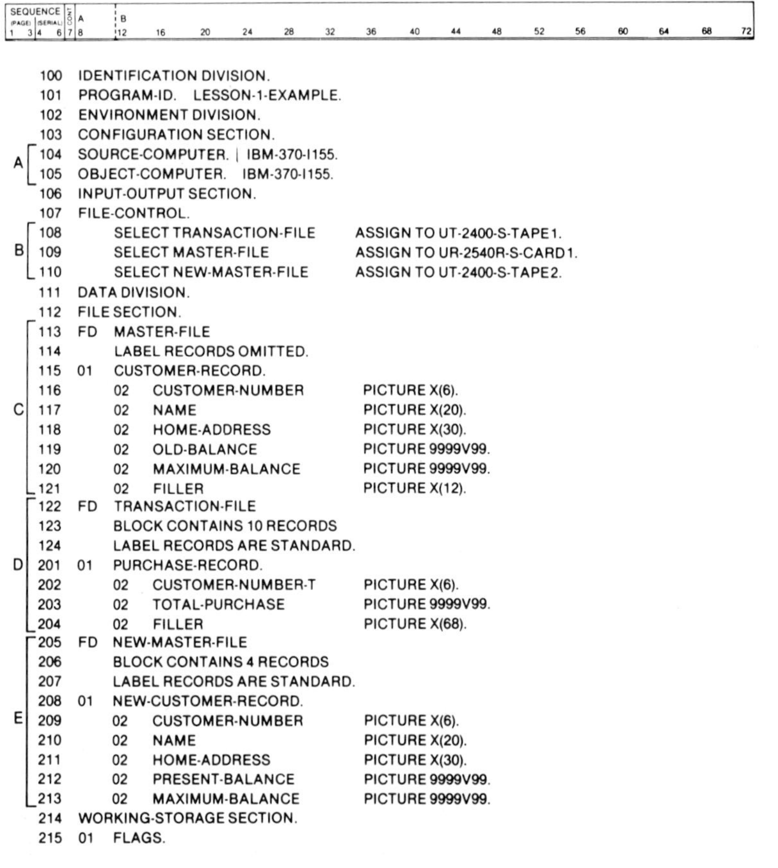

```
SEQUENCE (PAGE) (SERIAL) CONT.  A   B
1  3 4  6 7 8   12  16  20  24  28  32  36  40  44  48  52  56  60  64  68  72

     100  IDENTIFICATION DIVISION.
     101  PROGRAM-ID.   LESSON-1-EXAMPLE.
     102  ENVIRONMENT DIVISION.
     103  CONFIGURATION SECTION.
A  [ 104  SOURCE-COMPUTER. | IBM-370-I155.
   [ 105  OBJECT-COMPUTER.   IBM-370-I155.
     106  INPUT-OUTPUT SECTION.
     107  FILE-CONTROL.
   [ 108       SELECT TRANSACTION-FILE     ASSIGN TO UT-2400-S-TAPE1.
B  [ 109       SELECT MASTER-FILE          ASSIGN TO UR-2540R-S-CARD1.
   [ 110       SELECT NEW-MASTER-FILE      ASSIGN TO UT-2400-S-TAPE2.
     111  DATA DIVISION.
     112  FILE SECTION.
   [ 113  FD   MASTER-FILE
   [ 114       LABEL RECORDS OMITTED.
   [ 115  01   CUSTOMER-RECORD.
   [ 116       02   CUSTOMER-NUMBER         PICTURE X(6).
C  [ 117       02   NAME                    PICTURE X(20).
   [ 118       02   HOME-ADDRESS            PICTURE X(30).
   [ 119       02   OLD-BALANCE             PICTURE 9999V99.
   [ 120       02   MAXIMUM-BALANCE         PICTURE 9999V99.
   [ 121       02   FILLER                  PICTURE X(12).
   [ 122  FD   TRANSACTION-FILE
   [ 123       BLOCK CONTAINS 10 RECORDS
   [ 124       LABEL RECORDS ARE STANDARD.
D  [ 201  01   PURCHASE-RECORD.
   [ 202       02   CUSTOMER-NUMBER-T       PICTURE X(6).
   [ 203       02   TOTAL-PURCHASE          PICTURE 9999V99.
   [ 204       02   FILLER                  PICTURE X(68).
   [ 205  FD   NEW-MASTER-FILE
   [ 206       BLOCK CONTAINS 4 RECORDS
   [ 207       LABEL RECORDS ARE STANDARD.
   [ 208  01   NEW-CUSTOMER-RECORD.
E  [ 209       02   CUSTOMER-NUMBER         PICTURE X(6).
   [ 210       02   NAME                    PICTURE X(20).
   [ 211       02   HOME-ADDRESS            PICTURE X(30).
   [ 212       02   PRESENT-BALANCE         PICTURE 9999V99.
   [ 213       02   MAXIMUM-BALANCE         PICTURE 9999V99.
     214  WORKING-STORAGE SECTION.
     215  01   FLAGS.
```

Figure 3.4 Sample COBOL Program.

Figure 3.4 Continued

```
    216       02  MORE-DATA-FLAG              PICTURE XXX  VALUE 'YES'.
    217           88  MORE-DATA                            VALUE 'YES'.
    218           88  NO-MORE-DATA                         VALUE 'NO'.
    301  PROCEDURE DIVISION.
    302  MAIN-ROUTINE.
    303       OPEN  INPUT  MASTER-FILE,
    304                    TRANSACTION-FILE,
    305             OUTPUT NEW-MASTER-FILE.
    306       READ MASTER-FILE
    307            AT END MOVE 'NO' TO MORE-DATA-FLAG.
    308       PERFORM PROCESS-ROUTINE
    309            UNTIL NO-MORE-DATA.
    310       CLOSE  MASTER-FILE,
    311              TRANSACTION-FILE,
    312              NEW-MASTER-FILE.
    313       STOP RUN.
    314 *
    315  PROCESS-ROUTINE.
    316       MOVE CORRESPONDING CUSTOMER-RECORD TO
    317            NEW-CUSTOMER-RECORD.
 F┌ 318       ADD TOTAL-PURCHASE TO OLD-BALANCE GIVING
  └ 319            PRESENT-BALANCE.
  ┌ 320       IF PRESENT-BALANCE IS GREATER THAN MAXIMUM-BALANCE
  │ 321            OF NEW-CUSTOMER-RECORD,
  │ 322       THEN
 G│ 323            COMPUTE MAXIMUM-BALANCE OF
  │ 324                  NEW-CUSTOMER-RECORD = PRESENT-BALANCE
  │ 325       ELSE
  │ 326            NEXT SENTENCE.
  └ 327       WRITE NEW-MASTER-RECORD.
    328       READ MASTER-FILE
    329            AT END MOVE 'NO' TO MORE-DATA-FLAG.
```

WRITING THE FIRST COBOL PROGRAM

In the study of any programming language, it is important to write programs as soon as possible. COBOL is based on English; it uses English words and certain syntax rules derived from English. However, because it is a computer language, it is more precise than English. The programmer must, therefore, learn the rules that govern COBOL and follow them exactly. These rules are detailed later, beginning in the next chapter. Since COBOL source programs are written in an English-like language using paragraphs, sentences, statements, verbs, etc., it is possible to write a program using only the minimal requirements in each division. This should give one a general picture of how a COBOL program is put together and how it is compiled and executed.

Each of the divisions that constitute a COBOL source program (Identification, Environment, Data, and Procedure) is written in its stated order. Each division is placed in its logical sequence, each has its necessary logical function in the program, and each uses information developed in the divisions preceding it.

Coding Form

The COBOL coding form provides a standard method for writing COBOL source programs and aids the programmer in writing his source program for the subsequent punching of the program into cards (fig. 3.5). Only the minimum necessary entries to write a "file-free" program will be illustrated at this time.

The program sheet, despite its necessary restrictions, is relatively free-form. The programmer should note, however, that the rules for using it are precise and must be followed exactly.

COBOL PROGRAM SHEET

System		Punching Instructions		Sheet of
Program	Graphic		Card Form # *	Identification
Programmer	Date	Punch		73 80

SEQUENCE (PAGE) (SERIAL) 1 3 4 6 | CONT. 7 | A 8 | B 12 16 20 24 28 32 36 40 44 48 52 56 60 64 68 72

Figure 3.5 COBOL Coding Form.

The general format for the COBOL coding form is as follows:

Card Columns	Description
1–6	Represents the sequence number area. The sequence number has no effect in the source program and need not be written. The sequence number provides a check on the order of the cards in case they are scattered or misplaced.
7	is the continuation column. Explained in a later chapter.
8–11	*represent Area A } These columns are used for writing the COBOL source program.
12–72	represent Area B }
73–80	are used to identify the program.

*Area A, columns 8 through 11, is reserved for the beginning of division headers, section-names, paragraph-names, level indicators, and certain level numbers. All other entries start at Area B, columns 12 through 72.

To illustrate how a COBOL program is written, we will create a simplified procedure whereby we will pose a question and receive a response from the operator console. The program will *not* use files for input or output.

Identification Division

A name must be assigned to the program, presenting the information like this:

```
IDENTIFICATION DIVISION.
PROGRAM-ID. DISPLAY-ACCEPT-EXAMPLE.
```

PROGRAM-ID has informed the compiler that the unique name DISPLAY-ACCEPT-EXAMPLE was chosen for the program.

In addition to the name of the program, optionally, the name of the programmer, the date the program was written and other information that would serve to document the program could have been written in the Identification Division.

Environment Division

Although COBOL is to a large degree machine-independent, there are some aspects of any program that depend on a particular computer being used and on its associated input and output devices. In the Environment Division, the characteristics of the computer used may be identified.

In the Configuration Section, the source computer (the one that the compiler uses) and the object computer (the one that the object program uses) are described as follows:

```
ENVIRONMENT DIVISION.
CONFIGURATION SECTION.
SOURCE-COMPUTER. XEROX-530.
OBJECT-COMPUTER. XEROX-530.
```

These statements inform us that both computers will be the Xerox-530. If files were being used in the program, the location of each file referenced and how each would be used would have to be described. The identity of the files to be used in the program and the assignment of them to specific hardware devices would be specified in the Input-Output Section.

Data Division

The Data Division of the COBOL program gives a detailed description of all the data to be used in the program—whether to be read into the machine, used in intermediate processing, or written as output. Since no files will be used in the program, the File Section entry is not necessary. The Working-Storage Section must be constructed. This section describes the records and data items that are not part of files, but that are used during the processing of the object program.

For our simplified program, only one entry will be necessary. The Data Division would appear as follows:

```
DATA DIVISION.
WORKING-STORAGE SECTION.
77  RESPONSE   PICTURE X(40).
```

The level number 77 informs the compiler that RESPONSE is a noncontiguous data item—that this item has no relationship to any other data item described in the Working-Storage Section. RESPONSE is the unique data name, and PICTURE X(40) specifies that there are 40 alphameric characters written for the item. X represents any character in the COBOL character set and 40 represents the number of characters within the data item.

Procedure Division

The Procedure Division contains the instructions needed to solve the problem. To accomplish this, several types of COBOL statements are used. Since the program will be only using the operator console, the ACCEPT and DISPLAY statements will be the only ones used.

The ACCEPT statement with the FROM CONSOLE option enables the program to obtain information from the console operator during the running of the program. It allows the entry of low-volume data, up to one line of information, such as information needed to initialize program switches, balance totals, or serial numbers. It is not used to read files of data.

The DISPLAY statement with the UPON CONSOLE option enables low-volume data, up to one line of information, to be written on the console. It is used to write such information as exception records, control totals, or messages. It is not used to write files of data.

The Procedure Division would appear as follows:

```
PROCEDURE DIVISION.
INIT. DISPLAY 'WHAT DOES COBOL STAND FOR' UPON CONSOLE.
     ACCEPT RESPONSE FROM CONSOLE.
     DISPLAY 'THE CORRECT RESPONSE IS. . . . . .' UPON CONSOLE.
     DISPLAY 'CO-MMON B-USINESS O-RIENTED L-ANGUAGE'
          UPON CONSOLE.
     STOP RUN.
```

The first DISPLAY statement is used to type a one-line message to the operator in the form of an inquiry. Everything between the apostrophes (') will be printed.

The ACCEPT statement allows the entry of up to 40 columns of data from the operator as defined by the RESPONSE statement in the Working-Storage Section of the Data Division.

The last two DISPLAY statements will display two lines of messages on the console. There is an automatic space with each DISPLAY statement.

The last statement STOP RUN is used to end the job and turn control of the system back to the operating system.

Exercises

Write your answer in the space provided (answer may be one or more words).

1. COBOL is a ______________ programming language designed primarily for programming ______________ applications for computers.
2. COBOL is similar to the English language in the use of ______________, ______________, and ______________.
3. The ______________ is a set of instructions that carry out the logic of the particular data processing application.
4. The compiler is an ______________ routine that converts the ______________ statements of COBOL into ______________ instructions.
5. ______________ packages affect a great saving in coding time.
6. The initial program that the compiler translates is called the ______________.
7. The machine-coded program produced from the translation is known as the ______________.
8. The ______________ program will be used to process the data at execution time.
9. Every COBOL source program is divided into ______________ divisions.
10. Each division must be placed in its ______________ sequence.
11. The Identification Division identifies the ______________ to the ______________.
12. The Identification Division must appear ______________ at compile time.
13. The Remarks paragraph usually contains the ______________, a brief description of the ______________ and the ______________.
14. The Environment Division is ______________ dependent and contains the necessary information relative to the ______________ that will be used to ______________ and ______________ the program.

15. The Environment Division describes the ______________ features of the ______________ as well as the ______________ computers.
16. Any ______________ names assigned to ______________ devices must be stipulated in the Environment Division.
17. The Data Division describes the ______________ and detailed ______________ of the ______________ and ______________ data to be processed by the ______________ program.
18. All files that are named in the ______________ Division must be described in the ______________ Division.
19. In addition to ______________ and ______________ descriptions of data, work areas and constants to be used must be described in the ______________ section of the Data Division.
20. The Procedure Division specifies the ______________ expected of the ______________ program to process the ______________ to achieve the ______________.
21. The Procedure Division indicates the ______________ order of the ______________ steps and also any ______________ paths necessitated by decisions encountered during processing.
22. The Procedure Division is usually written from the ______________ flowchart.
23. The main types of actions that may be specified are ______________, ______________, ______________, and ______________.
24. The COBOL ______________ provides a standard method for writing COBOL ______________.
25. Columns ______________ of the coding form represent Area A.
26. Columns ______________ of the coding form represent Area B.
27. Area A is reserved for ______________, ______________, ______________, ______________, and certain ______________; all other entries start in ______________.
28. In the Configuration Section, the ______________ and ______________ computers are described.
29. The Working-Storage Section describes ______________ and ______________ items that are not part of ______________ but are used during the processing of the ______________.
30. The ACCEPT statement allows the entry of ______________ data, up to ______________.
31. The DISPLAY statement cannot be used to update ______________ of data.

Answers

1. NEAR-ENGLISH, BUSINESS
2. WORDS, SENTENCES, PARAGRAPHS
3. SOURCE PROGRAM
4. INTERMEDIATE, ENGLISHLIKE, COMPUTER-ACCEPTABLE
5. SOFTWARE
6. SOURCE PROGRAM
7. OBJECT PROGRAM
8. OBJECT
9. FOUR
10. PROPER
11. PROGRAM, COMPUTER
12. FIRST
13. PURPOSE OF THE PROGRAM, THE PROCESSING TO BE PERFORMED, OUTPUTS DESIRED
14. MACHINE, EQUIPMENT, COMPILE, EXECUTE
15. HARDWARE, SOURCE, OBJECT
16. SPECIAL, HARDWARE
17. FORMATS, CHARACTERISTICS, INPUT, OUTPUT, OBJECT
18. ENVIRONMENT, DATA
19. FILE, RECORD, WORKING-STORAGE
20. ACTIONS, OBJECT, DATA, DESIRED RESULTS
21. SEQUENTIAL, PROCESSING, ALTERNATE
22. PROGRAM
23. INPUT AND OUTPUT, ARITHMETIC, DATA TRANSMISSION, SEQUENCE CONTROL
24. CODING FORM, SOURCE PROGRAMS
25. 8–11
26. 12–72
27. DIVISION HEADERS, SECTION-NAMES, PARAGRAPH-NAMES, LEVEL INDICATORS, LEVEL NUMBERS, AREA B
28. SOURCE, OBJECT
29. RECORD, DATA, FILES, OBJECT PROGRAM
30. LOW-VOLUME, ONE LINE OF INFORMATION
31. FILES

Questions for Review

1. Briefly describe the main characteristics of the COBOL programming languages.
2. What are the two basic elements of the COBOL system and what are their main purposes?
3. Briefly explain the important role of software in COBOL programming.
4. What is a compiler?
5. What is a source program? An object program?
6. What are the four elements that must be provided for in a source program?
7. What are the main functions of each of the COBOL divisions?

8. What are four types of program verbs? Give examples of each.
9. Describe briefly the general format of the COBOL coding form.
10. Briefly explain the use of the ACCEPT statement with the FROM CONSOLE option.
11. Briefly explain the use of the DISPLAY statement with the UPON CONSOLE option.

Problems

1. *Identify the following:*

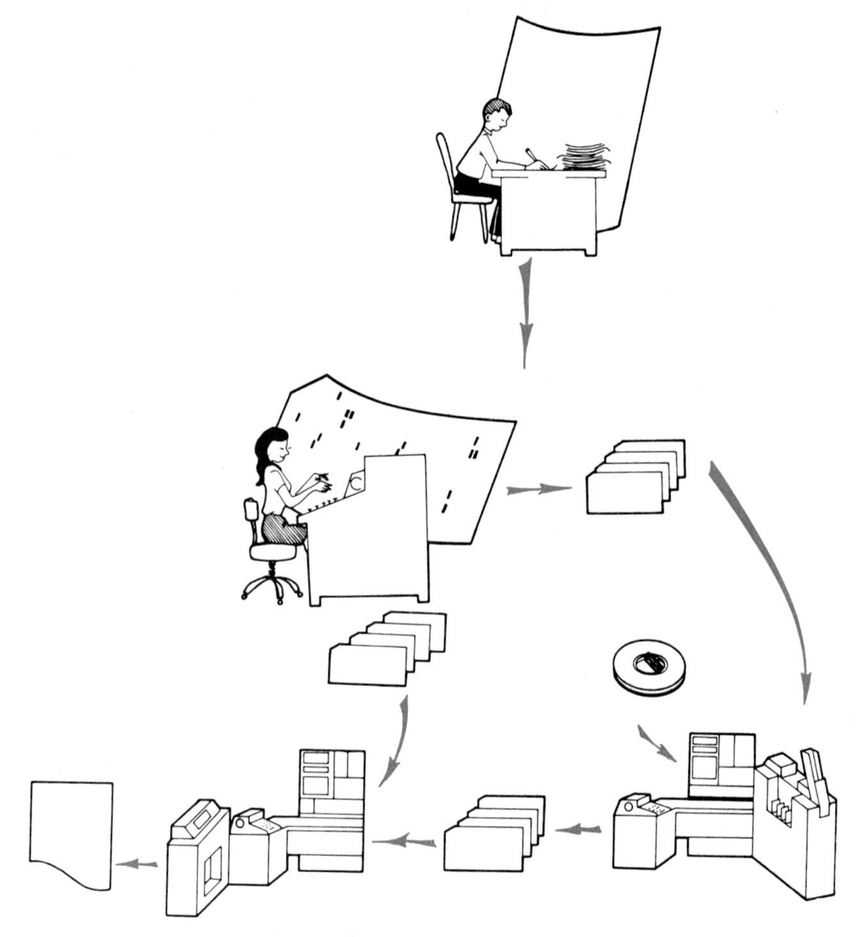

Compilation of a COBOL Source Program

2. *Match each item with its proper description.*

______ 1. MOVE	A.	Machine-coded program produced as a result of a translation of a source program.
______ 2. READ	B.	Near-English programming language designed primarily for programming business applications.
______ 3. Source Program	C.	Arithmetic verb.
______ 4. GO TO	D.	Intermediate routine that converts COBOL source statements into the computer-acceptable instructions.
______ 5. Software Packages	E.	Data transmission verb.
______ 6. COBOL	F.	Input and output verb.
______ 7. Object Program	G.	Set of instructions that carry out the logic of particular data processing applications.
______ 8. COMPUTE	H.	Perform many jobs undertaken by the programmer.
______ 9. Compiler	I.	Allows entry of low-volume data into the system.
______10. ACCEPT	J.	Sequence control verb.

3. *Number the following in their proper sequence of appearance in a COBOL source program.*

   ______a. DATA DIVISION.
   ______b. ENVIRONMENT DIVISION.
   ______c. PROCEDURE DIVISION.
   ______d. IDENTIFICATION DIVISION.

4. *Match each item with its description.*

______ 1. Procedure Division	A. Describes the computer to be used and the hardware features to be used in the program.
______ 2. Identification Division	B. Describes the characteristics of the information to be processed.
______ 3. Data Division	C. Identifies the program to the computer.
______ 4. Environment Division	D. Specifies the logical steps necessary to process the data.

5. *Identify each of the following types of program verbs as INPUT/OUTPUT, ARITHMETIC, DATA TRANSMISSION or SEQUENCE CONTROL.*

   ______________GO TO
   ______________MOVE
   ______________OPEN
   ______________ACCEPT
   ______________MULTIPLY
   ______________EXAMINE
   ______________ALTER
   ______________DISPLAY
   ______________COMPUTE
   ______________READ
   ______________STOP
   ______________SUBTRACT
   ______________PERFORM

6. *Write a simple procedure that will pose the question,*

   'WHAT ARE THE FOUR DIVISIONS OF COBOL'

   and receive a response from the operator console.
   Write messages on the console as follows:

   Line-1 'THE FOUR DIVISIONS OF COBOL ARE . . . .'
   Line-2 'IDENTIFICATION, ENVIRONMENT, DATA, AND PROCEDURE'

   The program should contain all four divisions of COBOL.

4 Writing COBOL Programs

COBOL PROGRAM SHEET FORMAT

The source program is written by the programmer on a COBOL Program Sheet Coding Form. The program sheet provides the programmer with a standard method of writing COBOL source programs (fig. 4.1). Despite the necessary restrictions, the program is written in rather free form. However, there are precise rules for using this form. Unless these rules are followed, especially in respect to spacing, many diagnostic errors will be generated unnecessarily.

The program sheet is so designed so that it can be readily keypunched. Each line on the form requires a separate card (fig. 4.2). The form provides for all columns of a card. Unnumbered boxes will not be punched.

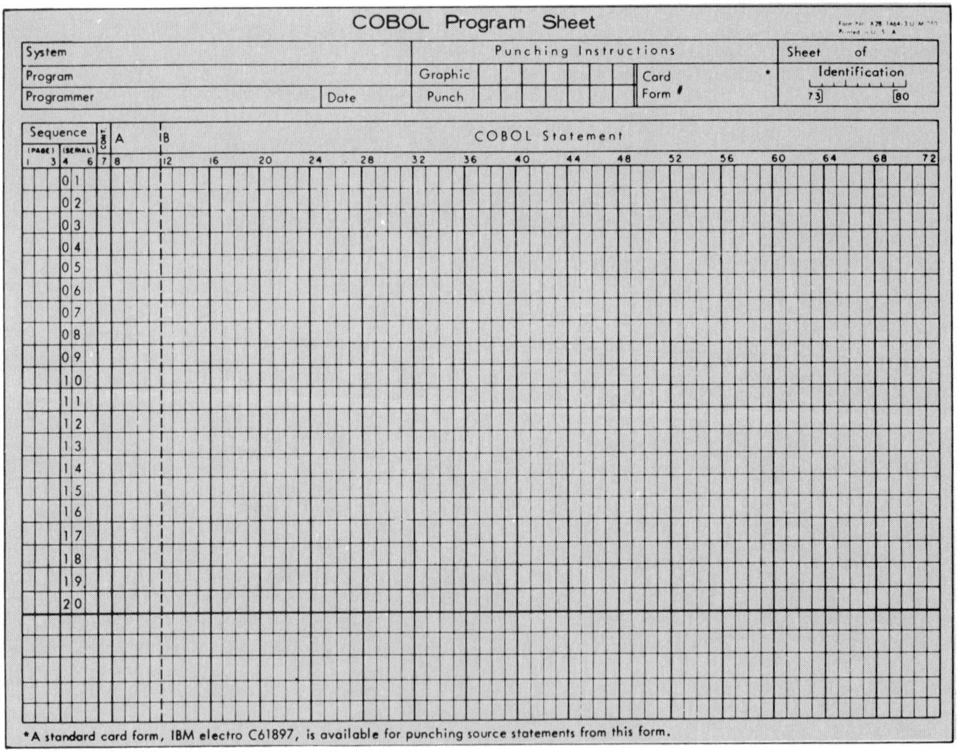

COBOL Program Sheet

System | Punching Instructions | Sheet of

Program | Graphic | Card Form * | Identification

Programmer | Date | Punch | 73 80

Sequence (PAGE) (SERIAL) 1 3 4 6 | CONT. 7 | A 8 | B 12 16 20 24 28 32 36 40 44 48 52 56 60 64 68 72 | COBOL Statement

01 02 03 04 05 06 07 08 09 10 11 12 13 14 15 16 17 18 19 20

*A standard card form, IBM electro C61897, is available for punching source statements from this form.

Figure 4.1
COBOL
Program Sheet.

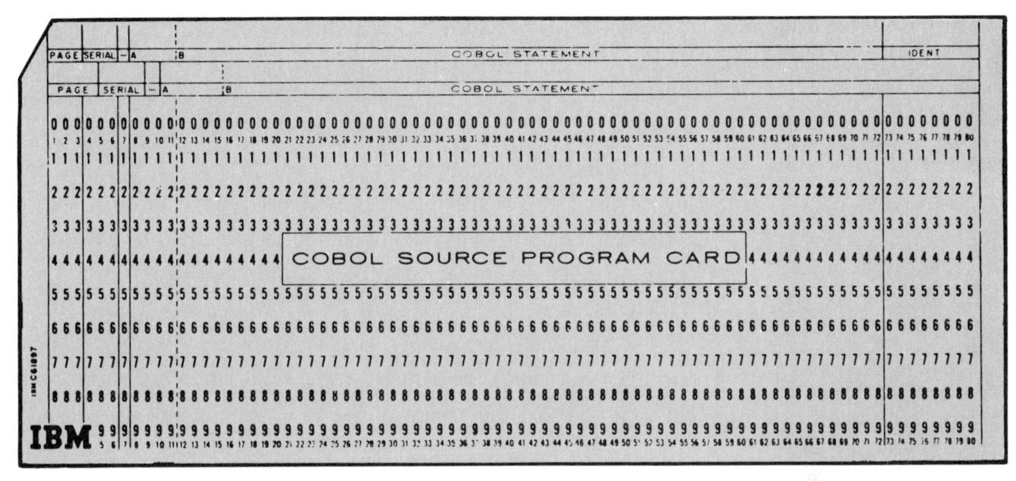

Figure 4.2 COBOL Program Card.

Care should be taken in the sequence of the instructions as the COBOL compiler will execute the program exactly as written. Punched cards from the program sheet serve as the initial input medium to the COBOL compiler. The compiler accepts the source program as written in the prescribed program sheet reference format and produces an output listing in the same format (fig. 4.3). Ample room in sequence numbers should be left for possible insertion of additional or "patch" instructions.

All characters should be written distinctly so that there aren't any questions as to the nature of the instructions. A keypunch operator will punch them exactly as written. Punctuation symbols, especially periods, can cause numerous errors when incorrectly used. Such characters as Z and 2, zero and O (letter), should be clearly indicated in the punching instructions of the form so that there will be no doubt as to what characters were intended.

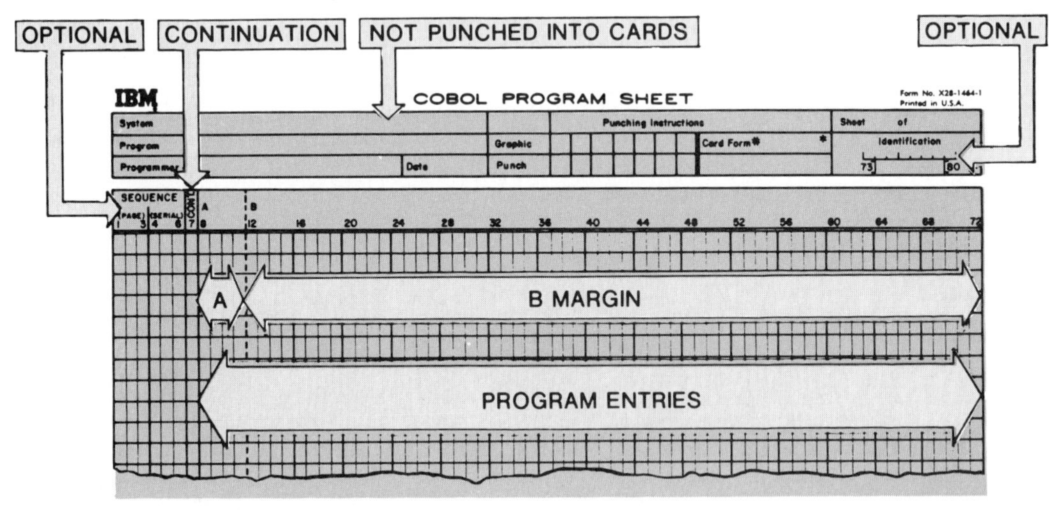

Figure 4.3 COBOL Program Sheet Entries.

Sequence Number (1–6)

The sequence numbers are written in columns 1 through 6 of the form.

The sequence number consisting of six digits is used to identify numerically each card to the COBOL compiler. The use of these sequence numbers is optional and has no effect on the object program. It is a good practice to use these sequence numbers, as they will provide a control on the sequence of the cards if the cards are scattered or if an insertion is to be made into the program. The single-character insertion permits one or more lines of coding to be inserted between existing lines. Column 6 normally contains a zero except where insertions are made.

If sequence numbers are used, they must be in ascending sequence. The compiler will check the sequence and indicate any sequence errors. No sequence check is made if the columns are left blank.

Continuation Indicator (7)

A coded statement may not extend beyond column 72 of the coding form. To continue a statement on a succeeding line, it is not necessary to use all spaces up to column 72 on the first line. Any excess spaces are disregarded by the compiler. If a statement must be continued on a succeeding line, the continuing word must begin at the B margin (column 12 or any column to the right of column 12). A continuation indicator is not necessary.

Words and Numeric Literals Continuation

When the splitting of a word or numeric literal is necessary, a hyphen is placed in column 7 of the continuation line to indicate that the first nonblank character follows the last nonblank character of the continued line without an intervening space (fig. 4.4). If the hyphen is omitted, the compiler will assume that the word or numeric literal was complete on the previous line and will insert an automatic blank character after it.

COBOL PROGRAM SHEET

System | Punching Instructions | Sheet of
Program | Graphic: N A N A N A | Card Form # | Identification
Programmer | Date | Punch: Ø 0 1 I 2 Z | 73 80

N-NUMERIC A-ALPHABETIC

SEQUENCE (PAGE) (SERIAL) 1 3 4 6 | CONT. 7 | A 8 | B 12 16 20 24 28 32 36 40 44 48 52 56 60 64 68 72

```
Ø27Ø5Ø A1. READ INVFILE AT END GO TO CLOSEOUT. MOVE INVENTORY-RECORD-
Ø27Ø6Ø-    MASTER TO NEW-MASTER.

      *THE CONTINUED COBOL WORD IS INVENTORY-RECORD-MASTER

Ø45Ø8Ø     ADD TOTAL-AMT1, TOTAL-AMT-2, GIVING NEW-TOTAL. MULTIPLY 741
Ø45Ø9Ø-    675.3 BY NEW GIVING NEW-AMT.

      *THE CONTINUED COBOL NUMERIC LITERAL IS 741675.3
```

Figure 4.4 Continued COBOL Word and Numeric Literal—Examples.

Nonnumeric Literal Continuation

Unlike the word or numeric literal, the nonnumeric literal, when continued, must be carried out to column 72, since spaces are considered part of the literal. To split a nonnumeric literal, the last character must be written in column 72, a hyphen placed in column 7 of the continuation (next) line, and a quotation mark placed in the B margin (Column 12 or anywhere to the right of column 12) and the literal continued. A final quotation mark at the end of the literal terminates the entry (figs. 4.5, 4.6).

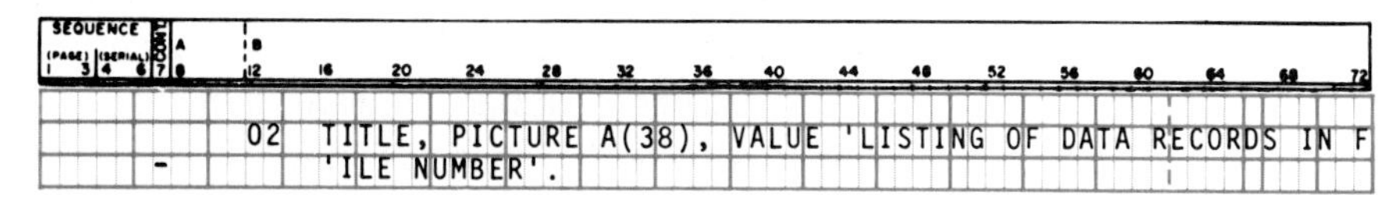

SEQUENCE
(PAGE) (SERIAL)
CONT.
A
B
02 TITLE, PICTURE A(38), VALUE 'LISTING OF DATA RECORDS IN F
- 'ILE NUMBER'.

Figure 4.5 Continuation Nonnumeric Literal.

```
       MULTIPLY AMOUNT OF CLIENT-PURCHASES BY TRADE-DISCOUNT, GIVING
```

```
      -REDUCTION.
        REDUCTION.
           REDUCTION.
      -        REDUCTION.
```

The third line is the only correct method to complete the entry. If an element ends in column 72, it is treated by the compiler as if it were followed by a space. Therefore, the continuation of such an entry may be written right at the beginning of the B-margin.

The first two lines show the continuation of the entry written in the A-margin, which is illegal. The first and fourth choices have hyphens in column 7, also illegal. The fourth line is correct, except for the hyphen in column 7.

```
       READ SERVICE-CALL; AT END, CLOSE SERVICE-CALLS-FILE, STOP RUN
```

The entry is incorrect.

The period used to end an entry *must not* be preceded by a space. Since the reserved word RUN ends in column 72, it is treated as if it were followed by a space—and that space precedes the period written on the next line. The simplest correction is to write the word RUN on the second line instead of the first, and write the period directly after it.

Figure 4.6 Continuation Entries—Examples.

Comment Line

An asterisk (*) in column 7 indicates that the entire line is a *comment line*. A comment line is printed on the compiler listing as documentation only. A comment line can appear as any line in the source program. Starting in column 8 of the comment line, any combination of characters from the COBOL character set may be used (fig. 4.7).

```
              A   B
016130            ADD QUANTITY TO TOTAL.
016140*             CALCULATE COMMISSION
016150            MOVE .015 TO RATE.
```

Figure 4.7 Comment Line—Example.

Program Statements (8–72)

These columns are used to write the source program entries, and are grouped as to margins. The *A* margin begins at column 8 and continues through column 11. Any information between these two columns is considered to be written at the A margin. The B margin begins at column 12 and continues through column 72. Any information written between these two columns is considered to be written at the B margin. A blank is assumed to appear in the column following column 72 except where the continuation indicator (hyphen in column 7 on succeeding line) is present. The assumed blank will terminate a word ending in column 72. If it is necessary to place a period beyond column 72 to terminate a sentence, a period should be written on the succeeding line at the B margin with a hyphen in column 7.

Rules for Margin Entries

1. The division header must begin at the A margin and be the first line in a division. The name of the division must be on a line by itself followed by a space then the word DIVISION and a period.
2. The section header must begin at the A margin followed by a space then the word SECTION and a period. If program segmentation is used, a space and priority number may follow the word SECTION; otherwise, no other text may appear on the same line as the section header except the USE or COPY sentence.
3. Paragraph headers must begin at the A margin and are followed by a period and a space. These headers need not be on a line by themselves. Statements may start on the same line at the B margin. Succeeding statements must be written starting at the B margin (fig. 4.8).
4. The level indicator FD (two-letter reserved word for file description) found in the Data Division must be written at the A margin, but the remainder of the file description entry must be written at the B margin (fig. 4.9).
5. Level numbers 01 and 77 found in the Data Division must be written at the A margin, but the remainder of the entry must begin at the B margin.
6. Special declarative headers in the Procedure Division, DECLARATIVES, and END DECLARATIVES must be written at the A margin on a line by themselves.

All other entries must be written at the B margin (except as noted in the succeeding statement).

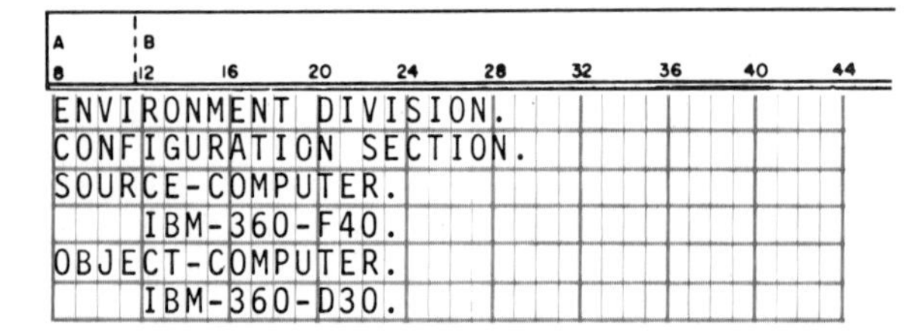
```
ENVIRONMENT DIVISION.
CONFIGURATION SECTION.
SOURCE-COMPUTER.
         IBM-360-F40.
OBJECT-COMPUTER.
         IBM-360-D30.
```

Figure 4.8 Environment Division—Sample Entries.

```
FD    SALES-MASTER-FILE,
      BLOCK CONTAINS 30 RECORDS,
      LABEL RECORDS ARE STANDARD,
      DATA RECORD IS SALES-MASTER-RECORD.
```

Figure 4.9 File Description Entry—Sample Entries.

Level numbers (01–49, 66, 77, 88) of the data description entries in the Data Division may begin at the A margin if so desired. Usually 01 and 77 level numbers, which are required to be written at the A margin, are written at the A margin with the other level numbers indented at the B margin to improve readability.

An entry that is required to start at the A margin must begin on a new line.

If an entry is too long to be completed on one line and must be continued on another line, the continuation of the entry is written at the B margin of the next line (see continuation indicator rules).

Good practice dictates the writing of short entries, leaving the remainder of a line blank. Individual statements on separate lines aid in the debugging of the COBOL program and increase the readability of the program. Insertions and corrections of programs are simplified by using short entry statements on single lines.

Identification Code (73–80)

These columns are used for the names of the program for identification purpose. Any character, including blanks, in the COBOL character set may be used. These codes are particularly useful in keeping programs separate from each other. A group of COBOL program cards could be sight-checked in colums 73–80 to assure that there are no other program cards in the deck. This code has no effect on the object deck or compilation, and may be left blank.

Blank Lines

A blank line is one that contains nothing but spaces from columns 7 through 72 inclusive. These blank lines are usually used to separate segments of a program and may appear anywhere in the source program, except immediately preceding a continuation line.

PROGRAM STRUCTURE

COBOL programs are arranged in a series of entries that comprise divisions, sections, and paragraphs. A division is composed of a series of sections, while a section is made up of paragraphs. Paragraphs are composed of a series of sentences containing statements (fig. 4.10).

Divisions

Four divisions are required in every COBOL program. They are Identification, Environment, Data, and Procedure. These divisions must always be written in the foregoing sequence. A fixed name header consisting of the division name followed by a space, the word DIVISION, and a period must appear on a line by itself.

Sections

All divisions do not necessarily contain sections. The Environment and Data Divisions always contain sections with fixed names. Sections are never found in the Identification Division, while the Procedure Division sections are optional and are created by the programmer if needed.

The beginning of each section is preceded by the name of the section followed by a space, the word SECTION, and a period. The header must appear on a line by itself unless otherwise noted. (See rules for margin entries.)

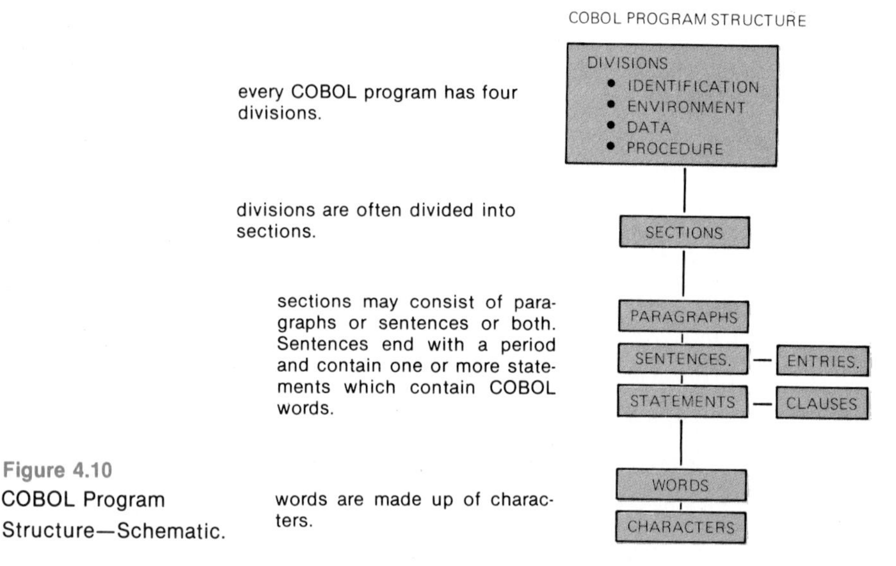
COBOL PROGRAM STRUCTURE
every COBOL program has four divisions.
DIVISIONS
• IDENTIFICATION
• ENVIRONMENT
• DATA
• PROCEDURE
divisions are often divided into sections.
SECTIONS
sections may consist of paragraphs or sentences or both. Sentences end with a period and contain one or more statements which contain COBOL words.
PARAGRAPHS
SENTENCES.
ENTRIES.
STATEMENTS
CLAUSES
words are made up of characters.
WORDS
CHARACTERS

Figure 4.10
COBOL Program Structure—Schematic.

Paragraphs

All divisions except the Data Division contain paragraphs. In the Identification and Environment Divisions, the paragraphs all have fixed names. The paragraph names in the Procedure Division are supplied by the programmer.

Each paragraph is identified by a paragraph name followed by a period and a space. Paragraph headers *do not* contain the word PARAGRAPH. The paragraph header need not appear on a line by itself; however, it must be the first entry, and can be followed on the same line by a series of entries.

A paragraph header entry consists of either a reserved word or a data-name and a period.

Entries

Entries (sentences) consist of a series of statements terminated by a period and a space. These statements must follow precise format rules as to sequence.

In order to write a COBOL program, the programmer should familiarize himself with the basic components of COBOL programming. There are many terms, rules, entry formats, and program structures to be learned before any attempt at COBOL programming is made (fig. 4.11).

TERMS

Source Program. The problem-solving program written in the COBOL language which will later be compiled and translated into the machine language of the particular computer.

Object Program. The machine language program that resulted from the compilation of the COBOL source program which will be used to process the data.

Compiler. A program supplied by a computer manufacturer that will translate the COBOL source program into the machine-language object program.

Source Computer. The computer that is used to compile the source program. Usually the same computer is used for the object computer.

Object Computer. The computer upon which the machine language program will be processed.

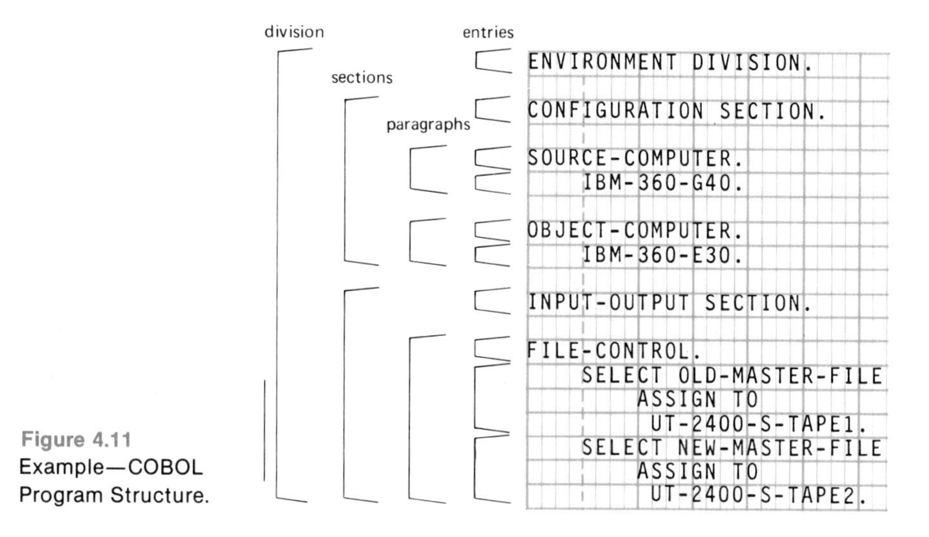

Figure 4.11 Example—COBOL Program Structure.

Character Set. The complete set of COBOL characters consists of fifty-one characters. These are the characters (alphabetic, numeric, and special characters or symbols) the manufacturer has included in the COBOL programming package (fig. 4.12).

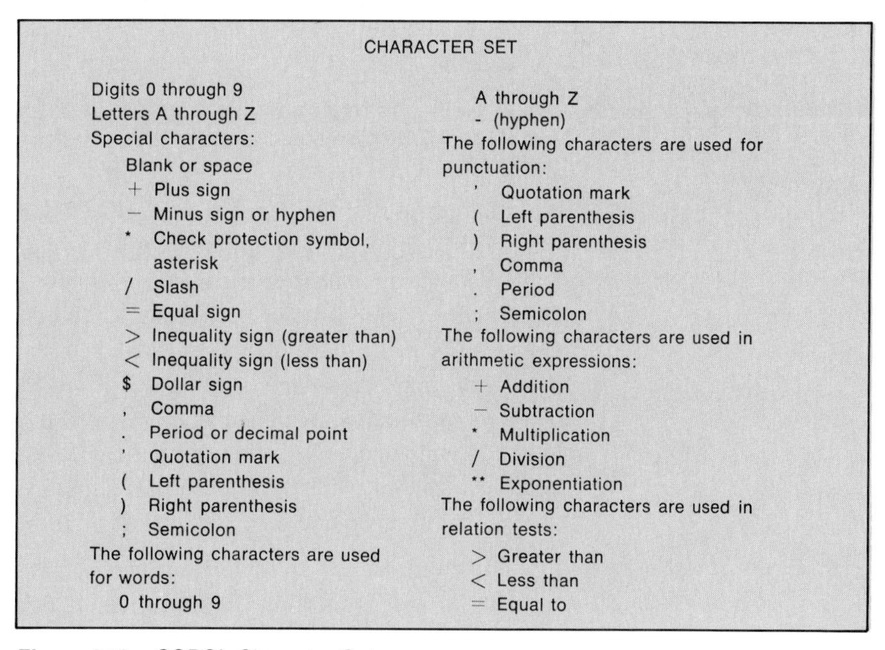

CHARACTER SET

Digits 0 through 9
Letters A through Z
Special characters:
- Blank or space
- + Plus sign
- − Minus sign or hyphen
- * Check protection symbol, asterisk
- / Slash
- = Equal sign
- > Inequality sign (greater than)
- < Inequality sign (less than)
- $ Dollar sign
- , Comma
- . Period or decimal point
- ' Quotation mark
- (Left parenthesis
-) Right parenthesis
- ; Semicolon

The following characters are used for words:
- 0 through 9
- A through Z
- - (hyphen)

The following characters are used for punctuation:
- ' Quotation mark
- (Left parenthesis
-) Right parenthesis
- , Comma
- . Period
- ; Semicolon

The following characters are used in arithmetic expressions:
- + Addition
- − Subtraction
- * Multiplication
- / Division
- ** Exponentiation

The following characters are used in relation tests:
- > Greater than
- < Less than
- = Equal to

Figure 4.12 COBOL Character Set.

NAMES

Names are a means of establishing words to identify certain data within a program. A symbolic name is attached to an item that is being used in the program. All reference to the item will be through that name, although the value may change many times throughout the execution of the program. The name must be unique or identified with the particular group of which it is a part.

Rules for the Assignment of Names

1. Names may range from one to thirty characters in length.
2. No spaces (blanks) may appear within a name.
3. Names may be formed from the alphabet, numerals, and the hyphen. No special characters may appear in a name except the hyphen.
4. Although the hyphen may appear in a name, no name may begin or end with a hyphen.
5. The procedure name may consist entirely of numerals, but all other names must have at least one alphabetic character.
6. Names which are identical must be qualified with a higher level name (see qualification rules).

Types of Names

Data-Names

A grouping of contiguous characters treated as data is called a *data item*. Data-names are words that are assigned by the programmer to identify data items in the COBOL program (fig. 4.13). For example, TRAN-AMOUNT could be the data-name of an item containing the amount of a transaction.

Data-names are devised and used by the programmer as needed in a program (fig. 4.14). The programmer must name and define each data item

DISBURSEMENTS
QUANTITY
RECORD-DATE
DOCUMENT-NUMBER
CLASS-STOCK
STOCK-NUMBER
UNIT-PRICE
AMOUNT
UNIT
DESCRIPTION

Figure 4.13 Data-Names—Examples.

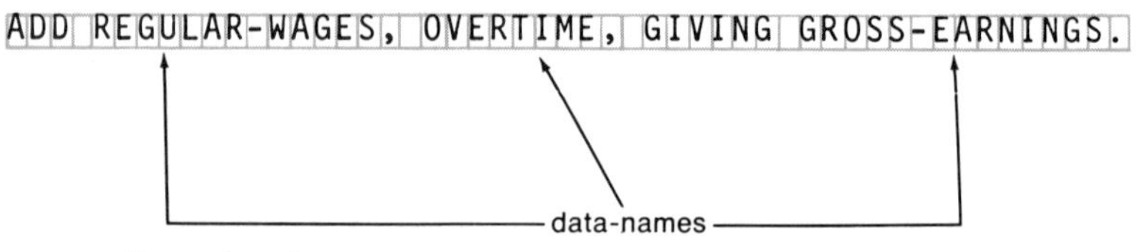

Figure 4.14 Example—Data-Names.

in a program so that the item can be accessed during the execution of the program. For example, if a program is written to process incoming accounts, one of the data items in the record would probably be the account number. The programmer assigns a name such as ACCOUNT-NUMBER to this data item and uses the name ACCOUNT-NUMBER each time the item is referred to.

If the same data-name is assigned to more than one data item then the data item must be made unique when referred to in the Procedure Division. A nonunique name must be followed by the necessary words that make it unique (see Qualification of Data-Names). All data items in the Data Division must be identified by a unique name or qualified data-names.

Procedure-Names

Procedure-names are symbolic names that are attached to the various segments of the Procedure Division (fig. 4.15). These names are used for reference by the program in a decision-making operation. The basic concept of computer programming is the ability of the program to leave the sequential order to another part of the program for further processing. A procedure-name may be either a paragraph-name or a section-name. The procedure-name may consist entirely of numerals (fig. 4.16).

```
B067.
PARAGRAPH-1.
456.
FICA-ROUTINE.
```

Figure 4.15 Procedure-Names—Examples.

```
PROCEDURE-NAME

INSERT-NEW-RECORD.
    MOVE CARD-RECORD TO OUTPUT-RECORD.
    WRITE OUTPUT-RECORD.
    READ CARD-FILE
        AT END MOVE 'NO' TO MORE-DATA-FLAG.
```

Figure 4.16 Example—Procedure-Name.

Condition-Names

Condition-names are assigned to an item that may have various values (fig. 4.17). The data item itself is called a condition variable and may assume a specific value, set of values, or a range of values. Condition-names are often used in the Procedure Division to specify certain conditions for branching to another part of the program.

Special-Names

Special-names are the mnemonic-names that are assigned to various components in the Environment Division (fig. 4.18). The term special-name refers to the mnemonic-name that is associated with a function-name. Function-names are fixed for each different type of computer. In the Procedure Division, the special-name can be written in place of the associated function-name in any format where substitution is valid.

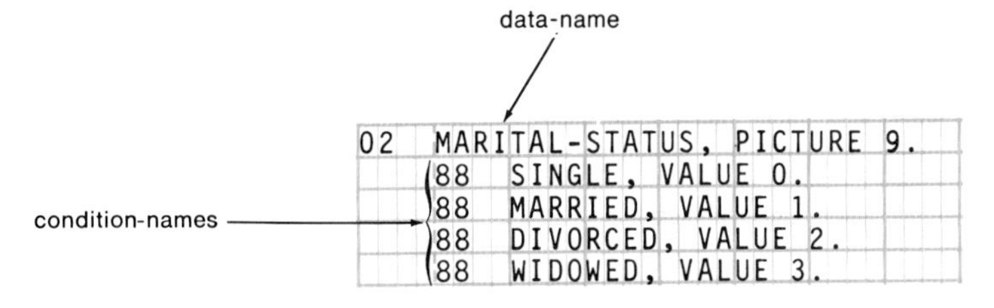

Figure 4.17 Example—Condition-Names.

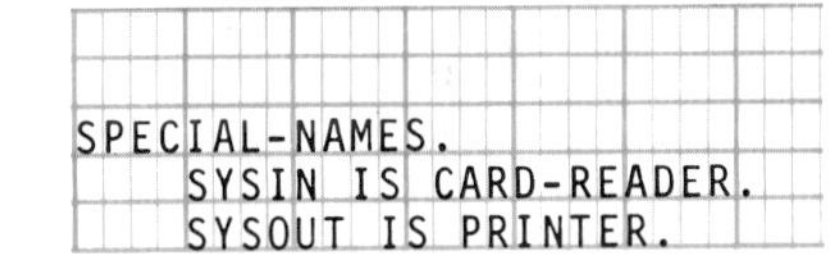

Figure 4.18 Example—Special-Names.

Qualification of Data-Names

Every data-name used in a COBOL program must be unique. Qualification is required where a single data-name has been used to name more than one item. Qualification is the process by which such a name is made unique (fig. 4.19). This is accomplished by placing a data-name or a paragraph-name, one or more phrases, each composed of the qualifier preceded by IN or OF (IN and OF are logically equivalent). Thus, if an item in MASTER-RECORD is called STOCK-NUMBER, and an item in DETAIL-RECORD is also called STOCK-NUMBER, then the data-name must appear in a qualified form, since the data-name STOCK-NUMBER is not unique. All references to this name must appear as STOCK-NUMBER OF MASTER-RECORD or STOCK-NUMBER IN DETAIL-RECORD, whichever name is intended.

Rules for Qualification of Data-Names

1. If any data-name, paragraph-name, or condition-name is assigned to more than one item in a program, it must be qualified whenever it is referred to in the Environment, Data, or Procedure Division.

```
01 OLD-MASTER                  01 NEW-MASTER
   02 CURRENT-DATE                02 CURRENT-DATE
      03 MONTH                       03 MONTH
      03 DAY                         03 DAY
      03 YEAR                        03 YEAR
```

The above example illustrates nonunique data-names. The names can only be referred to through qualification.

```
MOVE MONTH OF CURRENT-DATE IN NEW-MASTER . . .
MOVE MONTH IN CURRENT-DATE OF OLD-MASTER . . .
MOVE MONTH OF OLD-MASTER . . .
MOVE MONTH OF NEW-MASTER . . .
```

The above example illustrates a number of points.

1. Either connector (IN or OF) may be used.
2. The order in which the qualifiers appear must proceed from lower level to higher level.
3. It is not necessary to include all intermediate levels of qualification unless a data-name could not be uniquely determined if such intermediate levels were omitted.

Figure 4.19
Example—
Qualification of Names.

2. The name may be qualified by writing IN or OF after it, followed by the name of the group which contains the item being qualified.
3. A qualifier must be of a higher level of the group which contains the item being qualified.
4. The name must not appear at two levels in the same hierarchy in such a manner that it would appear to be qualifying itself.
5. The name of an item to which condition names have been assigned may be used to qualify any of its condition-names. For example, SINGLE OF STATUS is permitted where STATUS is the item and SINGLE is the condition-name associated with it.
6. Qualifications when not needed are permitted.
7. The highest-level qualifier that is permitted to qualify a data-name is a file-name.
8. No matter what qualification is available, a data-name cannot be the same as a procedure-name.
9. A procedure-name may be qualified only by a section-name. When this is done, SECTION must not appear as part of the qualifier.
10. A procedure-name must not be duplicated within the same section.
11. Qualifiers must not be subscripted, but the entire qualified name may be subscripted.
12. The highest-level qualifier must be a unique name.
13. No duplicate section names are allowed.
14. Similar data-names are not permitted where the data-name cannot be made unique by qualification.

WORDS

A COBOL word consists of one or more COBOL characters chosen from the character set. A word is followed by a space or by a period, right parenthesis, comma, or semicolon.

Reserved Words

Reserved words have preassigned meanings and must not be altered, misspelled, or changed in any manner from the specific purpose of the word. Each of these COBOL reserved words has a special meaning to the compiler; hence, it should not be used out of context. COBOL reserved words may appear in nonnumeric literals (enclosed in quotation marks). When appearing in this form, they lose their meanings as reserved words; therefore they violate no syntactical rules. (See figure 4.20.)

A list of reserved words is available for all computers and must be checked before attempting to program, since there are slight differences in the lists for different computers.

Interpretation of Words Used in COBOL Statements

The foregoing are used in describing the format of the COBOL statements and are not used in the actual programming of the statements. Following is the notation to be used in the remainder of the text to show the general forms of COBOL statements (fig. 4.21).

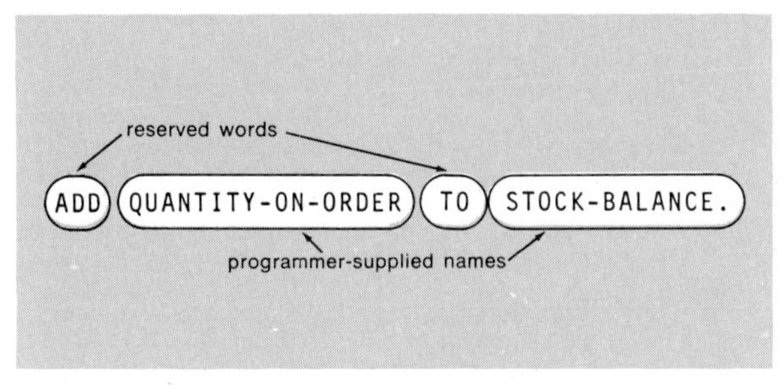

Figure 4.20 Example—Reserved Words and Programmer Supplied Names.

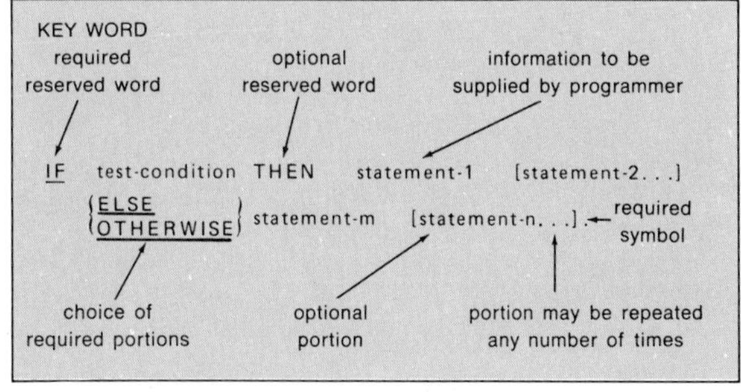

Figure 4.21 COBOL Words.

Reserved Words. Reserved words are printed entirely in capital letters.

Key Words. A key word is a reserved word that is required in a COBOL entry. The use of reserved words is essential to the meaning and structure of the COBOL statement. All key words are underlined and must be included in the program.

Optional Words. Reserved words become optional words in the format in which they appear if they are not underlined. These words appear at the user's option to improve the readability. The presence or absence of an optional word does not affect the compilation of the program. However, misspelling of an optional word or its replacement by another word is prohibited.

Lower Case Letter Words. Words printed in lower-case format represent information that must be supplied by the programmer.

Bracketed Words. [ ]. Words appearing in brackets must be included or omitted depending on the requirements of the program.

Braced Words { }. Braces enclosing words indicate that at least one of the words enclosed must be included.

Ellipsis (. . . . .). Ellipsis points immediately following the format indicate that the words may appear any number of times.

CONSTANT

A constant is an actual value of data that remains unchanged during the execution of the program. The value for the constant is supplied by the programmer at the time the program is loaded into storage.

There are two types of constants used in COBOL programming—literals and figurative constants. (See figure 4.22.)

Literals

Literals are composed of a string of characters wherein the value is determined by the set of characters of which the literal is a part. A literal may be named or unnamed (fig. 4.23). A named literal has a data-name assigned to it with a fixed value stipulated in the Working-Storage Section of the Data Division.

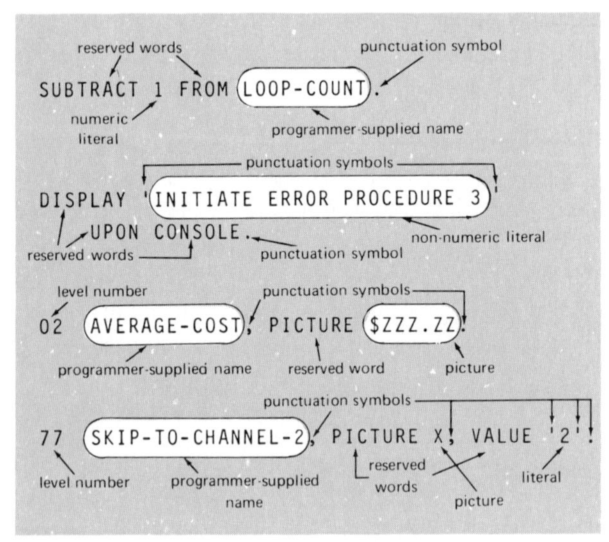

Figure 4.22 COBOL Symbols and Words.

Example 1. 77 FICA-RATE PICTURE V999 VALUE .052.
MULTIPLY GROSS BY FICA-RATE GIVING FICA-TAX.

Example 2. MULTIPLY GROSS BY .052 GIVING FICA-TAX.

In example 1, we are using a named numeric literal which must be described in the Working-Storage Section of the Data Division before it can be used in the Procedure Division.

In example 2, we are using an unnamed numeric literal which need not be described anywhere as it is used exactly as written.

One of the principal advantages of using a named numeric literal is that it can be changed easily. For example, if the rate of FICA-TAX changes, as it often does in a payroll application, all that need be changed is the description in the Working-Storage Section. If an unnamed numeric literal was used, it would have to be changed in every procedural statement that it appeared which may be rather cumbersome.

Figure 4.23 Examples—Named and Unnamed Literals.

An unnamed literal has the actual value specified at the time it is being used in the Procedure Division and does not require any separate definition in the Data Division.

There are two type of literals—numeric and nonnumeric.

Numeric Literal

A numeric literal is composed of a string of characters chosen from the digits 0–9, the plus or minus sign, and the decimal point. The value of the literal is implicit in the characters themselves. Thus, 842 is both the literal as well as its value. (See figure 4.24.)

Rules for Numeric Literals

1. A numeric literal may contain from 1 to 18 digits.
2. It may contain a sign only in the leftmost character position. If no sign is indicated, the compiler will assume the value to be positive. No space is permitted between the sign and the literal.
3. It may contain a decimal point anywhere in the literal except as the rightmost character. Integers may be written without decimal points. The decimal point is treated as an assumed decimal point.
4. It may contain only one sign and/or one decimal point.
5. It must not be enclosed in quotation marks.

(See figure 4.25.)

VALID NUMERIC LITERALS

−857394867.9842
+7583902.87
−.0006
5849245
205

INVALID NUMERIC LITERALS

2,678.56	Incorrect because it contains a comma.
− 294.84.	Incorrect because of the space between the minus sign and the first digit.
−30984378953592.87485	Incorrect because it contains too many digits.

Figure 4.24 Examples—Valid and Invalid Numeric Literals.

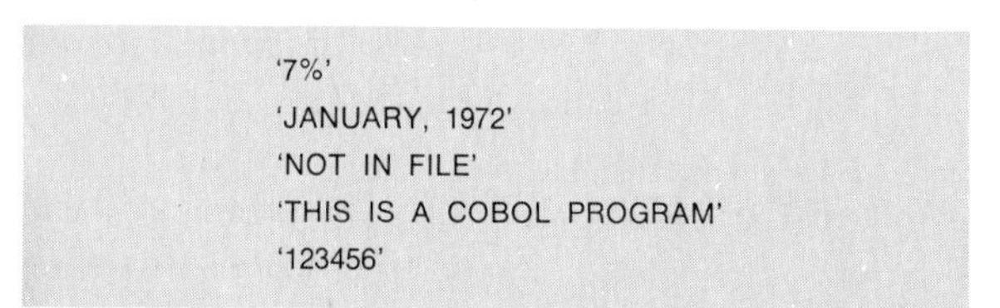

Figure 4.25 Examples—Usage of Numeric Literals.

'7%'
'JANUARY, 1972'
'NOT IN FILE'
'THIS IS A COBOL PROGRAM'
'123456'

Figure 4.26 Examples—Valid Nonnumeric Literals.

Nonnumeric Literal

A nonnumeric literal is composed of a string of any character in the computer's character set except the quotation marks (fig. 4.26). Non-COBOL characters may be included.

Rules for Nonnumeric Literals

1. A nonnumeric literal must be enclosed in quotation marks.
2. It can be used only for display purposes. The literal must not be used for computation. Only numeric literals may be used in computation.
3. It may contain from 1 to 120 characters.
4. It may contain any character in the character set of the particular computer, including blanks, special characters (except quote marks), and reserved words (fig. 4.27).

(*Note:* Signs and/or decimal points are not included in the size count of a numeric literal but are counted in the size of a nonnumeric literal. Quotation marks are not considered part of a nonnumeric literal and therefore are not included in the size count. A figurative constant may be used in place of a literal wherever a literal appears in the format.)

Figurative Constant

A figurative constant is a reserved word that has a predefined value recognized by the COBOL compiler (fig. 4.28). These words are frequently used in

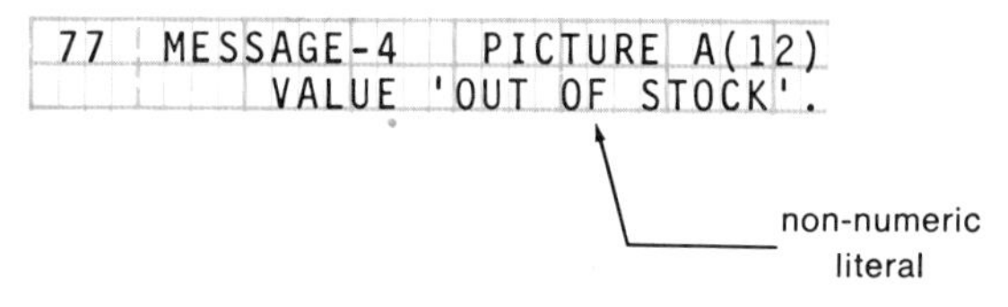

Figure 4.27 Example—Nonnumeric Literal.

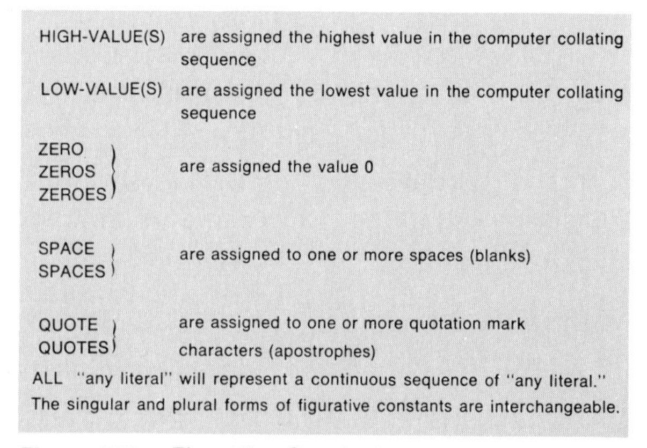

HIGH-VALUE(S)	are assigned the highest value in the computer collating sequence
LOW-VALUE(S)	are assigned the lowest value in the computer collating sequence
ZERO ZEROS ZEROES	are assigned the value 0
SPACE SPACES	are assigned to one or more spaces (blanks)
QUOTE QUOTES	are assigned to one or more quotation mark characters (apostrophes)

ALL "any literal" will represent a continuous sequence of "any literal."
The singular and plural forms of figurative constants are interchangeable.

Figure 4.28 Figurative Constants.

programming so that the programmer is relieved of the responsibility of assigning names for commonly used constants. A MOVE ZEROS TO WORK statement will fill the entire area WORK with zeros. Similarly MOVE SPACES TO OUT will blank out the entire area OUT. ZERO, ZEROS, ZEROES may be used interchangeably when singular or plural forms are desired (fig. 4.29). Usage of singular or plural forms does not affect the execution of the statement. They are only used to improve the readability of the statement.

OPERATORS

Operators are used in the COBOL language to specify some sort of action or relationship between items in the program. Symbols are special characters that have a specific meaning to the compiler. The type of operators and their symbolic forms are as follows:

Arithmetic Expression Operators

The characters used in arithmetic expressions are as shown in figure 4.30.

Assume that GRAND-TOTAL is defined as an 8 character data area. If we use the MOVE verb (to be explained fully in another section), which results in a transfer, the results will be:

COBOL STATEMENT	GRAND-TOTAL WILL CONTAIN
MOVE ZERO TO GRAND-TOTAL.	00000000
MOVE SPACES TO GRAND-TOTAL.	(all spaces)
MOVE QUOTE TO GRAND-TOTAL.	""""""""
MOVE ALL "9" TO GRAND-TOTAL.	99999999
MOVE ALL "ZERO" TO GRAND-TOTAL.	ZEROZERO

Figure 4.29 Example—Figurative Constants.

Meaning	Character	Name
ADDITION	+	Plus
SUBTRACTION	−	Minus
MULTIPLICATION	*	Asterisk (times)
DIVISION	/	Slash (divided by)
EXPONENTIATION	**	Double asterisk (raise to the power of)
EQUAL	=	Make equivalent to
PARENTHESIS	( )	To control sequence of calculations

Figure 4.30 Arithmetic Expression Operators and Meanings.

Sequence Rules for Processing Arithmetic Expressions

1. The innermost sets of imbedded parentheses are processed first, then the outermost pair.
2. Exponentiation of data is processed next.
3. Multiplication and division calculations are then processed from left to right.
4. Addition and subtraction calculations are then processed from left to right.

Arithmetic expression operators are used in the COMPUTE statement and in relational conditions (fig. 4.31). The uses and examples of these operators are discussed in greater detail in the Procedure Division section of the text.

Relational Expression Operators

The logical flow of a program frequently depends on the ability to make comparisons of the current value of the data-name and/or to compare this value with another or predetermined value. The expression can be reduced to a true or false statement. If the statement is true, the remainder of the statement is executed. If the statement is false, the program is directed to the next sentence unless an alternative action is specified. Figure 4.32 gives the symbols used in relational expressions together with their meanings.

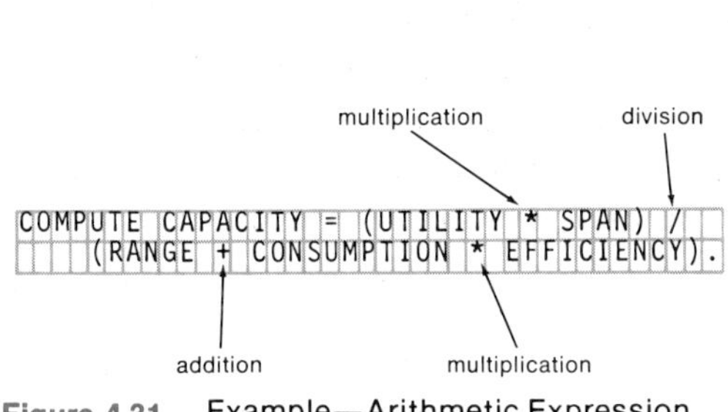
multiplication
division
COMPUTE CAPACITY = (UTILITY * SPAN) /
(RANGE + CONSUMPTION * EFFICIENCY).
addition
multiplication

Figure 4.31 Example—Arithmetic Expression Operators.

Meaning	Symbol	Name
EQUAL	=	"IS EQUAL TO"
GREATER THAN	>	"IS GREATER THAN"
LESS THAN	<	"IS LESS THAN"
PARENTHESIS	( )	To control sequence of statements to be evaluated.

```
IF PAYMENT < PREVIOUS-BAL GO TO PART-PAYMENT.
IF PAYMENT > PREVIOUS-BAL GO TO OVER-PAYMENT.
IF PAYMENT = PREVIOUS-BAL GO TO PROCESS.
```

Figure 4.32 Example—Relational Operators.

Figure 4.33 indicates how these symbols are used.

Relational expression operators may be used in place of their names in relation conditions. The uses and examples of these operators are discussed in greater detail in the Procedure Division section of the text.

Logical Expression Operators

These three logical operators (and, or, or not) are used to combine simple statements in the same expression for the purpose of testing the condition of the expression. The following are the operators used in logical operations together with their meanings. (See figure 4.34.)

```
IF PAYMENT < PREVIOUS-BAL
THEN
    PERFORM PART-PAYMENT
ELSE
    IF PAYMENT > PREVIOUS-BAL
    THEN
        PERFORM OVER-PAYMENT
    ELSE
        IF PAYMENT = PREVIOUS-BAL
        THEN
            PERFORM PROCESS
        ELSE
            NEXT SENTENCE.
```

Figure 4.33 Example—Relational Operators.

Operator	Meaning
AND	Used to evaluate both statements.
OR	Used to evaluate either or both statements.
NOT	Used to negate a positive condition.
( )	Used to control the sequence of enclosed statements.

```
IF A > B OR A = C AND D IS POSITIVE, PERFORM PROC-1.
```

Figure 4.34 Example—Logical Operators.

Sequence Rules for Processing Logical Expression

1. Parenthetical expressions are evaluated first, from the innermost pair to the outermost pair.
2. AND expressions are evaluated next, starting at the left of the expression and proceeding to the right.
3. OR expressions are evaluated last, starting at the left of the expression and proceeding to the right.

Logical expression operations are discussed in greater detail in the Procedure Division portion of the text.

Punctuation Symbols

Punctuation symbols are important to the successful execution of a COBOL program. Unless the correct usage of symbols are used, many diagnostic errors can be generated during the compilation phase. Figure 4.35 shows the symbols and their meanings as used to punctuate entries.

Rules for the Use of Punctuation Symbols

The following general rules apply to the use of punctuation in writing COBOL entries.

1. There must be at least one space between two successive words or parenthetical expressions. More than one space will not affect the execution of the program.
2. A period, comma, or semicolon, when used, must not be preceded by a space but must be followed by a space.
3. When any punctuation is indicated in the format, it is required in the program.
4. An arithmetic, relational, or logical operator must always be preceded and followed by a space.
5. Each sentence must be terminated by a period and a space.
6. Parentheses should be used to resolve potential ambiguity in a statement.
7. A left parenthesis must not be immediately followed by a space. A right parenthesis must not be immediately preceded by a space.
8. Semicolons or commas may be used to separate a series of clauses. Semicolons and commas are not required for correct COBOL execution of a program. They are included only to improve the readability of a program.
9. When a quotation mark (") double or (') single (permitted for IBM computers) is used, all entries between quotation marks are considered part of the entry, even including spaces.

STATEMENTS

A statement is a syntactically valid combination of words and symbols written in the Procedure Division used to express a thought in COBOL. The statement combines COBOL reserved words together with programmer-defined operands.

A COBOL statement may be either a simple or compound expression (figs. 4.36, 4.37). A *simple* statement would specify one action while a *compound* statement, usually joined by a logical operator, will specify more than one form of action.

The statement may be either imperative or conditional. An *imperative* statement directs the program to perform a particular operation under *all* conditions, while a *conditional* statement specifies the operation to be performed only if the condition is satisfied (true) or not (false).

Name	Symbol	Meaning
PERIOD	.	Used to terminate entries.
COMMA	,	Used to separate operands, clauses in a series of entries.
SEMICOLON	;	Used to separate clauses and statements.
QUOTATION MARK	'	Used to enclose nonnumeric literals.
PARENTHESIS	( )	Used to enclose subscripts.

Figure 4.35 Punctuation Symbols and Meanings.

```
PERFORM READ-RECORD.
```

Figure 4.36 Example—Simple Imperative Statement.

```
IF SCORE > 84
      AND SCORE < 93
THEN
      MOVE 'B' TO GRADE
ELSE
      PERFORM C-ROUTINE.

IF CONTRIBUTION = 100.00
      OR CONTRIBUTION > 100.00,
THEN
      DISPLAY 'GOOD SHOW' UPON CONSOLE
ELSE
      DISPLAY 'WE MUST TRY HARDER' UPON CONSOLE.
```

Figure 4.37 Example—Compound Conditional Statement.

WRITING A COBOL PROGRAM USING FILES

To illustrate how a COBOL program is written using files, a simplified procedure will be created that will list all records that have been read by a card reader and output on a line printer.

Identification Division

A name must be assigned to the program and, optionally, other information that will serve to document the program.

The Identification Division will appear as follows:

```
IDENTIFICATION DIVISION.
PROGRAM-ID. LISTING.
REMARKS. THIS PROGRAM WILL LIST ALL CARDS THAT ARE BEING READ.
```

PROGRAM-ID informs the compiler that the unique name LISTING was chosen for the program.

The REMARKS paragraph serves to state the purpose of the program. Other information, such as the name of the programmer, the date of the program, etc., may also be added.

Environment Division

The purpose of the Environment Division is to describe the hardware configuration of the compiling computer (source computer) and the computer on which the object program is run (object computer). It also describes the relationship between the files and input/output media.

The Environment Division will appear as follows:

```
ENVIRONMENT DIVISION.
CONFIGURATION SECTION.
SOURCE-COMPUTER. XEROX-530.
OBJECT-COMPUTER. XEROX-530.
INPUT-OUTPUT SECTION.
FILE-CONTROL.
    SELECT FILE-IN ASSIGN TO READER.
    SELECT FILE-OUT ASSIGN TO PRINTER.
```

The Configuration Section specifies that the computer to be used for the compilation and the computer to be used for the execution of the program will be one and the same—Xerox 530.

Next—the files to be used in the program must be identified and assigned to specific input/output devices. This is done in the Input-Output Section. The Input-Output Section gives the necessary information to control the transmission of data between the external media and the object program. The entry is written in the File-Control paragraph which associates the files with the external media. The Select clause is used to name each file in the user's COBOL program. File-In and File-Out are the file names supplied by the programmer. Every file name must have a File Description entry in the Data Division.

The Assign clause permits the files to be assigned to particular hardware devices. Thus File-In, the input file, will be associated with the card reader (*reader* is the implementor name for the card reader). File-Out, the output file, will be associated with the line printer (*printer* is the implementor name for the line printer).

Data Division

The Data Division describes data that the object program accepts as input in order to manipulate, create, or produce as output. Data falls into two categories:

External data, that is contained in files and enters or leaves the internal memory of the computer from specified areas.

Data developed or stored internally. This type of data is placed into storage within the computer's internal memory.

The Data Division is divided into three sections: File Section, Working-Storage Section, and Linkage Sections. (These sections will be discussed in greater detail further along in the text). In this problem, only the File Section will be used. The File Section defines the data stored on external files; all such data must be described here before it can be processed by the COBOL program. Each file is defined by a file description entry and one or more record description entries. Record description entries are written individually following the file description entry.

The Data Division will appear as follows:

```
DATA DIVISION.
FILE SECTION.
FD   FILE-IN
     LABEL RECORDS OMITTED.
01   CARD-IN                          PICTURE X(80).
FD   FILE-OUT
     LABEL RECORDS OMITTED.
01   RECORD-OUT.
     02  CARRIAGE-CONTROL             PICTURE X.
     02  PRINT-OUT                    PICTURE X(132).
```

The file description entry (FD) begins at Area A and is followed by at least one space. The name of the file is entered in Area B. The remainder of the file description entry is made up of independent clauses describing the file. The level indicator FD introduces File-In itself and informs the compiler that each entry written within File-In will be referred to as Card-In. Level number 01 identifies the record, Card-In. The level number must begin in Area A and the name of the record must begin in Area B. The Label Records clause is required in all FD entries. Since no labels are used in this program, the OMITTED option is used.

The second FD entry describes the output file, File-Out. The same rules apply as the first FD entry with the exception that Record-Out has a subordinate level. The concept of levels is a basic attribute of COBOL. The highest level is **FD**, the next highest level is 01. Level numbers 02–49 may subdivide the record, and the subdivisions themselves can be further subdivided if need be. *The smaller the subdivision, the larger the level number must be.*

Since the first print position is reserved for the carriage control character, no printing will take place in the first position. We wish to single space our listing, so we will leave our carriage control position blank. Later in the text, we will learn more about variable spacing. Level number 02 will begin in Area B with one or more intervening spaces followed by the data-name, Carriage-Control, and the Picture clause which indicates a length of one character. The 02 Print-Out indicates the actual print area of 132 positions for one line of printing.

Procedure Division

The Procedure Division contains the COBOL statements that give the computer step-by-step instructions for handling the data to be processed by the program. Procedures are logically successive instructions that process the data. Execution of the object program begins with the first statement in the procedures and continues in logical sequence. The end of the Procedure Division and the physical end of the COBOL source program is the physical position in the source program after which no procedures appear.

The Procedure Division will appear as follows:

```
PROCEDURE DIVISION.
INIT. OPEN INPUT FILE-IN OUTPUT FILE-OUT.
BEGIN. MOVE SPACES TO RECORD-OUT.
       READ FILE-IN AT END GO TO FINISH.
       MOVE CARD-IN TO PRINT-OUT.
       WRITE RECORD-OUT AFTER ADVANCING 1 LINE.
       GO TO BEGIN.
FINISH. CLOSE FILE-IN FILE-OUT.
        STOP RUN.
```

Each paragraph in the Procedure Division must have a procedure-name. INIT is designated as the name of the first paragraph.

The Open statement does the following: makes the records contained in File-In and File-Out available for processing; establishes a line of communication with each file; checks to make sure that each is available for use, brings the first record of File-In into special areas of internal storage known as buffers, and does other housekeeping chores. Each file must be defined as Input or Output or both (I-O). Each file named in the Open statement must have been described in the Data Division. An input file must be opened to enable the Read statement to obtain records for processing. The files can now be accessed.

The next statement MOVE SPACES TO RECORD-OUT will clear out the print area by moving blanks to the record in the output file. SPACES is a COBOL reserved word (explained in detail in the Figurative Constants section) that will be moved to the designated area.

The READ FILE-IN AT END GO TO FINISH statement makes the first record of the file, File-In available for processing. (Note that the AT END phrase is necessary in this sentence to branch to the FINISH paragraph which contains the necessary procedures taken when the end of the file is reached and the last record has been processed.)

The MOVE CARD-IN TO PRINT-OUT statement moves the necessary data from the input file to the output file ready for printing. The Move statement does not mean an actual physical movement of data. Instead it means that the data items from Card-In are copied into Print-Out. Data items within Card-In are not destroyed when a Move statement is executed and is still available for further processing if necessary.

The WRITE RECORD-OUT AFTER ADVANCING 1 LINE statement causes the record to be recorded on an output device specified for the file in the Environment Division; its format would be determined by the Data Division description of the file. The AFTER ADVANCING option indicates that the report is to be single-spaced. Notice that the format requires the name of a *record,* not that of a file. The file must have been opened before the record can be written into it.

The procedure-names provide a means for controlling the processing of successive items in File-In. If, for example, the processing is complete for the record and we wish to process the next record, control must be transferred to the procedure named BEGIN. Each procedure name indicates the beginning of a paragraph or a section within a program, and each indicates a reference point for programmer-specified transfer of control. When a procedure is entered, each logical successive instruction is processed in turn. The transfer to the procedure name BEGIN is accomplished through the use of the COBOL verb GO TO, which transfers control to the procedure indicated, as in the statement GO TO BEGIN. Processing will begin with the first sentence following the procedure name BEGIN.

The program must be terminated after all the records have been processed. In the Read statement, control of the program was directed to a procedure name FINISH after the last record had been read and processed. The FINISH paragraph consists of a few instructions necessary to terminate the program. The Close statement terminates the processing of the files and releases the storage areas that were used as buffers for those files. It is not necessary to stipulate whether the files were used for input or output or both. Any file that was previously opened must be closed before the run is stopped.

The STOP RUN statement terminates execution of the object program. Ending COBOL procedures are initiated, and control of the computer is returned to the operating system.

(*Note:* The above program is written in an "unstructured" form. Later in the text, you will learn to write programs, using the latest techniques of structured programming.)

Exercises

Write your answer in the space provided (answer may be one or more words).

1. The ______________ is written by a programmer on a COBOL Program Sheet Coding form.
2. Each line of the coding form must be punched on a ______________.
3. ______________ in sequence numbers should be allowed for the possible insertion of additional instructions.
4. The use of sequence numbers is ______________ and has no effect on the object program.
5. The compiler will check the ______________ of the program instructions.
6. A coded statement may not extend beyond column ______________ of the coding form.
7. If a statement must be continued on a succeeding line, it must begin at the ______________ margin.
8. A ______________ in column ______________ is used when a statement must be continued on a succeeding line.
9. The continuation indicator is used when it is necessary to ______________, ______________, or ______________.
10. An ______________ in column ______________ indicates that the entire line is a comment.
11. The A margin begins at column ______________ and continues through column ______________.
12. A ______________ is assumed to appear in the column following column ______________ unless a ______________ in column ______________ appears on the succeeding line.
13. A ______________, ______________, or ______________ header must begin at the ______________ margin followed by a ______________ and a ______________.
14. Level numbers ______________ and ______________ are required to begin at the A margin.
15. An entry that is required to start at the ______________ margin must begin on a new line.
16. Identification codes are punched in columns ______________ through ______________ and are particularly useful to keep programs separate from each other.
17. A ______________ is composed of a series of sections.
18. A section is composed of a series of ______________.
19. The machine coded program produced from the translation of a source program is known as the ______________.
20. ______________ are a means of establishing words to identify certain data within a program.
21. The different types of names used in COBOL programming are ______________, ______________, ______________, and ______________ names.
22. All items used in the Data Division must be identified by a ______________ or ______________ name.
23. A procedure name may either be a ______________ or ______________ name.
24. Names may range from ______________ to ______________ characters in length.
25. Names which are identical must be ______________ with a higher level name.
26. An identical name is made unique by preceding the qualifier by the words ______________ or ______________.
27. A ______________ name must not be duplicated within the same paragraph.
28. The highest level qualifier must be a ______________ name.
29. A COBOL word is followed by a ______________, ______________, ______________, ______________, or ______________.
30. Reserved words are printed entirely in ______________.
31. A ______________ word is a reserved word that is required in a COBOL entry.
32. An actual value of data that remains unchanged during the execution of a program is known as a ______________.

33. A numeric literal is composed of a string of characters chosen from ________________, ________________ or ________________, and the ________________.
34. A numeric literal may not exceed ________________ digits.
35. A numeric literal may contain only ________________ and/or ________________.
36. A nonnumeric literal must be enclosed in ________________.
37. A nonnumeric literal may contain ________________ characters in the character set of the particular computer and may be up to ________________ characters in length.
38. A ________________ is a reserved word that has a predefined value recognized by the COBOL compiler.
39. Operators are used in the COBOL language to specify some ________________ or ________________ in a program.
40. The types of operators used in COBOL programs are ________________, ________________, and ________________ operators.
41. Arithmetic expression operators are used in the ________________ statement and in ________________ conditions.
42. Relational expression operators reduce an expression to a ________________ or ________________ statement.
43. Logical expression operators are used to ________________ simple statements in the same expression for the purpose of ________________ the condition of the expression.
44. The incorrect use of ________________ can generate many diagnostic errors during the ________________ phase of the program.
45. Each operator must be preceded and followed by at least ________________.
46. Each COBOL sentence must be terminated by ________________ and ________________.
47. A ________________ is a syntactically valid combination of words and symbols written in the Procedure Division used to express a thought in COBOL.
48. A COBOL statement may be either a ________________ or ________________ expression usually joined by a ________________.
49. A statement that directs the program to perform a particular procedure under all conditions is known as a ________________ statement while a ________________ statement specifies the operation to be performed only if the condition is satisfied (true) or not (false).
50. The ________________ the subdivision, the ________________ the level number.
51. The ASSIGN clause permits ________________ to be assigned to particular ________________ devices.
52. The File Section defines data stored on ________________ files.
53. Each file is defined by a ________________ entry and one or more ________________ entries.
54. Each paragraph in the Procedure Division must have a ________________.

Answers

1. SOURCE PROGRAM
2. SEPARATE CARD
3. AMPLE ROOM
4. OPTIONAL
5. SEQUENCE
6. 72
7. B
8. HYPHEN, 7
9. SPLIT A WORD, CONTINUE A NUMERIC LITERAL, CONTINUE A NONNUMERIC LITERAL
10. ASTERISK, 7
11. 8, 11
12. BLANK, 72, HYPHEN, 7
13. DIVISION, SECTION, PARAGRAPH, A, PERIOD, SPACE
14. 01, 77
15. A
16. 73, 80
17. DIVISION
18. PARAGRAPHS
19. OBJECT PROGRAM
20. NAMES
21. DATA, PROCEDURE, CONDITION, SPECIAL
22. UNIQUE, QUALIFIED
23. PARAGRAPH, SECTION
24. ONE, THIRTY
25. QUALIFIED
26. OF, IN
27. PROCEDURE
28. UNIQUE
29. SPACE, PERIOD, RIGHT PARENTHESIS, COMMA, SEMICOLON
30. CAPITALS
31. KEY
32. CONSTANT
33. DIGITS 0-9, PLUS SIGN, MINUS SIGN, DECIMAL POINT
34. 18
35. ONE SIGN, ONE DECIMAL POINT
36. QUOTATION MARKS
37. ANY, 120
38. FIGURATIVE CONSTANT
39. ACTION, RELATIONSHIP BETWEEN ITEMS
40. ARITHMETIC EXPRESSIONS, RELATIONAL EXPRESSIONS, LOGICAL EXPRESSIONS
41. COMPUTE, RELATIONAL
42. TRUE, FALSE

43. COMBINE, TESTING
44. PUNCTUATION SYMBOLS, COMPILATION
45. ONE SPACE
46. A PERIOD, A SPACE
47. STATEMENT
48. SIMPLE, COMPOUND, LOGICAL OPERATOR
49. IMPERATIVE, CONDITIONAL
50. SMALLER, LARGER
51. FILES, HARDWARE
52. EXTERNAL
53. FILE DESCRIPTION, RECORD DESCRIPTION
54. PROCEDURE-NAME

Questions for Review

1. Explain the purpose of a COBOL program coding sheet.
2. What is the importance of the sequence number on a programming coding sheet?
3. What is the purpose of the continuation indicator? What are some of the principal uses?
4. What is a comment line and how is it written on a coding form?
5. List the rules for A and B margin entries.
6. Explain the major program structure of COBOL.
7. Define the terms, Source Program, Object Program and Compiler and explain how they are used in COBOL programming.
8. What should the programmer familiarize himself with before attempting to write a program in COBOL?
9. Explain the different types of names used in COBOL programming and their expressed purpose.
10. What are the rules for assigning names?
11. What is a Reserved Word and how is it used?
12. What are the rules for the use of words?
13. Name seven rules for qualification of data names.
14. What is a literal?
15. What are the rules for the use of numeric literals?
16. What are the rules for the use of nonnumeric literals?
17. What is a figurative constant? What is its main advantage to a programmer? Give an example of a figurative constant.
18. What is an operator and how is it used in COBOL programming?
19. What are Relational Expression operators?
20. What is the purpose of Logical Expression operators?
21. What are the important rules of punctuation?
22. What is the difference between an imperative and conditional statement?

Problems

1. *Identify the purposes and uses of numbered items.*

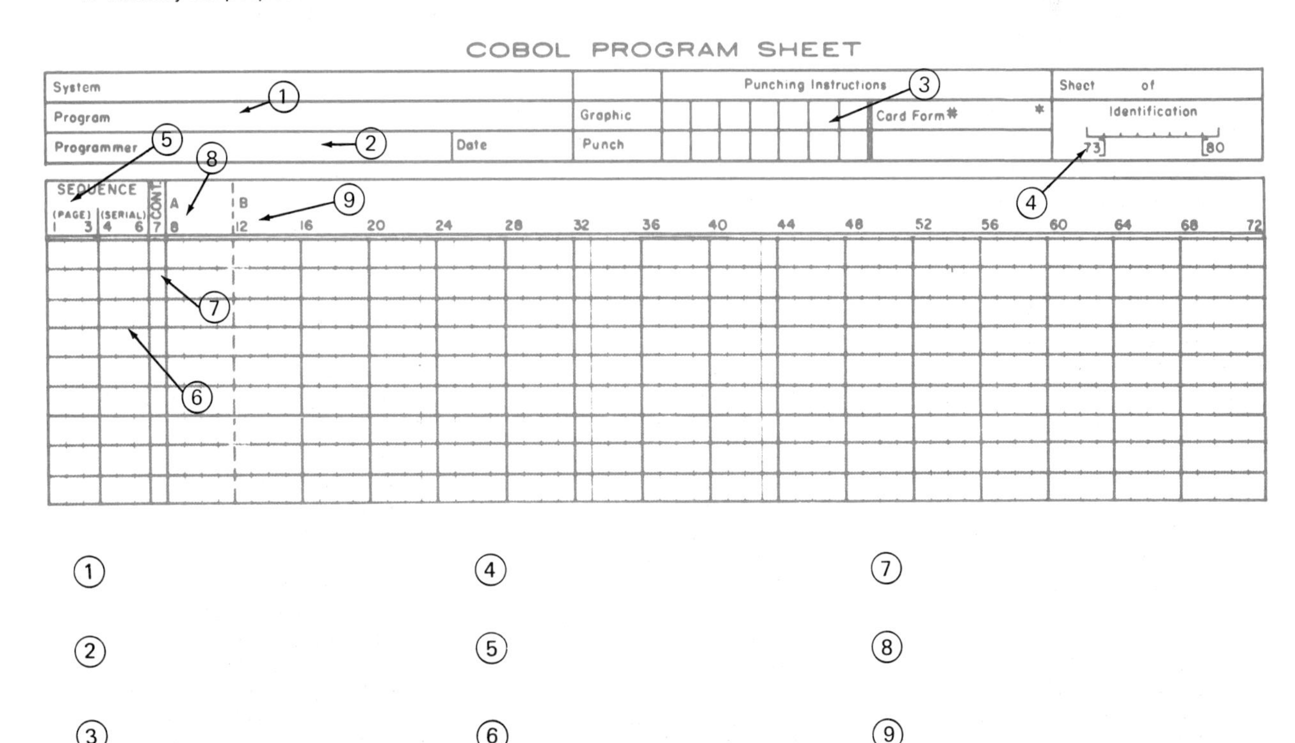

2. *Match each item with its proper description.*

_____ 1. Constant	A. Preassigned meaning.
_____ 2. Comment Line	B. Provides programmer with a standard method of writing COBOL source programs.
_____ 3. B Margin	C. Series of statements, terminated by a period and a space.
_____ 4. Special-Names	D. Hyphen in column 7.
_____ 5. Names	E. Columns 8 through 11.
_____ 6. Program Sheet	F. Words to identify certain data within a program.
_____ 7. Reserved Words	G. Columns 12 through 72.
_____ 8. Continuation Indicator	H. Mnemonic names assigned to various components in the Environmental Division.
_____ 9. A Margin	I. An asterisk in column 7.
_____10. Entries	J. Unchanging value during execution of the program.

3. *Match each item with its proper description.*

_____ 1. Statement	A. Specifies some sort of action or relationship between items in the program.
_____ 2. Key Word	B. Must be enclosed in quotation marks.
_____ 3. Figurative Constant	C. Value is determined by the set of characters.
_____ 4. Relational Expression Operator	D. Required in COBOL entry.
_____ 5. Arithmetic Expression Operator	E. Used to combine simple statements in the same expression for the purpose of testing condition of the expression.
_____ 6. Word	F. Used to reduce a statement to a true or false condition.
_____ 7. Operator	G. One or more COBOL characters chosen from the character set.
_____ 8. Literal	H. Has a predefined value.
_____ 9. Logical Expression Operator	I. Syntactically valid combination of words and symbols written in the Procedure Division used to express a thought.
_____10. Nonnumeric Literal	J. Used only in COMPUTE statement.

4. *Indicate the* incorrect *numeric literals in the following list.*

   a. – 8573956894183456.98
   b. – .00015
   c. 2,900.56
   d. – 192.85
   e. .5
   f. SIX

5. *Rewrite the following entry correctly.*

   COMPUTE GROSS,ROUNDED = (HOURS*RATE) + OTPAY.

6. *Correct the following data-names.*

   a. SPACE
   b. JOB 1
   c. 12345*
   d. LEVEL – 1
   e. – 10T4
   f. TAX/RATE

7. *Identify the following operators.*

	Operator	Name of Operator	Type of Operator
a.	=	____________	____________
b.	>	____________	____________
c.	*	____________	____________
d.	**	____________	____________
e.	AND	____________	____________
f.	/	____________	____________

8.

SEQUENCE (PAGE) 1 3 (SERIAL) 4 6	CONT. 7	A 8	B 12 ... 72
			MULTIPLY AMOUNT OF CLIENT-PURCHASES BY TRADE-DISCOUNT, GIVING

Which of the lines below shows a correct way of completing this entry?

```
a.        -REDUCTION.
b.           REDUCTION.
c.               REDUCTION.
d.        -            REDUCTION.
```

9. *Number the following in an ascending to descending order.*

_____a. Data-name.
_____b. Division headers.
_____c. Paragraph-name.
_____d. Section-name.

10. *Write a COBOL program to list 80-column cards on the printer, double space. Be sure to code all four divisions.*

5 Identification and Environment Divisions

IDENTIFICATION DIVISION

The Identification Division is the first and simplest of all four divisions to write, and must be included in every COBOL program (fig. 5.1).

Function. The function of this Division is to identify both the source program and the resultant output listing. In addition, the user may include any other information that he considers vital to his program, such as the date the program was written, the date of the compilation of the source program, etc. (See figure 5.2.)

The required entries are:

1. Division Header—Identification Division.
2. Program-ID.
3. Program-Name. The program-name is used to identify the object name to the control program and must be given in the first paragraph.

Rules for the Formation of the Program-Name are

1. The name *must not* be enclosed in quotation marks.
2. The name must conform to the rules for the formation of a procedure-name, such as
   a. A unique data-name.
   b. No special characters permitted.
   c. No reserved words, etc.

```
IDENTIFICATION DIVISION.
PROGRAM-ID. program-name.
[AUTHOR. [comment-entry]...]
[INSTALLATION. [comment-entry]...]
[DATE-WRITTEN. [comment-entry]...]
[DATE-COMPILED. [comment-entry]...]
[SECURITY. [comment-entry]...]
[REMARKS. [comment-entry]...]
```

Figure 5.1 Identification Division Format.

The Identification Division may contain all of these paragraphs.

```
IDENTIFICATION DIVISION.
PROGRAM-ID. PAYROLL-REPORT.
AUTHOR. S. D. HAMLIN.
INSTALLATION. EASTERN DIVISION OF ABC COMPANY.
DATE-WRITTEN. MAY 15, 1978.
DATE-COMPILED.
SECURITY. NONE.
REMARKS. THIS PROGRAM WILL BE RUN EVERY THURSDAY.
```

To generate an object program, the COBOL compiler needs only these entries.

```
IDENTIFICATION DIVISION.
PROGRAM-ID. PAYROLL-REPORT.
```

Figure 5.2 Examples—Identification Division Entries.

The other paragraphs are optional and may be included at the user's option. The six additional fixed-named paragraphs are AUTHOR, INSTALLATION, DATE-WRITTEN, DATE-COMPILED, SECURITY, and REMARKS. These optional paragraphs, if used, must be presented in the order shown in the format. The sentences may contain any numerals, letters, blanks, special characters, or reserved words organized to conform to sentence and paragraph structure. (See figures 5.3, 5.4.)

The programmer has complete freedom in what he chooses to write in each of these optional paragraphs. Entries in the REMARKS paragraph usually stipulate the purpose and what the program is to accomplish (fig. 5.5).

(*Note:* Regardless of what date is placed in the DATE-COMPILED paragraph, the actual compilation date will be printed on the source program listing.)

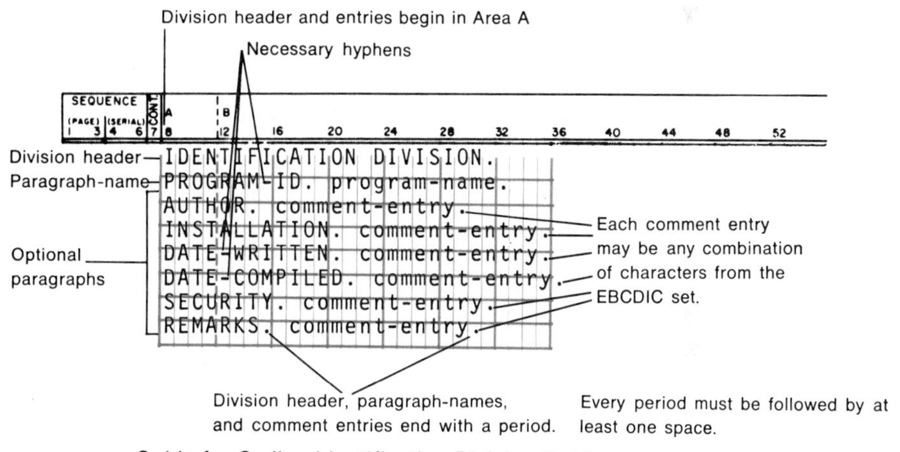

Figure 5.3 Guide for Coding Identification Division Entries.

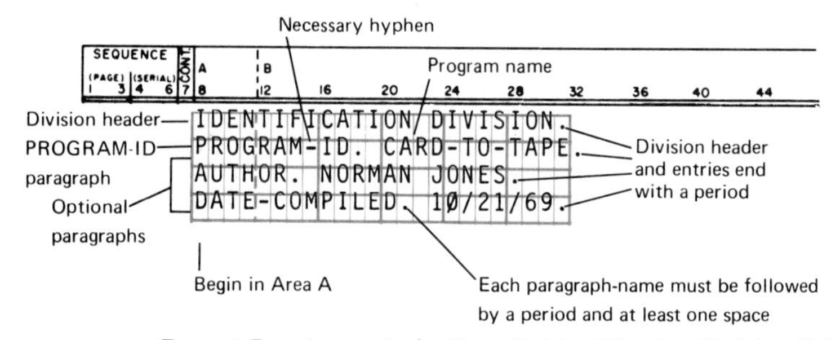

Figure 5.4 Format Requirements for Sample Identification Division Entries.

ENVIRONMENT DIVISION

All aspects of a data processing problem that depend upon the physical characteristics of a specific computer are expressed in the Environment Division. This division is the one division of COBOL that is machine-dependent, and the programmer must familiarize himself with the characteristics and special name of the machine upon which the particular source program is to be run. Any changes in a computer requires many changes in this division. A link is

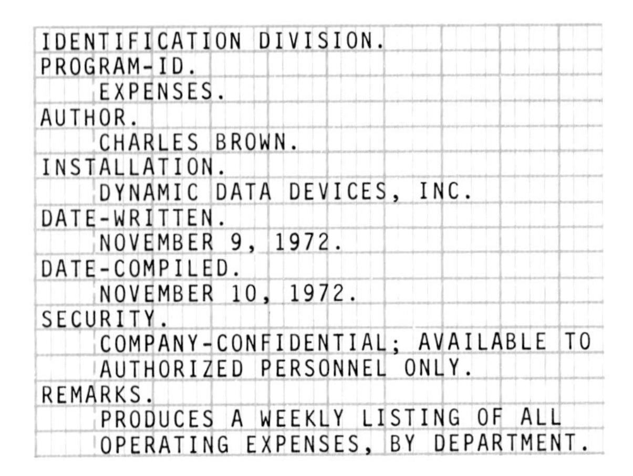

```
IDENTIFICATION DIVISION.
PROGRAM-ID.
      EXPENSES.
AUTHOR.
      CHARLES BROWN.
INSTALLATION.
      DYNAMIC DATA DEVICES, INC.
DATE-WRITTEN.
      NOVEMBER 9, 1972.
DATE-COMPILED.
      NOVEMBER 10, 1972.
SECURITY.
      COMPANY-CONFIDENTIAL; AVAILABLE TO
      AUTHORIZED PERSONNEL ONLY.
REMARKS.
      PRODUCES A WEEKLY LISTING OF ALL
      OPERATING EXPENSES, BY DEPARTMENT.
```

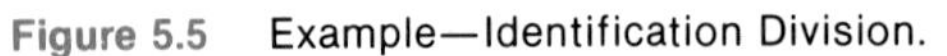

Figure 5.5 Example—Identification Division.

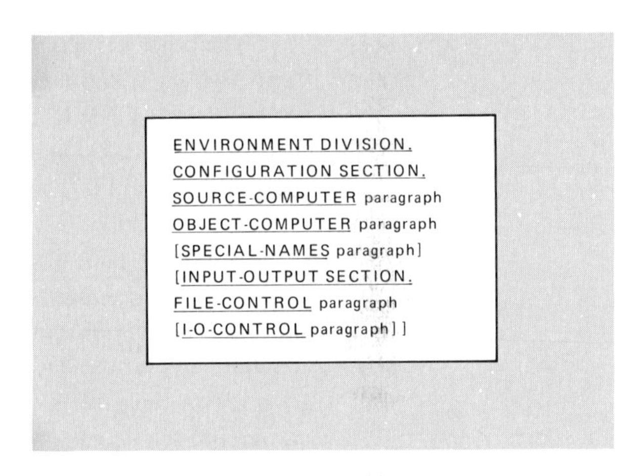

ENVIRONMENT DIVISION.
CONFIGURATION SECTION.
SOURCE-COMPUTER paragraph
OBJECT-COMPUTER paragraph
[SPECIAL-NAMES paragraph]
[INPUT-OUTPUT SECTION.
FILE-CONTROL paragraph
[I-O-CONTROL paragraph]]

Figure 5.6 Environment Division Format.

provided between the logical concept of the files and the physical aspects of the devices upon which the files will be processed and stored (fig. 5.6).

Function. The function of the Environment Division is to specify the configuration of the computer that will be used to compile the source program, as well as the configuration of the computer that will execute the object program. In addition, all input and output files will be assigned to individual hardware devices. Any special input and output techniques that will be used in the processing of data will be defined here (fig. 5.7).

```
ENVIRONMENT DIVISION.
CONFIGURATION SECTION.
SOURCE-COMPUTER.  NCR-CENTURY-200.
OBJECT-COMPUTER.  NCR-CENTURY-200  MEMORY SIZE 32000
    CHARACTERS.
SPECIAL-NAMES. SWITCH-5 ON STATUS IS PROCESS-FLIGHTFILE,
    OVERHAUL-FILE IS OVFILE.
.
INPUT-OUTPUT SECTION.
FILE-CONTROL.
.   FILE SPECIFICATIONS WORKSHEETS FOR ALL FILES
    SELECT TRANFILE   ASSIGN TO NCR-TYPE-41.
    SELECT FLIGHTFILE   ASSIGN TO NCR-TYPE-00.
    SELECT AIRFILE   ASSIGN TO NCR-TYPE-41.
    SELECT NEWAIRFILE   ASSIGN TO NCR-TYPE-41.
    SELECT OVERHAUL-FILE   ASSIGN TO NCR-TYPE-25.
.
I-O-CONTROL.
    RERUN EVERY END OF REEL OF NEWAIRFILE.
```

In the above, the SELECT entries in the Input-Output Section make the following hardware assignments:

- TRANFILE is assigned to a magnetic tape unit.
- FLIGHTFILE is assigned to the card reader.
- AIRFILE is assigned to a magnetic tape unit.
- NEWAIRFILE is assigned to a magnetic tape unit. This file has a rescue dump taken at the end of every reel.
- OVERHAUL-FILE is assigned to the printer.

Figure 5.7 Examples—Environment Division Entries.

The Environment Division must be included in every COBOL source program. This Division must begin at the A margin, with the heading ENVIRONMENT DIVISION followed by a period on a line by itself (fig. 5.8).

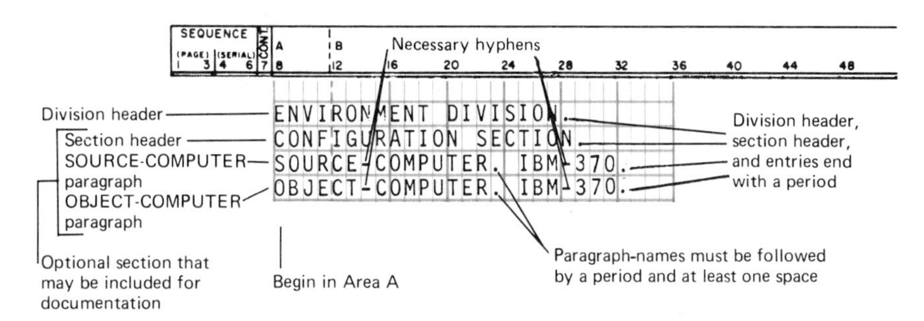

Figure 5.8 Format Requirements for Sample Environment Division Entries.

The Environment Division is divided into two sections, the Configuration Section and the Input-Output Section. The sections and paragraphs, when used, must appear in the sequence as shown in the format.

Configuration Section

The Configuration Section specifies the overall characteristics of the computer involved in the compilation and execution of a COBOL program. The section is divided into three paragraphs: SOURCE-COMPUTER, OBJECT-COMPUTER, and the SPECIAL-NAMES paragraphs (fig. 5.9).

The SOURCE-COMPUTER paragraph describes the computer upon which the compilation of the source program is to take place.

The OBJECT-COMPUTER paragraph describes the computer upon which the object compiled program is to be executed. Also, the specifications as to size and range of main storage to be used may be stipulated.

In both the SOURCE-COMPUTER and the OBJECT-COMPUTER paragraphs, the computer name must include the model number and conform to the rules for COBOL names (figs. 5.10, 5.11).

The SPECIAL-NAMES paragraph is used to equate user-specified mnemonic-names with function-names used by the compiler. This paragraph may also be used to exchange the functions of the comma and period in PICTURE character strings and numeric literals. If a currency symbol other than the $ is used in a PICTURE clause, the user must specify the currency substitute character in this paragraph (fig. 5.12). (Discussed in greater detail further along in the text.)

```
CONFIGURATION SECTION.
SOURCE-COMPUTER. source-computer-entry
OBJECT-COMPUTER. object-computer-entry
[SPECIAL-NAMES. special-names-entry]
```

Figure 5.9 Configuration Section—Format.

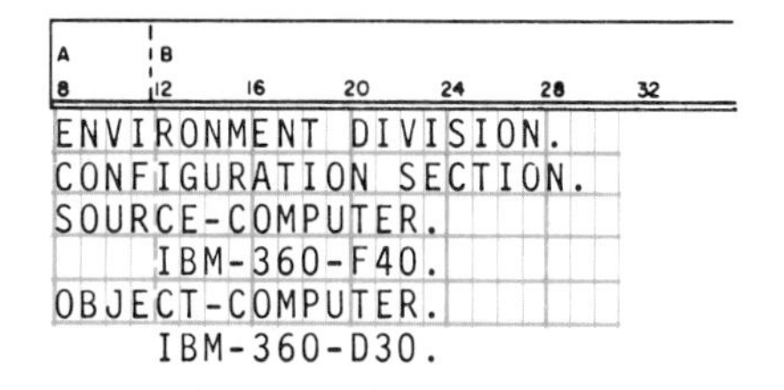

Figure 5.10 Example—Configuration Section.

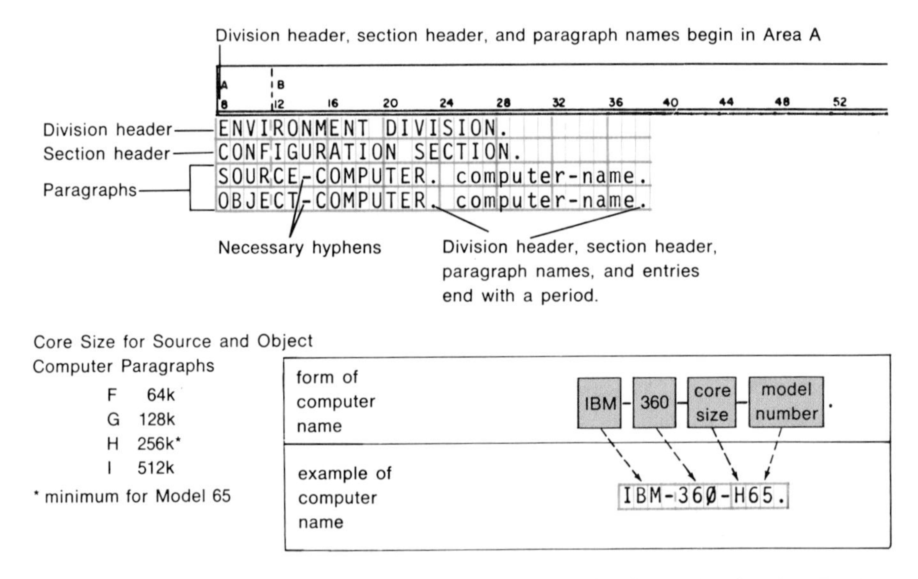

Figure 5.11 Guide for Coding Environment Division—Configuration Section Entries.

Figure 5.12 Examples—Special-Names Entries.

```
SPECIAL-NAMES.  TRANSACTION-FILE IS TRANFILE.
```

In the above, TRANSACTION-FILE is equated to TRANFILE and within the source program, both files can be used interchangeably.

```
SPECIAL-NAMES.  CURRENCY SIGN IS 'R'.
```

In the above, the character R is used as the currency symbol in the PICTURE clause.

The SOURCE-COMPUTER and the OBJECT-COMPUTER paragraphs are required in all COBOL programs (optional in 1974 ANS COBOL), while the SPECIAL-NAMES paragraph is optional and is included only when necessary in the program.

Input-Output Section

The Input-Output Section must be included in any COBOL source program if there are any input and output files required. As most programs involve the processing of files, this section is required in most programs.

The Input-Output Section is concerned with the definition of the input and output devices as well as the most efficient method of handling data between the devices and the object program.

This section is divided into two paragraphs: the File-Control paragraph, which names each file used in the program and identifies the media on which each file is located; and the I-O-Control paragraph, which specifies any special Input/Output control techniques to be used in object program. The individual clauses that make up these paragraphs may appear in any sequence within their respective sentences or paragraphs, but must begin at the B margin (fig. 5.13).

File-Control Paragraph

The File-Control paragraph names and associates files with external media. The names of files given in the file description entries in the Data Division are assigned to input/output devices. There is a relationship between the

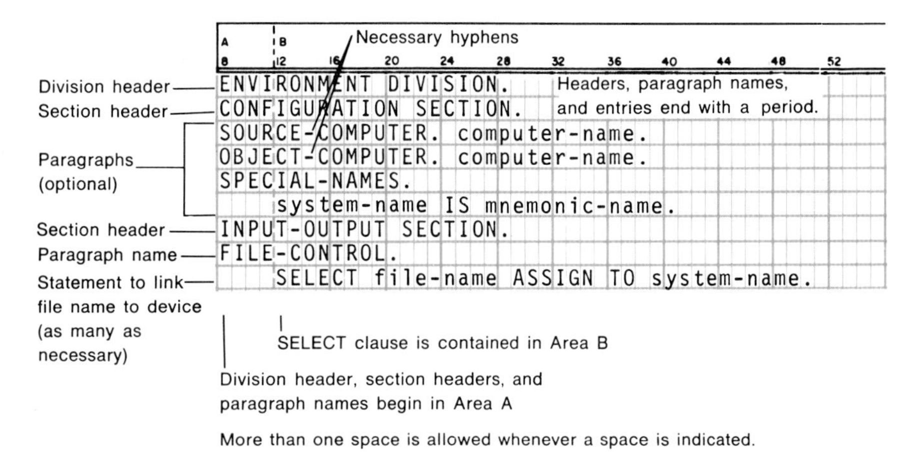

Figure 5.13 Guide for Coding Environment Division—Configuration Section and Input-Output Section Entries.

file entries in the three divisions. The Data Division entries specify the characteristics and the structure of the data within these files. The Procedure Division will specify the READ and WRITE entries for these files. The input/output device that will be used to read or write will be determined by the Environment Division entry which names the input/output devices assigned to the particular file.

Select Clause

The Select clause is used to name files within a COBOL source program. The Select entry must begin with the word SELECT followed by the file-name, and must be given for each file referred to by the COBOL source program (fig. 5.14). A separate SELECT clause is required for each file name in the Data Division.

File-Name. The unique name of the file assigned in the file description entry in the Data Division of the source program. This name will also be used in entries in the Procedure Division.

Optional. A key word that may be specified only for input files accessed sequentially, yet may not be present each time the object program is executed. If this option is used and the file is not present at object time, the first READ statement causes the control to be passed to the imperative statement following the key words AT END.

Assign Clause

The Assign clause is used to assign files to external media (fig. 5.15).

```
SELECT [OPTIONAL] file-name
```

Figure 5.14 Select Clause—Format.

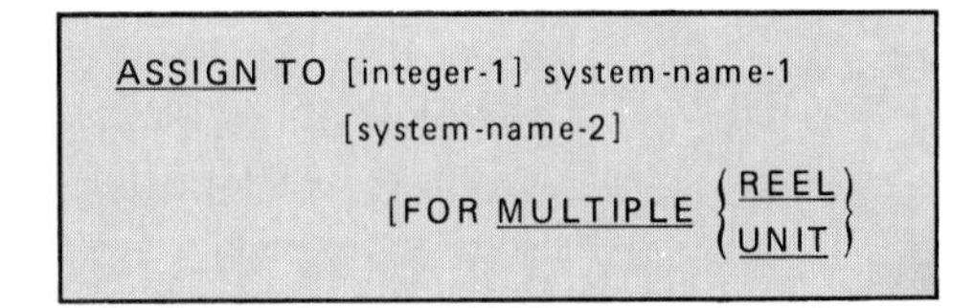

Figure 5.15 Assign Clause—Format.

Integer. Indicates the number of input/output devices that may be used for a given medium assigned to file-name. Since the number of units is automatically assigned by the operating system, this integer is rarely used.

System Name. Identifies the device class, the particular input/output device, the organization of the data upon the device, and the external name of the file. The external name is a 1-to-8-character field by which the file is known to the system.

Device Class. Two-character field that specifies the particular category of devices to be used. Each file must be assigned to a device class.

DA. The *Direct-Access* class is composed of mass storage devices that can read and write records randomly—disks, drums, data cell devices. The same devices may also appear in the Utility class.

UT. The *Utility* class is composed of devices that can read and write records sequentially. They include such devices as magnetic tape, disk, drum, and data cells.

UR. The *Unit-Record* class is composed of devices that can read and provide data, such as printers and card read/punch devices.

(*Note:* Files that are assigned to UT or UR device class must have a standard sequential organization, and records can only be accessed sequentially. Files that are assigned to DA devices may have both standard sequential or direct organization. When the organization is specified as direct, access may be either sequential or random.)

Device Number. A four- or five-character field used to specify a particular device within a device class. If device independence is specified, the device class must be UT; therefore, no device number is given, and no END-OF-PAGE class is associated with the file. At execution time, such a file may be assigned to any device, including unit record devices.

The allowable device numbers for the IBM 360 and 370 computers are:

Direct-Access	DA	2301, 2302, 2303, 2311, 2314, 2321.
Utility	UT	2301, 2302, 2311, 2314, 2321, 2340, 2400.
Unit-Record	UR	1403, 1404, 1442R, 1442P, 1443, 1445, 2501, 2520P, 2520R, 2540P, 2540R. (R indicates reader; P indicates punch)

Organization. A one-character field indicating the type of organization used for a particular file.

S. Used for files with standard sequential organization. When standard sequential organization is specified, the logical records of a file are positioned and read sequentially, in the order in which they were created. If the REVERSED option is specified, these records within these files may be sequentially reversed for read and write operations. The type *S* organization must be used for magnetic tape or unit record files and may be used for files assigned to direct-access (mass storage) devices. No actual keys are associated with records on a sequentially organized file.

D. Used for files with direct organization. A direct-organization file specifies that the records are organized in a random sequence. The addressing scheme is determined by the programmer whereby the positioning of the logical records within the file is determined by keys supplied by the actual user in the Environment Division. The ACTUAL KEY is used to locate records within the file.

Name. The external file-name by which the file is known to the control program. The name may consist of from one to eight characters. (See figures 5.16, 5.17.)

DOS ASSIGN Clause Guide‡

Device name	Class indicator*	Device number	Organization indicator†
card reader	UR	2540R	S
card punch	UR	2540P	S
printer	UR	1403	S
tape drive	UT	2400	S
disk drive	UT or DA	2311, 2314	S I
data cell	UT or DA	2321	S I

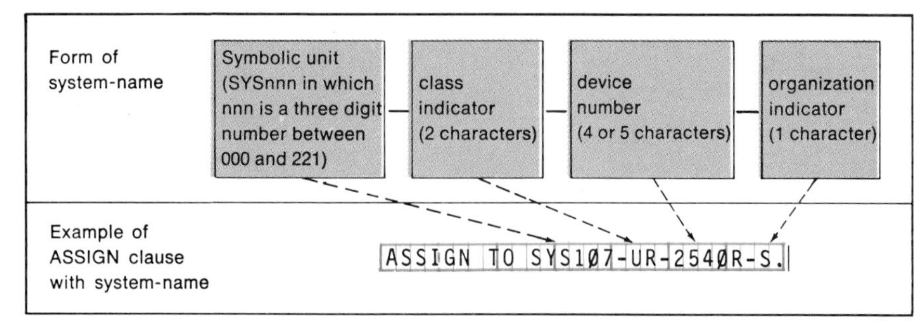

* Class indicator UR stands for Unit-Record, UT for Utility, and DA for Direct Access.
† Organization indicator S stands for Sequential and I for Indexed sequential.
‡ This Guide is applicable to both the complete and subset ANS COBOL compilers.

Figure 5.16 IBM 360/370 DOS Assign Clause Guide.

I-O-Control Paragraph

The I-O-Control paragraph defines special control techniques to be used in the object program. It may specify certain conditions in an object program, such as which checkpoints are to be established, which storage areas to be shared by different files, the location of files on multiple-file reels, and the optimization techniques.

If special techniques or conditions need to be defined in the program, the I-O-Control paragraph is used; otherwise, the entire paragraph and its associated clauses may be omitted.

The features of the I-O-Control paragraph and other features of the Environment Division are discussed later in the text. (See also figure 5.18.)

OS ASSIGN Clause Guide

Device name	Class indicator*	Device number	Organization indicator†
card reader	UR	2540R	S
card punch	UR	2540P	S
printer	UR	1403	S
tape drive	UT	2400	S
disk drive	UT or DA	2302, 2311, 2314	S
			I
drum	UT or DA	2301	S
		2301, 2303	I
data cell	UT or DA	2321	S
			I

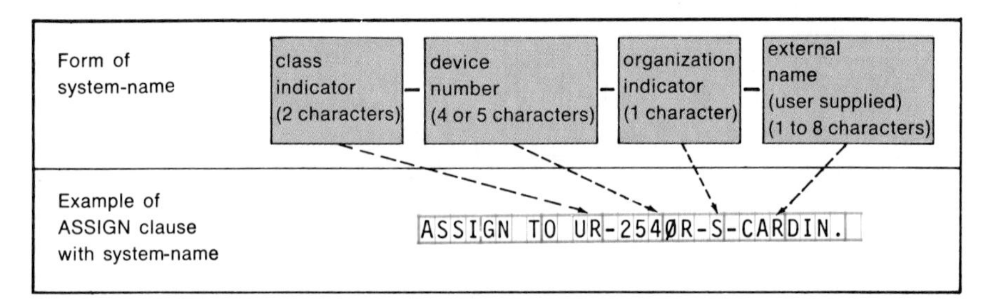

* Class indicator UR stands for Unit-Record, UT for Utility, and DA for Direct Access.
† Organization indicator S stands for Sequential and I for Indexed sequential.

Figure 5.17 IBM 360/370 OS Assign Clause Guide.

SAMPLE ENVIRONMENT DIVISION ENTRIES

BURROUGHS

```
ENVIRONMENT DIVISION.
CONFIGURATION SECTION.
SOURCE-COMPUTER.   B-5500.
OBJECT-COMPUTER.   B-5500.
INPUT-OUTPUT SECTION.
FILE-CONTROL.
    SELECT CUSTOMER ASSIGN TO READER.
    SELECT BILL ASSIGN TO PRINTER.
```

HONEYWELL

```
ENVIRONMENT DIVISION.
CONFIGURATION SECTION.
SOURCE-COMPUTER.   H-200.
OBJECT-COMPUTER.   H-200     MEMORY SIZE 16384 CHARACTERS.
INPUT-OUTPUT SECTION.
FILE-CONTROL.
    SELECT OMAST ASSIGN TO TAPE-UNIT DA.
    SELECT NMAST ASSIGN TO TAPE-UNIT HB.
    SELECT INV-CARD   TO CARD-READER E.
    SELECT INV-LIST ASSIGN TO PRINTER B.
```

IBM

```
ENVIRONMENT DIVISION.
CONFIGURATION SECTION.
SOURCE-COMPUTER.   IBM-370-I155.
OBJECT-COMPUTER.   IBM-370-I155.
INPUT-OUTPUT SECTION.
FILE-CONTROL.
    SELECT PURCHASING-FILE   ASSIGN TO SYS107-UR-2540R-S.
    SELECT PURCHASE-REPORT-FILE   ASSIGN TO SYS109-UR-1403-S.
```

NCR

```
ENVIRONMENT DIVISION.
CONFIGURATION SECTION.
SOURCE-COMPUTER.   NCR-CENTURY-200.
OBJECT-COMPUTER.   NCR-CENTURY-200 MEMORY SIZE 32000 WORDS.
INPUT-OUTPUT SECTION.
FILE-CONTROL.
    SELECT WHFILE ASSIGN TO NCR-TYPE-32.
    SELECT INVFILE ASSIGN TO NCR-TYPE-30.
```

XEROX

```
ENVIRONMENT DIVISION.
CONFIGURATION SECTION.
SOURCE-COMPUTER.   XEROX-530.
OBJECT-COMPUTER.   XEROX-530.
INPUT-OUTPUT SECTION.
FILE-CONTROL.
    SELECT   FILE-IN   ASSIGN TO READER.
    SELECT   FILE-OUT   ASSIGN TO PRINTER.
```

Figure 5.18 Sample Environment Division Entries.

Exercises

Write your answer in the space provided (answer may be one or more words).

The following relate to the Identification Division.

1. The Identification Division is the ____________ and ____________ of all four divisions to write.
2. This division must be included in every COBOL program and is used to identify both the ____________ and the resultant ____________.
3. The required entries are ____________, ____________, and ____________.
4. The program name must not be enclosed in ____________ and must conform to the rules for the formation of a ____________.
5. The six optional named paragraphs in order of sequence are ____________, ____________, ____________, ____________, ____________, and ____________.
6. The entries in the REMARKS paragraph usually stipulate the ____________.

The following relate to the Environment Division.

7. This division is the one division of COBOL that is ____________ dependent.
8. The programmer must familiarize himself with the ____________ and ____________ names of the machine upon which the particular ____________ is to be run.
9. A link is provided between the ____________ aspects of the files and the ____________ characteristics of the devices upon which the files will be processed and stored.
10. The configuration of the computer that will be used to ____________ the object program must be specified in this division.
11. All input and output files must be assigned to individual ____________ devices.
12. The Environment Division is divided into ____________ sections.
13. The ____________ section specifies the overall characteristics of the computer involved in the compilation and execution of a COBOL program.
14. The ____________ paragraph is used to equate user specified ____________ names with ____________ names used by the compiler.
15. The Input-Output section is concerned with the definitions of the ____________ and ____________ devices as well as the most efficient method of handling data between the ____________ and the ____________ program.
16. The File-Control paragraph ____________ and ____________ files with external media.
17. The ____________ clause names ____________ within a COBOL source program.
18. The ASSIGN clause assigns ____________ to ____________.
19. The System-Name identifies the ____________, the particular ____________ devices, the ____________ of the data and the ____________ name of the file.
20. The I-O-Control paragraph specifies specific ____________ to be used in the object program.

Answers

1. FIRST, SHORTEST
2. SOURCE PROGRAM, OUTPUT LISTING
3. DIVISION HEADER, PROGRAM-ID, PROGRAM NAME
4. QUOTATION MARKS, PROCEDURE-NAME
5. AUTHOR, INSTALLATION, DATE-WRITTEN, DATE-COMPILED, SECURITY, REMARKS
6. PURPOSE OF THE PROGRAM
7. MACHINE
8. CHARACTERISTICS, SPECIAL, SOURCE PROGRAM
9. LOGICAL, PHYSICAL
10. COMPILE
11. HARDWARE
12. TWO
13. CONFIGURATION
14. SPECIAL-NAMES, MNEMONIC, FUNCTION
15. INPUT, OUTPUT, DEVICES, OBJECT
16. NAMES, ASSOCIATES
17. SELECT, FILES
18. FILES, EXTERNAL MEDIA
19. DEVICE-CLASS, INPUT/OUTPUT, ORGANIZATION, EXTERNAL
20. CONTROL TECHNIQUES

Questions for Review

1. What is the function of the Identification Division?
2. What are the required entries in the Identification Division?
3. What is the importance of the REMARKS paragraph in the Identification Division?
4. What are six additional optional paragraphs in sequence that may be included in the Identification Division?
5. What is the importance of the Environment Division and what is its main function?
6. What are the sections of the Environment Division and what is the main function of each of these sections?
7. What are the main functions of the SELECT and ASSIGN clauses?
8. What are the components of the System-Name and what is the function of each of its segments?
9. What is the purpose of the I-O-Control paragraph and what is it customarily used for?

Problems

1. *In the following entries, which one is correctly written?*

   a. IDENTIFICATION DIVISION.
      PROGRAM-ID. SALES-ANALYSIS.
   b. IDENTIFICATION DIVISION
      PROGRAM-ID. CARD-TO-TAPE.
   c. IDENTIFICATION DIVISION.
      PROGRAM-ID. PAYROLL-MASTER.

2. *Which of the following PROGRAM-ID names is written correctly?*

   a. DATA
   b. 'INVENTORY-CONTROL-REPORT'
   c. INVENTORY-MASTER
   d. PAYROLL*

3. *In the following list of Identification Division paragraph names, which one is incorrect?*

   AUTHOR, TITLE, INSTALLATION, SECURITY, REMARKS

4. *Match each item with its proper paragraph name.*

______ 1. Program-Id.	A. Non-Military.
______ 2. Author.	B. May 17, 1977.
______ 3. Date-Written.	C. J Morse.
______ 4. Date-Compiled.	D. This program calculates payroll.
______ 5. Installation.	E. Payroll 04.
______ 6. Security.	F. District Office.
______ 7. Remarks.	

5. *In the Environment Division, indicate with a check mark which of the following must be written at the* A *margin.*

______SOURCE-COMPUTER.	______SELECT Clause.
______CONFIGURATION SECTION.	______FILE-CONTROL.
______SPECIAL-NAMES.	______ASSIGN Clause.

6. *Match each item with its proper device class.*

______Utility	A. Magnetic disks and data cells.
______Unit Record	B. Magnetic tape and magnetic drum.
______Direct Access	C. Card Readers, Card Punches and Printers.

7. *Write the Identification Division, using all the required and optional entries for the following:*

   The Acme Manufacturing Company is initiating an inventory control system. You are the programmer assigned to writing the program. The program is restricted to production control personnel and is to be run at the Boston Center.

8. *Write the Environment Division for the following systems flowchart. The program will be compiled and executed on an IBM 370 model 155 computer with the following hardware assignments:*

SYS005	1403	Printer
SYS009	2540R	Card Reader
SYS006	2540P	Card Punch
SYS011	2400	Magnetic Tape
SYS012	2400	Magnetic Tape

SYSTEMS FLOWCHART

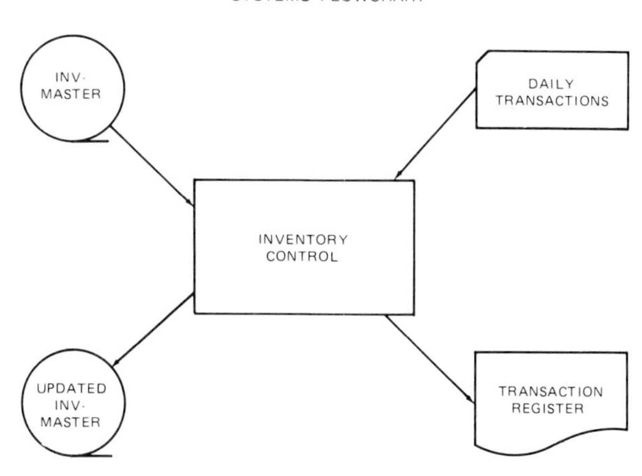

9. *Write the Environment Division for the following systems flowchart using the same hardware assignments as problem 8.*

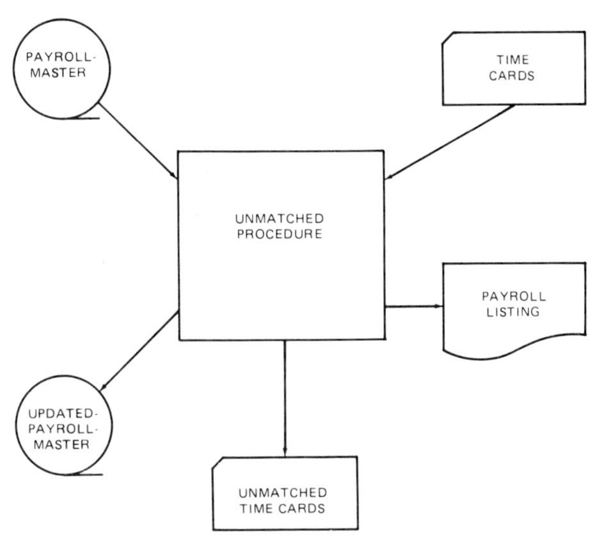

6 Data Division

INTRODUCTION

The Data Division of a COBOL source program describes the characteristics of the information to be processed by the object program. The separation of divisions provides the programmer with flexibility as the Procedure Division is interwoven with the Data Division. The manner in which data is organized and stored has a major effect upon the efficiency of the object program. Data to be processed falls into three categories.

1. The data in the files that are entering or leaving the internal storage areas of the computer.
2. The data in the work areas of the computer that have been developed internally by the program.
3. Constant data that is to be used by the program.

The structure of each record within a file is usually shown with the items described in the sequence in which they appear in the record.

The Data Division begins with the header DATA DIVISION at the A margin followed by a period on a line by itself. *Each of the sections within the Data Division has a fixed name.* The sections are followed by the word SECTION and a period, and are on a line by themselves. These sections consist of entries rather than paragraphs. Each entry must contain:

1. A level indicator or level number.
2. A data-name or other name (FILLER).
3. A series of clauses defining the data that may be separated by commas. The clauses may be written in any sequence by the programmer (except the REDEFINES clause). Each entry must be terminated by a period and a space. (See figure 6.1.)

UNITS OF DATA—TERMINOLOGY

Data in COBOL source programs are referred to by various names. They are:

Item. An item is considered a field and is an area used to contain a particular type of information.

Elementary Item. The smallest item available that is not divided into smaller units.

```
DATA DIVISION.
FILE SECTION.
FD  TRANSACTION; BLOCK CONTAINS 25
    RECORDS; LABEL RECORDS ARE STANDARD;
    DATA RECORD IS TRANSACTION-RECORD.
01  TRANSACTION-RECORD.
    02  ACCOUNT-NUMBER, PICTURE X(10).
    02  TRANSACTION-CODE, PICTURE X.
    02  AMOUNT, PICTURE 9(6)V99.
WORKING-STORAGE SECTION.
77  PREVIOUS-NUMBER, PICTURE X(10).
01  MESSAGE.
    02  ACCOUNT, PICTURE X(10).
    02  FILLER, PICTURE XX, VALUE SPACE.
    02  TOTAL, PICTURE $ZZZ, ZZZ.99.
    02  FILLER, PICTURE XX, VALUE SPACE.
    02  COMMENTS, PICTURE X(30).
```

Figure 6.1 Example—Data Division.

Group Item. A larger item that is composed of a named sequence of one or more elementary items. A referral to a group item applies to the entire area of elementary items.

Independent Item. An elementary item appearing in the Working-Storage Section of the Data Division that is not a record or a part of a record. These items are usually used as work areas or to contain constant data. (See figure 6.2.)

Data Record. The data record is usually considered to be the group item comprising several related items. It is also referred to as the "logical record." A *logical record* is one unit of information in a file of like units; for example, an item of inventory or an employee's record. One or more logical records can be included in a *physical record* or the logical record may itself be a physical record. It may be contained within a single physical unit or it may extend beyond the physical unit. The logical record is normally the unit of each program that processes input and output operations.

File. A file is composed of a series of related data records. The data records may have the same or varying lengths.

Block. A block is referred to as the "physical record" consisting of a series of logical records. A *physical record* is a group of characters or records which

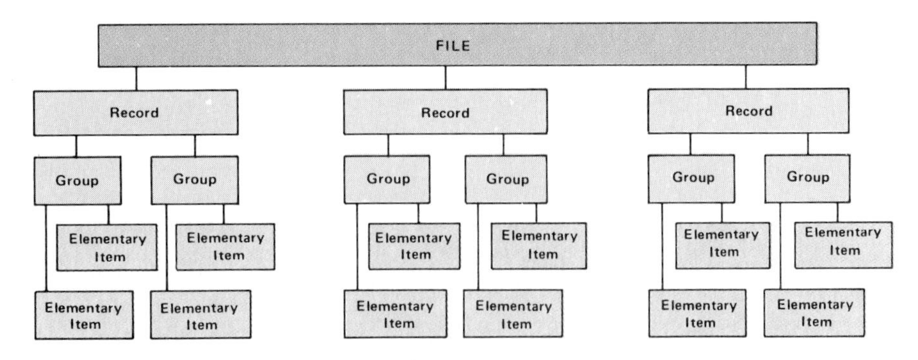

Figure 6.2 File Organization—Schematic.

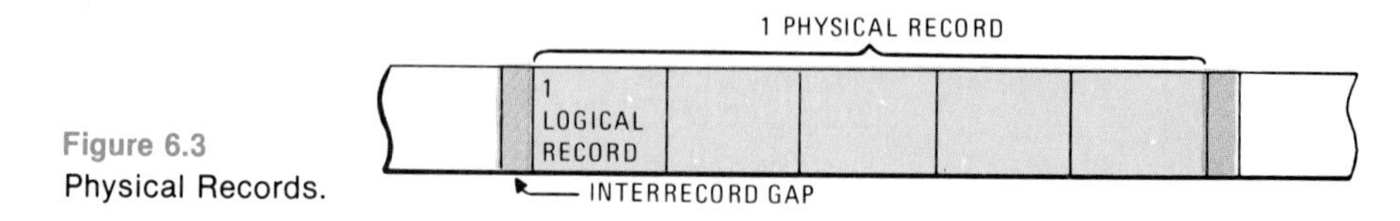

Figure 6.3
Physical Records.

is treated as an entity when moved into and out of main storage (fig. 6.3). When data is stored on magnetic tape or direct-access devices, the logical records are grouped in blocks. Each read or write operation may transfer an entire block of data to or from main storage at one time and to or from an input/output device. Each logical record within the block is then processed separately (fig. 6.4).

COBOL source language statements provide a means of describing the relationship between physical and logical records. Once this relationship is estalished, only logical records are made available to the program.

Label Records. Label records are normally used for files that are stored on magnetic tape or direct-access devices. The record usually contains information relative to the file (fig. 6.5). Card files do not contain any label records.

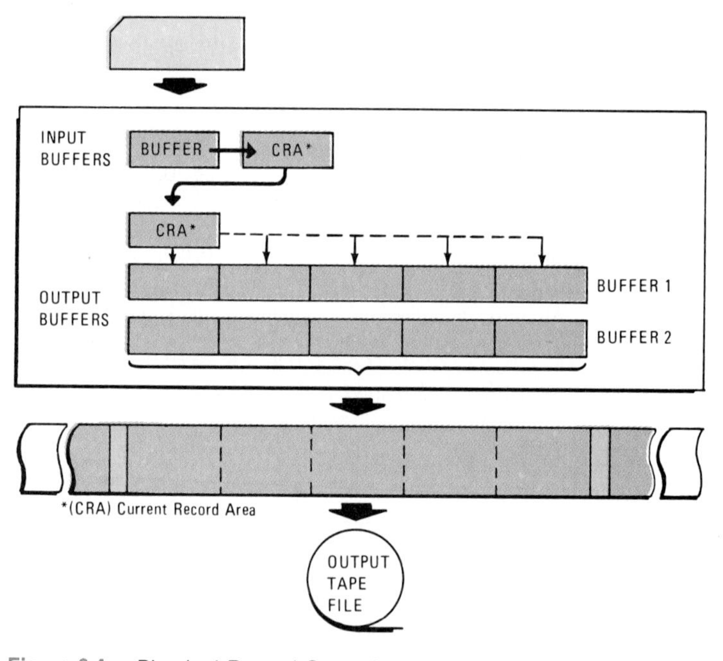

Figure 6.4 Physical Record Operation—Input/Output Buffer.

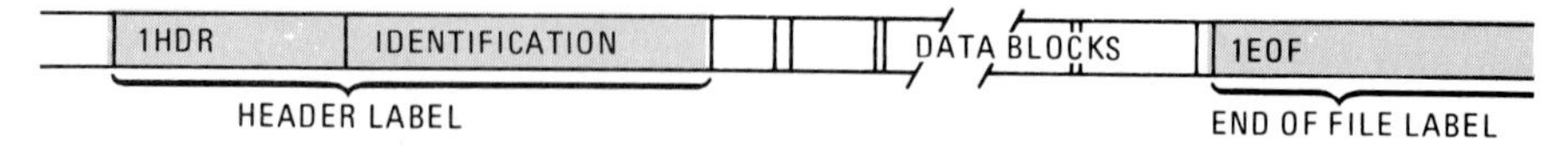

Figure 6.5 Label Records.

ORGANIZATION

Data Description

In the discussion of data description, a distinction must first be made between the record's external description and its internal content.

External description refers to the physical aspect of a file, such as the way the file appears on an external medium. For example, the number of logical records per physical record describe the grouping of records in the file. The physical aspects of a file are specified in the file description entries.

A COBOL record usually consists of groups of related information that are treated as an entity. The explicit description of the contents of each record defines its internal characteristics. For example, the type of data to be contained within each field of a logical record is an internal characteristic. This type of information about each field of a particular record is grouped into a record description entry.

The Data Division is divided into three fixed sections: File, Working-Storage, and Report (fig. 6.6). The File Section defines the contents of the data files that are stored on the external medium. The Working-Storage Section describes a record or noncontiguous data items which are not part of the external data files but which are developed and processed internally, or data items whose values are assigned in the source program and do not change during the execution of the program. Both logical records and noncontiguous items may be specified in the Working-Storage Section. The Report Section describes the content and format of all reports that are generated by the Report Writer Feature. The Report Section is discussed and described in greater detail later in the text.

The sections must appear in the above-mentioned sequence. If the section is not required in the source program, it may be omitted along with its name. (See figures 6.7, 6.8.)

```
DATA DIVISION.
FILE SECTION.
{file description entry
{record description entry}...}...
WORKING-STORAGE SECTION.
[data item description entry]...
[record description entry]...
REPORT SECTION.
{report description entry
{report group description entry}...}...
```

Figure 6.6 Structure of the Data Division.

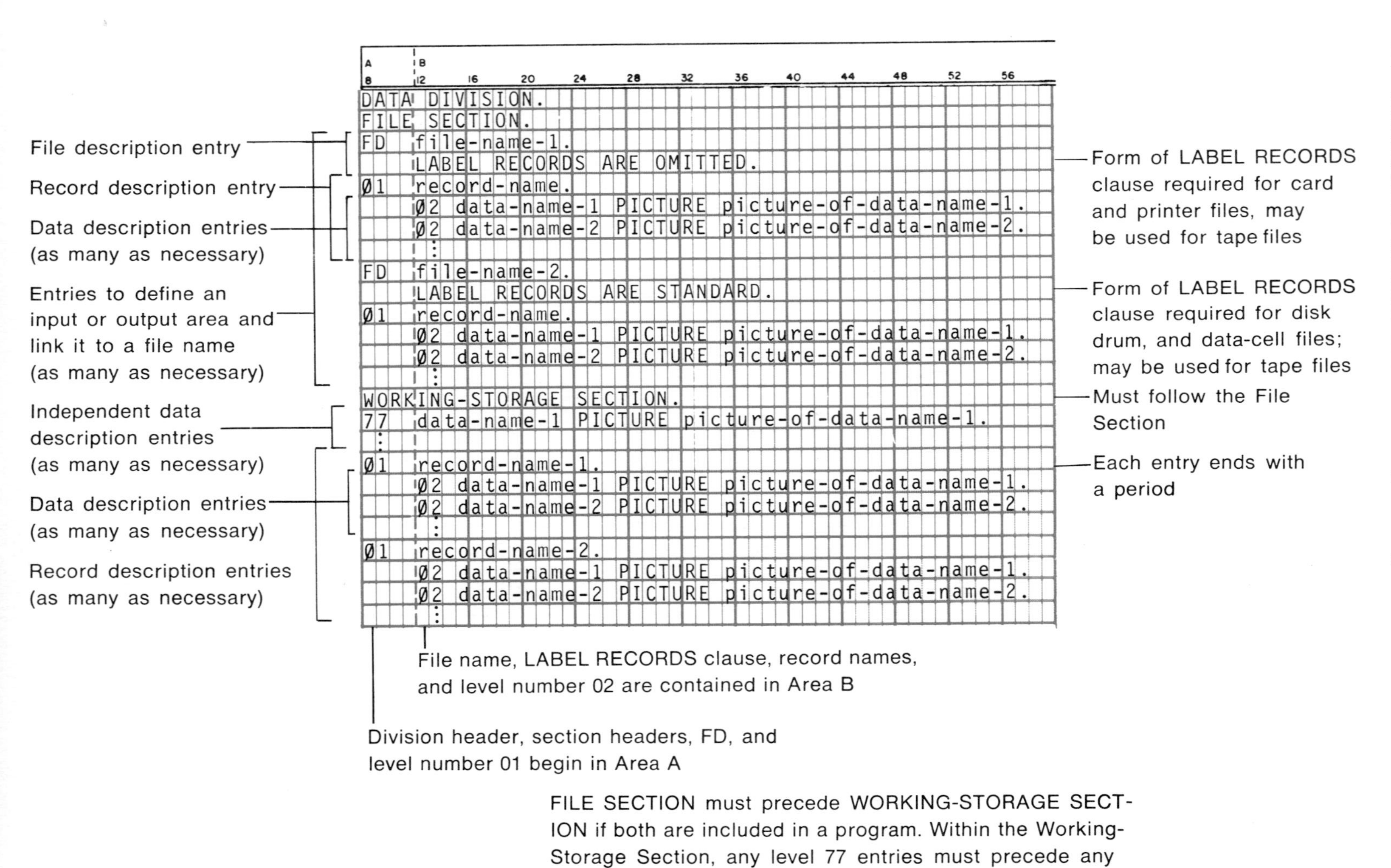

A
B
8
12
16
20
24
28
32
36
40
44
48
52
56
DATA DIVISION.
FILE SECTION.
FD file-name-1.
LABEL RECORDS ARE OMITTED.
Ø1 record-name.
Ø2 data-name-1 PICTURE picture-of-data-name-1.
Ø2 data-name-2 PICTURE picture-of-data-name-2.
FD file-name-2.
LABEL RECORDS ARE STANDARD.
Ø1 record-name.
Ø2 data-name-1 PICTURE picture-of-data-name-1.
Ø2 data-name-2 PICTURE picture-of-data-name-2.
WORKING-STORAGE SECTION.
77 data-name-1 PICTURE picture-of-data-name-1.
Ø1 record-name-1.
Ø2 data-name-1 PICTURE picture-of-data-name-1.
Ø2 data-name-2 PICTURE picture-of-data-name-2.
Ø1 record-name-2.
Ø2 data-name-1 PICTURE picture-of-data-name-1.
Ø2 data-name-2 PICTURE picture-of-data-name-2.
File description entry
Record description entry
Data description entries (as many as necessary)
Entries to define an input or output area and link it to a file name (as many as necessary)
Independent data description entries (as many as necessary)
Data description entries (as many as necessary)
Record description entries (as many as necessary)
Form of LABEL RECORDS clause required for card and printer files, may be used for tape files
Form of LABEL RECORDS clause required for disk drum, and data-cell files; may be used for tape files
Must follow the File Section
Each entry ends with a period
File name, LABEL RECORDS clause, record names, and level number 02 are contained in Area B
Division header, section headers, FD, and level number 01 begin in Area A
FILE SECTION must precede WORKING-STORAGE SECTION if both are included in a program. Within the Working-Storage Section, any level 77 entries must precede any record description entries.

Figure 6.7 Guide for Coding the Data Division with the File Section and the Working-Storage Section.

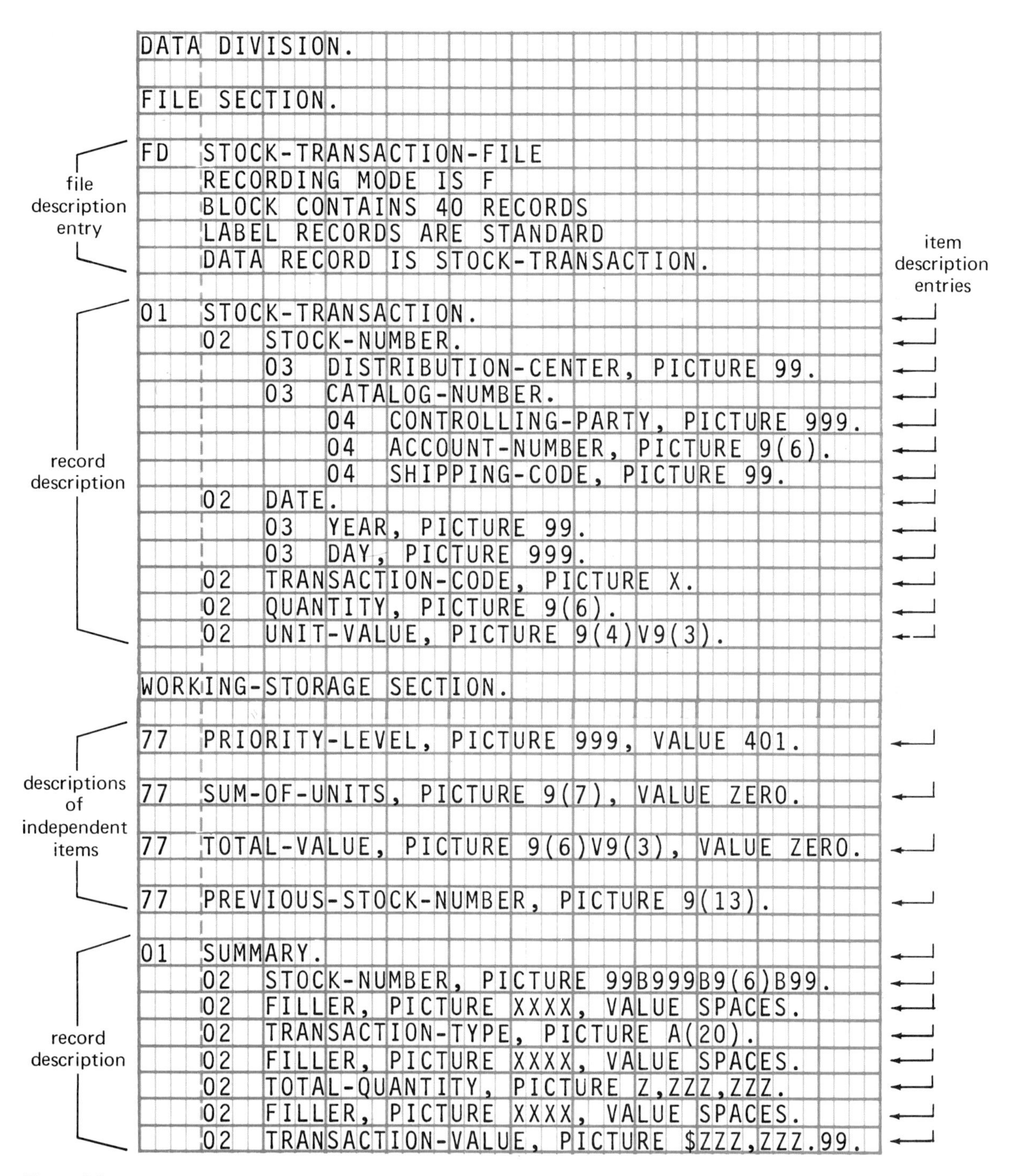

```
DATA DIVISION.

FILE SECTION.

FD  STOCK-TRANSACTION-FILE
    RECORDING MODE IS F
    BLOCK CONTAINS 40 RECORDS
    LABEL RECORDS ARE STANDARD
    DATA RECORD IS STOCK-TRANSACTION.

01  STOCK-TRANSACTION.
    02  STOCK-NUMBER.
        03  DISTRIBUTION-CENTER, PICTURE 99.
        03  CATALOG-NUMBER.
            04  CONTROLLING-PARTY, PICTURE 999.
            04  ACCOUNT-NUMBER, PICTURE 9(6).
            04  SHIPPING-CODE, PICTURE 99.
    02  DATE.
        03  YEAR, PICTURE 99.
        03  DAY, PICTURE 999.
    02  TRANSACTION-CODE, PICTURE X.
    02  QUANTITY, PICTURE 9(6).
    02  UNIT-VALUE, PICTURE 9(4)V9(3).

WORKING-STORAGE SECTION.

77  PRIORITY-LEVEL, PICTURE 999, VALUE 401.

77  SUM-OF-UNITS, PICTURE 9(7), VALUE ZERO.

77  TOTAL-VALUE, PICTURE 9(6)V9(3), VALUE ZERO.

77  PREVIOUS-STOCK-NUMBER, PICTURE 9(13).

01  SUMMARY.
    02  STOCK-NUMBER, PICTURE 99B999B9(6)B99.
    02  FILLER, PICTURE XXXX, VALUE SPACES.
    02  TRANSACTION-TYPE, PICTURE A(20).
    02  FILLER, PICTURE XXXX, VALUE SPACES.
    02  TOTAL-QUANTITY, PICTURE Z,ZZZ,ZZZ.
    02  FILLER, PICTURE XXXX, VALUE SPACES.
    02  TRANSACTION-VALUE, PICTURE $ZZZ,ZZZ.99.
```


Figure 6.8 Example—Data Division Entries.

FILE SECTION

Every program that processes input or output files is required to have a File Section. Since most programs employ files in the processing, a File Section appears in most programs. The File Section describes the characteristics of the file, and the record descriptions contained in those files (fig. 6.9).

For every file named in the SELECT clause in the Environment Division, a file description entry must appear in the Data Division.

File Description Entry

The file description entry describes:

1. The name of the file.
2. How the information is recorded.
3. The size of the blocks and records in the file.
4. Information about the label records.
5. The names of the data records within the file.

The file description entry consists of a level indicator (FD) followed by the file-name, followed by a series of independent clauses. The entry is terminated by a period (fig. 6.10).

```
Format  FD      file-name

        [RECORDING MODE IS   mode]

                                    {CHARACTERS}
        [BLOCK CONTAINS  integer-1  {RECORDS   }]

        [RECORD CONTAINS  [integer-2 TO]  integer-3  CHARACTERS]

        {LABEL RECORD IS   }  {STANDARD }
        {LABEL RECORDS ARE }  {OMITTED  }
                              {data-name}
          [VALUE OF Clause]

        {DATA RECORD IS   }
        {DATA RECORDS ARE }  record-name . . .
```

Figure 6.9 Format—File Description.

```
FD   OPINION-SURVEY,
     RECORDING MODE IS V,
     BLOCK CONTAINS 5 RECORDS,
     RECORD CONTAINS 120 TO 200
             CHARACTERS,
     LABEL RECORDS ARE STANDARD,
     DATA RECORD IS RESPONSE.
```

Figure 6.10 Example—File Description Entry.

Level Indicator

The file description entry always begins with the level indicator "FD," which is a reserved word. The indicator must be written at the A margin.

File-Name

The file-name always follows the level indicator. The name is supplied by the programmer and must be the same as stipulated in the SELECT clause in the Environment Division. The name must begin at the B margin.

The clauses that follow the name of the file are optional in many cases, and the order of their appearance is not important to the program.

Block Contains Clause

The Block Contains clause specifies either the number of records in a block or the number of characters in a block. When the number of characters per block is given, the clause specifies the largest number of characters that the longest block in storage will occupy. If both Integer-1 and Integer-2 are shown, they refer to the minimum and maximum size of the physical records respectively.

The Block Contains clause states the number of logical records or characters per physical record. The clause may be omitted when there is only one logical record per block. In all other instances, the clause is required (fig. 6.11).

```
BLOCK CONTAINS [integer-1 TO] integer-2 {CHARACTERS}
                                        {RECORDS   }
```

```
BLOCK CONTAINS 20 RECORDS
```

Figure 6.11 Block Contains Clause Format and Example.

Record Contains Clause

The Record Contains clause specifies the number of characters contained in the file (fig. 6.12). If the record does not range in size, this clause will specify how many characters will appear in the longest record. If the record has a range of record sizes, Integer-1 will indicate the size of the smallest record, and Integer-2 will indicate the size of the largest record.

```
RECORD CONTAINS [integer-1 TO] integer-2 CHARACTERS
```

```
RECORD CONTAINS 80 CHARACTERS
```

Figure 6.12 Record Contains Clause Format and Example.

Regardless of whether or not this clause is included, the record lengths are determined by the compiler from the record description entries. Since the size of each record is completely defined within the record description entry, this clause is never required.

Recording Mode Clause

The Recording Mode Clause is used to specify the format of logical records within a file. If the computer has more than one recording mode for data, this clause must be written (fig. 6.13). The IBM/360 and IBM/370 computers use the following recording modes.

```
RECORDING MODE IS mode
```

```
RECORDING MODE IS F;
```

Figure 6.13 Recording Mode Clause Format and Example.

F (Fixed Length). F recording mode assumes all records to be of the same length, and each is wholly contained within the block. No length or block length fields are necessary in this mode. The size of all logical records is fixed, and the logical records are not preceded by any control words. When these records are blocked, there is usually more than one record per block. The number of records per block is usually also fixed.

V (Variable Length). V recording mode assumes that records may be either fixed or variable in length and are preceded by a control word that specifies the length of the record. This is the only mode in which blocks of two or more variable length records may be handled. Each record contains a control word length field and each block contains a block length field. The control word is not described in any record entry in the Data Division and cannot be referred to by the program.

U (Undefined Length). U recording mode assumes that all records may be either fixed or variable in length. There are no record or block length fields, and there is only one record per block. Unlike the recording mode *V,* there are no control words preceding the logical record indicating the size of the record. The COBOL compiler considers the files with recording mode *U* as containing one record per block. The READ statement in the Procedure Division makes only one block available for processing (one record).

S (Spanned). In the *S* recording mode, the recording may be fixed or variable and may be larger than the block. If the record is larger than the block, a segment of the record is written in the block and the remainder is stored in the next block or blocks. Only complete records are made available to the user since each segment of a record in a block contains a control word, and each block contains a control word. These control words are automatically provided by the compiler and need no descriptive entries in the Data Division. These words cannot be referenced by the user.

(Note: If the Recording Mode clause is omitted, the default option is determined by an algorithm which does not always give *V* [variable length] recording mode).

Label Records Clause

The Label Records clause specifies whether labels are present and, if present, identifies the labels (fig. 6.14). Usually, magnetic tape files are labeled at the beginning to identify the file and the tape unit, and another label at the end of the file to provide a control to signal the end or indicate if there are more tapes in the file.

```
LABEL RECORDS ARE OMITTED;
```

Figure 6.14 Label Records Clause Format and Example.

The Label Records clause specifies the presence of standard or nonstandard labels in a file, or the absence of labels. This clause is required to appear in every file description entry. The clause may indicate that the label records are omitted or standard, or it may give a name for the label record.

Omitted. This option is used where there are no explicit labels for the file or where the existing file labels are nonstandard. The Omitted option is used for files assigned to unit record devices. This option may also be specified with nonstandard labels that the user wishes not to be processed by a label declarative.

Standard. The Standard option is used for labels that exist for a file and have the standard label format for the particular computer.

Data-Name. The Data-Name option indicates either the presence of user labels in addition to the standard labels or the presence of nonstandard labels. Data-Name is a programmer-supplied name of a storage area in which the labels will be processed. Data-Name will be defined in the File Section of the Data Division, where it must be associated with the appropriate FD entry.

Value Of Clause

The Value Of clause particularizes the description of an item in the label records associated with the file and serves only as documentation. To specify the required value of identifying data items in the label record for the file, the programmer must use the Value Of clause (fig. 6.15).

```
VALUE OF data-name-1 IS { literal-1   }
                        { data-name-2 }
         [data-name-3 IS { literal-2   } ] . . .
                         { data-name-4 }
```

```
VALUE OF IDENT IS DESC
```

Figure 6.15 Value of Clause Format and Example.

Data Records Clause

The Data Records clause serves only as documentation and informs the compiler what the name or names of each of the records in a file are (fig. 6.16).

```
DATA { RECORD IS   } data-name-1 [data-name-2] . . .
     { RECORDS ARE }
```

```
DATA RECORDS ARE EXPENSE-DETAIL,
     DEPARTMENT-TOTAL.
```

Figure 6.16 Data Record Clause Format and Example.

The name of each record is the data-name supplied by the programmer. There must be at least one record in each file so that this clause will appear in every file description entry. The presence of more than one data-name indicates that the file has more than one data record. Two or more record descriptions may occupy the same storage area for a given file. These records need not have the same description, and may be of differing sizes, differing formats, etc. When records of differing sizes are defined, the size of each record written is equal to the length of the largest record defined. The order in which these records are written is not significant.

Conceptually, all data records within a file share the same area. This is in no way altered by the presence of more than one type of record within a file.

Below the file description entry, each record-name must also appear in the level 01 entry in the record description clauses. This Data Records clause is never required in the file description entry.

Report Clause

The Report clause is used in conjunction with the Report Writer Feature. A complete description and information regarding use of this clause can be found in the "Report Writer Feature" section.

RECORD DESCRIPTION ENTRY

At least one record description entry is found below each file description entry. A record description entry specifies the characteristics of each item in a data record. Every entry must be described in the same order as that in which the item appears in the record, and must indicate whether the items are related to each other. Each record description entry consists of a level number, a data-name or FILLER, and a series of clauses separated by spaces. The entry must be terminated by a period.

Some records may be divided into smaller units as follows:

1. Each entry must be given a level number beginning with a 01 for the record, and succeeding entries are given higher-level numbers.
2. In subdividing an entry, the level numbers need not be consecutive. Level numbers 01–49 may be used for entries in the File Section.

Elementary Items are not further subdivided. Elementary items may be part of a group, or may be an independent item (not part of a group).

Group Items consist of all items under it until a level number equal to or less than the group number is reached.

Indentation. Item descriptions are usually indented to show the reader the relationship of the items within the group. Indenting is not required. If one entry at a given level is indented, then all similar entries should be indented for consistency.

Level Numbers

Level numbers are used to structure a logical record to satisfy the need to specify subdivision of data record for the purpose of data reference. The system of level numbers shows the organization of elementary and group items (fig. 6.17).

Level numbers are the first items of a record description entry.

1. Level number 01 and 77 must begin at the A margin followed by data-names and associated clauses beginning at the B margin. All other level numbers may begin at the A or B margin, with the data-names and associated clauses beginning at the B margin.
2. At least one space must separate a level number from its data-name.
3. Separate entries are written for each level number.
4. A single-digit level number may be written as a space followed by a digit or as a zero followed by a digit.

01 Level number indicates that the item is a record. Since records are the most inclusive data items, the level number for a record must be 1 or 01. A

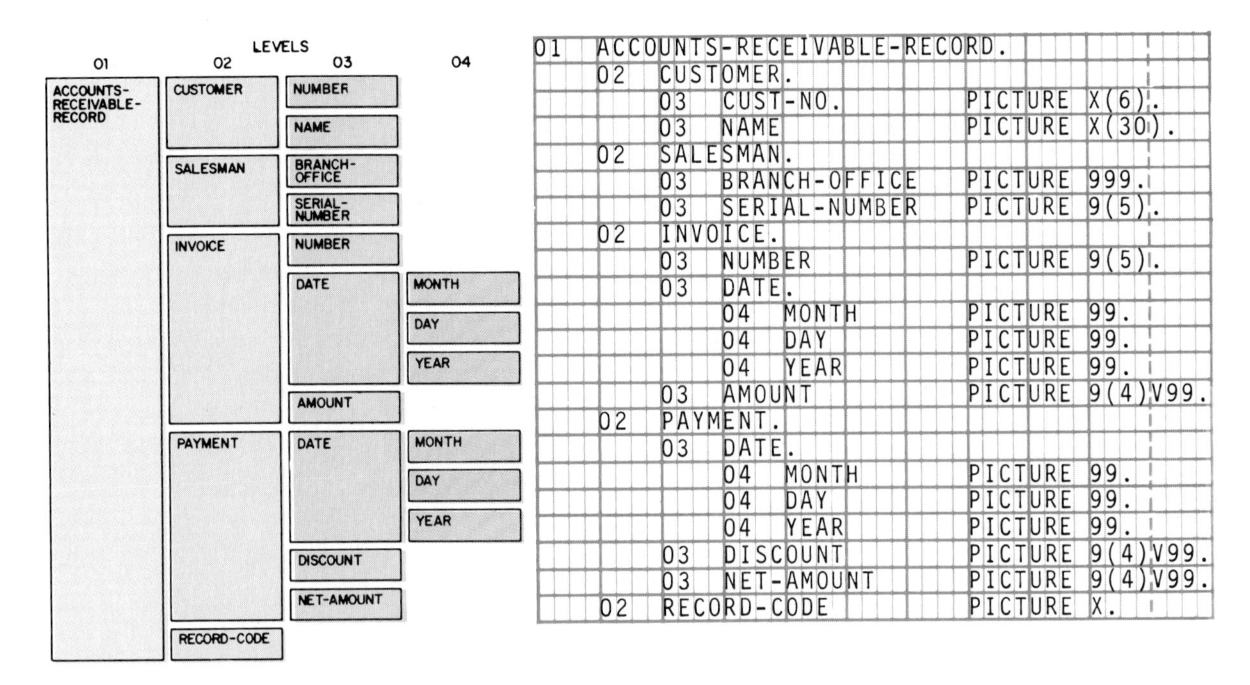

Figure 6.17 Example—Record Description Entries and Levels.

record is usually composed of related elementary items, but may be an elementary item itself.

Level numbers 02–49 are used for subdivisions of group-related record items (not necessarily in consecutive order). See figure 6.18.

There are several special level numbers used within data items where there is no real concept of level. These level numbers are:

66 Level number is used for names of elementary items or groups described by a RENAMES clause for the purpose of regrouping data items (see RENAMES clause section for an example of the function of the clause).

77 Level number is used to identify an independent elementary item in the Working-Storage Section. The item is not part of any record and is not related to any other item. Level 77 is usually used for noncontiguous items to define a work area or to store constant data (fig. 6.19).

88 Level number designates a condition entry and is used to assign values to particular items during execution time. A name is furnished to values that the preceding item assumes. It does not reserve any storage area. Level 88 is associated only with elementary items (fig. 6.20). (See Condition-Name clause section for examples of the use of level 88 entries.)

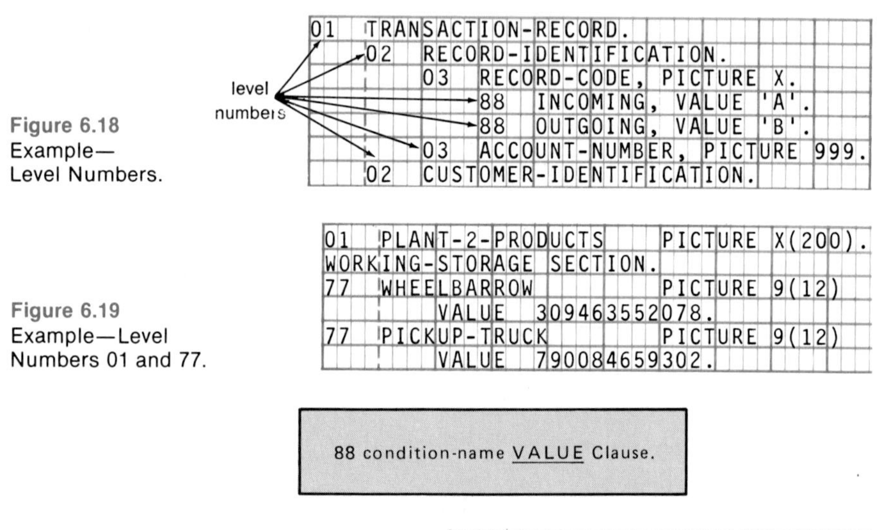

01 TRANSACTION-RECORD.
02 RECORD-IDENTIFICATION.
03 RECORD-CODE, PICTURE X.
88 INCOMING, VALUE 'A'.
88 OUTGOING, VALUE 'B'.
03 ACCOUNT-NUMBER, PICTURE 999.
02 CUSTOMER-IDENTIFICATION.
level numbers
01 PLANT-2-PRODUCTS PICTURE X(200).
WORKING-STORAGE SECTION.
77 WHEELBARROW PICTURE 9(12)
VALUE 309463552078.
77 PICKUP-TRUCK PICTURE 9(12)
VALUE 790084659302.
88 condition-name VALUE Clause.

Figure 6.18 Example—Level Numbers.

Figure 6.19 Example—Level Numbers 01 and 77.

```
02  MARITAL-STATUS, PICTURE 9.
    88  SINGLE, VALUE 0.
    88  MARRIED, VALUE 1.
    88  DIVORCED, VALUE 2.
    88  WIDOWED, VALUE 3.
```

Figure 6.20 Condition-Name Clause Format and Example.

Data-Name or FILLER

Data-Name. Each item in the record description entry must contain either a data-name or the reserved word FILLER immediately following the level number beginning at the B margin. The data-name permits the programmer to refer to items individually in procedural statements. The data-name must be unique (not a reserved word) or must be properly qualified if not unique. The data-name can be made unique by either spelling the data-name differently from any other data-name used in the program, or through qualification of a nonunique name (see Qualification of Names section).

The data-name refers to the name of the storage area that contains the data, not to a particular value; the item referred to may assume numerous values during the execution of the program. (See figure 6.21.)

In addition to the rules mentioned earlier in the Qualification of Names section, the following rules apply to data-names in the Data Division.

1. The highest possible qualifier would be the name of the file; thus it is possible for two records to have the same name.
2. The highest possible qualifier in the Working-Storage Section would be the record-name; thus all record-names in this section must be unique. The data-names for all independent items (level 77) must be unique since they can never be qualified.

FILLER. The reserved word FILLER may be used in place of a data-name. The name cannot be referenced by any procedural statements. Its primary use is in the description of items that will not be referred to because the information contained is not necessary for the processing of the program. FILLER *does not always represent a blank area.*

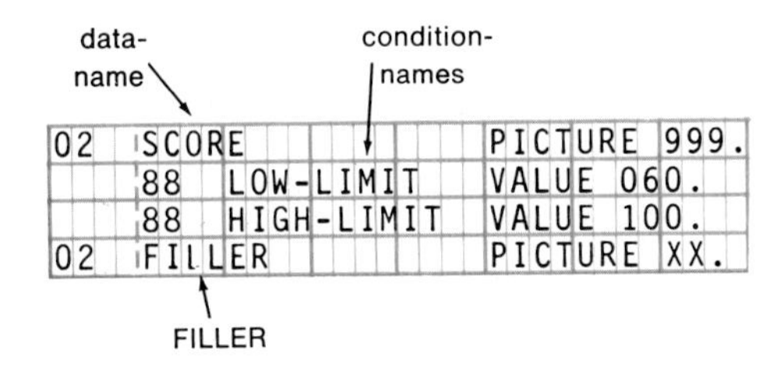

Figure 6.21 Example—Data-Names, Condition-Names, and Filler.

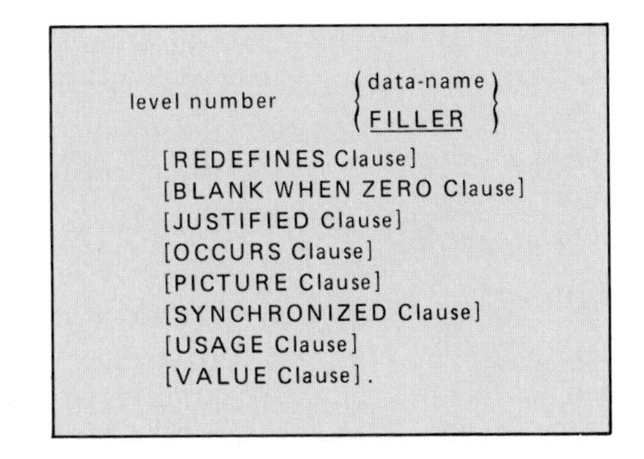

Figure 6.22 Record Description Format.

Independent Clauses

Each record description entry may consist of one or more clauses that provide information about the data item. The most commonly used clauses are the USAGE, PICTURE, and VALUE clauses (fig. 6.22).

Usage

The USAGE clause describes the form in which the data is stored in the computer's main storage (fig. 6.23).

Rules for USAGE clauses are

1. The clause may be written at any level.
2. If the clause is written at a group level, it applies to each elementary item within the group (fig. 6.24).
3. The usage of an elementary item must not contradict the explicit usage of the group item of which the item is a part. (See figure 6.25.)

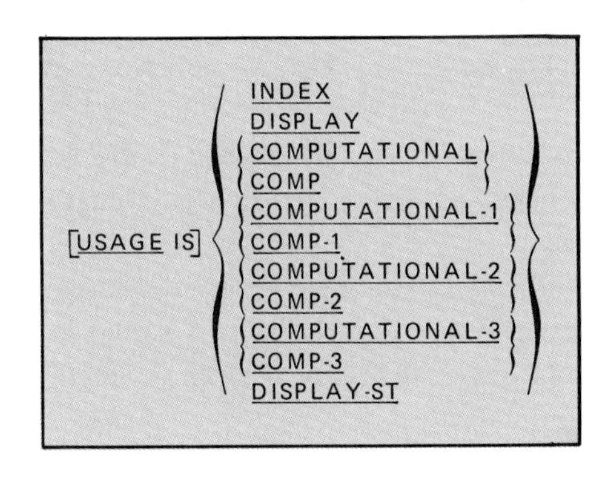

Figure 6.23 Usage Clause Format.

```
02  PAYMENT, COMPUTATIONAL-3.
    03  AMOUNT-DUE, PICTURE S9(6)V99.
    03  AMOUNT-PAID, PICTURE S9(6)V99.
```

Figure 6.24 Example—Group Usage Clause.

```
01  STOCK-TRANSFER.
    02  STOCK-NUMBER      PICTURE X(7).
    02  DESCRIPTION       PICTURE X(15).
    02  UNITS-OF-STOCK    PICTURE S9(8)
              COMPUTATIONAL.
```

Figure 6.25 Example—Elementary Usage Clause.

4. The usage of an elementary item is assumed DISPLAY if some other usage clause is not specified at the group or elementary level.

DISPLAY. This option specifies that one character is stored in each byte of the item. This corresponds to the form in which the information is represented for initial card input or for final printed or punched output. If the item is used to store numeric data, the rightmost byte may contain an operational sign in addition to the digit.

COMPUTATIONAL CLAUSE. All items in a computational clause represent values to be used in arithmetic operations, and must be numeric. If a group USAGE clause is used, it is only the elementary items that have that usage, since the group item cannot be used in computation.

The SYNCHRONIZED clause must be added to all definitions of COMPUTATIONAL clauses that require alignment or the level number 01 descriptions that contain such items. The SYNCHRONIZED clause assures proper alignment of an elementary item on computer memory boundaries. The intent of the clause is to efficiently align data items on integral storage boundaries.

COMPUTATIONAL. This option is specified for binary data items. One binary digit is stored in each bit of the item except the leftmost, which will contain the operational sign. Such items have the decimal equivalent consisting of the decimal digits 0 through 9 plus the operational sign.

The amount of storage to be occupied by a binary item depends on the number of digits in its PICTURE clause.

Digits in PICTURE clause	Storage Occupied
1 through 4	2 bytes (halfword)
5 through 9	4 bytes (fullword)
10 through 18	8 bytes (2 fullwords not necessarily a doubleword)

The PICTURE clause of an item having COMPUTATIONAL usage may contain only 9s, the operational sign character S, the implied decimal point V and one or more Ps. An operational sign character S must appear in COMPUTATIONAL usage items.

Items are aligned at the nearest halfword, fullword, or doubleword boundary.

COMPUTATIONAL-1. This option specifies that the item is stored in short precision internal floating-point format. Such items are 4 bytes in length and aligned on the next fullword boundary.

The data code is internal floating point, short (fullword) format.

COMPUTATIONAL-2. This option specifies that the item is stored in long precision internal floating-point format. Such items are 8 bytes in length and are aligned on the next double-word boundary.

The data code is internal floating-point long (doubleword) format.

Both COMPUTATIONAL-1 and COMPUTATIONAL-2 options have special formats designed for floating-point arithmetic operations. Part of the item may be stored in binary form, and part may be stored in hexadecimal format.

No PICTURE clauses are associated with internal floating-point items.

COMPUTATIONAL-3. This option is specified for an item that is stored in packed decimal format (2 digits per byte) with the low-order 4 bits of the rightmost byte containing the operational sign.

The PICTURE clause of COMPUTATIONAL-3 usage may contain only 9s, the operational sign S, the assumed decimal point V, and one or more Ps.

The data code is internal decimal (packed decimal format). (See figure 6.26.)

INDEX. This option is discussed in the Table-Handling Section of the text. (See figure 6.27.)

Item	Value	Description	Internal Representation*
External Decimal	−1234	DISPLAY PICTURE 9999	Z1 | Z2 | Z3 | F4 (byte)
		DISPLAY PICTURE S9999	Z1 | Z2 | Z3 | D4 (byte) Note that, internally, the Dr, which represents −4, is the same bit configuration as the EBCDIC character M.
Binary	−1234	COMPUTATIONAL PICTURE S9999	1111 | 1011 | 0010 | 1110 (S; byte) Note that, internally, negative binary numbers appear in two's complement form.
Internal Decimal	+1234	COMPUTATIONAL-3 PICTURE 9999	01 | 23 | 4F (byte)
		COMPUTATIONAL-3 PICTURE S9999	01 | 23 | 4C (byte)
External Floating-point	+12.34E+2	DISPLAY PICTURE +99.99E-99	+ | 1 | 2 | . | 3 | 4 | E | b | 0 | 2 (byte)
Internal Floating-point		COMPUTATIONAL-1	S | Exponent | Fraction (0 1 7 8 31)
		COMPUTATIONAL-2	S | Exponent | Fraction (0 1 7 8 63)

*Codes used in this column are as follows:

Z = zone, equivalent to hexadecimal F, bit configuration 1111

Hexadecimal numbers are their equivalent meanings are:

F = non-printing plus sign (treated as an absolute value)
C = internal equivalent of plus sign, bit configuration 1100
D = internal equivalent of minus sign, bit configuration 1101

S = sign position of a numeric field; internally,
1 in this position means the number is negative
0 in this position means the number is positive

b = a blank

Figure 6.26 Internal Representation of Numeric Items.

If the usage is	Then the data code is	Which means that
display	external decimal—also called BCD (binary-coded decimal), or EBCDIC (extended binary coded decimal interchange code)	one character is stored in each byte of the item; if the item is used to store a number, the rightmost byte may contain an operational sign in addition to a decimal digit
computational	binary	one binary digit is stored in each bit of the item, except the leftmost bit, in which the operational sign is stored
computational-1	internal floating-point, short (fullword) format	the item has a special format designed for floating-point arithmetic operations; part of the item is stored in binary code, and part in hexadecimal code
computational-2	internal floating-point, long (doubleword) format	
computational-3	internal decimal—also called packed decimal	two decimal digits are stored in each byte of the item, except the rightmost byte, in which one digit and the operational sign are stored

Figure 6.27 IBM 360/370 Chart—Usage Clause.

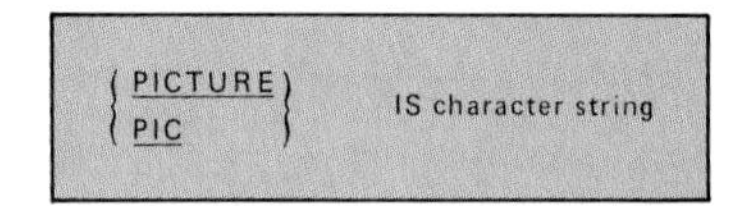

Figure 6.28 Picture Clause Format.

PICTURE Clause

The PICTURE clause specifies the general characteristics and the detail description of an elementary item (fig. 6.28).

Rules for the Use of PICTURE Clause are

1. This clause is required in the description of every elementary item except those whose usages are COMPUTATIONAL-1 or COMPUTATIONAL-2. (Floating-point items have definite storage formats.)
2. The clause tells how many characters will be stored and describes the types of characters through the use of various symbols.
3. The clause is forbidden at the group level.
4. Numeric literals enclosed in parenthesis is a shorthand method of expressing the repetitively consecutive occurrence of a character. For example, X(10) is another way of writing XXXXXXXXXX.
5. All characters except P, V, and S are counted in the total size of any item.
6. CR and DB occupy two character positions in storage, and both may not appear in the same PICTURE clause.
7. A maximum of 30 characters is permitted in the clause. For example, PICTURE X(60) consists of 5 PICTURE characters, since only the actual characters appearing in the PICTURE clause are included in the count.
8. The characters S, V, CR, and DB may appear only once in a clause.

(See figures 6.29, 6.30.)

If the picture contains	and also (possibly)	For example . . .	Then the item is called	And will be used to store
one or more Xs		XXX	alphanumeric	characters of any kind; letters, digits, special characters, or spaces
one or more As		A(35)	alphabetic	only letters or spaces
one or more 9s, but no editing symbols	S V P	S9(7)V99	numeric	only digits, and possibly an operational sign
one or more editing symbols; Z * $. , DB CR + − 0 B	9 V P	$ZZ,ZZZ.99	report	numeric data that is edited with spaces or certain special characters when the data is moved into the item
an E, in addition to 9s	+ − . V	+.9(8)E+99	external floating-point	a decimal quantity in an edited floating-point format that includes spaces or certain special characters

Figure 6.29
Identification of an Item by its Picture.

X	Each X stands for one character of any kind—a letter, digit, special character, or space. The picture X(12) indicates that the item will contain twelve characters, but gives no indication of what characters they will be; all twelve could be spaces, or all could be digits, or there could be a mixture of various kinds of characters.
A	Each A stands for one letter or space.
9	Each 9 stands for one decimal digit. Numbers are always described in terms of the *decimal* digits they are the equivalent of—even when the data code is *binary*.
S	S indicates that the number has an operational sign. An "operational" sign tells the computer that the number is negative or positive; it is *not* a separate character that will print as "+" or "−".
V	V shows the location of an assumed decimal point in the number. An "assumed" decimal point is *not* a separate character in storage.
P	Each P stands for an assumed zero. Ps are used to position the assumed decimal point away from the actual number. For example, an item whose actual value is 25 will be treated as 25000 if its picture is 99PPPV; or as .00025 if its picture is VPPP99.

Figure 6.30
Meaning of Some Common Picture Characters.

Categories of Data

The categories of data that can be described with a PICTURE clause are

1. Alphabetic.
2. Numeric.
3. Alphanumeric.
4. Alphanumeric Edited.
5. Numeric Edited.

(See figure 6.31.)

Level of Item	Class	Category
Elementary	Alphabetic	Alphabetic
	Numeric	Numeric
	Alphanumeric	Alphanumeric Alphanumeric Edited Numeric Edited
Group	Alphanumeric	Alphabetic Numeric Alphanumeric Alphanumeric Edited Numeric Edited

Figure 6.31
Class and Category of Elementary and Group Data Items.

Alphabetic. An alphabetic item may contain any combination of the 26 letters of the alphabet and the space. No special characters or numerics are permitted in an alphabetic item. The permissible character in an alphabetic picture is A.

Numeric. A numeric item may contain any combination of the numerals 0–9; the item may have an operational sign. Permissible characters in a numeric picture are 9, V, P, and S.

Alphanumeric. An alphanumeric item may contain any combination of characters in the COBOL character set. A permissible character in an alphanumeric picture is X.

Alphanumeric Edited. An alphanumeric edited item is one whose picture clause is restricted to certain combinations of the following characters: A, X, 9, B, 0. To qualify as an alphanumeric edited item, one of the following conditions must exist. The PICTURE clause must contain at least:

1. One B and at least one X.
2. One 0 and at least one X.
3. One 0 and at least one A.

Numeric Edited. A numeric edited item is one whose PICTURE clause is restricted to certain combinations of the following characters: B, P, V, Z, 0, 9, ., *, +, −, CR, DB, $. The maximum number of digits in a numeric edited picture is 18. (See figure 6.32.)

Picture Characters

Nonedited PICTURE Clauses. A nonedited PICTURE clause may contain a combination of the characters as shown in figure 6.33. For examples, see figure 6.34.

Edited PICTURE Clauses. An edited PICTURE clause is used to describe items to be output on the printer (alphanumeric and numeric edited items).

Rules Governing the Use of Edited PICTURE Clauses

1. There must be at least one digit position character in the clause.
2. If a fixed or floating string of plus or minus insertion characters are used, no other sign control character may be used.

	Source Area		Receiving Area	
	PICTURE	Data Value	PICTURE	Edited Data
1.	S99999	12345	-ZZ,ZZ9.99	12,345.00
2.	S99999V	00123	$ZZ,ZZ9.99	$ 123.00
3.	S9(5)	00100	$ZZ,ZZ9.99	$ 100.00
4.	S9(5)V	00000	-ZZ,ZZ9.99	0.00
5.	9(5)	00000	$ZZ,ZZZ.99	$ .00
6.	9(5)	00000	$ZZ,ZZZ.ZZ	
7.	999V99	12345	$ZZ,ZZ9.99	$ 123.45
8.	V99999	12345	$ZZ,ZZ9.99	$ 0.12
9.	9(5)	12345	$**,**9.99	$12,345.00
10.	9(5)	00123	$**,**9.99	$***123.00
11.	9(5)	00000	$**,***.99	$******.00
12.	9(5)	00000	$**,***,**	**********
13.	99V999	12345	$**,**9.99	$****12.34
14.	9(5)	12345	$$$,$$9.99	$12,345.00
15.	9(5)	00123	$$$,$$9.99	$123.00
16.	9(5)	00000	$$$,$$9.99	$0.00
17.	9(4)V9	12345	$$$,$$9.99	$1,234.50
18.	V9(5)	12345	$$$,$$9.99	$0.12
19.	S99999V	-12345	-ZZZZ9.99	-12345.00
20.	S9(5)V	12345	-ZZZZ9.99	12345.00
21.	S9(5)	-00123	-ZZZZ.99	- 123.00
22.	S99999	12345	ZZZZ9.99	12345.00
23.	S9(5)	-12345	ZZZZ9.99-	12345.00-
24.	S9(5)	00123	------.99	123.00
25.	S9(5)	-00001	------.99	-1.00
26.	S9(5)	12345	+ZZZZZ.99	+12345.00
27.	S9(5)	-12345	+ZZZZZ.99	-12345.00
28.	S9(5)	12345	ZZZZZ.99+	12345.00+
29.	S9(5)	-12345	ZZZZZ.99	12345.00
30.	S9(5)	00123	++++++.99	+123.00
31.	S9(5)	00001	------.99	1.00
32.	9(5)	00123	++++++.99	+123.00
33.	9(5)	00123	------.99	123.00
34.	9(5)	12345	BB999.00	345.00
35.	9(5)	12345	00099.00	00045.00
36.	S9(5)	-12345	ZZZZZ.99CR	12345.00CR
37.	S9(5)	12345	$$$$$$.99CR	$12345.00

Figure 6.32 Examples—Editing Applications of Picture Clauses.

Character	Meaning
9	The character 9 indicates that the position contains one decimal digit. Numbers are always described in terms of the decimal digits that they are equivalent to, even when the data is binary.
X	The character X indicates that the position can contain any type of character in the COBOL character set; a letter, digit, or special character.
V	The character V indicates the presence of an assumed decimal point. Since a numeric nonedited item may not contain an actual decimal point, an assumed decimal point provides the compiler with information concerning decimal alignment involved in computations. An "assumed decimal point" is not counted in the size of an elementary item and does not reserve any storage space.
P	The character P indicates the presence of an assumed zero. The Ps are used to position the assumed decimal point away from the actual number. For example, the actual value in storage is 15. It would be treated as 15000 if the PICTURE clause is 99PPP, or as .00015 if the PICTURE clause is VPPP99. The character V may be used or omitted when using the P character. When the V is used, it must be placed in the position of the assumed decimal point, to the left or the right of P or Ps that have been specified. The scaling position character P is not counted in the size of the data item.
S	The character S indicates the presence of an operational sign to the computer, either a positive or a negative number. It is not a separate character that will be printed as + or −. If used, S must be written as the leftmost character of the PICTURE clause. The presence of S is required where the USAGE clause is indicated as COMPUTATIONAL, since a sign appears in all binary numbers. The absence of S in a PICTURE clause will indicate a positive value. The operational sign is not counted in the size of an item.
A	The character A indicates the presence of a letter or space in an item. No special characters are permitted in a PICTURE clause with A picture.

Figure 6.33 Nonedited Picture Characters and Their Meanings.

3. The character to the left of an actual or assumed decimal point in the PICTURE clause (excluding the floating string of characters) are subject to the following restrictions.

   a. A Z may not follow 9, a floating string, or *.
   b. * may not follow 9, Z, or a floating string.

4. A floating string must begin with two consecutive characters (+, −, or $).
5. There may be only one type of floating string characters.
6. If the PICTURE clause does not contain 9s, BLANK WHEN ZERO is implied unless all the numeric positions contain asterisks. If the PICTURE clause does contain asterisks, and the area is zero, the area will be filled with asterisks.

A V indicates the location of the assumed decimal point.

If the data item contains:	and the PICTURE clause is:	then the data is interpreted as:
3492	99V99	34.92
169	9V99	1.69
175	99V9	17.5
254	V999	.254
36985	9V9999	3.6985
45694	9999V9	4569.4
98745	99999	98745.

A P indicates as assumed decimal scaling position and specifies the location of an assumed decimal point when the point is not within the data item.

If the data item contains:	and the PICTURE clause is:	then the data is interpreted as:
246	999PPP	246000.
387	PP999	.00387
487	999P	4870.

Figure 6.34 Nonedited Picture Examples.

7. The following restrictions apply to the characters to the right of the decimal point up to the end of the PICTURE (excluding insertion characters of +, −, CR, DB, if present).
   a. Only one type of digit character is permissible.
   b. If any of the characters appearing to the right of the decimal point is represented by +, −, Z, *, or $, then all the numeric characters in the PICTURE must be represented by the same characters.
8. The PICTURE character 9 can never appear to the left of the floating string or replacement character.
9. There cannot be a mixture of floating or replacement characters in an editing picture. They may appear as follows:
   a. An * or Z may appear with a fixed $.
   b. An * or Z may appear with a fixed leftmost + or fixed leftmost −.
   c. An * or Z may appear with a fixed rightmost + or fixed rightmost −.
   d. $ (fixed or floating string) may appear with fixed rightmost + or −.

The characters and meanings of allowable editing characters in edited PICTURE clauses are shown in figures 6.35, 6.36. (See also figures 6.37, 6.38.)

Character	Meaning
Z	The character Z represents a digit-suppression character. 1. Each character Z represents a digit position. 2. All leading zeros appearing in positions represented by Zs are suppressed, leaving the positions blank. 3. Zero suppression is terminated when the actual or assumed decimal point is encountered. 4. A Z may appear to the right of the decimal point point only if all positions to the right are represented by Zs. 5. If all digit positions are represented by Zs, and the value of the data is zero, the entire area will be filled with blanks. 6. A Z character may not appear anywhere to the right of a 9 character. 7. Each Z is counted in the size of the item.
•	The character (.) represents an actual decimal point to be inserted in the printed output. 1. The decimal point is actually printed in the position indicated. 2. The source data is decimal aligned. 3. The character that appears to the right of the actual decimal point must consist of characters of one type (Z, *, 9, +, $, or −). 4. The character is counted in the size of a data item. 5. The actual decimal point may not be the last character in the PICTURE clause.
*	The asterisk (*) in the edited PICTURE clause is primarily used for protection of the amount in the printing of checks. 1. Each asterisk represents a digit position. 2. Leading nonsignificant zeros are replaced by asterisks. 3. Each field so defined will be replaced by asterisks until an actual or assumed decimal point is encountered. 4. An asterisk may appear to the right of the decimal point only if all digit positions are represented by asterisks. 5. If all digit positions are zero, the entire area will be filled with asterisks, except the actual decimal point. 6. The BLANK WHEN ZERO clause does not apply to any item having an asterisk (*) in its PICTURE. 7. An asterisk is counted in the size of an item.
CR DB	These character symbols are used as editing sign control symbols. These character symbols are printed *only if an item is negative*. They are called credit and debit symbols. 1. They may appear only at the right end of a PICTURE. 2. A positive value will blank out the characters, and only spaces will appear. 3. These symbols occupy 2 character positions and are counted in determining the size of an item.

Figure 6.35 Edited Characters and Their Meanings.

Picture character	Data type	Specification	Additional explanation
X	alphanumeric	The associated position in the value will contain any character from the COBOL character set.	
A	alphabetic	The associated position in the value will contain an alphabetic character or a space.	
9	numeric or numeric edited	The associated position in the value will contain any digit.	
V	numeric	The decimal point in the value will be assumed to be at the location of the V. The V does not represent a character position.	
	numeric edited	The associated position in the value will contain a point or a space.	A space will occur if the entire data item is suppressed.
$	numeric edited	a. (simple insertion) The associated position in the value will contain a dollar sign. b. (floating insertion) The associated position in the value will contain a dollar sign, a digit, or a space.	The leftmost $ in a floating string does not represent a digit position. If the string of $ is specified only to the left of a decimal point, the rightmost $ in the picture corresponding to a position that precedes the leading nonzero digit in the value will be printed. A string of $ that extends to the right of a decimal point will have the same effect as a string to the left of the point unless the value is zero; in this case blanks will appear. All positions corresponding to $ positions to the right of the printed $ will contain digits; all to the left will contain blanks.
	numeric edited	The associated position in the value will contain a comma, space, or dollar sign.	A comma included in a floating string is considered part of the floating string. A space or dollar sign could appear in the position in the value corresponding to the comma.
S	numeric	A sign (+ or −) will be part of the value of the data item. The S does not represent a character position.	

Figure 6.36 Picture and Edit Characters.

Each Z in a character-string represents a leading numeric character position that is replaced by a space character when the content of that character position is zero.

If data moved to data-name is:	and the PICTURE clause is:	then the contents of data-name after move is:
00000	ZZZZZ	
39052	ZZZZZ	39052
00006	ZZZZZ	6
00295	ZZZZZ	295
00005	ZZZ99	05

Each asterisk (*) in the character-string represents a leading numeric character position into which an * is placed when the content of that position is zero.

If data moved to data-name is:	and the PICTURE clause is:	then the contents of data-name after move is:
00000	*****	*****
00820	*****	**820
00858	***99	**858
00075	**999	**075
78963	*****	78963

Figure 6.37 Edited Picture—Examples.

Editing Symbol in PICTURE Character String	Result	
	Data Item Positive or Zero	Data Item Negative
+	+	−
−	space	−
CR	2 spaces	CR
DB	2 spaces	DB

Figure 6.38 Editing Sign Control Symbols and Their Results.

Insertion Characters

Insertion characters are counted in determining the size of an item and represent the position into which the character will be inserted.

The characters and meanings of the insertion characters are indicated in figure 6.39. (See also figures 6.40, 6.41, 6.42.)

Character	Meaning
,(comma) B(space) 0(zero)	1. The insertion character does not represent a digit position 2. Zero Protection (Z) and Check Protection (*) indicates the replacement of insertion characters with spaces or asterisks if a significant digit or decimal point has not been encountered. 3. The comma, blank, or zero may appear with floating strings.

Figure 6.39 Insertion Characters and their Meanings.

PICTURE	Value of Data	Edited Result
99,999	12345	12,345
9,999,000	12345	2,345,000
99B999B000	1234	01 234 000
99B999B000	12345	12 345 000
99BBB999	123456	23 456

Figure 6.40 Examples—Simple Insertion Editing.

PICTURE	Value of Data	Edited Result
999.99	1.234	001.23
999.99	12.34	012.34
999.99	123.45	123.45
999.99	1234.5	234.50

Figure 6.41 Examples—Special Insertion Editing.

PICTURE	Value of Data	Edited Result
999.99+	+6555.556	555.55+
+9999.99	-5555.555	-5555.55
9999.99-	+1234.56	1234.56
$999.99	-123.45	$123.45
-$999.99	-123.456	-$123.45
$9999.99CR	+123.45	$0123.45
$9999.99DB	-123.45	$0123.45DB

Figure 6.42 Examples—Fixed Insertion Editing.

Floating Strings

Floating strings are a series of continuous characters of either $, or +, or −, or a string composed of one or a repetition of one, such characters may be interrupted by one, or more insertion characters (comma, 0 , B) and/or V, or an actual decimal point (fig. 6.43). Floating strings begin with at least two consecutive occurrences of the characters to be floated.

```
$$,$$$,$$$
++++
--,---,--
$$$B$$$
+(8)V++
$$,$$$.$$
```

Figure 6.43
Example—Floating Strings.

1. The floating string characters are inserted immediately to the left of the digit position indicated.
2. Blanks are placed in all positions to the left of the singly floating string character after insertion.
3. The presence of an actual or assumed decimal point in a floating string is treated as if all digit positions to the right of the decimal point where indicated by the PICTURE character 9 and BLANK WHEN ZERO clause were written for them.
4. A floating string need not constitute the entire picture.
5. When B (blank) or , (comma) or 0 (zero) appears to the right of the floating string, the character floats there to be as close to the leading digit as possible.
6. A comma may not be the last character in a PICTURE clause. (See figure 6.44.)

PICTURE	VALUE	Edited Result
$$$$.99	12.34	$12.34
$$$$.99	1234	$234.00
$$$$.99	.1234	$.12
- - - - .99	+12.34	12.34
- - - - .99	- 1.234	- 1.23
$$99.99	1.234	$01.23

Figure 6.44 Example—Floating Strings Insertion Editing.

Character	Meaning
B(blank)	The character B indicates that an imbedded blank is to appear in the indicated position unless the position immediately precedes a nonsignificant zero. Embedded blanks need not be single characters.
,(comma) 0(zero)	The characters comma and zero operate in the same manner as the blank except that the character themselves appear in the output instead of blanks.
$(Dollar Sign) + (plus) – (minus)	These characters may appear in an edited PICTURE clause either in a floating string or singly as a fixed character. 1. As a fixed sign character, the + or – must appear as the first or last character (not both). 2. The plus sign (+) indicates that the sign of an item may be either plus or minus, depending on the algebraic values of the item. The plus or minus sign will be placed in the output area. 3. The minus sign (–) indicates that a minus sign for items will only be placed in the output area. If the item is positive, a blank will replace the minus sign. 4. As a fixed insertion character, the character $ may appear only once in a PICTURE clause. 5. Each character symbol is used in determining the size of the item.

Figure 6.45
Floating String Characters and Their Meanings.

The characters and meanings of floating strings are detailed in figure 6.45. (See also examples shown in figure 6.46.)

Relationship Between PICTURE and USAGE Clauses

The usage of an item must be compatible with the PICTURE clause. The following kinds of items can have only DISPLAY usage: alphabetic, alphanumeric, alphanumeric edited, numeric edited, external decimal, and external floating point. Digits may have any usage: DISPLAY, COMPUTATIONAL, COMPUTATIONAL-1, -2, or -3, or INDEX. DISPLAY items may have any PICTURE clause; other than DISPLAY usage can have only numeric PICTURE clauses. (See figure 6.47.)

The following examples illustrate a fixed insertion and a floating currency symbol using the symbol $.

If data moved to data-name is:	and the PICTURE clause is:	then, the contents of data-name after move is:
0000	$9999	$0000
6794	$9999	$6794
0008	$9999	$0008
0015	$ZZZZ	$ 15
003	$Z99	$ 03
0005	$$$$$	$5
1575	$$$$$	$1575
0004	$$$99	$04
00000	$****9	$****0
00225	$*****	$**225
4579	$999	$579

Each comma in the character-string represents a position into which a comma is inserted. This character position is counted in the size of the item. The comma cannot be the last character in the character-string. If the zero to the left of the symbol position has been suppressed, then the comma is not inserted at this position; the character which is replacing the zeros is inserted instead.

If data moved to data-name is:	and the PICTURE clause is:	then, the contents of data-name after move is:
00000	99,999	00,000
00000	ZZ,ZZZ	
00345	ZZ,ZZZ	345
02466	ZZ,ZZZ	2,466
02466	$ZZ,ZZZ	$ 2,466
00000	$ZZ,ZZZ	
00000	$**,***	*******
00000	$**,*99	$****00

Each period appearing in the character-string represents the decimal point for alignment purposes; it also represents a position into which the character . is inserted. The period is counted in the size of the item. It cannot be the last character in the PICTURE character-string. If the entire data item is suppressed to space characters, a space will replace the period. If the entire data item is suppressed to asterisks, an asterisk will not replace the period.

If data moved to data-name is:	and the PICTURE clause is:	then, the contents of data-name after move is:
‸000135	.999999	.000135
0013‸59	9999.99	0013.59
0004‸28	99.9999	04.2800
3 9	999.99	003.90
1375	$$,$$$.99	$1,375.00
‸1375	$$,$$$.99	$.13
24675	$$,$$$.99	$4,675.00
00000	ZZZZZ.ZZ	
00000	***** **	** ***,**
‸05	$$.$$	$.05
‸00	$$.$$	
235‸07	$$$.$$	$35.07

^ represents assumed decimal point

Each B in the character-string represents a position into which the space character is inserted by the object program when data is placed in the field. The B is counted in the size of the field.

If data moved to data-name is:	and the PICTURE clause is:	then, the contents of data-name after move is:
A4892	XBXXXX	A 4892
CITYSTATE	XXXXBXXXXX	CITY STATE
FMLASTNAME	XBXBXXXXXXXX	F M LASTNAME
556086543	999B99B9999	556 08 6543
234‸45	999BV99	234 45

Each zero in the character-string represents a position into which the numeral zero is inserted by the object program when data is placed in the field. The 0 is counted in the size of the item.

If data moved to data-name is:	and the PICTURE clause is:	then, the contents of data-name after move is:
246	999000	246000
745	00999	00745
ABCXYZ	XXX0XXX	ABC0XYZ

When the minus sign (−) appears at either end of the character-string, the object program inserts a character describing the data as either positive or negative.

The minus sign itself will appear in the edited field only if the data being moved into the field is negative. If the data moving into the field is positive, a space character is inserted into the position indicated by the minus sign.

The minus sign can also be floated from the left end of the PICTURE by placing it in each leading numeric position to be suppressed. In the edited field each leading zero is replaced with a space or a minus sign (depending on whether the sending data is negative or positive) will appear adjacent to the leftmost significant digit.

If data moved to data-name is:	and the PICTURE clause is:	then, the contents of data-name after move is:
9687	-9999	9687
-9687	-9999	-9687
4756	9999-	4756
-0756	9999-	0756-
12345	----99	12345
00045	----99	45
-00045	----99	-45
-00045	------	-45
000	-----	
-000	-----	

When the plus sign (+) appears at either end of the character-string, the object program inserts into the position either a plus or minus sign, describing the data as either positive or negative.

If the data moving into the edited field is positive, then a plus sign is inserted. If the data is negative, then a minus sign is inserted.

The plus sign can also be floated from the left end of the PICTURE by placing it in each leading numeric position to be suppressed. Each leading zero is replaced with a space, and either a plus sign or minus sign (depending on whether the sending data is positive or negative) will appear adjacent to the leftmost significant digit.

If data moved to data-name is:	and the PICTURE clause is:	then, the contents of data-name after move is:
5500	+9999	+5500
-5500	+9999	-5500
7689	9999+	7689+
-0987	9999+	0987-
12345	++++99	+12345
00045	++++99	+45
-00045	++++99	-45
000	++++	
-0000	++++	

Figure 6.46 Examples—Insertion and Floating Strings.

Figure 6.46 Continued.

When the symbol CR is written at the righthand end of the character-string, the object program inserts either CR or space characters, depending on whether the data is negative or positive.

If the data moving into the field is negative, a CR is inserted. If the data is positive, two space characters are inserted.

If data moved to data-name is:	and the PICTURE clause is:	then, the contents of data-name after move is:
123‸45	$999.99CR	$123.45
−123‸45	$999.99CR	$123.45CR
−123‸45	$***.99BCR	$123.45 CR
−45‸6	$$,$$$.99CR	$45.60CR
4820‸33	$$,$$$.99CR	$4,820.33
−‸00	$$,$$$.99CR	$.00CR
0‸00	$$,$$$.99CR	$.00
0‸00	$*,***.99CR	$******.00

‸ represents assumed decimal point

When the symbol DB is written into the righthand end of the character-string, the object program inserts either DB or space characters, depending on whether the data is negative or positive.

If the data moving into the field is negative, a DB is inserted; if the data is positive, two space characters are inserted.

If data moved to data-name is:	and the PICTURE clause is:	then, the contents of data-name after move is:
123‸45	$999.99DB	$123.45
−123‸45	$999.99DB	$123.45DB
−123‸45	$***.99BDB	$123.45 DB
45‸6	$$,$$$.99DB	$45.60DB
4820‸33	$$,$$$.99DB	$4,820.33
−‸00	$$,$$$.99DB	$.00DB
0‸00	$$,$$$.99DB	$.00
0‸00	$*,***.99DB	$*****.00

USAGE clause — PICTURE clause

```
02  Y-T-D-DEMAND, COMPUTATIONAL, PICTURE S9(6)V99.
```

Figure 6.47 Example—Picture and Usage Clauses.

Value Clause

VALUE IS literal

Figure 6.48 Value Clause Format.

The VALUE clause defines the condition-name values in the initial value of an item in the Working-Storage Section (fig. 6.48). This clause is used mainly in the Working-Storage Section to assign values to constants at the elementary level. The value assigned to the item remains constant during the execution of the object program unless it is changed by a procedure in the program.

Rules Governing the Use of the Value Clause

1. The size of the literal in the VALUE clause must be less than or equal to the size of an item as given in the PICTURE clause. All leading or trailing zeros reflected by Ps in a PICTURE clause must be included in the VALUE clause.
2. This clause is not permitted in the description of data items in the File Section other than condition-name entries at the 88 level.
3. When an initial value is not specified for an item in the Working-Storage Section, no assumption should be made regarding the original contents of the item.
4. A numeric literal must be used if the PICTURE clause designates a numeric item (fig. 6.49).
5. A nonnumeric literal should be used if the item is alphabetic or alphanumeric (fig. 6.50).

```
77  DISCOUNT, USAGE COMPUTATIONAL,
    PICTURE SV99, VALUE .02.
```

VALUE clause

Figure 6.49 Example—Value Clause—Numeric Literal.

```
03  SUBCRIPTION-BASIS, PICTURE X.
    88  REGULAR, VALUE '1'.
```

VALUE clause

Figure 6.50 Example—Value Clause—Conditional Name—Nonnumeric Literal.

6. A Figurative Constant ZERO may be used in place of numeric or nonnumeric literal. The number of zeros generated will be the same as the size specified in the PICTURE clause.
7. A Figurative Constant SPACE may be used as an initial value in the place of a nonnumeric literal. The number of blanks generated will depend upon the size of the item in the PICTURE clause.
8. The VALUE clause can only be specified for elementary items.
9. The VALUE clause must not be specified for any item whose size, explicit or implicit, is variable.
10. The VALUE clause must not be written in a record description entry that contains an OCCURS clause or REDEFINES clause, or an entry that is contained in an OCCURS or REDEFINES clause.
11. If the VALUE clause is written in an entry at the group level, the literal must be a Figurative Constant or nonnumeric literal, and the group area is initialized without consideration for the usage of the individual elementary or group items contained within this group. The VALUE clause cannot be specified at subordinate levels within this group.

Condition-Name Clause

A condition-name is a name assigned by the programmer to a particular value that may be assumed by a data item (fig. 6.51).

```
88 condition-name VALUE IS literal .
```

Figure 6.51
Format—Condition-Name Clause.

Rules Governing the Use of Condition-Name Clauses

1. The condition-name is the name of the value of an item, not the name of the item itself. The item description entry complete with a PICTURE clause is required to describe the item.
2. A level 88 must be used in any condition-name entry.
3. A condition-name may pertain to an elementary item in a group item, with these exceptions: a level 66 item; a group containing items with descriptions which include JUSTIFIED, SYNCHRONIZED, or USAGE other than DISPLAY; or an index data item.
4. The condition-name is used in the Procedure Division simple relational test statements.
5. The VALUE clause is the only clause required in a condition-name entry.
6. The condition-name must immediately follow the item with which it is associated.
7. The type of literal used must be consistent with the data type of the condition variable. For example, a numeric literal must be used if the item has a numeric picture, or a nonnumeric literal for alphabetic or alphanumeric pictures. The Figurative Constant zero may be used in place of a numeric or nonnumeric literal. (See figure 6.52.)

SALESMAN

REGION
A = EASTERN
B = CENTRAL
C = WESTERN

OFFICE-NUMBER
999

BADGE-NUMBER
9999

INDUSTRY
20 = PETROLEUM
21 = METALS
26 = CHEMICALS
32 = UTILITIES

```
02  SALESMAN.
    03  REGION, PICTURE A.
        88  EASTERN, VALUE 'A'.
        88  CENTRAL, VALUE 'B'.
        88  WESTERN, VALUE 'C'.
    03  OFFICE-NUMBER, PICTURE 999.
    03  BADGE-NUMBER, PICTURE 9999.
    03  INDUSTRY, PICTURE 99.
        88  PETROLEUM, VALUE 20.
        88  METALS, VALUE 21.
        88  CHEMICALS, VALUE 26.
        88  UTILITIES, VALUE 32.
```

Figure 6.52 Example—Condition-Name Clauses.

Other Independent Clauses

Redefines Clause

The REDEFINES clause specifies that the same area is to contain different data items (fig. 6.53). The entry gives another name and description to an item previously described. That is, the REDEFINES clause specifies the redefinition of a storage area, not of the items occupying the area.

level number data-name-1 <u>REDEFINES</u> data-name-2

Figure 6.53
Redefines Clause Format.

The same area may be called by different names during the processing of the data. The area may contain different types of information and may be processed in a different manner under changing conditions. (See figure 6.54.)

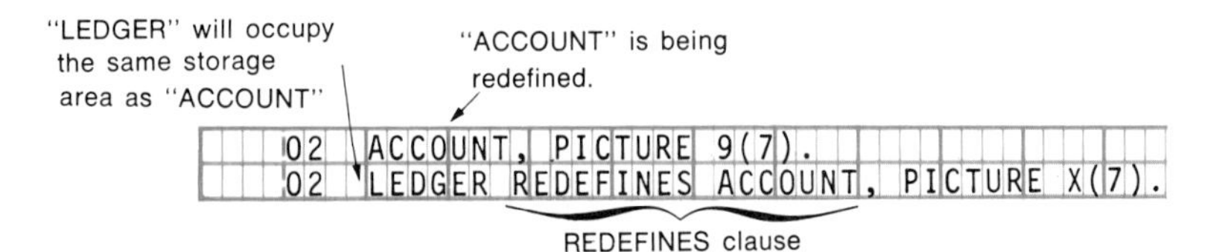
"LEDGER" will occupy the same storage area as "ACCOUNT"
"ACCOUNT" is being redefined.
02 ACCOUNT, PICTURE 9(7).
02 LEDGER REDEFINES ACCOUNT, PICTURE X(7).
REDEFINES clause

Figure 6.54 Example—Redefines Clause.

Rules Governing the Use of the Redefines Clause

1. The word REDEFINES must be written right after the data-name followed by the name of the item being redefined.
2. The level numbers of the two entries sharing the same area must be the same.
3. Redefines may be used at the 01 level in the working-storage section, but not in the file section.
4. The usage of data within the area cannot be redefined.
5. The redefinition starts at data-name-2 and ends when a level number that is less than or equal to that of data-name-2 is encountered. Between the data description of data-name-2 and data-name-1, there may be no entries having lower-level numbers than data-name-1 or -2.
6. A new storage area is not set aside by the redefinition. All descriptions of the area remain in effect.

7. The entries giving new descriptions of the area must immediately follow the entry that is being redefined.
8. A REDEFINES clause may be used for items subordinate to items who are themselves being redefined.
9. This entry should not contain any VALUE clauses.

(See figures 6.55, 6.56, 6.57.)

Blank When Zero Clause

The BLANK WHEN ZERO clause is used when an item is to be filled with spaces when the value of the item is zero. The clause may be specified only for an item whose PICTURE is numeric at the elementary level. The clause may not be specified for level 66 or 88 items (figs. 6.58, 6.59).

```
01  RECORDI.
    02  ACCOUNT-NUMBER PICTURE 9(6).
    02  AMOUNT   PICTURE 9(5)V99.
    02  PERCENT   PICTURE V999.
    02  PERCENT-OUT   REDEFINES PERCENT PICTURE 99V9.
    02  PRICE   PICTURE 9V999.
```

In the above example, a three character data item originally defined as PERCENT has been redefined as PERCENT-OUT. The redefinition with one decimal position enables the percentage to be printed as 36.5% instead of the decimal .365 which is used for internal calculation.

Figure 6.55 Example—Redefines Clause.

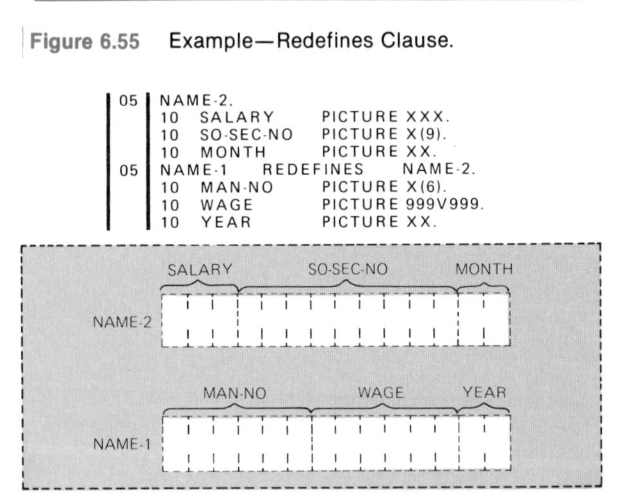
05 NAME-2.
10 SALARY PICTURE XXX.
10 SO-SEC-NO PICTURE X(9).
10 MONTH PICTURE XX.
05 NAME-1 REDEFINES NAME-2.
10 MAN-NO PICTURE X(6).
10 WAGE PICTURE 999V999.
10 YEAR PICTURE XX.
SALARY
SO-SEC-NO
MONTH
NAME-2
MAN-NO
WAGE
YEAR
NAME-1

Figure 6.56 Example—Redefines Clause.

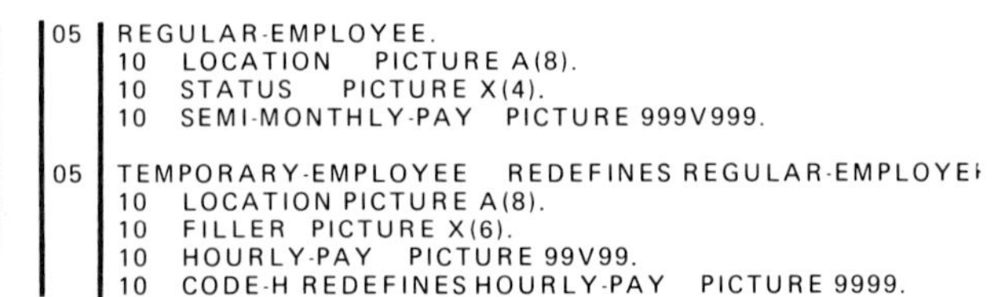

```
05  REGULAR-EMPLOYEE.
    10  LOCATION    PICTURE A(8).
    10  STATUS    PICTURE X(4).
    10  SEMI-MONTHLY-PAY   PICTURE 999V999.

05  TEMPORARY-EMPLOYEE   REDEFINES REGULAR-EMPLOYEE
    10  LOCATION PICTURE A(8).
    10  FILLER  PICTURE X(6).
    10  HOURLY-PAY   PICTURE 99V99.
    10  CODE-H REDEFINES HOURLY-PAY   PICTURE 9999.
```

Figure 6.57 Example—Redefines Clause.

<u>BLANK</u> WHEN <u>ZERO</u>

Figure 6.58 Blank When Zero Clause Format.

If data moved to data-name is:	and the PICTURE clause is:	then, the contents of data-name is:
00000	99,999 BLANK WHEN ZERO	
00020	99,999 BLANK WHEN ZERO	00,020
00000	99,999	00,000
00000	ZZ,ZZZ	
00000	$99,999 BLANK WHEN ZERO	
00000	$ZZ,ZZZ BLANK WHEN ZERO	
000000	$9,999.99 BLANK WHEN ZERO	
000000	$Z,ZZZ,ZZ BLANK WHEN ZERO	
000000	$*,***.** BLANK WHEN ZERO	****** **

Figure 6.59 Example—Blank When Zero Clause.

Justified Clause

The JUSTIFIED clause is used to override the normal positioning of alphabetic or alphanumeric data when it is moved to a larger area (fig. 6.60). If an item is moved to a location that is larger than itself, it may be necessary to specify the position that the data is to occupy in the new area. In the absence of the JUSTIFIED clause, normal justification will be performed on the movement of data as follows:

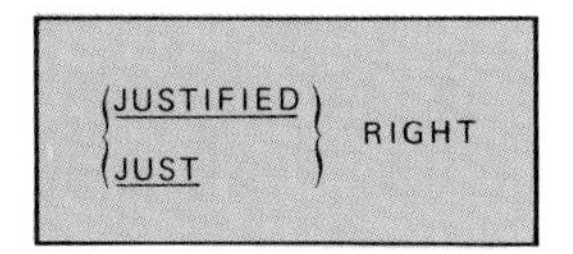

Figure 6.60
Justified Clause Format.

Numeric data will be *right justified* after decimal alignment, and any unused positions at the right or left will be filled with zeros. The rightmost character will be placed in the rightmost positions in the new area (if no decimals are involved), and zeros will be supplied to the unused positions at the left.

Alphabetic and *Alphanumeric* data will be *left justified* after the move, and any unused character positions at the right will be filled with blanks. If the sending field is larger than the receiving area, excess characters at the right will be truncated.

If the programmer *wishes to alter* the normal justification of alphabetic or alphanumeric items, he could do so with the JUSTIFIED clause. If the JUSTIFIED clause is used, it affects the positioning of the receiving area, as follows:

1. If the data being sent is longer than the receiving area, the leftmost characters will be truncated.
2. If the data being sent is shorter than the receiving area, the unused positions at the left will be filled with blanks.

The JUSTIFIED clause may be used only at the elementary level, and must not be specified for level 66 or level 88 data items (fig. 6.61).

Figure 6.61 Example—Justified Right Clause.

Sending: JOE-JONES → Receiving: JOE-JONES

Synchronized Clause

Some computer memories are organized in such a way that there are natural addressing boundaries in the computer memory (e.g., word boundaries, halfword boundaries, doubleword boundaries, byte boundaries). The manner in which data is stored is determined by the object program and need not respect these natural boundaries.

However, certain uses of data (e.g., in arithmetic operations or in subscripting) may be facilitated if the data is stored so as to be aligned on these natural boundaries. Specifically, additional machine operations in the object program may be required for the accessing and storage of data if portions of two or more data items appear between adjacent natural boundaries, or if certain natural boundaries divide a single data item into two branches.

Data items which are aligned in natural boundaries in such a way as to avoid additional machine operations are defined as being synchronized. A synchronized item is assumed to be introduced and carried in that form; conversion to synchronized form occurs only during the execution of a procedure (other than READ or WRITE) which stored data in the item.

Synchronization can be accomplished in two ways:

1. By use of the SYNCHRONIZED clause.
2. By recognizing the appropriate natural boundaries and organizing the data suitable without the use of the SYNCHRONIZED clause.

The SYNCHRONIZED clause is used to specify alignment of an elementary item in the natural boundaries of the computer memory (fig. 6.62). This clause specifies that the COBOL processor, in creating the internal format for this item, must arrange the item in contiguous units of memory in such a way that no other data item appears in any of the memory units between the right and left natural boundary delimiting these data items. If the size of the item is such that it does not itself utilize all of the storage area between the delimiting natural boundaries, the unused storage positions (or portion thereof) may not be used for any other data item.

Rules Governing the Use of the Synchronized Clause

1. SYNCHRONIZED not followed by RIGHT or LEFT specifies that an elementary item is to be positioned between the natural boundaries in such a way as to effect utilization of the elementary data items. The specific positions are determined by the implementor.
2. If SYNCHRONIZED LEFT is specified, the leftmost character will occupy the leftmost position in the contiguous memory area. The right-hand positions of the area will be unoccupied.
3. If SYNCHRONIZED RIGHT is specified, the rightmost character will occupy the right-hand position in the contiguous memory area with the leftmost positions of the area unoccupied.
4. Whenever a SYNCHRONIZED item is referenced in the source program, the original size of the items, as shown in the PICTURE clause, is used in determining any action that depends on size, such as justification, truncation, or overflow.
5. In the data description for an item, the sign appears in the normal operational size position regardless of whether the item is SYNCHRONIZED LEFT or SYNCHRONIZED RIGHT.
6. This clause is hardware-dependent and, in addition to the rules stated above, the implementor must specify how elementary items associated with this clause are handled. The user should consult individual reference manuals for particular computers for further information relative to this clause.

(See figure 6.63.)

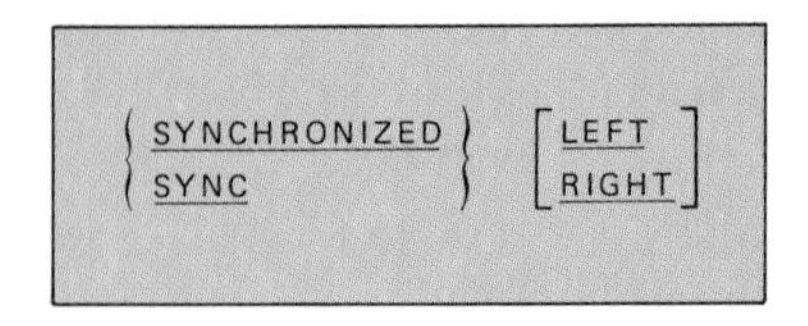

Figure 6.62
Synchronized Clause Format.

```
01  WORK-RECORD.
    05  WORK-CODE                  PICTURE X.
    05  COMP-TABLE OCCURS 10       TIMES.
        10  COMP-TYPE              PICTURE X.
    [   10  IA-Slack-Bytes         PICTURE XX. Inserted by compiler]
        10  COMP-PAY               PICTURE S9(4)V99 COMP SYNC.
        10  COMP-HRS               PICTURE S9(3)  COMP SYNC.
        10  COMP-NAME              PICTURE X(5).
    [   10  IE-Slack-Bytes         PICTURE XX. Inserted by compiler]
```

Figure 6.63 Synchronized Clause—Example.

Occurs Clause (See Table Handling Section)

The OCCURS clause is used to define tables and other homogeneous sets of data whose elements can be referred to by subscripting or indexing. The clause specifies the number of times that an item is repeated with no change in its USAGE or PICTURE clauses. The clause is used primarily in defining related sets of data such as tables, lists, matrixes, etc. (See figures 6.64, 6.65.)

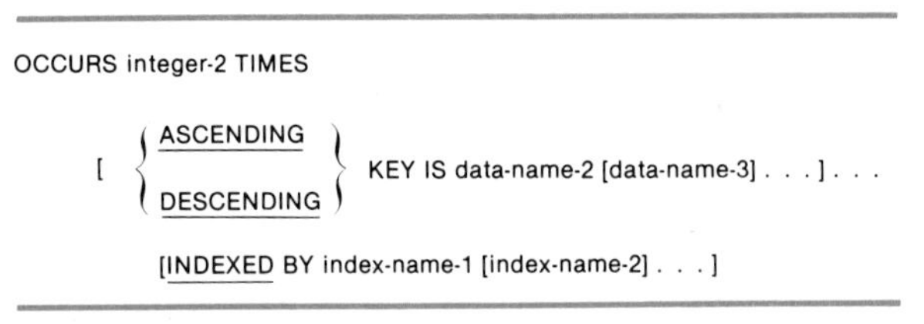

OCCURS integer-2 TIMES
[{ ASCENDING / DESCENDING } KEY IS data-name-2 [data-name-3] . . .] . . .
[INDEXED BY index-name-1 [index-name-2] . . .]

Figure 6.64 Occurs Clause Format.

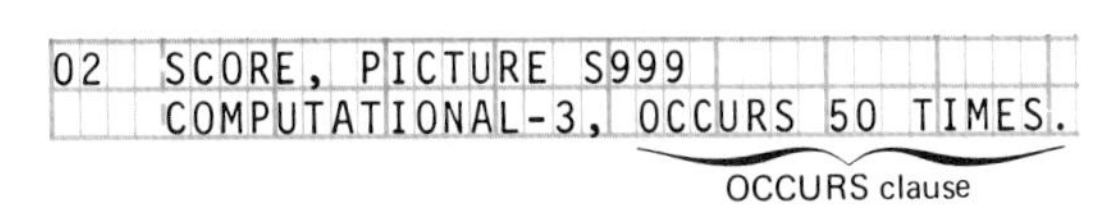

Figure 6.65 Example—Occurs Clause.

Rules Governing the Use of the Occurs Clause

1. Record description clauses associated with an item that contains an OCCURS clause, apply to each repetition of the item being described.
2. Whenever the OCCURS clause is used, the data-name that is the defining name of the entry must be subscripted whenever used in the Procedure Division.
3. If the data-name is the name of a group item, then all data-names belonging to the group must be subscripted whenever used.
4. The OCCURS clause may not be used in the 01, 77, or 88 level of a record description entry.
5. The clause cannot describe an item whose size is variable.

(See figures 6.66, 6.67.)

PRODUCT NUMBER	QUANTITY				
9 9 9	9 9 9 ^ 9	9 9 9 ^ 9	9 9 9 ^ 9	9 9 9 ^ 9	9 9 9 ^ 9

```
02  PRODUCT-NUMBER   PICTURE 999.
02  QUANTITY   PICTURE 999V9 OCCURS 5 TIMES.
```

The definition of the first QUANTITY also applies to each of the four repetitions of QUANTITY. Thus, all five fields consist of four numeric characters, including one decimal position.

Figure 6.66 Example—Occurs Clause.

PRODUCT NUMBER	WAREHOUSE-DATA								
	LOC	BIN	QUANTITY						
9 9 9 9	1 7	6 2 9	9 9 9 9 9	1 8	1 5 8	9 9 9 9 9	1 8	1 6 0	9 9 9 9 9

```
02  PRODUCT-NUMBER   PICTURE 9999.
02  WAREHOUSE-DATA   OCCURS 3 TIMES ASCENDING KEYS ARE
      LOCATION, BIN-NUMBER.
      03  LOCATION   PICTURE 99.
      03  BIN-NUMBER   PICTURE 999.
      03  QUANTITY-ON-HAND   PICTURE 9(5).
```

In the example above of data descriptions and data structure, the three occurrences of WAREHOUSE-DATA appear in ascending order according to the major key LOCATION and its minor key BIN-NUMBER.

Figure 6.67 Example—Occurs Clause.

Subscripting

The need arises to have tables of information accessible to a source program for referencing. Subscripting provides the facility for referring to data items in a list or a table that has been assigned individual values.

Like all data tables, the tables must be described in the Data Division with an OCCURS clause to indicate the number of appearances of a particular item. The subscripts are used in the Procedure Division to reference a particular item in the table. If subscripting were not used, each item would have to be described in a separate entry.

Rules Governing the Use of Subscripts

1. A subscript must always have a positive nonzero integral value whose value determines which item is being referenced within a table or list.
2. The subscript must be represented either by a numeric literal or a data-name that has an integral value.
3. Subscripts are enclosed in parentheses to the right of the subscripted data-name with an intervening space. If the subscripted data-name is qualified, the subscripts must appear to the right of all qualifiers.
4. If more than one level of subscript is present, the subscripts are separated by commas and arranged from right to left in increasing order of inclusiveness of the grouping within the table. Multiple subscripts are written with a single pair of parentheses separated by commas and followed by a space. A space should also separate the data-name from the subscripted expressions. A *maximum of three levels of subscripts is permitted.*
5. A subscripted data-name must be qualified when it is not unique in accordance with rules for qualification.
6. A subscript must always be used to reference an item that has an OCCURS clause or belongs to a group having an OCCURS clause. *Subscription may not be used with any undescribed data-name using an OCCURS clause.*
7. A programmer may refer to blocks (sets) of data within a table. The data-name is written, followed by the subscript of the particular block plus any other subscript necessary to locate it. A complete table may be referenced just by using the name of the table.
8. Subscripts may not themselves be subscripted.
9. A data-name may not be subscripted when it is being used as
   a. A subscript or qualifier.
   b. A defining name of a record description entry.
   c. Data-name-2 in a REDEFINES clause.
   d. A data-name in a LABEL RECORDS clause.
   e. A data-name in the DEPENDING ON option of the OCCURS clause.

(See figures 6.68, 6.69, 6.70.)

Additional examples of subscripting will be found in the Procedure Division PERFORM statements and the Table Handling Section of the text.

Qualification and subscripting are entirely different. Qualification involves appending additional data-names to a name which has been used to represent different items. Subscripting is used to refer to one item among a group organized into a table or list.

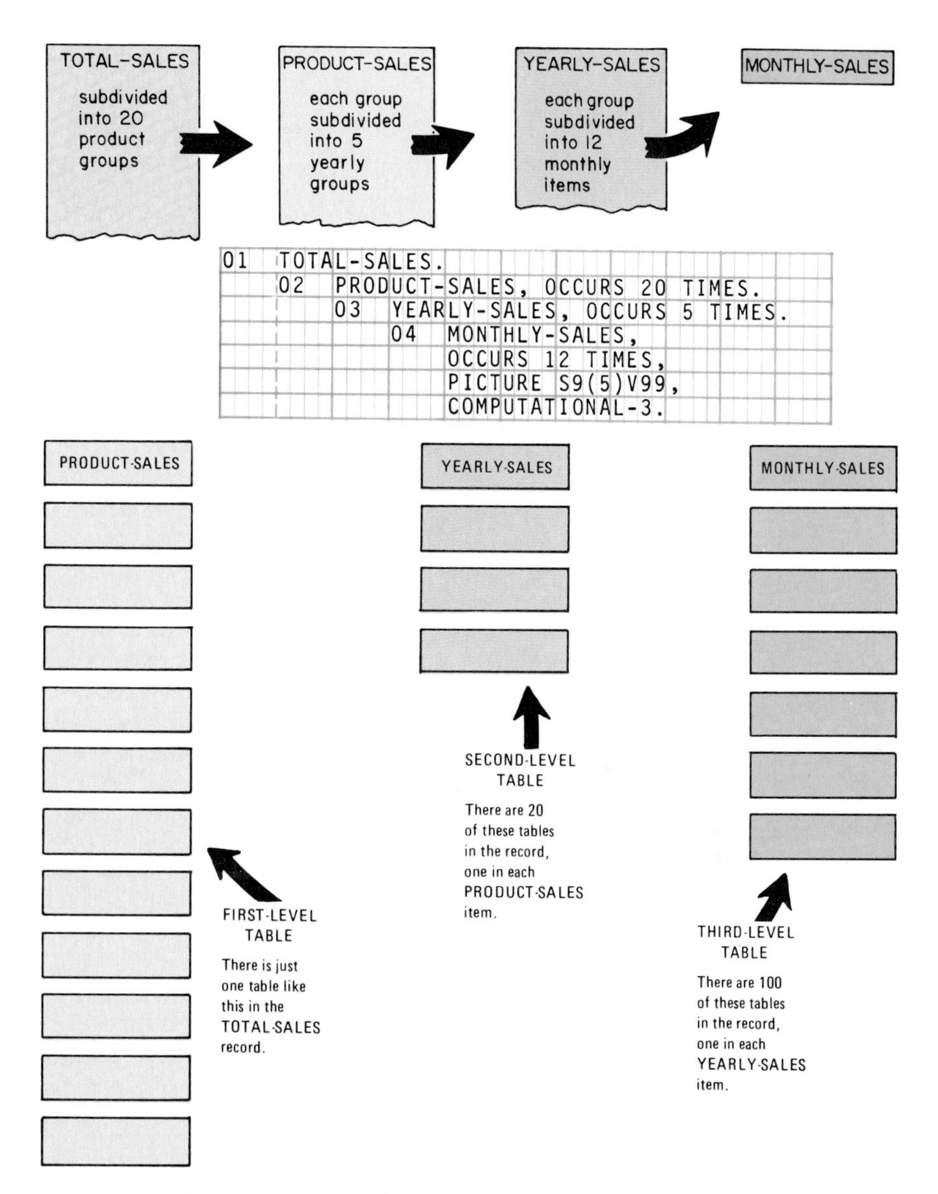

Figure 6.68 Example—Subscripting.

```
01    GROUP.
  02    ARRAY OCCURS 2 TIMES.
    03    VECTOR OCCURS 2 TIMES.
      04    ELEMENT OCCURS 3 TIMES USAGE
            COMPUTATIONAL PICTURE S9(9).
```

Figure 6.69
Example—Subscripting 3 Levels.

```
ADD ELEMENT (2, 1, 3) TO SUM-1.
```

Figure 6.70 Example—Procedure Division Entry.

WORKING-STORAGE SECTION

The Working-Storage Section may contain descriptions of records which are not part of external data files but which are developed and processed internally. This section is used to describe areas wherein intermediate results are stored temporarily at object time. The section is used also for descriptions of data to be used in the program. The section may be omitted if there aren't any constants or work areas needed in the program.

The Working-Storage Section is often used to provide headings for a report. Since this is the only section that is permitted to have VALUE clauses (outside of condition-name entries), report headings can be designed and moved to an output file description prior to a write operation. Output formats for detail lines to be printed can also be provided for in this section. In many programs, the Working-Storage Section is the largest unit in the Data Division.

Structure

The Working-Storage Section must begin with the header WORKING-STORAGE SECTION followed by a period and a space on a line by itself. The section contains data entries for independent (noncontiguous) items and record items in that order (fig. 6.71).

```
WORKING-STORAGE SECTION.

[77 noncontiguous item description entry] . . .

[01 record description entry] . . .
```

Figure 6.71 Working-Storage Section Format.

Independent Items are items in the Working-Storage Section that bear no hierarchical relationship to one another and need not be grouped into records, provided that they do not need to be further subdivided (fig. 6.72).

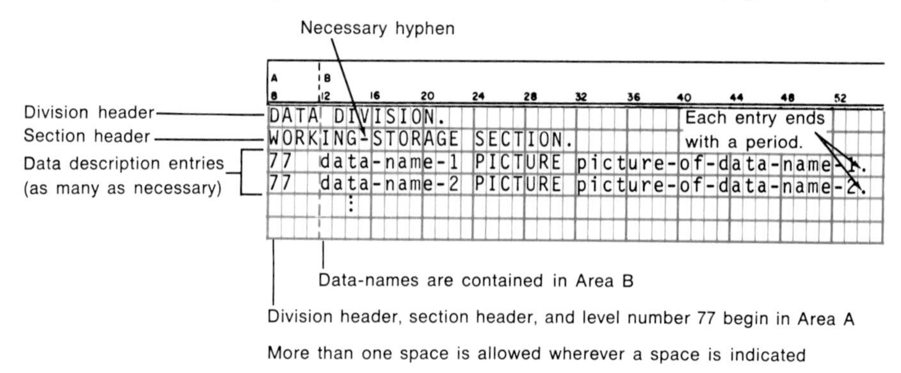

Figure 6.72 Guide for Coding Level 77 Entries in the Working-Storage Section.

1. These items must not be subdivided or themselves be a subdivision of another item.
2. These entries must precede record item entries (level number 01–49).
3. Each item must be defined in a separate record description as follows: Level number 77, Data Name, USAGE Clause (optional), VALUE clause (optional), and a PICTURE clause (required).
4. An OCCURS clause must not be used in describing an independent item.

These items are used primarily as temporary storage of an item pending the completion of a calculation, or to define a constant to be used in the program (fig. 6.73).

```
WORKING-STORAGE SECTION.

77  OLD-NUMBER                  PICTURE 9(12).
77  QUANTITY-TOTAL              PICTURE 9(7), VALUE ZERO.
77  PURCHASE-COST-TOTAL         PICTURE 9(7)V99,
                                          VALUE ZERO.
```

Figure 6.73 Example—Level 77 Entries—Working-Storage Section.

Record Items are data elements in the Working-Storage Section among which there is definite hierarchical relationship; they must be grouped together into records according to the rules for the formation of record descriptions (fig. 6.74).

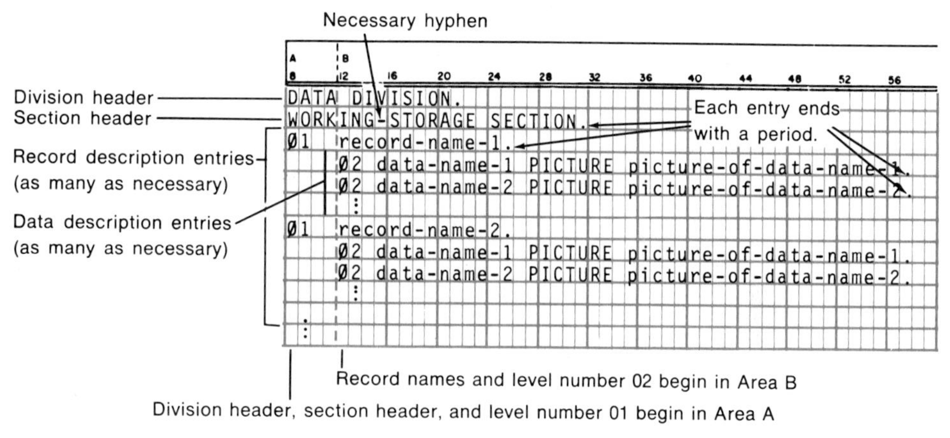

Figure 6.74 Guide for Coding Level 01 and 02 Entries in the Working-Storage Section.

1. These items are subdivided into smaller units and bear a definite relationship to each other.
2. The entries used to describe these record items are identical to those used to describe a record in the File Section. However, one difference exists between records in the File Section and the entries in the Working-Storage Section. The entries in the File Section may be elementary items, but entries at the record level in the Working-Storage Section must be group items.

Record entries are often used for output heading and detail formats because of the ability to use the VALUE clause in the Working-Storage Section.

An internal value of an item in the Working-Storage Section may be specified by using the VALUE clause. The value is assumed by the item at the time of the execution of the program. *No assumption can be made of the initial value of an item that has not been defined with a VALUE clause.* (See figures 6.75, 6.76.)

Additional Data Division clauses, entries, examples, and uses will be found further along in the text.

```
DATA DIVISION.
WORKING-STORAGE SECTION.
01  INPUT-DATA.
      02  FIELD-1, PICTURE X(10).
      02  FIELD-2, PICTURE X(5).
01  OUTPUT-DATA.
      02  FIELD-A, PICTURE X(5).
      02  FILLER, PICTURE X(19), VALUE SPACES.
      02  FIELD-B, PICTURE X(10).
```

Figure 6.75 Example—Level 01 and 02 Entries in Working-Storage Section.

```
DATA DIVISION.
WORKING-STORAGE SECTION.
77  TOTAL, PICTURE 9(8), VALUE ZERO.
01  NUMBERS.
      02  SMALLER, PICTURE 9999.
      02  LARGER, PICTURE 9999.
```

Figure 6.76 Example—Level 77, 01 and 02 Entries in Working-Storage Section.

ACCOUNTS RECEIVABLE PROBLEM

INPUT

Field	Card Columns
Entry Date	1–5
Entry	6–7
Customer Name	8–29
Invoice Date	30–33
Invoice Number	34–38
Customer Number	39–43
Location	44–48
Blank	49–62
Discount Allowed	63–67
Amount Paid	68–73
Blank	74–80

CALCULATIONS TO BE PERFORMED

1. Calculate Accounts Receivable = Amount Paid + Discount Allowed.
2. Final Totals for Accounts Receivable, Discount Allowed and Amount Paid.

OUTPUT

Print a report as follows:

ACCOUNTS RECEIVABLE REGISTER

CUST. NO. CUST. NAME INV. NO. ACCTS. REC. DISCT. ALLOW. AMT. PAID

TOTALS

ACCOUNTS RECEIVABLE REGISTER

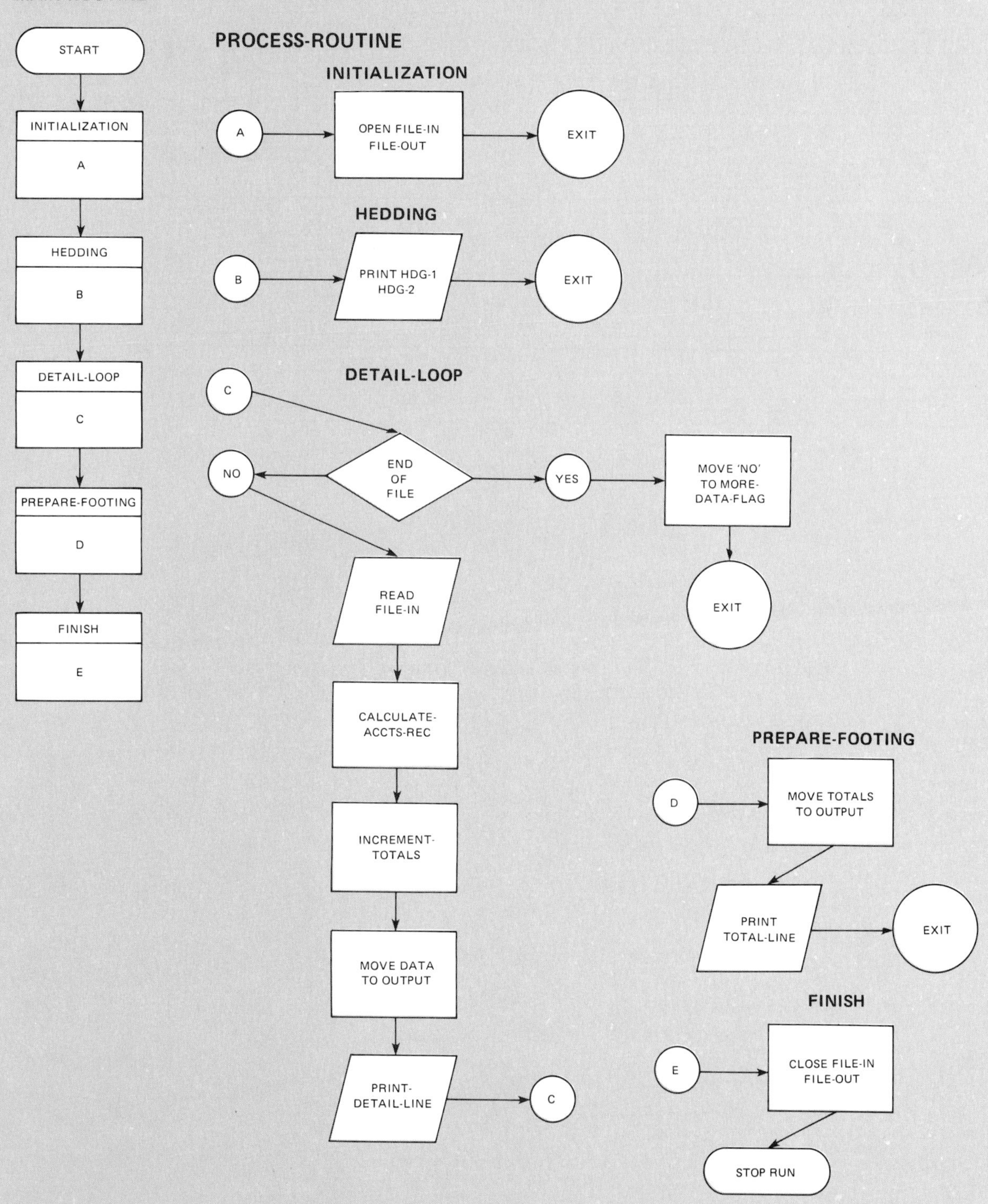

Line							
H 6				ACCOUNTS RECEIVABLE REGISTER			
H 9	CUST. NO.	CUST. NAME	INV. NO.	ACCTS. REC.	DISCT. ALLOW.	AMT. PAID	
O 11	ZZZZZ	A————A	ZZZZZ	$$$$,$$$.99	$$,$$$.99	$$$$,$$$.99	
T 15			TOTALS	$$,$$$,$$$.99	$$$,$$$.99	$$,$$$,$$$.99	**

```
PP 5734-CB2 V4 RELEASE 1.3 01AUG75        IBM OS AMERICAN NATIONAL STANDARD COBOL        DATE AUG  4,1977

    1

00001    00100   IDENTIFICATION DIVISION.
00002
00003    00110   PROGRAM-ID.
00004    00115       ACCOUNTS-RECEIVABLE-REGISTER.
00005    00120   AUTHOR.
00006    00125       DOROTHY R HENNESSEY.
00007    00130   INSTALLATION.
00008    00135       WEST LOS ANGELES COLLEGE.
00009    00140   DATE-WRITTEN.
00010    00145       17 MARCH 1977.
00011    00155   DATE-COMPILED. AUG  4,1977.
00012    00165   REMARKS.
00013    00170       THIS PROGRAM PREPARES AN ACCCUNTS RECEIVABLE REGISTER.
00014
00015
00016    00204   ENVIRONMENT DIVISION.
00017
00018    00210   CONFIGURATION SECTION.
00019    00215   SOURCE-COMPUTER.
00020    00220       IBM-370.
00021    00225   OBJECT-COMPUTER.
00022    00230       IBM-370.
00023    00235   SPECIAL-NAMES. C01 IS SKIP-TO-ONE.
00024    00242   INPUT-OUTPUT SECTION.
00025    00250   FILE-CONTROL.
00026    00255       SELECT FILE-IN       ASSIGN TO UR-2540R-S-CARDIN.
00027    00265       SELECT FILE-OUT      ASSIGN TO UR-1403-S-PRINT.
```

```
2

00029   01004  DATA DIVISION.
00030
00031   01010  FILE SECTION.
00032   01015  FD  FILE-IN
00033   01020      RECORDING MODE IS F
00034   01025      LABEL RECORDS ARE OMITTED
00035   01030      DATA RECORD IS CARDIN.
00036   01036  01  CARDIN.
00037   01040      03  ENTRY-DATE       PICTURE X(5).
00038   01045      03  ENTRI            PICTURE 99.
00039   01050      03  CUSTOMER-NAME    PICTURE X(21).
00040   01055      03  INVOICE-DATE     PICTURE X(5).
00041   01060      03  INVOICE-NUMBER   PICTURE 9(5).
00042   01065      03  CUSTOMER-NUMBER  PICTURE 9(5).
00043   01070      03  LOCATION         PICTURE 9(5).
00044   01075      03  FILLER           PICTURE X(14).
00045   01080      03  DISCOUNT         PICTURE 999V99.
00046   01085      03  AMOUNT-PAID      PICTURE 9999V99.
00047   01090      03  FILLER           PICTURE X(7).
00048   01096  FD  FILE-OUT
00049   01100      RECORDING MODE IS F
00050   01105      LABEL RECORDS ARE OMITTED
00051   01110      DATA RECORD IS PRINT.
00052   01116  01  PRINT                PICTURE X(133).
00053   01122  WORKING-STORAGE SECTION.
00054   01130  77  ACCT-REC-WS          PICTURE 9(5)V99 VALUE ZEROES.
00055   01135  77  ACCT-REC-TOT-WS      PICTURE 9(6)V99 VALUE ZEROES.
00056   01140  77  DISCOUNT-TOT-WS      PICTURE 9(5)V99 VALUE ZEROES.
00057   01145  77  AMT-PAID-TOT-WS      PICTURE 9(6)V99 VALUE ZEROES.
00058   01146  01  FLAGS.
00059   01147      03  MORE-DATA-FLAG   PICTURE XXX        VALUE 'YES'.
00060   01148          88  MORE-DATA                       VALUE 'YES'.
00061   01149          88  NO-MORE-DATA                    VALUE 'NO '.
00062   01151  01  HDG-1.
00063   01155      03  FILLER           PICTURE X(53)      VALUE SPACES.
00064   01160      03  FILLER           PICTURE X(30)      VALUE 'ACCOUNTS RECEIVABL
00065   01161  -                                             'E REGISTER'.

3

00067   01166  01  HDG-2.
00068   01170      03  FILLER           PICTURE X(17)      VALUE SPACES.
00069   01175      03  FILLER           PICTURE X(9)       VALUE 'CUST. NO.'.
00070   01180      03  FILLER           PICTURE X(10)      VALUE SPACES.
00071   01185      03  FILLER           PICTURE X(10)      VALUE 'CUST. NAME'.
00072   01190      03  FILLER           PICTURE X(11)      VALUE SPACES.
00073   01195      03  FILLER           PICTURE X(8)       VALUE 'INV. NO.'.
00074   01200      03  FILLER           PICTURE X(7)       VALUE SPACES.
00075   01205      03  FILLER           PICTURE X(11)      VALUE 'ACCTS. REC.'.
00076   01210      03  FILLER           PICTURE X(8)       VALUE SPACES.
00077   01215      03  FILLER           PICTURE X(13)      VALUE 'DISCT. ALLOW.'.
00078   01220      03  FILLER           PICTURE X(5)       VALUE SPACES.
00079   01225      03  FILLER           PICTURE X(9)       VALUE 'AMT. PAID'.
00080   01231  01  DETAIL-LINE.
00081   01235      03  FILLER           PICTURE X(19)      VALUE SPACES.
00082   01240      03  CUSTOMER-NUMBER  PICTURE Z(5).
00083   01245      03  FILLER           PICTURE X(7)       VALUE SPACES.
00084   01250      03  CUSTOMER-NAME    PICTURE X(21).
00085   01255      03  FILLER           PICTURE X(7)       VALUE SPACES.
00086   01260      03  INVOICE-NUMBER   PICTURE Z(5).
00087   01265      03  FILLER           PICTURE X(8)       VALUE SPACES.
00088   01270      03  ACCT-REC-OUT     PICTURE $$$,$$$.99.
00089   01275      03  FILLER           PICTURE X(13)      VALUE SPACES.
00090   01280      03  DISCOUNT         PICTURE $$$$.99.
00091   01285      03  FILLER           PICTURE X(7)       VALUE SPACES.
00092   01290      03  AMOUNT-PAID      PICTURE $$,$$$.99.
00093   01301  01  TOTAL-LINE.
00094   01305      03  FILLER           PICTURE X(59)      VALUE SPACES.
00095   01310      03  FILLER           PICTURE X(6)       VALUE 'TOTALS'.
00096   01315      03  FILLER           PICTURE X(6)       VALUE SPACES.
00097   01320      03  ACCT-REC-T-OUT   PICTURE $$$$,$$$.99.
00098   01325      03  FILLER           PICTURE X(10)      VALUE SPACES.
00099   01330      03  DISCOUNT-T-OUT   PICTURE $$$,$$$.99.
00100   01335      03  FILLER           PICTURE X(5)       VALUE SPACES.
00101   01340      03  AMT-PAID-T-OUT   PICTURE $$$$,$$$.99.
00102   01345      03  FILLER           PICTURE XXX        VALUE ' **'.
```

4

```
00104   02004  PROCEDURE DIVISION.
00105
00106   02010  MAIN-ROUTINE.
00107   02015      PERFORM INITIALIZATION.
00108   02019      PERFORM HEDDING.
00109              READ FILE-IN
00110                  AT END MOVE 'NO ' TO MORE-DATA-FLAG.
00111              PERFORM DETAIL-LOOP
00112                  THRU DETAIL-LOOP-EXIT
00113                  UNTIL NO-MORE-DATA.
00114   02021      PERFORM PREPARE-FOOTING.
00115   02022      PERFORM FINISH.
00116
00117          PROCESS-ROUTINES.
00118
00119   02062  INITIALIZATION.
00120              OPEN INPUT  FILE-IN
00121                   OUTPUT FILE-OUT.
00122
00123   02086  HEDDING.
00124   02090      WRITE PRINT FROM HDG-1
00125   02091          AFTER ADVANCING SKIP-TO-ONE LINES.
00126   02095      WRITE PRINT FROM HDG-2
00127   02096          AFTER ADVANCING 3 LINES.
00128   02100      MOVE SPACES TO PRINT.
00129   02101      WRITE PRINT
00130   02102          AFTER ADVANCING 1 LINES.
00131
00132   02104  DETAIL-LOOP.
00133   02107      PERFORM CALCULATE-ACCTS-REC.
00134   02108      PERFORM INCREMENT-TOTALS.
00135   02109      PERFORM MOVE-DATA-TO-OUTPUT.
00136   02110      PERFORM PRINT-DETAIL-LINE.
00137   02115      READ FILE-IN
00138   02116          AT END MOVE 'NO ' TO MORE-DATA-FLAG.
00139   02117  DETAIL-LOOP-EXIT.  EXIT.
```

5

```
00141   02121  CALCULATE-ACCTS-REC.
00142   02125      COMPUTE ACCT-REC-WS = DISCOUNT OF CARDIN
00143                  + AMOUNT-PAID OF CARDIN.
00144
00145   02131  INCREMENT-TOTALS.
00146   02135      ADD ACCT-REC-WS TO ACCT-REC-TOT-WS.
00147   02140      ADD DISCOUNT OF CARDIN TO DISCOUNT-TOT-WS.
00148   02145      ADD AMOUNT-PAID OF CARDIN TO AMT-PAID-TOT-WS.
00149
00150   02151  MOVE-DATA-TO-OUTPUT.
00151   02155      MOVE CORRESPONDING, CARDIN TO DETAIL-LINE.
00152   02170      MOVE ACCT-REC-WS TO ACCT-REC-OUT.
00153
00154   02181  PRINT-DETAIL-LINE.
00155   02185      WRITE PRINT FROM DETAIL-LINE
00156   02190          AFTER ADVANCING 1 LINES.
00157
00158   02196  PREPARE-FOOTING.
00159   02200      MOVE ACCT-REC-TOT-WS TO ACCT-REC-T-OUT.
00160   02205      MOVE DISCOUNT-TOT-WS TO DISCOUNT-T-OUT.
00161   02210      MOVE AMT-PAID-TOT-WS TO AMT-PAID-T-OUT.
00162   02215      WRITE PRINT FROM TOTAL-LINE
00163   02216          AFTER ADVANCING 4 LINES.
00164
00165   02221  FINISH.
00166   02225      CLOSE FILE-IN
00167   02226            FILE-OUT.
00168   02230      STOP RUN.
```

ACCOUNTS RECEIVABLE REGISTER

CUST. NO.	CUST. NAME	INV. NO.	ACCTS. REC.	DISCT. ALLOW.	AMT. PAID
67451	ACME MFG CO	345	$697.17	$13.67	$683.50
67452	AMERICAN STEEL CO	342	$1,398.93	$27.43	$1,371.50
67453	TAIYO CO LTD	447	$1,211.25	$23.75	$1,187.50
67454	ALLIS CHALMERS CO	451	$2,307.75	$45.25	$2,262.50
67455	XEROX CORP	435	$163.71	$3.21	$160.50
67456	GLOBE FORM CO	436	$229.50	$4.50	$225.00
67457	WATSON MFG CO	428	$113.73	$2.23	$111.50
67458	CALCOMP CORP	429	$165.75	$3.25	$162.50
67459	SHOP--RITE MARKETS	433	$168.30	$3.30	$165.00
67460	MICROSEAL CORP	440	$5.61	$.11	$5.50
67461	MITSUBISHI LTD	420	$2,305.20	$45.20	$2,260.00
67462	MARK KLEIN & SONS	431	$1,393.32	$27.32	$1,366.00
67463	HONEYWELL CORP	432	$11.73	$.23	$11.50
67464	SPERRY RAND CORP	449	$2,345.49	$45.99	$2,299.50
67465	WESTINGHOUSE CORP	460	$3,047.25	$59.75	$2,987.50
67466	GARRETT CORP	399	$184.62	$3.62	$181.00
67467	NANCY DOLL TOY CO	400	$22.95	$.45	$22.50
67468	RAMONAS FINE FOODS	430	$3,557.25	$69.75	$3,487.50
67469	EL CHOLOS	434	$1,795.71	$35.21	$1,760.50
67470	DATAMATION INC	437	$2,247.00	$374.50	$1,872.50
67471	MICROFICHE CORP	441	$2,555.10	$50.10	$2,505.00
67472	REALIST INC	389	$2,872.32	$56.32	$2,816.00
67473	EASTMAN KODAK CO	401	$2,311.32	$45.32	$2,266.00
67474	UNIVAC INC	410	$3,348.15	$65.65	$3,282.50
67475	AVCO CO	411	$5,015.85	$98.35	$4,917.50
67476	TRW SYSTEMS GROUP	412	$2,311.32	$45.32	$2,266.00
67477	BELL HELICOPTER CO	413	$2,878.95	$56.45	$2,822.50
67478	BOEING AEROSPACE CORP	414	$2,328.15	$45.65	$2,282.50
		TOTALS	$46,993.38	$1,251.88	$45,741.50 **

Exercises

Write your answer in the space provided (answer may be one or more words).

1. The Data Division of a COBOL program describes the ______________ of the information to be processed by the ______________ program.
2. The manner in which data is ______________ and ______________ has a major effect upon the efficiency of the object program.
3. The categories of data to be processed are data in ______________, data in ______________ and ______________ data.
4. Each Data Division entry must contain a ______________, a ______________ or ______________ and a series of ______________.
5. An item is considered a ______________.
6. An item that is not divided into smaller units is known as a ______________ item.
7. An item appearing in the Working-Storage Section that is not part of a record or record itself is known as a ______________ item.
8. The ______________ record is considered a group item comprising several related items.
9. A ______________ is composed of a series of related data records.
10. Blocks of ______________ records are referred to as the ______________ record.
11. Once the relationship is established between ______________ and ______________ records, only the ______________ records are made available to the program.
12. Label records are normally used for files that are stored on ______________ or ______________ devices.
13. The external description refers to the ______________ aspect of a file.
14. The explicit description of the contents of each record defines its ______________.

15. The Data Division is divided into three fixed sections in the following sequence: ________________, ________________, and ________________.
16. The ________________ Section defines the contents of data files stored on the external medium.
17. The ________________ Section describes records or noncontiguous data items which are not part of the data files.
18. For every file named in the ________________ clause in the ________________ Division, a file description entry must be written.
19. The file description entry describes the ________________ of the file, how the data is ________________, the size of the ________________ and ________________ in the file, information about ________________ records, and the names of the ________________ in the file.
20. The file description entry always begins with the level indicator ________________ followed by the ________________ name and a series of ________________.
21. The Block Contains clause specifies the number of ________________ or ________________ in a block.
22. The Record Contains clause is a ________________ clause that specifies the size of the ________________ records within a ________________.
23. The IBM 360 or 370 computer has the following recording formats: ________________ length, ________________ length, ________________ length and ________________.
24. The Label Records clause is ________________ in every file description entry and specifies whether the labels are present or ________________.
25. The ________________ clause informs the compiler what the record(s) in the file are.
26. The Report clause is used in conjunction with the ________________ feature.
27. A record description entry specifies the ________________ of each item in the data record.
28. Each record description entry consists of a ________________, a ________________ or ________________ and a series of ________________ clauses.
29. The level number ________________ is used for the record itself.
30. In subdividing a level, level numbers need not be ________________.
31. Group items consist of all items under it until a level number ________________ or ________________ than the group level number is reached.
32. Indenting of items is ________________ required.
33. Level numbers ________________ through ________________ are used for subdivisions of group items.
34. Separate entries are written for each ________________.
35. The ________________ permits the programmer to refer to items individually in procedure statements.
36. The highest qualifier of a data-name permitted is a ________________.
37. FILLER does not always represent a ________________ area.
38. The Usage clause describes the form that ________________ is stored in ________________.
39. If the Usage clause is written at a ________________ level, it applies to each elementary item in the group.
40. The type of Usage clauses permitted in the IBM 360 or 370 computer are ________________, ________________, ________________, ________________, and ________________.
41. The Picture clause specifies the general ________________ and the ________________ descriptions for a ________________ item.
42. The Picture clause is forbidden at the ________________ level.
43. The categories of data that can be described with a Picture clause are ________________, ________________, ________________, ________________, and ________________.
44. An ________________ item may only contain the letters of the alphabet and a space.
45. An ________________ item may contain any combination of characters in the COBOL character set.
46. A nonedited Picture clause may contain a combination of the following characters: ________________, ________________, ________________, ________________, ________________, and ________________.
47. The character ________________ is used to represent an assumed decimal point.
48. The character ________________ indicates that the position contains a decimal digit.

49. An ______________ Picture clause is used to describe items to be output on a printer.
50. When zero suppression is used, a ______________ character may not follow a ______________ character.
51. A floating string must begin with ______________ consecutive characters.
52. The character ______________ represents an actual decimal point to be inserted in the output.
53. The DB character symbols may appear only at the ______________ end of a Picture clause and will print only if the item is ______________.
54. ______________ are a continuous string of characters and inserted immediately to the ______________ of the digit position indicated.
55. The usage of an item must be compatible with the ______________ clause.
56. The ______________ clause may be used to assign an initial value to data in the ______________ section.
57. A ______________ is a name assigned by the programmer to a particular value that may be assumed by a data item.
58. To use the same area to contain different data items, a ______________ clause is used.
59. In the absence of the Justified clause, when an item is moved to a location larger than itself, numeric data will be ______________ justified after decimal alignment with any unused positions being filled with ______________ while nonnumeric data will be ______________ justified with unused positions being filled with ______________.
60. The Occurs clause is used to ______________ tables and other ______________ sets of data whose elements can be referred to by ______________ and ______________.
61. The ______________ clause is used to specify alignment of an elementary item in the natural boundaries of the computer memory.
62. ______________ provides the facility to refer to data items in a list or a table that has been assigned individual values.
63. A ______________ must always have a positive ______________ integral value whose value determines which item is being ______________ in the table or list.
64. The Working-Storage Section contains descriptions of records which are developed and processed ______________. It is also used for ______________ work areas and for ______________ to be used in the program.
65. ______________ items in the Working-Storage Section must precede record items.

Answers

1. CHARACTERISTICS, OBJECT
2. ORGANIZED, STORED
3. FILES, WORKING AREAS, CONSTANT
4. LEVEL INDICATOR, DATA-NAME, OTHER NAME, CLAUSES
5. FIELD
6. ELEMENTARY
7. INDEPENDENT
8. DATA
9. FILE
10. LOGICAL, PHYSICAL
11. LOGICAL, PHYSICAL, LOGICAL
12. MAGNETIC TAPE, DIRECT ACCESS
13. PHYSICAL
14. INTERNAL CHARACTERISTICS
15. FILE, WORKING-STORAGE, REPORT
16. FILE
17. WORKING-STORAGE
18. SELECT, ENVIRONMENT
19. NAME, RECORDED, BLOCKS, RECORDS, LABEL, RECORDS
20. FD, FILE, CLAUSES
21. RECORDS, CHARACTERS
22. OPTIONAL, LOGICAL, FILE
23. FIXED, VARIABLE, UNDEFINED, SPANNED
24. REQUIRED, OMITTED
25. DATA RECORDS
26. REPORT WRITER
27. CHARACTERISTICS
28. LEVEL NUMBER, DATA-NAME, FILLER, INDEPENDENT
29. 01
30. CONSECUTIVE
31. EQUAL TO, LESS THAN
32. NOT
33. 02, 49
34. LEVEL NUMBER
35. DATA-NAME
36. FILE-NAME
37. BLANK
38. DATA, MAIN STORAGE
39. GROUP
40. DISPLAY, COMPUTATIONAL, COMPUTATIONAL-1, COMPUTATIONAL-2, COMPUTATIONAL-3
41. CHARACTERISTICS, DETAIL, ELEMENTARY
42. GROUP
43. ALPHABETIC, NUMERIC, ALPHANUMERIC, ALPHA-NUMERIC EDITED, NUMERIC EDITED
44. ALPHABETIC
45. ALPHANUMERIC
46. 9, X, V, P, S, A
47. V
48. 9

49. EDITED
50. Z, 9
51. TWO
52. .
53. RIGHT, NEGATIVE
54. FLOATING STRINGS, LEFT
55. PICTURE
56. VALUE, WORKING-STORAGE
57. CONDITION-NAME
58. REDEFINES
59. RIGHT, ZEROS, LEFT, SPACES
60. DEFINE, HOMOGENOUS, SUBSCRIPTING, INDEXING
61. SYNCHRONIZED
62. SUBSCRIPTING
63. SUBSCRIPT, NONZERO, REFERENCED
64. INTERNALLY, INTERMEDIATE, CONSTANTS
65. INDEPENDENT

Questions for Review

1. Why is it important that data be stored properly?
2. What must each entry contain?
3. What is the relationship between an item, elementary item, group item, and independent item?
4. What is the relationship between a data record, file, and block?
5. What is the distinction between the record's external description and its internal content?
6. What is the function of the File Section?
7. What does the file description entry describe?
8. Briefly differentiate between the types of recording modes used in a system 360 or 370 computer.
9. When is the Block Contains clause used?
10. What clauses are required in the file description entry?
11. Why are the items indented in the record description entries?
12. How are the various level numbers used?
13. How is 'FILLER' used and does it always represent a blank area?
14. What are the rules governing the use of the Usage clauses? Explain the various types of Usage clauses in the System 360 or 370 computer.
15. Describe the use of a Picture clause.
16. Describe the relationship between the Picture and Usage clauses.
17. What is the Value clause and how is it used?
18. What is the Occurs clause and how is it used?
19. What is the purpose of the Redefines clause?
20. What is the Blank When Zero clause used for?
21. What is the purpose of Editing Picture clauses?
22. What are Floating Strings and what is their purpose?
23. How does the COBOL compiler perform automatic justification?
24. How is the Synchronized clause used?
25. What is a condition-name and how is it used?
26. What are the principal uses of the Working-Storage Section?
27. What is subscripting and how is it used?

Problems

1. *Match each item with its proper description.*

______	1. Group Item	A. Series of related data records.
______	2. Data Record	B. Named sequence of one or more elementary items.
______	3. Item	C. Physical record.
______	4. Block	D. Smallest item available.
______	5. Independent item	E. Item appearing in Working-Storage Section that is not a record or part of a record.
______	6. File	F. Field.
______	7. Elementary Item	G. Logical record.

2. *Match each clause with its proper description.*

______	1. Recording Mode	A. Presence or absence of labels.
______	2. Record Contains	B. Format of logical record.
______	3. Block Contains	C. Particularizes description of item in Label Records clause.
______	4. Label Records	D. Size of logical record.
______	5. Value Of	E. Used with Report Writer feature.
______	6. Data Records	F. Number of records or characters in block.
______	7. Report	G. Name of record(s) in file.

3. *Match each level number group with its proper classification.*

______	1. 01	A. Subdivisions of group record items.
______	2. 02–49	B. Condition entry.
______	3. 66	C. Designates item as a record.
______	4. 77	D. Used with RENAMES clause.
______	5. 88	E. Independent elementary item in Working-Storage Section.

4. *Match each USAGE clause with its proper description.*

_____ 1. Display	A. Short precision internal floating point format.
_____ 2. Computational	B. Packed decimal format.
_____ 3. Computational-1	C. One character per byte.
_____ 4. Computational-2	D. Binary data items.
_____ 5. Computational-3	E. Long precision internal floating point format.

5. *Match each category of data with its proper description.*

_____ 1. Alphabetic	A. Numerals.
_____ 2. Numeric	B. Combination of numerals and editing characters.
_____ 3. Alphanumeric	C. Letters and space.
_____ 4. Alphanumeric Edited	D. Any combination of characters in COBOL character set.
_____ 5. Numeric Edited	E. Combination of any character plus the characters B or 0.

6. *Match each non-editing character with its proper description.*

_____ 1. 9	A. Assumed decimal point.
_____ 2. X	B. One decimal digit.
_____ 3. V	C. Operational sign.
_____ 4. P	D. Letter or space.
_____ 5. S	E. Any type of character in the COBOL character set.
_____ 6. A	F. Assumed zero.

7. *Match each editing character with its proper description.*

_____ 1. Z	A. Credit symbol.
_____ 2. .	B. Digit position.
_____ 3. *	C. Credit symbol.
_____ 4. CR	D. Actual decimal point.
_____ 5. DB	E. Check protection.

8. *Which of the following is a floating strings or insertion character? Use* F *for floating strings and* I *for insertion characters.*

_____ $
_____ B
_____ –
_____ 0
_____ +
_____ ,

9. *Match each clause with its proper description.*

_____ 1. Picture	A. Same area to contain different data items.
_____ 2. Value	B. Item to be filled with spaces when value is zero.
_____ 3. Condition-Name	C. Alignment of item on natural boundaries of computer memory.
_____ 4. Redefines	D. Initial value of an item.
_____ 5. Blank When Zero	E. Override normal positioning of alphabetic data when moved to a larger area.
_____ 6. Justified	F. Particular value that may be assigned by a data item.
_____ 7. Synchronized	G. General characteristics and description of an item.
_____ 8. Occurs	H. Define tables and other homogenous sets of data.

10. *List the group and elementary items in the following.*

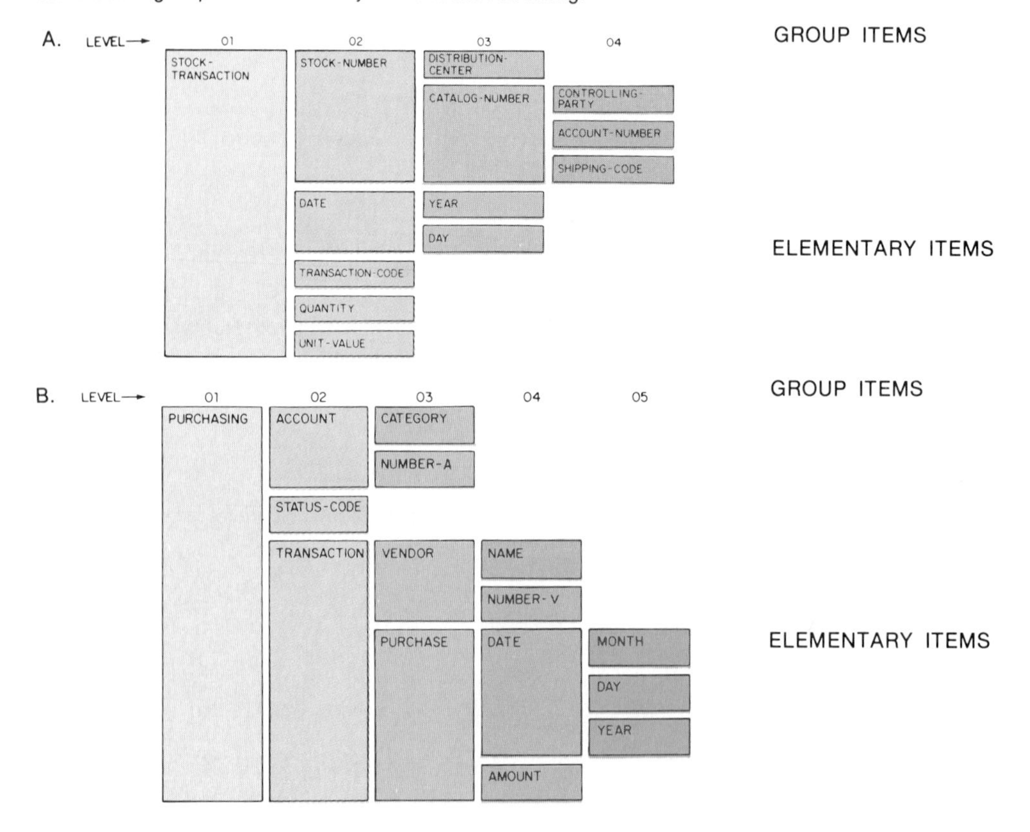

11. *Write the file description entry for a file whose name is EXPENSE-FILE.*
    The file is on magnetic tape, has standard label records and one type of data record called EXPENSE-RECORD. All records are fixed length and each record is preceded by a record length control field. There are twenty records per block.

12. *Write the record description entry for the following record. We are not concerned with the USAGE and PICTURE clauses of these items, just the level numbers, data-names or FILLER.*

STUDENT-RECORD
STUDENT
ATTENDEE
LAST-NAME
INITIALS
EMPLOYER
NUMBER
NAME
GRADES
APTITUDE TEST
FINAL EXAM
DATE
M D Y
SCORE
QUIZ AVERAGE
DAYS ATTENDED
COURSE
NAME
SUBJECT
TOPIC
CLASS
SERIAL
REGION
DURATION
START
M D Y
FINISH
M D Y

13. *Write the complete record description entry for the following input record including level numbers, data-names or FILLER, together with all necessary PICTURE and VALUE clauses.*

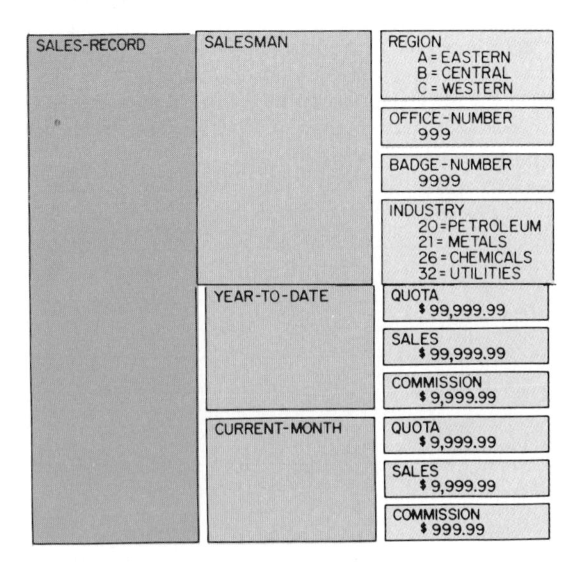

14. *In the Working-Storage Section, write the entries to set up the following:*

   a. A work area large enough to hold 25 alphanumeric characters.
   b. An independent item called DIFFERENCE to contain a sign, five digits, and stored in packed decimal format.
   c. A constant called LIMIT, whose value is 600 and is stored in a binary format.
   d. An alphanumeric constant which is to serve as a title of a report. The contents of the item are to be DEPRECIATION SCHEDULE and the item is to be named TITLE.
   e. A record to be called ADDRESS composed of STREET (20 alphanumeric characters), CITY (20 alphanumeric characters), STATE (5 Alphanumeric characters) and ZIPCODE (5 digits).

15. *Write the necessary entries in the Working-Storage Section for the following headings:*

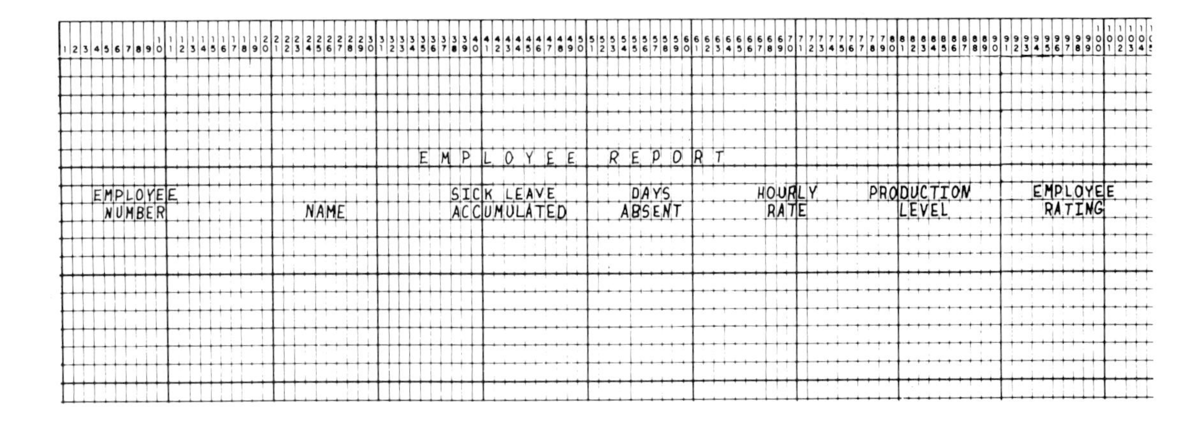

16. *Set up the Working-Storage Section for the following:*

  a. A percentage value of 25% to be used as the multiplier in a multiplication operation.
  b. A 4-digit counter containing an initial value of 1000.
  c. A 5-position field to be used as a temporary storage area.
  d. A 3-digit page number which will be incremented for each new page printed. The initial value equals 001.
  e. A record containing six fields of 15 characters each with the initial values equal to the names of the six New England States (Connecticut, Massachusetts, Maine, New Hampshire, Rhode Island and Vermont).
  f. Headings for a report as described in the following:

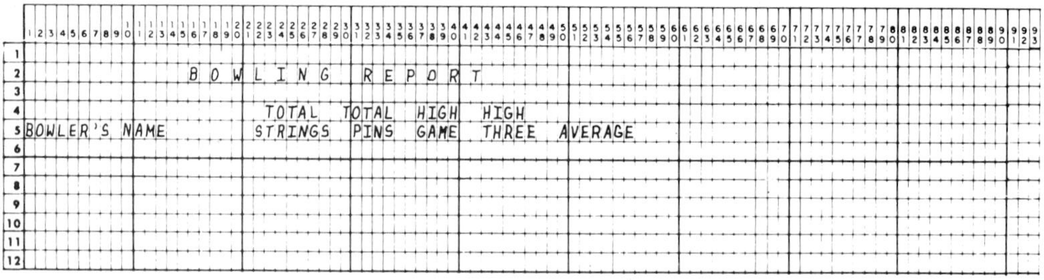

```
 2                    B O W L I N G   R E P O R T

 4                           TOTAL   TOTAL  HIGH    HIGH
 5 BOWLER'S NAME             STRINGS  PINS  GAME    THREE   AVERAGE
```

17. *How will the following source data be interpreted?*

	Source Data	Picture	Interpretation
a.	123	9V99	________
b.	13$\bar{2}$	S999	________
c.	15671	999V999	________
d.	4071629	9(5)V99	________
e.	9263	V999	________
f.	6$\overset{+}{1}$	S99	________

18. *Specify the actions that will take place when the source area data is moved to the corresponding receiving area.*

	Source Area		Receiving Area	
	Picture	Data Value	Picture	Data Value Interpretation
a.	9999	1234	9(6)	________
b.	99V99	1234	999V99	________
c.	99V99	1234	99V999	________
d.	99V99	1234	9(4)V9(4)	________
e.	9(4)	1234	999	________
f.	999V9	1234	99V99	________
g.	999V9	1234	999	________
h.	999V999	123456	99V99	________
i.	99V999	12345	9(4)V99	________
j.	999V99	12345	99V999	________

19. *Specify the actions that will take place when the source area data is moved to the corresponding receiving area.*

	Source Area		Receiving Area	
	Picture	Data Value	Picture	Edited Data
a.	999V9	1234	999.9	________
b.	9(5)V99	0001234	9(5).99	________
c.	9(5)V99	0001234	Z(5).99	________
d.	9(5)V99	0000123	ZZZ99.99	________
e.	9(5)V99	0123456	99,999.99	________
f.	9(5)V99	0123456	ZZ,999.99	________

g.	9(5)V99	0000000	ZZ,ZZZ.ZZ	____________
h.	9(5)V99	1234567	$99,999.99	____________
		+		____________
i.	S9(6)	123456	+ 9(6)	____________
		−		____________
j.	S999	123	− 999	____________
		+		____________
k.	S9(4)	0012	ZZZ9 +	____________
		−		____________
l.	S99V99	1234	99.99CR	____________
		−		____________
m.	S99V99	1234	99.99DB	____________
		+		____________
n.	S99V99	1234	99.99DB	____________
o.	9(5)V99	0234567	$$$,$$9.99	____________
p.	9(6)V99	00001234	***,***.99	____________
		+		____________
q.	S9999	0123	+ + + + 9	____________
		−		____________
r.	S9999	0123	+ + + + 9	____________
		+		____________
s.	S9999	0001	− − − − 9	____________
t.	9(9)	123456789	999B99B9999	____________

20. *Write the Identification, Environment and Data divisions for the following:*

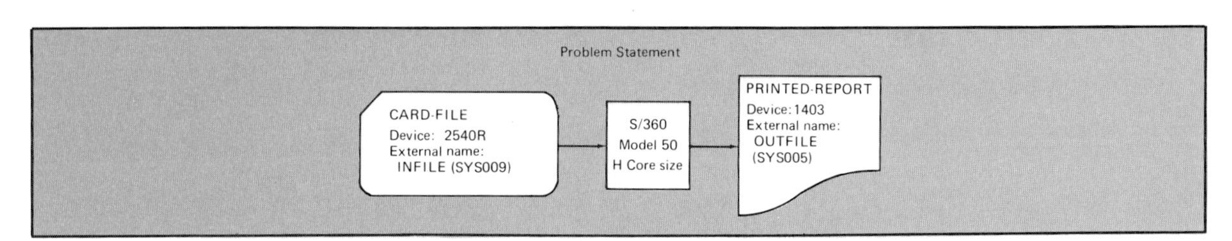

The system flowchart above shows the files and equipment to be used in this program. The forms of records in CARD-FILE and PRINTED-REPORT are illustrated below.

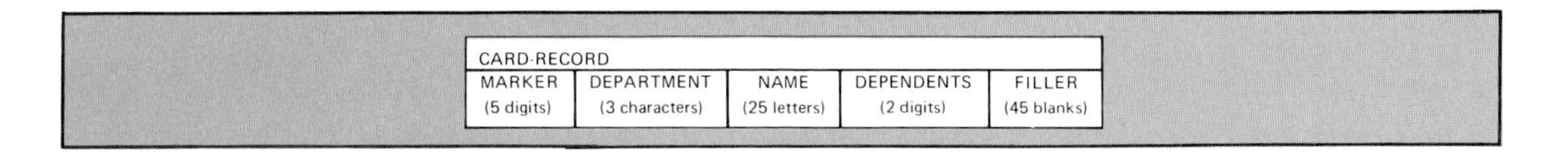

CARD-RECORD				
MARKER (5 digits)	DEPARTMENT (3 characters)	NAME (25 letters)	DEPENDENTS (2 digits)	FILLER (45 blanks)

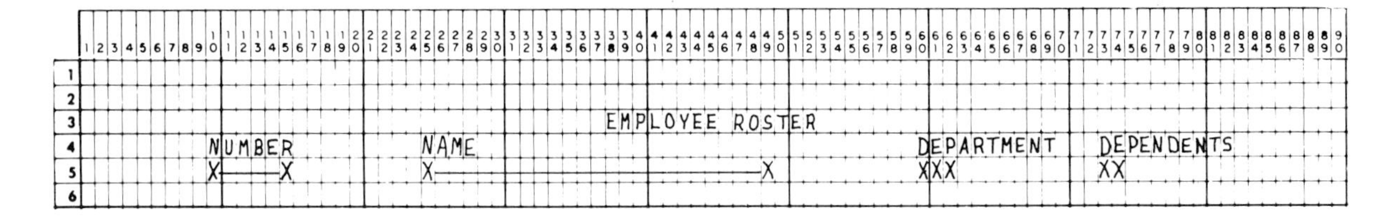

The file PRINTED-REPORT is to consist of a title and headings followed by a listing of the records in CARD-FILE with the data items rearranged as shown in the Printer Spacing Chart. Use the variables CARD-RECORD, PRINT-RECORD, WORK-RECORD-1, WORK-RECORD-2, and WORK-RECORD-3.

21. *Write the Identification, Environment and Data divisions for the following:*

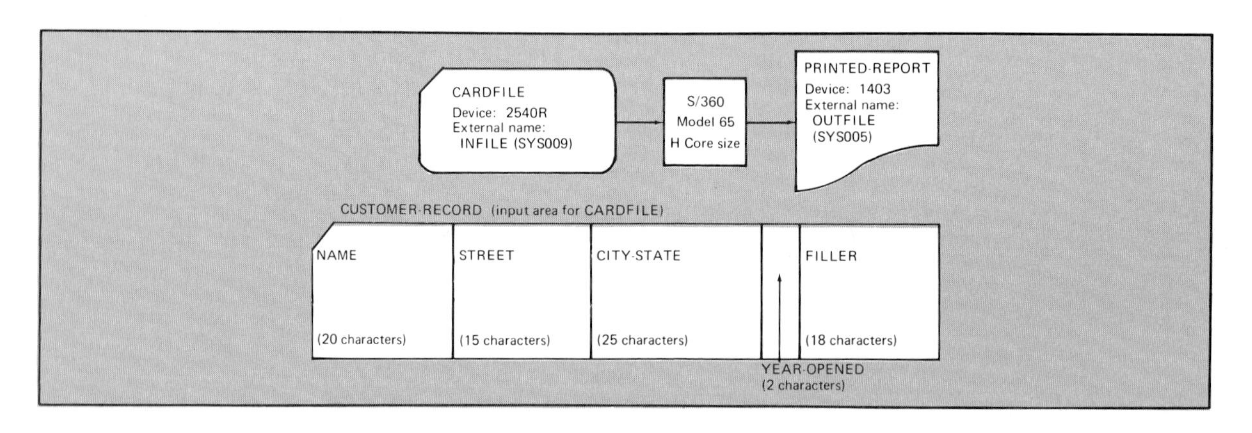

PRINT-RECORD (output area for PRINTED-REPORT)

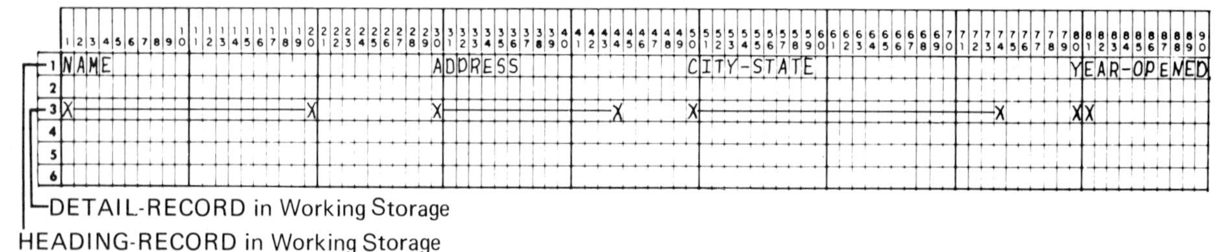

22. *Write the Identification, Environment and Data divisions for the following:*

Hardware

Computer	—	IBM 370 Model 155 with H Core size
Input	—	Magnetic Tape Unit Model 2400 External name RECBLES (SYS012).
Output	—	Printer Model 1403 External name ACCTLIST (SYS005).

Input File—RECEIVABLE

There are 50 records per block and there are standard labels.

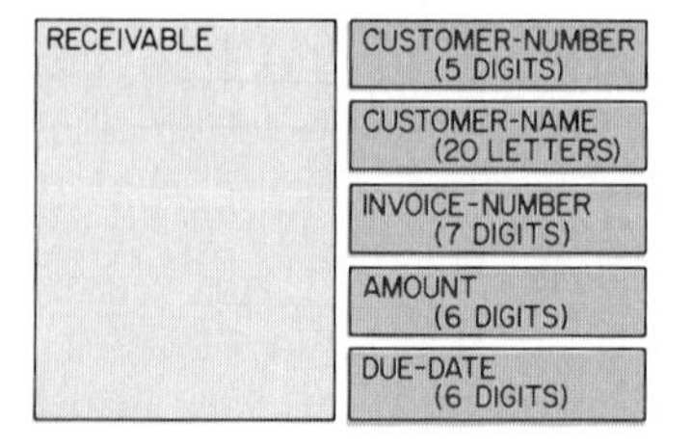

Output File—ACCOUNT

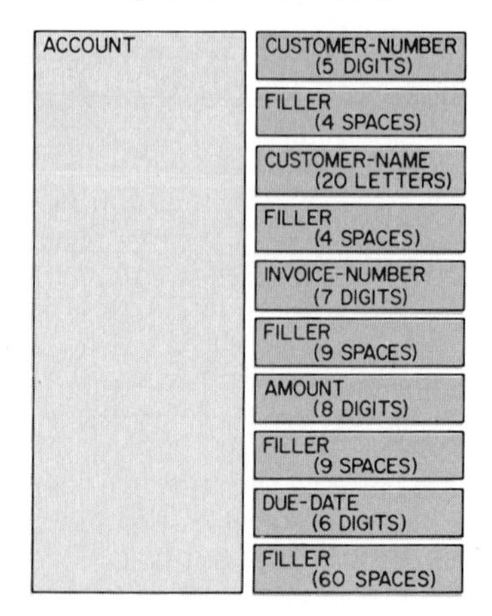

23. *Write the Identification, Environment and Data divisions for the following:*

Companies have master files of information that require changes and constant updating. Such a change may be the deletion of discontinued items from the master file and the subsequent listing of the deleted items. In order to make these changes, a maintenance program should be written.

The following processing is involved in writing the maintenance program.

1. *Card-to-tape-conversion*—Read a deck of delete cards and write the deleted items on a transaction file tape.
2. *File update*—Update the master file by passing the transaction file tape against the old master file deleting items according to the item number.
3. *Print*—Print those items which are deleted from the master file.

Hardware—

Computer—IBM 370	*Model 155 H Core size*
4 Magnetic Tape Units	*Model 2400 (Tape-I-1, Tape-I-2, Tape-O-1, Tape-O-2).*
1 Card Reader	*Model 2540 (SYS009).*
1 Printer	*Model 1403 (SYS005).*

The following systems flowchart illustrates the two phases of the program.

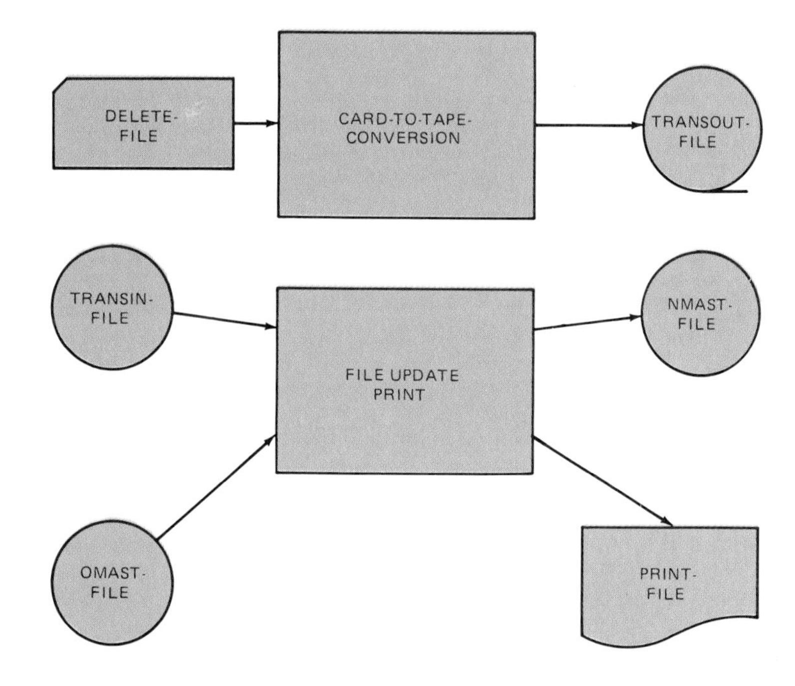

Data Division

File Section

DELETE-FILE	*Read from Card Reader*	*Record name is DELETE-RECORD*

Field Name	**Card Columns**	**Field Class**
Item number	1–5	Numeric
Delete date	6–11	Numeric
Description	12–24	Alphabetic

TRANSOUT-FILE	*Tape Unit SYS012*	*Record name is TRANSOUT-RECORD*

FORMAT SAME AS DELETE-FILE

TRANSIN-FILE	*Tape Unit SYS013*	*Record name is TRANSIN-RECORD*

FORMAT SAME AS TRANSOUT-FILE

OMAST-FILE	*Tape Unit SYS014*	*Record name is OMAST-RECORD*

Field Name	**Card Columns**	**Field Class**
Item number	1–5	Numeric
Item description	6–18	Alphabetic
Quantity	19–22	Numeric
Balance	23–28	Numeric XXXX.XX

NMAST-FILE	*Tape Unit SYS015*	*Record name is NMAST-RECORD*

FORMAT SAME AS OMAST-FILE
ALL TAPE RECORDS IN BLOCKS OF 4 RECORDS AND HAVE STANDARD LABELS.

PRINT-FILE	*LISTED REPORT 'DELETED ITEMS REPORT'*	*Record Name is PRINT-RECORD*

Field Name	**Print Positions**	**Field Class**
Delete Date	4–9	Numeric
Item number	14–18	Numeric
Item description	23–35	Alphabetic
Quantity	40–43	Numeric
Balance	49–56	Numeric Edited ($9999.99)

Working-Storage Section

Field Name	**Size**	**Field Class**
TOTAL	7	Numeric XXXXX.XX

Header Information

Constant Value	**Print Positions**
DATE	5–8
ITEM-NO	13–19
DESCRIPTION	24–34
QUANTITY	39–46
BALANCE	50–56

7 Structured Programming

INTRODUCTION

Structured programming applies a theoretical concept to the coding of computer programs. The main element of the concept is the use of very basic structures to create complex programs. To apply structured programming in a commercial data processing environment requires a change in program design methods and rigid application of standards to insure the best results. In addition, further difficulties arise from the use of COBOL, which does not provide all the basic structures directly.

The concept of structured programming and related aspects are currently getting wide coverage and attention in the computer field. The concepts, to be useful, must be distilled into a practical approach that will fit the environment of the organization wishing to use structured programs.

A structured program tends to be much easier to understand than programs written in other styles. This capability of being more easily understood facilitates code checking, which in turn may reduce program testing and debugging time. This is true partly because structured programming concentrates on one of the most error-prone factors in programming, the logic.

A program that is easy to read and which is composed of well-defined segments tends to be simpler, faster, and less expensive to maintain. These benefits derive in part from the fact that since the program is to a significant extent its own documentation, the documentation is always up-to-date. This is seldom true with conventional methods.

Structured programming offers these benefits, but it should not be thought of as a panacea. Program development is still a demanding task requiring skill, effort, and creativity (fig. 7.1).

BENEFITS

- **Encourages Programming Discipline**
- **Fewer Errors**
- **More easily modified and maintained**
- **More nearly self-documenting**
- **Programmer can control larger amount of code**

Figure 7.1 Structured Programming Benefits.

OBJECTIVES

The main objectives of structured programming is to provide a single generalized method of program design to provide the following:

1. Program development by segment
2. Ease of coding
3. Speed of coding
4. Ease of debugging
5. Ease of maintenance

STRUCTURE

Structured programming is a style of programming in which the structure of a program (that is, the interrelationship of its parts) is made as clear as possible by restricting control logic to just three structures (fig. 7.2).

1. *Simple Sequence (Sequence)*—In the absence of instructions to the contrary, statements are executed in the order in which they are written.
2. *IFTHENELSE (Selection)*—combine with statement brackets (begin and end) so that groups of statements can be included in the THEN and ELSE clauses. In fact, the THEN and ELSE clauses may themselves contain any of the three structures.
3. *DOWHILE or DOUNTIL (Repetition)*—A loop control mechanism.

In a structured program, any program function can be performed using any of the above three basic control logic structures; simple sequence, selection, or repetition. Any kind of processing, any combination of decisions, any sort of logic, can be accommodated with any one of these control structures or a combination of them. Each structure is characterized by a simple and single point of transfer of control into the structure, and a single point of transfer of control out of the structure. These structures can be combined to form a program that is simple, in the sense that control flows from top to bottom or from beginning to end. There is no back tracking. The control structures can be nested as shown in figure 7.3, but they retain their characteristic of single/entry—single/exit.

Reducing program complexity can be thought of as a process of removing things from a program, e.g., obscure structure, complicated control paths, redundant and obsolete code, meaningless notes, etc. Improving program clarity can be thought of as a process of adding material features to a program, e.g., self-explanatory labels, meaningful notes, code layout, and indentation that has information content for the reader, more levels of modularity, etc. (fig. 7.4).

Much of a program's complexity arises from the fact that the program contains many jumps to other parts of a program—jumps both forward and backward in the code. These jumps make it difficult to follow the logic of the program and moreover, make it difficult to ascertain at any given point of the program what existing conditions may be (such as, what the start of the variables is, and what other paths of the program have already been executed, etc.). Furthermore, as a program undergoes change during its development period, as it gets further debugged during its maintenance period, as it gets modified in subsequent new projects, its complexity grows alarmingly. New jumps are inserted, increasing that complexity. In some cases, new code is added because the pro-

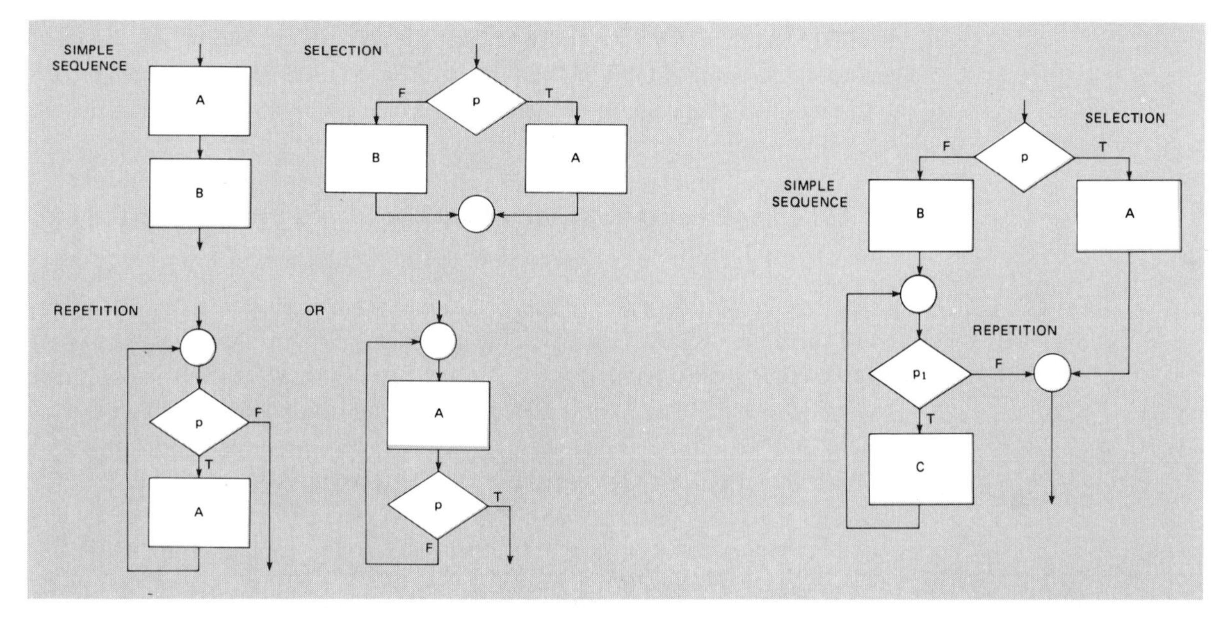

Figure 7.2 Control Structures.

Figure 7.3 Nesting of Control Structures.

COBOL CODING RULES

ALL PARAGRAPHS EXPLICITLY PERFORMED

NO 'PERFORM A THRU Z'
NO PERFORM OF SECTION

NO EXTRA PARAGRAPH NAMES

PARAGRAPHS IN ORDER OF EXECUTION

MINIMIZE COMMENTS

DON'T REPEAT THE CODE

USE MEANINGFUL NAMES AND CLEAR CODE

REMARKS AND OVERALL COMMENTS GOOD

NEW PAGE FOR DIVISIONS, SECTIONS, SEGMENTS

SEPARATE PARAGRAPHS WITH BLANK LINES

ONE STATEMENT PER LINE

INDENT CONTINUATIONS AND IF STATEMENTS

Figure 7.4 COBOL Coding Rules.

grammer cannot find existing code that performs the desired functions, or isn't sure how the existing code works, or is afraid to disturb the existing code for fear of undoing another desirable function. The result, after many modifications, is a program that is nearly unintelligible. The program is by now shopworn; the time has come when it is better to throw the entire program out and start afresh.

Easy program readability requires that it not be necessary to turn a lot of pages in order to understand how something works. A structured program is composed of *segments,* which may range from a few statements to a page of coding. Each segment is allowed to have just one entry and one exit. A practical rule is that a segment should not exceed a page of code, about fifty lines. In

COBOL terms, a segment can be a paragraph, section, subroutine, or code incorporated with a COPY. (The term segment as used here has nothing to do with the different meanings of the term in connection with the functions of the operating system.)

Such a segment, assuming it has no infinite loops and no unreachable code, is called a *proper program*. When proper programs are combined using the three basic control logic structures, the result is a proper program (fig. 7.5).

GO TO-LESS PROGRAMMING

Structured programming has occasionally been referred to as "GO TO-less Programming." A well-structured program gains an important part of its easy-to-read quality from the fact that it can be read in sequence, without "skipping around" from one part of the program to another. The "sequential ability," or "top-down readability," is beneficial because there is a definite limit to the amount of detail the human mind can encompass at one time. It is far easier to grasp completely what a statement does if its function can be understood in terms of just a few other statements, all of which are adjacent physically. The trouble with GO TO statements is that they generally defeat this purpose; in extreme cases, they can make a program essentially incomprehensible. (See figure 7.6.)

There are uncommon situations where the use of GO TOs may improve readability compared with other ways of expressing a procedure. Such examples are exceptional, however, and do not usually occur in everyday programming. The impact of deviations from installation guidelines, such as using GO TOs in other than prescribed ways, should be given careful consideration before such deviations are permitted.

The blatant use of GO TO statements results in unnecessary and complex flow patterns leading to difficult debugging effort on the part of the programmer. No special effort is really required to "eliminate GO TOs," which sometimes has been misunderstood as the goal of structured programming. They just never occur when the standard control logic structures are used.

The fundamental difficulty with the GO TO statement is that it distracts the reader from the program by forcing him to examine the program in an unnatural way.

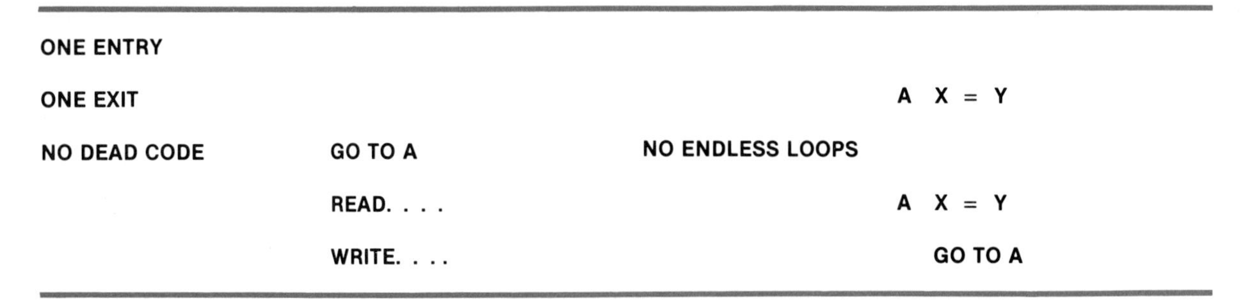

Figure 7.5 Proper Program.

```
PAR-1.
     IF A IS GREATER THAN .20, GO TO PAR-3.
     IF A IS GREATER THAN .10, GO TO PAR-2.
     MOVE 5 TO FLD-X.
     GO TO PAR-4.
PAR-2.
     MOVE 6 TO FLD-X.
     GO TO PAR-4.
PAR-3.
     MOVE 4 TO FLD-X.
PAR-4.
     ...............
```

To understand what is going on in the above program, the programmer would consider the conditional statements in the sequence presented. First, if the statement, "A IS GREATER THAN .20" is true, the program continues execution at PAR-3. Next, if the statement "A IS GREATER THAN .20" is false but the statement "A IS GREATER THAN .10" is true, the program continues execution at PAR-2. Only if both conditionals fail, does the program execute the next statement "MOVE 5 TO FLD-X" and then pass control to PAR-4.

To a COBOL-experienced eye, this does not seem to be a very difficult code structure, but if the eventual target, PAR-4, were several pages away (and if PAR-2 and PAR-3 were not so conveniently located) there would be a considerable amount of page flipping in order to discern the intended meaning of the program.

The problem would be eliminated or at least greatly simplified, if the program were organized so that there was greater "locality." In a sense, achieving this kind of correspondence between static placement of statements and dynamic flow depends on the vague concept of "program style." To illustrate how this program segment would look in GO TO-free form, it can be rewritten as follows:

```
IF A IS GREATER THAN .20
THEN
     MOVE 4 TO FLD-X
ELSE
     IF A IS GREATER THAN .10
     THEN
          MOVE 6 TO FLD-X
     ELSE
          MOVE 5 TO FLD-X.
............
```

The resulting value of FLD-X, after the set of checks on the value of A, is evident by the organization of the statements. Because there are no GO TO statements there are no labels and there is a direct correspondence between the static form of the program and the dynamic flow during execution.

Figure 7.6 GO TO-Less Programming—Illustration.

TOP-DOWN DESIGN

The mere removal of all of the GO TO statements will not of itself "structure" the program; in fact, even though GO TO-free programs are intrinsically easier to read and debug than their counterparts (programs with GO TO statements), the form and style of the expression of the algorithms is not explicitly changed by the avoidance of the GO TO statements. Structured programming is also concerned with ways of developing complicated program structures in an orderly fashion.

The design of the program should proceed from top to bottom. Individual program segments should be as short as possible, preferably no longer than one page of machine output, to facilitate the partitioning of the logic into individual chunks that are easy to debug.

There are two major advantages of a strictly hierarchical form for a program. First, adhering to the hierarchical constraints forces the organization of the program along "natural" algorithm boundaries; individual program segments can be organized so that each performs some specific function. The result is that each segment is easier to debug and so the entire program is easier to debug.

Structured programming is an aggregation of three basic ideas.

1. The beneficial properties of GO TO-less programming.
2. The application of management techniques to process from top to bottom design.
3. The idea of each program segment being specifically related to each of the others.

Each level of programming is supported by the next lower level of the hierarchy. The program at the lower level manipulates the data and at the same time participates in generating a higher level of information for the next higher level of the hierarchy to manipulate.

For example, assume that a program is to be written to retrieve information from a file. A file may be considered a collection of records with each record residing on some section of the disk. Each record is, in turn, composed of words, each word is composed of bytes, and each byte is composed of bits. The hierarchy of resources needed to retrieve a single bit from the file is as follows:

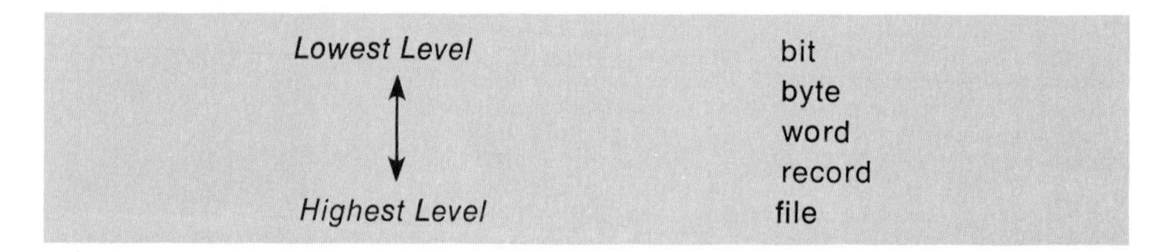

A set of instructions would be written to manipulate files, in this case, to locate and extract particular records. The next level of instructions would operate solely on records, extracting words from records, retrieved by manipulating the program. In turn, the next level would be represented by a set of instructions which would extract bytes from words; at the top would be the program to extract bits from bytes.

INDENTATION

The indentation of the program statements will make the structure more obvious to the reader of the program. This may make the program more difficult to code but it will greatly simplify the reading. When a program has to be maintained, it is the reading that becomes crucial to its success. Following such a practice also makes it much easier for another programmer to check a program for correctness.

The use of indentation is important because consistent indentation enhances readability so that the finished program exhibits in a pictorial way the relationships among statements. The basic idea is that all statements controlled by a control logic structure should be indented by a consistent amount, to show the scope of the control of the structure. In COBOL this means that the statements between the IF and the ELSE should be indented a consistent amount, and similarly for the statements between the ELSE and the next sentence (fig. 7.7).

Indentation can be a major benefit.

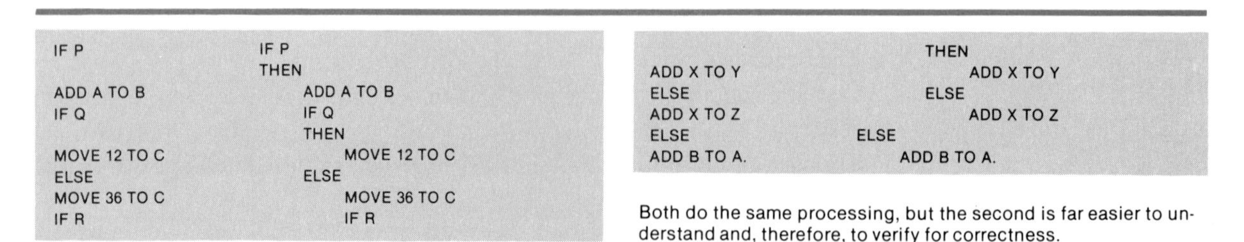

```
IF P
ADD A TO B
IF Q
MOVE 12 TO C
ELSE
MOVE 36 TO C
IF R
ADD X TO Y
ELSE
ADD X TO Z
ELSE
ADD B TO A.
```

```
IF P
THEN
    ADD A TO B
    IF Q
    THEN
        MOVE 12 TO C
    ELSE
        MOVE 36 TO C
        IF R
        THEN
            ADD X TO Y
        ELSE
            ADD X TO Z
ELSE
    ADD B TO A.
```

Both do the same processing, but the second is far easier to understand and, therefore, to verify for correctness.

Figure 7.7 Nested IF Statements With and Without Indentation.

ADVANTAGES

Use of the three classical structures of structured programming in their pure form results in inefficiency in two situations. This inefficiency is avoided through the use of a variant of the selection structure and a slight relaxation of the single/exit rule. The first situation is handled by computed GO TOs or switches; the CASE structure, where only one of a series of functions is to be performed, depending on the value of a variable. This is really a generalization of the selection process (IFTHENELSE) from a two-valued to a multivalued operation (fig. 7.8). The second situation arises when the programmer wishes to terminate a repetition block abnormally and the language does not explicitly allow this (fig. 7.9). Although such an abnormal termination violates the single/entry—single/exit rule of structured programming, it may produce a significant savings in space and time. If properly flagged, this practice maintains the spirit of structured programming.

An example where GO TO is useful is the handling of the INVALID KEY situation.

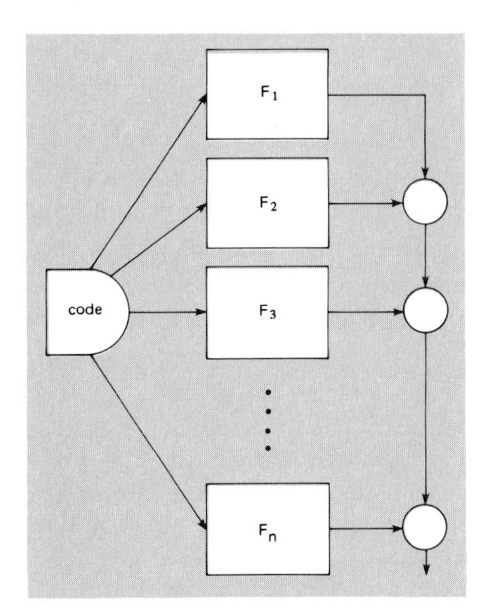

Figure 7.8
From a Two-Valued to a Multi-Valued Operation.

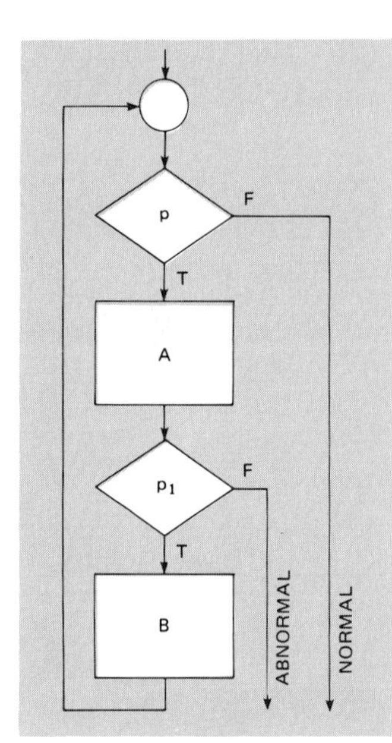

Figure 7.9
An Abnormal Termination of a Repetition Block.

In addition to the use of restricted control structures, many other refinements and attributes have been attached to structured programming, such as,

1. Segment length should be limited to a manageable size, usually expressed as a number of lines. This size restriction helps limit programs to a comprehensible unit.
2. Since the segment has a single entry and a single exit and no arbitrary jumps to other parts of the program, there is little need for page-turning or for holding several places in a listing to which one can refer constantly.
3. Careful indention of coding to show nesting levels also gives increased clarity to the code.

Structured programming combined with some traditional coding practices, such as good annotation, descriptive labels, and judicious spacing in the source code, greatly clarifies source coding. This increased clarity, and the reduced complexity of structured programs are responsible for another advantage of structured programming: its correctness is more easily verified than that of unstructured programs. There are two senses in which this is true. First, because the flow of control is simpler in structured programs, the development and execution of test cases to adequately debug the program is simpler. Second, since the program is more understandable, its correctness is more easily verified by reading, that is, desk-checking. Compared to unstructured programs, structured programs are very easy to read and verify for correctness. The use of structured programming and of more desk-checking will, therefore, improve the quality of programs and reduce the cost of their development.

COBOL IMPLEMENTATION OF STRUCTURED PROGRAMMING

Once the principles of structured programming are understood, writing structured programs in COBOL is a matter of habitually following a few simple rules. The structure theorem states that any *proper program* can be written using only the control logic structure of *sequence, selection,* and *iteration (loop mechanism).* A proper program is defined as one that meets the following two requirements:

1. It has exactly one entry point and one exit point for program control.
2. There are paths from the entry to the exit that lead through every part of the program; this means that there are no infinite loops and no unreachable code. This requirement is, of course, no restriction but simply a statement that the structure theorem applies only to meaningful programs.

All control logic structures can be expressed in COBOL although in some cases not directly.

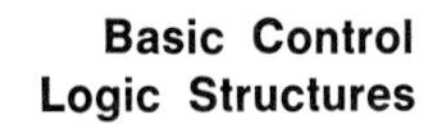

Basic Control Logic Structures

Sequence is simply a formularization of the idea that unless otherwise stated, program statements are executed in the order in which they appear in the program (fig. 7.10). This is true of all commonly used programming languages; it is not always realized that sequence is in fact a control logic structure.

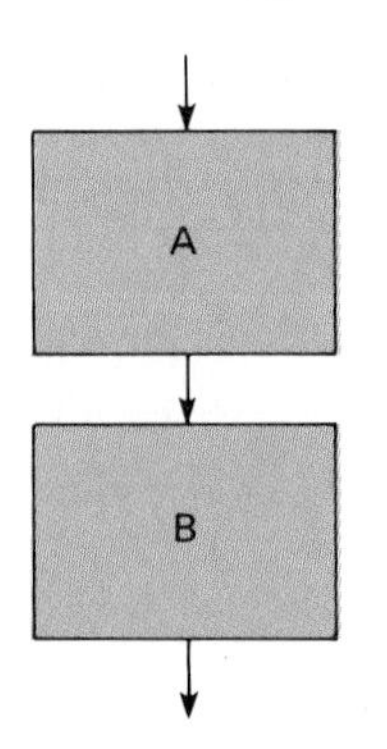

Figure 7.10
Flowchart for Sequence Program Structure.

Sequence is implemented in the COBOL language by simply writing statements in succession. (See figure 7.11.)

Selection is the choice between two actions based on a condition (predicate); this is called the IFTHENELSE structure.

In COBOL it is implemented with the IF statement and the conditional clauses.

The *iteration* structure is used for repeated execution of code while a condition is true (or false). It is also referred to as the loop control. The iteration structure is called the DOWHILE structure.

In COBOL, the DOWHILE structure is implemented with the PERFORM verb with the UNTIL option.

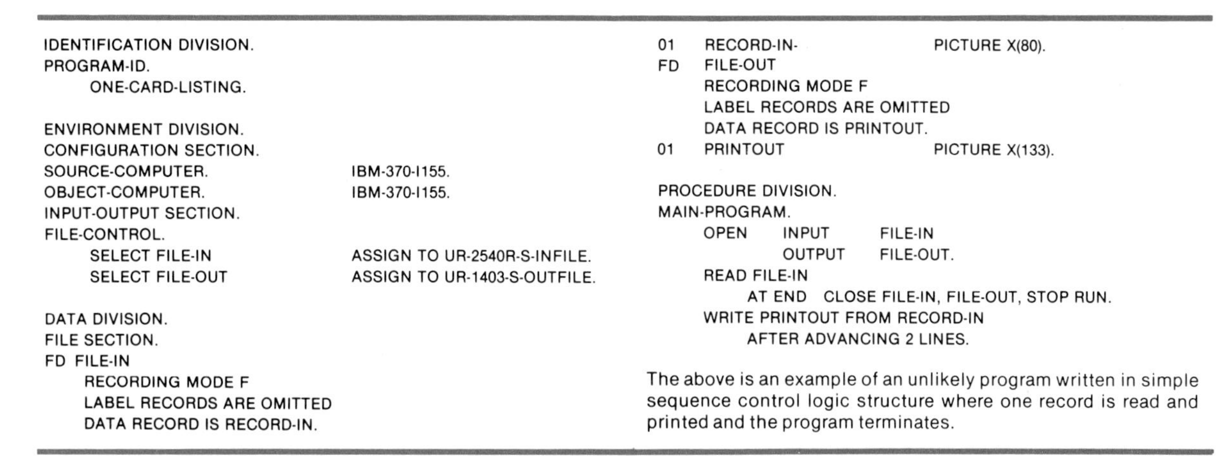

```
IDENTIFICATION DIVISION.
PROGRAM-ID.
       ONE-CARD-LISTING.

ENVIRONMENT DIVISION.
CONFIGURATION SECTION.
SOURCE-COMPUTER.                    IBM-370-I155.
OBJECT-COMPUTER.                    IBM-370-I155.
INPUT-OUTPUT SECTION.
FILE-CONTROL.
       SELECT FILE-IN               ASSIGN TO UR-2540R-S-INFILE.
       SELECT FILE-OUT              ASSIGN TO UR-1403-S-OUTFILE.

DATA DIVISION.
FILE SECTION.
FD  FILE-IN
      RECORDING MODE F
      LABEL RECORDS ARE OMITTED
      DATA RECORD IS RECORD-IN.
01    RECORD-IN-                PICTURE X(80).
FD    FILE-OUT
      RECORDING MODE F
      LABEL RECORDS ARE OMITTED
      DATA RECORD IS PRINTOUT.
01    PRINTOUT                  PICTURE X(133).

PROCEDURE DIVISION.
MAIN-PROGRAM.
      OPEN    INPUT      FILE-IN
              OUTPUT     FILE-OUT.
      READ FILE-IN
          AT END   CLOSE FILE-IN, FILE-OUT, STOP RUN.
      WRITE PRINTOUT FROM RECORD-IN
          AFTER ADVANCING 2 LINES.
```

The above is an example of an unlikely program written in simple sequence control logic structure where one record is read and printed and the program terminates.

Figure 7.11 Example—Simple Sequence Control Logic Structure Program.

The basic iteration structure is DOWHILE, but there is a closely related structure called DOUNTIL that is sometimes used, depending on the procedure to be expressed and on the availability of appropriate language features.

The difference between the DOWHILE and DOUNTIL structures is that with the DOWHILE, the condition is tested *before* executing the function; if the condition is false, the function is not executed at all. With the DOUNTIL, the condition is tested *after* executing the function; the function will always be executed at least once, regardless of whether the condition is true or false.

It is sometimes helpful—from both readability and efficiency standpoints—to have some way to express a multiway branch commonly referred to as the CASE structure. For example, it is necessary to execute appropriate routines based on a two-digit decimal code, it is certainly possible to write 100 IF statements, or a compound statement with 99 ELSE IF statements, but common sense suggests that there is no reason to adhere so rigidly to the three basic structures.

The CASE structure uses the value of the variable to determine which of several routines is to be executed. Efficiency and convenience dictate reasonable use of language elements that may carry out logic functions in ways slightly different from those of the three basic structures.

Two of the standard programming structures—IFTHENELSE and DOWHILE—are implemented directly through COBOL statements. The DOUNTIL and CASE structures are simulated by conventional COBOL statements in particular structural forms.

The IFTHENELSE Program Structure

The IFTHENELSE structure tests a single condition (predicate) to determine which of two function blocks will be executed. Alternately, the ELSE section may be replaced with ELSE NEXT SENTENCE when there isn't any function block to be executed.

The flowchart for this structure is detailed in figure 7.12.

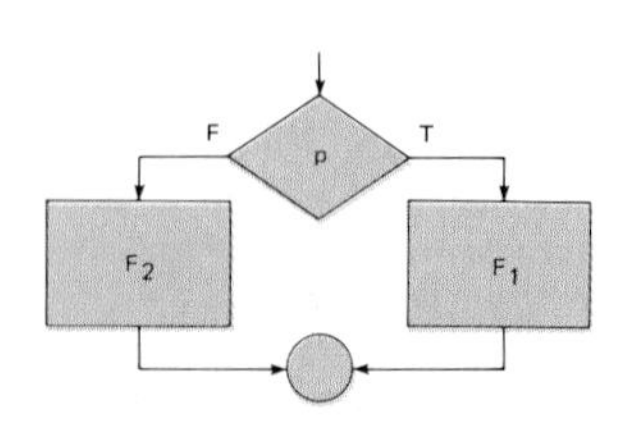

Figure 7.12
Flowchart for
The IFTHENELSE
Program Structure.

The COBOL format is:

```
IF      (condition)
THEN
        code for F1
ELSE
        code for F2
```

(See figure 7.13.)

Alternately, the structure may be coded as:

```
IF      (condition)
THEN
        code for F1
ELSE
        NEXT SENTENCE.
```

(See figure 7.14.)

Note: THEN is an optional word and will be omitted where not included in compiler.

```
IF STOCK-ON-HAND = ZERO
THEN
    PERFORM STOCK-BAL-ZERO
ELSE
    MOVE STOCK-ON-HAND TO QUANTITY-OUT.
```

Figure 7.13 Example—IFTHENELSE Program Structure.

```
IF STOCK-ON-HAND IS LESS THAN QUANTITY-ORD
THEN
    PERFORM BACK-ORDER
ELSE
    NEXT SENTENCE.
```

Figure 7.14 Example—IFTHENELSE Program Structure.

Statements controlled by THEN and ELSE parts of the IF statement are indented to exhibit the span of control of the structure. The words THEN and ELSE are started in the same columns as IF. The word THEN is an optional extension of ANSI COBOL and should be included, where available, to improve readibility of the program.

The DOWHILE Program Structure

The DOWHILE structure provides the basic loop capability. A COBOL paragraph or section is repeated while some condition is satisfied. The condition is tested prior to each execution of the paragraph or section, including the first.

The verb PERFORM with the UNTIL option is the COBOL implementation of the DOWHILE structure. Because the execution of the COBOL UNTIL option terminates on a "true" condition and the DOWHILE terminates on a "false" condition, it is necessary to code the inverse of the condition to loop while the condition is true. In the example below, this is shown (NOT condition).

The flowchart for the DOWHILE structure is shown in figure 7.15.

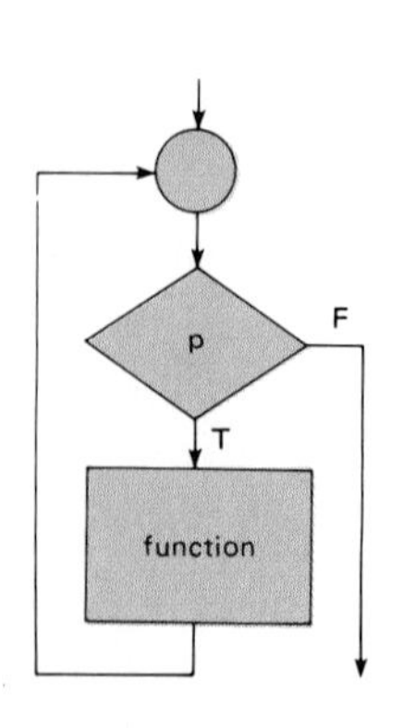

Figure 7.15
Flowchart—for
DOWHILE Program
Structure.

The COBOL format is:

```
PERFORM paragraph-name
        UNTIL (NOT condition)
```

(See figure 7.16.)

With indexing, the structure may be coded as:

```
PERFORM paragraph-name
        VARYING identifier-1   FROM identifier-2   BY identifier-3
        UNTIL (NOT condition)
```

(See figure 7.17.)

```
PERFORM MAIN-PROCESS-ROUTINE
    UNTIL MASTER-FILE-EOF.
```

Figure 7.16 Example—DOWHILE Program Structure.

```
PERFORM ACCEPT-TABLE-RTN
    VARYING TABLE-INDEX FROM 1 BY 1
    UNTIL TABLE-INDEX > TABLE-SIZE.
```

Figure 7.17 Example—DOWHILE Program Structure.

It should be emphasized that the PERFORM with the UNTIL option carries out the test of the condition *before* executing the controlled function. If the condition in the UNTIL phrase is true when first encountered, the paragraph or section in the PERFORM statement is not executed at all.

The DOUNTIL Program Structure

The DOUNTIL structure provides essentially the same loop capability as the DOWHILE, differing from it in two respects:

1. The test on the condition is reversed. The DOWHILE terminates when the condition is *false;* the DOUNTIL terminates when the condition is *true*.
2. In the DOUNTIL, the condition may be tested after each execution of the function, so that the function is always executed at least once.

The flowchart for the DOUNTIL structure is detailed in figure 7.18.

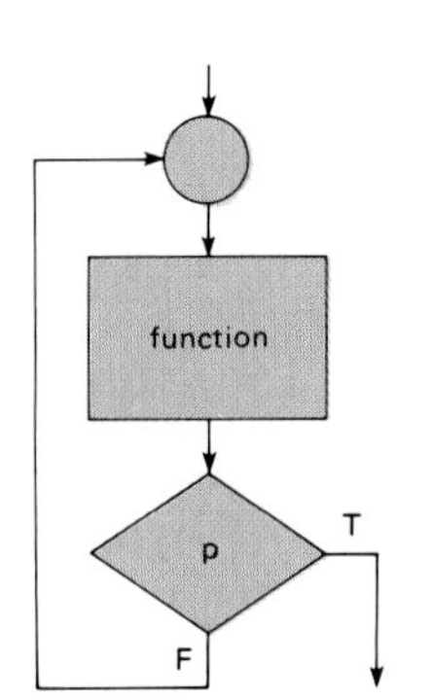

Figure 7.18 Flowchart for DOUNTIL Program Structure.

Since this structure is not directly implemented by any COBOL statement, it is easily simulated with the PERFORM verb in either of two ways. The first way starts with an unconditional PERFORM statement to guarantee that the function is carried out at least once.

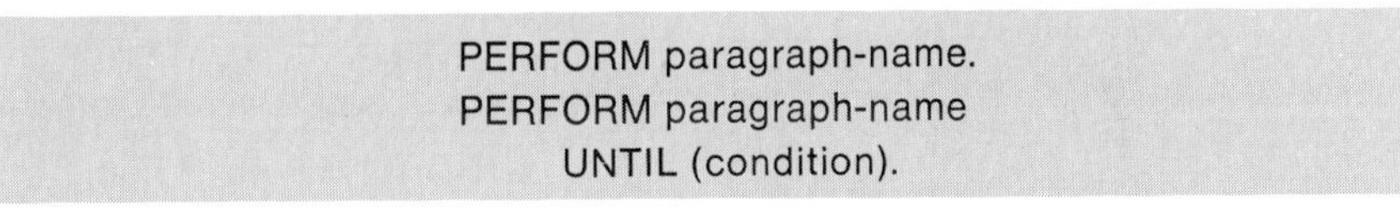

```
PERFORM paragraph-name.
PERFORM paragraph-name
    UNTIL (condition).
```

(See figure 7.19.)

```
PERFORM PROCEDURE-1.
PERFORM PROCEDURE-1
    UNTIL END-OF-TRANS.
```

Figure 7.19 Example—DOUNTIL Program Structure.

The second (cumbersome) way is to set a flag to a value before executing the PERFORM statement, and then conditionally changing it within the named paragraph.

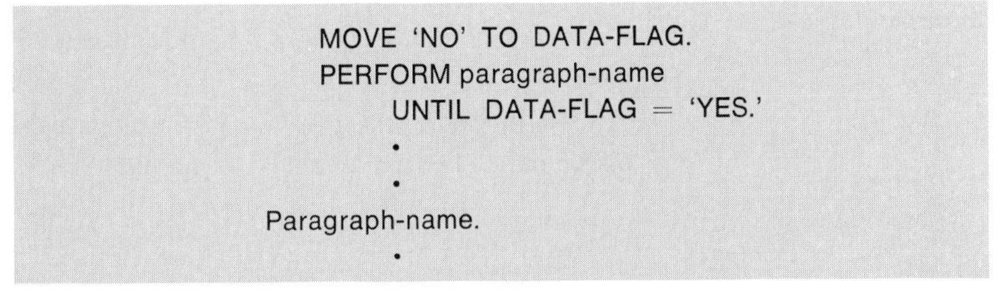

```
MOVE 'NO' TO DATA-FLAG.
PERFORM paragraph-name
    UNTIL DATA-FLAG = 'YES.'
        •
        •
Paragraph-name.
        •
```

The CASE Program Structure

The CASE structure is used to select one of a set of functions for execution depending on the value of the integer-identifier whose range is from 1 to the number of procedure-names listed in the statement. For any value outside of the defined range, the statement is ignored. The use of the CASE structure is recommended when there are more than three (or any other number decided on for use with the IFTHENELSE statement) conditions. Three or less conditions are usually handled with nested IF statements.

COBOL offers no direct implementation of the CASE structure, but it may be simulated by using a combination of PERFORM and the GO TO statements with the DEPENDING ON option. The PERFORM statement must be written with the THRU option because the paragraphs that carry out the individual functions all end with a GO TO-paragraph-EXIT. Without the THRU option, the EXIT paragraph would not be branched to correctly.

Where the "processing statements" are shown, it would also be quite possible to have PERFORM statements invoking out-of-line segments to do the processing.

The flowchart for the CASE structure is shown in figure 7.20.

The CASE statement is not part of conventional COBOL. It must be simulated as follows:

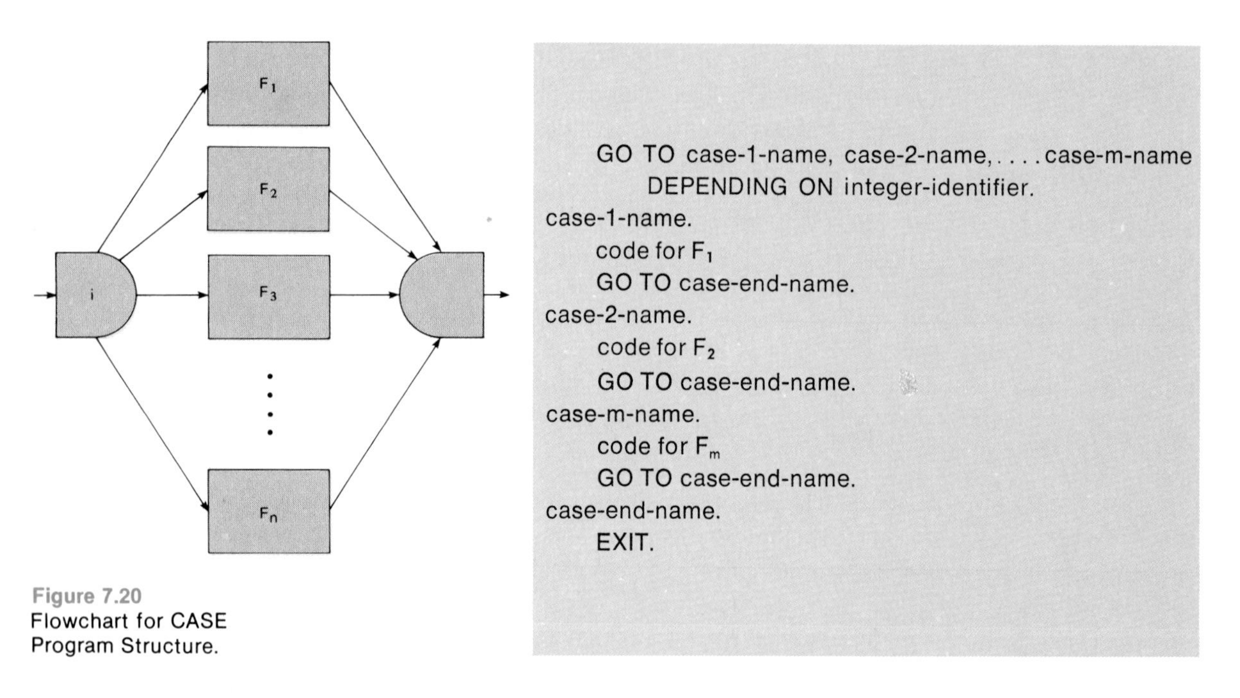


```
        GO TO case-1-name, case-2-name,. . . .case-m-name
              DEPENDING ON integer-identifier.
case-1-name.
        code for F1
        GO TO case-end-name.
case-2-name.
        code for F2
        GO TO case-end-name.
case-m-name.
        code for Fm
        GO TO case-end-name.
case-end-name.
        EXIT.
```

Figure 7.20
Flowchart for CASE Program Structure.

Furthermore, the paragraph implementing the CASE statement must be placed out-of-line and executed via an inline PERFORM such as the following:

```
PERFORM case-paragraph THRU case-end-name.
This is the only allowable use of the THRU option.
```

The CASE statement cannot be implemented in COBOL without using the GO TO statement, but the problems associated with the GO TO statement are minimized by using it only in a highly controlled way. The implementation of the CASE structure is assured that transfer will be limited to paragraphs within this structure, and the processing paragraphs all transfer to the CASE EXIT. The problems associated with "skipping around" while reading a program are thereby avoided. (See figure 7.21.)

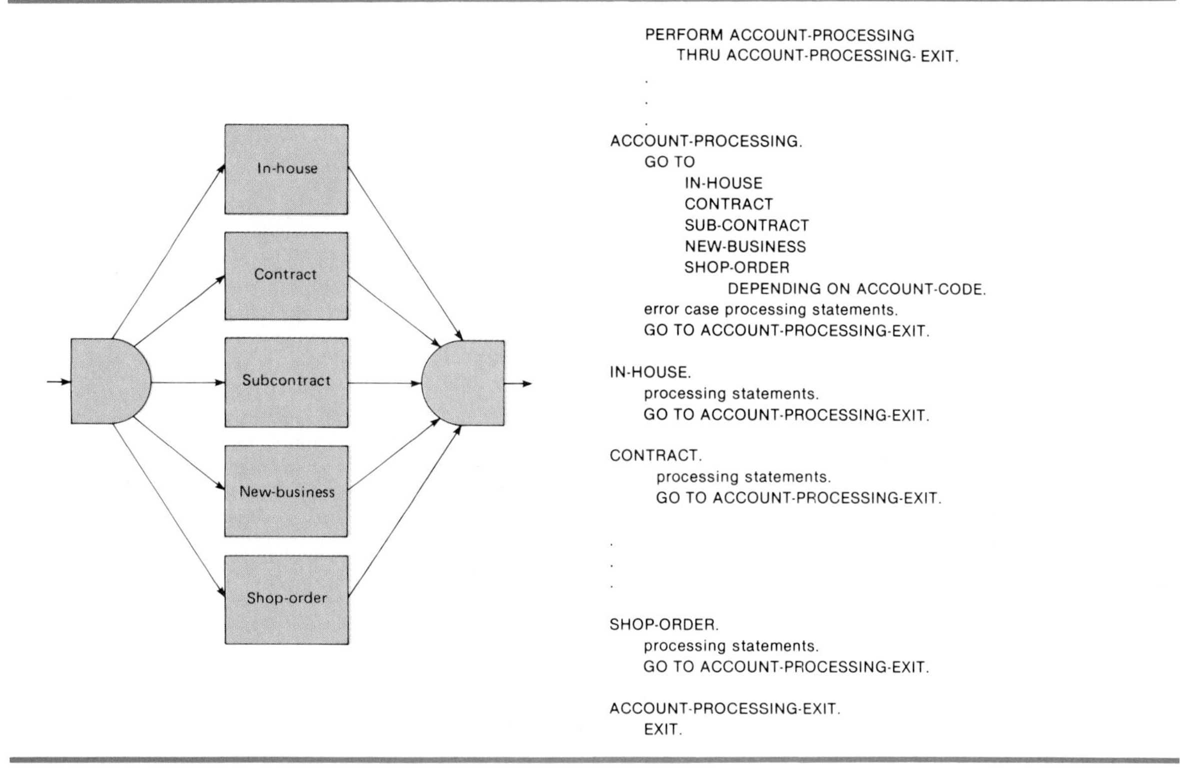


```
          PERFORM ACCOUNT-PROCESSING
              THRU ACCOUNT-PROCESSING- EXIT.
          .
          .
          .
      ACCOUNT-PROCESSING.
          GO TO
              IN-HOUSE
              CONTRACT
              SUB-CONTRACT
              NEW-BUSINESS
              SHOP-ORDER
                   DEPENDING ON ACCOUNT-CODE.
          error case processing statements.
          GO TO ACCOUNT-PROCESSING-EXIT.

      IN-HOUSE.
          processing statements.
          GO TO ACCOUNT-PROCESSING-EXIT.

      CONTRACT.
           processing statements.
           GO TO ACCOUNT-PROCESSING-EXIT.
      .
      .
      .
      SHOP-ORDER.
          processing statements.
          GO TO ACCOUNT-PROCESSING-EXIT.

      ACCOUNT-PROCESSING-EXIT.
          EXIT.
```

Figure 7.21 Example—CASE Program Structure.

CODING STANDARDS

The standards described here represent recommendations based on experience to date in writing structured programs in COBOL. They are all devised to satisfy the same basic goals; to produce programs that are easy to read, easy to understand, and easy to maintain and modify.

Program Organization

These standards provide for the organization of a COBOL source program into a set of segments that are easily linked for compilation. (This use of the term *segment* refers to a block of code that becomes part of a program for compilation directly as input to the computer, is COPYed into the program, or is a linked part of the COBOL compiler's primary data set. It is specifically not a reference to the COBOL reserved word "SEGMENT").

Any COBOL program requires a certain ordering of statements within the program. This ordering is further restricted to that below for the sake of readability, clarity, and consistency. The first segment of code in the program must contain the Identification and Environment Divisions. The second segment should contain the Data Division, which may include COPY statements for detailed data specifications and is the only segment in the program that may exceed one page in the source program listing, about fifty lines, including blank lines. Succeeding segments should contain the Procedure Division which is made up of segments consisting of sections and paragraphs.

The construction of a segment of the Procedure Division is as follows: At the head of the first segment of the Procedure Division, the programmer will place the main line coding for the program. The code must be complete within this segment. If the main line coding contains any PERFORMs, the placement of the PERFORMed routine is determined by the reason for coding the PERFORM.

1. If the PERFORM is used to invoke a large block of code, or code too complex to be physically inserted at the location of the PERFORM, then the PERFORMed paragraph will be placed in the next lower level segment.
2. If the PERFORM statements are being used to implement a structured element (IFTHENELSE, DOWHILE or DOUNTIL), then the PERFORMed paragraph is considered to be at the same logical level as the statements in the current segment, and therefore, the PERFORMed paragraph will be placed in the current segment after the main body of code.

If these paragraphs (the ones PERFORMed to implement structured elements) themselves PERFORM other paragraphs, then the "next-lower level" paragraphs will be placed in the current or lower segment based upon the above paragraph placement rules. Thus, all code for a single logical level will be in the same segment, and subordinate code will be in lower level segments. Hence, the number of segments in a program will indicate the level of complexity. Additionally, any PERFORMed paragraph will be found close to the PERFORMing statement and behind it, thereby continuing the philosophy that programs should be readable from the top down.

The segments of the Procedure Division should appear in some logical sequence, such as the order in which they are invoked, or a sequence that reflects their position in the logical hierarchy of the program.

CODING LINE SPACING

The following are *only suggestions.* No standardization of spacing and indentation has been developed as yet, and there seems to be little pressure for such standardization so long as consistent standards are followed within any one organization.

The key idea in devising helpful indentation is the production of programs in which the visual layout of the program elements aids the reader in understanding program relationship and functioning.

1. PERFORMed paragraphs will be separated from the main body of code above by blank lines(s).

2. Logically noncontiguous paragraphs (other than those used in the CASE structure) will be separated by a blank line.

The free use of blank lines can exhibit more clearly that relationships exist among items so grouped.

Segment Size

Each segment of the Procedure Division should consist of no more than fifty lines of coding (approximately one page of the compiler listing).

Indenting and Formatting Standards

The purpose of these standards is to make all structured programs similar in format and to highlight the structuring graphically.

1. Paragraph names will begin in column 8 (except for the CASE variation) and will be the only coding on the line.
2. The first statement in every paragraph will begin in column 12. (Paragraph and section names are easier to locate if they are always written on a separate line, that is, if the first statement following a paragraph name begins on the next line.)
3. Any statement not subject to indentation rules will start in the same columns as the statements above it.
4. Only one statement per line should be written. Verbs are easier to locate, and programs are easier to correct and modify, if two statements are never written on the same line.
5. Any statements requiring more than one line should be indented on successive lines. Statements are easier to pick out if second and following lines of a continued statement are indented by some consistent amount of spaces, such as four spaces.
6. Options used with normal verbs will be coded separately on successive lines with indentation. These options are

```
GIVING           AT END         UNTIL
ON SIZE ERROR    INVALID KEY    VARYING
DEPENDING ON     UPON           BEFORE }
AFTER            TALLYING       AFTER  }   ADVANCING
TIMES            REPLACING
```

All Sort Keys and Sort Options should be indented on separate lines. Important clauses of certain COBOL verbs are better set off if they are written on separate lines, indented by whatever amount of space is chosen as standard for continuation lines. For example, when the UNTIL and/or VARYING options are used with the PERFORM's, they can be written on separate lines and indented.

```
PERFORM PRINTOUT
    VARYING LINE-NUMBER FROM 1 BY 1
    UNTIL LINE-NUMBER IS GREATER THAN 50.
```

The VARYING, AT END, and WHEN clauses of a SEARCH verb can be written on separate lines and indented:

```
SEARCH RATE-TABLE
    VARYING MAN-IDENTIFICATION
    AT END MOVE T TO NO-FIND-SW
    WHEN RATE-MF = RATE-TABLE (RATE-INDEX)
```

7. Various statements are easier to read if appropriate portions of the statements are aligned vertically. For example,

```
OPEN  INPUT    TAPE-IN
               FILE-IN
      OUTPUT   TAPE-OUT
               FILE-OUT.
```

8. For multiple field operations (MOVE ZEROS TO FLD-A, FLD-G), place each successive field on a separate line directly below the preceding field. For example,

```
MOVE ZEROS TO FLD-A,
              FLD-G.
```

The IFTHENELSE Structure

1. Statements within the IF should be indented.
2. The ELSE should be coded on a line by itself in the same column as the IF.

```
IF    (condition)
THEN
      statement-1
      •
      statement-n
ELSE
      statement-1
      •
      statement-n.
```

3. When there are four or more statements (or any other number decided on) as a function of the IF, the following form should be used:

```
IF    (condition)
THEN
      PERFORM paragraph-1
ELSE
      PERFORM paragraph-2.
```

4. The condition portion of the IF will be enclosed in parentheses.

```
NESTED IFS

IF (condition-1)
THEN
    IF (condition-2)
    THEN
        statement(s)
    ELSE
        statement(s)
ELSE
    statement(s)
```

Figure 7.22
Nested IF Format.

5. When there are no statements to follow the ELSE, ELSE NEXT SENTENCE will be coded fully.
6. A limit of levels of nested IF statements should be established (fig. 7.22).
7. If the IF or ELSE clause should include a nested IF which is followed by one or more statements that should be executed without regard to evaluation of the nested IF (e.g., "statement-2" in the following example), then the nested IF must be PERFORMed.

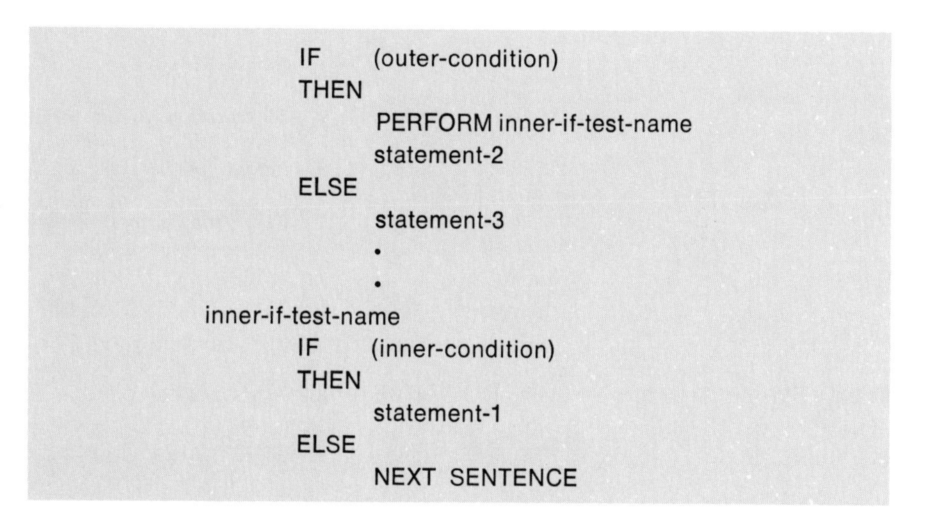

```
        IF     (outer-condition)
        THEN
               PERFORM inner-if-test-name
               statement-2
        ELSE
               statement-3
               .
               .
inner-if-test-name
        IF     (inner-condition)
        THEN
               statement-1
        ELSE
               NEXT SENTENCE
```

The statements controlled by the THEN and ELSE parts of an IF statement are shown to be subordinate to the logic by indenting these statements by a consistent amount. If one tends to use nested IF statements sparingly, an indentation unit of four spaces would be reasonable. If one writes deeply nested IF statements, however, the indentation would need to be three or two positions, to avoid running out of space in the line (see figs. 7.23, 7.24).

8. When the same field or condition is consecutively tested for more than three numeric values (or any other number decided on) using nested IF state-

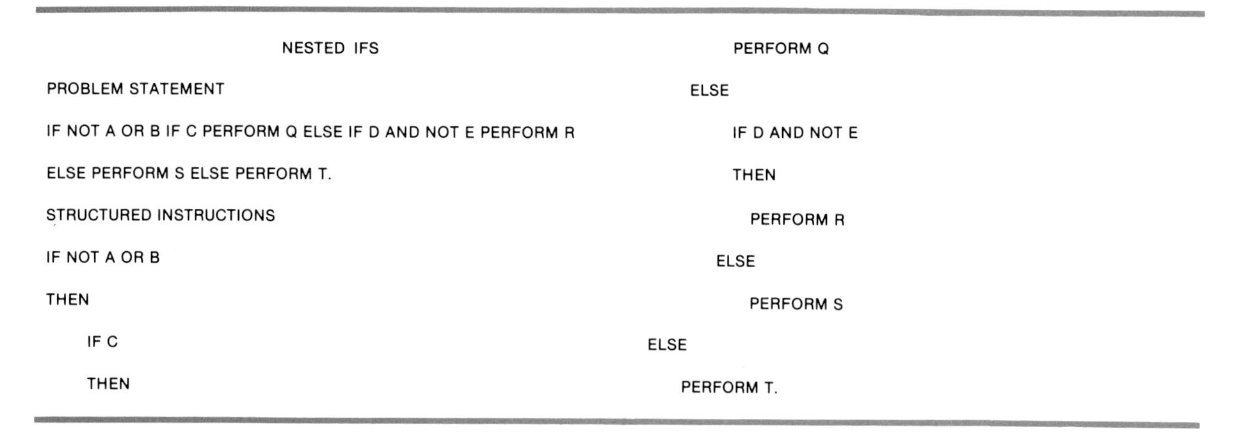

```
NESTED IFS

PROBLEM STATEMENT

IF NOT A OR B IF C PERFORM Q ELSE IF D AND NOT E PERFORM R
ELSE PERFORM S ELSE PERFORM T.

STRUCTURED INSTRUCTIONS

IF NOT A OR B
THEN
    IF C
    THEN
        PERFORM Q
    ELSE
        IF D AND NOT E
        THEN
            PERFORM R
        ELSE
            PERFORM S
ELSE
    PERFORM T.
```

Figure 7.23 Example—Nested IFs.

HOW TO SIMPLIFY NESTED IF'S

Keep compounds to a minimum

- **Avoid NOT's**
- **Avoid implied subjects and operators**

Restrict depth

No GO TO's

All on one page (use PERFORM's if necessary)

Indent

Use readable condition names

Figure 7.24
How to Simplify Nested IFs.

ments, the CASE structure should be used. In designing condition field values, the use of the CASE structure should be considered.

The DOWHILE Structure

The DOWHILE structure should be coded as follows:

```
PERFORM paragraph-name
      UNTIL (NOT condition).
```

1. With indexing the structure should be coded:

```
PERFORM paragraph-name
      VARYING identifier-1    FROM....BY....
      UNTIL (NOT condition).
```

2. Only single paragraphs will be performed (i.e., the THRU option should not be used).
3. The condition portion of the DOWHILE will be enclosed in parentheses.

The DOUNTIL Structure

1. The COBOL simulation of the DOUNTIL structure will be coded as follows:

```
PERFORM   paragraph-name.
PERFORM   paragraph-name
      UNTIL (condition).
```

2. The condition portion of the DOUNTIL will be enclosed in parentheses.

The CASE Structure

The COBOL simulation of the CASE structure is shown in the example below.

1. If the names of the paragraphs to be executed do not fit on one line, indent each continued line.
2. The DEPENDING ON clause will be coded on a separate line and will be indented.

3. The GO TO case-end-name statement, required to handle the situation where the value of the variable is greater than the greatest expected value or is less than one, will be coded on the next line and should be indented.
4. Each paragraph to be executed should be indented and must end with a GO TO statement that transfers control to the common end point of the CASE structure.
5. The case-end paragraph will follow the last of the function paragraphs and contain just the EXIT statement.

```
          PERFORM ACCOUNT-PROCESSING
                THRU ACCOUNT-PROCESSING-END.
          •
          •
      ACCOUNT-PROCESSING
          GO TO IN-HOUSE, CONTRACT, SUB-CONTRACT,
                ESTIMATES, SHOP-ORDER
                DEPENDING ON ACCOUNT-CODE.
                GO TO ACCOUNT-PROCESSING-END.
      IN-HOUSE.
          •
          GO TO ACCOUNT-PROCESSING-END.
      CONTRACT.
          •
          •
          •
          GO TO ACCOUNT-PROCESSING-END.
          •
          •
      SHOP-ORDER.
          •
          •
          GO TO ACCOUNT-PROCESSING-END.
      ACCOUNT-PROCESSING-END.
          EXIT.
```

(*Note:* The CASE structure is the only allowable variance to the rule for using the THRU option of the PERFORM.)

CONCLUSIONS

The first few programs using structured programming techniques may be somewhat difficult. However, once the learning phase is over, the coding should be faster using structured programming and associated techniques (e.g., top-down design, etc.) than with previous programming methods. Proper use of the structured programming coding standards is the responsibility of the individual programmer with enforcement being provided through subsequent review of the programs. (See figures 7.25, 7.26.)

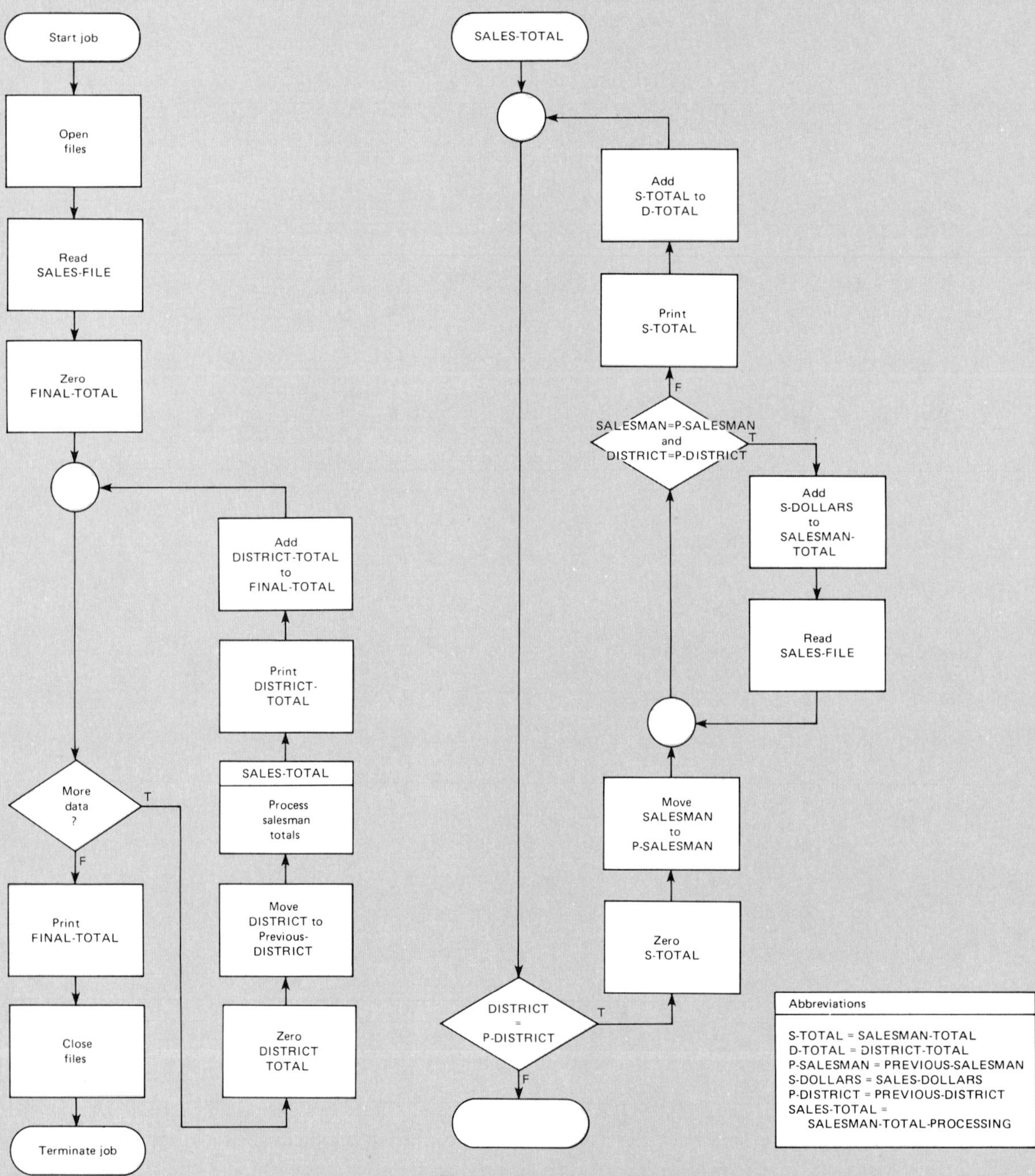

Flowchart for the mainline processing portion of a two-level control total processing application

Flowchart for the record processing portion of a two-level control total processing application

Figure 7.25 Illustrative Structured Program.

Figure 7.25 Continued.

```
IDENTIFICATION DIVISION.
PROGRAM-ID.
    TWOLEVEL.

ENVIRONMENT DIVISION.
INPUT-OUTPUT SECTION.
FILE-CONTROL.
    SELECT SALES-FILE ASSIGN TO SYS012-UR-2540R-S.
    SELECT REPORT-FILE ASSIGN TO SYS009-UR-1403-S.

DATA DIVISION.
FILE SECTION.

FD  SALES-FILE
    RECORD CONTAINS 80 CHARACTERS
    RECORDING MODE IS F
    LABEL RECORDS ARE OMITTED
    DATA RECORD IS SALES-RECORD.
01  SALES-RECORD.
    05  SALESMAN                PICTURE X(5).
    05  DISTRICT                PICTURE XXX.
    05  SALES-DOLLARS           PICTURE 9(5)V99.

FD  REPORT-FILE
    RECORD CONTAINS 133 CHARACTERS
    RECORDING MODE IS F
    LABEL RECORDS ARE OMITTED
    DATA RECORD IS REPORT-RECORD.
01  REPORT-RECORD.
    05  CARRIAGE-CONTROL        PICTURE X.
    05  SALESMAN-OUT            PICTURE ZZZZ9.
    05  FILLER                  PICTURE XXX.
    05  SALESMAN-TOTAL-OUT      PICTURE $$$,$$$,$$9.99.
    05  FILLER                  PICTURE X(8).
    05  DISTRICT-OUT            PICTURE ZZ9.
    05  FILLER                  PICTURE XXX.
    05  DISTRICT-TOTAL-OUT      PICTURE $$$,$$$,$$9.99.
    05  FILLER                  PICTURE X(8).
    05  FINAL-TOTAL-OUT         PICTURE $$$,$$$,$$9.99.

WORKING-STORAGE SECTION.

01  FLAGS.
    05  MORE-DATA-FLAG          PICTURE XXX     VALUE 'YES'.
        88  MORE-DATA                           VALUE 'YES'.
        88  NO-MORE-DATA                        VALUE 'NO'.

01  SAVE-ITEMS.
    05  PREVIOUS-SALESMAN       PICTURE X(5).
    05  PREVIOUS-DISTRICT       PICTURE XXX.

01  TOTALS COMPUTATIONAL-3.
    05  SALESMAN-TOTAL          PICTURE S9(8)V99 VALUE ZERO.
    05  DISTRICT-TOTAL          PICTURE S9(8)V99 VALUE ZERO.
    05  FINAL-TOTAL             PICTURE S9(8)V99 VALUE ZERO.

PROCEDURE DIVISION.
PREPARE-SALES-REPORT.
    OPEN INPUT   SALES-FILE
         OUTPUT  REPORT-FILE.
    READ SALES-FILE
        AT END MOVE 'NO' TO MORE-DATA-FLAG.
    PERFORM DISTRICT-TOTAL-PROCESSING
        UNTIL NO-MORE-DATA.
    MOVE SPACES TO REPORT-RECORD.
    MOVE FINAL-TOTAL TO FINAL-TOTAL-OUT.
    WRITE REPORT-RECORD.
    CLOSE  SALES-FILE
           REPORT-FILE.
    STOP RUN.

DISTRICT-TOTAL-PROCESSING.
    MOVE ZERO TO DISTRICT-TOTAL.
    MOVE DISTRICT TO PREVIOUS-DISTRICT.
    PERFORM SALESMAN-TOTAL-PROCESSING
        UNTIL DISTRICT IS NOT EQUAL TO PREVIOUS-DISTRICT
        OR NO-MORE-DATA.
    MOVE SPACES TO REPORT-RECORD.
    MOVE PREVIOUS-DISTRICT TO DISTRICT-OUT.
    MOVE DISTRICT-TOTAL TO DISTRICT-TOTAL-OUT.
    WRITE REPORT-RECORD.
    ADD DISTRICT-TOTAL TO FINAL-TOTAL.

SALESMAN-TOTAL-PROCESSING.
    MOVE ZERO TO SALESMAN-TOTAL.
    MOVE SALESMAN TO PREVIOUS-SALESMAN.
    PERFORM PROCESS-AND-READ
        UNTIL DISTRICT IS NOT EQUAL TO PREVIOUS- DISTRICT
        OR SALESMAN IS NOT EQUAL TO PREVIOUS- SALESMAN
        OR NO-MORE-DATA.
    MOVE SPACES TO REPORT-RECORD.
    MOVE PREVIOUS-SALESMAN TO SALESMAN-OUT.
    MOVE SALESMAN-TOTAL TO SALESMAN-TOTAL-OUT.
    WRITE REPORT-RECORD.
    ADD SALESMAN-TOTAL TO DISTRICT-TOTAL.

PROCESS-AND-READ.
    ADD SALES-DOLLARS TO SALESMAN-TOTAL.
    READ SALES-FILE
        AT END MOVE 'NO' TO MORE-DATA-FLAG.
```

Structured program for a two-level control total processing application (2 of 2)

```
  41    $203.37
  52    $110.00
  69    $134.65
                      1     $448.02
  18    $207.69
  32    $185.60
                      2     $393.29
  36    $194.15
  39    $121.40
  50     $51.80
                      3     $367.35     $1,208.66
```

Illustrative output from the two-level control total program of Figure 22

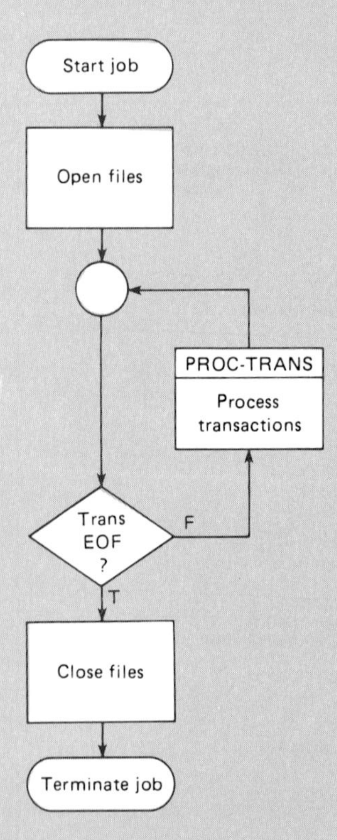

Flowchart of the mainline processing for an inquiry response application

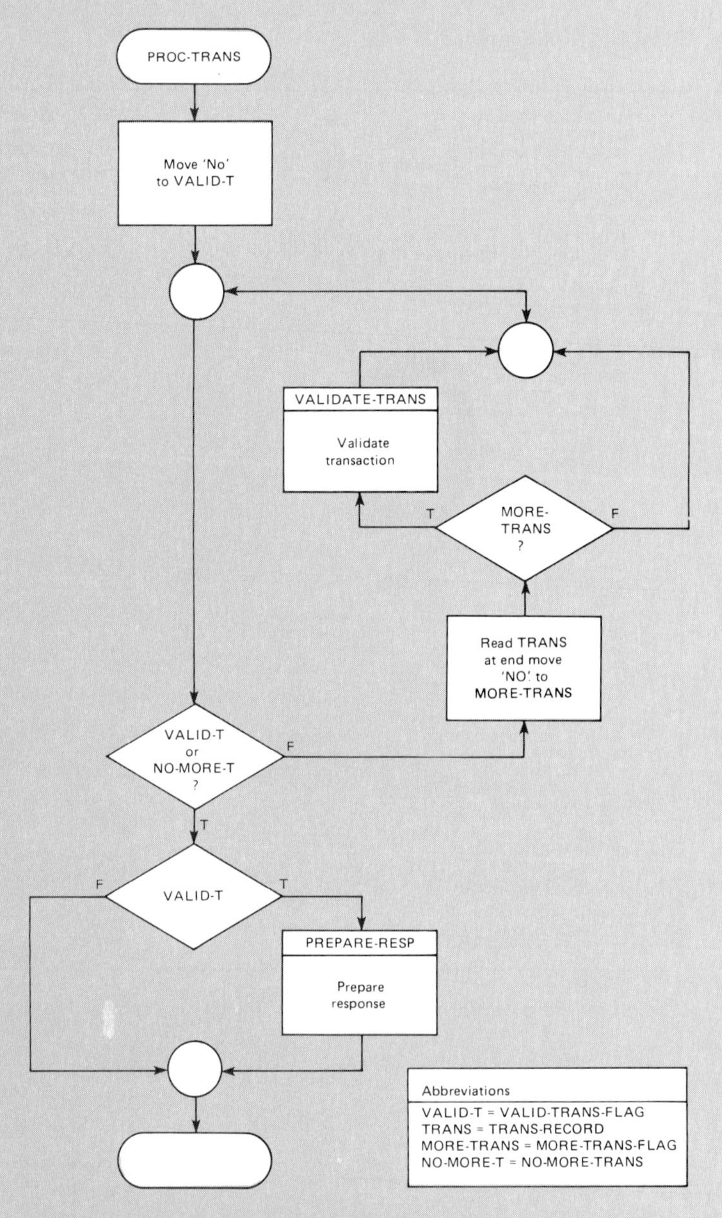

Flowchart of the transaction processing logic for an inquiry response application

Figure 7.26 Illustrative Structured Program.

Figure 7.26 Continued.

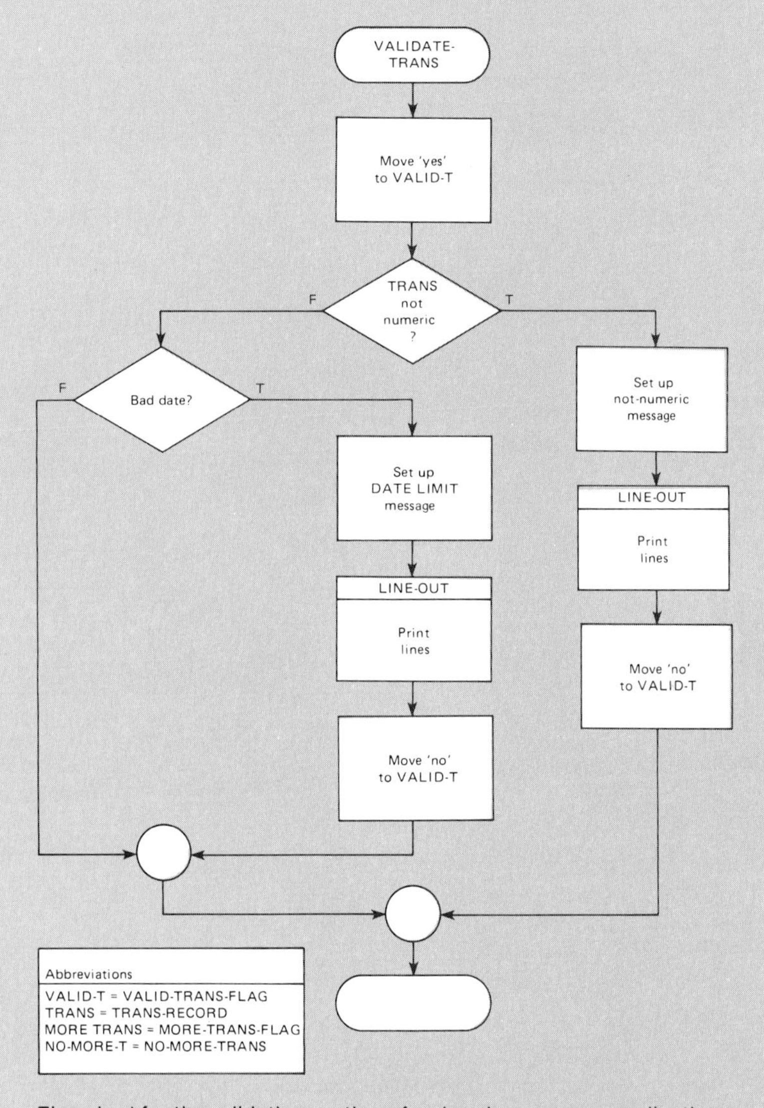

Flowchart for the validation portion of an inquiry response application

Figure 7.26 Continued.

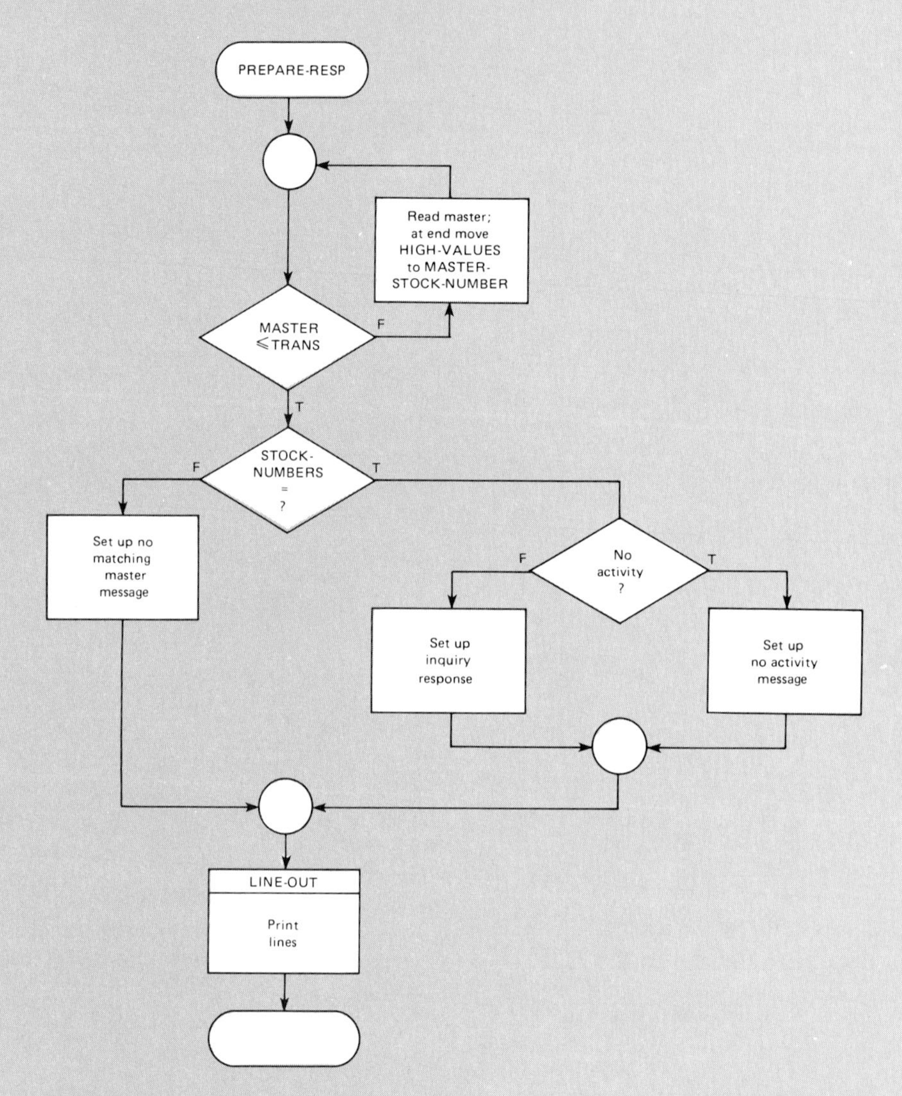

Flowchart of the logic for preparing a response in an inquiry response application

Figure 7.26 Continued.

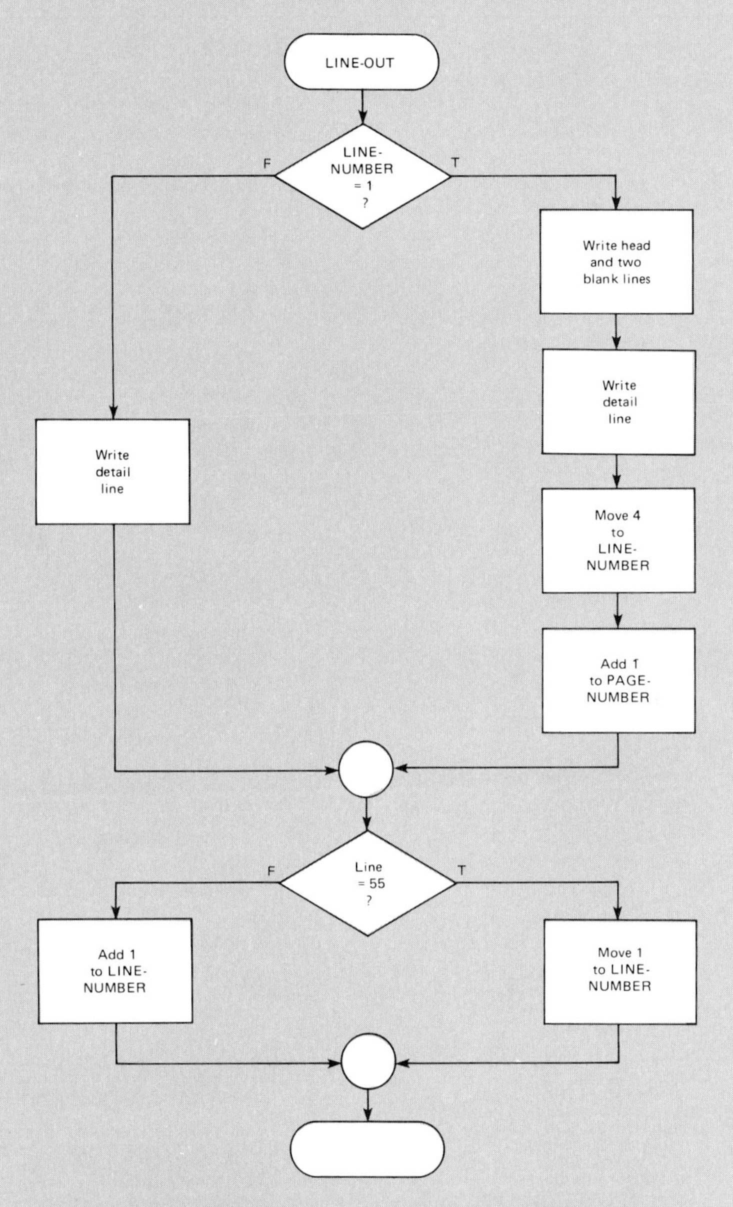

Flowchart of the logic for printing heading and detail lines in an inquiry response application

Figure 7.26 Continued.

```
IDENTIFICATION DIVISION.
PROGRAM-ID.
        STRUCTURED PROGRAMMING EXAMPLE.

ENVIRONMENT DIVISION.
INPUT-OUTPUT SECTION.
FILE-CONTROL.
        SELECT MASTER-FILE ASSIGN TO SYS011-UT-2400-S.
        SELECT TRANS-FILE   ASSIGN TO SYS012-UR-2540R-S.
        SELECT REPORT-FILE ASSIGN TO SYS009-UR-1403-S.

DATA DIVISION.

FILE SECTION.

FD  MASTER-FILE
        RECORD CONTAINS 80 CHARACTERS
        RECORDING MODE IS F
        LABEL RECORDS ARE OMITTED
        DATA RECORD IS MASTER-RECORD-BUFFER.
01      MASTER-RECORD-BUFFER            PICTURE X(80).

FD  TRANS-FILE
        RECORD CONTAINS 80 CHARACTERS
        RECORDING MODE IS F
        LABEL RECORDS ARE OMITTED
        DATA RECORD IS TRANS-RECORD.
01      TRANS-RECORD.
        05   STOCK-NUMBER               PICTURE X(6).
        05   DATE-LIMIT.
             10   YEAR                  PICTURE XX.
             10   DAY-OF-YEAR           PICTURE XXX.
        05   FILLER                     PICTURE X(69).
FD  REPORT-FILE
        RECORDING MODE IS F
        LABEL RECORDS ARE OMITTED
        DATA RECORD IS REPORT-RECORD.
01      REPORT-RECORD.
        05  FORMS-CONTROL               PICTURE X.
        05  FILLER                      PICTURE X (132).

WORKING-STORAGE SECTION.

01      MASTER-RECORD.
        05   STOCK-NUMBER               PICTURE X(6) VALUE LOW-VALUES.
        05   DESCRIPTION                PICTURE X(20).
        05   UNIT-PRICE                 PICTURE 9(5)V99.
        05   QUANTITY-ON-HAND           PICTURE 9(5)V99.
        05   LAST-ACTIVITY-DATE.
             10   YEAR                  PICTURE XX.
             10   DAY-OF-YEAR           PICTURE XXX.
        05   FILLER                     PICTURE X(35).

01   FLAGS.
     05   MORE-TRANS-FLAG            PICTURE XXX   VALUE 'YES'.
          88   MORE-TRANS                          VALUE 'YES'.
          88   NO-MORE-TRANS                       VALUE 'NO'.
     05   VALID-TRANS-FLAG           PICTURE XXX.
          88   VALID-TRANS                         VALUE 'YES'.

01   MESSAGES.
     05   NO-ACTIVITY-MSG               PICTURE X(47)
          VALUE 'NO ACTIVITY FOR THIS ITEM SINCE DATE IN INQUIRY'.
     05   NO-MATCHING-MASTER-MSG   PICTURE X(36)
          VALUE 'NO MASTER FOR THIS STOCK NUMBER'.
     05   NOT-NUMERIC-MSG               PICTURE X(50)
          VALUE 'ALL ITEMS IN INQUIRY MUST BE NUMERIC'.
     05   DATE-LIMIT-MSG                PICTURE X(50)
          VALUE 'DATE-LIMIT MUST NOT BE LESS THAN 70001'.

01   LINE-AND-PAGE-COUNTERS.
     05   LINE-NUMBER                PICTURE 99    VALUE 1.
     05   PAGE-NUMBER                PICTURE 999   VALUE 1.
01   OUTPUT-LINE.
     05   FORMS-CONTROL              PICTURE X.
     05   FILLER                     PICTURE X(132).

01   INQUIRY-RESPONSE.
     05   FORMS-CONTROL              PICTURE X.
     05   STOCK-NUMBER               PICTURE X(6).
     05   FILLER                     PICTURE XXX   VALUE SPACES.
     05   DATE-LIMIT                 PICTURE X(5).
     05   FILLER                     PICTURE X(5)  VALUE SPACES.
     05   DESCRIPTION                PICTURE X(20).
     05   FILLER                     PICTURE XXX   VALUE SPACES.
     05   UNIT-PRICE                 PICTURE $$$,$$9.99.
     05   FILLER                     PICTURE XXX   VALUE SPACES.
     05   QUANTITY-ON-HAND           PICTURE Z(4)9.99.
     05   FILLER                     PICTURE XXX   VALUE SPACES.
     05   COST-OF-STOCK              PICTURE $$,$$$,$$9.99.
     05   FILLER                     PICTURE XXX   VALUE SPACES.
     05   LAST-ACTIVITY-DATE.
          10     CENTURY             PICTURE XX    VALUE '19'.
          10     YEAR                PICTURE XX.
          10     FILLER              PICTURE XX    VALUE SPACES.
          10     DAY-OF-YEAR         PICTURE XXX.
     05   FILLER                     PICTURE X(9)  VALUE SPACES.

01   MESSAGE-RECORD.
     05   FORMS-CONTROL              PICTURE X.
     05   STOCK-NUMBER               PICTURE X(6).
     05   FILLER                     PICTURE XXX   VALUE SPACES.
     05   DATE-LIMIT                 PICTURE X(5).
     05   FILLER                     PICTURE X(5)  VALUE SPACES.
     05   MESSAGE-FIELD              PICTURE X(82).

01   HEADING-LINE.
     05   FORMS-CONTROL              PICTURE X.
     05   FILLER                     PICTURE X(46)
          VALUE ' TRANSACTION              DESCRIPTION         UNIT'.
     05   FILLER                     PICTURE X(44)
          VALUE ' PRICE      QOH       TOTAL COST   LAST ACTIVITY'.
     05   PAGE-NUMBER-OUT            PICTURE Z(6)9.

PROCEDURE DIVISION.
ANSWER-INQUIRIES.
     OPEN INPUT MASTER-FILE
                    TRANS-FILE
          OUTPUT REPORT-FILE.
     PERFORM PROCESS-TRANSACTIONS
          UNTIL NO-MORE-TRANS.
     CLOSE MASTER-FILE
          TRANS-FILE
          REPORT-FILE.
     STOP RUN.

PROCESS-TRANSACTIONS.
     MOVE 'NO' TO VALID-TRANS-FLAG.
     PERFORM GET-VALID-TRANSACTION
          UNTIL VALID-TRANS
          OR NO-MORE-TRANS.
     IF VALID-TRANS
     THEN
          PERFORM PREPARE-RESPONSE.

GET-VALID-TRANSACTION.
     READ TRANS-FILE
          AT END MOVE 'NO' TO MORE-TRANS-FLAG.
     IF MORE-TRANS
     THEN
          PERFORM VALIDATE-TRANSACTION.

VALIDATE-TRANSACTION.
     MOVE 'YES' TO VALID-TRANS-FLAG.
     IF STOCK-NUMBER OF TRANS-RECORD IS NOT NUMERIC
          OR DATE-LIMIT OF TRANS-RECORD IS NOT- NUMERIC
     THEN
          MOVE CORRESPONDING TRANS-RECORD TO MESSAGE-RECORD
          MOVE NOT-NUMERIC-MSG TO MESSAGE-FIELD
          MOVE MESSAGE-RECORD TO OUTPUT-LINE
          PERFORM LINE-OUT
          MOVE 'NO' TO VALID-TRANS-FLAG
     ELSE
          IF DATE-LIMIT OF TRANS-RECORD IS LESS THAN '70001'
          THEN
               MOVE CORRESPONDING TRANS-RECORD TO
                    MESSAGE-RECORD
               MOVE DATE-LIMIT-MSG TO MESSAGE-FIELD
               MOVE MESSAGE-RECORD TO OUTPUT-LINE
               PERFORM LINE-OUT
               MOVE 'NO' TO VALID-TRANS-FLAG.

PREPARE-RESPONSE.
     PERFORM READ-MASTER
          UNTIL STOCK-NUMBER OF MASTER-RECORD
                    IS EQUAL TO STOCK-NUMBER OF TRANS-RECORD
               OR STOCK-NUMBER OF MASTER-RECORD
                    IS GREATER THAN STOCK-NUMBER OF TRANS-RECORD.
```

Structured program for an inquiry response application

Figure 7.26 Continued.

```
IF STOCK-NUMBER OF TRANS-RECORD
    IS EQUAL TO STOCK-NUMBER OF MASTER-RECORD
THEN
    IF DATE-LIMIT OF TRANS-RECORD IS NOT LESS THAN
        LAST-ACTIVITY-DATE OF MASTER-RECORD
    THEN
        MOVE CORRESPONDING TRANS-RECORD TO
            MESSAGE-RECORD
        MOVE NO-ACTIVITY-MSG TO MESSAGE-FIELD
        MOVE MESSAGE-RECORD TO OUTPUT-LINE
    ELSE
        MOVE CORRESPONDING MASTER-RECORD TO
            INQUIRY-RESPONSE
        MOVE DATE-LIMIT OF TRANS-RECORD TO
            DATE-LIMIT OF INQUIRY-RESPONSE
        MULTIPLY UNIT-PRICE OF MASTER-RECORD
            BY QUANTITY-ON-HAND OF MASTER- RECORD
            GIVING COST-OF-STOCK
        MOVE INQUIRY-RESPONSE TO OUTPUT-LINE
ELSE
    MOVE CORRESPONDING TRANS-RECORD TO MESSAGE-RECORD
    MOVE NO-MATCHING-MASTER-MSG TO MESSAGE- FIELD
    MOVE MESSAGE-RECORD TO OUTPUT-LINE.
PERFORM LINE-OUT.

READ-MASTER.
    READ MASTER-FILE INTO MASTER-RECORD
        AT END MOVE HIGH-VALUES TO STOCK-NUMBER
            OF MASTER-RECORD.

LINE-OUT.
    IF LINE-NUMBER = 1
    THEN
        MOVE PAGE-NUMBER TO PAGE-NUMBER-OUT
        WRITE REPORT-RECORD FROM HEADING-LINE
            AFTER POSITIONING 0
        WRITE REPORT-RECORD FROM OUTPUT-LINE
            AFTER POSITIONING 2
        MOVE 4 TO LINE-NUMBER
        ADD 1 TO PAGE-NUMBER
    ELSE
        WRITE REPORT-RECORD FROM OUTPUT-LINE
            AFTER POSITIONING 1.
    IF LINE-NUMBER = 55
    THEN
        MOVE 1 TO LINE-NUMBER
    ELSE
        ADD 1 TO LINE-NUMBER.
```

Structured program for an inquiry response application

```
000108DESK                   0018500000160075010
000115CHAIR, FOLDING         0001810001270075100
000180LAMP, FLOOR            0003750000120075180
000181LAMP, DESK             0002200001170075093
000200TYPEWRITER STAND       0002490000400074350
000309BOOKCASE, 5 SHELF      0004125000200075105
000310BOOKCASE, 4 SHELF      0003650000310075090
000311BOOKCASE, 3 SHELF      0002800000170075110
000480FILE CABINET, 4 DWR    0006180001000075130
000481FILE CABINET, 2 DWR    0003990000500075150
010684WASTEBASKET, GREEN     0000417000120075190
010686WASTEBASKET, GRAY      0000417001900075120
010687WASTEBASKET, BLUE      0000417000570075182
021732SOFA, LEATHER, BROWN0035620000290075070
021739SOFA, LEATHER, RED     0035620000370075040
```

Illustrative master file for the inquiry response program

```
00010875001
00018075001
00020075001
00025075001
000310 75001
00031075001
00048075140
00048175140
010.68575140
01069075150
02173975030
03194075150
```

Illustrative transaction file for the inquiry response program

TRANSACTION		DESCRIPTION	UNIT PRICE	QOH	TOTAL COST	LAST ACTIVITY	
000108	75001	DESK	$185.00	16.00	$2,960.00	1975	010
000180	75001	LAMP, FLOOR	$37.50	12.00	$450.00	1975	180
000200	75001	NO ACTIVITY FOR THIS ITEM SINCE DATE IN INQUIRY					
000250	75001	NO MASTER FOR THIS STOCK NUMBER					
000310	7500	ALL ITEMS IN INQUIRY MUST BE NUMERIC					
000310	75001	BOOKCASE, 4 SHELF	$36.50	31.00	$1,131.50	1975	090
000480	75140	NO ACTIVITY FOR THIS ITEM SINCE DATE IN INQUIRY					
000481	75140	FILE CABINET, 2 DWR	$39.90	50.00	$1,995.00	1975	150
010.68	57514	ALL ITEMS IN INQUIRY MUST BE NUMERIC					
010690	75150	NO MASTER FOR THIS STOCK NUMBER					
021739	75030	SOFA, LEATHER, RED	$356.20	37.00	$13,179.40	1975	040
031940	75150	NO MASTER FOR THIS STOCK NUMBER					

Output of the program

Exercises

Write your answer in the space provided (answer may be one or more words).

1. Structured programming applies a __________ to the coding of computer programs.
2. COBOL does not provide all the __________ directly for structured programming.
3. Structured programming concentrates on one of the most error-prone factors in programming, __________.
4. The main objectives of structured programming are the following; __________, __________, __________, __________, and __________.
5. Structured programming is a style of programming in which the __________ of a program is made as clear as possible by restricting __________ to just __________.
6. In a sequence structure, in absence of instructions to the contrary, statements are executed in the __________ that they are __________.
7. The IFTHENELSE statements are used for the __________ structure.
8. The DOWHILE or DOUNTIL is a __________ structure that is used for __________ mechanism.
9. In a structured program, any program function can be performed by using any of the basic control structures: __________, __________, and __________.
10. Each structure is characterized by a __________ and __________ of transfer of control into the structure, and a __________ of transfer of control out of the structure.
11. The structures can be combined to form a program where the control flows from __________ to __________ or from __________ to __________.
12. Control structures can be __________ but they retain their characteristic of __________ and __________.
13. Reduced program complexity can be thought of as a process of removing things from a program, such as __________, __________, __________, __________, and __________.
14. Improving program clarity can be thought of as a process of adding things to a program, such as __________, __________, __________, and __________.
15. A structured program is composed of __________ which may range from a __________ up to a __________.
16. A well-structured program gains an important part of its easy readability from the fact that there is no __________ from one part of the program to another.
17. The blatant use of GO TO statements results in unnecessary __________ flow patterns leading to difficult __________ effort on the part of the programmer.
18. The design of a program should proceed from __________ to __________.
19. Individual program segments should be as __________ as possible.
20. Each level of programming is supported by the next __________ level of the hierarchy.
21. The indentation of the program will make the __________ more __________ to the reader.
22. The use of indentation is important because __________ indentation enhances __________ so that the finished program exhibits in a __________ way the __________ among __________.
23. In a CASE structure, only one series of __________ is to be performed depending on the value of a __________.
24. Segment length should be limited to a __________ size.
25. Careful __________ of coding to show __________ levels also gives increased __________ to __________.
26. The basic requirements for a proper program are that it has exactly __________ point and __________ point for __________ and there are paths from __________ to __________ that lead through every part of the program.
27. Sequence is implemented in the COBOL language by simple writing statements in __________.
28. Selection is a choice between __________ based on a __________.
29. Iteration is used for __________ execution of a __________ while a __________ is __________ or __________.

30. In COBOL, the DOWHILE structure is implemented with the ______________ verb with the ______________ option.
31. In the DOWHILE structure, the condition is tested ______________ executing the function and in the DOUNTIL structure, the condition is tested ______________ executing the function.
32. The CASE structure uses the ______________ of the variable to determine which of several ______________ is to be ______________.
33. ______________ and ______________ structures are implemented directly through COBOL statements.
34. In the IFTHENELSE structure, the ELSE section may be replaced with ______________ when there isn't any function block to be executed.
35. Statements controlled by THEN and ELSE parts of the ______________ statement are ______________ to exhibit the ______________ of the structure.
36. The DOWHILE structure provides the ______________ capability.
37. In a DOWHILE structure, the ______________ is tested ______________ to the execution of each ______________ or ______________, including the ______________.
38. The verb ______________ with the ______________ option is the COBOL implementation of the DOWHILE structure.
39. Because the execution of the COBOL UNTIL option terminates on a ______________ condition and the DOWHILE structure terminates on a ______________ condition, it is necessary to code the ______________ of the condition to loop while the condition is ______________.
40. If the condition in the UNTIL phrase is true when first encountered, the ______________ or ______________ in the ______________ statement is ______________ executed at all.
41. The DOWHILE structure terminates when the condition is ______________ and the DOUNTIL structure terminates when the condition is ______________.
42. In the CASE structure, for any value ______________ of the defined range, the statement is ______________.
43. The CASE structure is implemented in COBOL by a combination of ______________ and the ______________ statements with the ______________ option.
44. In the CASE structure, the PERFORM statement must be written with the ______________ option because the paragraphs that carry out individual ______________ all end with a ______________.
45. The first segment of code in a program must contain the ______________ and ______________ Divisions.
46. The second segment of the code in a program should contain the ______________.
47. The Procedure Division is made up of ______________ consisting of ______________ and ______________.
48. At the head of the first segment of the Procedure Division, the programmer will place the ______________ coding of the ______________.
49. All code for a single logical level will be in the ______________ segment and subordinate code will be in ______________ segments.
50. The segments of the Procedure Division should appear in some ______________ sequence, such as the order in which they are ______________, or a sequence that reflects their ______________ in the logical ______________ of the program.
51. The free use of ______________ can exhibit more clearly that ______________ exist among ______________ so grouped.
52. Each segment of the Procedure Division should consist of no more than ______________ lines of coding.
53. Paragraph and section names are easier to locate if they are always written on a ______________ line.
54. Verbs are easier to ______________ and paragraphs are easier to ______________ and ______________, if ______________ statements are never written on the same line.
55. Important clauses of certain COBOL verbs are better set off if they are written on ______________ lines, ______________ by whatever amount of space is chosen as standard for ______________ lines.
56. Statements with the IFs should be ______________ and the ELSE should be coded on a ______________ line.
57. A limit of levels of ______________ statements should be established.
58. In the DOWHILE structure, only ______________ paragraphs should be performed and the ______________ option should not be used.

59. The CASE structure is the allowable variance to the rule for using the ______________ option of the ______________ verb.

60. Proper use of structured programming coding standards is the responsibility of ______________ with information provided through ______________ of the program.

Answers

1. THEORETICAL CONCEPT
2. BASIC STRUCTURES
3. THE LOGIC
4. PROGRAM DEVELOPMENT BY SEGMENTS, EASE OF CODING, SPEED OF CODING, EASE OF DEBUGGING, EASE OF MAINTENANCE
5. STRUCTURE, CONTROL LOGIC, THREE STRUCTURES
6. ORDER, WRITTEN
7. SELECTION
8. REPETITION, LOOP CONTROL
9. SEQUENCE, SELECTION, REPETITION
10. SIMPLE, SINGLE POINT, SINGLE POINT
11. TOP, BOTTOM, BEGINNING, END
12. NESTED, SINGLE/ENTRY, SINGLE/EXIT
13. OBSCURE STRUCTURE, COMPLICATED CONTROL PATHS, REDUNDANT AND OBSOLETE CODE, MEANINGLESS NOTES
14. SELF-EXPLANATORY LABELS, GOOD NOTES, CODE LAYOUT, INDENTATION
15. SEGMENTS, FEW STATEMENTS, PAGE OF CODING
16. SKIPPING AROUND
17. COMPLEX, DEBUGGING
18. TOP, BOTTOM
19. SHORT
20. LOWER
21. STRUCTURE, OBVIOUS
22. CONSISTENT, READABILITY, PICTORIAL, RELATIONSHIP, STATEMENTS
23. FUNCTIONS, VARIABLE
24. MANAGEABLE
25. INDENTING, NESTING, CLARITY, CODE
26. ONE ENTRY, ONE EXIT, PROGRAM CONTROL, ENTRY, EXIT
27. SUCCESSION
28. TWO ACTIONS, CONDITION
29. REPEATED, CODE, CONDITION, TRUE, FALSE
30. PERFORM, UNTIL
31. BEFORE, AFTER
32. VALUE, ROUTINES, EXECUTED
33. IFTHENELSE, DOWHILE
34. ELSE NEXT SENTENCE
35. IF, INDENTED, SPAN OF CONTROL
36. BASIC LOOP
37. CONDITION, PRIOR, PARAGRAPH, SECTION, FIRST
38. PERFORM, UNTIL
39. TRUE, FALSE, INVERSE, TRUE
40. PARAGRAPH, SECTION, PERFORM, NOT
41. FALSE, TRUE
42. OUTSIDE, IGNORED
43. PERFORM, GO TO, DEPENDING ON
44. THRU, FUNCTIONS, GO TO-PARAGRAPH-EXIT
45. IDENTIFICATION, ENVIRONMENT
46. DATA DIVISION
47. SEGMENTS, SECTIONS, PARAGRAPHS
48. MAIN LINE, PROGRAM
49. SAME, LOWER LEVEL
50. LOGICAL, INVOKED, POSITION, HIERARCHY
51. BLANK LINES, RELATIONSHIPS, ITEMS
52. 50
53. SEPARATE
54. LOCATE, CORRECT, MODIFY, TWO
55. SEPARATE, INDENTED, CONTINUATION
56. INDENTED, SEPARATE
57. NESTED IF
58. SINGLE, THRU
59. THRU, PERFORM
60. INDIVIDUAL PROGRAMMER, SUBSEQUENT REVIEWS

Questions for Review

1. What is a structured program and what difficulties does it create in COBOL programming?
2. What are the principal benefits of structured programming?
3. What are the main objectives of structured programming?
4. What is structured programming?
5. Briefly, what are the three basic structures of structured programming?
6. How are the basic structures combined into a program?
7. What is meant by "reduced program complexity" and "improving program clarity"?
8. What causes "jumps" within a program and what are the problems involved therein?
9. What is a segment?
10. What is a proper program?
11. What are the problems with using GO TO statements?
12. Why is it important that a program should proceed from top to bottom?
13. What are the three main ideas of structured programming?
14. What is the importance of indentation in a program?
15. What are the refinements and attributes attached to structured programming?
16. What are the two requirements for a proper program?
17. What is the Sequence control structure and how is it implemented in COBOL?
18. What is the Selection control structure and how is it implemented in COBOL?

19. What is the Iteration control structure and how is it implemented in COBOL?
20. What is the basic difference between the DOWHILE and DOUNTIL structures?
21. What is the CASE structure?
22. What is the function of the IFTHENELSE structure and how is it coded?
23. What is the function of the DOWHILE structure and how is it coded?
24. What is the function of the DOUNTIL structure and how is it coded?
25. What is the function of the CASE structure and how is it coded?
26. What is the ordering of statements in a structured COBOL program?
27. What are the rules for construction of the Procedure Division?
28. How are blank lines used in coding line spacing?
29. List the rules for indenting and formatting structured programs.
30. What are the rules for indenting and formatting IFTHENELSE structures?
31. What are the rules for indenting and formatting DOWHILE structures?
32. What are the rules for indenting and formatting DOUNTIL structures?
33. What are the rules for indenting and formatting CASE structures?

Problems

1. *Match each item with its proper description.*

______ 1. Top Down Design	A. Choice between two actions based on a condition.
______ 2. GO TO Statements	B. Control logic restricted to three structures.
______ 3. Selection	C. Used for repeated execution of code while a condition is true (false).
______ 4. Structured Programming	D. Exhibits in a pictorial way the relationships among statements.
______ 5. Segment	E. Uses the value of a variable to determine which of several routines is to be executed.
______ 6. CASE Structure	F. A segment that has no infinite loops and no unreachable code.
______ 7. Sequence	G. A paragraph, section, subroutine, or code incorporated with a COPY.
______ 8. Indentation	H. Single point of transfer into or out of a structure.
______ 9. Proper Program	I. Formularization of the idea that unless otherwise stated, statements are executed in the order in which they appear in a program.
______10. Iteration	J. Distracts reader from program by forcing him to examine program in an unnatural way.

2. *Rewrite the following statements using structured programming indenting and formatting standards.*

   A. IF TCODE IS EQUAL TO 20, ADD 7 TO AMOUNT, MOVE AMOUNT TO WORK-AMOUNT, ELSE SUBTRACT 10 FROM AMOUNT, MOVE AMOUNT TO EXTRA-WORK DIVIDE BALANCE BY 2 GIVING NEW-BAL.
   B. IF TRCODE IS EQUAL TO 50 NEXT SENTENCE ELSE ADD 100 TO NEW-AMT. MOVE NEW-AMT TO WORK-AMT.
   C. PERFORM DAILY-SALES-PROCESS VARYING DAY FROM 1 BY 1 UNTIL DAY IS GREATER THAN 7.
   D. OPEN INPUT QUANFILE, OUTPUT CALFILE, TFILE.

3. *Rewrite the following statements using structured programming indenting and formatting standards.*

   A. IF ACCT-NO = MACCOUNT-NO
   IF TCODE = 15
   ADD TRAN-AMT TO BALANCE
   ELSE SUBTRACT TRAN-AMT FROM BALANCE
   ELSE MOVE TRAN-FILE TO OUT-TRAN-FILE.
   B. MOVE SPACES TO FLD-A, FLD-B, FLD-C.
   C. IF TCODE = 9 AND AMT IS GREATER THAN 10 OR BALANCE IS LESS THAN ZERO ADD 700 TO EXCESS ELSE SUBTRACT 300 FROM OVER-DUE. MULTIPLY BALANCE BY 10 GIVING NEW-BALANCE.

4. *Place the number in the second column (COBOL Implementation) that correctly identifies the program structure.*

Program Structure	*COBOL Implementation*
1. DOWHILE	_____Writing statements in succession.
2. IFTHENELSE	_____PERFORM with UNTIL option.
3. SEQUENCE	_____IF statements.
4. CASE	_____PERFORM
	_____PERFORM with UNTIL option.
5. DOUNTIL	_____PERFORM with VARYING option.
	_____PERFORM with THRU option
	_____GO TO with DEPENDING ON option.

5. *Rewrite the following statement in correct structured programming form.*

   PERFORM EDIT-ROUTINE UNTIL TRAN-NO IS EQUAL TO MAST-NO.

   What type of program structure is being used?

6. *In the following statement, what is wrong with indentation? How can the statement be corrected to give the action represented by the indentation?*

```
IF SIZE-CODE = 'M'
THEN
    IF TOTAL-PRICE IS GREATER THAN 500
    THEN
        MOVE CHECK-PRICE-MESSAGE TO OUTPUT-LINE
        PERFORM PRINT-ROUTINE
ELSE
    MOVE BAD-SIZE-CODE-MESSAGE TO OUTPUT-LINE
    PERFORM PRINT-ROUTINE.
```

7. *Write a program segment using the CASE program structure based on the following information. Indicate each paragraph of the structure.*

```
IF TRAN-CODE = 1 GO TO DEPOSIT.
IF TRAN-CODE = 2 GO TO WITHDRAWAL.
IF TRAN-CODE = 3 GO TO INTEREST.
IF TRAN-CODE IS LESS THAN 1 OR GREATER THAN 3 GO TO NEXT SENTENCE.
```

8 Procedure Division

INTRODUCTION

The Procedure Division of a source program specifies the actions necessary to solve a given problem (fig. 8.1). These steps (input/output, logical decisions, computations, etc.) are required to process the data and to control the sequence in which these actions are to be carried out. Statements similar to English are used to denote the processing to be performed. *A statement is a syntactical valid combination of words and symbols beginning with a COBOL verb.* Verbs are used in statements in a source program to specify the steps the object program is to perform. (See figures 8.2, 8.3)

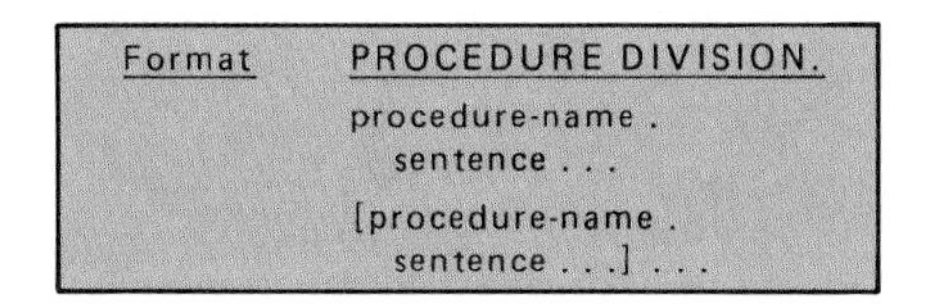

```
Format   PROCEDURE DIVISION.
         procedure-name .
           sentence . . .
         [procedure-name .
           sentence . . .] . . .
```

Figure 8.1 Format—Procedure Division.

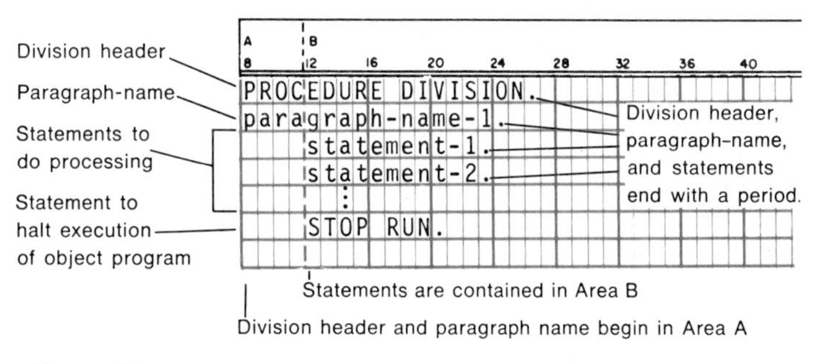

Figure 8.2 Guide for Coding Procedure Division Entries.

UNITS OF DATA

The following are units of expressions that constitute the Procedure Division. These units may be combined to form larger units.

Statement

A statement expresses an action to be taken during the execution of the object program. It begins with a verb and is completed by a combination of words that designate the data to be acted upon and that, at times, amplify the instructions. The portion of the statement following the verb consists of key words, optional words, and operands.

Each statement has its own particular format that specifies the types of words required and the organization of the statement. It is necessary to include all words required in a statement and to present them to the compiler according to the prescribed structure. If these two requirements are not fulfilled, the compiler will not interpret the statements.

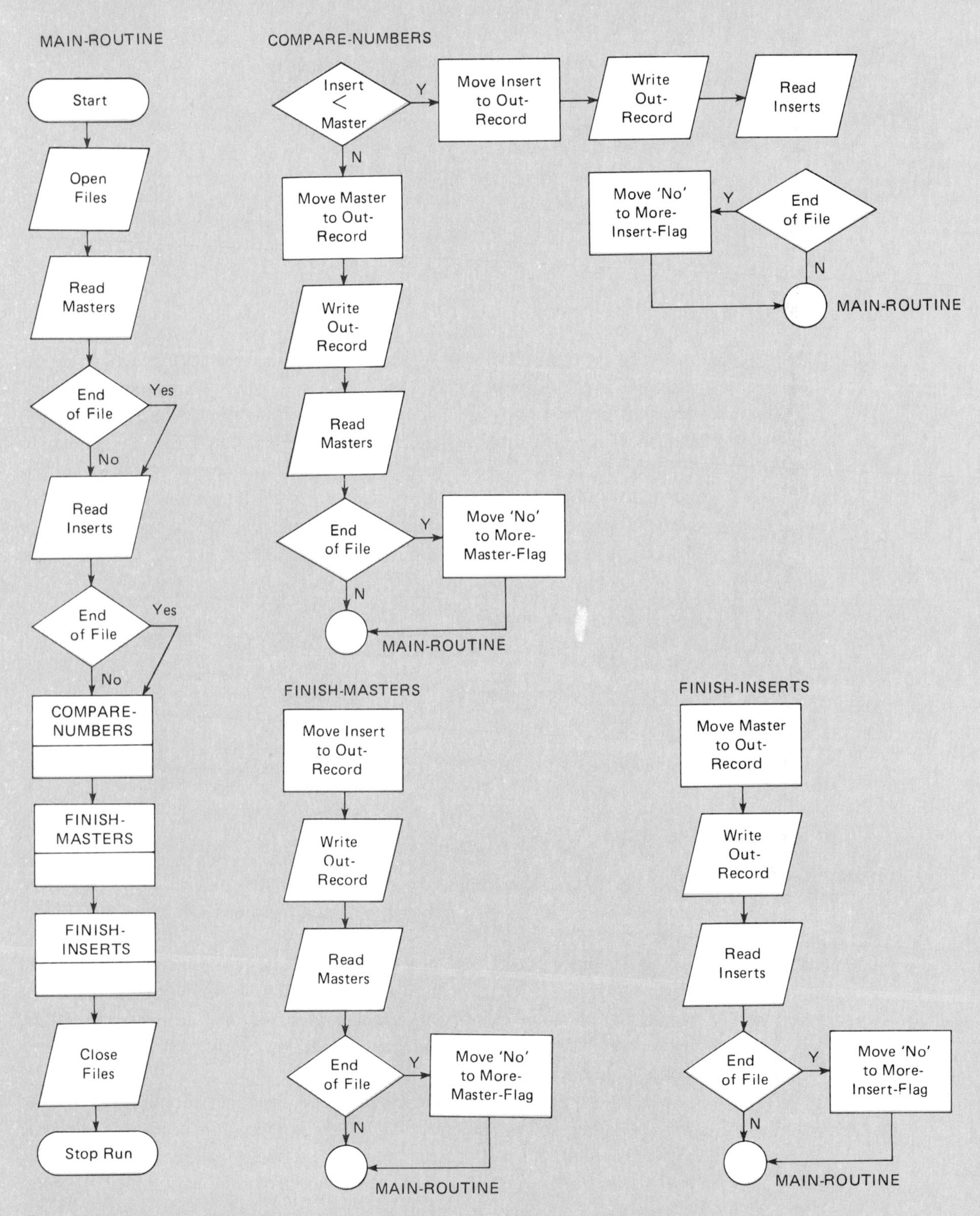

Figure 8.3 Example—Procedure Division Coding.

Figure 8.3 Continued

```
PROCEDURE DIVISION.
MAIN-ROUTINE.
    OPEN    INPUT    MASTERS,
                     INSERTS,
            OUTPUT   UPDATED-MASTERS.
        READ MASTERS
            AT END NEXT SENTENCE.
        READ INSERTS
            AT END NEXT SENTENCE.
        PERFORM COMPARE-NUMBERS
            UNTIL NO-MORE-INSERTS
               OR NO-MORE-MASTERS.
        PERFORM FINISH-MASTERS
            UNTIL NO-MORE-MASTERS.
        PERFORM FINISH-INSERTS
            UNTIL NO-MORE-INSERTS.
        CLOSE MASTERS
            INSERTS.
        STOP RUN.
    •
   COMPARE-NUMBERS.
        IF NUMBER OF INSERT < NUMBER OF MASTER
        THEN
            MOVE INSERT TO OUT-RECORD
            WRITE OUT-RECORD
            READ INSERTS
                AT END MOVE 'NO' TO MORE-INSERT-FLAG
        ELSE
            MOVE MASTER TO OUT-RECORD
            WRITE OUT-RECORD
            READ MASTERS,
                AT END MOVE 'NO' TO MORE-MASTER-FLAG.
    FINISH-MASTERS.
        MOVE MASTER TO OUT-RECORD.
        WRITE OUT-RECORD.
        READ MASTERS
            AT END MOVE 'NO' TO MORE-MASTER-FLAG.
    FINISH-INSERTS.
        MOVE INSERT TO OUT-RECORD.
        WRITE OUT-RECORD.
        READ INSERTS
            AT END MOVE 'NO' TO MORE-INSERT-FLAG.
```

The statement is the basic unit of the Procedure Division. A statement consists of a COBOL verb or the words IF or ON followed by the appropriate operands (file-names, literals, data-names, etc.) and other COBOL words that are essential to the completion of the statement. COBOL statements may be compared to clauses in the English language. The statement may be one of three types: imperative, conditional, or compiler-directing.

Imperative

An imperative statement consists of one or more unconditional "commands" to be performed by the object program. A simple imperative statement consists of one COBOL verb and its associated operands, excluding compiler-directing statements and conditional statements (fig. 8.4). An imperative statement may also consist of a series of imperative statements.

Imperative statements direct the computer to perform certain specified actions. These actions are specified and unequivocal, and the computer does not have the option of not performing them. An example of an imperative statement: SUBTRACT DEDUCTIONS FROM GROSS GIVING NET-PAY.

Conditional

A conditional statement is a statement that is to be tested, and the evaluation of the conditional expression will determine which of the alternate paths the program will follow (fig. 8.5). The modification of an imperative statement

```
MOVE CATALOG-NUMBER TO CONTROL-ITEM.
```

Figure 8.4 Example—Imperative Statement.

```
IF A EQUALS B
THEN
    MOVE A TO J
ELSE
    MOVE A TO M.

ADD C, D
    ON SIZE ERROR PERFORM PARA-3.

READ EMPLOYEE-FILE
    AT END PERFORM END-ROUTINE.
```

Figure 8.5 Example—Conditional Statements.

permits the computer to perform an operation under certain conditions. If the programmer attaches one or more conditional statements to an imperative statement, then the entire statement becomes a conditional statement.

In a conditional statement, the stated action is performed only if the specified conditions are present. Some examples of conditional statements are found in a READ statement, an arithmetic statement with the ON SIZE ERROR option, and IF statements. These and other statements involving conditions will be discussed later in the chapter. An example of a conditional statement: IF AGE IS LESS THAN 21 PERFORM MINOR.

Compiler-Directing

A compiler-directing statement directs the computer to take certain action at compile time (fig. 8.6). The statement contains one of the compiler-directing

```
ENTER PARA-A
```

Figure 8.6 Example—Compiler Directing Statement.

verbs (COPY, ENTER, NOTE) and its associated operands. These statements will be discussed in greater detail further along in the chapter.

Sentence

A sentence is composed of one or more statements specifying action to be taken terminated by a period and followed by a space (fig. 8.7). Commas, semicolons, or the word THEN may be used as separators between statements. When separators are used, they must be followed by a space. Separators improve readibility, but their absence or presence has no effect upon the compilation of the object program. An example of a sentence: ADD EARNINGS TO GROSS, PERFORM FICA-PROC. (See also figure 8.8.)

Paragraph

COBOL sentences may be combined into a logical entity called a paragraph. *A paragraph may be composed of one or more successive sentences* (fig. 8.9).

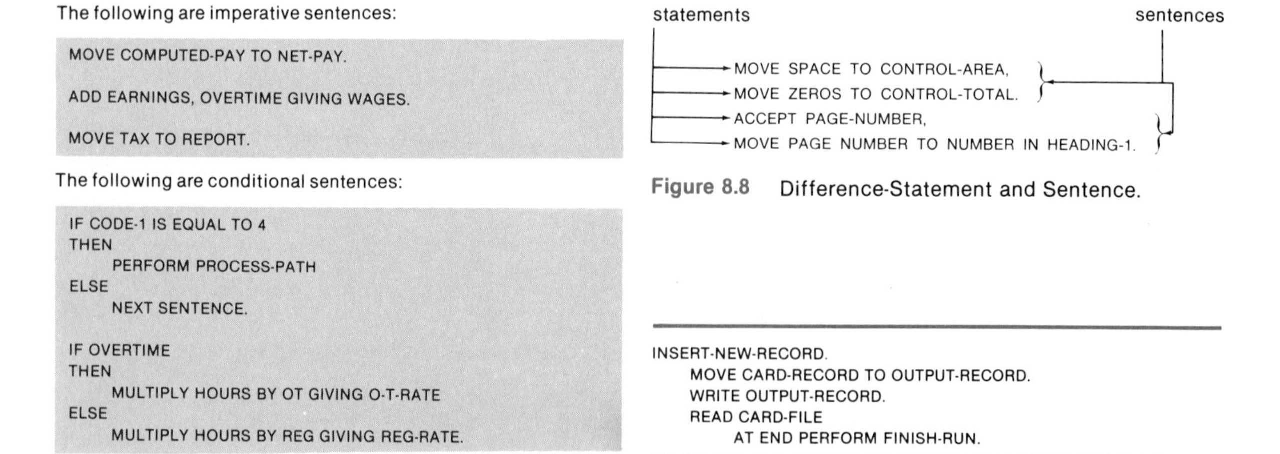

The following are imperative sentences:

```
MOVE COMPUTED-PAY TO NET-PAY.

ADD EARNINGS, OVERTIME GIVING WAGES.

MOVE TAX TO REPORT.
```

The following are conditional sentences:

```
IF CODE-1 IS EQUAL TO 4
THEN
    PERFORM PROCESS-PATH
ELSE
    NEXT SENTENCE.

IF OVERTIME
THEN
    MULTIPLY HOURS BY OT GIVING O-T-RATE
ELSE
    MULTIPLY HOURS BY REG GIVING REG-RATE.
```

Figure 8.7 Examples—Imperative and Conditional Sentences.

statements
sentences
MOVE SPACE TO CONTROL-AREA,
MOVE ZEROS TO CONTROL-TOTAL.
ACCEPT PAGE-NUMBER,
MOVE PAGE NUMBER TO NUMBER IN HEADING-1.

Figure 8.8 Difference-Statement and Sentence.

```
INSERT-NEW-RECORD.
    MOVE CARD-RECORD TO OUTPUT-RECORD.
    WRITE OUTPUT-RECORD.
    READ CARD-FILE
        AT END PERFORM FINISH-RUN.
```

Figure 8.9 Example—Paragraph.

1. Each paragraph must begin with a procedure-name. Statements may be written on the same line as the procedure-name.
2. A procedure-name must not be duplicated within the same section.
3. Procedure-names follow the same rules as data-names with one exception: a procedure-name may be made up entirely of numerals.
4. A paragraph ends immediately before the next procedure-name or section-name, or at the end of the Procedure Division. If declaratives are used, the key words END DECLARATIVES will terminate the paragraph.

Sections

The section in a COBOL program is the largest unit to which a procedure-name may be assigned.

1. A section is composed of one or more successive paragraphs.
2. A section must begin with a section header (a procedure-name) followed by a space and the word SECTION followed by a period. The section header must appear on a line by itself, except in the DECLARATIVES portion of the Procedure Division, where it may be followed after an intervening space by a USE statement.
3. The Procedure Division need not be broken down into sections. Section usage is at the discretion of the programmer.
4. A section ends immediately before the next section name or at the end of the Procedure Division. If declaratives are used, the key words END DECLARATIVES will terminate the section.

ORGANIZATION

The Procedure Division consists of instructions that are written in statement form which may be combined to form sentences. Groups of sentences form paragraphs.

The Procedure Division generally consists of a series of paragraphs which may be optionally grouped into programmer-created sections. Each paragraph has a data-name and may consist of a varying number of entries (fig. 8.10).

The Procedure Division may contain both declaratives and procedures.

The *Declaratives Section* must be grouped at the beginning of the Procedure Division, preceded by the key word DECLARATIVES, and followed by a period or space. Declarative sections are concluded by the key words END DECLARATIVES and followed by a period and a space (fig. 8.11). For a complete discussion, including examples and uses of declarative statements, see the section "Declaratives" later in the text.

Procedures are composed of paragraphs, groups of successive paragraphs, a section, or a group of successive sections within the Procedure Division (fig. 8.12). Paragraphs need not be grouped into sections. Execution begins with the first statement of the Procedure Division, *excluding the declaratives*. Statements are then executed in the sequence in which they are written, unless altered by the program.

The end of the Procedure Division is the physical end of a COBOL program after which no further procedures may appear.

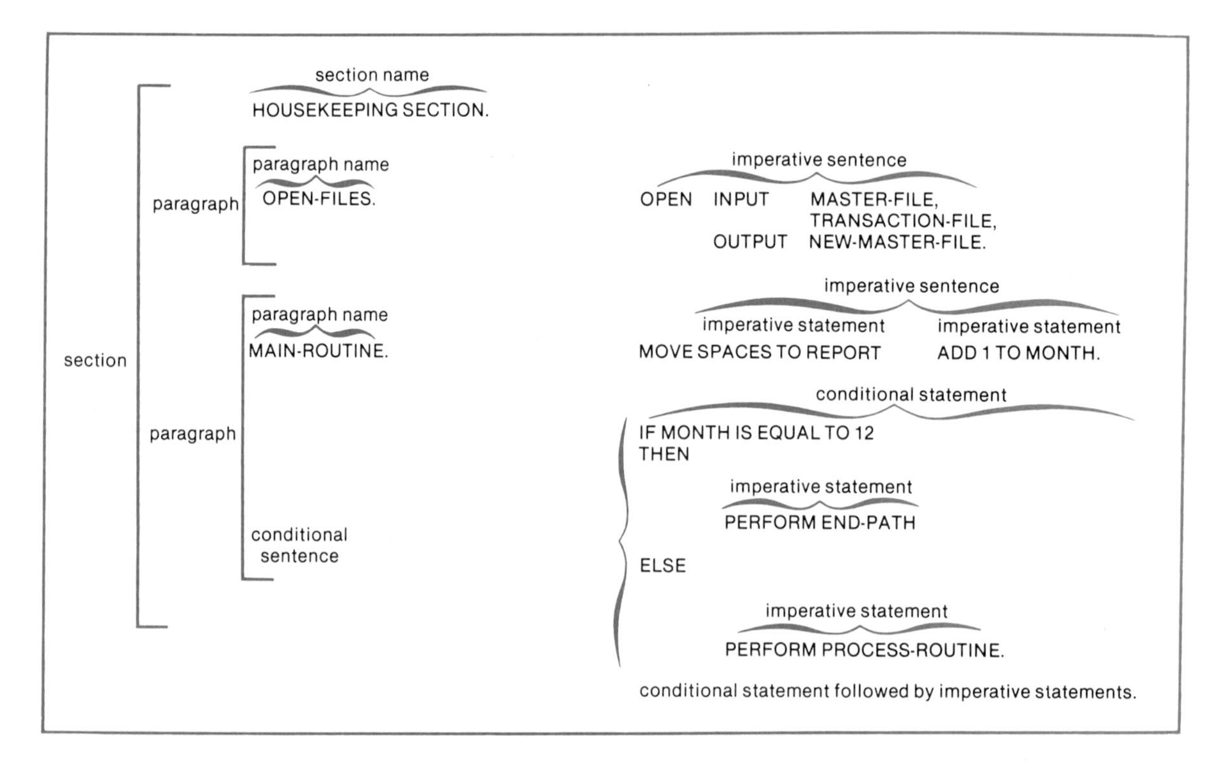

Figure 8.10 Procedure Division Structure.

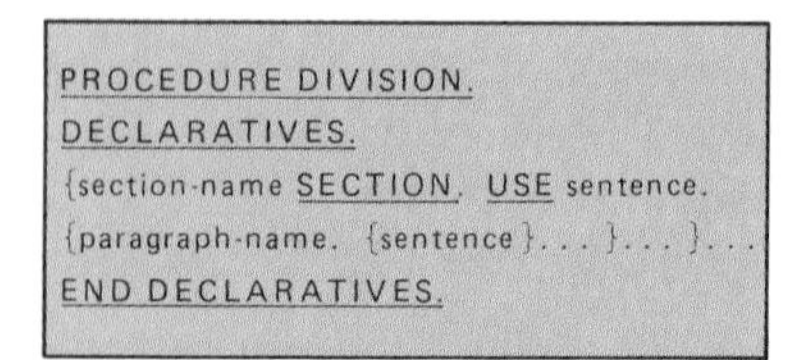

PROCEDURE DIVISION.
DECLARATIVES.
{section-name SECTION. USE sentence.
{paragraph-name. {sentence}. . . }. . . }. . .
END DECLARATIVES.

Figure 8.11 Declaratives Format.

```
PROCEDURE DIVISION.
MAIN-ROUTINE.
      OPEN   INPUT     CUSTOMER-FILE,
             OUTPUT    PRINT-FILE.
      PERFORM HEADING-LINE.
      READ CUSTOMER-FILE
            AT END MOVE 'NO' TO MORE-DATA-FLAG.
      PERFORM LISTING-ROUTINE
            UNTIL NO-MORE-DATA.
      CLOSE CUSTOMER FILE.
             PRINT-FILE.
      STOP RUN.

HEADING-LINE.
      MOVE HEADING TO PRINT-RECORD.
      WRITE PRINT-RECORD AFTER ADVANCING PAGE.
LISTING-ROUTINE.
      MOVE CORRESPONDING CUSTOMER TO LIST-RECORD.
      WRITE PRINT-RECORD FROM LIST-RECORD AFTER ADVANCING
            2 LINES
            AT EOP PERFORM HEADING-LINE.
      READ CUSTOMER-FILE
            AT END MOVE 'NO' TO MORE-DATA-FLAG.
```

Figure 8.12 Example—Procedures.

COBOL VERBS

COBOL Verbs are the basis of the Procedure Division of a source program. The organization of the division is based on the classification of COBOL verbs to be found in figure 8.13. (Other COBOL verbs will be discussed further along in the text.)

Each of the verbs causes some event to take place either at compile time or at program-execution time.

Input/Output Verbs

In data processing operations, the flow of data through a system is governed by an input/output system. The COBOL statements discussed in this section are used to initiate the flow of data that is stored on an external media, such as punched cards or magnetic tape, and to govern the flow of low-volume information that is to be obtained from or sent to an input/output device, such as a console typewriter.

The programmer is concerned only with the use of individual records. The input/output system provides for operations such as the movement of data into buffers and/or internal storage, validity checking, and unblocking and blocking of physical records.

One of the important advantages of COBOL programming is the use of pretested input and output statements to get data into and out of data processing systems. The COBOL input and output verbs provide the means of storing data on an external device (magnetic tape, disk units, as well as on card readers, punches, and printers) and extracting such data from these external devices.

Four verbs, OPEN, READ, WRITE, and CLOSE are used to specify the flow of data to and from files stored in an external media. ACCEPT and DISPLAY are used in conjunction with low-volume data that has to be obtained or sent to a card reader, console typewriter, or printer. (See figure 8.14.)

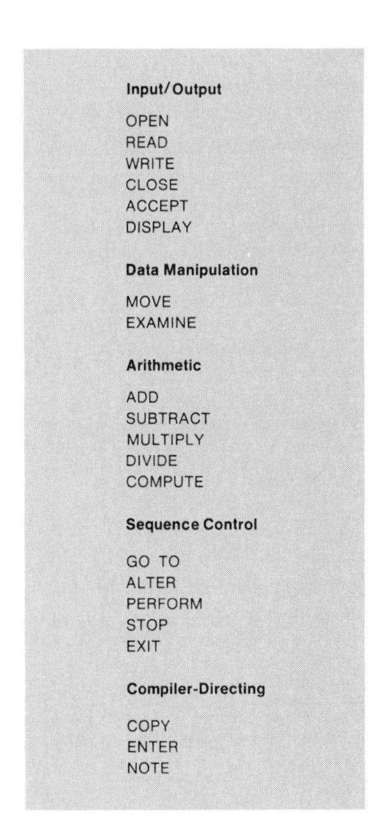

Input/Output
OPEN
READ
WRITE
CLOSE
ACCEPT
DISPLAY
Data Manipulation
MOVE
EXAMINE
Arithmetic
ADD
SUBTRACT
MULTIPLY
DIVIDE
COMPUTE
Sequence Control
GO TO
ALTER
PERFORM
STOP
EXIT
Compiler-Directing
COPY
ENTER
NOTE

Figure 8.13 COBOL Verbs.

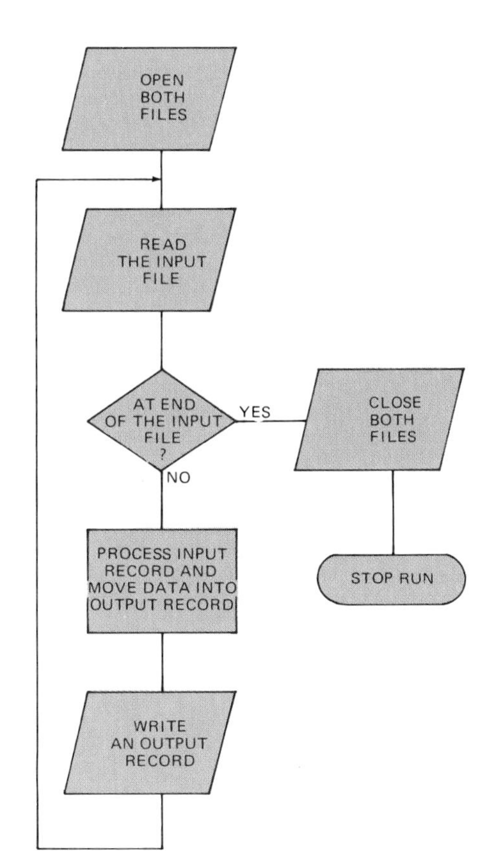

READ THE INPUT FILE
AT END OF THE INPUT FILE ?

Figure 8.14
Overall Logic Input and Output (Sequential Files).

Open

The OPEN statement makes one or more input or output files ready for reading or writing, checks or writes labels if needed, and prepares the storage areas to receive or send data (figs. 8.15, 8.16).

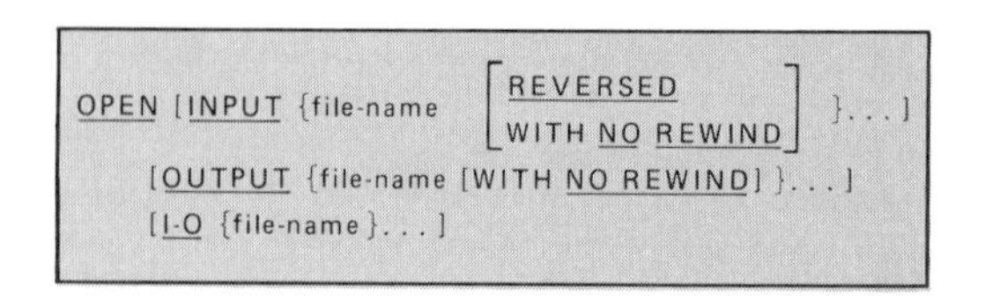

Figure 8.15 Format—Open Statement.

```
OPEN      INPUT     MASTER-FILE.
OPEN      OUTPUT    ERROR-LISTING.
OPEN      INPUT     MASTER-FILE,
                    TRANSACTION-FILE.
OPEN      INPUT     MASTER-FILE,
                    TRANSACTION-FILE,
          OUTPUT    NEW-MASTER-FILE,
                    DELETIONS,
                    ADDITIONS.
```

Figure 8.16 Examples—Open Statement.

Rules Governing the Use of the Open Statement

1. An OPEN statement must be specified for all files used in a COBOL program. The file must be designated as either INPUT, OUTPUT, or I-O (mass storage files).
2. An OPEN statement must be executed prior to any other input or output statement for a particular file.
3. If the file has been closed during the processing, a second OPEN statement must be executed before the file can be used again.
4. The OPEN statement does not make input records available for processing nor release output records to their respective devices. A READ or WRITE statement respectively is required to perform these functions.
5. When a file is opened, such actions as checking and creating beginning file labels are done automatically for those files requiring such action.
6. An OPEN statement can name one or all the files to be processed by the program.
7. Each file that has been opened must be defined in the file description entry in the Data Division as well as the SELECT entry in the Environment Division.
8. At least one of the three optional clauses (INPUT, OUTPUT, or I-O) must be written.
9. The I-O option permits the opening of a mass storage file for both input and output operations. Since this option implies the existence of a file, it cannot be used if the mass storage file is being initially created.

Read

The READ statement makes a data record from a sequential input file (magnetic tape and card files) available for processing and allows the performance of one or more specified statements when the end of the file is detected (fig. 8.17).

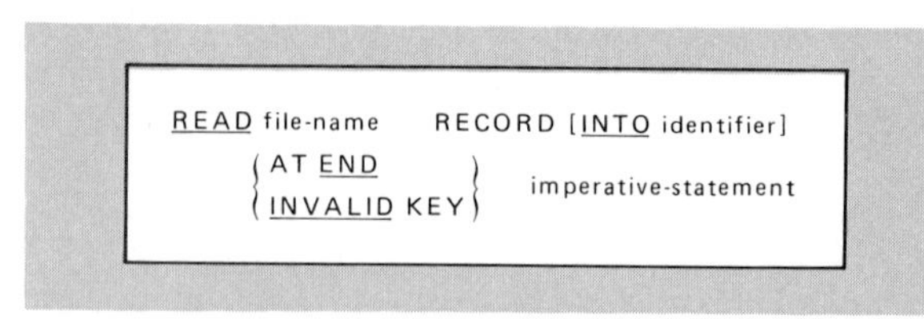

Figure 8.17
Format—Read Statement.

The READ statement may also make a specific record from a mass storage available for random file processing and give control to a specified imperative statement if the contents of the associated ACTUAL KEY data item are found to be invalid.

Rules Governing the Use of the Read Statement

1. The data records are made available in the input block one at a time. The record is available in the input until the next READ or a CLOSE statement is executed.
2. If the file contains more than one type of record for a file being executed, the next record is made available regardless of type. If more than one record description is specified in the FD entry, it is the programmer's responsibility to recognize which record is in the input block at any one time, since these records automatically share the same storage area, one that is equivalent to an implicit redefinition of the area. The programmer cannot specify the type of record to be read because the format of the READ statement requires the name of a file, not a record.
3. The file must be opened before it can be read.
4. When the end of a volume is reached for multivolume files, such as tape files, volumes are automatically switched, the tape is rewound, and the next reel is read. All normal header and trailer labels are checked. (See figure 8.18.)
5. The INTO option converts the READ statement into a READ and MOVE statement (fig. 8.19). The identifier must be the name of a Working-Storage

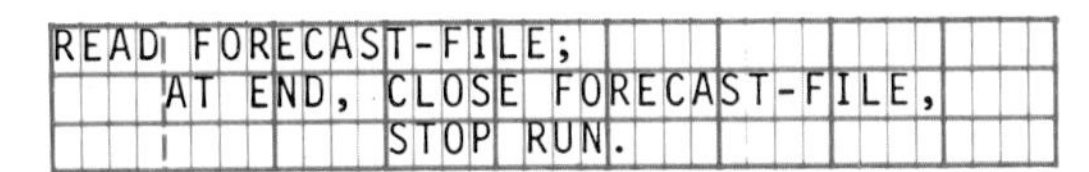
READ FORECAST-FILE;
AT END, CLOSE FORECAST-FILE,
STOP RUN.

Figure 8.18 Example—Read Statement.

```
READ MASTER-FILE RECORD INTO MASTER-WORK
    AT END PERFORM END-DATA-MASTER.
```

Figure 8.19 Example—Read Statement—Into Option.

Section entry or a previously opened output record. The current record is now available in the input area as well as the area specified by the identifier. If the format of the INTO area is different than the input area, the data is moved into that area in accordance with the rules for the MOVE statement *without* the CORRESPONDING option.

(*Note:* The largest record may be described in any 01 level entry; it need not be the first level 01 entry. Using the INTO option, data is moved using the size of the largest record specified in the file description (FD) entry as the sending field size.)

6. An AT END clause must be included in all READ statements for sequential input files. The statements following the records AT END up to the period are taken to be the end of file conditions. When the AT END clause is encountered, the last data record of the file has already been read.
7. Once the imperative statements in the AT END clause have been executed for a file, any later referral to the file will constitute an error unless subsequent CLOSE and OPEN statements for that file are executed.

8. The INVALID KEY option must be specified for mass storage files in the random access mode. Control of the program will be processed according to the imperative statements following the INVALID KEY when the contents of the ACTUAL KEY are invalid.

Write

The WRITE statement releases a data record for insertion in an output file (figs. 8.20, 8.21). Format-1 is used for standard sequential files. Format-2 is used for processing mass storage files. See chapter 12.

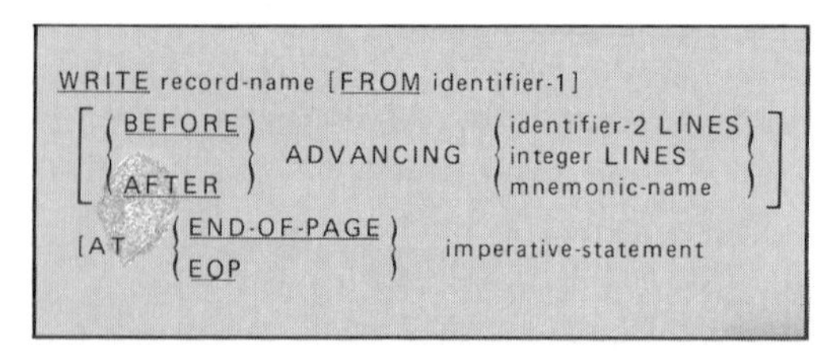

Figure 8.20 Format—Write Statement—Advancing Option.

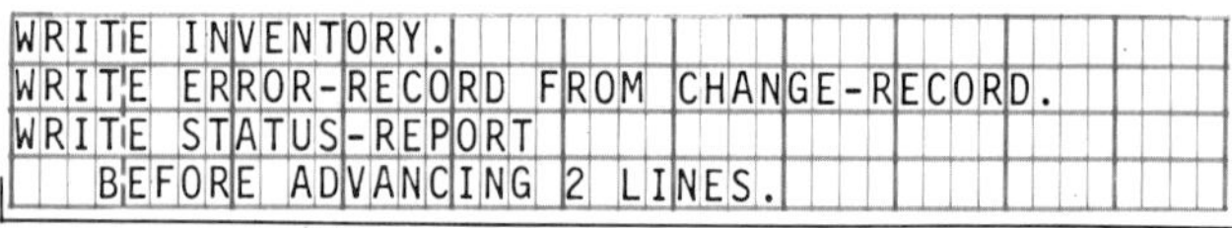

Figure 8.21 Examples—Write Statement.

Rules Governing the Use of the Write Statement

1. If the records are blocked, the actual transfer of the data to the output block may not occur until later in the processing cycle when the output block is filled with the number of records specified in the file description entry in the Data Division.
2. When an end of volume is reached for multivolume files, such as magnetic tape files, volumes are switched, the tape is rewound, and the next reel is written. All normal standard header and trailer labels are written.
3. An OPEN statement must be executed prior to the execution of the first WRITE statement.
4. After the record has been released, the logical record named by the record-name is no longer available for processing. All necessary processing of a record must be done prior to the WRITE statement.
5. The file associated with the record-name must be defined in the FD entry in the Data Division of the program. When a WRITE statement is executed, the record-name record is released to the output device.
6. The format requires a *record-name* rather than a file-name.
7. When the FROM option is used, Identifier-1 must not be the name of an item in the file containing the record-name. The FROM option converts the WRITE statement into a MOVE and WRITE statement. Identifier-1 must be the name of an item defined in the Working-Storage Section or in another FD. Moving takes place according to the rules specified for the MOVE statement without the CORRESPONDING option. After the execution of the WRITE statement with the FROM option, the information is still available in Identifier-1 although it is no longer available in the record-name area.
8. The ADVANCING options allow control of the vertical position of each record on a printed page of a report.

a. If the ADVANCING option is used with a WRITE statement, every WRITE statement for records associated with the same file must also contain one of these options. Automatic spacing is overriden by the ADVANCING option.
b. If the ADVANCING option is not used, automatic advancing will be provided by the implementor so as to cause single spacing.
c. When the ADVANCING option is used, the first character in each logical record of a file must be reserved by the user for control characters. In a printed report, if 132 characters are to be printed, PICTURE X(133) should be specified to allow for the control character. The compiler will generate instructions to insert the appropriate control character as the first character of a record. If the records are to be punched, the first character is used for pocket selection. PICTURE X(81) should be specified for punched output. It is the user's responsibility to see that the proper carriage-control tape is mounted on the printer prior to the execution of the program.

9. *ADVANCING option*

Identifier-2. If the Identifier-2 option is specified, the printer page is advanced the number of lines equal to the current value of Identifier-2. If the identifier is used, it must be the name of a nonnegative numeric elementary item (less than 100) described as an integer.

Integer. If the Integer option is specified, the printer page is advanced the number of lines equal to the value of the integer. The integer must be a nonnegative amount less than 100.

Mnemonic-Name. If the Mnemonic-Name option is specified, the printer page is advanced according to the rules specified by the implementor for that hardware device. The mnemonic-name must be defined as function-name in the SPECIAL-NAMES paragraph of the Environment Division. It is used to skip to channels 1–12 and to suppress spacing. It is also used for pocket selection for punched-card output files. (See figures 8.22, 8.23.)

Action Taken for Function-names—ADVANCING Option

Function-name	Action Taken
CSP	Suppress spacing.
C01 through C09	Skip to channel 1 through 9, respectively.
C10 through C12	Skip to channel 10, 11, and 12, respectively.
S01, S02	Pocket select 1 or 2 on the IBM 1442, and P1 or P2 on the IBM 2540.

Figure 8.22 Advancing Option—Function-Names.

Before Advancing. If the BEFORE ADVANCING option is used, the record is written *before* the printer page is advanced according to the preceding rules (figs. 8.24, 8.25).

```
ENVIRONMENT DIVISION.
CONFIGURATION SECTION.
SOURCE-COMPUTER. IBM-370-I155.
OBJECT-COMPUTER. IBM-370-I155.
SPECIAL-NAMES.
    CØ1 IS TO-FIRST-LINE.
```

```
    WRITE OUTPUT-RECORD
        AFTER ADVANCING TO-FIRST-LINE.
```

Figure 8.23 Example—Writing Record after Skipping to Next Page.

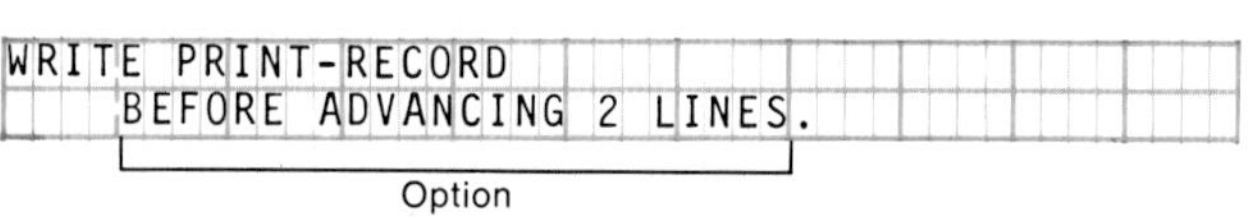


```
WRITE PRINT-RECORD
    BEFORE ADVANCING 2 LINES.
```

Option

Figure 8.24 Before Advancing Option.

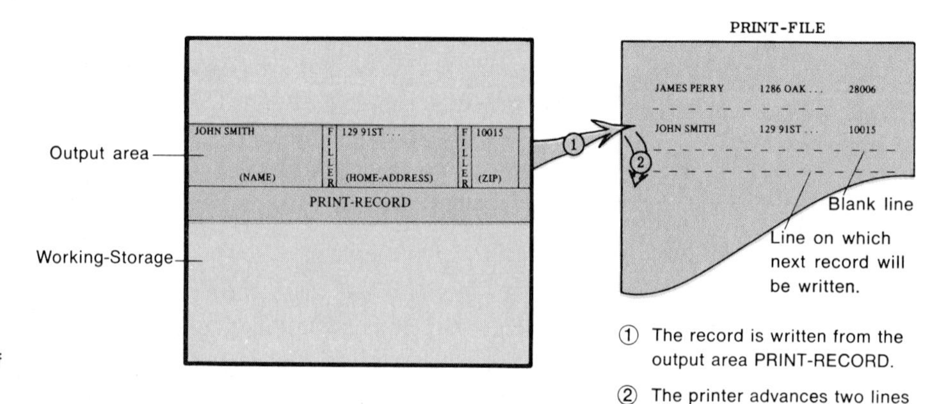

Figure 8.25 Execution of Write Statement—Before Advancing Option.

After Advancing. If the AFTER ADVANCING option is used, the record is written *after* the printer page is advanced according to the preceding rules (figs. 8.26, 8.27).

```
WRITE record-name
    AFTER ADVANCING integer LINES.
```

Option

Figure 8.26 After Advancing Option.

```
WRITE OUTPUT-RECORD
    AFTER ADVANCING 3 LINES.
```

Figure 8.27 Example—After Advancing Option.

10. *END-OF-PAGE.* The END-OF-PAGE option can be used to test for channel 12 on an on-line printer (fig. 8.28). When the end of a page is reached, the imperative statement following END-OF-PAGE or EOP is executed. The writing and spacing operations are completed before the END-OF-PAGE imperative is executed. If the ADVANCING and END-OF-PAGE options are used together, the page is advanced before the END-OF-PAGE test.

```
WRITE OUTPUT-RECORD
    AFTER ADVANCING 1 LINE
    AT END-OF-PAGE PERFORM HEADING-SEQUENCE.
    (END-OF-PAGE may be abbreviated EOP.)
```

Figure 8.28 Example—End-of-Page Option.

11. The INVALID KEY phrase must be specified for a file that resides on a mass storage device. The imperative statement is executed when the mass storage file is specified as in sequential-access mode, the last segment of the file has been reached, and an attempt is made to execute a WRITE statement.

    Control is passed to the imperative statement if the access is random and a record is not found. See chapter 12 for an example.
12. *AFTER POSITIONING* is used with IBM/360-/370 computers (fig. 8.29). In the AFTER POSITIONING option, Identifier-2 must be described as a one-character alphanumeric item, that is, with PICTURE X. The table shows the valid values that Identifier-2 may assume and their interpretations (fig. 8.30). If the integer option is used, the integer must be unsigned, and it must be the value 0, 1, 2, or 3 (figs. 8.31, 8.32). The values assume the meanings as shown in the table.

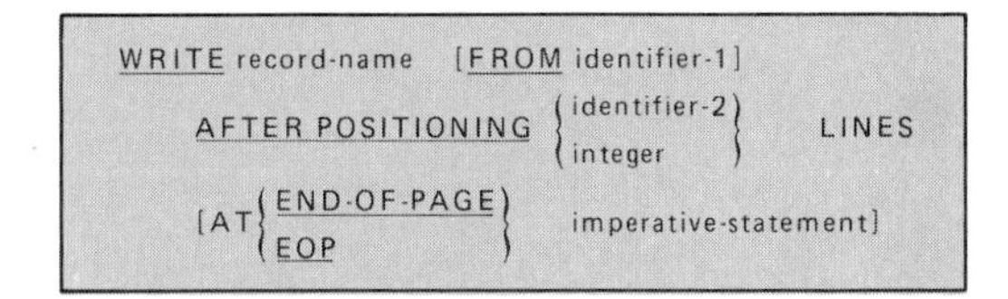
```
WRITE record-name  [FROM identifier-1]
    AFTER POSITIONING {identifier-2} LINES
                      {integer     }
    [AT {END-OF-PAGE} imperative-statement]
        {EOP        }
```

Figure 8.29
Format—Write Statement—Positioning Option.

Value of Identifier-2	Interpretation
b (blank)	Single-spacing
0	Double-spacing
—	Triple-spacing
+	Suppress spacing
1 - 9	Skip to channel 1 - 9, respectively
A, B, C	Skip to channel 10, 11, 12, respectively
V, W	Pocket select 1 or 2, respectively, on the IBM 1442, and P1 or P2 on the IBM 2540.

Figure 8.30 Positioning Option—Values of Identifier-2 and Interpretation.

Value of Integer	Interpretation
0	Skip to channel 1 of next page (carriage control "eject")
1	Single-spacing
2	Double-spacing
3	Triple-spacing

Figure 8.31 Positioning Option—Values of Integers and Interpretation.

```
PROCEDURE DIVISION.
MAIN-ROUTINE.
    OPEN  INPUT  CARDFILE
          OUTPUT PRINTED-REPORT.
    PERFORM HEADING-ROUTINE.
    READ CARDFILE
        AT END MOVE 'NO' TO MORE-DATA-FLAG.
    PERFORM MAIN-SEQUENCE
        UNTIL NO-MORE-DATA.
    CLOSE  CARDFILE
           PRINTED-REPORT.
    STOP RUN.

HEADING-ROUTINE.
    MOVE HEADING-RECORD-1 TO PRINT-RECORD.
    WRITE PRINT-RECORD
        AFTER POSITIONING 0.
    MOVE HEADING-RECORD-2 TO PRINT-RECORD.
    WRITE PRINT-RECORD
        AFTER POSITIONING 2.
MAIN-SEQUENCE.
    MOVE CORRESPONDING STUDENT-RECORD TO DETAIL-RECORD.
    MOVE DETAIL-RECORD TO PRINT-RECORD.
    WRITE PRINT-RECORD
        AFTER POSITIONING 2
        AT END-OF-PAGE PERFORM HEADING-ROUTINE.
    READ CARDFILE
        AT END MOVE 'NO' TO MORE-DATA-FLAG.
```

(Remember that you cannot mix ADVANCING and POSITIONING options within a program)

Figure 8.32 Example—Positioning Option.

Close The CLOSE statement terminates the processing of one or more data files, releases areas that serve as buffers, and optionally reverses and/or lock tape files where applicable (fig. 8.33).

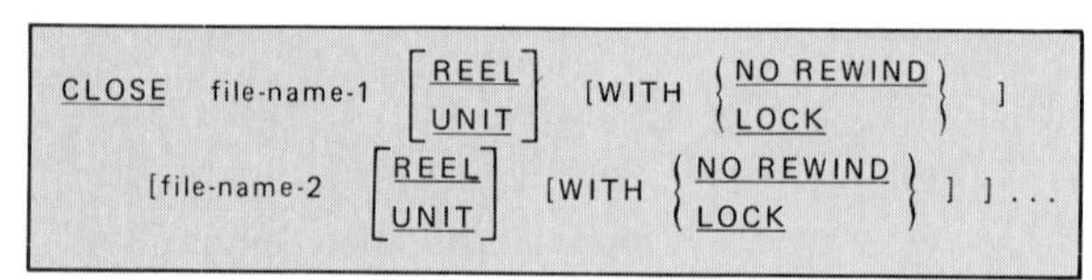

```
CLOSE  file-name-1 [REEL|UNIT] [WITH {NO REWIND|LOCK} ]
      [file-name-2 [REEL|UNIT] [WITH {NO REWIND|LOCK} ] ] ...
```

Figure 8.33 Format—Close Statement.

Rules Governing the Use of the Close Statement

1. A CLOSE statement may be executed only for files that have been previously opened.
2. The file-name is the name of a file upon which the CLOSE statement is to operate. The file-name must be defined by a FD entry in the Data Division.
3. The REEL and WITH NO REWIND options apply only to files stored on magnetic tape devices and other devices to which these terms are applicable.
4. The UNIT option is applicable to mass storage devices in the sequential-access mode.
5. The LOCK option insures that the file cannot be opened during the execution of the object program.
6. The optional clauses (INPUT and OUTPUT) are not written for files closed.

(See figures 8.34, 8.35.)

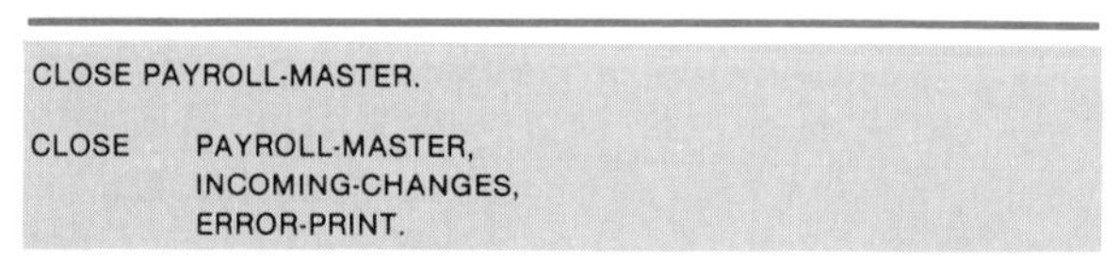

```
CLOSE PAYROLL-MASTER.

CLOSE    PAYROLL-MASTER,
         INCOMING-CHANGES,
         ERROR-PRINT.
```

This will close all three files as specified under the CLOSE file-name-1 statement.

```
CLOSE    INVENTORY WITH NO REWIND.

CLOSE    INVENTORY WITH NO REWIND,
         INVENTORY-CHANGES.

CLOSE    NEW-INVENTORY-MASTER WITH LOCK,
         INVENTORY-MASTER WITH NO REWIND,
         INVENTORY-CHANGES.
```

Figure 8.34 Examples—Close Statement.

INPUT-RECORD			
NAME		HOME-ADDRESS	
SUR (12 characters)	GIVEN (8 characters)	(40 characters)	(blank) (20 characters)

OUTPUT-RECORD	
LAST-NAME (12 characters)	ADDRESS-O (40 characters)

The program flowchart shows the order of operations for the Procedure Division.

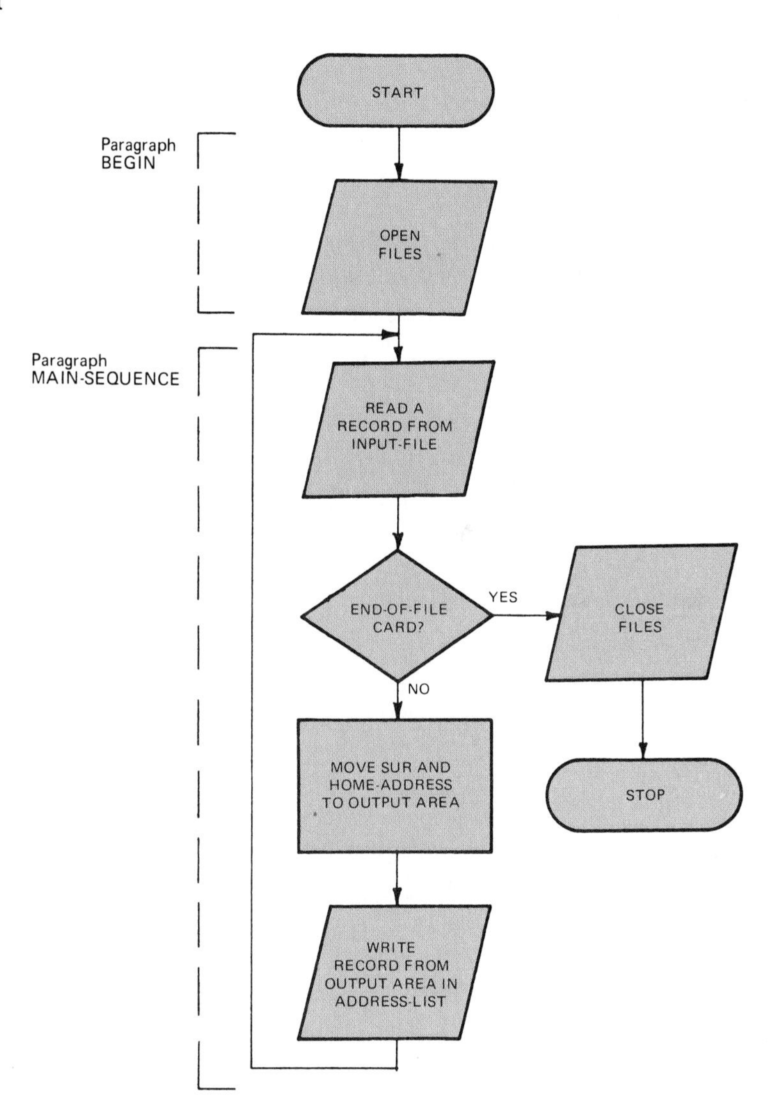

Figure 8.35 Example—Read and Write Statements.

Figure 8.35 Continued.

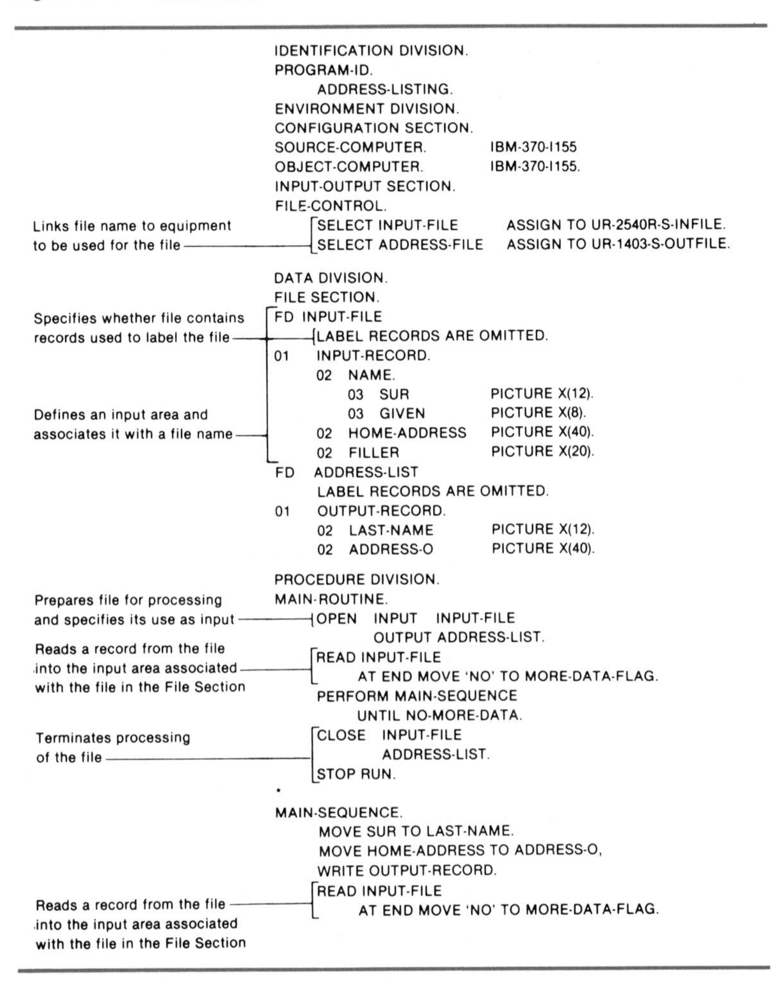

```
IDENTIFICATION DIVISION.
PROGRAM-ID.
       ADDRESS-LISTING.
ENVIRONMENT DIVISION.
CONFIGURATION SECTION.
SOURCE-COMPUTER.          IBM-370-I155
OBJECT-COMPUTER.          IBM-370-I155.
INPUT-OUTPUT SECTION.
FILE-CONTROL.
    SELECT INPUT-FILE       ASSIGN TO UR-2540R-S-INFILE.
    SELECT ADDRESS-FILE     ASSIGN TO UR-1403-S-OUTFILE.

DATA DIVISION.
FILE SECTION.
FD  INPUT-FILE
    LABEL RECORDS ARE OMITTED.
01  INPUT-RECORD.
    02  NAME.
        03  SUR           PICTURE X(12).
        03  GIVEN         PICTURE X(8).
    02  HOME-ADDRESS      PICTURE X(40).
    02  FILLER            PICTURE X(20).
FD  ADDRESS-LIST
    LABEL RECORDS ARE OMITTED.
01  OUTPUT-RECORD.
    02  LAST-NAME         PICTURE X(12).
    02  ADDRESS-O         PICTURE X(40).

PROCEDURE DIVISION.
MAIN-ROUTINE.
    OPEN   INPUT   INPUT-FILE
           OUTPUT ADDRESS-LIST.
    READ INPUT-FILE
         AT END MOVE 'NO' TO MORE-DATA-FLAG.
    PERFORM MAIN-SEQUENCE
         UNTIL NO-MORE-DATA.
    CLOSE  INPUT-FILE
           ADDRESS-LIST.
    STOP RUN.

MAIN-SEQUENCE.
    MOVE SUR TO LAST-NAME.
    MOVE HOME-ADDRESS TO ADDRESS-O,
    WRITE OUTPUT-RECORD.
    READ INPUT-FILE
         AT END MOVE 'NO' TO MORE-DATA-FLAG.
```


Accept

The ACCEPT statement obtains low-volume data from the system's logical input device, or from the console (fig. 8.36).

Rules Governing the Use of the Accept Statement

1. If the same input/output device is specified for the READ and ACCEPT statements, the results may be unpredictable.
2. The identifier will be described in the Working-Storage Section of the Data Division. The ACCEPT statement will cause the transfer of data from the hardware device specified to the area specified by the identifier. This data replaces the previous contents of the area. (See figure 8.37.)
3. The *mnemonic-name* must be specified in the SPECIAL-NAMES paragraph of the Environment Division. The mnemonic-name may be either the system logical input device, a card reader with an assumed input record size of 80 characters, or the CONSOLE which must not exceed 255 characters. If the FROM option is not specified, the systems logical input device is assumed.

```
ACCEPT identifier [FROM {CONSOLE      } ]
                        {mnemonic-name}
```

Figure 8.36 Format—Accept Statement.

EMPLOYEE-RECORD		
NAME (20 characters)	HOME-ADDRESS (30 characters)	EMPLOYEE-NUMBER (5 characters)

```
DATA DIVISION.
WORKING-STORAGE SECTION.
Ø1  EMPLOYEE-RECORD.
      Ø2 NAME PICTURE X(2Ø).
      Ø2 HOME-ADDRESS PICTURE X(3Ø).
      Ø2 EMPLOYEE-NUMBER PICTURE X(5).
```

```
      ACCEPT EMPLOYEE-RECORD FROM CONSOLE.
```

This statement will allow values of elementary variables in EMPLOYEE-RECORD to be entered into Working-Storage through the console typewriter.

Figure 8.37 Example—Accept Statement—Console Option.

4. When the FROM CONSOLE option is used,
   a. A message code is automatically displayed followed by the literal "AWAITING REPLY." The operation is suspended until the operator types the same message code and the necessary information for the continuance of the program. The message code serves as a key in the control program to correlate the console input with the proper program.
   b. As many records as necessary are read to exhaust the operand, up to 255 characters.
5. If the hardware device specified is capable of transferring data of the same size as the receiving area, the transferred data is stored in the receiving data item. If the hardware device is not capable of transferring data of the same size as the receiving area item, then the following takes place:
   a. If the size of receiving area is greater than the transferred data, the transferred data is stored in the left portion of the receiving area, and additional data is requested.
   b. If the size of the receiving area is less than the transferred data, only the leftmost characters will be moved until the area is filled, with the excess character positions at the right being truncated.
6. If the mnemonic-name is associated with the logical input device, up to 80 characters can be obtained. The data to be moved will come from the leftmost positions of the input block. The data must be punched into a card and entered, together with the job-control cards, at the time the object program is to be executed. (See figure 8.38.)

CUSTOMER-RECORD		
NAME (25 characters)	HOME-ADDRESS (30 characters)	BALANCE (5 characters)

Figure 8.38 Example—Accept Statement—Mnemonic Name Option.

Figure 8.38 Continued.

```
DATA DIVISION.
WORKING-STORAGE SECTION.
Ø1  CUSTOMER-RECORD.
    Ø2 NAME PICTURE X(25).
    Ø2 HOME-ADDRESS PICTURE X(3Ø).
    Ø2 BALANCE PICTURE X(5).

PROCEDURE DIVISION.
SEQUENCE-1.
    ACCEPT CUSTOMER-RECORD FROM CONSOLE.
    STOP RUN.
```

```
    ACCEPT CUSTOMER-RECORD FROM SYSIN.
```

The ACCEPT statement may be used for low-volume input from a card reader as well as from a console typewriter. The card reader, when it is the system logical input device, is referred to in a COBOL statement as SYSIN. The statement would transmit values to a Working-Storage variable from a punched card through a card reader.

Display

The DISPLAY statement causes the writing of low-volume data on an output device (fig. 8.39).

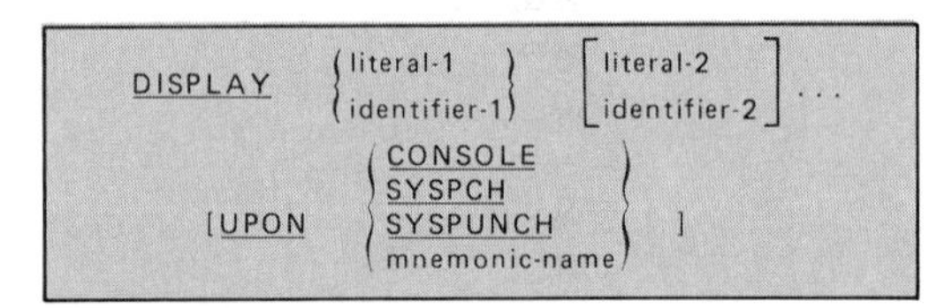

Figure 8.39
Format—Display Statement.

Rules Governing the Use of the Display Statement

1. A maximum logical record is assumed for each hardware device.

   a. Systems logical printing device—120 characters.
   b. Systems logical console device—100 characters.
   c. Systems logical punch device—72 characters with columns 73–80 reserved for identification purposes. If fewer than 72 characters are required, the remaining positions up to column 73 are filled with spaces.

   (See figure 8.40.)

2. If the same input/output device is used with both the WRITE and DISPLAY statement, the output resulting from the statements may not be in the same sequence as that in which the statements were encountered.
3. The mnemonic-name is associated with a hardware device in the SPECIAL-NAMES paragraph in the Environment Division.
4. The identifier may be either an elementary or a group item.
5. When a DISPLAY statement contains more than one operand, the size of the sending item is the sum of the sizes associated with the operands, and the value of the operands are transferred in the same sequence as that in which the operands are encountered.

6. Numeric or nonnumeric literals may be used.
7. Figurative constants, except ALL, may be used in DISPLAY statements. If a figurative constant is used as one of the operands, only a single occurrence of the figurative constant is displayed.
8. Any number of identifiers, literals, and figurative constants may be combined into one statement, but they must not exceed the specified maximum limit size. When more than one item is displayed, any spaces desired between multiple operands must be explicitly specified, either with designated spaces included in the literal or with the figurative constant SPACE between operands.
9. If the hardware device is capable of receiving the data of the same size being transferred, then the data is moved; otherwise, the following applies:
   a. If the size of the data item being transferred exceeds the size of the data that the hardware device is capable of receiving in a single transfer, the data, beginning with the leftmost character, is stored aligned to the left in the receiving hardware device, and additional data is requested.
   b. If the size of the data item that the hardware device is capable of receiving exceeds the size of the data being transferred, the transferred data is stored aligned to the left in the receiving hardware device.
10. If the UPON option is not used, the systems logical display device is assumed (fig. 8.41).

```
DISPLAY 'ENTER PROGRAM-NAME.'
        UPON CONSOLE.
```

The message ENTER PROGRAM-NAME will be written on the console typewriter.

```
DISPLAY PROGRAM-NAME
        UPON CONSOLE.
```

The value of the variable PROGRAM-NAME will be written on the console typewriter.

Figure 8.40
Examples—
Display Statement.

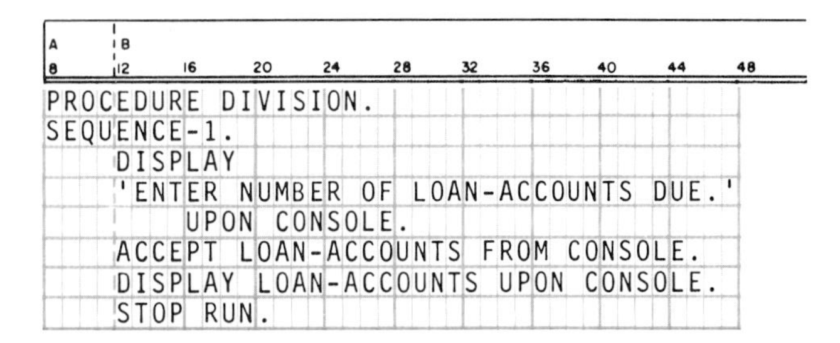

```
PROCEDURE DIVISION.
SEQUENCE-1.
        DISPLAY
        'ENTER NUMBER OF LOAN-ACCOUNTS DUE.'
                UPON CONSOLE.
        ACCEPT LOAN-ACCOUNTS FROM CONSOLE.
        DISPLAY LOAN-ACCOUNTS UPON CONSOLE.
        STOP RUN.
```

The above entries will cause the following:
1. Write the message ENTER NUMBER OF LOAN-ACCOUNTS DUE on the console typewriter.
2. Allow the number of loan accounts that are due to be keyed into LOAN-ACCOUNTS.
3. Write the value of LOAN-ACCOUNTS on the console typewriter.
4. The STOP RUN statement will halt execution of the program.

Figure 8.41
Examples—Display
and Accounts.

Data-Manipulation Verbs

The data-manipulation verbs move data from one storage area to another within the computer, and the inspection of the data is explicit in the functioning of several of the COBOL words. The MOVE verb has as its primary function the transmission of information from one storage area to another, while the EXAMINE verb inspects the data with or without the movement of data.

Move

The MOVE statement is used to move data from one area in main storage to one or more areas within the computer. (See figures 8.42, 8.43, 8.44.)

```
MOVE {identifier-1} TO identifier-2 [identifier-3] . . .
     {literal     }
```

Figure 8.42
Format—Move Statement.

Sending variable \ Receiving variable	Group	Alphabetic	Alphanumeric	External decimal	Packed decimal	Edit
Group	A	A	A	AU	AU	I
Alphabetic	A	A	A	I	I	I
Alphanumeric	A	A	A	N*	N*	E*
External decimal	AU	I	A*	N	N	E
Packed decimal	AU	I	A*	N	N	E
Edit	A	I	A	I	I	I

A Alphanumeric move
E Edit move
AU Alphanumeric move (value of receiving field is unpredictable)
N Numeric move
* Integers only
I Invalid

Figure 8.43
Type of Moves.

Type of move	Receiving item	Compiler action during move	Alignment	Padding if necessary	Truncation if necessary
Alphanumeric	Group	none	at left of value	on right with spaces	on right
	Alphabetic or alphanumeric	any necessary conversion	at left of value	on right with spaces	on right
Numeric	External decimal or packed decimal	any necessary conversion	at decimal point	on left and right with zeros	on left and right
Edit	Edited	editing and any necessary conversion	at decimal point	on left and right with zeros (unless suppressed)	on left and right

Figure 8.44 Effects of Types of Moves.

Rules Governing the Use of the Move Statement

1. Source data can be transferred to any number of receiving items (fig. 8.45).

```
MOVE ZEROS TO HIGH-SCORE,
              LOW-SCORE,
              AVERAGE-SCORE.
```

Figure 8.45
Example—Move Statement.

The above statement will fill the entire area of HIGH-SCORE, LOW-SCORE, and AVERAGE-SCORE with zeros.

2. When a group item is involved in a move, the data is moved without any regard for the level structure of the group items involved, and without editing. Thus, when a group item is present, the data being moved is treated as simply as a sequence of alphanumeric characters and is placed in the receiving area in accordance with the rules for moving elementary nonnumeric items. If the size of the group item differs, the compiler will produce a warning message when the statement is encountered. Normally, when a group item is involved in a move, it is a group transfer, and the descriptions of the two items are the same. (See figure 8.46.)

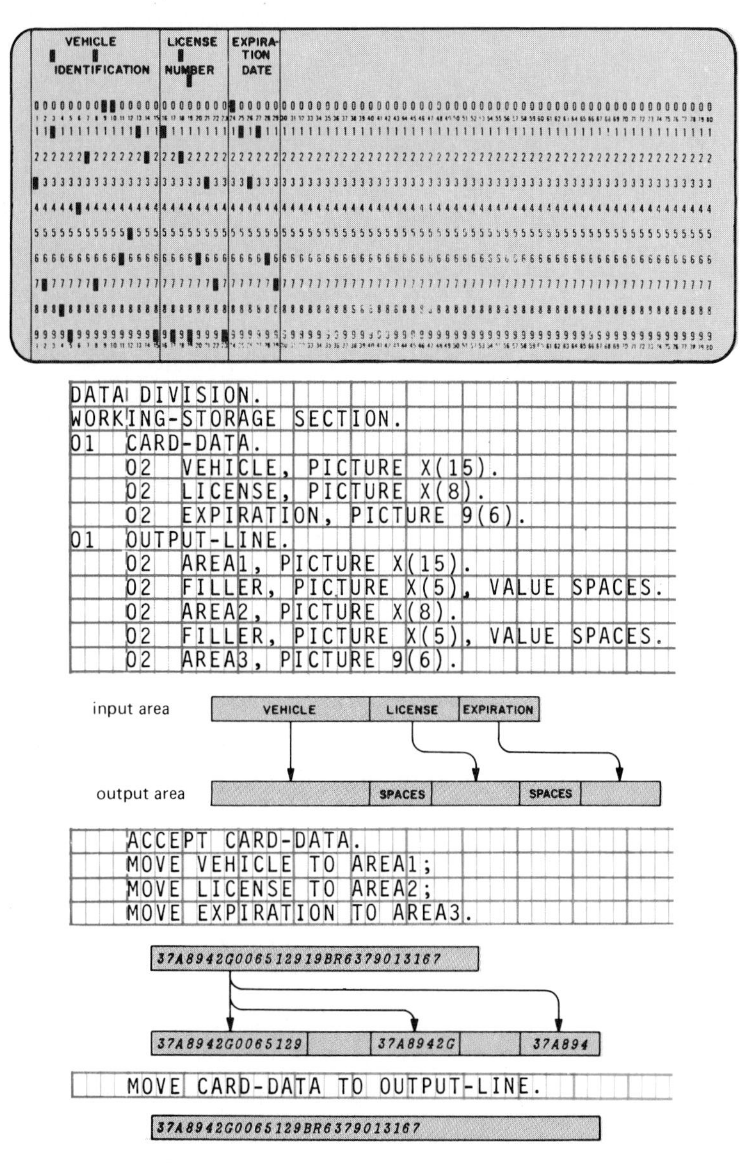

Figure 8.46 Example—Individual Move and Group Move.

3. When both the source and receiving areas are elementary items, editing appropriate to the format of the receiving area takes place automatically after the MOVE instruction is executed. The type of editing depends upon whether the item is numeric or nonnumeric. (See figure 8.47.)
4. Numeric literals and the figurative constant ZERO belong to the numeric category (fig. 8.48). Nonnumeric literals and the figurative constant SPACE belong to the nonnumeric category.

Source Field		Receiving Field		
PICTURE	Value	PICTURE	Value before MOVE	Value after MOVE
99V99	1234	99V99	9876	1234
99V99	1234	99V9	987	123
9V9	12	99V999	98765	01200
XXX	A2B	XXXXX	Y9X8W	A2Bbb
9V99	123	99.99	87.65	01.23
AAAAAA	REPORT	AAA	JKL	REP

Figure 8.47 Example—Data Movement.

Numeric Data

1. The data from the source area is aligned with respect to the decimal point (assumed or actual) in the receiving area. This alignment may result in the loss of leading or low-order digits (or both) if the source area is larger than the receiving area. Excess positions in the receiving area at either end will be filled with zeros. (See figure 8.49.)

MOVE ZEROES TO FIELD-B.

	Picture	Before execution	After execution
FIELD-B	9999	0 1 2 3	0 0 0 0

Figure 8.48 Example—Move Statement—Numeric Literal.

MOVE 125.7 TO DOLLARS.

	Picture	Before execution	After execution
DOLLARS	9999V99	1 2 3 4 ‸ 6 6	0 1 2 5 ‸ 7 0

Figure 8.49 Example—Move Statement—Numeric Data.

2. If the USAGE clause of the source and receiving fields differ, conversion to the representation specified in the receiving area takes place (fig. 8.50).
3. If the receiving area specifies editing, zero suppression, insertion of dollar signs, commas, decimal points, etc., and decimal point alignment, all will take place in the receiving area (figs. 8.51, 8.52).
4. If no decimal point is specified, and the receiving area is larger than the sending area, right justification will take place with the blank left positions being filled with zeros.

(See figures 8.53,8.54.)

MOVE FIELD-6 TO FIELD-7.

	Picture	Before execution	After execution
FIELD-6	999	1 5 7	1 5 7
FIELD-7	XXXX	A B 2 4	1 5 7

Figure 8.50 Example—Move Statement—Numeric Data.

MOVE AMOUNT TO AMOUNT-PR.

	Picture	Before execution	After execution
AMOUNT	9999V99	1 2 5 8 ‸ 3 9	1 2 5 8 ‸ 3 9
AMOUNT-PR	$9,999.99	$ 3 , 3 3 3 . 3 3	$ 1 , 2 5 8 . 3 9

Figure 8.51 Example—Move Statement—Numeric Data.

MOVE AMOUNT-1 TO AMOUNT-1-OUT.

	Picture	Before execution	After execution
AMOUNT-1	9999V99	0 0 0 0 0 3	0 0 0 0 0 3
AMOUNT-1-OUT	$$,$$$.99	$ 2 2 1 9	$. 0 3

Figure 8.52 Example—Move Statement—Numeric Data.

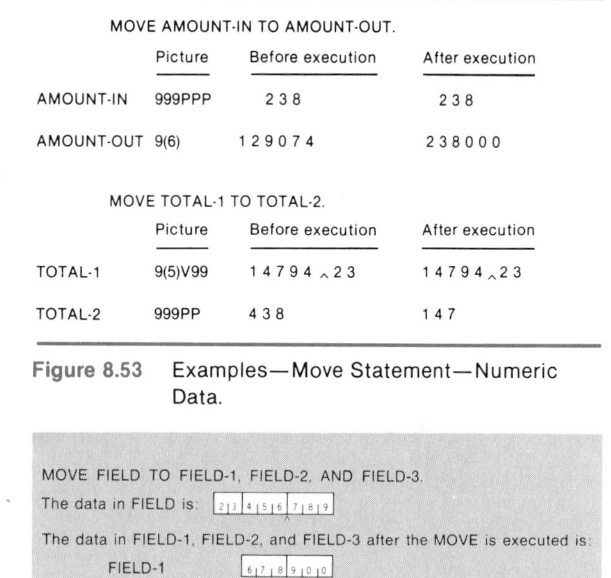

MOVE AMOUNT-IN TO AMOUNT-OUT.

	Picture	Before execution	After execution
AMOUNT-IN	999PPP	2 3 8	2 3 8
AMOUNT-OUT	9(6)	1 2 9 0 7 4	2 3 8 0 0 0

MOVE TOTAL-1 TO TOTAL-2.

	Picture	Before execution	After execution
TOTAL-1	9(5)V99	1 4 7 9 4 ‸ 2 3	1 4 7 9 4 ‸ 2 3
TOTAL-2	999PP	4 3 8	1 4 7

Figure 8.53 Examples—Move Statement—Numeric Data.

MOVE FIELD TO FIELD-1, FIELD-2, AND FIELD-3.

The data in FIELD is: 2 3 4 5 6 7 8 9

The data in FIELD-1, FIELD-2, and FIELD-3 after the MOVE is executed is:

FIELD-1 6 7 8 9 0 0

FIELD-2 2 3 4 5 6

FIELD-3 0 2 3 4 5 6 7 8

‸ Indicates Assumed Decimal Point

Figure 8.54 Example—Move Statement—Numeric Data.

Nonnumeric Data

1. The data from the source area is placed in the receiving area beginning at the left and continuing to the right, unless the field is specified as JUSTIFIED RIGHT, in which case the source data is placed in the right positions of the receiving area.
2. If the receiving area is not completely filled with data, the remaining positions are filled with spaces at the right or left for justified right items.
3. If the source field is longer than the receiving area, the move is terminated as soon as the receiving area is filled. Excess characters are truncated when the receiving area is filled. (See figures 8.55–8.60.)

MOVE FIELD-1 TO FIELD-2

	Picture	Before execution	After execution
FIELD-1	XXX	A B C	A B C
FIELD-2	XXXX	X Y W K	A B C

Figure 8.55 Example—Move Statement—Nonnumeric Data.

MOVE NAME TO NAME-1.

The data in NAME is: J O H N

The data in NAME-1 after the MOVE is executed is: J O H N

Figure 8.56 Example—Move Statement—Nonnumeric Data.

MOVE NAME TO FIELD-A AND FIELD-B.

Data in NAME J O H N B R O W N 1 5 0 2

Data in FIELD-A and FIELD-B after the MOVE is executed.

FIELD-A J O H N FIELD-B J O H N B R O W N

Figure 8.57 Example—Move Statement—Nonnumeric Data.

MOVE FIELD-3 TO FIELD-4.

	Picture	Before execution	After execution
FIELD-3	XXX	A B C	A B C
FIELD-4	XX	X Y	A B

Figure 8.58 Example—Move Statement—Nonnumeric Data.

MOVE '123' TO FIELD-5.

	Picture	Before execution	After execution
FIELD-5	XXXX	A B C D	1 2 3

Figure 8.59 Example—Move Statement—Nonnumeric Data.

MOVE ACCOUNT-NO TO ACCT-NO-PR.

	Picture	Before execution	After execution
ACCOUNT-NO	XXXX	A 1 2 3	A 1 2 3
ACCT-NO-PR	XBXXX	A C D E	A 1 2 3

Figure 8.60 Example—Move Statement—Nonnumeric Data.

Rules Governing Elementary Items Move Statements

1. A numeric edited, alphanumeric edited, the figurative constant SPACE, or any alphabetic data item must not be moved to a numeric edited data item.
2. A numeric literal, the figurative constant ZERO, a numeric data item, or a numeric edited data item must not be moved to an alphabetic item.
3. A numeric literal or a numeric data item whose implicit decimal point is not immediately to the right of the least significant digit must not be moved to an alphanumeric or alphanumeric edited item.
4. All other elementary moves are permissible and are performed in accordance with the rules mentioned previously.

Move Corresponding

When a MOVE CORRESPONDING statement is executed at object time, selected items within the source area are moved with any required editing to selected areas within the receiving area (fig. 8.61).

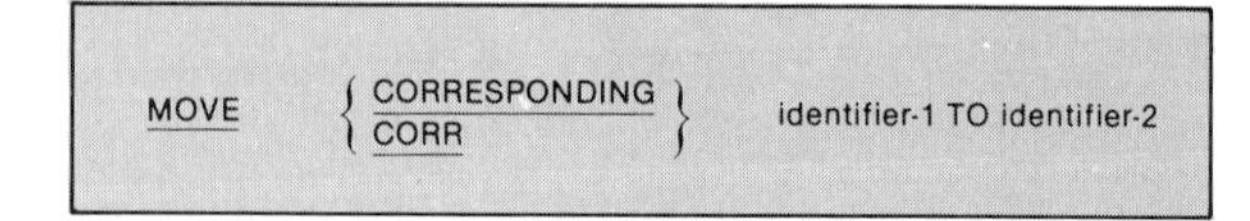

Figure 8.61 Format—Move Corresponding Statement.

Rules Governing the Use of the Move Corresponding Statement

1. Only corresponding data items having the same name and qualification as Identifier-1 and Identifier-2 are moved (fig. 8.62).
2. At least one of the items of the pair of matching items must be an elementary item.
3. The effect of a MOVE CORRESPONDING statement is equivalent to a series of simple MOVE statements.
4. Identifier-1 and Identifier-2 must be group items.
5. An item subordinate to Identifier-1 or Identifier-2 is not considered corresponding if:

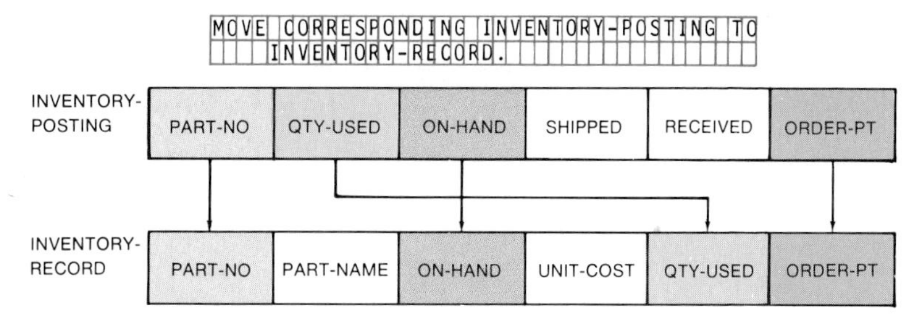

Figure 8.62 Data Movement Affected by Move Corresponding Statements.

a. It is an item identified by the word FILLER and any items subordinate to it, or
b. An item identified by a REDEFINES, OCCURS, RENAMES, or USAGE IS INDEX clauses and any items subordinate to it.

6. Either identifier may have REDEFINES or OCCURS clauses in its description or may be subordinate to a data item described with these clauses. If either identifier is described with an OCCURS clause, then the items must be subscripted; each data item that corresponds will also have to be subscripted by the computer.
7. Data-names with level number 66, 77, or 88 (RENAMES clause, independent item clause, or condition-names) cannot be referenced by Identifier-1 or Identifier-2.
8. Each matched source item is moved in conformity with the description of the receiving item.

(See figures 8.63.)

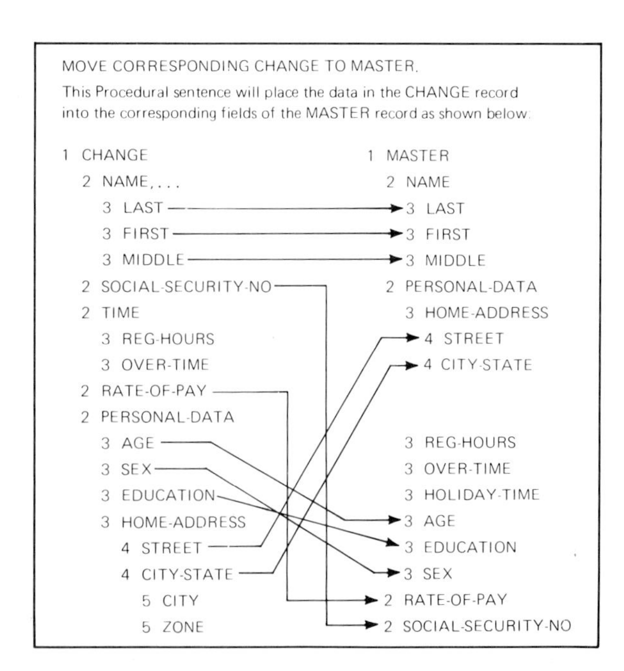

Figure 8.63 Example—Move Corresponding Statement.

Examine

The EXAMINE statement is used to replace a given character and/or to count the number of times it appears in a data item (fig. 8.64).

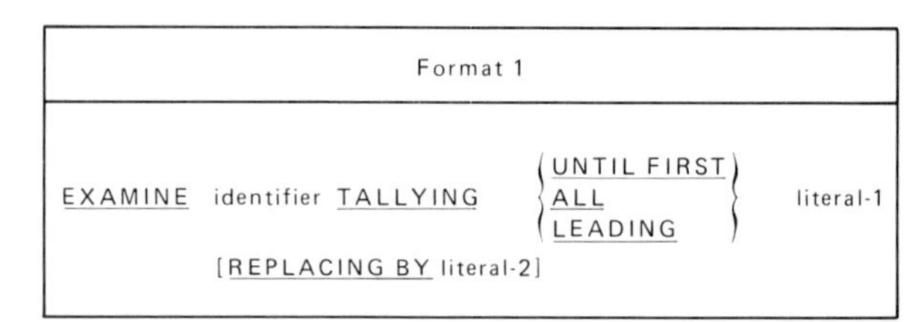

Figure 8.64
Format—Examine Statement—Format 1.

Rules Governing the Use of the Examine Statement

1. The EXAMINE statement may be only applied to an item whose USAGE IS DISPLAY.
2. Any literal used in the EXAMINE statement must be a number of characters associated with the class specified for the identifier. For example, if the class of the identifier is numeric, each literal in the statement must be numeric and may possess an operational sign. All must be single characters.
3. Nonnumeric literals must be single characters enclosed in quotation marks.
4. The examination of the data item begins with the first (leftmost) character of the data set and proceeds to the right. If the data item is numeric, any operational sign associated with it will be ignored.
5. Figurative constants may be used in place of Literal-2, with the exception of the figurative constant ALL.
6. *Tallying Option.* A count is made of the number of occurrences of certain characters in the identifier, and the count replaces the value of a special register called TALLY, whose length is five decimal digits. TALLY may be used in other procedural statements. The COBOL compiler sets up this special register. The count in TALLY at object time depends upon which of the following options is used.
   a. *UNTIL FIRST.* If the UNTIL FIRST option is specified, the TALLY count represents the number of characters other than Literal-1 encountered prior to the first occurrence of Literal-1 (fig. 8.65).
   b. *ALL.* If the ALL option is specified, all occurrences of Literal-1 are counted and accumulated in the TALLY register (fig. 8.66).
   c. *LEADING.* If the LEADING option is specified, the TALLY count represents the number of occurrences of Literal-1 prior to encountering other than Literal-1 data (fig. 8.67).
7. *Replacing Option.* The REPLACING option may be used with or without the TALLYING option. The replacement of characters depends upon which of the options is employed (fig. 8.68).
   a. *ALL.* If the ALL option is specified, Literal-2 is substituted for all occurrences of Literal-1 (fig. 8.69).
   b. *LEADING.* If the LEADING option is specified, Literal-2 is substituted for Literal-1. The substitution is terminated either when a character other

than Literal-1 is encountered or when the right-hand boundary of the data item is reached (fig. 8.70).

c. *FIRST*. If the FIRST option is specified, only the first occurrence of Literal-1 is replaced by Literal-2 (fig. 8.71).

d. *UNTIL FIRST*. If the UNTIL FIRST option is specified, the substitution of Literal-2 terminates as soon as the first Literal-1 is encountered or until the right-hand boundary of the data item is reached (fig. 8.72).

EXAMINE ACCOUNT TALLYING UNTIL FIRST "A".

Data in ACCOUNT before EXAMINE	Data in ACCOUNT after EXAMINE	Contents of TALLY after EXAMINE
1296AA10	no change	000004
AA123456		000000
ØØØA123	in	000003
FRTS987123		000010
+A1234	data	000001

Figure 8.65 Example—Examine Statement—Format 1.

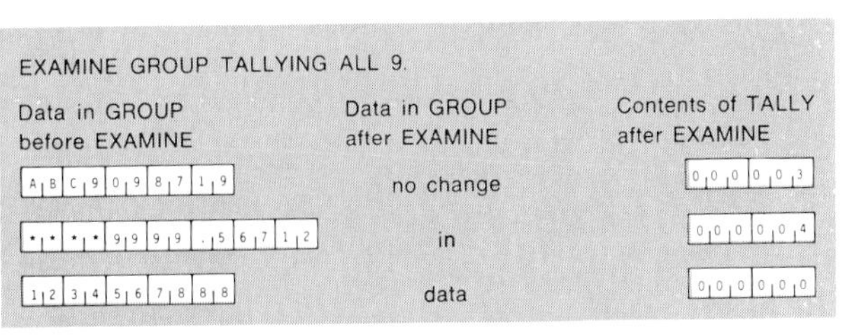

EXAMINE GROUP TALLYING ALL 9.

Data in GROUP before EXAMINE	Data in GROUP after EXAMINE	Contents of TALLY after EXAMINE
ABC9098719	no change	000003
••••9999.56712	in	000004
1234567888	data	000000

Figure 8.66 Example—Examine Statement—Format 1.

EXAMINE PART-NUMBER TALLYING LEADING "Z".

Data in PART-NUMBER before EXAMINE	Data in PART-NUMBER after EXAMINE	Contents of TALLY after EXAMINE
ZZZZ90982	no change	000004
ZZ123@ZZ		000002
ZJOHNDOE	in	000001
YOUZZZZZQ		000000
12345	data	000000

Figure 8.67 Example—Examine Statement—Format 2.

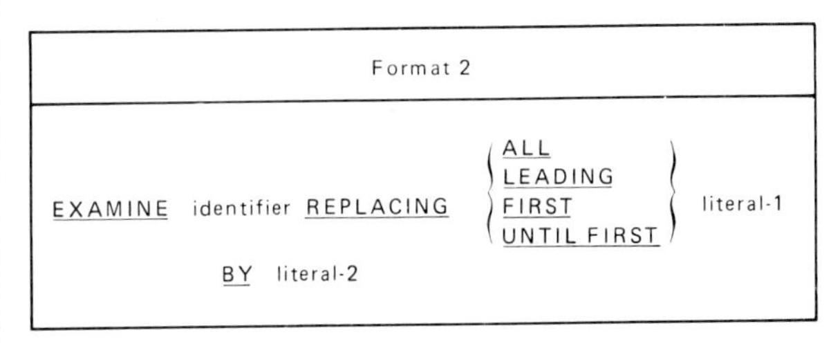

Format 2

EXAMINE identifier REPLACING { ALL / LEADING / FIRST / UNTIL FIRST } literal-1

BY literal-2

Figure 8.68 Format—Examine Statement—Format 2.

EXAMINE GROUP TALLYING ALL 9 REPLACING BY ".".

Data in GROUP before EXAMINE	Data in GROUP after EXAMINE	Contents of TALLY after EXAMINE
987612345	•87612345	000001
9999999	•••••••	000007
90909090	•0•0•0•0	000004

Figure 8.69 Example—Examine Statement—Format 1.

EXAMINE PART-NUMBER TALLYING LEADING "M" REPLACING BY "T".

Data in PART-NUMBER before EXAMINE	Data in PART-NUMBER after EXAMINE	Contents of TALLY after EXAMINE
MMMM9876	TTTT9876	000004
MM@12@45MM	TT212@45MM	000002
MJOHNDOE	TJOHNDOE	000001
YOUMMM	YOUMMM	000000
909876	909876	000000

Figure 8.70 Example—Examine Statement—Format 1.

EXAMINE LOCATION REPLACING FIRST "E" BY ",".

Data in LOCATION before EXAMINE	Data in LOCATION after EXAMINE	TALLY after EXAMINE
EEENODATA	,EENODATA	not
BILLEMARY	BILL,MARY	
SUEEBILLEMARY	SU,EBILLEMARY	affected

Figure 8.71 Example—Examine Statement—Format 2.

EXAMINE ACCOUNT TALLYING UNTIL FIRST "A" REPLACING BY "Z".

Data in ACCOUNT before EXAMINE	Data in ACCOUNT after EXAMINE	Contents of TALLY after EXAMINE
09876AB@	ZZZZZAB@	000005
A8761AWE	A8761AWE	000000
JOHN01ZFFF	ZZZZZZZZZZ	000010

Figure 8.72 Example—Examine Statement—Format 1.

Arithmetic Statements

Five arithmetic verbs are provided for in COBOL to perform the necessary arithmetic functions: ADD, SUBTRACT, MULTIPLY, DIVIDE, and COMPUTE. ADD, SUBTRACT, MULTIPLY, and DIVIDE are arithmetic verbs used to perform individual arithmetic operations. The fifth verb, COMPUTE, permits the programmer to combine arithmetic operations into arithmetic expressions in a formula style using the various arithmetic operators.

Arithmetic expressions can be composed of an identifier of a numeric elementary item, a numeric literal, those identifiers and literals separated by arithmetic operators, or an arithmetic expression enclosed in parentheses.

Rules Governing the Use of Arithmetic Statements

1. All identifiers used in arithmetic statements must represent elementary numeric items that are defined in the Data Division.
2. The identifier that follows GIVING may contain editing symbols if it is not itself involved in the computation.
3. All literals used in arithmetic statements must be numeric.
4. The maximum size of a numeric literal or identifier is 18 decimal digits.
5. The maximum size of a result of a computation is 18 decimal digits after decimal alignment.
6. Decimal alignment is supplied automatically throughout the computation in accordance with individual PICTURE clauses of the results and operands.
7. The GIVING option applies to all arithmetic statements except the COMPUTE statement.

Giving

If the GIVING option is used, the value of the identifier following the word GIVING will be made equal to the calculated value of the arithmetic expression. This identifier may be an edited numeric item but must not be involved in the computation. *If the GIVING option is not used, the replaced operand in the arithmetic calculation must not be a literal.* (See figure 8.73.)

Rounded

The ROUNDED option is used when the number of places in the calculated result exceeds the number of places allocated for the sum, difference, product, quotient, or computed result (fig. 8.74).

Rules Governing the Use of the Rounded Statement

1. Truncation (dropping of excess digits) is determined by the identifier associated with the result.
2. The least significant digit in the result is increased by 1 if the most significant digit of the excess is greater than or equal to 5.
3. If the option is not specified, truncation occurs without rounding after decimal alignment.
4. Rounding of a computed negative result occurs by rounding the absolute value of the computed result and making the final result negative (fig. 8.75).

```
ADD A, B, C, D, GIVING R.
SUBTRACT A, B, C, FROM D, GIVING R.
MULTIPLY A BY B, GIVING R.
DIVIDE A INTO B, GIVING R.
```

Figure 8.73 Example—Giving Option.

```
MULTIPLY QUANTITY BY PRICE,
    GIVING AMOUNT, ROUNDED.
```

Figure 8.74 Example—Rounded Option.

	Item to Receive Calculated Result		
Calculated Result	PICTURE	Value After Rounding	Value After Truncating
-12.36	S99V9	-12.4	-12.3
8.432	9V9	8.4	8.4
35.6	99V9	35.6	35.6
65.6	99V	66	65
.0055	V999	.006	.005

Figure 8.75 Rounding or Truncation of Calculations.

Size Error

The SIZE ERROR option is used where the computed result after decimal alignment exceeds the number of integral places in the format of the identifier associated with the result (fig. 8.76).

Figure 8.76 Example—On Size Error Option.

```
ADD OVERTIME TO REGULAR-EARNINGS
    ON SIZE ERROR, GO TO OVERSIZE-SUM.
```

Rules Governing the Use of the Size Error Statement

1. If the ROUNDED option is specified, rounding takes place before checking for size error.
2. If the SIZE ERROR option is not specified, and a size error condition arises, the result is unpredictable.
3. If the SIZE ERROR option is specified, and a size error condition does arise, the value of the result is not altered, and a series of imperative statements specified for the condition will be executed.
4. An arithmetic statement written with the SIZE ERROR option becomes a conditional statement and is prohibited in context where only imperative statements are allowed.
5. Division by zero always causes a size error condition.
6. The SIZE ERROR option applies only to final results, and no assumption of an answer can be made if the size error condition arises in an intermediate result.

(See figure 8.77.)

Corresponding

The CORRESPONDING option allows the user to perform computations on elementary items of the same name simply by specifying the group item to which they belong. This option can be used with the ADD or SUBTRACT verb whereby elementary data items within are added to or subtracted from elementary items of the same name in another group. (See fig. 8.83.)

The rules for MOVE CORRESPONDING apply to the ADD and SUBTRACT CORRESPONDING statements. When the SIZE ERROR option is used with the CORRESPONDING option, the size error test is made after the completion of all add or subtract operations. If any of the additions or subtrac-

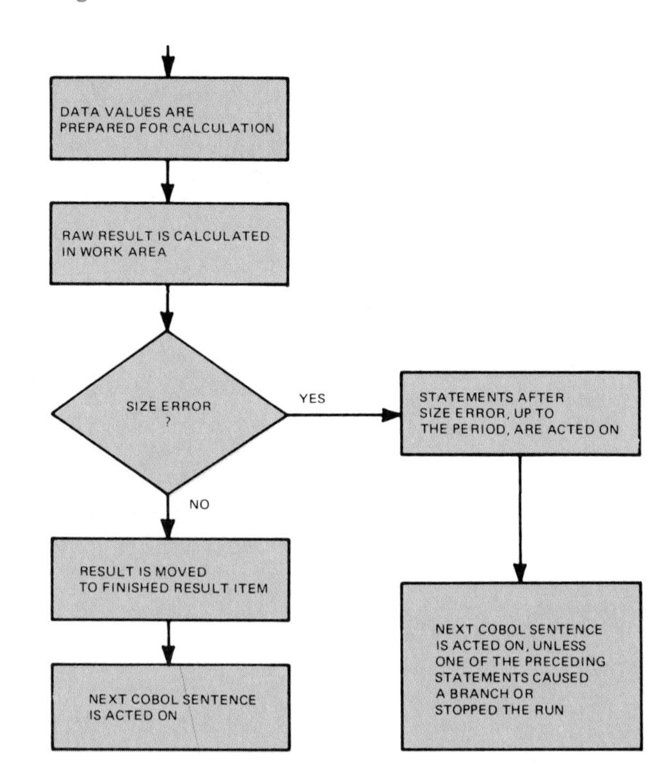

Figure 8.77
Flow of Control Through Arithmetic Statements that Control on Size Error.

tions produce a size error condition, the resultant field remains unchanged for that particular item, and the imperative statement specified in the SIZE ERROR option is executed.

Add

The ADD statement is used to specify the addition of the numeric values of two or more items and to substitute the sum for the current value of an item (fig. 8.78).

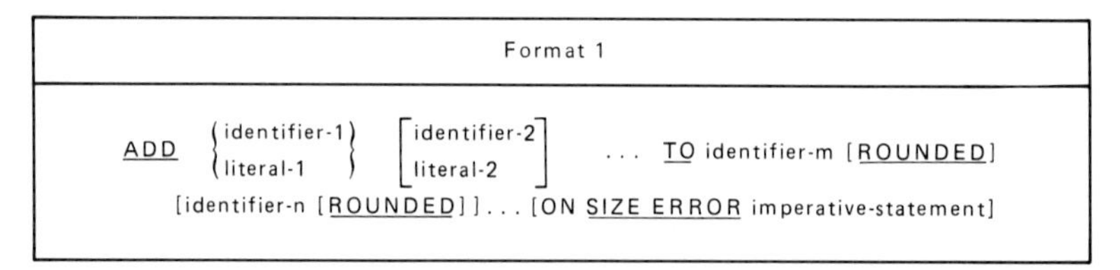

Format 1

ADD {identifier-1 | literal-1} [identifier-2 | literal-2] . . . TO identifier-m [ROUNDED]
[identifier-n [ROUNDED]] . . . [ON SIZE ERROR imperative-statement]

Figure 8.78 Format—Add Statement—Format 1.

Rules Governing the Use of the Add Statement

1. An ADD statement must contain at least two addends (elementary numeric items).
2. When the TO option is used, the values of the operands (literals and identifiers) preceding the word TO are added together. The sum is then added to the current value in each Identifier-m, Identifier-n, etc., and the resultant sum replaces the current values of Identifier-m, Identifier-n, etc. (fig. 8.79).
3. The resultant sums are not edited with the TO option.
4. The word GIVING may not be written in the same statement as TO.

5. When the GIVING option is used, there must be at least two operands (literals and/or identifiers) preceding the word GIVING. The sum may be edited according to the items PICTURE and may be either an elementary numeric item or an elementary numeric edited item. (See figures 8.80, 8.81, 8.82.)

```
ADD 5 TO FIELD-A.
The data in FIELD-A before the ADD is executed:          0 0 1 5 0 5
The data in FIELD-A after the ADD has been executed:     0 0 1 5 5 5

    ADD 125.25, FIELD-B, TO FIELD-C.
    Data in the fields before the ADD is executed:
    FIELD-B          6 8 5
    FIELD-C    0 1 0 0 0 0
    Data in the fields after the ADD has been executed:
    FIELD-B    Unchanged
    FIELD-C    0 2 9 3 7 5
```

Figure 8.79 Example—Add Statement—Format 1.

```
                              Format 2

ADD  {identifier-1}  {identifier-2}  [identifier-3]  ...
     {literal-1   }  {literal-2   }  [literal-3   ]
   GIVING identifier-m [ROUNDED]  [ON SIZE ERROR imperative-
                                                    statement]
```

Figure 8.80 Format—Add Statement—Format 2—Giving Option.

```
ADD FIELD-E, 5, FIELD-F GIVING FIELD-G.
Data in the fields before the ADD is executed:
FIELD-E        0 0 0 2 5 0
FIELD-F              1 0 0
FIELD-G        - 0 0 5 0 0
Data in the fields after the ADD has been executed:
FIELD-E        Unchanged
FIELD-F        Unchanged
FIELD-G        0 0 0 6 0 0
```

Figure 8.81 Example—Add Statement—Format 2.

```
ADD AMOUNT, DISCOUNT, SUB-TOTAL GIVING GRAND-TOTAL
ROUNDED, ON SIZE ERROR PERFORM ERROR-ROUTINE.
Data in the fields before the ADD is executed:
AMOUNT           0 5 9 5 4
DISCOUNT           - 0 5 9
SUB-TOTAL      1 0 2 0 0 1
GRAND-TOTAL    0 0 5 0 0 6
Data in the fields after the ADD has been executed:
AMOUNT         Unchanged
DISCOUNT       Unchanged
SUB-TOTAL      Unchanged
GRAND-TOTAL    0 1 0 7 9 0
```

Figure 8.82 Example—Add Statement—Format 2.

6. Decimal points are aligned in all ADD operations.
7. When the CORRESPONDING option is used, elementary data items within Identifier-1 are added to and stored in the corresponding data items in Identifier-2. Identifier-1 and Identifier-2 must be group items (figs. 8.83, 8.84).
8. When the SIZE ERROR option is specified for CORRESPONDING option items, the test is made after the completion of all add operations. If any of the additions produce a size error condition, the resultant field for that addition remains unchanged, and the imperative statement specified in the SIZE ERROR option is executed.

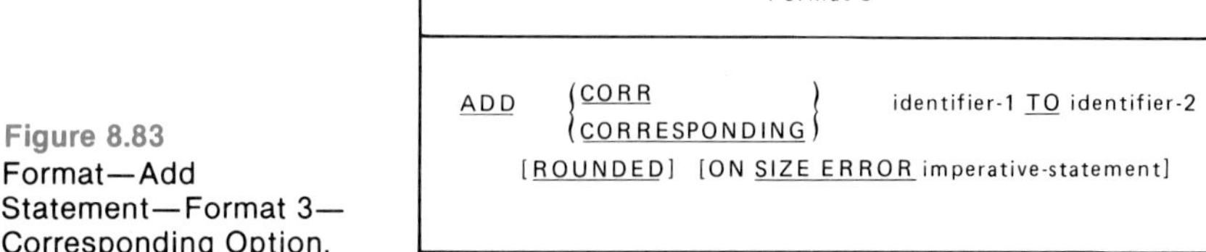

Format 3
ADD
CORR
CORRESPONDING
identifier-1 TO identifier-2
[ROUNDED] [ON SIZE ERROR imperative-statement]

Figure 8.83 Format—Add Statement—Format 3—Corresponding Option.

```
ADD CORRESPONDING EMPLOYEE-RECORD TO PAYROLL-CHECK.

01 EMPLOYEE-RECORD                 01 PAYROLL-CHECK
   02 EMPLOYEE-NUMBER                 02 EMPLOYEE-NUMBER
      03 PLANT-LOCATION                  03 CLOCK-NUMBER
      03 CLOCK-NUMBER                    03 FILLER
         04 SHIFT-CODE                02 DEDUCTIONS
         04 CONTROL-NUMBER               03 FICA-RATE
   02 INCOME                             03 WITHHOLDING-TAX
      03 HOURS-WORKED                    03 PERSONAL-LOANS
      03 PAY-RATE                     02 INCOME
   02 FICA-RATE                          03 HOURS-WORKED
   02 DEDUCTIONS                         03 PAY-RATE
                                      02 NET-PAY
                                      02 EMPLOYEE- NAME
                                         03 SHIFT-CODE
```

According to the ADD CORRESPONDING rules, the following operations would take place:

1. HOURS-WORKED of INCOME. } WOULD BE ADDED TO
2. PAY-RATE of INCOME. } CORRESPONDING ITEMS.

The following items would not be added for the reasons stated:

1. EMPLOYEE-NUMBER (item is not elementary in both groups.)
2. PLANT-LOCATION of EMPLOYEE-NUMBER (Name does not appear in PAYROLL-CHECK.)
3. CLOCK-NUMBER of EMPLOYEE-NUMBER (Item is not elementary in one group.)
4. SHIFT-CODE of CLOCK-NUMBER of EMPLOYEE-NUMBER of EMPLOYEE-RECORD (Qualification is not identical in PAYROLL-CHECK.)
5. CONTROL-NUMBER of CLOCK-NUMBER of EMPLOYEE-NUMBER of EMPLOYEE-RECORD (Name does not appear in PAYROLL-CHECK.)
6. INCOME (Item is not elementary in both groups.)
7. FICA-RATE of EMPLOYEE-RECORD (Qualification is not identical in PAYROLL-CHECK.)
8. DEDUCTIONS (Item is not elementary in both groups.)

Figure 8.84 Example—Add Statement—Format 3—Corresponding Option.

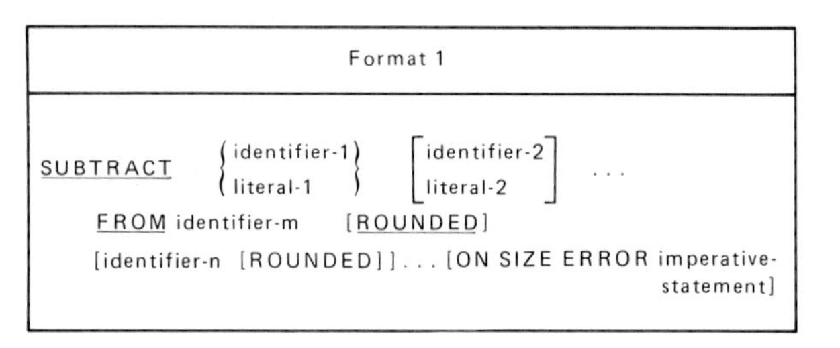

Format 1
SUBTRACT {identifier-1 / literal-1} [identifier-2 / literal-2] . . .
FROM identifier-m [ROUNDED]
[identifier-n [ROUNDED]] . . . [ON SIZE ERROR imperative-statement]

Figure 8.85 Format—Subtract Statement—Format 1.

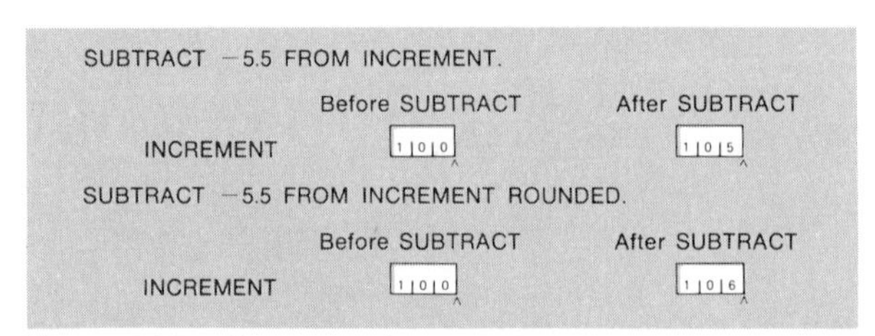

Figure 8.86 Example—Subtract Statement—Format 1.

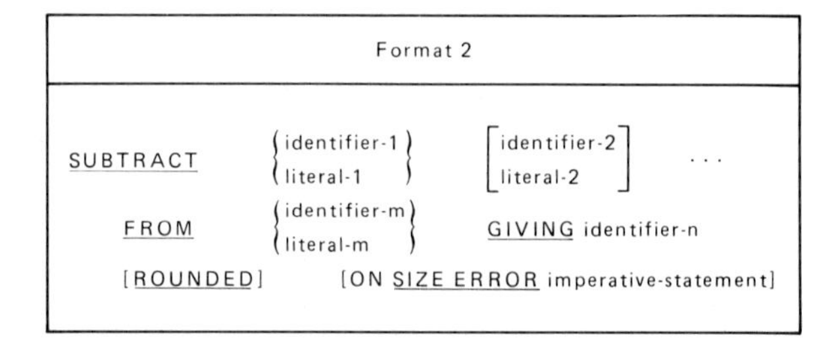

Format 2
SUBTRACT {identifier-1 / literal-1} [identifier-2 / literal-2] . . .
FROM {identifier-m / literal-m} GIVING identifier-n
[ROUNDED] [ON SIZE ERROR imperative-statement]

Figure 8.87 Format—Subtract Statement—Format 2—Giving Option.

Subtract

The SUBTRACT statement is used to specify the subtraction of one or more numeric values from one or more specified items and to set the values of one or more items equal to the results (fig. 8.85).

Rules Governing the Use of the Subtract Statement

1. All operands (literals and identifiers) must be elementary numeric items.
2. All values of the operands (literals and identifiers) preceding the word FROM are added together, and this total is subtracted from the value of the Identifier or Literals-m, -n, etc.; the difference replaces the value of Identifier-m or Identifier-n, etc., if GIVING option is used. (See figure 8.86.)
3. Each identifier must refer to an elementary numeric item, except the identifier following the word GIVING, which may be a numeric edited item. (See figures 8.87, 8.88.)
4. When the CORRESPONDING option is used, elementary data items within Identifier-1 are subtracted from corresponding data items in Identifier-2, and the differences are stored in the corresponding Identifier-2 data items. (See figure 8.89.)

SUBTRACT TOTAL-DEDUCTIONS FROM GROSS-PAY GIVING NET-PAY.

	Before SUBTRACT	After SUBTRACT
TOTAL-DEDUCTIONS	02 1^22	Unchanged
GROSS-PAY	011 4^76	Unchanged
NET-PAY	-89 4^00	009 3^54

Figure 8.88 Example—Subtract Statement—Format 2—Giving Option.

```
Format 3

SUBTRACT {CORR         } identifier-1 FROM identifier-2
         {CORRESPONDING}
    [ROUNDED]  [ON SIZE ERROR imperative-statement]
```

Figure 8.89 Format—Subtract Statement—Format 3—Corresponding Option.

5. When the CORRESPONDING option is used in conjunction with the SIZE ERROR option and a size error condition arises, the result for SUBTRACT is analogous to that of ADD.

Multiply

The MULTIPLY statement is used to specify the multiplication of two numeric values and to substitute the resulting product for the current value of an item (fig. 8.90).

Rules Governing the Use of the Multiply Statement

1. All operands (literals and identifiers) must be elementary numeric items.
2. The value of Identifier-1 or Literal-1 is multiplied by the value of Identifier-2 or Literal-2. The value of Identifier-2 is replaced by the product (fig. 8.91).
3. Each identifier must refer to an elementary numeric item, except the identifier following the word GIVING, which may be a numeric edited item. (See figures 8.92, 8.93.)

```
Format 1

MULTIPLY {identifier-1} BY identifier-2  [ROUNDED]
         {literal-1   }
    [ON SIZE ERROR imperative-statement]
```

Figure 8.90 Format—Multiply Statement—Format 1.

MULTIPLY BASE BY RATE.

	Before MULTIPLY	After MULTIPLY
BASE	-50 0^00	Unchanged
RATE	000 0^06	-03 0^00

Figure 8.91 Example—Multiply Statement—Format 1.

```
Format 2

MULTIPLY {identifier-1} BY {identifier-2} GIVING identifier-3
         {literal-1   }    {literal-2   }
    [ROUNDED]       [ON SIZE ERROR imperative-statement]
```

Figure 8.92 Format—Multiply Statement—Format 2—Giving Option.

MULTIPLY BASE BY RATE GIVING NEW-BASE.

	Before MULTIPLY	After MULTIPLY
BASE	-50 0^00	Unchanged
RATE	000 0^06	Unchanged
NEW-BASE	-70 0^00	-03 0^00

Figure 8.93 Example—Multiply Statement—Format 2—Giving Option.

Divide

The DIVIDE statement specifies the division of one numeric data item by another and uses the result to replace the value of an item (fig. 8.94).

Rules Governing the Use of the Divide Statement

1. All operands (literals and identifiers) must represent elementary numeric items. (See figure 8.95.)
2. When the INTO option is used without the GIVING option, the value of Identifier-1 (or Literal-1) is divided into the value of Identifier-2. The value of the dividend (Identifier-2) is replaced by the quotient.

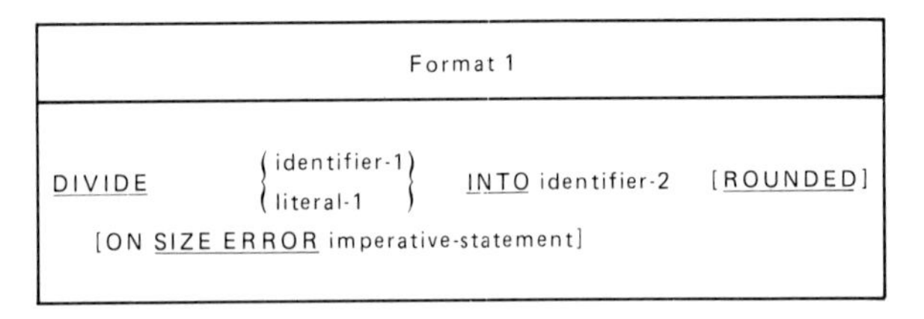

Figure 8.94 Format—Divide Statement—Format 1.

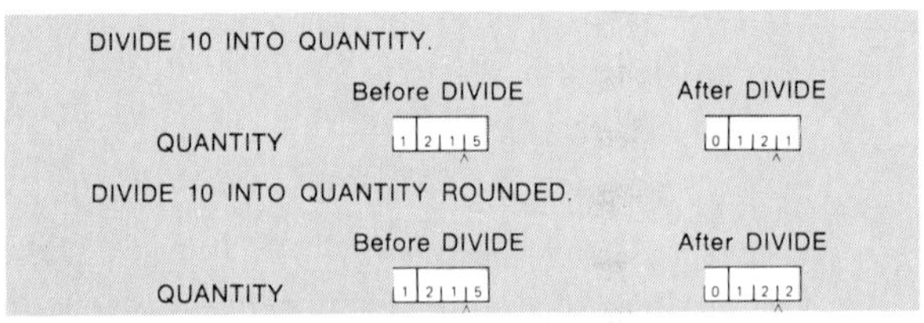

Figure 8.95 Example—Divide Statement—Format 1.

3. When the BY or INTO option is used in conjunction with the GIVING and REMAINDER options, the value of Identifier-1 or Literal-1 is divided into or by Identifier-2 or Literal-2, and the quotient is stored in Identifier-3, with the remainder being optionally stored in Identifier-4. A remainder is the result of subtracting the product of the quotient and the divisor from the dividend. If the ROUNDED option is specified, the quotient is rounded after the remainder is determined. (See figure 8.96.)
4. Each identifier must refer to an elementary numeric item, except the identifier following the word GIVING, which may be a numeric edited item (fig. 8.97).
5. A division by zero will always result in a size error condition.

(*Note:* In all four arithmetic statements, if the word GIVING is not used, the last operand [one replaced by the result] must not be a literal.)

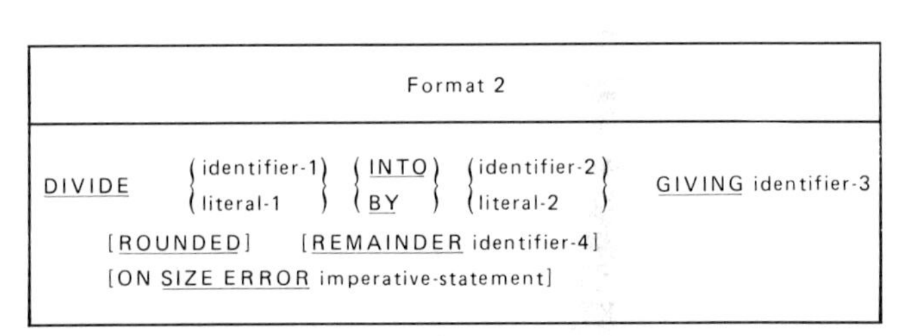

Figure 8.96 Format—Divide Statement—Format 2—Giving Option.

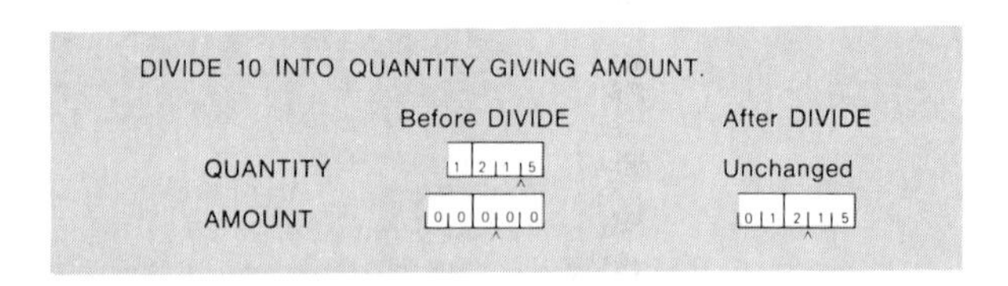

Figure 8.97 Example—Divide Statement—Format 2—Giving Option.

Compute

The COMPUTE statement specifies the use of an arithmetic expression, an identifier, or a numeric literal for a series of arithmetic operations (fig. 8.98).

Rules Governing the Use of the Compute Statement

1. The arithmetic expression option permits the use of a meaningful combination of identifier, numeric literal, figurative constant ZERO, joined by the operators in figure 8.99.

   This permits the user to combine arithmetic operations without the restrictions imposed by the arithmetic statements ADD, SUBTRACT, MULTIPLY, and DIVIDE. (See figures 8.100, 8.101.)

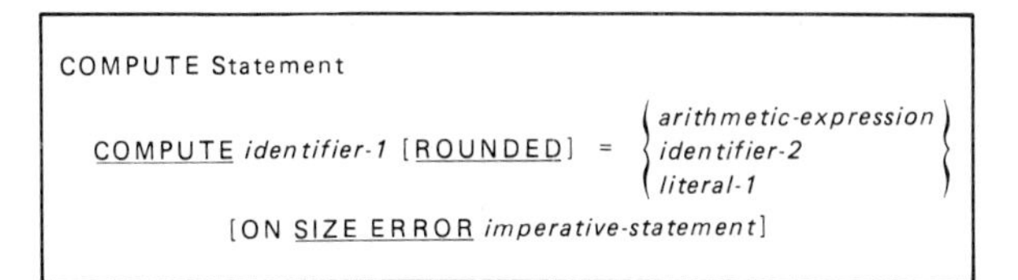

COMPUTE Statement
COMPUTE identifier-1 [ROUNDED] = { arithmetic-expression / identifier-2 / literal-1 }
[ON SIZE ERROR imperative-statement]

Figure 8.98 Format—Compute Statement.

Operator	Function
+	Addition
−	Subtraction
*	Multiplication
/	Division
**	Exponentiation

Figure 8.99 Arithmetic Operators and Their Functions.

Arithmetic Operator	Meaning	COBOL Example	Arithmetic Example
+	Addition	AMT1 + AMT2	2 + 4 = 6
−	Subtraction	BAL − AMT	7 − 5 = 2
*	Multiplication	AMT * 4	16 x 4 = 64
/	Division	BAL / 4	'32 / 4 = 8
−	The effect of multiplication by −1.	− QUAN	− 16 = 16x −1 = −16

Figure 8.100 Arithmetic Operators, Meanings, and Examples.

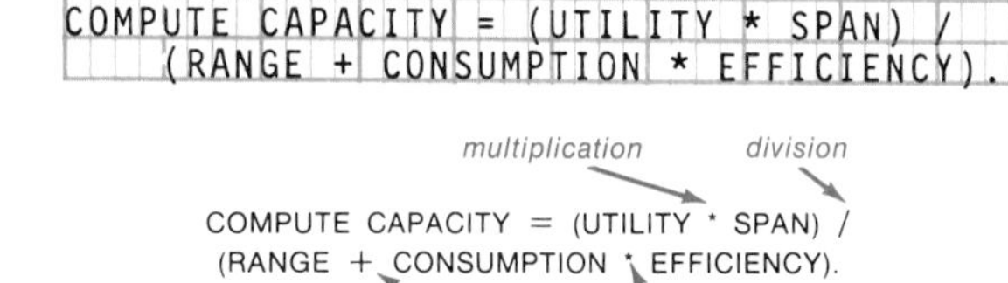

COMPUTE CAPACITY = (UTILITY * SPAN) /
(RANGE + CONSUMPTION * EFFICIENCY).

COMPUTE CAPACITY = (UTILITY * SPAN) /
(RANGE + CONSUMPTION * EFFICIENCY).

Figure 8.101 Example—Compute Statement.

2. Operators must be preceded and followed by one or more spaces.
3. If exponentiation is desired, the COMPUTE statement must be used.
4. When the Identifier-2 or Literal-1 option is used, the result is the same as a MOVE operation. The value of Identifier-1 is made equal to the value of Identifier-2 or Literal-1.
5. The value of the result of the calculation must be written to the left of the equal sign as the item represented by Identifier-1.
6. Identifier-1 must be an elementary numeric item or an elementary numeric edited data item. The calculated value is placed here and is edited according to the Identifier-1 item picture.
7. The ROUNDED and SIZE ERROR options apply also to the COMPUTE statement.

The COMPUTE statement permits most arithmetic operations in much the same manner as specifying arithmetic verbs. (See figure 8.102.)

COMPUTE YEARS = MONTHS / 12.

	Before execution	After execution
MONTHS	1 0 0 3	Unchanged
YEARS	0 1 2	0 8 3

COMPUTE YEARS ROUNDED = MONTHS / 12.

	Before execution	After execution
MONTHS	1 0 0 3	Unchanged
YEARS	0 0 1	0 8 4

Figure 8.102
Example—Compute Statement.

Rules for the Sequence of Calculation of Arithmetic Expressions Containing a Combination of Operators

1. Parenthetical arithmetic expressions are calculated first.
2. All exponentiations are performed next.
3. Multiplication and division operations are then calculated, from left to right.
4. Addition and subtraction operations are performed last, from left to right. (See figure 8.103.)

Arithmetic Statements / Allowable Options	GIVING variable-name	variable-name ROUNDED	SIZE ERROR statement	REMAINDER variable-name
ADD {identifier-1 / numeric-literal-1} [identifier-2 / numeric-literal-2] . . . TO identifier-m	X*	X	X	
SUBTRACT {identifier-1 / numeric-literal-1} [identifier-2 / numeric-literal-2] . . . FROM identifier-m	X	X	X	
MULTIPLY {identifier-1 / numeric-literal-1} BY identifier-2	X	X	X	
DIVIDE {identifier-1 / numeric-literal-1} INTO identifier-2	X	X	X	X
COMPUTE identifier-1 = {identifier-2 / numeric-literal-1 / arithmetic-expression}		X	X	

*The reserved word TO is omitted when the GIVING option is specified.

Figure 8.103 Summary of Arithmetic Statements and Their Options.

Sequence Control Statements

The SEQUENCE CONTROL statements are designed to specify the sequence in which the various source program instructions are to be executed. Statements, sentences, and paragraphs of the Procedure Division are executed normally in the sequence in which they are written, except when one of these sequence control verbs is encountered. The five verbs that follow are used for procedure branching operations. The GO TO and PERFORM verbs interrupt the normal sequence and transfer control to another point in the program. The ALTER verb is used in conjunction with the GO TO verb to modify a branch instruction. The EXIT verb is used with the PERFORM verb in conditional exits from paragraphs. The STOP verb is used to halt execution of the program.

Go To

The subject of GO TO-less programming still stirs considerable controversy. Most structured program supporters claim that GO TO statements lead to difficulty in debugging, modifying, understanding, and proving programs. GO TO advocates argue that this statement, used correctly, need not lead to problems, and that it provides a natural, straightforward solution to common programming procedures.

No special effort is required to "eliminate GO TOs," which sometimes has mistakenly been considered the goal of structured programming. There are indeed good reasons for not wanting GO TOs, but no extra effort is required to "avoid" them; they just never occur when standard control logic structures are used. Naturally, if the chosen program language lacks essential control logic structures, they have to be simulated, and that does involve GO TOs. But this can be accomplished by carefully controlled means, such as the use of GO TO with the DEPENDING ON option in CASE structure problems.

Many experienced COBOL programmers suggest that the use of the ALTER statement in connection with the GO TO verb is so difficult to document adequately, and so likely to lead to serious difficulties in program test and maintenance, that the ALTER verb should be used sparingly or not at all. To accomplish the same purpose, there are alternate means that are easier to understand and simpler to maintain.

There are situations in which the use of GO TOs may improve reading, compared with other ways of expressing a procedure. The GO TOs may be used to disable error routines that prevent further processing when certain types of interrupts occur. Good judgment should be used to determine whether the maintainability of the program is improved by using GO TOs in the particular situation.

The GO TO statement provides a means of transferring control conditionally or unconditionally from one part of the program to another (fig. 8.104).

Format 1
GO TO procedure-name-1

Figure 8.104 Format Go To Statement—Format 1.

The PERFORM statement also causes a branch out of normal sequence, but in addition provides a return to the program.

Rules Governing the Use of Go To Statements

Unconditional

1. A procedure-name (name of paragraph or section) in the Procedure Division must follow the GO TO statement.
2. If the procedure-name is omitted, a paragraph-name must be assigned. The paragraph-name must be the only name in a paragraph, and must be modified by an ALTER statement prior to the execution of the GO TO statement.
3. If the procedure-name is omitted, and the GO TO sentence is not preset by an ALTER statement prior to its execution, erroneous processing will occur.
4. The GO TO statement can be used only as the final sentence in the sequence in which it appears.
5. A procedure-name may be the name of the procedure of which the GO TO statement is part. It is permissible to branch from a point in the paragraph back to the beginning of the procedure. (See figure 8.105.)

Figure 8.105
Examples—Go To Statement—Format 1.

```
GO TO DETERMINE-TYPE-RECORD.
GO TO DEV-MR.
```

Depending On (Conditional Go To)

The DEPENDING ON option permits the multiple-branch type of operations according to the value of the current value of the identifier (fig. 8.106).

1. The identifier must have a positive integral value.
2. Control is passed to the 1st, 2nd. . . . . nth procedure-name as the value name of the identifier is 1, 2 . . . . (See figure 8.107.)

Figure 8.106
Format—Go To Statement—Format 2—Depending On Option.

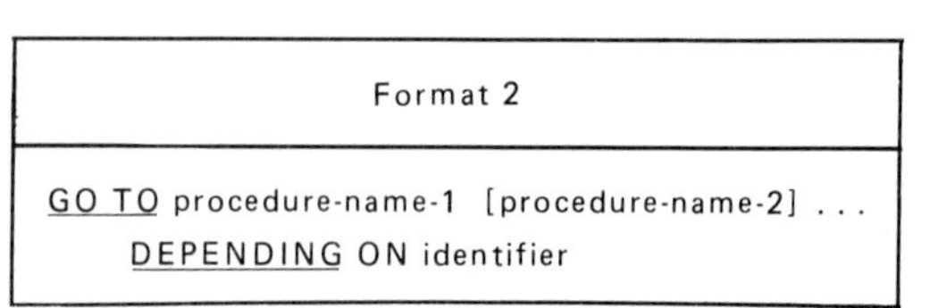

Format 2

GO TO procedure-name-1 [procedure-name-2] . . .
DEPENDING ON identifier

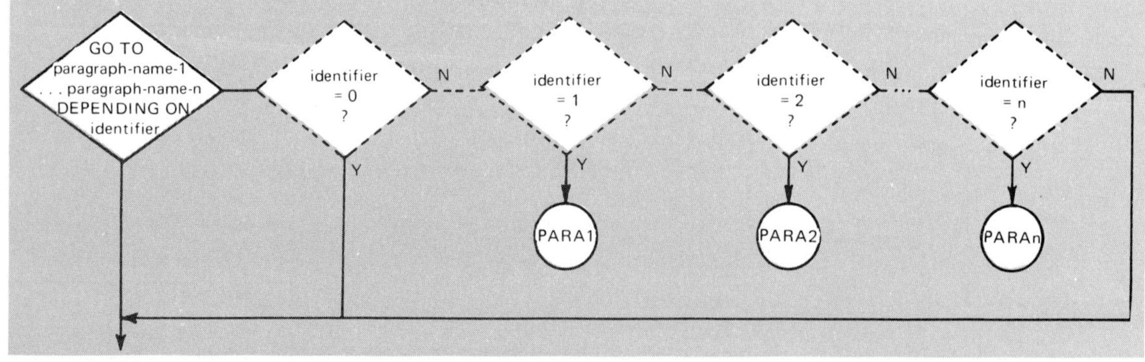

Figure 8.107 Flow of Logic—Go To Statement—Format 2—Depending On Option.

3. The identifier must have a range of values starting at 1 and continuing successively upward.
4. If the value of the identifier is outside of the range of 1 through n, no branch occurs, and control passes to the first statement after the GO TO statement. (See figures 8.108, 8.109.)

In a branching operation, after the transfer to the particular point in the program is executed, normal flow of control is resumed at the beginning of the particular procedure.

```
GO TO   RECEIPTS
        SHIPMENTS
        CUSTOMER-ORDERS
            DEPENDING ON TRANSACTION-CODE.
```

In the object program, if TRANSACTION-CODE contains the value 3, control will be transferred to CUSTOMER-ORDERS (the third procedure name in the series.)

Figure 8.108 Example—Go To Statement—Format 2—Depending On Option.

```
GO TO   DEPOSIT
        WITHDRAWAL
        INTEREST
            DEPENDING ON TRAN-CODE.
```

The above GO TO statement transfers control to DEPOSIT, WITHDRAWAL, or INTEREST depending on the number in TRAN-CODE. The following statements are equivalent to the above GO TO..DEPENDING ON statement.

```
IF TRAN-CODE = 1,  GO TO DEPOSIT.
IF TRAN-CODE = 2,  GO TO WITHDRAWAL.
IF TRAN-CODE = 3,  GO TO INTEREST.
IF TRAN-CODE IS LESS THAN 1
    OR GREATER THAN 3,
THEN
    GO TO the next sentence
ELSE
    continue processing.
```

Figure 8.109 Example—Go To Statement—Format 2—Depending On Option.

Alter The ALTER statement is used to modify the effect of the unconditional GO TO statement elsewhere in the program thus changing the sequence of operations to be performed (fig. 8.110).

Rules Governing the Use of the Alter Statement

1. A GO TO statement to be altered must be written as a single paragraph consisting solely of the unconditional GO TO statement preceded by a paragraph-name (figs. 8.111, 8.112).

```
ALTER procedure-name-1 TO [PROCEED TO] procedure-name-2
    [procedure-name-3 TO [PROCEED TO] procedure-name-4] . . .
```

Figure 8.110 Format—Alter Statement.

Format 3
GO TO.

Figure 8.111 Format—Go To Statement—Format 3.

```
MODIFIER-GO. GO TO.
```

Figure 8.112 Example—Go To Statement—Format 3. Giving Option.

2. The ALTER statement replaces the procedure-name specified in the GO TO statement (if any) by the procedure-name specified in the ALTER statement (fig. 8.113).

Perform The PERFORM statement provides a method of temporarily transferring control from the normal sequence of procedure execution in order to execute

```
PARAGRAPH-1.
         GO TO BYPASS-PARAGRAPH.
PARAGRAPH-1A.
         |
BYPASS-PARAGRAPH.
         |
         ALTER PARAGRAPH-1 TO PROCEED TO PARAGRAPH-2.
         |
PARAGRAPH-2.
```

Before the ALTER statement is executed, when control reaches PARAGRAPH-1, the GO TO statement transfers control to BYPASS- PARAGRAPH. After execution of the ALTER statement, however, when control reaches PARAGRAPH-1, the GO TO statement transfers control to PARAGRAPH-2.

Figure 8.113 Example—Alter Statement.

some other procedure a specified number of times or until a specified condition is satisfied (fig. 8.114). At the conclusion of the execution of the procedures, control is transferred back to the statement immediately following the point from which the transfer was made.

PERFORM has several different formats which vary in complexity. In its simplest form, the procedure referred to is executed once each time the PERFORM is encountered. Other formats provide repetitive execution using one or more optional controls to control the "looping."

Rules Governing the Use of the Perform Statement

1. When a procedure is performed, the PERFORM statement transfers the sequence control to the first statement in procedure name-1 and also provides for the return of the control. The point at which the control is returned to the main program depends on the structure of the procedure being executed (fig. 8.115).
2. If procedure-name-1 is a paragraph-name, and procedure-name-2 is not specified, control is returned after the execution of the last statement of procedure-name-1 paragraph.

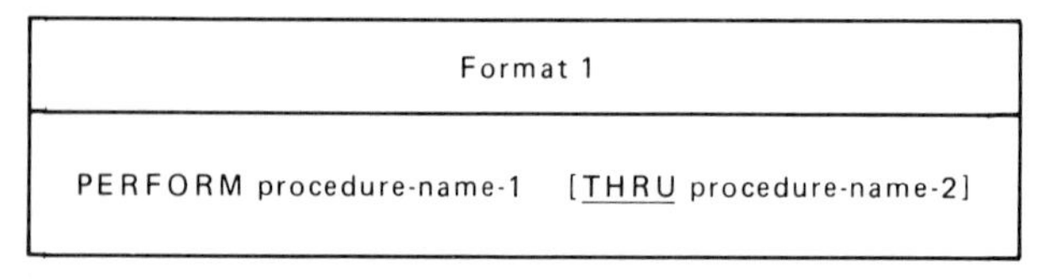
Format 1

PERFORM procedure-name-1 [THRU procedure-name-2]

Figure 8.114 Format—Perforn Statement—Format 1—Thru Option.

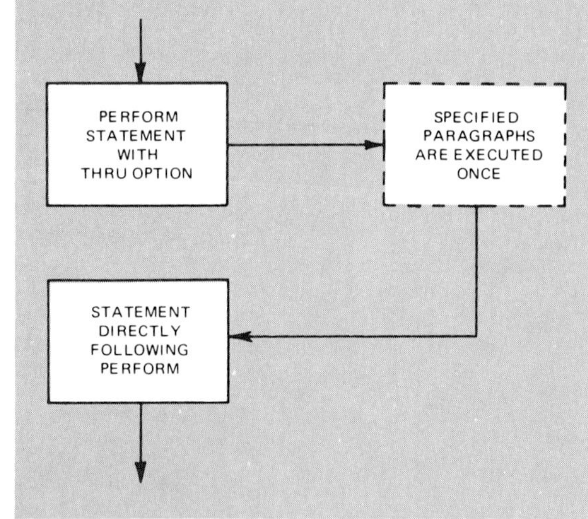

Figure 8.115 Flow of Logic—Perform Statement—Format 1—Thru Option.

3. If procedure-name-1 is the name of a section, and procedure-name-2 is not specified, control is returned after the last statement of the last paragraph in procedure-name-1 section.
4. If procedure-name-2 is specified, control is transferred after the last statement of the procedure-name-2 paragraph.
5. If procedure-name-2 is specified and is a section, control is transferred after the last statement of the procedure-name-2 section.
6. When procedure-name-2 is specified, the relationship between procedure-name-1 and procedure-name-2 must exist. Execution must proceed from procedure-name-1 throughout the last statement of procedure-name-2. GO TO and PERFORM statements are permitted between procedure-name-1 and the last statement in procedure-name-2 provided that the sequence ultimately returns prior to the final statement in procedure-name-2.
7. The last sentence in the procedure (referred to in 6 above) must not be an unconditional GO TO sentence. If the logic of the procedure requires a conditional exit prior to the last sentence, the EXIT verb is used to satisfy the requirement. An EXIT statement consists solely of a paragraph-name and the word EXIT (see exit statement).
8. The procedure-name may be either a paragraph or section-name. The word SECTION is not required.
9. The procedure-name must not be the name of a procedure of which the PERFORM statement is a part.
10. A procedure-name can be referenced by more than one PERFORM statement.
11. Procedures to be performed can be outside the main program or can be part of the main routine so that they can be executed in line.
12. A referenced procedure may itself contain other PERFORM statements.
13. All procedures should be arranged in the order in which they are to be performed.

Simple Perform

In a simple PERFORM statement, the procedure referenced is executed once, and control then passes to the first statement following the PERFORM statement (figs. 8.116, 8.117). All statements in paragraphs or sections named in procedure-name-1 (through procedure-name-2) constitute the range and are executed before control is returned. (See figure 8.118.)

Times

The TIMES option provides a means for performing a procedure or section a repetitive number of times, and for then returning control to the next statement after PERFORM. (See figures 8.119, 8.120.)

Rules Governing the Use of the Times Option

1. The number of times the procedure is to be performed is specified as a number or identifier.

```
PERFORM COMPUTE-FICA
```

The paragraph COMPUTE-FICA will be executed; that is, PERFORMed and control will then pass to the statement following the PERFORM verb.

```
PERFORM COMPUTE-FICA THRU COMMON-CHECK-PRINT
    .  .
    .  .
    .  .
COMPUTE-FICA. . . .
COMPUTE-NET-PAY. . . .
COMMON-CHECK-PRINT. . . .
```

The range of procedures specified above is executed and control is passed to the statement following the PERFORM.

Figure 8.116 Example—Perform Statement—Format 1—Thru Option.

```
PAR-1.      MULTIPLY AMOUNT BY 300 GIVING TOTAL-AMOUNT.
            PERFORM CALCULATE.
            ADD 100 TO TOTAL.
            .
            .
CALCULATE.
            ADD 10 TO TOTAL.
            MOVE TOTAL TO NEW-TOTAL.
            SUBTRACT TOTAL FROM OLD-TOTAL.
MOVE-DATE.
            IF DATA IS EQUAL TO TODAY-DATE, MOVE TODAY-DATE TO
            OUTPUT-DATE.
            .
            .
```

In the above example, the statement PERFORM CALCULATE executes the three statements contained in the paragraph CALCULATE. The instructions are executed in the following sequence:

```
MULTIPLY AMOUNT BY 300 GIVING TOTAL-AMOUNT.
ADD 10 TO TOTAL.                  ┐
MOVE TOTAL TO NEW-TOTAL.          │
SUBTRACT TOTAL FROM OLD-TOTAL.    ┘—PERFORM CALCULATE
ADD 100 TO TOTAL.
.
.
```

Figure 8.117 Example—Perform Statement—Format 1—Thru Option.

```
PAR-1.      MOVE NEW-ACCT-NO TO ACCT-NO.
            PERFORM MOVEMENT THRU COMPUTATION.
            WRITE RECORD-OUT.
            .
            .
TEST-EQUALITY.
            IF TCODE = 1,  PERFORM ROUTINE-1.
            IF TCODE = 2,  PERFORM ROUTINE-2.
            IF TCODE = 3,  PERFORM ROUTINE-3.
MOVEMENT.   MOVE BALANCE TO NEW-BALANCE
            MOVE TCODE TO NEW-CODE.
NEW-RECORD.
            MOVE 2 TO NEW-LL-CODE.
            MOVE DATE-1 TO NEW-DATE.
COMPUTATION.
            ADD AMOUNT TO BALANCE.
            SUBTRACT 35 FROM LOWER-LIMIT.
OUTPUT-ROUTINE.
            WRITE NEW-RECORD FROM OLD-AREA.
            ADD 1 TO COUNTER-1.
```

In the above example, the statement PERFORM MOVEMENT THRU COMPUTATION executes the six statements in the paragraphs entitled MOVEMENT, NEW-RECORD, and COMPUTATION. The instructions are executed in the following sequence:

```
MOVE NEW-ACCT-NO TO ACCT-NO.
MOVE BALANCE TO NEW-BALANCE.     ┐
MOVE TCODE TO NEW-CODE.          │
MOVE 2 TO NEW-LL-CODE.           ├—PERFORM MOVEMENT
MOVE DATE-1 TO NEW-DATE.         │  THRU COMPUTATION.
ADD AMOUNT TO BALANCE.           │
SUBTRACT 35 FROM LOWER-LIMIT.    ┘
WRITE RECORD-OUT.
```

Figure 8.118 Example—Perform Statement—Format 1—Thru Option.

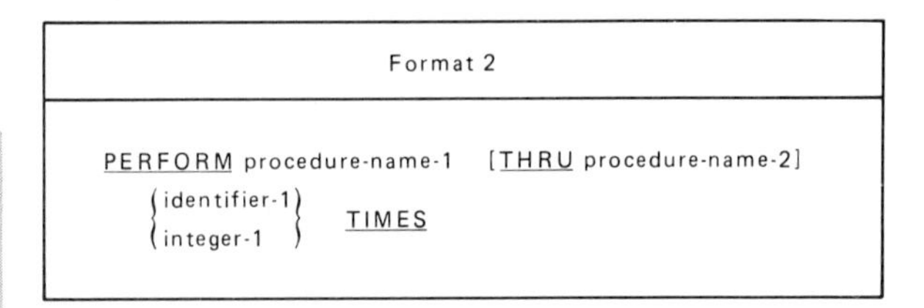
Format 2

PERFORM procedure-name-1 [THRU procedure-name-2]
{identifier-1 / integer-1} TIMES

Figure 8.119 Format—Perform Statement—Format 2—Times Option.

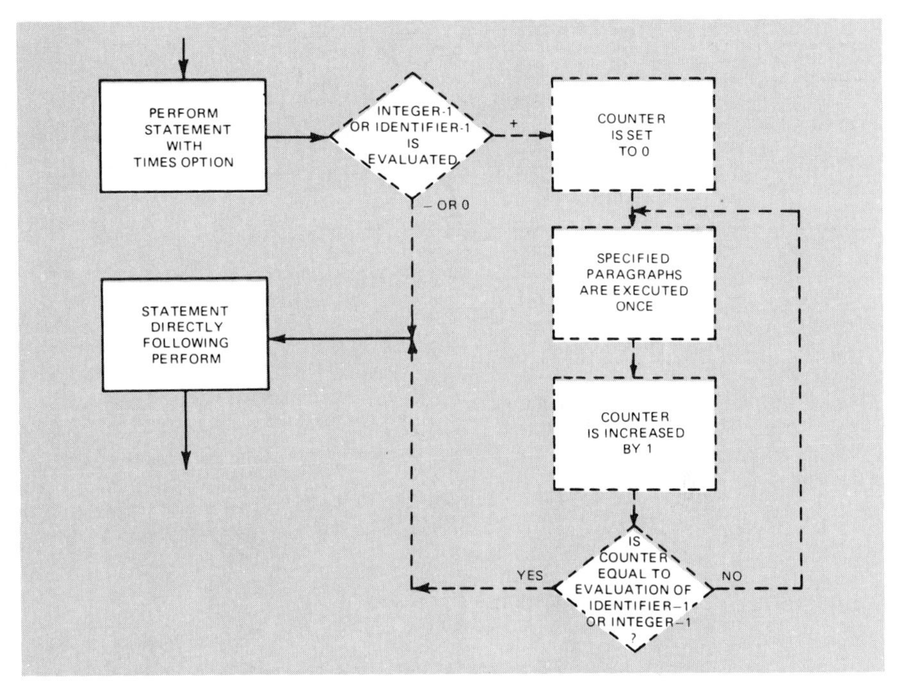

Figure 8.120 Flow of Logic—Perform Statement—Format 2—Times Option.

2. If an identifier is used, it must have an integral value.
3. The identifier or the number must have a positive value. If the value of the identifier is zero or negative, control is transferred immediately to the next statement following the PERFORM statement.
4. When the TIMES option is used, a counter is set up, and this counter is tested against the specified number of execution (times) before control is transferred to procedure-name-1. The counter is incremented by one after each execution, and the process is repeated until the value of the counter is equal to the number of times specified. At that point, control is passed to the next statement following the PERFORM statement. An initial value of zero will cause no execution of procedure-name-1. (See figure 8.121.)

```
PERFORM ITERATION Q-NUMBER TIMES.
```

The procedure ITERATION will be performed exactly the number of times as specified by the numeric quantity in the field Q-NUMBER, each time the PERFORM verb is executed. If the numeric quantity in the field Q-NUMBER is zero, then the procedure is not executed. Keep in mind that the numeric quantity in the field Q-NUMBER can be changed from time to time.

```
PERFORM A THROUGH E DETERMINED-NUMBER-OF TIMES.
```

The entire series of procedures beginning with A and ending with E will be executed exactly the number of times as specified by the numeric quantity in the field DETERMINED-NUMBER-OF, each time the PERFORM verb is executed.

```
PERFORM ITERATION 3 TIMES.
```

The procedure ITERATION will be executed exactly 3 times each time the PERFORM verb is executed.

Figure 8.121 Example—Perform Statement—Format 2—Times Option.

Until

The UNTIL option operates in the same manner as the Times option, except that no counting takes place, and the PERFORM causes an evaluation of a specified test condition instead of testing the value of a counter against a specified number of executions. (See figures 8.122, 8.123.)

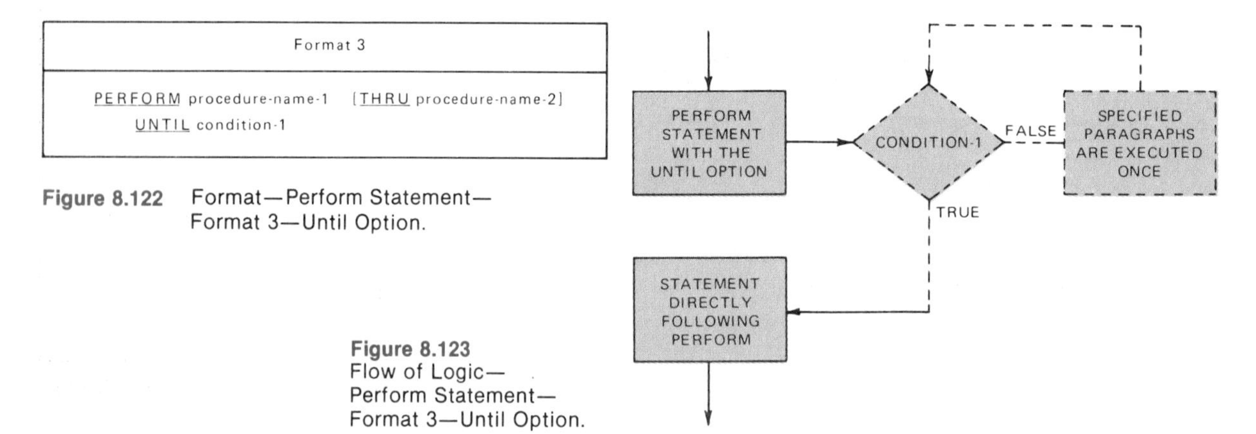

Figure 8.122 Format—Perform Statement—Format 3—Until Option.

Figure 8.123 Flow of Logic—Perform Statement—Format 3—Until Option.

Rules Governing the Use of the Until Option

1. Condition-1 may be a simple or compound expression.
2. Condition-1 is evaluated before the specified procedures are executed.
3. If condition-1 is true, control passes to the next statement after the PERFORM statement. The specified procedure is not executed.
4. If condition-1 is not true, control transfers to procedure-name-1.
5. The process is repeated until condition-1 is detected to be true.

(See figure 8.124.)

Varying

The VARYING option is used to PERFORM a procedure repetitively, increasing or decreasing the value of one or more identifiers or index names once for each repetition until a specified condition is satisfied (fig. 8.125).

Rules Governing the Use of the Varying Option

1. The option may be used to increase or decrease the value of one or more identifiers or index names depending upon whether the BY value is positive or negative. (See figure 8.126.)
2. The specified test condition may be a simple or compound expression.
3. The identifier, index name, or literal is set to the specified initial value (FROM) when commencing the PERFORM statement. Then condition-1 is evaluated (UNTIL).
4. If condition-1 is true, control passes to the next statement immediately following the PERFORM statement, and no execution of the procedures take place.

```
PERFORM EDIT-ROUTINE
    UNTIL KODE IS EQUAL TO '5'.
```

The procedure EDIT-ROUTINE will be executed until the data in the field KODE compares equal to 5.

```
PERFORM EDIT-ROUTINE
    UNTIL TRANSACTION-NUMBER IS GREATER THAN '12345'
    OR IS EQUAL TO KEY-NUMBER.
```

The procedure EDIT-ROUTINE will be executed until the data in the field TRANSACTION-NUMBER compares greater than 12345 or until the data compares equal to the data in the field KEY-NUMBER.

```
PERFORM A THROUGH G
    UNTIL V IS EQUAL TO M
    OR EQUAL TO N
    OR EQUAL TO P.
```

Figure 8.124 Format—Perform Statement—Format 3—Until Option.

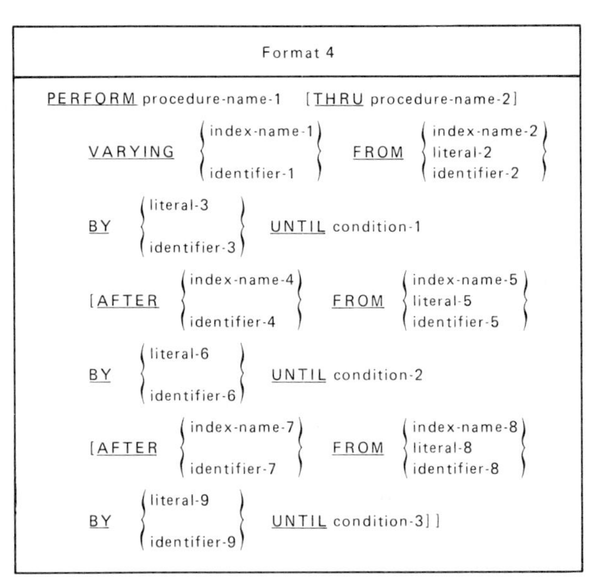

Format 4

```
PERFORM procedure-name-1  [THRU procedure-name-2]

         {index-name-1}        {index-name-2}
VARYING  {            }  FROM  {literal-2   }
         {identifier-1}        {identifier-2}

     {literal-3   }
BY   {            }  UNTIL condition-1
     {identifier-3}

         {index-name-4}        {index-name-5}
[AFTER   {            }  FROM  {literal-5   }
         {identifier-4}        {identifier-5}

     {literal-6   }
BY   {            }  UNTIL condition-2
     {identifier-6}

         {index-name-7}        {index-name-8}
[AFTER   {            }  FROM  {literal-8   }
         {identifier-7}        {identifier-8}

     {literal-9   }
BY   {            }  UNTIL condition-3]]
     {identifier-9}
```

Figure 8.125 Format—Perform Statement—Format 4—Varying Option.

If the statement is false, the procedure specified in procedure-name-1 through procedure-name-2 is executed once. The BY value is added to the index name or identifier and again causes condition-1 to be evaluated. This process continues until the conditional expression is found to be true, whereupon control passes to the next statement after the PERFORM statement.

5. The items used in the BY and FROM must represent numeric value but need not be integers; such values may be positive, negative, or zero. (See figure 8.127.)
6. When more than one identifier is varied, the value of each identifier goes through the complete cycle (FROM, BY, UNTIL) each time that Identifier-1 is altered by its BY value.

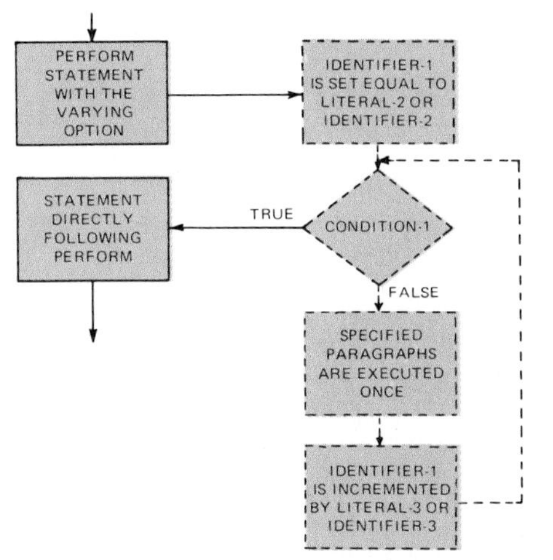

Figure 8.126 Flow of Logic—Perform Statement—Format 4—Varying Option.

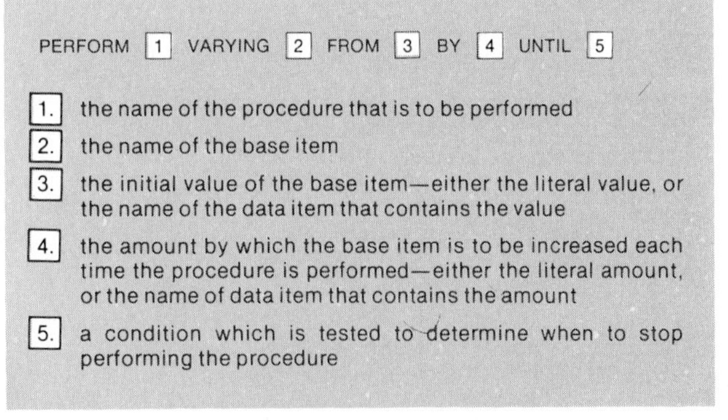

PERFORM [1] VARYING [2] FROM [3] BY [4] UNTIL [5]

1. the name of the procedure that is to be performed
2. the name of the base item
3. the initial value of the base item—either the literal value, or the name of the data item that contains the value
4. the amount by which the base item is to be increased each time the procedure is performed—either the literal amount, or the name of data item that contains the amount
5. a condition which is tested to determine when to stop performing the procedure

Figure 8.127 Using Perform Statement—Format 4—Varying Option.

7. Regardless of the number of identifiers being varied, as soon as test condition-1 is satisfied, control is transferred to the next statement after the PERFORM statement. (See figures 8.128, 8.129, 8.130.)

   (Note: The PERFORM verb is used extensively in structured programming. The COBOL implementation of the DOWHILE structure involves the PERFORM verb with an UNTIL phrase. If indexing is used, the DOWHILE structure may be carried out using the PERFORM verb with the VARYING option.

   COBOL has no implementation of DOUNTIL structure, but it is rather

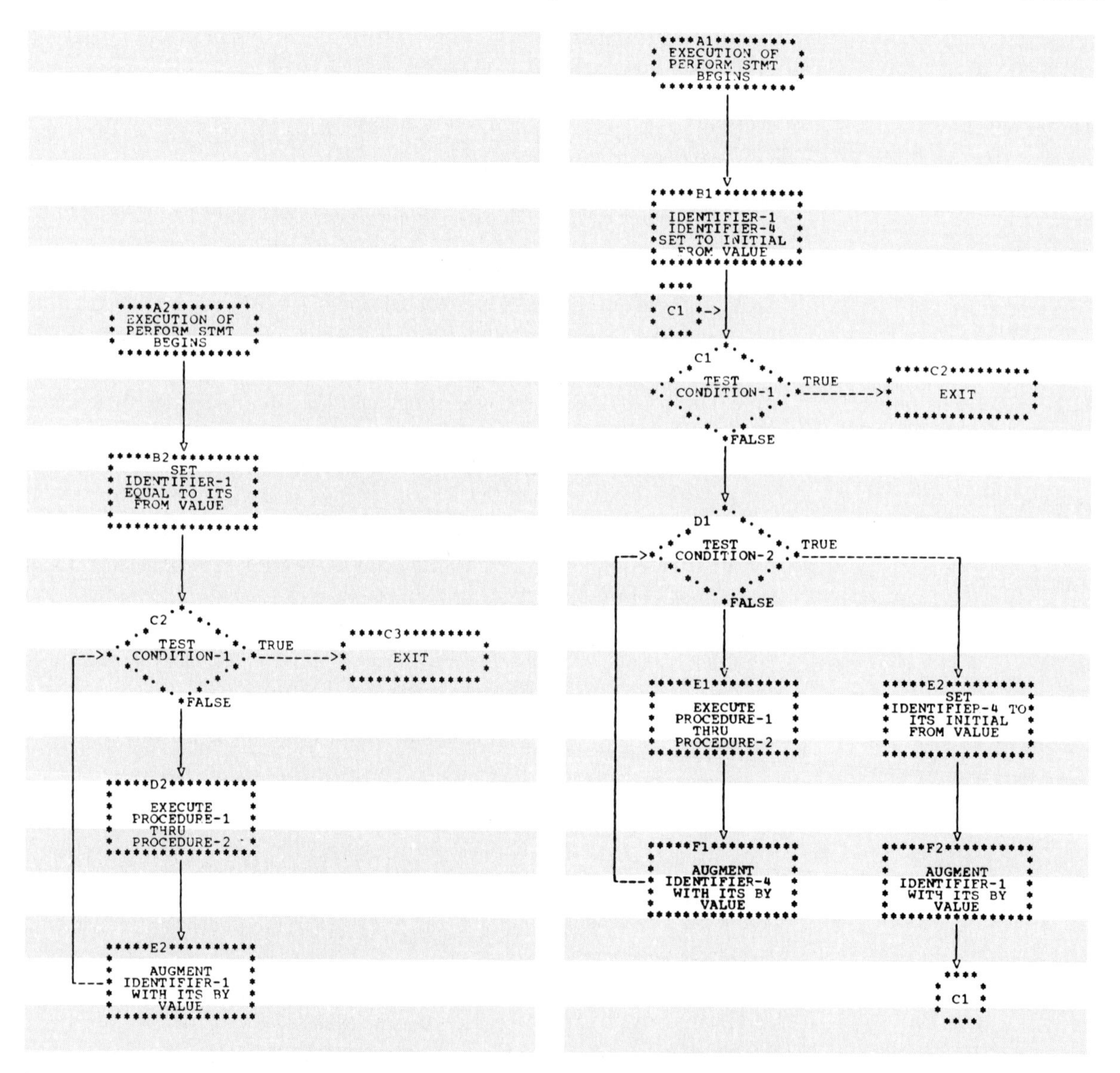

Figure 8.128 Logical Flow—Perform Statement—Format 4—Varying 1 Identifier.

Figure 8.129 Logical Flow—Perform Statement—Format 4—Varying 2 Identifiers.

```
PERFORM COMPUTE-FICA THRU CHECK-PRINT
    VARYING EMPLOYEE-COUNT FROM 0 BY 1
    UNTIL EMPLOYEE-COUNT IS EQUAL TO TOTAL-EMPLOYEE.
```

The previous example illustrates an item (EMPLOYEE-COUNT) that is set to 0 and incremented by 1 each time procedures COMPUTE-FICA through CHECK-PRINT are PERFORMed. If the item TOTAL-EMPLOYEE value is 5, the procedures are performed five times; that is, until EMPLOYEE-COUNT is incremented to 5.

Figure 8.130
Example—Perform Statement—
Format 4—Varying Option.

easily implemented with the PERFORM statement in either of two ways. The first simply starts with an unconditional PERFORM to guarantee that the function is carried out at least once. The second, more cumbersome, is to set a flag to a value before executing the PERFORM and then unconditionally changing it within the named paragraph.

The PERFORM statement must be written with the THRU option when using the CASE structure because the paragraphs that carry out the individual functions all end with a GO TO paragraph-EXIT; without the THRU option, the EXIT paragraph would not be correctly interpreted.)

(See Structured Programming section (chap. 7) for a more comprehensive discussion of structured programming.)

Stop

The STOP statement permits the programmer to specify a temporary or permanent halt to the program (fig. 8.131).

```
STOP  {RUN    }
      {literal}
```

Figure 8.131
Format—Stop Statement.

```
STOP RUN.
```

Figure 8.132
Example—Stop Run Statement.

Stop Run

Rules Governing the Use of the Stop Run Statement

1. The STOP RUN statement terminates the execution of a program.
2. Because of its terminal effect, the STOP RUN statement can be only used as a final statement in the sequence in which it appears; otherwise, the succeeding statements will never be executed.
3. All files should be closed before a STOP RUN statement is issued.
4. The actions following the execution of a STOP RUN statement depend upon the particular installation and/or a particular computer.

(See figure 8.132.)

Stop 'Literal'

Rules Governing the Use of the Stop 'Literal' Statement

1. The STOP 'LITERAL' statement is used by the programmer to specify a temporary halt to the program.
2. When this statement is used, the specified literal will be displayed on the console at the time the stop occurs.
3. The program may be resumed only by operator intervention. A reply must be keyed in on the console to resume execution of the program.
4. Following the execution of the STOP 'LITERAL' statement, continuation of the object program begins with the next sentence in sequence.

5. The literal may be numeric or nonnumeric, or it may be a figurative constant except ALL.

(See figure 8.133.)

Figure 8.133
Example—Stop Literal Statement.

```
      STOP 'HALT 350 -- CONSULT RUN BOOK'.
```

Exit

The EXIT statement is used when it is necessary to provide a common end point for a series of procedures (fig. 8.134). The EXIT statement is also used when it is necessary to provide an ending point for a procedure that is executed by a PERFORM statement possibly having one or more conditional exits prior to the last sentence. The EXIT verb serves as an ending point common to all paths. (See figure 8.135.)

```
paragraph-name. EXIT.
```

Figure 8.134
Format—Exit Statement.

Rules Governing the Use of the Exit Statement

1. The statement must appear in a source program as a one-word paragraph preceded by a paragraph-name.
2. In a PERFORM statement, the EXIT paragraph-name may be given as the object of the THRU option.
3. If the THRU option is used for the EXIT paragraph, a statement in the range of the PERFORM being executed may transfer to an EXIT paragraph, bypassing the remainder of the statements in the PERFORM range.
4. If the control reaches an EXIT paragraph and no associated PERFORM statement is used, control passes through the exit point to the first sentence of the next paragraph.

(See figure 8.136.)

```
    PERFORM ANALYSIS-ROUTINE
            .
            .
            .
ANALYSIS-ROUTINE.
    COMPUTE RETURNS-RATIO = RETURNS/(ORDERS - FILLED +
        BACK-ORDERS-RETURNS).
    IF RETURNS-RATIO IS LESS THAN 20
    THEN
        GO TO FINISH-ANALYSIS
    ELSE
        IF RETURNS-RATIO IS LESS THAN .33,
        THEN
            ADD 1 TO HIGH-RATIO-COUNTER
            GO TO FINISH-ANALYSIS.
    PERFORM HIGH-RATIO-REPORT.
FINISH-ANALYSIS.
    EXIT.
```

Figure 8.135 Example—Exit Statement.

```
PERFORM-1.     PERFORM TEST-CODE THRU EXIT-POINT.
SENTENCE-1.    .
               .
               .
TEST-CODE.
               IF TCODE = 12, GO TO PATH-B.
               IF TCODE = 15, GO TO PATH-C.
               IF TCODE = 16 GO TO EXIT-POINT.
PATH-A.
               .
               .
               GO TO EXIT-POINT.
PATH-B.
               .
               .
               GO TO EXIT-POINT.
PATH-C.
               .
               .
EXIT-POINT.    EXIT.
```

In the above example, the original PERFORM statement named PERFORM-1 executes the set of procedures from TEST-CODE through EXIT-POINT. In these procedures, the testing of TCODE, results in the execution of one of four paths: PATH-A, PATH-B, PATH-C, and EXIT-POINT.

Figure 8.136 Example—Exit Statement.

Compiler-Directing Statements

Compiler-directing statements are special instructions for the COBOL compiler. They cause the compiler to take certain specific action at compile time. The compiler-directing statements are COPY, ENTER, and NOTE.

Copy

The COPY statement is used to include prewritten source program entries in a COBOL program at compile time. Thus, an installation can utilize standard file descriptions, record descriptions, or procedures, without having to repeat programming them. These entries and procedures are contained in user-created libraries. They are included in a source program by means of a COPY statement.

The use of the COPY statement together with its formats, functions, and examples are explained later in the text.

Enter

The ENTER statement enables the programmer to use other programming languages in the same COBOL source program (fig. 8.137). The ENTER statement serves as documentation and is intended to provide a means of allowing the use of more than one source language in the same source program (fig. 8.138).

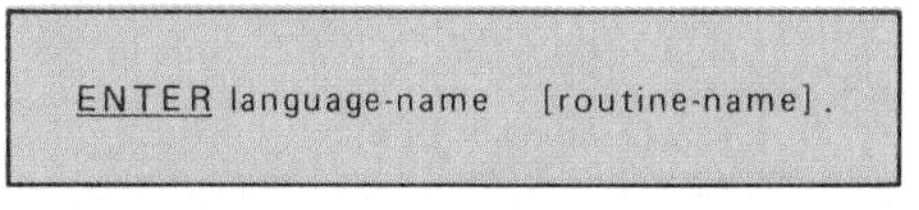
ENTER language-name [routine-name].

Figure 8.137 Format—Enter Statement.

```
ENTER FORTRAN.
```

Figure 8.138 Example—Enter Statement.

Rules Governing the Use of the Enter Statement

1. The language-name informs the compiler as to what other language is desired at this point in the program. The publication for the various compilers will specify what languages may be entered.
2. The language entered must be written in a series of statements immediately following the ENTER statement.
3. If the statements in the entered language cannot be written in line, a routine-name is given to identify the portion of the other language coding to be executed at this point in the procedure sequence. A routine-name is a COBOL word, and it may be referenced to only in an ENTER statement.
4. If the other language statements can be written in-line, the routine-name option is not used.
5. An ENTER COBOL statement must follow the last other language statement in order to indicate to the compiler the point at which a return to COBOL source language takes place (fig. 8.139).
6. Each ENTER statement constitutes a separate paragraph in the source program.
7. The other language statements between the ENTER statement of the language and the following ENTER COBOL statement are executed in the object program as if they had been compiled in the object program following the ENTER statement. These other language statements must conform to the rules of the particular named language.

```
MOVE TCODE TO WCODE.
ENTER NEAT/3.
      INSERTMFILE,WORKI
      ENTER COBOL
PERFORM READ-FILEIN.
```

In the above example, the NEAT/3 instructions must be written in the NEAT/3 programming format. The ENTER COBOL statement must be written on a separate line immediately following the last NEAT/3 instruction in order to tell the compiler when to return to COBOL.

Figure 8.139 Example—Enter Statement.

8. Implementors will specify all details on how the other languages are to be written for their compilers.

Note

The NOTE statement permits the programmer to insert comments and statements into the source program to explain or annotate the procedures being defined (fig. 8.140).

```
NOTE    any words, numbers, or symbols .
              [any sentence . . . ]
```

Figure 8.140 Format—Note Statement.

Rules Governing the Use of the Note Statement

1. A NOTE statement will be printed in the source program listing but will not be compiled into an object instruction.
2. Any combination of COBOL characters can follow the word NOTE.
3. If NOTE is the first word of a paragraph, the entire paragraph must be devoted to note(s). The paragraph must be named and must follow the format rules for paragraph structure (fig. 8.141).
4. If a NOTE sentence appears as other than the first sentence of a paragraph, the commentary must be terminated by a period followed by a space (fig. 8.142).

(*Note:* Many compilers, including IBM, include an extension for writing comments in a program. Explanatory comments may be inserted on any line within a source program by placing an asterisk (*) in column 7 (continuation column) of that line. Any combination of the characters from the EBCDIC set may be included at the A and B margins of that line. The asterisk and the

```
COMMENT-6.
    NOTE THAT JUST THE FIRST 3 DIGITS OF
    THE CLASSIFICATION CODE ARE
    ANALYZED. THIS IS DONE IN ORDER TO
    CHOOSE AMONG 8 SPECIALIZED SUB-
    ROUTINES WHICH ANALYZE THE REMAINING
    DIGITS. IF THE FIRST 3 DIGITS ARE
    ILLEGAL, THE RECORD IS DUMPED OUT ON
    THE CHECKPOINT TAPE AND THE NEXT
    RECORD IS READ.
```

Figure 8.141 Example—Note Statement Paragraph.

```
NOTE -- CONTROL AREA IS FILLED WITH
    9'S AFTER LAST RECORD OF MASTER
    FILE HAS BEEN READ.
```

Figure 8.142 Example—Note Statement Sentence.

characters will be produced on the source listing but will serve no other purpose.)

This feature is implemented in the ANS 1974 COBOL.

CONDITIONAL EXPRESSIONS

A conditional expression is an expression containing one or more variables whose value may change during the course of the program. The conditional expression can be reduced to a single value that can be tested to determine to which of the alternate paths the program flow is to be taken. *A test condition is an expression that, taken as a whole, may be either true or false, depending on the circumstances existing when the expression is evaluated* (fig. 8.143). Although

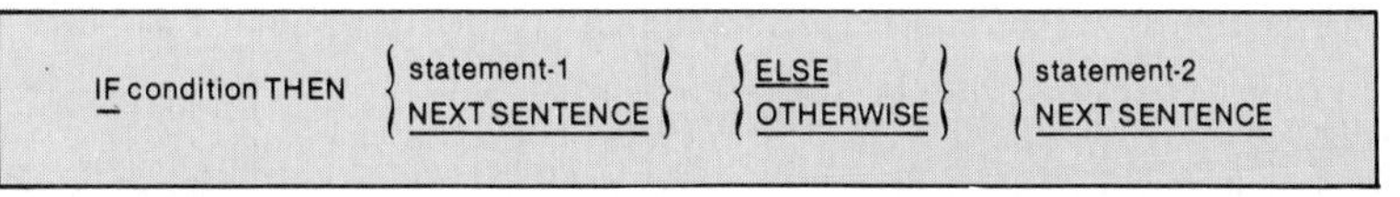

Figure 8.143
Format—Conditional Statement.

Note: THEN and OTHERWISE are not standard ANS 1974 COBOL statements.

IF is not a verb in the grammatical sense, it is regarded as such in COBOL. IF statements are used to evaluate test conditions. There are four types of test conditions; relation conditions, class conditions, sign conditions, and condition-name conditions.

When the test condition is evaluated, the following action will take place.

1. If the condition is true, the statements immediately following the conditional expression are executed. Control then passes to the next sentence.
2. If the condition is false, the statements immediately following ELSE or OTHERWISE are executed, or the next sentence, if the ELSE or OTHERWISE clause is omitted (figs. 8.144, 8.145).
3. An IF statement must be terminated by a period and a space.

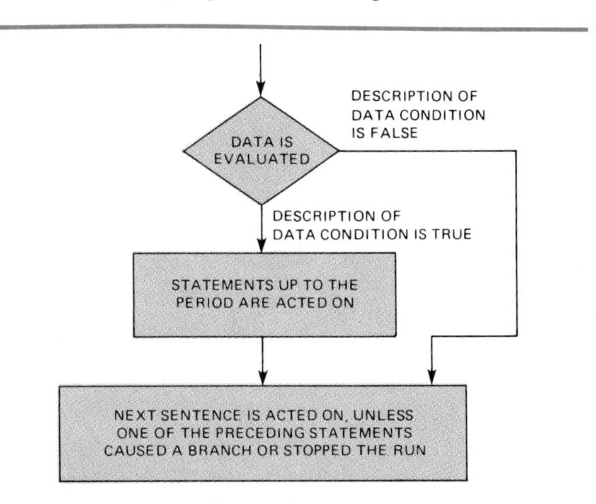


```
IF BALANCE IS NEGATIVE,
THEN
     MOVE 'BALANCE IS IN YOUR FAVOR' TO MESSAGE-AREA.
WRITE CUSTOMER-BILL-RECORD.
```

Figure 8.144
Flow of Control Through an If Statement that Does Not Contain an Else or Otherwise Statement.

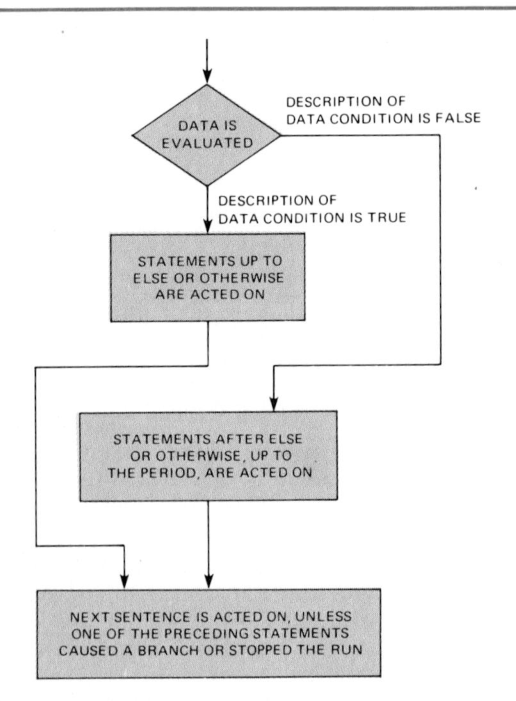


```
IF RECOVERY IS NOT GREATER THAN SCRAP-POINT,
THEN
      MOVE ITEM-NUMBER TO SCRAP-ITEM,
      ADD RECOVERY TO SCRAP;
ELSE
      WRITE CONSERVATION-RECORD.
ADD 1 TO TRANSACTION-COUNT.
```

Figure 8.145 Flow of Control Through an If Statement that Contains an Else or Otherwise Statement.

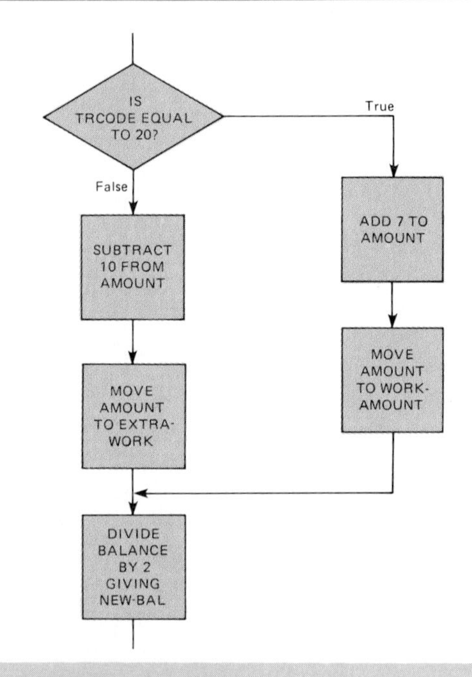


```
IF TRCODE IS EQUAL TO 20,
THEN
      ADD 7 TO AMOUNT,
      MOVE AMOUNT TO WORK-AMOUNT
ELSE
      SUBTRACT 10 FROM AMOUNT,
      MOVE AMOUNT TO EXTRA-WORK.
DIVIDE BALANCE BY 2 GIVING NEW-BAL.
```

In the above example, when TRCODE is equal to 20, the imperative statements ADD 7 TO AMOUNT and MOVE AMOUNT TO WORK-AMOUNT are executed. Control then returns to the sentence after the IF sentence. In this example, DIVIDE BALANCE BY 2 GIVING NEW-BAL is executed next.

When the IF statement is executed and the condition is found to be false, then the statements immediately following the word ELSE are executed. In this example, SUBTRACT 10 FROM AMOUNT and MOVE AMOUNT TO EXTRA-WORK are executed when TRCODE is not equal to 20. Control then transfers the sentence immediately following the IF sentence (DIVIDE BALANCE BY 2 GIVING NEW-BAL).

Figure 8.146 Example—If Statement.

4. Any number of statements may follow the test condition. These statements are acted upon if the condition exists and skipped over if the condition does not exist or to the statements following ELSE or OTHERWISE. (See figure 8.146.)
5. In a series of imperative statements executed when the condition is true, only the last statement may be an unconditional GO TO or STOP RUN statement; otherwise, the series of statements would contain statements into which control cannot flow. It is the programmer's responsibility to write the program steps in a logical sequence for execution.

Relation Condition

A Relation Condition test involves the comparison of two data values, either of which may be an identifier, a literal, or an arithmetic expression (fig. 8.147). Either the relational operator symbol or relational operator may be used in the test. If the symbols are used, they must be preceded and followed by a space. NOT is used to specify the opposite of the expression. (See figure 8.148.)

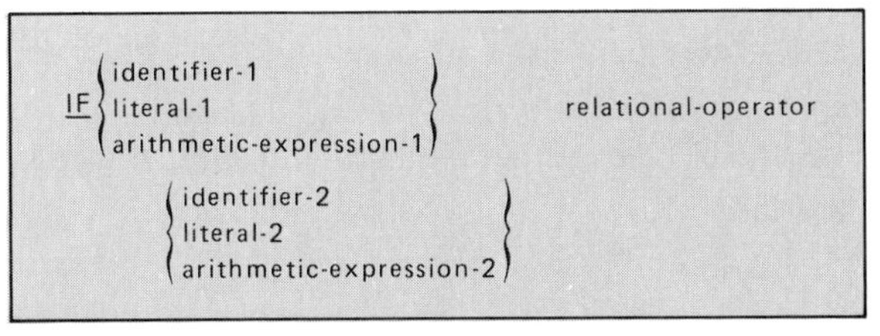

```
   { identifier-1            }
IF { literal-1               }   relational-operator
   { arithmetic-expression-1 }

      { identifier-2            }
      { literal-2               }
      { arithmetic-expression-2 }
```

Figure 8.147 Format—Relation Condition.

Relational-operator	Meaning
IS [NOT] GREATER THAN IS [NOT] >	Greater than or not greater than
IS [NOT] LESS THAN IS [NOT] <	Less than or not less than
IS [NOT] EQUAL TO IS [NOT] =	Equal to or not equal to

Figure 8.148 Relational Operators and Their Meanings.

Rules Governing Relation Condition Tests

1. The first operand is called the subject of the condition; the second operand is called the object of the condition. The subject and object may not both be literals.
2. Both operands must have the same USAGE, except when two numeric operands are involved.

(See figure 8.149.)

First Operand \ Second Operand		Group	Elementary			
			Alphanumeric	Alphabetic	Numeric	Literal
Group		C	C	C	C	C
Elementary	Alphanumeric	C	C	C	C	C
	Alphabetic	C	C	C	I	C
	Numeric	C	C	I	N	C
	Literal	C	C	C	C	I

C Compared logically (one character at a time, according to collating sequence of a particular computer)
N Compared algebraically (numeric values are compared)
I Invalid comparison

Example:

IF TOTAL GREATER THAN MAXIMUM PERFORM MESSAGE.

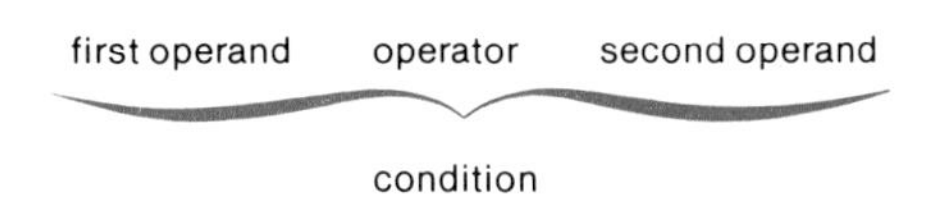

Explanation:

To use this chart find the data type (determined by the picture) of the first operand in the column headed First Operand. Then find the data type of the second operand across the top of the figure opposite Second Operand. Extend imaginary lines into the figure from the data types of the first and second operands. In the block where these two lines intersect is a letter that tells you how the values are compared.

Figure 8.149 Types of Valid Comparisons.

Numeric Items Comparison Tests

1. The test determines that the value of one of the items is GREATER THAN, LESS THAN, or EQUAL TO the other item, regardless of the length of the operands.
2. The items are compared algebraically after decimal point alignment.
3. Zero is considered a unique value regardless of its length, sign, or implied decimal-point location.
4. Numeric operands that do not have signs are considered positive values for purposes of comparison.
5. Comparison of numeric operands is permitted regardless of the manner in which their USAGE is described.

(See figure 8.150.)

The following are examples of comparison of numeric operands:

Operand 1	Operand 2	Result of Comparison
1 2 5‸5 0	1 2‸4 0 1 0	125.50 is greater than 12.4010
1 5 1	5 1‸2 5	151 is greater than 51.25
1 0 0 0‸0	1 0 0 0‸1	1000.0 is less than 1000.1
1‸0 –	0 0‸0 +	–1.0 is less than +0.0
0 0 –	0 0 +	0 equals 0
3 4	0 1 2 8	34 is less than 0128
0 2 5	2 5	025 equals 25

Figure 8.150 Examples—Numeric Operands Comparisons.

Nonnumeric Comparison Tests

1. The test determines that one item is GREATER THAN, LESS THAN, or EQUAL TO the other item with respect to the specified collating sequence of characters for the particular computer. (In the IBM collating sequence, the numerals are in the highest category, followed by alphabetic characters, with the special characters the lowest of the group.)
2. Numeric and nonnumeric operands may be compared only if both items have the same USAGE, implicitly or explicitly.
3. The size of the operand is the total number of characters in the operand. All group items are considered in the nonnumeric operand group.
4. If both operands are of equal length in a nonnumeric comparison, the test proceeds from left (high-order position) to right (low-order position), and each character is compared to the corresponding character of the other item. The comparison of characters continues until an unequal condition is noted.

   If each individual character compared results in an equality, and the two items consist of the same number of characters, the items are considered equal.
5. If the operands are of differing lengths, the comparison proceeds as if the shorter item was filled with spaces until it is of the same length as the other operands.

(See figures 8.151, 8.152.)

The following are examples of comparison of nonnumeric operands:

Operand 1	Operand 2	Result of Comparison
A B 9 5 4	A B 9 5 4	AB954 is equal to AB954
A 4 5 0	9 4 5 0	A450 is less than 9450
9 5 0	9 5 J	950 is greater than 95J
A B C D	S T U V	ABCD is less than STUV
M N O P Q	A B C	MNOPQ is greater than ABC
A B C	M N O P Q	ABC is less than MNOPQ
J B H N A L C A R N	J B H N A L C A N	JBHNALCARN is greater than JBHNALCAN
J K H N ø ø A L	J K H N ø ø A M	JKHNøøAL is less than JKHNøøAM
D E F	D E F G	DEF is less than DEFG
D E F	D E F ø	DEF is equal to DEFø

Figure 8.151 Examples—Nonnumeric Operands Comparisons.

```
IF AMOUNT IS LESS THAN BALANCE
THEN
     MOVE SHIPMENT TO WORKSTORE
ELSE
     NEXT SENTENCE.
```

Note: The MOVE is not executed on an "equal" or "greater" condition.

```
IF DATE-IN OF MASTER IS EQUAL TO TODAYS-DATE,
THEN
     PERFORM REVIEW
ELSE
     PERFORM NEXT-DETAIL.
```

Note: On a "less" or "greater" condition the program performs NEXT-DETAIL.

Figure 8.152 Example—Relational Condition.

Class Condition

The Class Condition test is used to determine whether the data is numeric or alphabetic (figs. 8.153, 8.154).

Numeric data consists entirely of the numerals 0–9 with or without an operational sign. If the PICTURE clause of the record description of the identifier being tested does not contain an operational sign, the identifier is determined to be numeric only if the contents are numeric and an operational sign is not present.

Type of Identifier	Valid Forms of the Class Test	
Alphabetic	ALPHABETIC	NOT ALPHABETIC
Alphanumeric	ALPHABETIC NUMERIC	NOT ALPHABETIC NOT NUMERIC
External-Decimal	NUMERIC	NOT NUMERIC

Figure 8.153 Format—Class Condition.

```
IF identifier is [NOT] { NUMERIC    }
                       { ALPHABETIC }
```

Figure 8.154 Valid Forms of Class Tests.

A numeric test cannot be used with an item whose data descriptions describes items as alphabetic.

Alphabetic data consists of the characters A through Z plus the space character.

An alphabetic test cannot be used with an item whose record description is numeric. (See figures 8.155, 8.156.)

```
IF ACTIVITY-RATING IS ALPHABETIC
THEN
     PERFORM HIGH-ACTIVITY-ANALYSIS
ELSE
     NEXT SENTENCE.
```

Figure 8.155 Example—Class Test.

The following are examples of the class condition:

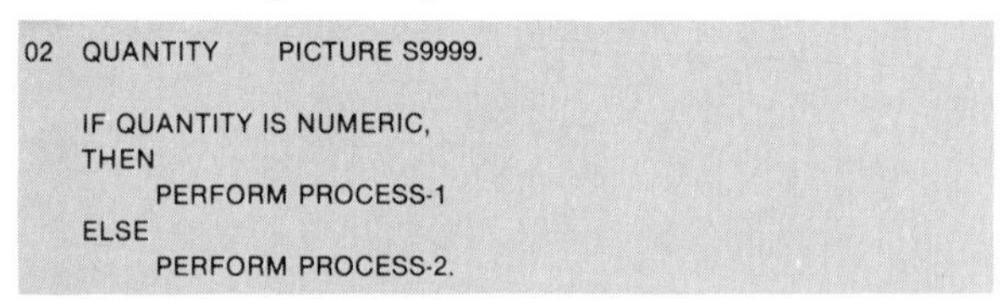

```
02  QUANTITY     PICTURE S9999.

    IF QUANTITY IS NUMERIC,
    THEN
         PERFORM PROCESS-1
    ELSE
         PERFORM PROCESS-2.
```

In the above, this class condition tests each character in QUANTITY for a numeric value and the rightmost character for a zone-sign.

```
02  TCODE        PICTURE A.

    IF TCODE IS ALPHABETIC,
    THEN
         MOVE MASTER TO NEW-MASTER
    ELSE
         NEXT SENTENCE.
```

In the above, this class condition tests each character in TCODE for a letter of the alphabet or a space character.

Figure 8.156 Examples—Class Tests.

Sign Condition

The Sign Condition test is used to determine whether the algebraic value of a numeric item is less than zero (NEGATIVE), greater than zero (POSITIVE), or zero (ZERO) (fig. 8.157).

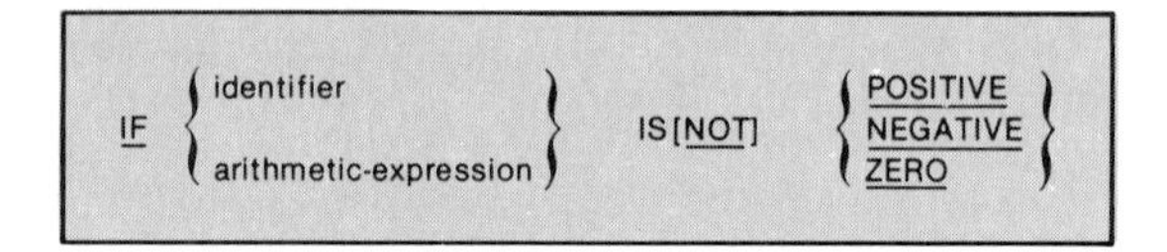

Figure 8.157
Format—Sign Condition.

1. The value of zero is considered neither positive nor negative.
2. If an identifier appears in a Sign Condition test, it must represent a numeric value. If the value is unsigned and not equal to zero, it is considered to be positive.

(See figures 8.158, 8.159.)

Condition-Name Condition

A Condition-Name Condition test is used to test a condition variable to determine whether or not its value is equal to one of the values assumed with its condition-name (fig. 8.160).

1. The condition-name must be defined in a level 88 entry in the Data Division associated with the condition-name.

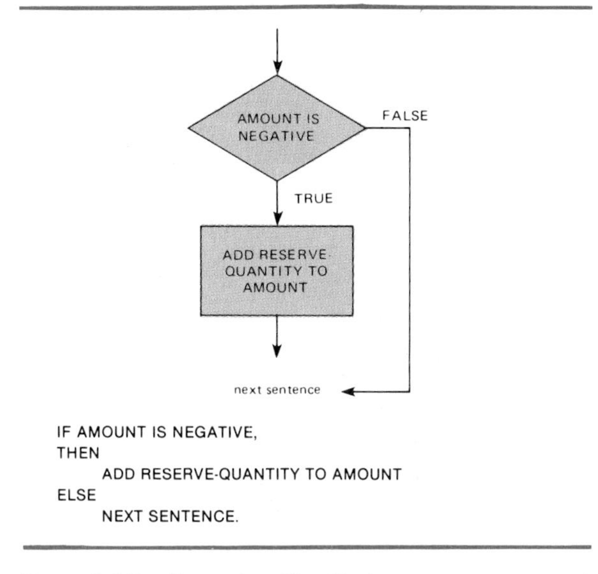


```
IF AMOUNT IS NEGATIVE,
THEN
    ADD RESERVE-QUANTITY TO AMOUNT
ELSE
    NEXT SENTENCE.
```

Figure 8.158 Example—Sign Test.

The following are examples of the sign condition:

```
IF AMOUNT IS NOT ZERO,
THEN
    MOVE NEW-AMOUNT TO AMOUNT IN MASTER-RECORD
ELSE
    NEXT SENTENCE.
```

In the above, the sign condition tests the field AMOUNT for a nonzero value.

```
IF A * B IS POSITIVE,
THEN
    PERFORM COMPUTATION
ELSE
    PERFORM ERROR-PROCEDURE.
```

In the above, the sign condition tests the product of A times B for a positive value.

```
IF BALANCE IS NEGATIVE,
THEN
    PERFORM OVERDRAFT-PROCEDURE
ELSE
    PERFORM REGULAR-PROCEDURE.
```

In the above, the sign condition tests the field BALANCE for a negative value.

Figure 8.159 Examples—Sign Test.

```
IF [NOT] condition-name
```

Figure 8.160 Format—Condition-Name Test.

2. The Condition-Name Condition test is an alternative way of expressing certain conditions that could be expressed by a simple relational condition. The rules for comparing a condition variable with a condition-name value are the same as those specified for relation conditions.
3. The test is true if the value corresponding to the condition-name equals the value of its associated condition variables.

(See figure 8.161.)

COMPOUND CONDITIONAL EXPRESSIONS

A compound Conditional Expression consists of two or more simple conditions combined with logical operators AND and OR. These conditions are linked by AND and OR in any sequence which would produce the overall desired result (fig. 8.162).

The logical operators must be preceded by a space and followed by a space.

Rules for the Formation of Compound Conditional Expressions

1. Two or more simple conditions combined by AND/OR make up a compound condition. (See figure 8.163.)
2. The word OR is used to mean either or both. Thus the expression A OR B is true if A is true or B is true or both A and B are true.
3. The word AND is used to mean that both expressions must be true. Thus the expression A AND B is true only if both A and B are true.

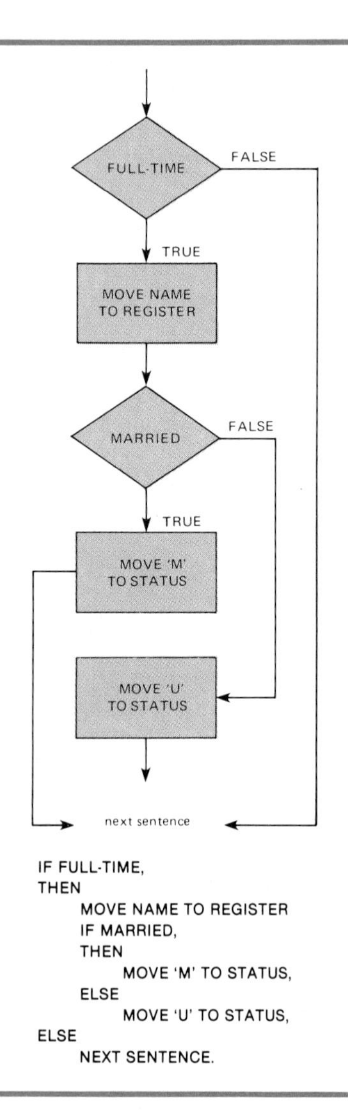


```
IF FULL-TIME,
THEN
      MOVE NAME TO REGISTER
      IF MARRIED,
      THEN
            MOVE 'M' TO STATUS,
      ELSE
            MOVE 'U' TO STATUS,
ELSE
      NEXT SENTENCE.
```

Figure 8.161 Example—Condition-Name Test.

Logical Operator	Meaning
OR	Logical inclusive (either or both are true)
AND	Logical conjunction (both are true)
NOT	Logical negation (not true)

Figure 8.162 Logical Operators and Their Meanings.

Type of Operation	Operator (operation symbol)	Operation
Relational	IS GREATER THAN (>)	is greater than
	IS LESS THAN (<)	is less than
	IS EQUAL TO (=)	is equal to
Logical	OR	logical inclusive OR (either or both are true)
	AND	logical conjunction (both are true)
	NOT	logical negation

Figure 8.163 Relational and Logical Operators.

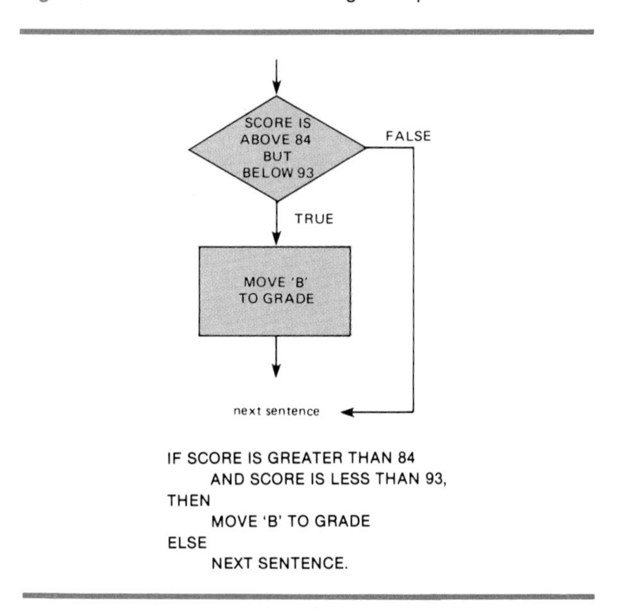


```
IF SCORE IS GREATER THAN 84
      AND SCORE IS LESS THAN 93,
THEN
      MOVE 'B' TO GRADE
ELSE
      NEXT SENTENCE.
```

Figure 8.164 Example—Compound Conditional Statements.

4. The word NOT may be used to specify the opposite of the compound expression. Thus NOT A AND B is true if both A and B are false, A is false, or if B is false.
5. Parentheses may be used to specify the sequence in which the conditions are to be evaluated. Parentheses must always appear as a pair. Logical evaluation begins with the innermost pair of parentheses and proceeds to the outermost pairs.

6. If the sequence of evaluation is not specified by parentheses, the expression is evaluated in the following manner.
   a. Arithmetic expressions.
   b. Relational operators.
   c. [NOT] conditions.
   d. Conditions surrounding all ANDs are evaluated first, starting at the left and proceeding to the right.
   e. OR and its surrounding conditions are then evaluated, also proceeding from left to right.

(See figures 8.164, 8.165, 8.166, 8.167.)

Thus the expression A IS GREATER THAN B OR A IS EQUAL TO C AND D IS POSITIVE would be evaluated as if it were parenthesized as follows.

(A IS GREATER THAN B) OR (A IS EQUAL TO C) AND (D IS POSITIVE).

(See figure 8.168.)

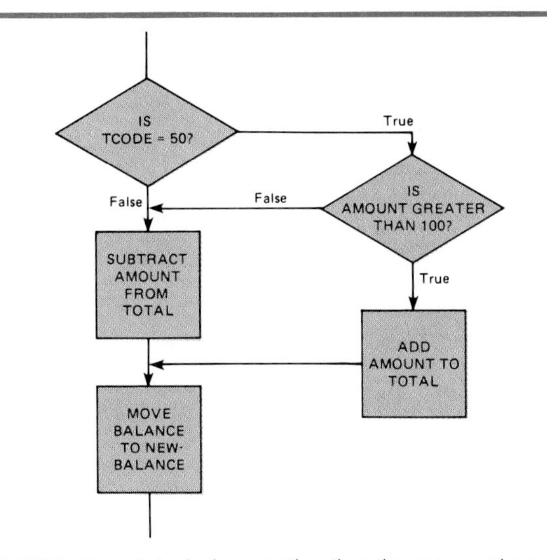

If AND is the only logical connective, then the compound condition is true only if each simple condition is true.

```
IF TCODE IS EQUAL TO 50
      AND AMOUNT IS GREATER THAN 100,
THEN
      ADD AMOUNT TO TOTAL,
ELSE
      SUBTRACT AMOUNT FROM TOTAL.
MOVE BALANCE TO NEW-BALANCE.
```

Figure 8.165 Example—Compound Conditional Statement.

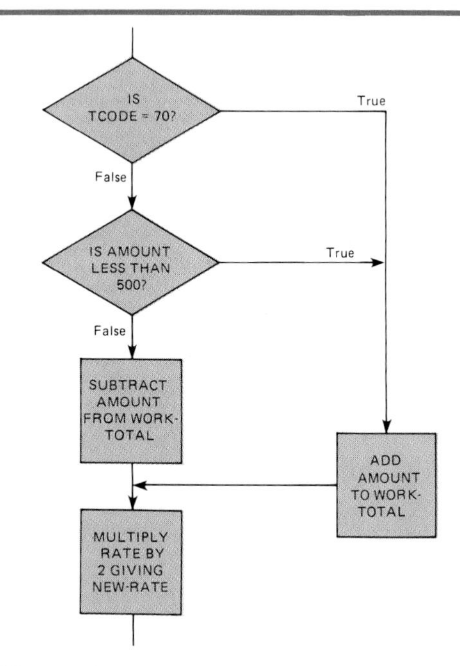

If OR is the only logical connective, then the compound condition is true if at least one of the simple conditions is true.

```
IF TCODE IS EQUAL TO 70,
      OR AMOUNT IS LESS THAN 500,
THEN
      ADD AMOUNT TO WORK-TOTAL,
ELSE
      SUBTRACT AMOUNT FROM WORK-TOTAL.
MULTIPLY RATE BY 2 GIVING NEW-RATE.
```

Figure 8.166 Example—Compound Conditional Statement.

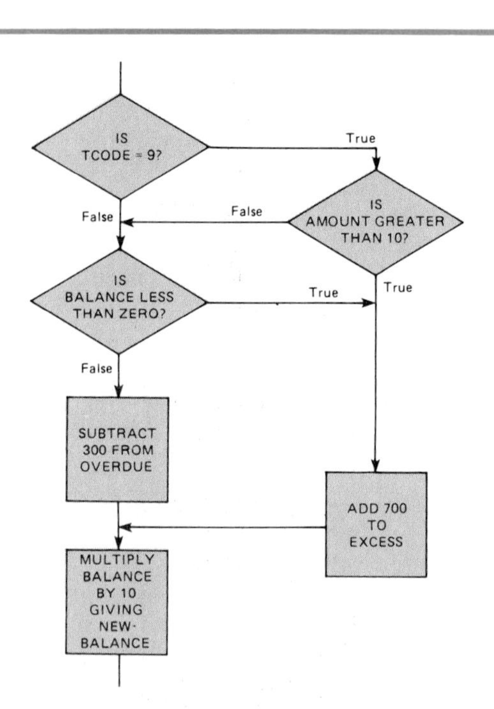

If both AND and OR are used, the conditions are evaluated proceeding left to right with each AND connective referring to the simple conditions before and after the word AND.

```
IF TCODE = 9,
      AND AMOUNT IS GREATER THAN 10,
      OR BALANCE IS LESS THAN ZERO,
THEN
      ADD 700 TO EXCESS,
ELSE
      SUBTRACT 300 FROM OVERDUE.
MULTIPLY BALANCE BY 10 GIVING NEW-BALANCE.
```

Figure 8.167 Example—Compound Conditional Statement.

The following table summarizes the true or false value resulting from conditions having the logical connective AND, the logical connective OR, and the logical operator NOT.

If Condition A is:	If Condition B is:	Then A AND B is:	Then A OR B is:	Then NOT A is:
True	True	True	True	False
True	False	False	True	False
False	True	False	True	True
False	False	False	False	True

Figure 8.168 And/Or Value Table.

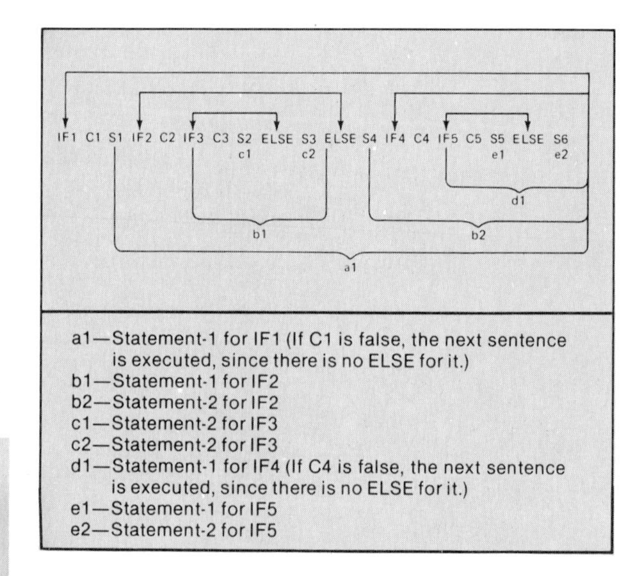

IF1 C1 S1 IF2 C2 IF3 C3 S2 ELSE S3 ELSE S4 IF4 C4 IF5 C5 S5 ELSE S6
c1
c2
e1
e2
d1
b1
b2
a1
a1—Statement-1 for IF1 (If C1 is false, the next sentence is executed, since there is no ELSE for it.)
b1—Statement-1 for IF2
b2—Statement-2 for IF2
c1—Statement-2 for IF3
c2—Statement-2 for IF3
d1—Statement-1 for IF4 (If C4 is false, the next sentence is executed, since there is no ELSE for it.)
e1—Statement-1 for IF5
e2—Statement-2 for IF5

Figure 8.169 Conditional Statements with Nested If Statements.

NESTED CONDITIONAL EXPRESSIONS

If a conditional statement appears as Statement-1 or as part of Statement-1, it is said to be nested. Nesting a statement is like specifying a subordinate arithmetic expression enclosed in parentheses combined in a larger arithmetic expression. IF statements contained within are considered paired, IF and ELSE or OTHERWISE combinations proceeding from left to right. Thus any ELSE or OTHERWISE statement encountered must be considered to apply to the immediately preceding IF, if it has not already been paired with an ELSE or OTHERWISE.

Certain compilers may place some restrictions on the number and types of conditionals that can be nested. (See figures 8.169, 8.170, 8.171.)

(Note: Selection is a structured program theorem that is implemented by the IFTHENELSE structure. In COBOL, it is implemented with the IF statement and a stated condition is tested. Nested IFs and compound conditions for nested IFs are also used in structured programming.)

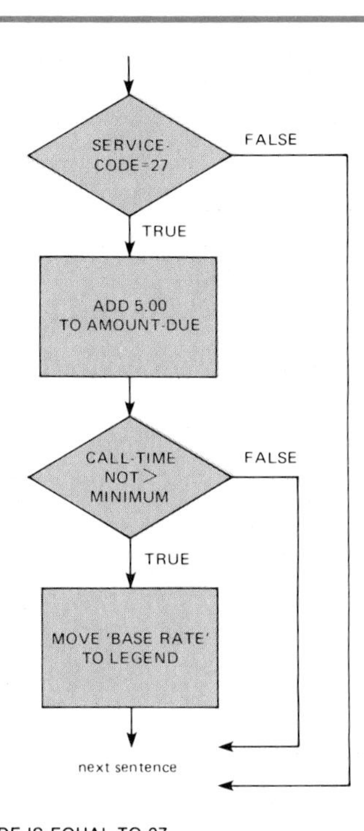


```
IF SERVICE-CODE IS EQUAL TO 27,
THEN
    ADD 5.00 TO AMOUNT-DUE
    IF CALL-TIME IS NOT GREATER THAN MINIMUM
    THEN
        MOVE 'BASE RATE' TO LEGEND
    ELSE
        NEXT SENTENCE
ELSE
    NEXT SENTENCE.
```

Figure 8.170 Example—Nested If Statement.

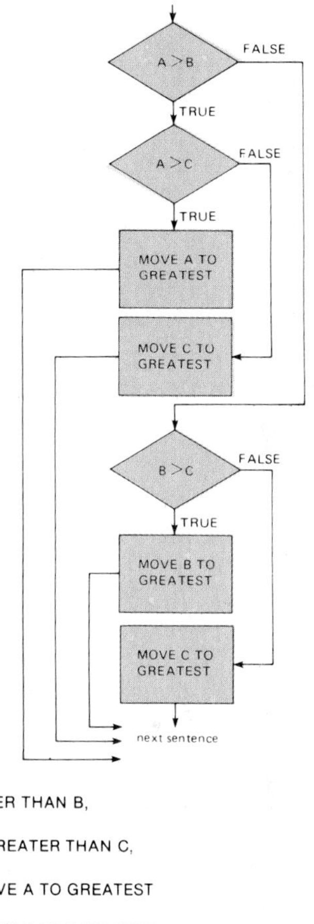


```
IF A IS GREATER THAN B,
THEN
    IF A IS GREATER THAN C,
    THEN
        MOVE A TO GREATEST
    ELSE
        MOVE C TO GREATEST
ELSE
    IF B IS GREATER THAN C,
    THEN
        MOVE B TO GREATEST
    ELSE
        MOVE C TO GREATEST
```

Figure 8.171 Example—Nested If Statement.

IMPLIED SUBJECTS

Many times a conditional expression will contain several simple relational conditions. These conditions may have the same subject. For example, IF A IS GREATER THAN B AND LESS THAN C, the second occurrence of A is implied. (See figure 8.172.)

```
. . . subject relational-operator   object

{ AND }
{ OR  }  [NOT]  relational-operator   object . . .
```

Figure 8.172
Format—Implied Subject.

Rules Governing the Use of Implied Subjects

1. Only conditional expressions written as simple relational conditions may have implied subjects. SIGN and CLASS condition tests can never have implied subjects.
2. The first series of simple relational conditions must always consist of a subject, operator, and operation, all of which must be explicitly stated.
3. The subject may be implied only in a series of simple relational conditions connected by the logical operators AND and/or OR.
4. When the subject of a simple relational condition is implied, the subject used is the first subject to the left which is explicitly stated. For example, IF A = B OR =C OR D= E AND = F, A is the implied subject for C and D, while D is the implied subject of F since D is the first subject to the left of F.
5. When NOT is used in conjunction with a relational operator and an implied subject, the NOT is treated as a logical operator. For example, A IS GREATER THAN B AND NOT EQUAL TO C AND D is equivalent to A IS GREATER THAN B AND NOT A IS EQUAL TO C AND A IS EQUAL TO D.

(See figure 8.173.)

```
A = B OR NOT > C        (The subject, A, is implied.)
A = B OR NOT A > C      (The subject, A, is explicit.)
```

Figure 8.173
Example—Implied Subject.

IMPLIED OPERATORS

Relational operators may be implied in a series of consecutive simple relational conditions in much the same way as that in which the subject can be implied. For example, IF A IS GREATER THAN B AND C OR D, PERFORM X. Not only is the subject implied (A), but also the relational operator (GREATER THAN). (See figure 8.174.)

```
                                          { AND }
... subject  relational-operator  object  {     }  [NOT]  object ...
                                          { OR  }
```

Figure 8.174
Format—Implied Subject and Operator.

Rules Governing the Use of Implied Operators

1. A relational operator may be implied only in a simple relational condition when the subject is implied. SIGN and CLASS conditions can never be implied (do not have operators).
2. When an operator is implied, it is assumed to be the operator of the nearest completed stated simple condition to the left.

(See figures 8.175, 8.176.)

A = B AND C	(Subject and relational-operator, A = , are implied.)
A = B AND A = C	(Subject and relational-operator, A = , are explicit.)

Figure 8.175 Example—Implied Subject and Operator.

A > B AND NOT < C AND D	(Subject, A, is implied in the second condition. Subject, A, and relational-operator, <, are implied in the third condition.)
A > B AND NOT A < C AND A < D	(Subject, A, and relational-operator, <, are explicit.)

Figure 8.176 Example—Implied Subject, and Subject and Operator.

PAYROLL REGISTER PROBLEM

INPUT

Field	**Card Columns**
Department	14–16
Serial	17–21
Gross Earnings	57–61
Insurance	62–65
Federal Withholding Tax	69–72
State Withholding Tax	73–75
Miscellaneous Deductions	76–79
Code (Letter E)	80

CALCULATIONS TO BE PERFORMED

1. FICA TAX = Gross Earnings × 0.0585 (Rounded to 2 decimals)
2. NET EARNINGS = Gross Earnings − Insurance − FICA Tax − Federal Withholding Tax − State Withholding Tax − Miscellaneous Deductions.
3. Department Earnings value is the sum of the Net Earnings for each employee.
4. The Total Net Earnings is the sum of the Net Earnings for each department.

OUTPUT

Print a report as follows:

WEEKLY PAYROLL REGISTER

Employee No. Dept.	Serial	Gross Earnings	Insurance	FICA Tax	Fed. Withhold.	St. Withhold.	Misc. Deduct.	Net Amt.

Dept. Earnings:
Total Net Earnings:

WEEKLY PAYROLL REGISTER

MAIN-ROUTINE

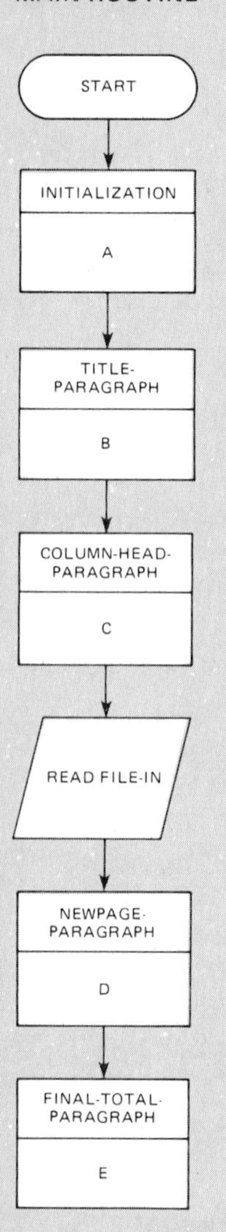

PROCESS-ROUTINES

INITIALIZATION

TITLE-PARAGRAPH

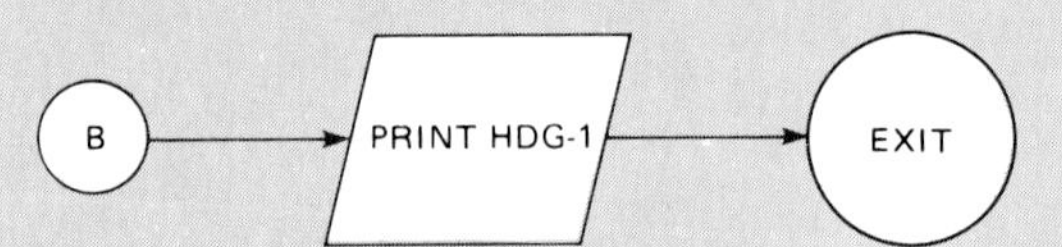

COLUMN-HEAD-PARAGRAPH

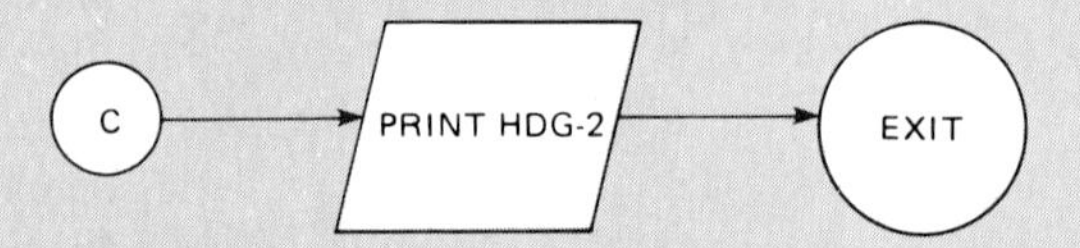

(PROCESS-ROUTINES, CONTINUED)

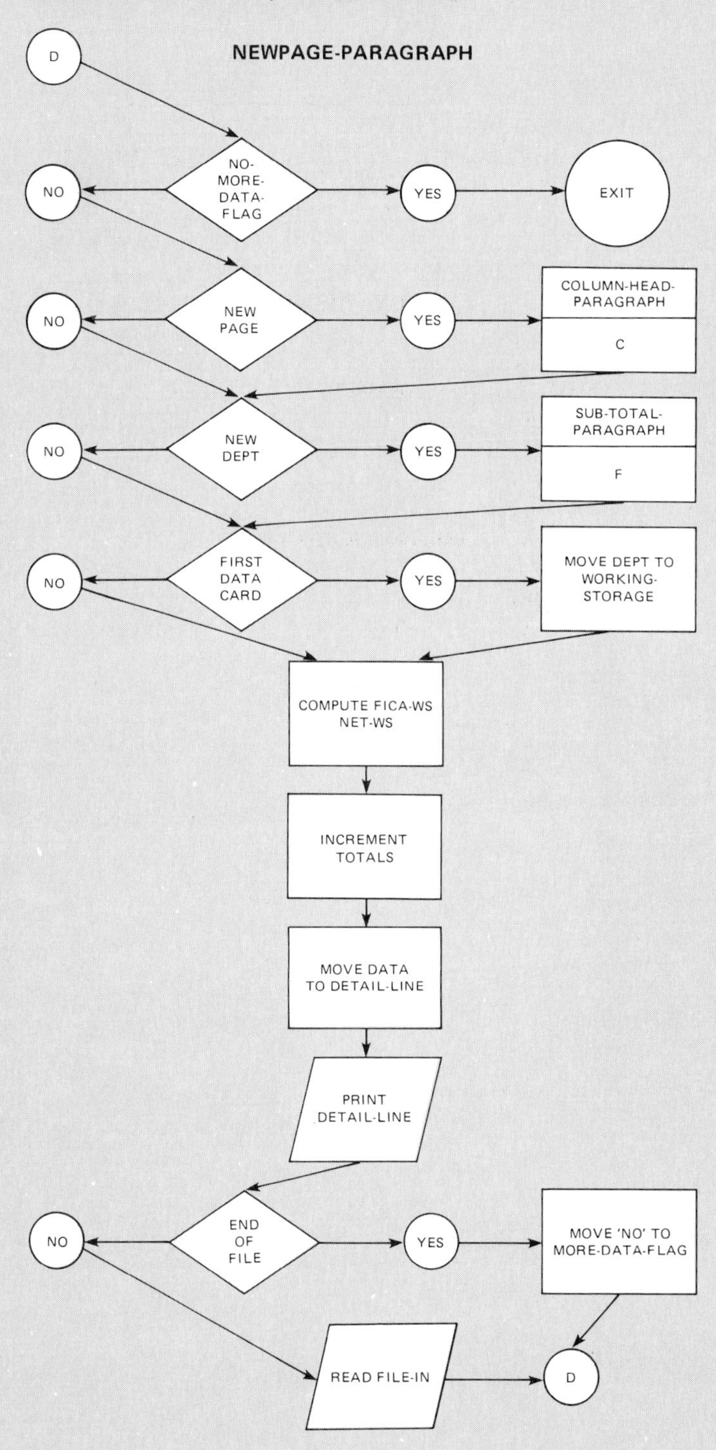

SUB-TOTAL-PARAGRAPH

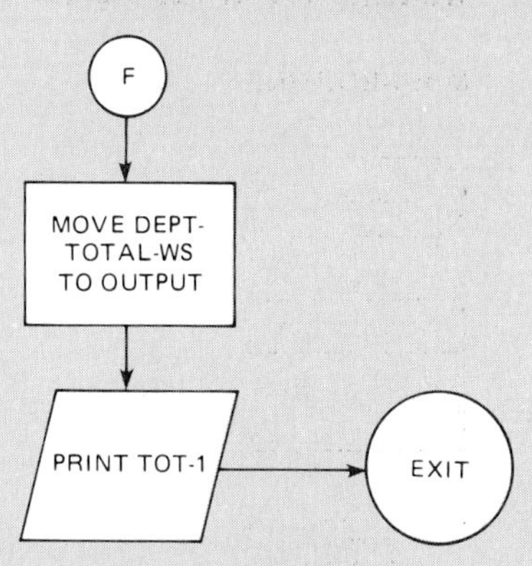

FINAL-TOTAL-PARAGRAPH

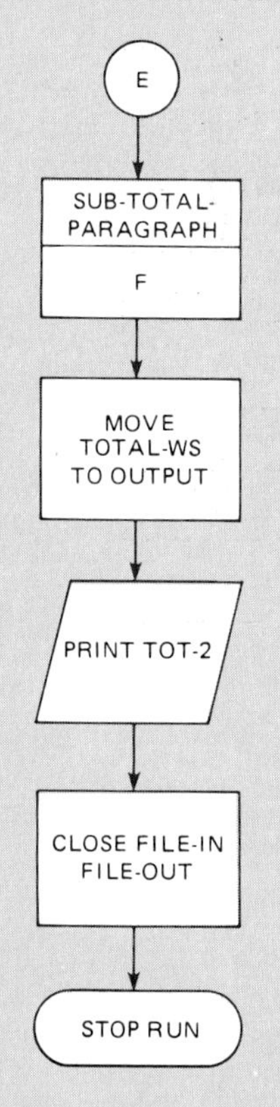

```
                                          WEEKLY PAYROLL REGISTER

           EMPLOYEE NO.         GROSS                          FICA        FEDERAL         STATE          MISC.           NET
         DEPT.     SERIAL      EARNINGS      INSURANCE         TAX       WITHHOLDING    WITHHOLDING      DEDUCT.        AMOUNT

          ZZZ      ZZZZZZ      $$$$.99        $$$.99         $$$$.99        $$$.99         $$.99          $$$.99        $$$$$.99

          ZZZ      ZZZZZZ      $$$$.99        $$$.99         $$$$.99        $$$.99         $$.99          $$$.99        $$$$$.99

                                                                                               DEPT. EARNINGS  $$,$$$.99 *

          ZZZ      ZZZZZZ      $$$$.99        $$$.99         $$$$.99        $$$.99         $$.99          $$$.99        $$$$$.99
          ZZZ      ZZZZZZ      $$$$.99        $$$.99         $$$$.99        $$$.99         $$.99          $$$.99        $$$$$.99

                                                                                               DEPT. EARNINGS  $$,$$$.99 *

                                                                                      TOTAL NET EARNINGS  $$$,$$$.99 **
```

```
PP 5734-CB2 V4 RELEASE 1.3 01AUG75         IBM OS AMERICAN NATIONAL STANDARD COBOL        DATE AUG  4,1977

    1

00001          IDENTIFICATION DIVISION.                                        4
00002
00003          PROGRAM-ID.                                                     5
00004              WEEKLY-PAYROLL-REGISTER.                                    6
00005          AUTHOR.                                                         7
00006              DOROTHY R HENNESSEY.                                        8
00007          INSTALLATION.                                                   9
00008              WEST LOS ANGELES COLLEGE.                                   10
00009          DATE-WRITTEN.                                                   11
00010              8 MAY 1977.                                                 12
00011          DATE-COMPILED. AUG  4,1977.                                     13
00012          REMARKS.                                                        14
00013              THIS PROGRAM PREPARES A WEEKLY PAYROLL REGISTER.            15
00014
00015
00016          ENVIRONMENT DIVISION.                                           16
00017
00018          CONFIGURATION SECTION.                                          17
00019          SOURCE-COMPUTER.                                                18
00020              IBM-360.                                                    19
00021          OBJECT-COMPUTER.                                                20
00022              IBM-360.                                                    21
00023          SPECIAL-NAMES.                                                  22
00024              C01 IS TO-TOP.                                              23
00025          INPUT-OUTPUT SECTION.                                           24
00026          FILE-CONTROL.                                                   25
00027              SELECT FILE-IN       ASSIGN TO UR-2540R-S-CARDIN.           26
00028              SELECT FILE-OUT      ASSIGN TO UR-1403-S-PRINT.             28
```

2

```
00030          DATA DIVISION.                                                                31
00031
00032          FILE SECTION.                                                                 32
00033          FD  FILE-IN                                                                   33
00034              RECORDING MODE IS F                                                       34
00035              LABEL RECORDS ARE OMITTED                                                 35
00036              DATA RECORD IS CARDIN.                                                    36
00037          01  CARDIN.                                                                   37
00038              03  FILLER             PICTURE X(13).                                     38
00039              03  DEPT               PICTURE XXX.                                       39
00040              03  SERIAL             PICTURE X(5).                                      40
00041              03  FILLER             PICTURE X(35).                                     41
00042              03  GROSS              PICTURE 999V99.                                    42
00043              03  INSURANCE          PICTURE 99V99.                                     43
00044              03  FILLER             PICTURE XXX.                                       44
00045              03  FED-WH-TAX         PICTURE 99V99.                                     45
00046              03  ST-WH-TAX          PICTURE 9V99.                                      46
00047              03  MISC-DEDUCT        PICTURE 99V99.                                     47
00048              03  KODE               PICTURE X.                                         48
00049          FD  FILE-OUT                                                                  49
00050              RECORDING MODE IS F                                                       51
00051              LABEL RECORDS ARE OMITTED                                                 52
00052              DATA RECORD IS PRINT.                                                     53
00053          01  PRINT                  PICTURE X(133).                                    54
00054          WORKING-STORAGE SECTION.                                                      55
00055          77  DEPT-WS                PICTURE XXX       VALUE SPACES.                    56
00056          77  FICA-WS                PICTURE 999V99    VALUE ZEROES.                    57
00057          77  NET-WS                 PICTURE 999V99    VALUE ZEROES.                    58
00058          77  DEPT-TOTAL-WS          PICTURE 999V99    VALUE ZEROES.                    59
00059          77  TOTAL-WS               PICTURE 9(5)V99   VALUE ZEROES.                    60
00060          77  LINE-COUNT-WS          PICTURE 99        VALUE ZEROES.                    61
00061          01  FLAGS.                                                                    61.2
00062              03  MORE-DATA-FLAG     PICTURE XXX       VALUE 'YES'.
00063                  88  MORE-DATA                        VALUE 'YES'.
00064                  88  NO-MORE-DATA                     VALUE 'NO '.
```

3

```
00066          01  HDG-1.                                                                    62
00067              03  FILLER   PICTURE X(55)    VALUE SPACES.                               63
00068              03  FILLER   PICTURE X(23)    VALUE 'WEEKLY PAYROLL REGISTER'.            64
00069              03  FILLER   PICTURE X(55)    VALUE SPACES.                               65
00070          01  HDG-2.                                                                    67
00071              03  FILLER   PICTURE X(23)    VALUE '             EMPLOYEE NO.'.          68
00072              03  FILLER   PICTURE X(16)    VALUE '           GROSS'.                   69
00073              03  FILLER   PICTURE X(25)    VALUE SPACES.                               70
00074              03  FILLER   PICTURE X(22)    VALUE 'FICA            FEDERAL  '.          71
00075              03  FILLER   PICTURE X(24)    VALUE '       STATE          MISC.'.        72
00076              03  FILLER   PICTURE X(23)    VALUE '               NET      '.          73
00077          01  HDG-3.                                                                    74
00078              03  FILLER   PICTURE X(20)    VALUE '            DEPT.     '.             75
00079              03  FILLER   PICTURE X(21)    VALUE 'SERIAL        EARNINGS'.             76
00080              03  FILLER   PICTURE X(23)    VALUE '      INSURANCE        '.            77
00081              03  FILLER   PICTURE X(24)    VALUE 'TAX          WITHHOLDING  '.         78
00082              03  FILLER   PICTURE X(23)    VALUE 'WITHHOLDING        DEDUCT.'.         79
00083              03  FILLER   PICTURE X(22)    VALUE '            AMOUNT       '.          80
00084          01  DETAIL-LINE.                                                              81
00085              03  FILLER             PICTURE X(9)      VALUE SPACES.                    82
00086              03  DEPT               PICTURE ZZZ.                                       83
00087              03  FILLER             PICTURE X(8)      VALUE SPACES.                    84
00088              03  SERIAL             PICTURE X(5).                                      85
00089              03  FILLER             PICTURE X(8)      VALUE SPACES.                    86
00090              03  GROSS              PICTURE $$$$.99.                                   87
00091              03  FILLER             PICTURE X(8)      VALUE SPACES.                    88
00092              03  INSURANCE          PICTURE $$$.99.                                    89
00093              03  FILLER             PICTURE X(8)      VALUE SPACES.                    90
00094              03  FICA-OUT           PICTURE $$$$.99.                                   91
00095              03  FILLER             PICTURE X(8)      VALUE SPACES.                    92
00096              03  FED-WH-TAX         PICTURE $$$.99.                                    93
00097              03  FILLER             PICTURE X(8)      VALUE SPACES.                    94
00098              03  ST-WH-TAX          PICTURE $$.99.                                     95
00099              03  FILLER             PICTURE X(8)      VALUE SPACES.                    96
00100              03  MISC-DEDUCT        PICTURE $$$.99.                                    97
00101              03  FILLER             PICTURE X(8)      VALUE SPACES.                    98
00102              03  NET-OUT            PICTURE $$$$.99.                                   99
00103              03  FILLER             PICTURE X(8)      VALUE SPACES.                    100
```

```
   4

00105          01  TOT-1.                                                      101
00106              03  FILLER          PICTURE X(100)  VALUE SPACES.           102
00107              03  FILLER          PICTURE X(16)   VALUE 'DEPT. EARNINGS  '.103
00108              03  DEPT-TOTAL-OUT  PICTURE $$,$$$.99.                      104
00109              03  FILLER          PICTURE X(8)    VALUE ' *      '.       105
00110          01  TOT-2.                                                      106
00111              03  FILLER          PICTURE X(96)   VALUE SPACES.           107
00112              03  FILLER          PICTURE X(10)   VALUE 'TOTAL NET '.
00113              03  FILLER          PICTURE X(9)    VALUE 'EARNINGS '.
00114              03  TOTAL-OUT       PICTURE $$$,$$$.99.                     110
00115              03  FILLER          PICTURE X(10)   VALUE ' **       '.
00116          01  EMPTY         PICTURE X(133)  VALUE SPACES.                 111.05
00117
00118
00119          PROCEDURE DIVISION.                                             112
00120
00121          MAIN-ROUTINE.                                                   112.2
00122              PERFORM INITIALIZATION-PARAGRAPH.                           114
00123              PERFORM TITLE-PARAGRAPH.                                    116
00124              PERFORM COLUMN-HEAD-PARAGRAPH.
00125              READ FILE-IN                                                154
00126                  AT END MOVE 'NO ' TO MORE-DATA-FLAG.                    1555
00127              PERFORM NEWPAGE-PARAGRAPH
00128                  THRU NEWPAGE-PARA-EXIT
00129                  UNTIL NO-MORE-DATA.                                     117.1
00130              PERFORM FINAL-TOTAL-PARAGRAPH.                              117.2
00131
00132          INITIALIZATION-PARAGRAPH.                                       133
00133              OPEN INPUT  FILE-IN
00134                   OUTPUT FILE-OUT.
00135
00136          TITLE-PARAGRAPH.                                                137
00137              WRITE PRINT FROM HDG-1                                      138
00138                  AFTER ADVANCING TO-TOP LINES.                           139
00139              WRITE PRINT FROM EMPTY                                      140
00140                  AFTER ADVANCING 3 LINES.                                141
00141              ADD 4 TO LINE-COUNT-WS.                                     142


   5

00143          COLUMN-HEAD-PARAGRAPH.                                          144
00144              WRITE PRINT FROM HDG-2                                      145
00145                  AFTER ADVANCING 1 LINES.                                146
00146              WRITE PRINT FROM HDG-3                                      147
00147                  AFTER ADVANCING 1 LINES.                                148
00148              WRITE PRINT FROM EMPTY                                      149
00149                  AFTER ADVANCING 1 LINES.                                150
00150              ADD 3 TO LINE-COUNT-WS.                                     151
00151
00152          NEWPAGE-PARAGRAPH.
00153              IF LINE-COUNT-WS IS > 37                                    117.11
00154              THEN
00155                  WRITE PRINT FROM EMPTY
00156                      AFTER ADVANCING TO-TOP LINES
00157                  MOVE 1 TO LINE-COUNT-WS
00158                  PERFORM COLUMN-HEAD-PARAGRAPH
00159              ELSE
00160                  NEXT SENTENCE.
00161
00162              IF DEPT OF CARDIN IS NOT = DEPT-WS
00163                  AND DEPT-WS IS NOT = '   '
00164              THEN
00165                  PERFORM SUB-TOTAL-PARAGRAPH
00166              ELSE
00167                  NEXT SENTENCE.
00168
00169              IF DEPT-WS IS = '   '
00170              THEN
00171                  MOVE DEPT OF CARDIN TO DEPT-WS
00172              ELSE
00173                  NEXT SENTENCE.
```

6

```
00175          COMPUTE FICA-WS ROUNDED = GROSS OF CARDIN              157.0
00176              * 0.0585.
00177          COMPUTE NET-WS = GROSS OF CARDIN                       158.0
00178              - INSURANCE OF CARDIN
00179              - FICA-WS
00180              - FED-WH-TAX OF CARDIN
00181              - ST-WH-TAX OF CARDIN
00182              - MISC-DEDUCT OF CARDIN.
00183          ADD NET-WS TO DEPT-TOTAL-WS, TOTAL-WS.                 161
00184          MOVE CORRESPONDING CARDIN TO DETAIL-LINE.              162
00185          MOVE FICA-WS TO FICA-OUT.                              162.1
00186          MOVE NET-WS TO NET-OUT.                                162.2
00187          WRITE PRINT FROM DETAIL-LINE                           163
00188              AFTER ADVANCING 1 LINES.                           164
00189          ADD 1 TO LINE-COUNT-WS.                                164.1
00190          READ FILE-IN
00191              AT END MOVE 'NO ' TO MORE-DATA-FLAG.
00192      NEWPAGE-PARA-EXIT.  EXIT.
00193
00194      SUB-TOTAL-PARAGRAPH.                                       166
00195          MOVE DEPT-TOTAL-WS TO DEPT-TOTAL-OUT.                  166.1
00196          WRITE PRINT FROM TOT-1                                 167
00197              AFTER ADVANCING 2 LINES.                           168
00198          WRITE PRINT FROM EMPTY                                 169
00199              AFTER ADVANCING 1 LINES.                           169.05
00200          ADD 3 TO LINE-COUNT-WS.                                169.06
00201          MOVE SPACES TO DEPT-WS.                                169.1
00202          MOVE ZEROES TO DEPT-TOTAL-WS.                          171
00203
00204      FINAL-TOTAL-PARAGRAPH.                                     172
00205          PERFORM SUB-TOTAL-PARAGRAPH.                           172.05
00206          MOVE TOTAL-WS TO TOTAL-OUT.                            172.1
00207          WRITE PRINT FROM TOT-2                                 173
00208              AFTER ADVANCING 1 LINES.                           174
00209          CLOSE FILE-IN                                          175
00210              FILE-OUT.                                          176
00211          STOP RUN.                                              177
```

WEEKLY PAYROLL REGISTER

EMPLOYEE NO. DEPT.	SERIAL	GROSS EARNINGS	INSURANCE	FICA TAX	FEDERAL WITHHOLDING	STATE WITHHOLDING	MISC. DEDUCT.	NET AMOUNT
9	16572	$115.78	$.00	$6.77	$8.90	$.84	$.00	$99.27
9	2A182	$84.17	$1.00	$4.92	$16.02	$1.19	$1.75	$59.29
9	970K5	$100.65	$3.00	$5.89	$12.57	$1.34	$.00	$77.85
9	63600	$116.40	$2.00	$6.81	$18.90	$1.85	$2.00	$84.84
							DEPT. EARNINGS	$321.25 *
10	3L715	$156.80	$2.00	$9.17	$22.46	$2.02	$.00	$121.15
10	38Y8J	$186.57	$.00	$10.91	$4.80	$.60	$.75	$169.51
10	W2923	$75.38	$1.25	$4.41	$9.75	$.78	$.00	$59.19
10	13E10	$84.17	$.00	$4.92	$8.90	$.84	$.00	$69.51
							DEPT. EARNINGS	$419.36 *
19	1P411	$100.65	$1.00	$5.89	$16.02	$1.19	$1.75	$74.80
19	64G12	$116.40	$3.00	$6.81	$12.57	$1.34	$.00	$92.68
19	13I43	$156.80	$7.00	$9.17	$18.90	$1.85	$7.00	$112.88
							DEPT. EARNINGS	$280.36 *
20	31B54	$186.57	$2.00	$10.91	$22.46	$2.02	$.00	$149.18
20	H1954	$67.20	$.00	$3.93	$4.80	$.60	$.75	$57.12
20	16X33	$75.38	$1.25	$4.41	$9.75	$.78	$.00	$59.19
20	137M5	$84.17	$.00	$4.92	$8.90	$.84	$.00	$69.51
							DEPT. EARNINGS	$335.00 *
29	V1388	$100.65	$1.00	$5.89	$16.02	$1.19	$1.75	$74.80
29	4C193	$116.40	$3.00	$6.81	$12.57	$1.34	$.00	$92.68
29	2N270	$156.80	$2.00	$9.17	$18.90	$1.85	$2.00	$122.88
							DEPT. EARNINGS	$290.36 *
30	52Q16	$186.57	$7.00	$10.91	$22.46	$2.02	$.00	$144.18

EMPLOYEE NO. DEPT.	SERIAL	GROSS EARNINGS	INSURANCE	FICA TAX	FEDERAL WITHHOLDING	STATE WITHHOLDING	MISC. DEDUCT.	NET AMOUNT
30	322D6	$67.70	$.00	$3.96	$4.80	$.60	$.75	$57.59
30	7U223	$75.18	$1.25	$4.40	$9.75	$.78	$.00	$59.00
30	25435	$84.17	$.00	$4.92	$8.90	$.84	$.00	$69.51
							DEPT. EARNINGS	$330.28 *
39	2R598	$100.64	$1.00	$5.89	$16.02	$1.19	$1.75	$74.79
39	226T3	$116.40	$3.00	$6.81	$12.57	$1.34	$.00	$92.68
							DEPT. EARNINGS	$167.47 *
							TOTAL NET EARNINGS	$2,144.08 **

DISTRICT SALES REPORT PROBLEM

INPUT

Field	Card Columns
Product Number	1–2
Sales Amount	3–7
Salesman Number	8–9
District Number	10

CALCULATIONS TO BE PERFORMED

Accumulate totals for sales amount as follows:

a. Minor Control—Salesman Number
b. Intermediate Control—District
c. Final Total—Monthly

OUTPUT

Prepare a salesman report per the following output format:

DISTRICT SALES REPORT

SALESMAN NUMBER	SALES	DISTRICT NUMBER	SALES

MONTHLY TOTAL SALES

DISTRICT SALES REPORT

MAIN-ROUTINE

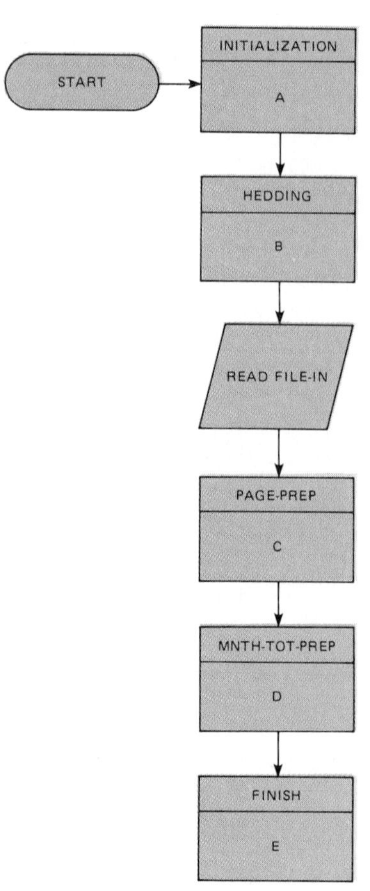

PROCESS-ROUTINES

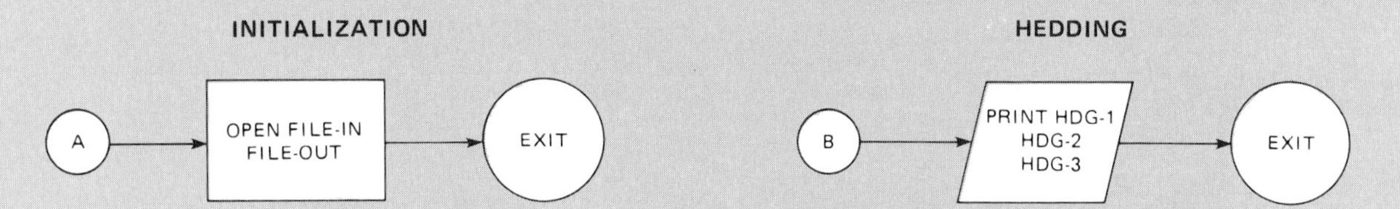

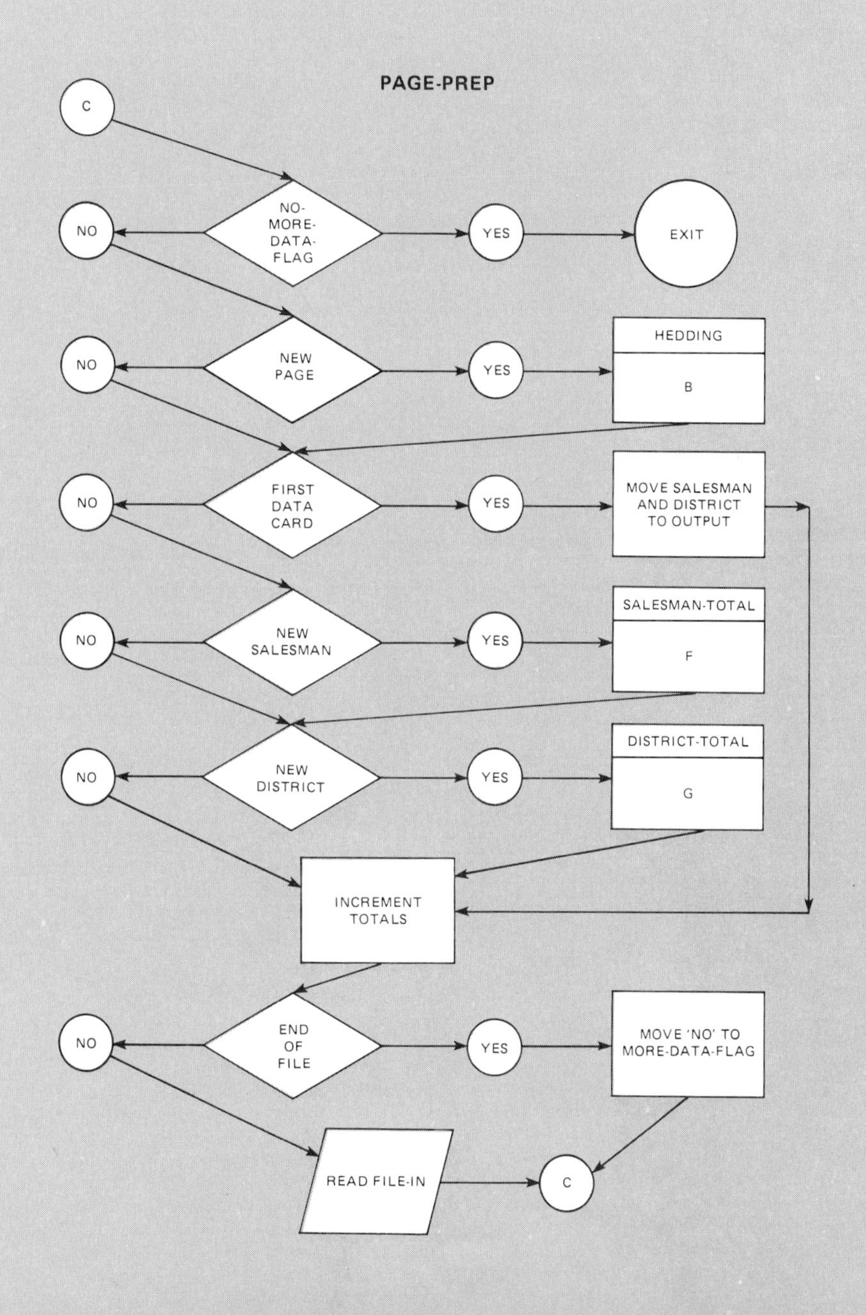

(PROCESS-ROUTINES, CONTINUED)

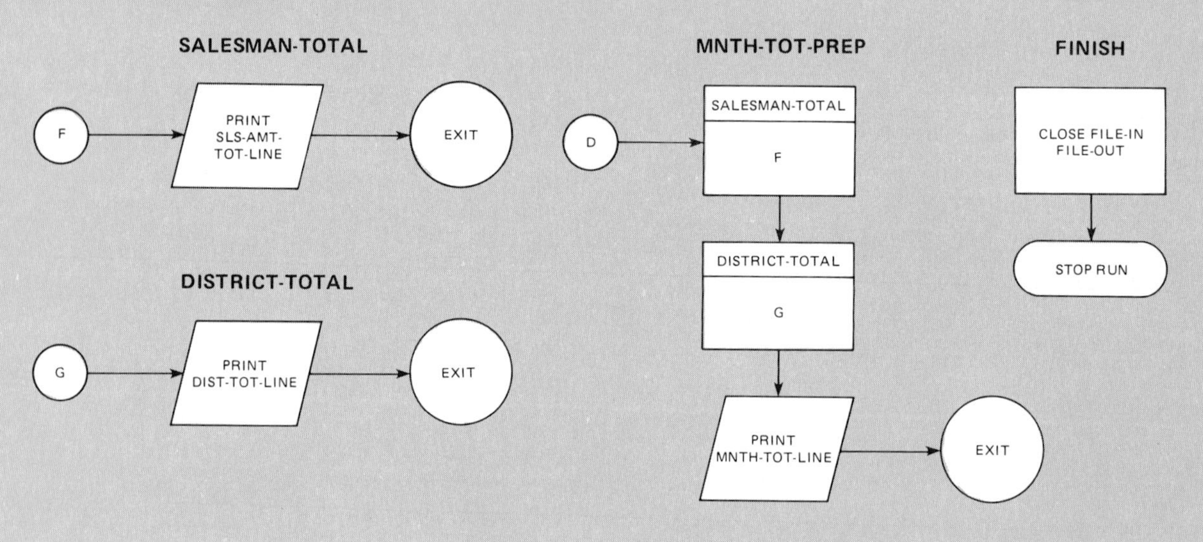

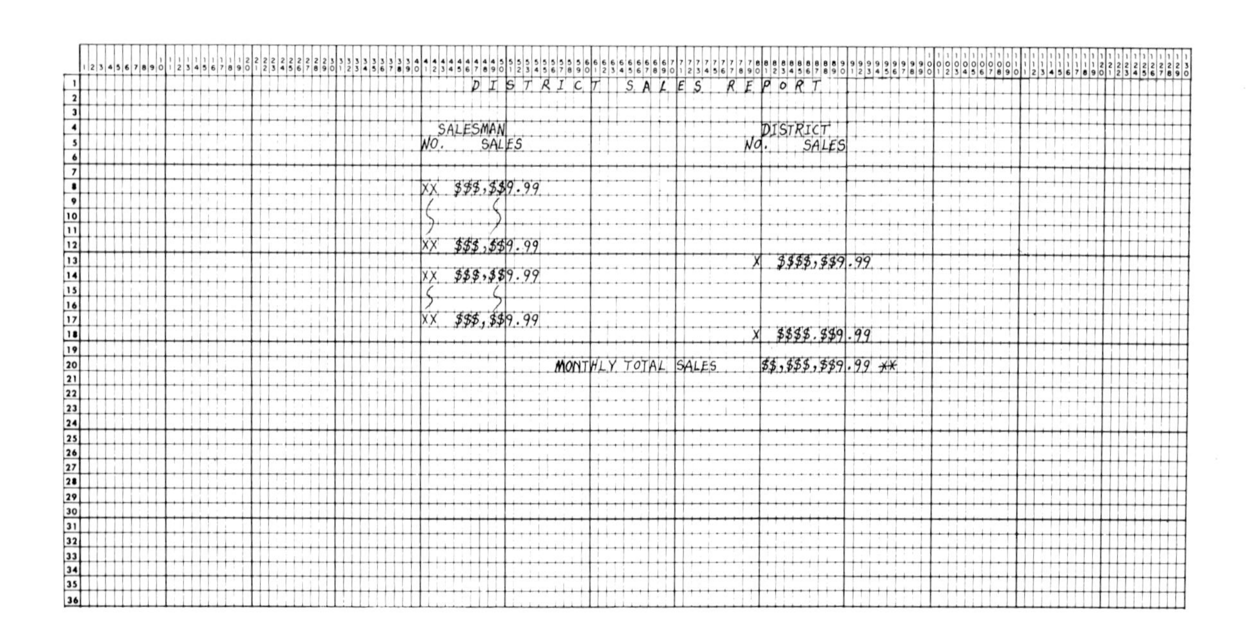

```
PP 5734-CB2 V4 RELEASE 1.3 01AUG75       IBM CS AMERICAN NATICNAL STANDARC COBOL       DATE AUG  4,1977

   1

00001          IDENTIFICATION DIVISION.                                          2
00002
00003          PROGRAM-ID.                                                       3
00004              DISTRICT-SALES-REPORT.                                        4
00005          AUTHOR.                                                           5
00006              DOROTHY R HENNESSEY.                                          6
00007          INSTALLATION.                                                     7
00008              WEST LOS ANGELES COLLEGE.                                     8
00009          DATE-WRITTEN.                                                     9
00010              16 JUNE 1977.                                                 10
00011          DATE-COMPILED. AUG  4,1977.                                       11
00012          REMARKS.                                                          12
00013              THIS PROGRAM PREPARES A REPORT OF AMOUNT OF SALES FOR EACH    13
00014                  SALESMAN AND DISTRICT WITH MONTHLY TOTALS.                14
00015
00016
00017          ENVIRONMENT DIVISION.                                             16
00018          CONFIGURATION SECTION.                                            17
00019          SOURCE-COMPUTER.                                                  18
00020              IBM-360.                                                      19
00021          OBJECT-COMPUTER.                                                  20
00022              IBM-360.                                                      21
00023          SPECIAL-NAMES. C01 IS SKIP-TC-CNE.                                22
00024          INPUT-OUTPUT SECTION.                                             23
00025          FILE-CCNTROL.                                                     24
00026              SELECT FILE-IN  ASSIGN TO UR-2540R-S-CARDIN.                  25
00027              SELECT FILE-OUT ASSIGN TO UR-1403-S-PRINT.                    26
```

```
   2

00029          DATA DIVISION.                                                    28
00030
00031          FILE SECTION.                                                     29
00032          FD  FILE-IN                                                       30
00033              RECORDING MODE IS F                                           31
00034              LABEL RECORDS ARE OMITTED                                     32
00035              DATA RECORD IS CARDIN.                                        33
00036          01  CARDIN.                                                       34
00037              03  PRODUCT       PICTURE XX.                                 35
00038              03  SALES-AMOUNT  PICTURE 999V99.                             36
00039              03  SALESMAN      PICTURE XX.                                 37
00040              03  DISTRICT      PICTURE X.                                  38
00041              03  FILLER        PICTURE X(70).                              39
00042          FD  FILE-OUT                                                      40
00043              RECORDING MODE IS F                                           41
00044              LABEL RECORDS ARE OMITTED                                     42
00045              DATA RECORD IS PRINT.                                         43
00046          01  PRINT                   PICTURE X(133).                       44
00047          WORKING-STORAGE SECTION.                                          45
00048          77  SLS-AMT-TOT-WS          PICTURE 9(5)V99 VALUE ZEROES.         48
00049          77  DIST-AMT-TOT-WS         PICTURE 9(6)V99 VALUE ZEROES.         49
00050          77  MNTH-TOT-WS             PICTURE 9(7)V99 VALUE ZEROES.         50
00051          77  PRNT-LINE-CNT           PICTURE 99      VALUE ZEROES.         51
00052          01  FLAGS.                                                        52
00053              03  MORE-DATA-FLAG  PICTURE XXX     VALUE 'YES'.              53
00054                  88  MORE-DATA                   VALUE 'YES'.              54
00055                  88  NO-MORE-DATA                VALUE 'NO '.              55
00056          01  HDG-1.                                                        56
00057              03  FILLER  PICTURE X(46)   VALUE SPACES.                     57
00058              03  FILLER  PICTURE X(24)   VALUE 'D I S T R I C T   S A L '.58
00059              03  FILLER  PICTURE X(17)   VALUE 'E S   R E P O R T'.        59
00060              03  FILLER  PICTURE X(46)   VALUE SPACES.                     60
00061          01  HDG-2.                                                        61
00062              03  FILLER        PICTURE X(42)       VALUE SPACES.           62
00063              03  FILLER        PICTURE X(8)        VALUE 'SALESMAN'.       63
00064              03  FILLER        PICTURE X(30)       VALUE SPACES.           64
00065              03  FILLER        PICTURE X(8)        VALUE 'DISTRICT'.       65
00066              03  FILLER        PICTURE X(45)       VALUE SPACES.           66
```

```
  3

00068         01  HDG-3.                                                        67
00069             03  FILLER        PICTURE X(40)        VALUE SPACES.          68
00070             03  FILLER        PICTURE X(12)        VALUE 'NO.     SALES'.  69
00071             03  FILLER        PICTURE X(26)        VALUE SPACES.          70
00072             03  FILLER        PICTURE X(21)        VALUE 'NO.     SALES'.  71
00073         01  SLS-AMT-TOT-LINE.                                             72
00074             03  FILLER        PICTURE X(40)        VALUE SPACES.          73
00075             03  SALESMAN      PICTURE XX           VALUE SPACES.          74
00076             03  FILLER        PICTURE XX           VALUE SPACES.          75
00077             03  SLS-AMT-TOT   PICTURE $$$,$$9.99   VALUE ZEROES.          76
00078             03  FILLER        PICTURE X(79)        VALUE SPACES.          77
00079         01  DIST-TOT-LINE.                                                78
00080             03  FILLER        PICTURE X(79)        VALUE SPACES.          79
00081             03  DISTRICT      PICTURE X            VALUE SPACES.          80
00082             03  FILLER        PICTURE XX           VALUE SPACES.          81
00083             03  DIST-AMT-TOT  PICTURE $$$$,$$9.99  VALUE ZEROES.          82
00084             03  FILLER        PICTURE X(40)        VALUE SPACES.          83
00085         01  MNTH-TOT-LINE.                                                84
00086             03  FILLER       PICTURE X(56)   VALUE SPACES.                85
00087             03  FILLER       PICTURE X(19)   VALUE 'MONTHLY TOTAL SALES'. 86
00088             03  FILLER       PICTURE X(5)    VALUE SPACES.                88
00089             03  MNTH-TOT     PICTURE $$,$$$,$$9.99 VALUE ZEROES.          89
00090             03  FILLER       PICTURE XXX     VALUE ' **'.                 90
00091             03  FILLER       PICTURE X(37)   VALUE SPACES.                91
00092         01  EMPTY            PICTURE X(133)  VALUE SPACES.                92
```

```
  4

00094         PROCEDURE DIVISION.                                               93
00095
00096         MAIN-ROUTINE.                                                     95
00097             PERFORM INITIALIZATION.                                       96
00098             PERFORM HEDDING.                                              96.1
00099             READ FILE-IN                                                  96.2
00100                 AT END MOVE 'NO ' TO MORE-DATA-FLAG.                      96.3
00101             PERFORM PAGE-PREP                                             97
00102                 THRU PAGE-PREP-EXIT
00103                 UNTIL NO-MORE-DATA.                                       98
00104             PERFORM MNTH-TOT-PREP.                                        99
00105             PERFORM FINISH.                                               100
00106
00107         PROCESS-ROUTINES.                                                 101
00108
00109         INITIALIZATION.                                                   109
00110             OPEN INPUT  FILE-IN
00111                  OUTPUT FILE-OUT.
00112
00113         HEDDING.                                                          113
00114             MOVE ZEROES TO PRNT-LINE-CNT.                                 114
00115             WRITE PRINT FROM HDG-1                                        115
00116                 AFTER ADVANCING SKIP-TO-ONE LINES.                        116
00117             WRITE PRINT FROM HDG-2                                        117
00118                 AFTER ADVANCING 3 LINES.                                  118
00119             WRITE PRINT FROM HDG-3                                        119
00120                 AFTER ADVANCING 1 LINES.                                  120
00121             WRITE PRINT FROM EMPTY                                        121
00122                 AFTER ADVANCING 2 LINES.                                  122
00123             ADD 7 TO PRNT-LINE-CNT.                                       123
00124
00125         PAGE-PREP.                                                        102
00126             IF PRNT-LINE-CNT > 39
00127             THEN
00128                 PERFORM HEDDING
00129             ELSE
00130                 NEXT SENTENCE.
```

5

```
00132           IF SALESMAN OF SLS-AMT-TOT-LINE IS = ' '
00133           THEN
00134               MOVE SALESMAN OF CARDIN
00135               TO SALESMAN OF SLS-AMT-TOT-LINE
00136               MOVE DISTRICT OF CARDIN
00137               TO DISTRICT OF DIST-TOT-LINE
00138           ELSE
00139               NEXT SENTENCE.
00140
00141           IF SALESMAN OF CARDIN IS NOT = SALESMAN OF SLS-AMT-TOT-LINE
00142           THEN
00143               PERFORM SALESMAN-TOTAL
00144           ELSE
00145               NEXT SENTENCE.
00146
00147           IF DISTRICT OF CARDIN IS NOT = DISTRICT OF DIST-TOT-LINE
00148           THEN
00149               PERFORM DISTRICT-TOTAL
00150           ELSE
00151               NEXT SENTENCE.
00152
00153           ADD SALES-AMOUNT TO SLS-AMT-TOT-WS
00154               DIST-AMT-TOT-WS
00155               MNTH-TOT-WS.
00156           READ FILE-IN                                              124
00157               AT END MOVE 'NO ' TO MORE-DATA-FLAG.                  125
00158       PAGE-PREP-EXIT.  EXIT.
00159
00160       SALESMAN-TOTAL.                                               146
00161           MOVE SLS-AMT-TOT-WS TO SLS-AMT-TOT                        146.1
00162           WRITE PRINT FROM SLS-AMT-TOT-LINE                         146.2
00163           AFTER ADVANCING 1 LINES.                                  146.3
00164           ADD 1 TO PRNT-LINE-CNT.                                   146.4
00165           MOVE ZEROES TO SLS-AMT-TOT-WS.                            146.5
00166           MOVE SALESMAN OF CARDIN                                   146.6
00167               TO SALESMAN OF SLS-AMT-TOT-LINE.                      146.7
```

6

```
00169       DISTRICT-TOTAL.                                               148.1
00170           MOVE DIST-AMT-TOT-WS TO DIST-AMT-TOT.                     148.2
00171           WRITE PRINT FROM DIST-TOT-LINE                            148.3
00172               AFTER ADVANCING 1 LINES.                              148.4
00173           ADD 1 TO PRNT-LINE-CNT.                                   148.5
00174           MOVE ZEROES TO DIST-AMT-TOT-WS.                           148.6
00175           MOVE DISTRICT OF CARDIN TO DISTRICT OF DIST-TOT-LINE.     145
00176
00177       MNTH-TOT-PREP.                                                150
00178           PERFORM SALESMAN-TOTAL                                    150.01
00179           PERFORM DISTRICT-TOTAL.                                   150.1
00180           MOVE MNTH-TOT-WS TO MNTH-TOT.                             155.4
00181           WRITE PRINT FROM MNTH-TOT-LINE                            157
00182               AFTER ADVANCING 2 LINES.                              158
00183           MOVE ZEROES TO MNTH-TOT-WS.                               159
00184
00185       FINISH.                                                       161
00186           CLOSE FILE-IN                                             162
00187                 FILE-OUT.                                           163
00188           STOP RUN.                                                 164
```

```
                 D I S T R I C T   S A L E S   R E P O R T

    SALESMAN                                      DISTRICT
NO.      SALES                                  NO.      SALES


01      $412.00
02      $632.75
03      $212.00
04      $959.50
                                                 A      $2,216.25
05      $523.00
06      $212.00
07      $662.00
08       $50.00
                                                 B      $1,447.00
09      $250.00
10      $650.00
11       $59.75
12      $412.00
14      $632.75
                                                 C      $2,004.50
15      $212.00
16      $959.50
17      $523.00
18      $212.00
                                                 D      $1,906.50
19      $662.00
20       $50.00
21      $250.00
22      $650.00
23       $59.75
24      $412.00
                                                 E      $2,083.75
25      $632.75
26      $212.00
27      $959.50
28      $523.00
                                                 F      $2,327.25

                 D I S T R I C T   S A L E S   R E P O R T

    SALESMAN                                      DISTRICT
NO.      SALES                                  NO.      SALES


29      $212.00
30      $662.00
31       $50.00
32      $250.00
33      $650.00
                                                 G      $1,824.00
34       $59.75
35      $412.00
36      $632.75
37      $212.00
                                                 H      $1,316.50
38      $959.50
39      $523.00
40      $212.00
41      $100.00
                                                 J      $1,794.50
42      $562.00
43       $50.00
44      $250.00
45      $650.00
46       $59.75
                                                 K      $1,571.75

                        MONTHLY TOTAL SALES           $18,492.00 **
```

Exercises

Write your answer in the space provided (answer may be one or more words).

1. The Procedure Division specifies the ________________ necessary to solve a given problem.
2. A statement is a syntactical valid combination of ________________ and ________________ beginning with a COBOL ________________.
3. A statement expresses an ________________ to be taken during the ________________ of the ________________ program.
4. Each statement has its own particular ________________ that specifies the types of ________________ required and the ________________ of the statements.
5. COBOL statements may be compared to ________________ in the English language.
6. The imperative statement consists of one or more ________________ commands to be performed by the object program.
7. A conditional statement is a statement that is to be ________________ and the ________________ of the conditional expression will determine which of the ________________ paths the program will follow.
8. A ________________ statement directs the computer to perform some action at compile time.
9. A sentence is composed of one or more ________________ specifying action to be taken and is terminated by a ________________ and a ________________.
10. Separators when used must be followed by at least one ________________.
11. COBOL sentences may be combined into a logical entity called a ________________.
12. A procedure-name must not be ________________ within the same section.
13. A ________________ is composed of one or more successive paragraphs.
14. The Procedure Division generally consists of a series of ________________ which may be optionally grouped by the programmer into ________________.
15. The Declaratives Section must be grouped at the ________________ of the Procedure Division.
16. Statements are executed in the sequence in which they are written unless altered by the ________________.
17. COBOL ________________ are the basis of the Procedure Division.
18. The programmer is concerned only with the use of ________________ records.
19. In data processing operations, the flow of data through a system is governed by the ________________ system.
20. The four verbs used to specify the flow of data to or from files stored on an external media are ________________, ________________, ________________, and ________________.
21. ________________ and ________________ verbs are used in conjunction with low volume data.
22. The ________________ statement makes one or more input/output files ready for reading or writing.
23. An ________________ statement must be specified for all ________________ in a COBOL program and must be executed prior to any other input/output statement for that file.
24. Each file that has been opened must be defined in the ________________ entry in the Data Division as well as in the ________________ clause in the Environment Division.
25. The ________________ statement makes a data record ready for processing.
26. The file must be ________________ before it is read.
27. The INTO option converts the READ statement into a ________________ and ________________ statement.
28. An ________________ clause must be included in all READ statements for sequential files.
29. The ________________ option must be specified for mass storage files in the random access mode.
30. The WRITE statement releases a ________________ for insertion in an ________________ file.
31. The format for a WRITE statement requires a ________________ rather than a file-name.
32. The ________________ option is used with the WRITE statement to allow vertical positioning up to 99 lines.
33. The ________________ option can be used to test for channel 12 on an online printer.
34. The CLOSE statement ________________ the processing of one or more files and must be written for all files that have been previously ________________.

35. The ACCEPT statement obtains ______________ data from the systems logical ______________ device or from the ______________.
36. The DISPLAY statement causes the ______________ of low volume data on an ______________ device.
37. The MOVE statement transmits data from ______________ storage area to ______________ or more ______________ areas within the computer.
38. When a group item is involved in a move, the data is moved without any regard to the ______________ of the group item.
39. When both the sending and receiving areas are elementary items, appropriate ______________ to the format of the ______________ area takes place.
40. The figurative constant ______________ belongs to the numeric category of data while the figurative constant ______________ belongs to the nonnumeric category.
41. In a numeric data MOVE statement, the source area is aligned with regard to the ______________ in the receiving area and any necessary ______________ will take place.
42. Nonnumeric data from a source area is placed in the receiving area beginning at the ______________ and continuing to the ______________ unless the field is specified as ______________.
43. In a MOVE CORRESPONDING statement execution, selected items within the ______________ area are moved with required ______________ to the selected ______________ area.
44. At least one of selected items in a MOVE CORRESPONDING statement must be an ______________ item.
45. The EXAMINE statement is used to replace a given ______________ and/or to ______________ the number of times it appears in the data item.
46. All identifiers used in arithmetic statements must represent elementary ______________ items.
47. The identifier that follows GIVING may contain ______________ characters.
48. The literals used in arithmetic statements must be ______________.
49. Decimal alignment is performed automatically throughout the computation in accordance with the ______________ clause of the ______________ and ______________.
50. The ______________ option is used to half adjust the result of an arithmetic operation.
51. The ______________ option is used when the computed result after ______________ alignment ______________ the number of positions identified with the result.
52. The CORRESPONDING option of the ADD and SUBTRACT verbs allows the user to perform ______________ on an elementary item by simply specifying the ______________ to which they belong.
53. The ADD statement must contain at least ______________ operands.
54. The ______________ must not be written in the same ADD statement as TO.
55. In a SUBTRACT statement, all values of operands preceding the word ______________ are added together and this total is subtracted from the value of the ______________.
56. The ______________ statement specifies the use of an arithmetic expression.
57. All operators used in a ______________ statement must be ______________ and ______________ by one or more spaces.
58. If exponentiation is desired, the ______________ statement must be used.
59. In an arithmetic expression, all ______________ expressions are performed first.
60. The GO TO statement provides a means of transferring control ______________ or ______________ from one part of the program to another.
61. The GO TO statement can only be used as the ______________ sentence in the sequence in which it appears.
62. If the DEPENDING ON option of the ______________ statement is used, the value of the identifier must have a range starting at ______________ and continuing successively upward.
63. The ______________ statement is used to modify the effect of a GO TO statement.
64. The PERFORM statement permits a ______________ transfer of sequence control and a ______________ back at the conclusion of the execution of a specified condition.
65. In its simplest form, the PERFORM statement provides control back to the ______________ statement after the PERFORM statement.

66. The last statement in a referenced procedure of a PERFORM statement must not contain an unconditional ________________ statement.
67. The ________________ statement is used to halt execution of a program.
68. The EXIT verb is used with the ________________ verb in ________________ exits from a paragraph.
69. The ________________ statements cause the compiler to take specific action at ________________ time.
70. The ________________ statement enables the programmer to use other programming languages in a COBOL source program.
71. The NOTE statement is used to insert ________________ in the source program.
72. If NOTE is the first word of a paragraph, the paragraph must be devoted to ________________.
73. A conditional expression is an expression containing one or more ________________ whose ________________ may change during the course of the program.
74. If a test condition is ________________, the statements immediately following the ________________ expression are executed.
75. A ________________ condition test involves the comparison of two data values.
76. A Class Condition test is used to determine whether an item is ________________ or ________________.
77. The Condition-Name Condition test is an alternate way of expressing certain conditions that could be expressed in a simple ________________ condition.
78. In a compound conditional expression, if either or both are true, the logical operator ________________ is used.
79. Nesting a conditional expression is like specifying a ________________ arithmetic expression enclosed in ________________ enclosed in a larger arithmetic expression.
80. If the operator is implied, then the ________________ must also be implied.
81. ________________ and ________________ conditions can never have implied subjects.
82. A Relational Operator may be implied only in a ________________ relational condition.

Answers

1. ACTIONS
2. WORDS, SYMBOLS, VERB
3. ACTION, EXECUTION, OBJECT
4. FORMAT, WORDS, ORGANIZATION
5. CLAUSES
6. UNCONDITIONAL
7. TESTED, EVALUATION, ALTERNATE
8. COMPILER-DIRECTING
9. STATEMENTS, PERIOD, SPACE
10. SPACE
11. PARAGRAPH
12. DUPLICATED
13. SECTION
14. PARAGRAPHS, SECTIONS
15. BEGINNING
16. PROGRAM
17. VERBS
18. INDIVIDUAL
19. INPUT/OUTPUT
20. OPEN, READ, WRITE, CLOSE
21. ACCEPT, DISPLAY
22. OPEN
23. OPEN, FILES
24. FILE DESCRIPTION, SELECT
25. READ
26. OPENED
27. READ, MOVE
28. AT END
29. INVALID KEY
30. RECORD, OUTPUT
31. RECORD NAME
32. ADVANCING
33. END-OF-PAGE
34. TERMINATES, OPENED
35. LOW VOLUME, INPUT, CONSOLE
36. WRITING, OUTPUT
37. ONE, ONE, STORAGE
38. LEVEL STRUCTURE
39. EDITING, RECEIVING
40. ZERO, SPACE
41. DECIMAL POINT, EDITING
42. LEFT, RIGHT, JUSTIFIED RIGHT
43. SOURCE, EDITING, RECEIVING
44. ELEMENTARY
45. CHARACTER, COUNT
46. NUMERIC
47. EDITING
48. NUMERIC
49. PICTURE, OPERANDS, RESULTS
50. ROUNDED
51. SIZE ERROR, DECIMAL, EXCEEDS
52. COMPUTATION, GROUP
53. TWO
54. GIVING OPTION
55. FROM, IDENTIFIER
56. COMPUTE
57. COMPUTE, PRECEDED, FOLLOWED
58. COMPUTE
59. PARENTHETICAL
60. CONDITIONALLY, UNCONDITIONALLY
61. FINAL
62. GO TO, ONE
63. ALTER
64. TEMPORARY, RETURN
65. NEXT
66. GO TO
67. STOP
68. PERFORM, CONDITIONAL
69. COMPILER-DIRECTING, COMPILE
70. ENTER
71. COMMENTS
72. NOTES
73. VARIABLES, VALUES

74. TRUE, CONDITIONAL
75. RELATIONAL
76. ALPHABETIC, NUMERIC
77. RELATIONAL
78. OR
79. SUBORDINATE, PARENTHESIS
80. SUBJECT
81. SIGN, CLASS
82. SIMPLE

Questions for Review

1. Define a statement and explain the three types of statements used in the Procedure Division.
2. Define a sentence, paragraph and section.
3. How is the Procedure Division organized?
4. What are the main functions of input and output verbs? Give examples.
5. What are the main function of the OPEN and READ statements and how are they related to each other?
6. What are the main functions of the WRITE and CLOSE statements?
7. Explain the use of the ADVANCING, POSITIONING and END-OF-PAGE options of the WRITE statement.
8. What are the functions of the ACCEPT and DISPLAY statements?
9. What are data manipulation verbs? Give purposes and examples of usage of each type of verb.
10. Explain the movement of numeric and nonnumeric data.
11. What is the function of a MOVE CORRESPONDING verb? What are the advantages and disadvantages of using this verb?
12. What is meant by "Rounding" and how is it used in arithmetic statements?
13. How is the ON SIZE ERROR option used in arithmetic statements?
14. Explain the use of the CORRESPONDING option of the ADD and SUBTRACT statements.
15. How is the COMPUTE statement used?
16. What are the rules for the sequence of arithmetic expression concerning a number of operators?
17. What are the purposes of sequence control verbs? Give function and examples of each type of verb.
18. What are the functions of the compiler-directing verbs? Give functions and examples of each type of verb.
19. What is a conditional expression? Explain the functions of the four test condition statements.
20. What is a compound conditional expression?
21. What is a nested conditional statement?
22. What is an implied subject? An implied operator?

Problems

1. *Match each term with its proper description.*

_____ 1. Statement
_____ 2. Sentence
_____ 3. Paragraph
_____ 4. Section
_____ 5. Procedures
_____ 6. Imperative
_____ 7. Conditional

A. One or more successive sentences.
B. Unconditional command.
C. Largest unit within the Procedure Division.
D. Basic unit of the Procedure Division.
E. Tested and evaluation.
F. One or more statements.
G. Groups of successive paragraphs.

2. *Match each verb with its proper description.*

_____ 1. Open
_____ 2. Read
_____ 3. Write
_____ 4. Close
_____ 5. Accept
_____ 6. Display

A. Terminates processing of files.
B. Causes writing of low volume data.
C. Makes record available for processing.
D. Obtains low volume data.
E. Prepares an input or output file for reading or writing.
F. Releases record for insertion in output file.

3. *Match each verb with its proper description.*

_____ 1. Move
_____ 2. Examine
_____ 3. Add
_____ 4. Subtract
_____ 5. Multiply
_____ 6. Divide
_____ 7. Compute

A. Product of two numeric items.
B. Specifies accumulation of numeric values.
C. Requires use of an arithmetic expression.
D. Replace and/or tally the occurrences of a given character.
E. Difference between two numeric values.
F. Shift data to another area.
G. Division of one numeric value by another.

4. *Match each verb with its proper description.*

____	1. Go To	A.	Causes transfer of control from normal execution and returns control back.
____	2. Alter	B.	Common end point for set of procedures.
____	3. Perform	C.	Temporary or permanent halt to the program.
____	4. Stop	D.	Include other programming languages in a COBOL program.
____	5. Exit	E.	Include prewritten procedures at compile time.
____	6. Copy	F.	Transfers control conditionally or unconditionally to another point in the program.
____	7. Enter	G.	Insert comments into a source program listing.
____	8. Note	H.	Modify the effect of an unconditional Go To statement.

5. *Match each statement at left with its proper category at right.*

____ 1. ADD X, Y; ON SIZE ERROR GO TO SPEC-ROUTINE.
____ 2. STOP RUN
____ 3. GO TO ROUNTINE-1.
____ 4. NOTE THE FOLLOWING ROUTINE. . . .
____ 5. READ FILE-IN; AT END GO TO FINISH.
____ 6. CLOSE FILE-IN, FILE-OUT.

A. Imperative.
B. Conditional.
C. Compiler-directing.

6. *The hierarchy of entries in the Procedure Division beginning with the largest unit are:*

1. ________________________.
2. ________________________.
3. ________________________.
4. ________________________.

7. *Write Procedure Division statements for the following:*

a. Prepare a READ statement for a card file named CARD-INPUT with a record named CARD-REC-DATA. The termination paragraph is named LAST-RECORD.
b. Prepare a WRITE statement for a printed report with a file name of PRINT-LIST and a record name of PRINT-IT, where each line is to be separated by two blank lines.
c. Type on the console typewriter, the literal VALUE EXCEEDS BALANCE and the field defined in the Working-Storage Section as REMAINING-BALANCE.
d. Prepare a CLOSE statement for the file named, INPUT-FILE, OUTPUT-FILE and PRINT-FILE.

8. *Write the necessary WRITE statement with the ADVANCING or POSITIONING option to accomplish the following:*

a. Advance the form three lines before a record is written.
b. Advance the form three lines after a record is written.
c. Double spacing.
d. Triple spacing.
e. Skipping to the first printing line of a new page before the line is printed.
f. Branching at the bottom of a page.

9. *Write the necessary Procedure Division entries for the following:*

a. Allow a value of CODE-DATA to be entered into Working-Storage through the console typewriter.
b. Write the value of CODE-DATA on the console typewriter.

10. *Match the result with the correct statement.*

	Statement		Result
____	1. DISPLAY DATE UPON CONSOLE.	A.	The word DATE will be written on the console typewriter.
____	2. DISPLAY 'DATE' UPON CONSOLE.	B.	The word 'DATE' will be written on the console typewriter.
		C.	The value of the variable DATE will be written on the console typewriter.

11. *Write the necessary Data Division and Procedure Division entries for the following:*

a. In the paragraph INIT, the following message is to be written out on the console typewriter ENTER OPERATION-CODE.
b. A value of OPERATION-CODE is to be entered into storage through the console keyboard. Values of OPERATION-CODE have a form such as 107B, 509X, or 879G.
c. The value of OPERATION-CODE is to be written on the console typewriter.

12.
```
    02  TRANSACTION-FLD.
        03  TRANS-CODE   PICTURE X   OCCURS 20 TIMES.
```

Write an instruction to clear data from the fifth TRANS-CODE item in the TRANSACTION-FLD.

13. *Using the partial Data Division entry that follows, write the necessary coding for the MOVE problem.*

```
    01  RECORD-1                    01  RECORD-2
        03  STOCK-NUMBER                03  STOCK-NUMBER
        03  UNITS                       03  DATA-MESSAGE
        03  VALUE                       03  QUANTITY
        03  WEIGHT                      03  UNITS
```

a. Transfer UNITS from RECORD-1 to RECORD-2.
b. Transfer all like named fields from RECORD-1 to RECORD-2.
c. Clear a field in Working-Storage Section called DATA-MESSAGE of previous field, move information to area above in RECORD-2.

14. *A name has been defined in the input record as follows:*

```
    02  NAME.
        03  LAST-NAME          PICTURE A(20).
        03  FIRST-INITIAL      PICTURE A.
        03  SECOND-INITIAL     PICTURE A.
```

a. Write the necessary entries to define a Working-Storage record called EDITED-NAME, with INITIAL-1 in the first position, followed by a period and space; then INITIAL-2 in the fourth position, another period and space; finally 20 positions called SURNAME.
b. Write the MOVE statement to put each part of the name into their proper places in the Working-Storage record.
c. Write a MOVE statement to transfer EDITED-NAME to CUSTOMER-NAME (Assume that CUSTOMER-NAME is a 26 position alphanumeric item in the output record which has been previously defined).

15. *Imagine you are preparing a customer invoice, at the bottom of which you want to print three lines that look like*

```
    GROSS AMOUNT     $205.30
        DISCOUNT     $   4.11
      NET AMOUNT     $201.19
```

Set up the data division entry for the item AMOUNT-ID to store the literals and appropriate MOVE statements to accomplish the above.

a. Use the JUSTIFIED RIGHT clause.
b. Without the use of the JUSTIFIED RIGHT clause.

16. *Given the following data record.*

I-RECORD	**W-RECORD**
EMPLOYEE-NO	EMPLOYEE-NO
EMPLOYEE-NAME	FILLER
RATE	EMPLOYEE-NAME
	FILLER
	RATE
	FILLER

Write the necessary procedural statements to move the I-RECORD items to the appropriate W-RECORD fields. Use two different methods.

17. *Indicate the RECEIVED DATA in the following:*

	Source Data	Source Picture	Receiving Picture	Received Data
1.	8736	9999	9999	____________
2.	8736	9999P	P9999	____________
3.	8736	99V99	99V99	____________
4.	8736	99V99	99.99	____________

5.	8736	P9999	99.99	______
6.	8736	9999P	99.99	______
7.	8736	P9999	99V99	______
8.	8736	9999P	99V99	______
9.	8736	9999	$99.99	______
10.	8736	99V99	$99.999	______
11.	8736	9999P	99999.99	______
12.	ERROR	XXXXX	XXXXX	______
13.	ERROR	AAAAA	AAAAA	______
14.	ERROR	AAAAA	AAAAAAAA	______

18. *For each of the following, fill in TALLY.*

a. EXAMINE FLD-A TALLYING UNTIL FIRST ZERO.

FLD-A | 1 | 0 | 3 | 5 | 0 | 0 | 1 | TALLY | | | | | |

b. EXAMINE FLD-B TALLYING ALL ZEROS.

FLD-B | 1 | 0 | 3 | 5 | 0 | 0 | 1 | TALLY | | | | | |

c. EXAMINE FLD-C TALLYING LEADING SPACES.

FLD-C | | | | | * | * | * | * | TALLY | | | | | |

For each of the following, fill in the resulting data.

d. EXAMINE FLD-D REPLACING ALL ZEROS BY QUOTES.

FLD-D | 0 | 0 | 0 | 3 | 4 | 5 | 0 | FLD-D | | | | | | | |

e. EXAMINE FLD-E REPLACING FIRST "0" BY "–".

FLD-E | 5 | 7 | 3 | 0 | 5 | 1 | 4 | FLD-E | | | | | | | |

f. EXAMINE FLD-F TALLYING LEADING ZEROS REPLACING BY SPACES.

FLD-F | | | | | | | | (result)

FLD-F | 0 | 0 | 0 | 0 | 6 | 7 | 0 | TALLY | | | | | |

19. *ARITHMETIC PROBLEMS*

Data-Name	Picture	Data Values Before Execution	Data Values After Execution
FLD-A	S999V99	+ 10000	______
FLD-B	S999V999	+ 045550	______
FLD-C	S999V99	– 12345	______
FLD-D	S9999	1234	______
FLD-E	S99V9999	+ 123456	______
FLD-F	S999V99	+ 90000	______
FLD-G	S999V9	+ 12345	______
FLD-H	S9V9	– 45	______

Data-Name	Picture	Data Values Before Execution	Data Values After Execution
FLD-I	S999V99	− 32045	________
FLD-J	S99V99	+ 0475	________
FLD-K	S9999V9999	+ 46250000	________
FLD-L	S999V9	+ 4259	________
FLD-M	S999V99	− 32007	________
FLD-N	S999V99	00000	________
FLD-O	S9999	4567	________
FLD-P	S9999V99	+ 123456	________

Arithmetic Statements

1. ADD FLD-A, FLD-B, GIVING FLD-C.
2. ADD FLD-A, FLD-B, FLD-H, FLD-I, FLD-J TO FLD-K.
3. SUBTRACT FLD-M FROM FLD-I ROUNDED.
4. MULTIPLY FLD-A BY FLD-B GIVING FLD-D ROUNDED.
5. DIVIDE FLD-A INTO FLD-B GIVING FLD-E ON SIZE ERROR GO TO ERROR-RT.
6. MULTIPLY FLD-A BY FLD-F GIVING FLD-G ROUNDED ON SIZE ERROR PERFORM FIX-IT.
7. DIVIDE FLD-N INTO FLD-A GIVING FLD-P.
8. ADD FLD-O to FLD-O.

Required: In the DATA VALUES AFTER EXECUTION column, write the results of the arithmetic statements shown above.

20. *Show the contents of each field after the calculation in the After Execution area.*

a. ADD FLD-A to FLD-B.

Data-Name	Picture	Before Execution	After Execution
FLD-A	S99V99	+ 1234	________
FLD-B	S99V99	− 1200	________

b. SUBTRACT FLD-A, FLD-B FROM FLD-C.

Data-Name	Picture	Before Execution	After Execution
FLD-A	S99V99	+ 1234	________
FLD-B	S9999V99	− 987654	________
FLD-C	S9999V99	+ 123456	________

c. MULTIPLY FLD-A BY FLD-B GIVING FLD-C.

Data-Name	Picture	Before Execution	After Execution
FLD-A	S99V99	+ 1234	______
FLD-B	S9999V99	+ 98765	______
FLD-C	S9(7)V9999	+ 1234567890	______

d. DIVIDE FLD-A INTO FLD-B ROUNDED.

Data-Name	Picture	Before Execution	After Execution
FLD-A	S99V99	+ 1234	______
FLD-B	S999V9	+ 9879	______

e. ADD FLD-A, FLD-B TO FLD-C ON SIZE ERROR GO TO ERROR-ROUTINE.

Data-Name	Picture	Before Execution	After Execution
FLD-A	S99V99	+ 1234	______
FLD-B	S99V99	+ 9876	______
FLD-C	S999V99	+ 98765	______

21. *Prepare the following arithmetic statements.*

   a. Add the fields of GIANT and CONTAINER and place the result in CONTAINER.
   b. Add the fields TOOL, TOTAL-NUMB, and NUMB, placing the result in TOTAL-NUMB.
   c. Add the fields DATA-IN, PROD, ROYALTY and place the result in GRAND-SUM.
   d. Subtract the field QUANTITY from TOTAL-BALANCE and place the result in TOTAL-BALANCE.
   e. Subtract the fields DATA-GIVEN, HOLD-DATA, CON-HOLD from TOTAL-HOLD and place the result in TOTAL-HOLD.
   f. Subtract the field BALANCE-B from CON-NUMB and place the result in NEW-NUMB.

22. *To permanently change the sequence of execution of instructions in the Procedure Division, you use the ______ verb. To change the sequence temporarily, you use the ______.*

23. *Prepare the following procedural statements using the GO TO verb.*

   a. A statement to transfer control to a paragraph called COMPUTE-PROCESS.
   b. A statement for the true condition of an equality comparison. The fields BALANCE and TOTAL are compared for equality. When the condition is true, control is transferred to COMPUTE-ROUTINE.
   c. A statement to transfer the program to PROC-1, PROC-2, PROC-3 or PROC-4 if the CLASS-CODE is 1, 2, 3 or 4. If the CLASS-CODE is out of this range, the program should branch to a paragraph called ERROR-ROUTINE.

24. *Assume that a numeric item has been defined and that it meets the requirements of the GO TO statement with the DEPENDING ON option. This series of IF statements is based on testing that item. Write the GO TO statements with the DEPENDING ON option which will cause the same branches as the series of IFs.*

   a.
```
IF JOB = 1 GO TO MANAGER.
IF JOB = 2 GO TO ANALYST.
IF JOB = 3 GO TO PROGRAMMER.
IF JOB = 4 GO TO OPERATOR.
```

b.

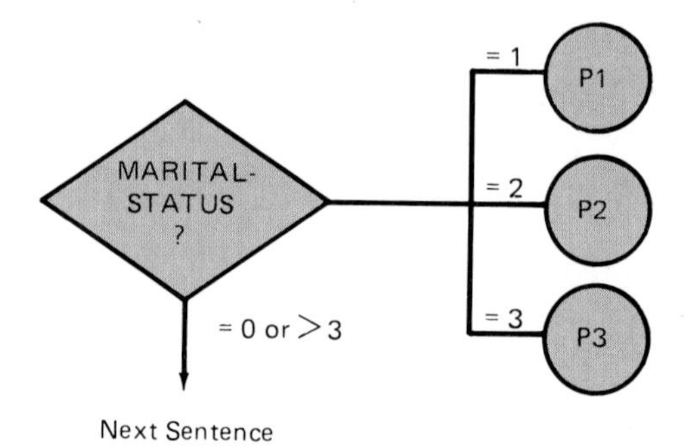

The diagram above shows the paragraph to which control is to be passed depending on the value of MARITAL-STATUS. Write the procedural statements to transfer control as specified in the diagram.

25. FIX. GO TO IN-STATE.

    a. ALTER FIX TO PROCEED TO OUT-STATE.
    b. ALTER IN-STATE TO PROCEED TO OUT-STATE.
    c. ALTER FIX TO PROCEED TO IN-STATE.

    Which of the above ALTER statements could alter the GO TO statement above so that the execution of the paragraph FIX would cause control to be transferred to paragraph OUT-STATE.

26. SWITCH-PARAGRAPH. GO TO PENALTY.
    ALTER-PARAGRAPH.

    Write a statement that will cause execution of SWITCH-PARAGRAPH to transfer control to DISCOUNT.

27. *You are writing the Procedure Division for PROGRAM-1. In the Data Division, there is a two character alphanumeric field called CDE. At various times during the execution of the program, CDE can contain a value ranging from 01 through 05 and from AA through AE.*

    Using the PERFORM statement, write the necessary entries to execute the following three separate points in the program depending upon the contents of CDE.

CDE Value	Action
01	ADD AMT-01 TO GRAND-TOTAL. MOVE A TO WORK-AREA.
02	ADD AMT-02 TO GRAND-TOTAL. MOVE B TO WORK-AREA.
.	
.	
AA	ADD AMT-AA TO GRAND-TOTAL. MOVE F TO WORK-AREA.
.	
.	
AE	ADD AMT-AE TO GRAND-TOTAL. MOVE H TO WORK-AREA.

28. *Using the PERFORM verb, write the necessary procedural statements for the following print routine.*

    a. Print a header line (HDG-1) at the top of each page.
    b. Print a maximum of 40 detail lines (PRINTOUT) per page.
    c. Before processing a new department (compare DEPT-NO with PREV-DEPT, which is set up in Working-Storage) print a total header line (DEPT-TOTAL-HDG) followed by a department total (DEPT-TOTAL-LINE). Skip to a new page and print HDG-1 before continuing.

29. *A program contains the following:*

```
TAX-DEDUCTION SECTION.
    TAX-PARA-1.
    TAX-PARA-2.
    TAX-PARA-3.
FICA-DEDUCTION SECTION.
    FICA-PARA-1.
    FICA-PARA-2.
    FICA-PARA-3.
TOTAL-DEDUCTIONS SECTION.
    TOTAL-PARA-1.
    TOTAL-PARA-2.
```

In the following PERFORM statement indicate after the *last Statement of which paragraph,* will control be transferred to the next statement after PERFORM.

a. PERFORM TAX-PARA-2, SUBTRACT TAX-FIGURED FROM TOTAL-TAX.
b. PERFORM TAX-DEDUCTION, SUBTRACT TAX-FIGURED FROM TOTAL-TAX.
c. PERFORM TAX-DEDUCTION THRU FICA-PARA-2, ADD TAX-FIGURED, FICA-FIGURED GIVING TAX-DEDUCTIONS.
d. PERFORM TAX-DEDUCTION THRU TOTAL-DEDUCTIONS.

30. *Which of the following COPY entries are correctly written?*

a. INPUT-OUTPUT SECTION.
   COPY IN-OUT-SECT.
b. SELECT COPY FILES.
c. FD CARD-FILE
   COPY FILE-DESCRIPTION-ENTRY.
d. 01 COPY CARD-RECORD.
e. 77 SAVE-NUMBER COPY LIBRARY-1.
f. BEGIN. COPY OPEN-PARAGRAPH.

31. *In the following NOTE statements indicate what will appear on the compilation output listing.*

a. EXPLANATION.
   NOTE WHEN THE FIELD LOW-QUANTITY CONTAINS 150 IT IS TIME TO REORDER. IF THIS IS A DISCONTINUED ITEM, A "D" WILL BE IN THE STATUS FIELD.
b. MOVE ORDER TO MASTER-LIST. NOTE THIS CUSTOMER IS ENTITLED TO A 2% DISCOUNT ON THIS ITEM. COMPUTE TOTAL-COST = (NO-ITEMS * COST-PER-ITEM) * .98.

32.

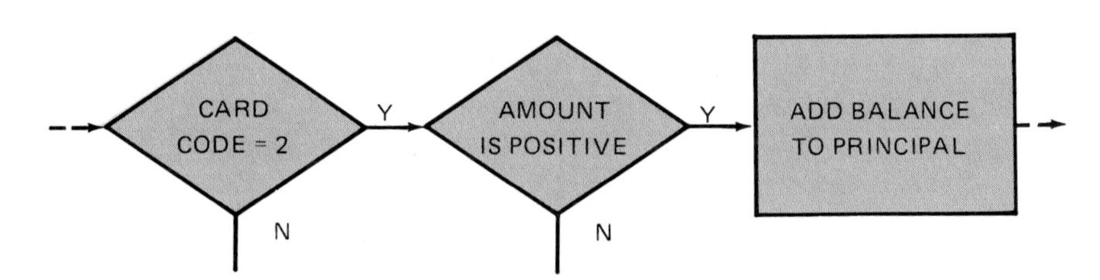

Select the proper coding for the procedures above.

a. IF CARD-CODE IS EQUAL TO 2 OR AMOUNT IS POSITIVE, ADD BALANCE TO PRINCIPAL.
b. IF CARD-CODE IS EQUAL TO 2 AND AMOUNT IS POSITIVE, ADD BALANCE TO PRINCIPAL.

33. *Write the conditional statements for the following:*

a. When the account number MAST-ACCT of the master file is not equal to the transaction account number, TRANS-ACCT, display 'SEQUENCE ERROR' TRANS-ACCT, on the console typewriter, otherwise continue regular processing.
b. When the stock of parts, TOTAL-UNITS, is below the minimum REORDER-POINT, transfer the stock number PART-NO to the reorder report, REORDER-NUMBER; otherwise continue processing.
c. When the line counter, LINE-COUNT, exceeds the specified number of lines per page, LINE-CONSTANT, control is passed to a page change routine, PAGE-CHANGE; otherwise a "1" (ONE-CONSTANT) is added to the line counter, LINE-COUNT.

34. *For a given application, it is necessary to select all records that contain an item number equal to 41571, 58001 through 59720 or 64225.*

Write the necessary procedural statements to accomplish the above.

35. *A percentage of sales is offered to each salesman as a commission. The percentage differs if the item sold is class A, B, C or D. If the item sold is not class A, B, C or D, no commission is calculated and the program proceeds to the NO-COMM-ROUTINE.*

Compute the sales commission based on the following rates.

CLASS A — equal to or less than 1,000, the commission is 6%.
— greater than 1,000 but less than 2,000, the commission is 7%.
— 2,000 or greater, the commission is 10%.
CLASS B — less than 1,000, the commission is 4%.
— 1,000 or greater, the commission is 6%.
CLASS C — all amounts, the commission is 4 1/2%
CLASS D — all amounts, the commission is 5%.

The result of the computation is stored in COMMISSION.
Upon completion of the computation, proceed to NEXT-ROUTINE.

Write the necessary procedural statements for the above.

36. *Using the following data values, solve the conditional statement by determining whether the tested condition is true or false.*

A1 = 5	B1 = 3	C1 = 8	D1 = 15
A2 = 4	B2 = 3	C2 = 4	D2 = 15

Statement:
IF A1 = B1 OR (C1 IS LESS THAN D1 AND (B2 = C2 OR A1 IS GREATER THAN A2) and A2 = C2) OR C1 is GREATER THAN D2 and D1 is EQUAL TO D2.

37. *Write the single IF sentence that corresponds to the flowchart at right. (Assume that all names have been defined. The decisions are condition-name tests.)*

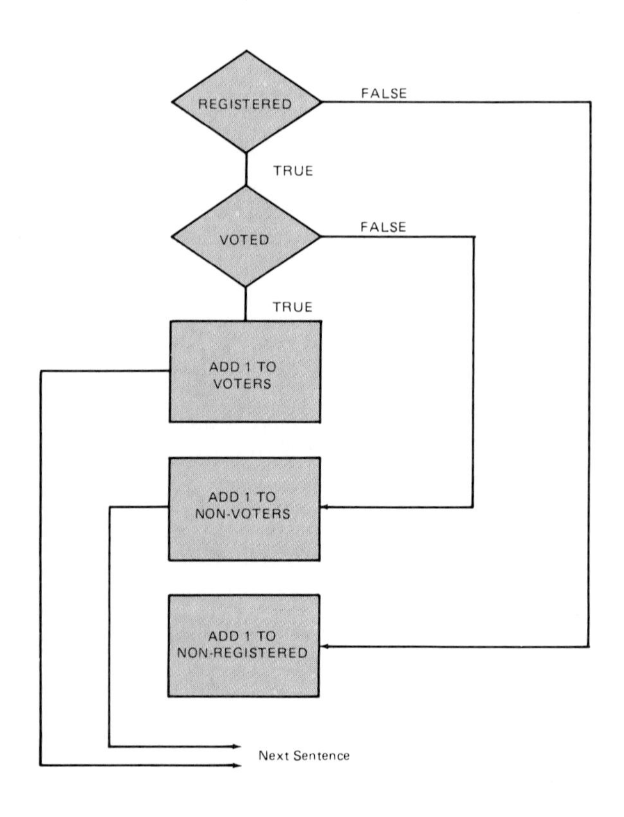

38. *Write the necessary procedural statements to accomplish the following:*

a.

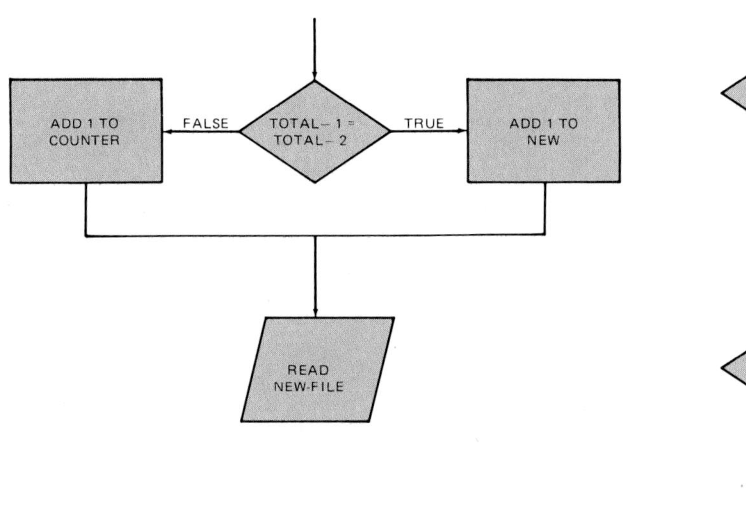

b.

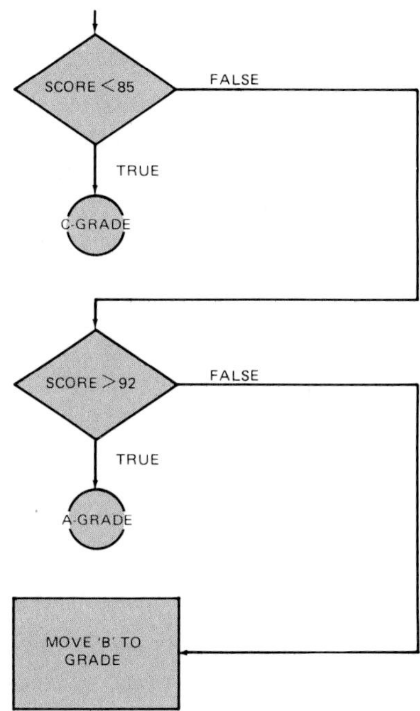

39. LOOP-1 SECTION.
SETUP-LOOP.

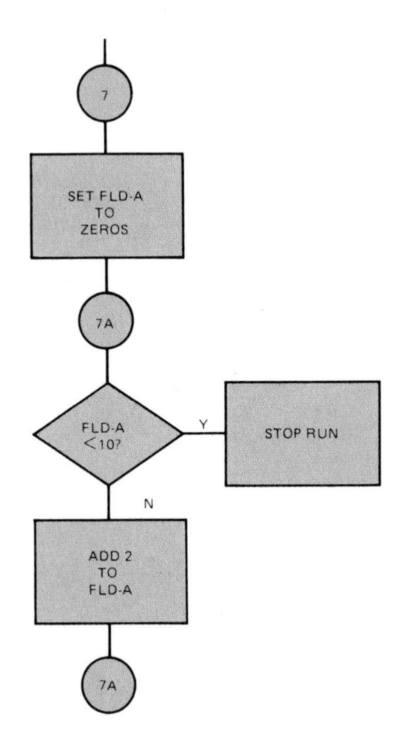

Write the Procedure Division entries for the above.

40. *LARGEST NUMBER PROBLEM*

There are three unequal numbers labeled NUM-A, NUM-B and NUM-C. Write the IF and MOVE statements necessary to move the largest number of the three to FLD-1, the next largest number to FLD-2 and the smallest number to FLD-3 and proceed to ROUT-X.

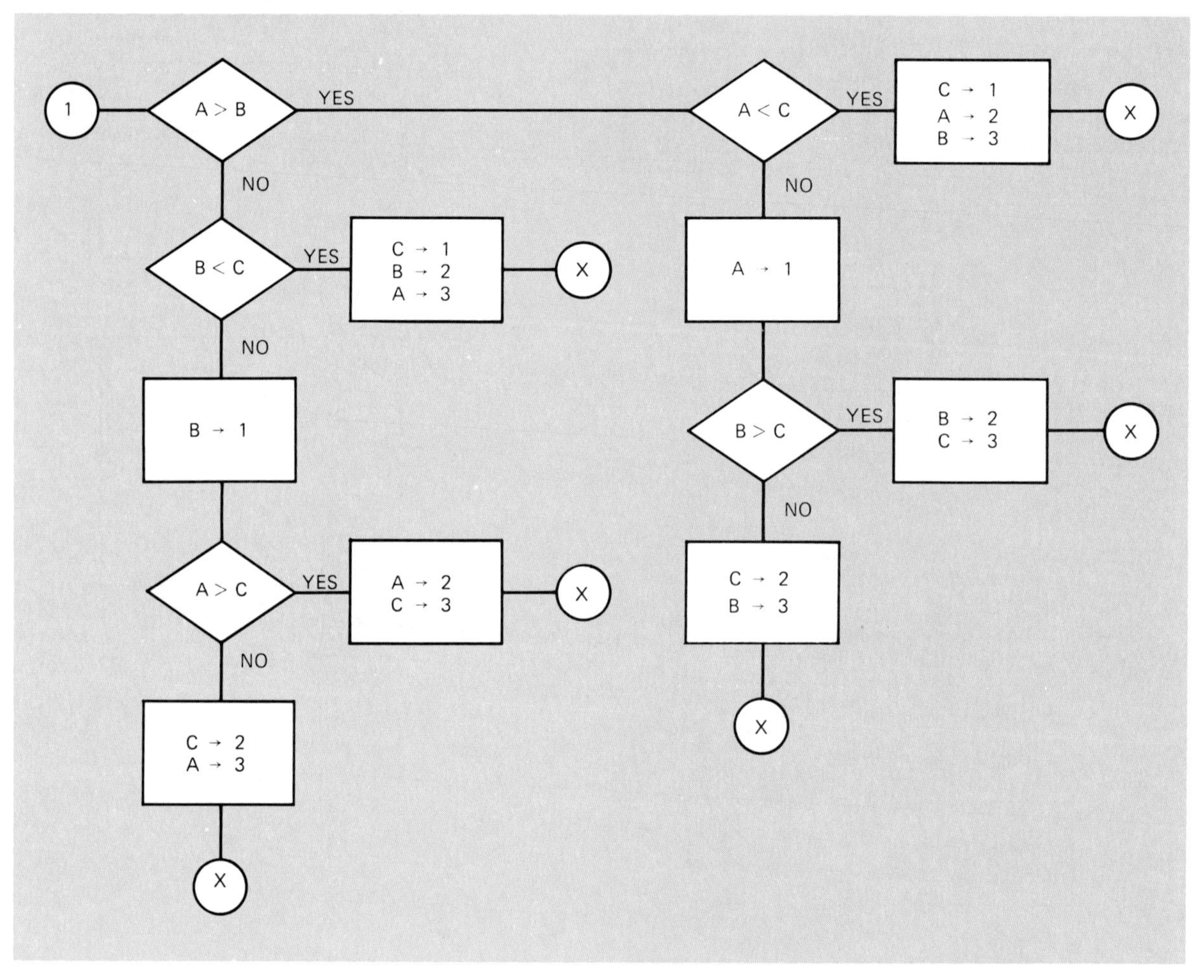

41. *LISTING*

This is a simple program whereby each card is read and a line printed for each card.

INPUT	*File-name*
OUTPUT	*File-name*
FILE-IN	*Record-name*
FILE-OUT	*Record-name*
RECORD-IN	
RECORD-OUT	

Write the procedural statements to accomplish the following:

PAR-1.

PAR-2.

FINISH.

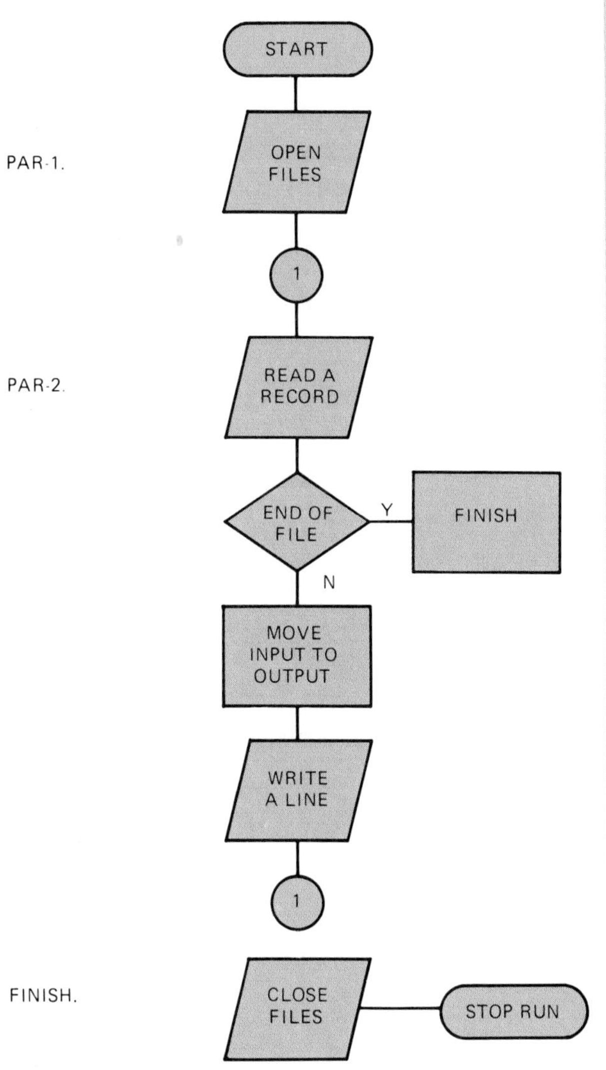

42. *PAYROLL PROBLEM*

One of the many problems in payroll is the computation of the tax deduction for Social Security. The law (for 1972) states that 5.2% of the gross pay is to be taken out of each pay check until a maximum of $468 (5.2% of $9,000) has been deducted. At this time, the deductions would cease. The following flowchart depicts a solution to this problem.

Write the Procedure Division statements to solve the following flowchart.

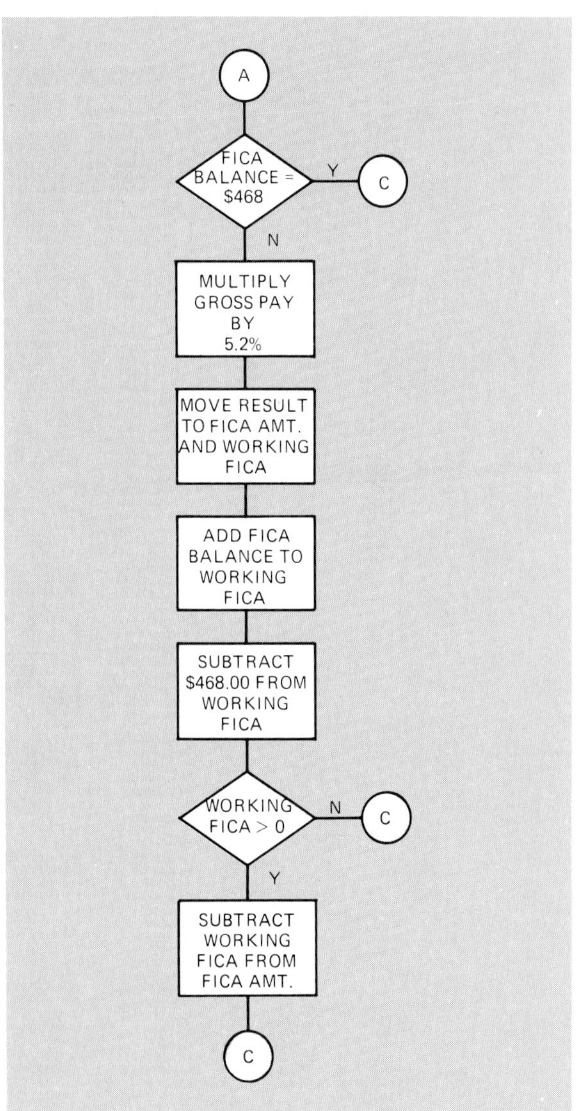

43.

Below you are given the COBOL Procedure Division statements necessary to code the flowchart below. Rearrange them in their correct sequence and enter them on a COBOL coding form.

STOP RUN.
READ FILE-IN;
OPEN INPUT FILE-IN:
OUTPUT FILE-OUT.
AFTER ADVANCING 2 LINES.
A-PARA.

B-PARA.
GO TO B-PARA.
AT END CLOSE FILE-IN, FILE-OUT;
WRITE E-RECORD.
PROCEDURE DIVISION.
ADD A, B, C, D, GIVING E.

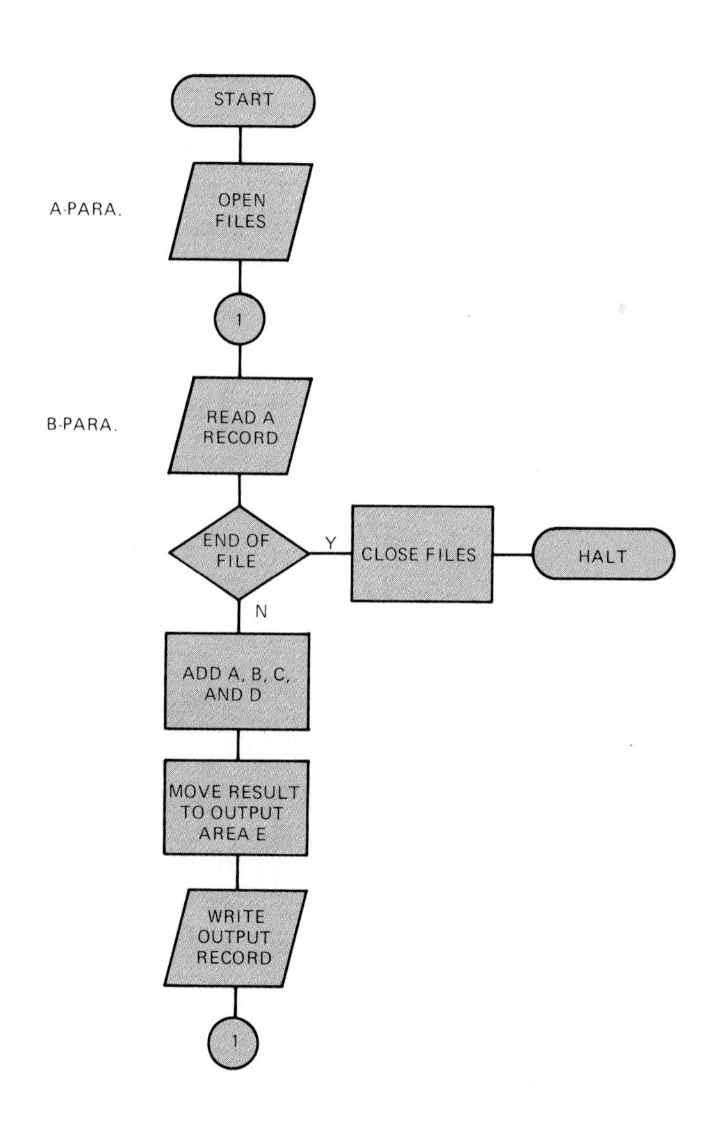

9 Table Handling

INTRODUCTION

Occasionally a program requires that one or more tables of values be stored in memory. These values can be referenced and used in computations or other procedural routines. For example, a tax percentage table might be useful for calculating tax deductions during a payroll run (fig. 9.1). Such a table might be set up in the Working-Storage section of the Data Division as shown in figure 9.2.

In the Procedure Division, the tax calculation routine might look like that shown in figure 9.3. Notice that many IF statements must be written and executed, as there are values in the table. For larger tables, this would require a tremendous amount of writing, memory space, and execution time.

COBOL provides a Table Handling module that has the capability of defining tables of contiguous data, and to access an item relative to its position in the table. In order to discuss table handling features, it is necessary to define a few standard terms.

TAX TABLE		
1 dependent	.175	(17.5%)
2 dependents	.164	
3 dependents	.159	
4 dependents	.153	
5 dependents	.147	
6 dependents	.141	
7 dependents	.127	
etc.		

Figure 9.1
Typical Tax Table—Diagram.

```
WORKING-STORAGE SECTION.

01  TAX-TABLE.
    02  DEP-1   PICTURE V999   VALUE  .175.
    02  DEP-2   PICTURE V999   VALUE  .164.
    02  DEP-3   PICTURE V999   VALUE  .159.
    02  DEP-4   PICTURE V999   VALUE  .153.
    02  DEP-5   PICTURE V999   VALUE  .147.
    02  DEP-6   PICTURE V999   VALUE  .141.
    02  DEP-7   PICTURE V999   VALUE  .127
```

Figure 9.2 Typical Tax Table—Working-Storage Entries.

```
TAX-CALC.
    IF NO-DEP IS EQUAL TO 0, MULTIPLY DEP BY .179 GIVING
        TAX-AMT.
    IF NO-DEP IS EQUAL TO 1, MULTIPLY DEP BY DEP-1 GIVING
        TAX-AMT.
    IF NO-DEP IS EQUAL TO 2, MULTIPLY DEP BY DEP-2 GIVING
        TAX-AMT.
    IF NO-DEP IS EQUAL TO 3, MULTIPLY DEP BY DEP-3 GIVING
        TAX-AMT.
           Etc.
```

Figure 9.3 Typical Tax Table—Procedure Division Entries.

A *table* consists of a number of contiguous data items that are of like format. The individual data items in a table are called *table elements*. Within the COBOL programming language specifications, table elements are accessed by either *indexing* or *subscripting* techniques.

Subscripting is a method of specifying an occurrence number and to affix a *subscript* to the data-name of the table element. A *subscript* is an integer whose value identifies a particular element in a table.

Indexing is another method of specifying an occurrence number and to affix an *index* to the data-name of the table element. An *index* is a computer storage position or register whose content corresponds to an occurrence number.

Table Value refers to any one of the values (i.e, the value of any cell) contained in a table. The table itself simply consists of all values (cells) contained in it. In the tax table example, each tax percentage is a table value.

Search Argument is any number used to locate one of the values within the table. In the tax calculation example, the number of dependents constitutes the search argument.

To locate a value within a table, the search argument is used to scan the table until a table element associated with the search argument is found. If, in the tax example, the search is for the tax percentage to be used for an employee with five dependents (search argument), the search would produce the table element .14 which is associated with five dependents. The operation is called *table search* or *table lookup.*

The Table Handling Feature enables the COBOL programmer to process tables or lists of repeated data conveniently. A table may be set up with three dimensions; for example, three levels of subscripting or indexing can be handled. This Table Handling module provides a capability for defining tables of contiguous data items and for accessing an item relative to its position in the table. Language facility is provided for specifying how many times an item can be repeated. Each item may be identified through the use of a subscript or an index-name. Such a case exists when a group item described with an OCCURS clause contains another group item with an OCCURS clause, which in turn contains another group with an OCCURS clause. To make reference to any element within such a table, each level must be subscripted or indexed.

The Table Handling Feature provides a capability for accessing items in three-dimensional variable-length tables. The feature also provides the additional facilities for specifying ascending or descending keys, and permits searching of a table for an item satisfying a specified condition.

TABLE DEFINITION

Tables composed of contiguous data items are defined in COBOL by including the OCCURS clause in their data description entries. This clause specifies that the data item is to be repeated as many times as stated. The item is considered to be a table element and its name and description apply to each repetition or occurrence (fig. 9.4). Since each occurrence of a table element does not have to be assigned by a unique data-name, reference to a desired

```
OCCURS
xx    data-name OCCURS y TIMES [PICTURE IS......]
      xx = level number (can be any level except 01)
       y = number of times the item is repeated (number of
           individual values within the table)
```

Figure 9.4
Occurs Clause—Example.

occurrence may be made simply by specifying the data-name of the table element together with the occurrence number of a desired table element.

To define a one-dimensional table, an OCCURS clause is used as part of the data description of the table element, but the OCCURS clause must not appear in the description of group items that contain the table element. An example of defining a one-dimensional table is:

```
01  TABLE-1.
    02  TABLE-ELEMENT OCCURS 20 TIMES.
        03  NAME....................
        03  SSNO....................
```

Defining a one-dimensional table within each occurrence of an element of another one-dimensional table calls for use of a two-dimensional table. To define a two-dimensional table, an OCCURS clause must appear in the group item that defines the table, and as part of an item which is subordinate to that group item. To define a three-dimensional table, an OCCURS clause should appear in the data description of the group item that defines the table and on two additional items (one of which is subordinate to the other). Note the following example of defining a three-dimensional table.

```
01  SALES-QUOTA-TABLE
    02  REGION-TABLE OCCURS 5 TIMES.
        03  REGION-NAME PICTURE X(10).
        03  DISTRICT-TABLE OCCURS 10 TIMES.
            04  DISTRICT-NAME PICTURE X(10).
            04  OFFICE-TABLE OCCURS 15 TIMES.
                05  OFFICE-NAME PICTURE X(10).
                05  SALES-QUOTA PICTURE 9(7).
```

REFERENCE TO TABLE-ITEMS

Whenever the user refers to a table element, the reference must indicate which occurrence of the element is intended. For access to a one-dimensional table, the occurrence number of the desired element provides complete information. For access to tables of more than one dimension, an occurrence number must be supplied for each dimension of the table accessed. Thus, in the above example of a three-dimensional table, a reference to the fourth REGION-NAME would be complete, whereas a reference to the fourth DISTRICT-NAME would not. To refer to DISTRICT-NAME, which is an element of a two-dimensional table, the user must refer to, for example, the fourth DISTRICT-NAME within the fifth REGION-TABLE.

SUBSCRIPTING

One method by which occurrence numbers may be specified is to append one or more subscripts to the data-name (fig. 9.5). (Refer to SUBSCRIPTING in the Data Division.) *A subscript is an integer whose value specifies the occurrence number of an element.* The subscript can be represented either by a literal that is an integer or by a data-name that is defined elsewhere as a numeric elementary item with no character positions to the right of the assumed decimal point. In either case, the subscript is enclosed in parentheses and written immediately following the name of the table element (see fig. 9.6); examples of each are REGION-NAME (5) and REGION-NAME (FIVE) where FIVE is a data item defined elsewhere as integer 5. A table reference must include as many subscripts as there are dimensions in the table whose element is being referenced. That is, there must be a subscript for each OCCURS clause in the hierarchy containing the data-name itself. In the three-dimensional table example above, reference to REGION-NAME requires only one subscript, reference to DISTRICT-NAME requires two, and reference to OFFICE-NAME and SALES-QUOTA requires three.

```
data-name (subscript[, subscript] [, subscript])
```

Figure 9.5 Format—Subscripting.

When more than one subscript is required, each is written in order of successively less inclusive dimensions of the data organization. When a data-name is used as a subscript, it may be used to refer to items in many different tables. These tables need not have elements of the same size. The data-name may also appear as the only subscript. It is also permissible to mix literal and data-names in subscripts; for example, DISTRICT-NAME (NEWKEY, 42). Notice from this example that when more than one subscript appears within a pair of parentheses, the subscripts must be separated by commas. A space must follow each comma, but no space may appear between the left parenthesis and the leftmost subscript or between the right parenthesis and the rightmost subscript.

Subscripts may be qualified, in which case, OF or IN follows the subscripting data-name (see fig. 9.7).

Subscripts are used only to refer to an individual element within a list or table of elements that have not been assigned individual data-names (see "Subscripting" in Data Division, chap. 6).

INDEXING

The number of times that a table element is repeated is specified by the OCCURS clause. The name and description of the table element applies to each repetition or *occurrence* of the table element. One method of specifying the desired occurrence is by subscripting. Another method of specifying an occurrence number is to affix an index to the data-name of the element. This technique is called *indexing.*

References can be made to individual elements within a table of elements by specifying indexing or by that reference. The INDEXED BY option allows both direct and relative referencing of tables of repeated data in a more efficient manner than subscripting.

PERCENT-TABLE (elementary variable)	(value)
PERCENT (1)	.90
PERCENT (2)	.94
PERCENT (3)	.96

common name subscript

```
01  PERCENT-TABLE     USAGE COMP-3.
    02  PERCENT OCCURS 3 TIMES     PICTURE V99.

DISCOUNT-ROUTINE.
    IF GOOD,
    THEN
        COMPUTE BILL ROUNDED = AMOUNT-WS * PERCENT (1)
    ELSE
        IF MEDIUM,
        THEN
            COMPUTE BILL ROUNDED = AMOUNT-WS * PERCENT (2)
        ELSE
            COMPUTE BILL ROUNDED = AMOUNT-WS * PERCENT (3).
```

Figure 9.6 Example—Subscripting.

```
data-name {OF|IN} data-name-1 [ {OF|IN} data-name-2] . . .

  (index-name [ {+|-} integer] [, index-name [ {+|-} integer] ]

  [, index-name [ {+|-} integer] ] )
```

Figure 9.7 Format—Qualification with Indexing.

An index is assigned to a given level of a table by using an INDEXED BY clause in the definition of the table. A name given in the INDEXED BY clause is known as an index-name and is used to refer to the assigned index. An index-name must be initialized by a SET statement before it can be used in a table reference. An index may be modified only by a SET, SEARCH, or PERFORM statement. Data items described by the USAGE IS INDEX clause permit storage of the values of index-names as data without conversion. Such data items are called index-data items.

Direct Indexing is specified by using an index-name in the form of a subscript; for example, ELEMENT (PRIME-INDEX).

Relative Indexing is specified when the terminal space of the data-name is followed by a parenthesized group of items: the index-name, followed by a space, followed by one of the operators + or −, followed by another space, followed by an unsigned integral numeric literal; for example, ELEMENT (PRIME-INDEX + 5).

Qualification may be used in conjunction with indexing, in which case OF or IN follows the data-name being indexed.

Rules Governing Subscripting, Indexing, and Qualification

Tables may have one, two, or three dimensions; therefore, reference to an element in a table may require up to three subscripts or indexes.

1. A data-name must not be subscripted or indexed when the data-name is itself being used as an index, subscript, or qualifier.
2. When subscripting, indexing, or qualification are required for a given data item, the indexes or subscripts are specified after all necessary qualifications have been given.
3. Subscripting and indexing must not be used together in a single reference.
4. Whenever subscripting is not permitted, indexing is not permitted.

5. The commas shown in the format for indexing and subscripting are required. (See figures 9.8, 9.9, 9.10, 9.11.)

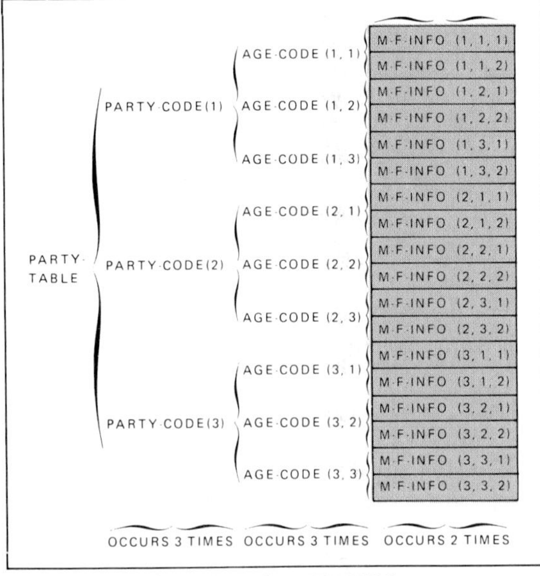

Storage Layout for PARTY-TABLE.

```
01  PARTY-TABLE REDEFINES TABLE.
    05  PARTY-CODE OCCURS 3 TIMES INDEXED BY PARTY.
        10  AGE-CODE OCCURS 3 TIMES INDEXED BY AGE.
            15  M-F-INFO OCCURS 2 TIMES INDEXED BY M-F
                PICTURE 9(7)V9 USAGE DISPLAY.
```

PARTY-TABLE contains three levels of indexing. Reference to elementary items within PARTY-TABLE is made by use of a name that is subscripted or indexed. A typical Procedure Division statement might be.

```
MOVE M-F-INFO (PARTY, AGE, M-F) TO M-F-RECORD.
```

Figure 9.8
Example—Subscripting and Indexing.

An example of a one-dimensional table is shown in the following.

The following is the record layout of each record in the file SALESFILE which is on a magnetic disk.

PNO1					Monday Qty			Tuesday Qty			Wednesday Qty			Thursday Qty			Friday Qty		
9	9	9	9	9	9	9	9	9	9	9	9	9	9	9	9	9	9	9	9
000	001	002	003	004	005	006	007	008	009	010	011	012	013	014	015	016	017	018	019

Saturday Qty			PRICE			
9	9	9	9	9	9	9
020	021	022	023	024	025	026

The contents of the fields within the SALESFILE record are as follows:

PNO1: Product number
QUANTITY: Quantity sold for each day
PRICE: Price per unit

Figure 9.9 Example—One-Dimensional Table Handling.

Figure 9.9 Continued.

Data from the records of SALESFILE is used to create records written onto SUMMFILE. The following is the record layout of each record in the file SUMMFILE.

PNO2					TOTAL QUANTITY				TOTAL-D-VALUE								Monday Dollar Value		
9	9	9	9	9	9	9	9	9	9	9	9	9	9	9	9	9	9	9	9
000	001	002	003	004	005	006	007	008	009	010	011	012	013	014	015	016	017	018	019

Monday Dollar Value				Tuesday Dollar Value							Wednesday Dollar Value							Thursday	
9	9	9	9	9	9	9	9	9	9	9	9	9	9	9	9	9	9	9	9
020	021	022	023	024	025	026	027	028	029	030	031	032	033	034	035	036	037	038	039

Thursday Dollar Value					Friday Dollar Value							Saturday Dollar Value						
9	9	9	9	9	9	9	9	9	9	9	9	9	9	9	9	9	9	9
040	041	042	043	044	045	046	047	048	049	050	051	052	053	054	055	056	057	058

The contents of the fields within the record of SUMMFILE are as follows.

PNO2: Product number
TOTAL-QUANTITY: Total quantity sold for one week.
TOTAL-D-VALUE: Dollar value of total quantity sold for one week.
D-VALUE: Dollar value of the quantity sold for each day.

The following flowchart illustrates the logic of the PERFORM VARYING statement in the following program.

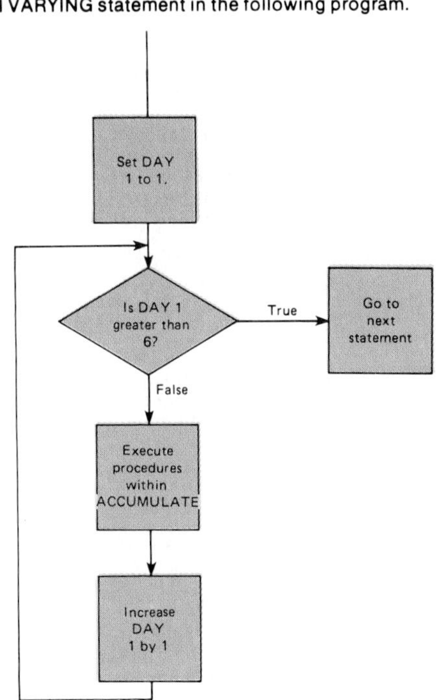


```
DATA DIVISION.
FILE SECTION.
FD  SALESFILE          BLOCK CONTAINS 18 RECORDS
                       RECORD CONTAINS 27 CHARACTERS
                       LABEL RECORD IS STANDARD
                       DATA RECORD IS SALESRECORD.

01  SALESRECORD.
    02  PNO1           PICTURE 9(5).
    02  QUANTITY       PICTURE 999 OCCURS 6 TIMES INDEXED BY DAY1.
    02  PRICE          PICTURE 99V99.

FD  SUMMFILE           BLOCK CONTAINS 8 RECORDS
                       RECORD CONTAINS 59 CHARACTERS
                       LABEL RECORD IS STANDARD
                       DATA RECORD IS SUMMRECORD.

01  SUMMRECORD.
    02  PNO2               PICTURE 9(5).
    02  TOTAL-QUANTITY     PICTURE 9999.
    02  TOTAL-D-VALUE      PICTURE 9(6)V99.
    02  D-VALUE   PICTURE 9(5)V99 OCCURS 6 TIMES INDEXED BY DAY2.

WORKING-STORAGE SECTION.
77  COUNTER      PICTURE 9.
01  FLAGS.
    02  MORE-DATA-FLAG      PICTURE XXX      VALUE 'YES'.
        88  MORE-DATA                        VALUE 'YES'.
        88  NO-MORE-DATA                     VALUE 'NO'.

PROCEDURE DIVISION.
MAIN-ROUTINE.
    OPEN    INPUT     SALESFILE,
            OUTPUT    SUMMFILE.
    READ SALESFILE
        AT END MOVE 'NO' TO MORE-DATA-FLAG.
    PERFORM READ-IN
        UNTIL NO-MORE-DATA.
    CLOSE   SALESFILE
            SUMMFILE.
    STOP RUN.

READ-IN.
    MOVE PNO1 TO PNO2.
    MOVE ZERO TO TOTAL-QUANTITY.
    PERFORM ACCUMULATE
        VARYING DAY1 FROM 1 BY 1
        UNTIL DAY1 IS GREATER THAN 6.
    MULTIPLY TOTAL-QUANTITY BY PRICE GIVING TOTAL-D-VALUE.
    WRITE SUMMRECORD.
    READ SALESFILE
        AT END MOVE 'NO' TO MORE-DATA-FLAG.

ACCUMULATE.
    ADD QUANTITY (DAY1) TO TOTAL-QUANTITY.
    SET DAY2 TO DAY1.
    MULTIPLY QUANTITY (DAY1) BY PRICE GIVING D-VALUE (DAY2).
```

An example of a two-dimensional table is shown in the following diagram.

	DEPARTMENT NUMBER	1st SALE	2nd SALE	3rd SALE	4th SALE	5th SALE
1st DEPARTMENT	1359	001293	029160	200015	025037	010035
2nd DEPARTMENT	1530	013000	000250	304000	002519	023410
3rd DEPARTMENT	2400	002459	002947	012044	002830	016625
4th DEPARTMENT	3594	000248	001590	002300	000122	003520

If the above two-dimensional table contains variable data, then the definition of the two-dimensional table would be as follows:

```
FILE SECTION.
01  TABLE-1.
    02  DEPARTMENT-ITEM  OCCURS 4 TIMES INDEXED BY DEPT.
        03  DEPARTMENT-NUMBER          PICTURE XXXX.
        03  SALE        PICTURE 9999V99  OCCURS 5 TIMES
                        INDEXED BY DAYI.
```

In the above definition of the two-dimensional table, SALE is an element of the two-dimensional table. SALE occurs five times within each DEPARTMENT-ITEM. DEPARTMENT-ITEM occurs four times within the table called TABLE-1. The INDEXED BY DEPT clause and INDEXED BY DAYI are only needed when the indexing technique accesses the elements of the table.

The procedural statements required to add all SALE items and place the result of the addition into the field TOTAL-SALES are shown below.

```
PROCEDURE DIVISION.

    PERFORM DAILY-SALES-ACCUMULATION,
        VARYING DEPT FROM 1 BY 1
        UNTIL DEPT IS GREATER THAN 4
        AFTER DAYI FROM 1 BY 1
        UNTIL DAYI IS GREATER THAN 5.

DAILY-SALES-ACCUMULATION.
    ADD SALE (DEPT, DAYI) TO TOTAL-SALES.
```

Figure 9.10 Example—Two-Dimensional Table Handling.

An example of a three-dimensional table is shown in the following diagram.

	DEPARTMENT						
	STORE-NO	DEPARTMENT-NO	1st SALE	2nd SALE	3rd SALE	4th SALE	5th SALE
1st STORE	123	1359	001293	029160	200015	025037	010035
		1530	013000	000250	304000	002519	023410
		2400	002459	002947	012044	002830	016625
		3594	000248	001590	002300	000122	003520
2nd STORE	530	1280	002510	020050	001257	011120	025910
		3320	100245	001945	027933	123450	300020
		7730	002460	001024	003913	029433	002930
		9250	002200	001050	023910	000200	000391

If the above three-dimensional table contains variable data, then the definition of the three-dimensional table would be as follows.

```
FILE SECTION.
01  TABLE-2.
    02  STORE-ITEM      OCCURS 2 TIMES     INDEXED BY STORE.
        03  STORE-NO       PICTURE XXX.
        03  DEPARTMENT-ITEM       OCCURS 4 TIMES INDEXED BY
                DEPT.
            04  DEPARTMENT-NO   PICTURE XXXX.
            04  SALE   PICTURE 9999V99  OCCURS 5 TIMES
                                    INDEXED BY DAYI.
```

In the above definition of the three-dimensional table, SALE is an element of the three-dimensional table. SALE occurs five times within each DEPARTMENT-ITEM. DEPARTMENT-ITEM occurs four times within each STORE-ITEM. STORE-ITEM occurs two times within the table called TABLE-2. The INDEXED BY STORE clause, the INDEXED BY DEPT clause, and the INDEXED BY DAYI clause are only needed when the indexing technique accesses the elements of the table.

The procedural statements required to add all SALE items and place the result of the addition into the field TOTAL-SALE are shown below.

```
    PERFORM DAILY-SALES-ACCUMULATION
        VARYING STORE FROM 1 BY 1
        UNTIL STORE IS GREATER THAN 2
        AFTER DEPT FROM 1 BY 1
        UNTIL DEPT IS GREATER THAN 4
        AFTER DAYI FROM 1 BY 1
        UNTIL DAYI IS GREATER THAN 5.

DAILY-SALES-ACCUMULATION.
    ADD SALE (STORE, DEPT, DAYI) TO TOTAL-SALES.
```

Figure 9.11 Example—Three-Dimensional Table Handling.

DATA DIVISION

The OCCURS and USAGE clause are included as part of the record description entry utilizing the Table Handling Feature.

Occurs Clause

The OCCURS clause eliminates the need for separate entries for repeated data since it indicates the number of times a series of items with an identical form is repeated. In addition, it also supplies the required information for the application of subscripts and indexes.

The OCCURS clause is used in defining tables and other homogenous sets of repeated sets of repeated data (fig. 9.12). Whenever the OCCURS clause is used, the data-name that is the subject of this entry must be either subscripted or indexed whenever it is referred to in a statement other than SEARCH. When subscripted, the subject refers to one occurrence within the table. When not subscripted (permitted only in a SEARCH statement), the subject represents the entire table element. (A table element consists of all occurrences of one level of a table.) Further, if the subject of this entry is the name of a group item, then all data-names belonging to the group must be subscripted or indexed whenever they are used as operands.

Format 1

```
OCCURS integer-2 TIMES
    [ {ASCENDING } KEY IS data-name-2 [data-name-3] . . . ] . . .
      {DESCENDING}
      [INDEXED BY index-name-1 [index-name-2] . . . ]
```

Format 2

```
OCCURS integer-1 TO integer-2TIMES [DEPENDING ON data-name-1]
    [ {ASCENDING } KEY IS data-name-2 [data-name-3] . . . ] . . .
      {DESCENDING}
      [INDEXED BY index-name-1 [index-name-2] . . . ]
```

Format 3

```
OCCURS integer-2TIMES [DEPENDING ON data-name-1]
    [ {ASCENDING } KEY IS data-name-2 [data-name-3] . . . ] . . .
      {DESCENDING}
      [INDEXED BY index-name-1 [index-name-2] . . . ]
```

Figure 9.12 Occurs Clause Formats.

Rules Governing the Use of the Occurs Clause

1. The OCCURS clause is optional in a data description entry and cannot be specified in a data description entry that
   a. Has an 01 or 77 level number.
   b. Describes an item whose size is variable. The size of an item is variable if the data description of any subordinate item contains an OCCURS clause with a DEPENDING ON option.
2. A record description entry that contains an OCCURS clause may not also contain a VALUE clause except for condition-name entries.

3. Integer-1 and Integer-2 must be positive integers. Where both are used, the value of Integer-1 must be less than the value of Integer-2. The value of Integer-1 may be zero, but Integer-2 may not be zero.
4. In Format-1, the value of Integer-2 represents the exact number of occurrences. In Format-2, the value of Integer-2 represents the maximum number occurrences.

Depending On Option

The DEPENDING ON option is used in Format-2. This option is only required when the end of the occurrences cannot otherwise be determined. This indicates that the subject has a variable number of occurrences. This does not mean that the subject is variable but rather that the number of times that the subject may be repeated is variable. The number of times is being controlled by the value of data-name-1 at object time.

Integer-1 represents the minimum number of occurrences while *Integer-2* represents the maximum number of occurrences.

Data-name-1 is the object of the DEPENDING ON option.

1. Must be described as a positive integer,
2. Must not exceed Integer-2 in value,
3. **May be qualified, for example, data-name-1 of group-name, when desired,**
4. Must not be subscripted (that is, itself the subject of or an entry within a table), and
5. Must, if it appears in the same record as the table it controls, appear before the variable portion of the record.

(See figure 9.13.)

Key Option

The KEY option is used in conjunction with the INDEXED BY option in the execution of a SEARCH ALL statement. The option is used to indicate that the repeated data is arranged in ASCENDING or DESCENDING order according to the values contained in data-name-2, data-name-3, etc. The data-names are listed in descending order of significance.

If data-name-2 is the subject of the table entry, it is the only key that may be specified for the table, otherwise

1. all of the items identified by the data-name in the KEY IS phrase must be within the group item which is the subject of this entry, and
2. none of the items identified by the data-name in KEY IS phrase can be described by an entry which either contains an OCCURS clause or is subordinate to an entry which contains an OCCURS clause.

Indexed By Option

The INDEXED BY option is required if the subject of this entry (the data name described by the OCCURS clause, or an item within this data-name, if it is a group item) is to be referred to by indexing. The index-name(s) identified by this clause is not defined elsewhere in the program since its allocation and format are dependent upon the system, and not being data, cannot be associated with any data hierarchy.

SALE-TABLE	
SALE-ITEM (1)	A21033
SALE-ITEM (2)	A21455
SALE-ITEM (3)	A22223
SALE-ITEM (4)	A23762
SALE-ITEM (5)	A24689
SALE-ITEM (6)	A24643
SALE-ITEM (7)	A29567
SALE-ITEM (8)	J33468
SALE-ITEM (9)	J24788
SALE-ITEM (10)	J67011

1. Working-Storage Section entries to define the variable:
   TABLE-SIZE which is to contain the number of sale items,
   TABLE-VALUE to which a table value will be transmitted prior to being moved to a specific table element,
   SUBSCRIPT to be used as a subscript in referring to table elements.
   SALE-TABLE whose size is to be determined by TABLE-SIZE, and

2. Procedure Division entries to accept the table size and the table values.

```
1.WORKING-STORAGE SECTION.
  77     TABLE-SIZE   PICTURE 99.
  77     TABLE-VALUE   PICTURE X(6).
  77     SUBSCRIPT   PICTURE 99.
  01     SALE-TABLE.
         02   SALE-ITEM   OCCURS 10 TIMES   DEPENDING ON TABLE-SIZE
                   PICTURE X(6).

2.PROCEDURE DIVISION.
  INITIAL-ROUTINE.
         ACCEPT TABLE-SIZE   FROM SYSIN.
         PERFORM ACCEPT-TABLE-VALUE
              VARYING SUBSCRIPT FROM 1 BY 1
              UNTIL SUBSCRIPT IS GREATER THAN TABLE-SIZE
                   .
                   .
   ACCEPT-TABLE-VALUE.
         ACCEPT TABLE-VALUE FROM SYSIN.
         MOVE TABLE-VALUE TO SALE-ITEM (SUBSCRIPT).
```

Figure 9.13 Example—Subscripting.

Rules Governing the Use of Index-Names

1. The number of index-names for a Data Division entry must not exceed 12.
2. An index-name must be initialized through a SET statement before it can be used.
3. Each index-name contains a binary value that represents an actual displacement from the beginning of the table that corresponds to an occurrence number in the table. The value is calculated as the occurrence number minus one, multiplied by the length of the entry that is indexed by the index-name.

(See figure 9.14.)

Usage Is Index Clause

The USAGE IS INDEX clause is used to specify the format of a data item in the computer storage. The clause permits the programmer to specify index-data items. (See figure 9.15.)

Rules Governing the Use of the Usage Is Index Clause

1. The USAGE clause may be written at any level. If the USAGE clause is written at a group level, it applies to each elementary item in the group. The USAGE clause at an elementary level cannot contradict the USAGE clause of a group to which the item belongs, unless the group USAGE clause is not stated.

PRODUCT-NUMBER	WAREHOUSE-DATA								
	LOC	BIN	QUANTITY						
9 9 9 9	1 7	6 2 9	9 9 9 9 9	1 8	1 5 8	9 9 9 9 9	1 8	1 6 0	9 9 9 9 9

```
02  PRODUCT-NUMBER          PICTURE 9999.
02  WAREHOUSE-DATA          OCCURS 3 TIMES
                            ASCENDING KEYS ARE LOCATION,
                            BIN-NUMBER INDEXED BY
                            WAREHOUSE-INDEX.
    03  LOCATION            PICTURE 99.
    03  BIN-NUMBER          PICTURE 999.
    03  QUANTITY-ON-HAND            PICTURE 9(5).
```

The data description entries enable each of the three WAREHOUSE-DATA items to be accessed by indexing. In the Procedure Division, the index-name identifying the table element is enclosed in parentheses immediately following the last space of the table element data-name. The following is an example of how the Procedure Division accesses the first WAREHOUSE-DATA item.

```
SET WAREHOUSE-INDEX TO 1.
ADD QUANTITY-ON-HAND (WAREHOUSE- INDEX) TO
    TOTAL-QUANTITY.
MOVE WAREHOUSE-DATA (WAREHOUSE-INDEX) TO PRINT-AREA.
```

In order to access the third WAREHOUSE-DATA item, the following Procedure Division statements are needed:

```
SET WAREHOUSE-INDEX TO 3.
ADD QUANTITY-ON-HAND (WAREHOUSE-INDEX) TO
    TOTAL-QUAN-TITY.
MOVE WAREHOUSE-DATA (WAREHOUSE-INDEX) TO PRINT-AREA.
```

Notice that both the ADD and MOVE statements are exactly the same. Thus, the programmer would probably place the two statements in a subroutine and change the above procedure to the following:

```
    SET WAREHOUSE-INDEX TO 1.
    PERFORM ADD-MOVE-ROUTINE.
    SET WAREHOUSE-INDEX TO 2.
    PERFORM ADD-MOVE-ROUTINE.

ADD-MOVE-ROUTINE.
    ADD QUANTITY-ON-HAND (WAREHOUSE-INDEX) TO
        TOTAL-QUAN-TITY.
    MOVE WAREHOUSE-DATA (WAREHOUSE-INDEX) TO PRINT-AREA.
```

The preceding example demonstrates the advantage of indexing when only one routine is written to process each element in a table.

Figure 9.14 Example—Indexed By, Key Options.

```
Format

[USAGE IS] INDEX
```

Figure 9.15 Format—Usage Is Index Clause.

The following data description entries describe a table containing five elements which are accesssd by an index called TABLE1-INDEX:

```
01  TABLE1.
    02  TABLE-ELEMENT OCCURS 5 TIMES INDEXED BY TABLE1-
            INDEX.
        03  ACCOUNT-NUMBER      PICTURE X(5).
        03  BALANCE             PICTURE 9999V99.
```

In order to store the contents of the index (TABLE1-INDEX) temporarily, the programmer must move the contents of the index to an index data item. The index data item is defined in the Working-Storage Section as follows:

```
77  INDEX-STORAGE      USAGE IS INDEX.
```

Since an index is accessed only by a SET, SEARCH, and PERFORM statement, a SET statement must be used to store the contents of the index (TABLE1-INDEX) into the index data item (INDEX-STORAGE).

```
SET INDEX-STORAGE TO TABLE1-INDEX.
```

Thus, if TABLE1-INDEX contains the address or offset for the third table element in TABLE1, then the SET statement places the offset or occurrence number of the third table element into the index data item (INDEX-STORAGE). When the programmer wants to reset the index (TABLE1-INDEX) to the content of the index data item (INDEX-STORAGE), the following SET statement would be executed.

```
SET TABLE1-INDEX TO INDEX-STORAGE.
```

Figure 9.16 Example—Usage Is Index Clause.

If the USAGE IS INDEX clause describes a group item, all the elementary items in the group are index data items. The group itself is not an index data item. In the following data descriptions, FIELD-A, FIELD-B, and FIELD-C are index data items.

```
01  GROUP-1   USAGE IS INDEX.
    02   FIELD-A.
    02   FIELD-B.
    03   FIELD-C.
```

Figure 9.17 Example—Group Usage Is Index Clause.

```
                                   {index-name-1}
SEARCH identifier-1  [VARYING      {identifier-2}  ]

    [AT END imperative-statement-1]

                         {imperative-statement-2}
    WHEN condition-1     {NEXT SENTENCE         }

                         {imperative-statement-3]}
    [WHEN condition-2    {NEXT SENTENCE          }   ] . . .
```

Figure 9.18 Format—Search Statement Format 1.

2. An elementary item described with the USAGE IS INDEX clause is called an index-data item. An index-data item is an elementary item (not necessarily connected with any table) that can be used to save index-name values for future reference. An index-data item must be assigned an index-name (e.g., [occurrence number-1] * entry length) through the SET statement. Such a value corresponds to an occurrence number in a table.
3. An index-data item can be referred to directly only in a SEARCH or SET statement or in a relation condition. An index-data item can be part of a group which is referred to in a MOVE or I/O statement, in which case no conversion will take place.

(See figures 9.16, 9.17.)

PROCEDURE DIVISION

The SEARCH and SET statements may be used to facilitate table handling. In addition, there are special rules involving Table-Handling elements when they are used in relation conditions.

Relation Conditions

Comparisons involving index-names and/or index-data items conform to the following rules:

1. The comparison of two index-names is actually the comparison of the corresponding occurrence numbers.
2. In the comparison of an index-name with a data item (other than an index-data item), or in the comparison of an index-name with a literal, the occurrence number that corresponds to the value of the index-name is compared with the data item or literal.
3. In the comparison of an index-data item with an index-name or another index-data item, the actual values are compared without conversion.

Search Statement

The SEARCH statement is used to search a table from a table-element that satisfies the specified condition and to adjust the value of the associated index-item to the occurrence number corresponding to the table-element (fig. 9.18).

Only one level of a table (a table element) can be referenced with one SEARCH. There are two formats for the SEARCH statement. Format 1, SEARCH, is used for a serial search. Format 2, SEARCH ALL, is used for a binary search.

Format 1 SEARCH statements perform a serial search of a table element. If the programmer knows that the "found" condition will come after some intermediate point or some table element, to speed up execution, the SET statement can be used to set the index-names at that point and search only part of the table elements. If the number of table elements are large and must be searched from the first occurrence to the last, the use of Format 2 (SEARCH ALL) is more efficient than Format 1, since it uses a binary search technique that is much faster; however, the table must be ordered.

In Format 2, the SEARCH ALL statement, the table must be ordered on the KEY(S) specified in the OCCURS clause.

Rules Governing the Use of the Search Statement

1. Identifier-1 must not be subscripted or indexed, but its description must contain an OCCURS clause and an INDEXED BY clause.
2. Identifier-2, when specified, must be described as USAGE IS INDEX, or as a numeric elementary item without any positions to the right of the assumed decimal point. Identifier-2 is incremented by the same amount, and at the same time, as the occurrence number.

Procedure of Search Statement—Format 1

1. Upon execution of a SEARCH statement, a serial search takes place, starting with the current index setting.
2. If at the start of the search the value of the index-name associated with Identifier-1 is not greater than the highest possible occurrence number for Identifier-1, the following action takes place:
   a. The conditions in the WHEN option are evaluated in the order in which they are written.
   b. If none of the conditions is satisfied, the index-number for Identifier-1 is incremented to reference the next table-element, and the search is repeated.
   c. If, upon evaluation, one of the WHEN conditions is satisfied, the search terminates immediately, and the imperative statement associated with that condition is executed. The index-name points to the table-element that satisfied the condition. (See figures 9.19, 9.20.)
   d. If the end of the table is reached without the WHEN condition being satisfied, the search is terminated immediately, and if the AT END option is specified, the imperative statement is executed. If the AT END option is omitted, control passes to the next sentence. (See figures 9.21, 9.22, 9.23.)

Condition-1 or Condition-2 may be any condition as follows: a relation condition, class condition, condition-name condition, or sign condition.

Varying Option

When the VARYING option is specified, one of the following applies:

1. If index-name-1 is one of the indexes for Identifier-1, index-name-1 is used for the search; otherwise, the first (or only) index-name is used for Identifier-1.
2. If index-name-1 is an index for another table entry, then when the index-name for Identifier-1 is incremented to represent the next occurrence of the table, index-name-1 is simultaneously incremented to represent the next occurrence of the table it indexes. (See figures 9.24, 9.25.)

Procedure for Search Statement—Format-2

1. A nonserial type of search operation takes place (fig. 9.26). When this occurs, the initial setting of the index-name for Identifier-1 is ignored, and its setting is varied during the search. At no time is it less than the value that corresponds to the first element of the table, nor is it ever greater than the value that corresponds to the last element of the table.

Simple SEARCH Sentence

Format — SEARCH identifier WHEN condition imperative-statement.

Explanation — Beginning with the table element whose occurrence number corresponds to the current value of the index defined for identifier, table elements with the common name-identifier are tested serially by index until an element that satisfies condition is found. The imperative statement is then executed and control moves either to the sentence directly following the SEARCH sentence or the paragraph specified in a GO TO statement within the imperative statement.

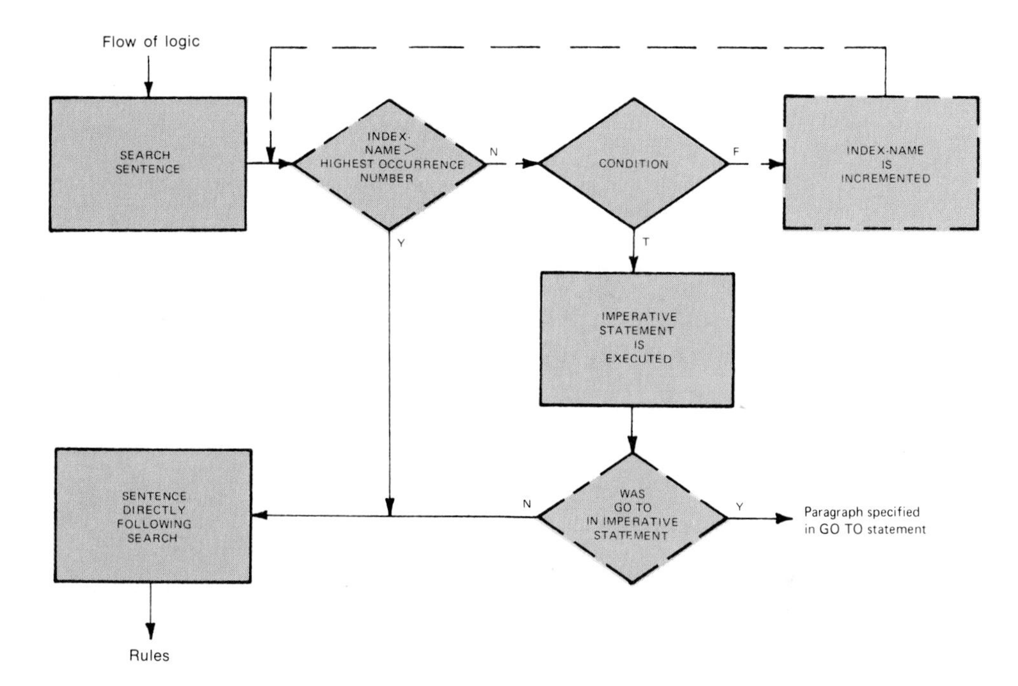

Rules

1. Identifier must be the common name of the table elements for which the OCCURS clause with the INDEXED BY option is specified.
2. Identifier must not be a level 01 variable.
3. Index-name (not specified in the SEARCH sentence but necessary for its execution) must be defined by the INDEXED BY option of the OCCURS clause in the data description entry for identifier.

(The dashed lines indicate logic that is done automatically by the compiler.)

Figure 9.19 Example—Simple Search Statement—Format 1.

```
01  SALE-TABLE.
    02  SALE-ITEM  OCCURS 10 TIMES
                   DEPENDING ON TABLE-SIZE
                   INDEXED BY S
                   PICTURE X(6).
    SET S TO 1.
    SEARCH SALE-ITEM
        WHEN ORDER-ITEM IS EQUAL TO SALE-ITEM (S)
        COMPUTE PRICE-WS ROUNDED = PRICE-WS * NINETY.
    COMPUTE AMOUNT = PRICE-WS * QUANTITY-WS.
```

Figure 9.20 Example—Search Statement—Format 1.

SEARCH Sentence with the AT END Option

Format

SEARCH identifier [AT END {imperative-statement-1 | NEXT SENTENCE}] WHEN condition imperative-statement-2.

Explanation

Beginning with the table element whose occurrence number corresponds to the current value of the index defined for identifier, table elements are tested serially by index until an element that satisfies condition is found. Imperative-statement-2 is then executed and control moves either to the sentence directly following the SEARCH sentence or the paragraph specified in a GO TO statement within imperative-statement-2. If no table element satisfies condition, imperative-statement-1 is executed and control moves either to the sentence directly following the SEARCH sentence or the paragraph specified in a GO TO statement within imperative-statement-1.

Flow of logic

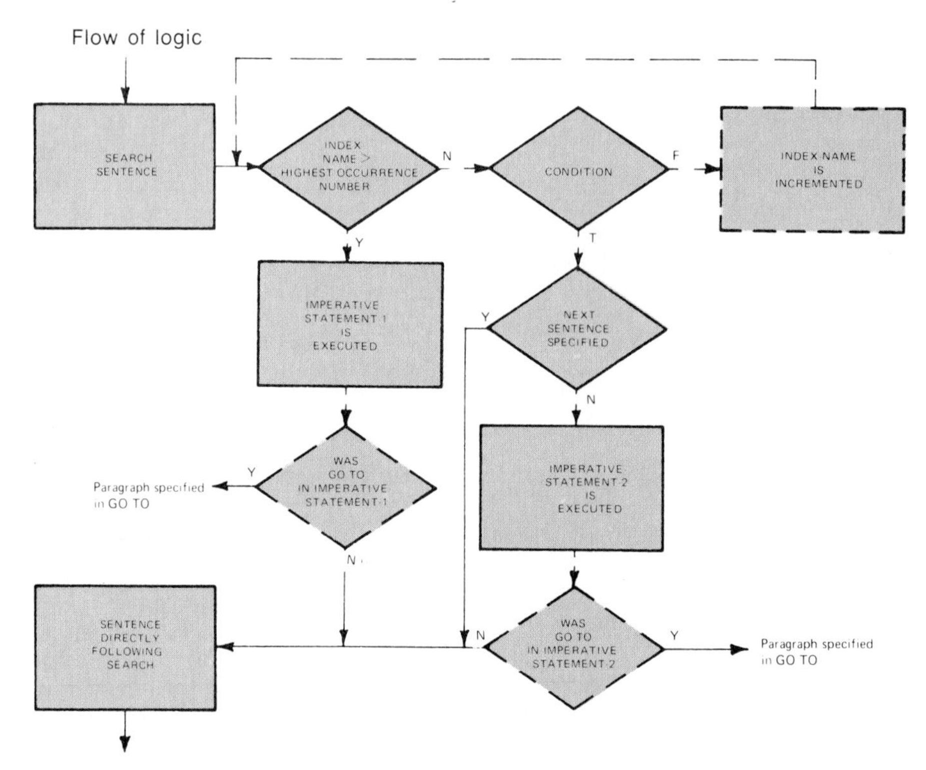

Rules

1. Identifier must be the common name of the table elements for which the OCCURS clause with the INDEXED BY option is specified.
2. Identifier must not be a level 01 variable.
3. Index-name (not specified in the SEARCH sentence but necessary for its execution) must be defined by the INDEXED BY option of the OCCURS clause in the data description entry for identifier.

(The dashed lines indicate logic that is done automatically by the compiler.)

Figure 9.21 Example—Search Statement—At End Option.

```
    SET D TO 1.
TEST.
    SEARCH DISCONTINUED-ITEM-NUMBER
        AT END PERFORM DETAIL-PROC
        WHEN ITEM-NUMBER EQUAL TO
            DISCONTINUED-ITEM-NUMBER (D)
        PERFORM DISCONTINUED-DETAIL.
EXIT-POINT.
    EXIT.
```

Figure 9.22 Example—Search Statement—At End Option.

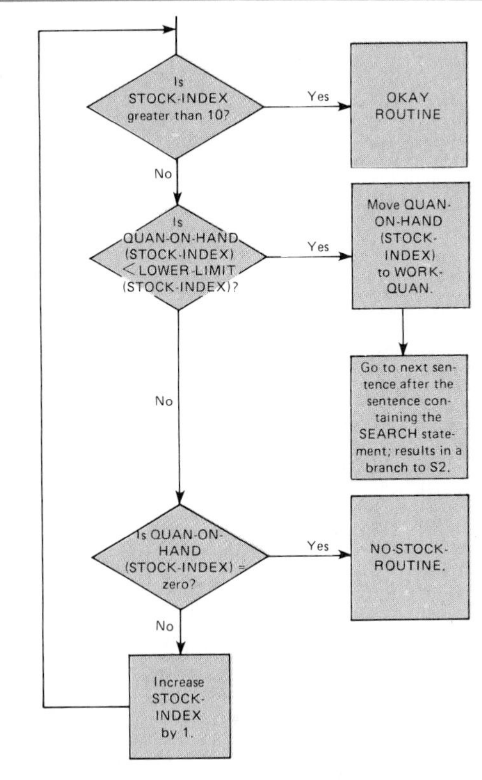

The table called STOCK-TABLE consists of 10 table elements (STOCK-ELEMENT). These are indexed by the index called STOCK-INDEX. SEARCH begins testing the fifth table element because the index associated with the table element was set to the occurrence number 5 before the SEARCH statement was executed. If the fifth table element does not meet either of the two conditions, the index (STOCK-INDEX) is incremented to the next occurrence number, 6. The search operation is then repeated, testing the fields in the sixth table element. The search continues with each successive table element, terminating when the end of the table is reached or a condition is met by the QUAN-ON-HAND field in one of the last, six table elements.

```
01  STOCK-TABLE.
    02  STOCK-ELEMENT  OCCURS 10 TIMES  INDEXED BY STOCK-
    INDEX.
        03  STOCK-NUMBER        PICTURE 9(5).
        03  PRICE               PICTURE 999V99.
        03  QUAN-ON-HAND        PICTURE 9999.
        03  BACK-ORDER-QUAN     PICTURE 9999.
        03  LOWER-LIMIT         PICTURE 999.

PROCEDURE DIVISION.
PAR-1.
    SET STOCK-INDEX TO 5.
    SEARCH STOCK-ELEMENT
        AT END PERFORM OKAY-ROUTINE,
        WHEN QUAN-ON-HAND (STOCK-INDEX) IS LESS THAN
            LOWER-LIMIT (STOCK-INDEX),
        MOVE QUAN-ON-HAND (STOCK-INDEX) TO WORK-QUAN
        WHEN QUAN-ON-HAND (STOCK-INDEX) EQUALS ZERO,
        PERFORM NO-STOCK-ROUTINE.
PAR-2.
    ADD 100 TO WORK-QUAN.
```

Figure 9.23 Example—Search Statement—At End Option.

SEARCH Sentence with the VARYING Option

Format

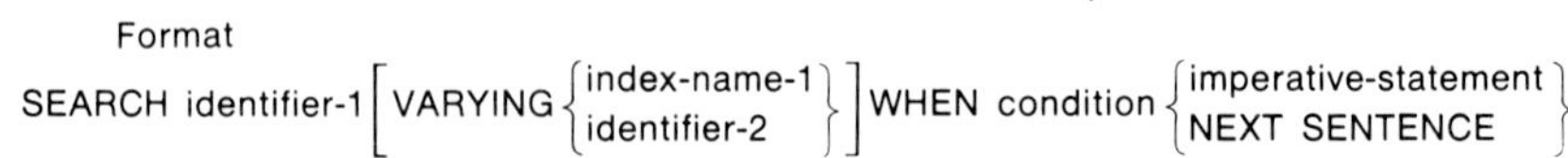

SEARCH identifier-1 [VARYING {index-name-1 / identifier-2}] WHEN condition {imperative-statement / NEXT SENTENCE}

Explanation Beginning with the table element whose occurrence number corresponds to the current value of the index defined for identifier-1, table elements with the common name identifier-1 are tested serially by index until an element that satisfies condition is found. Beginning with its value at the time the SEARCH sentence is executed the variable specified in the VARYING option is incremented each time the index defined for identifier-1 is incremented. When a table element satisfying condition is found, if NEXT SENTENCE is specified in the WHEN option, control moves to the sentence directly following the SEARCH sentence. Otherwise, the imperative statement is executed and control moves either to the sentence directly following the SEARCH sentence or the paragraph specified in a GO TO statement within the imperative statement.

Flow of Logic

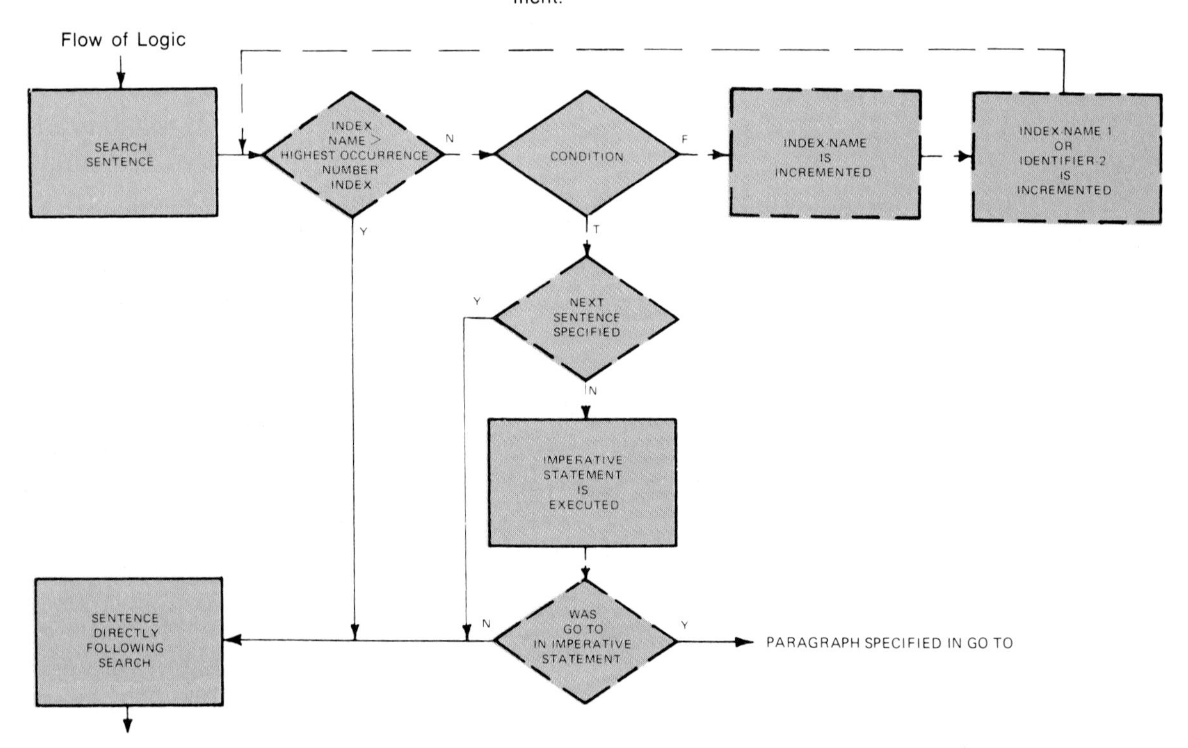

Rules

1. Identifier-1 must be the common name of the table elements for which the OCCURS clause with the INDEXED BY option is specified.
2. Identifier-1 must not be a level 01 variable.
3. Index-name (not specified in the SEARCH sentence but necessary for its execution) must be defined by the INDEXED BY option of the OCCURS clause in the data description entry for identifier-1.
4. Index-name-1 may be the index defined for identifier-1 or for another identifier.
5. Identifier-2 may be a variable defined with the USAGE IS INDEX clause or an elementary variable that will have integer values.

(The dashed lines indicate logic that is done automatically by the compiler.)

Figure 9.24 Example—Search Statement—Varying Option.

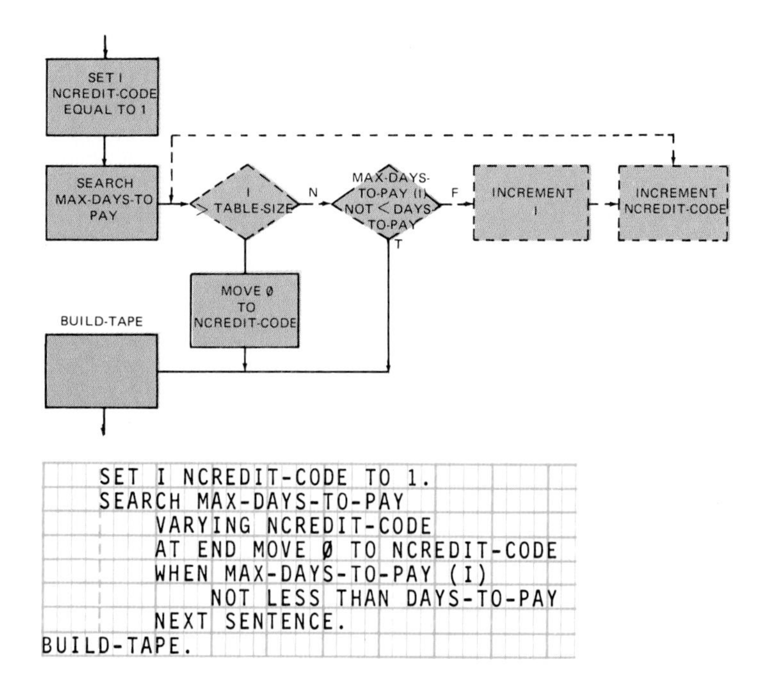


```
       SET I NCREDIT-CODE TO 1.
       SEARCH MAX-DAYS-TO-PAY
           VARYING NCREDIT-CODE
           AT END MOVE Ø TO NCREDIT-CODE
           WHEN MAX-DAYS-TO-PAY (I)
               NOT LESS THAN DAYS-TO-PAY
           NEXT SENTENCE.
BUILD-TAPE.
```

Figure 9.25 Example—Search Statement—Varying Option.

Figure 9.26
Format—Search Statement—Format 2.

```
SEARCH ALL identifier-1  [AT END imperative-statement-1]
      WHEN condition-1  { imperative-statement-2 }
                        { NEXT SENTENCE          }
```

a. If condition-1 cannot be satisfied for any setting of the index within the permitted range, control is passed to imperative statement-1 when the AT END option appears, or to the next sentence when this clause does not appear. In either case, the final setting of the index is unpredictable.
b. If the index indicates an occurrence that allows condition-1 to be satisfied, control passes to imperative statement-2.

2. The first index-name assigned to Identifier-1 will be used for the search.
3. The description of Identifier-1 must contain the KEY option in its OCCURS clause.

Condition-1 must consist of one of the following:

1. *Relation Condition*—One of the data-names must appear in the KEY clause.
2. *Condition-name Condition*—the conditional variable associated with condition-name must be one of the names that appear in the KEY clause of Identifier-1.
3. A compound condition from simple conditions of the types described above, with AND as the only connector.

Any data-name that appears in the KEY clause of Identifier-1 may be tested in condition-1. However, all data-names in the KEY clause preceding the one to be tested must also be tested on condition-1. No other tests can be made on condition-1. (See figure 9.27.)

The table called STOCK-TABLE consists of 10 elements (STOCK-ELEMENT). These are indexed by the index called STOCK-INDEX. The SEARCH ALL statement begins testing the first table element with the index STOCK-INDEX set to the occurrence number 1. If the first table element does not have STOCK-NUMBER equal to TRAN-STOCK-NO then STOCK-INDEX is incremented to the next occurrence number, 2. The search operation is then repeated with the testing of the STOCK-NUMBER in the second table element, continuing with each successive table element until the end of the table is reached or the condition is met.

```
01  STOCK-TABLE.
    02  STOCK-ELEMENT              OCCURS 10 TIMES ASCENDING KEY IS
        STOCK-NUMBER               INDEXED BY STOCK-INDEX.
        03  STOCK-NUMBER           PICTURE 9(5).
        03  PRICE                  PICTURE 999V99.
        03  QUAN-ON-HAND           PICTURE 9999.
        03  BACK-ORDER-QUAN        PICTURE 9999.
        03  LOWER-LIMIT            PICTURE 999.

PROCEDURE DIVISION.
PAR-1.
    SEARCH ALL STOCK-ELEMENT
        AT END PERFORM MISSING-ROUTINE
        WHEN STOCK-NUMBER (STOCK-INDEX) IS EQUAL TO
             TRAN-STOCK-NO
        MOVE STOCK-ELEMENT (STOCK-INDEX) TO WORK-AREA.
PAR-2.
    ADD QUANTITY TO TOTAL.
```

If an imperative-statement does not terminate with a GO TO or PERFORM statement, control passes to the sentence following the SEARCH statement. In the example above, the condition's imperative-statement does not terminate with a GO TO or PERFORM statement. Thus, after the move is executed, control passes to PAR-2 which is the next sentence after the SEARCH statement.

Figure 9.27 Example—Search Statement—Format 2.

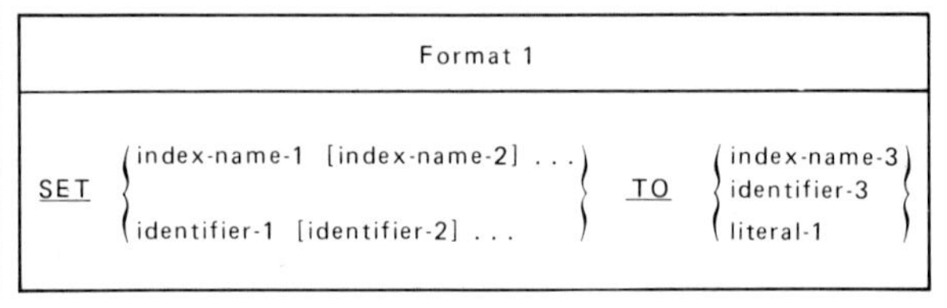

Format 2

SET index-name-4 [index-name-5] . . . {UP BY / DOWN BY} {identifier-4 / literal-2}

Figure 9.28 Format—Set Statement.

Set Statement The SET statement establishes reference points for table-handling operations by setting index-names associated with table-elements (fig. 9.28). The SET statement must be used when initializing index-name values before execution of a SEARCH statement. It may also be used to transfer values between index-names and other elementary data items.

The SET statement is used to assign values to index-names and to index-data names. An *index-name* is the name given to the table associated with a table of elements. An index is assigned by including the INDEXED BY option of the OCCURS clause in the definition of the table element. An *index-data name* is an elementary data item described with the USAGE IS INDEX clause in the Data Division.

When the SET statement assigns to an index-name the value of a literal, identifier, or an index-name from another table element, it is set to an actual displacement from the beginning of the table element that corresponds to an occurrence number of an element in the associated table.

Rules Governing the Use of the Set Statement

1. All identifiers must name either index-data items or elementary items described as integers, except that Identifier-4 must not name an index-data item. When a literal is used, it must be a positive integer. Index-names are considered related to a given table through the INDEXED BY option.

Format-1

When the SET statement is executed, one of the following actions takes place:

1. Index-name-1 is set to a value that corresponds to the same table-element to which either index-name-3, Identifier-3, or literal-1 corresponds. If Identifier-3 is an index-data item, or if index-name-3 is related to the same table as index-name-1, no conversion takes place.
2. If Identifier-1 is an index-data item, it may be set to equal either the contents of index-name-3 or Identifier-3 where Identifier-3 is also an index-data item. Literal-1 cannot be used in this case.
3. If Identifier-1 is not an index-data item, it may be set only to an occurrence number that corresponds to the value of index-name-3. Neither Identifier-3 nor literal-1 can be used in this case.

(See figures 9.29, 9.30.)

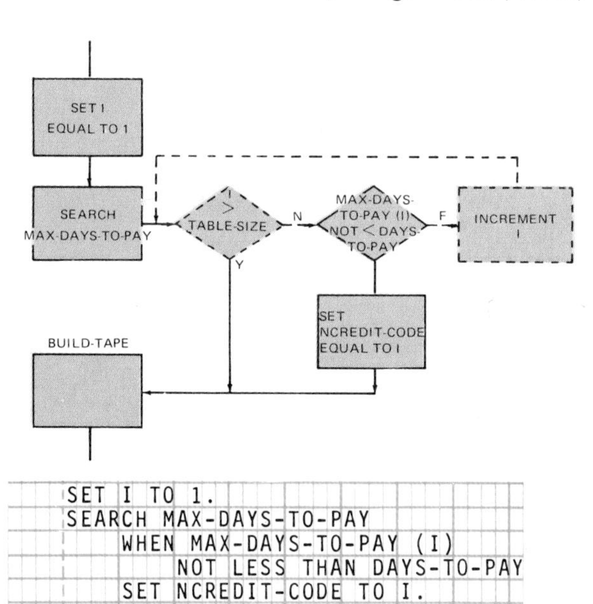


```
    SET I TO 1.
    SEARCH MAX-DAYS-TO-PAY
        WHEN MAX-DAYS-TO-PAY (I)
            NOT LESS THAN DAYS-TO-PAY
        SET NCREDIT-CODE TO I.
BUILD-TAPE.
```

Figure 9.29 Example—Set Statement.

INDEX-A indexes a table of 10 AMOUNT items. During the Procedure Division, the current contents of INDEX-A must be stored temporarily. Thus, the sentence SET INDEX-A-STORAGE TO INDEX-A is given. When it is time to restore INDEX-A to its original contents, the SET INDEX-A TO INDEX-A-STORAGE is given. Notice that the index-data-item (INDEX-A-STORAGE) is defined as a 77 level in the Working-Storage Section; the index (INDEX-A) is defined by the INDEXED BY clause associated with the table element (AMOUNT).

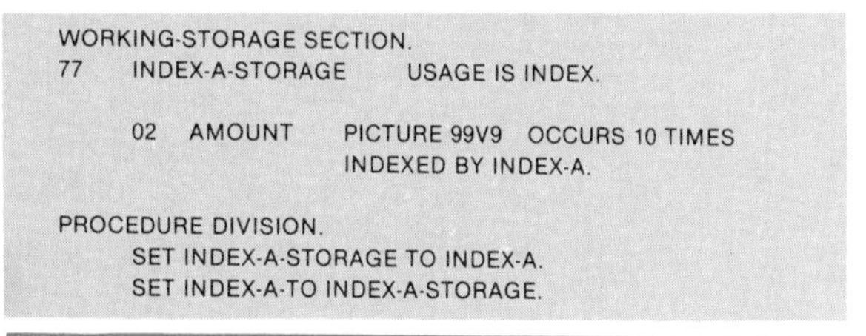

```
WORKING-STORAGE SECTION.
77    INDEX-A-STORAGE        USAGE IS INDEX.

      02    AMOUNT        PICTURE 99V9   OCCURS 10 TIMES
                          INDEXED BY INDEX-A.

PROCEDURE DIVISION.
      SET INDEX-A-STORAGE TO INDEX-A.
      SET INDEX-A-TO INDEX-A-STORAGE.
```

Figure 9.30 Example—Set Statement.

Format-2

When the SET statement is executed, the content of index-number-4 (index-name-5, etc), if present, is incremented (UP BY) or decremented (DOWN BY) a value that corresponds to the number of occurrence represented by the value of literal-2 or Identifier-4. (See figures 9.31, 9.32, 9.33, 9.34.)

The index Q2, is set to the value corresponding to the occurrence number represented by index Q1. If the index Q1 contains the offset for the second element in the table of QUANTITY elements, then the occurrence number of the index Q1 is 2. Thus, the SET statement changes the contents of the index Q2 to the offset for the second element in the table of LOWER-LIMIT elements. The offset contained in index Q2 now corresponds to the occurrence number of 2.

```
FILE SECTION.
    02  QUANTITY     PICTURE 999  OCCURS 5 TIMES INDEXED BY Q1.

    02  LOWER-LIMIT  PICTURE 99   OCCURS 5 TIMES INDEXED BY Q2.

PROCEDURE DIVISION.
    SET Q2 TO Q1.
    IF QUANTITY (Q1) IS LESS THAN LOWER-LIMIT (Q2)
    THEN
        PERFORM UNDER-ROUTINE
    ELSE
        NEXT SENTENCE.
```

Figure 9.31 Example—Set Statement.

The contents of index Q1 is incremented by 1. This permits the index to access each element in the table as the logic loops through the ADD statement five times.

```
FILE SECTION.

    02  QUANTITY     PICTURE 999   OCCURS 5 TIMES   INDEXED BY
            Q1.

PROCEDURE DIVISION.

    SET Q1 TO 1.
    PERFORM PAR-1
        UNTIL Q1 IS GREATER THAN 5.
PAR-1.
    ADD QUANTITY (Q1) TO TOTAL-QUANTITY.
    SET Q1 UP BY 1.
```

Figure 9.32 Example—Set Statement.

```
IDENTIFICATION DIVISION.
PROGRAM-ID.
ENVIRONMENT DIVISION.
CONFIGURATION SECTION.
SOURCE-COMPUTER.           IBM-370-I155.
OBJECT-COMPUTER.           IBM-370-I155.
SPECIAL-NAMES.    CONSOLE IS TYPEWRITER.
INPUT-OUTPUT SECTION.
FILE-CONTROL.
    SELECT INFILE          ASSIGN TO UT-2400-S-INTAPE.
    SELECT OUTFILE         ASSIGN TO UR-1403-S-PRTOUT.
    SELECT INCARDS         ASSIGN TO UR-2540R-S-ICARDS.

DATA DIVISION.
FILE SECTION.
FD  INFILE     LABEL RECORDS ARE OMITTED.
01  TABLE      PICTURE X(28200).
01  TABLE-2    PICTURE X(1800).
FD  OUTFILE    LABEL RECORDS ARE OMITTED.
01  PRTLINE    PICTURE X(133).
FD  INCARDS    LABEL RECORDS ARE OMITTED.
01  CARDS.
      05  STATE-NAME         PICTURE X(4).
      05  SEXCODE            PICTURE 9.
      05  YEARCODE           PICTURE 9(4).
      05  FILLER             PICTURE X(71).
WORKING-STORAGE SECTION.
01    PRTAREA-20.
      05  FILLER             PICTURE X         VALUE SPACES.
      05  YEARS-20           PICTURE 9(4).
      05  FILLER             PICTURE X(3)      VALUE SPACES.
      05  BIRTHS-20          PICTURE 9(7).
      05  FILLER             PICTURE X(3)      VALUE SPACES.
      05  DEATHS-20          PICTURE 9(7).
      05  FILLER             PICTURE X(108)    VALUE SPACES.
01    PRTAREA.
      05  FILLER             PICTURE X         VALUE SPACES.
      05  YEAR               PICTURE 9(4).
      05  FILLER             PICTURE X(3)      VALUE SPACES.
      05  BIRTHS             PICTURE 9(5).
      05  FILLER             PICTURE X(3)      VALUE SPACES.
      05  DEATHS             PICTURE 9(5).
      05    FILLER               PICTURE X(112)     VALUE SPACES.
01    CENSUS-STATISTICS-TABLE.
      05    STATE-TABLE     OCCURS 50 TIMES      INDEXED BY ST.
            10  STATE-ABBREV      PICTURE X(4).
            10  SEX     OCCURS 2 TIMES           INDEXED BY SE.
                15  STATISTICS    OCCURS 20 TIMES   ASCENDING KEY IS YEAR
                                  INDEXED BY YR.
                    20  YEAR      PICTURE 9(4).
                    20  BIRTHS    PICTURE 9(5).
                    20  DEATHS    PICTURE 9(5).
01    STATISTICS-LAST-20-YRS.
      05    SEX-20     OCCURS 2 TIMES     INDEXED BY SE-20.
            10  STATE-20     OCCURS 50 TIMES     INDEXED BY ST-20.
                15  YEARS-20      PICTURE 9(4).
                15  BIRTHS-20     PICTURE 9(7).
                15  DEATHS-20     PICTURE 9(7).
01    FLAGS.
      05    MORE-TABLE-FLAG            PICTURE XXX     VALUE 'YES'.
            88  MORE-TABLE                             VALUE 'YES'.
            88  NO-MORE-TABLE                          VALUE 'NO'.
      05    MORE-DATA-FLAG             PICTURE XXX     VALUE 'YES'.
            88  MORE-DATA                              VALUE 'YES'.
            88  NO-MORE-DATA                           VALUE 'NO'.
PROCEDURE DIVISION.
MAIN-ROUTINE.
    OPEN    INPUT     INFILE
                      INCARDS
            OUTPUT    OUTFILE.
    READ INFILE INTO CENSUS-STATISTICS-TABLE
        AT END MOVE 'NO' TO MORE-TABLE-FLAG.
    READ INFILE INTO STATISTICS-LAST-20-YRS
        AT END MOVE 'NO' TO MORE-TABLE-FLAG.
    PERFORM READ-TABLE
        UNTIL NO-MORE-TABLE.
    READ INCARDS
        AT END MOVE 'NO' TO MORE-DATA-FLAG.
    PERFORM PROCESS-DATA
        UNTIL NO-MORE-DATA.
    CLOSE   INFILE
            INCARDS
            OUTFILE.
    STOP RUN.
READ-TABLE.
```

Figure 9.33 Sample—Table Handling Program.

Figure 9.33 Continued.

```
    READ INFILE INTO CENSUS-STATISTICS-TABLE
        AT END MOVE 'NO' TO MORE-TABLE-FLAG.
    READ INFILE INTO STATISTICS-LAST-20-YRS
        AT END MOVE 'NO' TO MORE-TABLE-FLAG.

PROCESS-DATA.
    PERFORM DETERMINE-ST.
    PERFORM DETERMINE-YR.
    READ INCARDS
        AT END MOVE 'NO' TO MORE-DATA-FLAG.

DETERMINE-ST.
    SET ST ST-20 TO 1.
    SEARCH STATE-TABLE
        VARYING ST-20
        AT END DISPLAY 'INCORRECT STATE' STATE-NAME UPON
            TYPEWRITER
            PERFORM PROCESS-DATA
        WHEN STATE-NAME = STATE-ABBREV (ST)
            NEXT SENTENCE.
    SET SE SE-20 TO SEXCODE.

DETERMINE-YR.
    SEARCH ALL STATISTICS
        AT END DISPLAY 'INCORRECT YEAR' YEARCODE UPON
            TYPEWRITER
        WHEN YEAR OF STATISTICS (ST, SE, YR) = YEARCODE
        PERFORM WRITE-RECORD.
EXIT-POINT.
    EXIT.

WRITE-RECORD.
    MOVE CORRESPONDING STATISTICS (ST, SE, YR) TO PRTAREA.
    WRITE PRTLINE FROM PRTAREA
        AFTER ADVANCING 3.
    MOVE CORRESPONDING STATE-20 (SE-20, ST-20) TO PRTAREA-20.
    WRITE PRTLINE FROM PRTAREA-20
        AFTER ADVANCING 1.
```

The census bureau uses the program to compare:

1. The number of births and deaths that occurred in any of the 50 states in any of the past 20 years with
2. The total number of births and deaths that occurred in the same state over the entire 20-year period

The input, INCARDS, contains the specific information upon which the search of the table is to be conducted. INCARDS is formatted as follows:

STATE-NAME	a 4-character alphabetic abbreviation of the state name.
SEXCODE	1 = male; 2 = female
YEARCODE	a 4-digit field in the range 1950 through 1969

A typical run might determine the number of females born in New York in 1963 as compared with the total number of females born in New York in the past 20 years.

Problem Definition

The accessing of a table containing constant information is shown by the COBOL solution of a program processing data from a magnetic tape called READINGFILE. The following is the record layout of each record in the file READINGFILE.

The contents of the fields within a READINGFILE record are as follows:

RNO: Account number

BEGIN-METER: Beginning meter reading.

END-METER: Ending meter reading.

RSC: Service code.

Data from the records of READINGFILE is used to create records written onto BILLFILE. The following is the record layout of each record in the file BILLFILE.

RNO				BEGIN-METER				END-METER				RSC
9	9	9	9	9	9	9	9	9	9	9	9	X
000	001	002	003	004	005	006	007	008	009	010	011	012

The contents of the fields within a BILLFILE record are as follows:

BNO: Account number

BSC: Service code

QUANTITY: Quantity consumed

CROSS: Gross dollar amount

NET: Net dollar amount

The contents of the fields in a BILLFILE record are determined by the following formulas.

Quantity consumed = ending meter reading − beginning meter reading

Gross dollar amount = quantity consumed × charge rate (rounded)

Discount dollar amount = gross dollar amount × percent of discount

Net dollar amount = gross dollar amount − discount dollar amount.

The values of charge rate and percent of discount are determined from the following table.

Service Code	Quantity Consumed	Charge Rate	Percent of Discount
B	0100	$.055	01.5%
B	0500	$.045	05.3%
B	1000	$.032	05.3%
B	9999	$.027	05.3%
R	0050	$.060	01.5%
R	0100	$.058	03.2%
R	0200	$.051	03.2%
R	0500	$.047	03.2%
R	9999	$.040	03.2%

BNO				BSC	QUANTITY				GROSS					NET				
9	9	9	9	X	9	9	9	9	9	9	9 ^	9	9	9	9	9 ^	9	9
000	001	002	003	004	005	006	007	008	009	010	011	012	013	014	015	016	017	018

Figure 9.34 Example—Table Handling Program.

Figure 9.34 Continued.

If a zero results from subtracting beginning meter reading from ending meter reading or if the service code is not found within the table, then the READINGFILE record is to be output to the error file called BADFILE with the appropriate error code.

```
IDENTIFICATION DIVISION.
PROGRAM-ID.   METER.

ENVIRONMENT DIVISION.
CONFIGURATION SECTION.
SOURCE-COMPUTER.   NCR-CENTURY-200.
OBJECT-COMPUTER.   NCR-CENTURY-200   MEMORY SIZE 3200 WORDS.
*
INPUT-OUTPUT SECTION.
FILE-CONTROL.
    SELECT READNGFILE     ASSIGN TO NCR-TYPE-45.
    SELECT BILLFILE       ASSIGN TO NCR-TYPE-45.
    SELECT BADFILE        ASSIGN TO NCR-TYPE-45.

* EACH FILE IS ASSIGNED TO A MAGNETIC TAPE UNIT TYPE 45.

DATA DIVISION.
FILE SECTION.
FD READNGFILE           BLOCK CONTAINS 100 RECORDS
                        RECORD CONTAINS 13 CHARACTERS
                        LABEL RECORD IS STANDARD.
01     READRECORD.
       02   RNO            PICTURE 9999.
       02   BEGIN-METER    PICTURE 9999.
       02   END-METER      PICTURE 9999.
       02   RSC            PICTURE X.
*
FD BILLFILE                BLOCK CONTAINS 90 RECORDS
                           RECORD CONTAINS 19 CHARACTERS
                           LABEL RECORD IS STANDARD.
01     BILLRECORD.
       02   BNO            PICTURE 9999.
       02   BSC            PICTURE X.
       02   QUANTITY       PICTURE 9999.
       02   GROSS          PICTURE 999V99.
       02   NET            PICTURE 999V99.
*
FD BADFILE                 BLOCK CONTAINS 50 RECORDS
                           RECORD CONTAINS 14 CHARACTERS
                           LABEL RECORD IS STANDARD.
01     BADRECORD.
       02   BADATA         PICTURE X(13).
       02   ECODE          PICTURE X.

WORKING-STORAGE SECTION.
77     OPEN-CODE           PICTURE 9     VALUE IS ZERO.
       88   BADFILE-NOT-OPEN             VALUE IS ZERO.
       88   BADFILE-OPEN                 VALUE IS 1.
01     RATE-TABLE.
       02   FILLER         PICTURE X(11)    VALUE 'B0100055015'.
       02   FILLER         PICTURE X(11)    VALUE 'B0500045053'.
       02   FILLER         PICTURE X(11)    VALUE 'B1000032053'.
       02   FILLER         PICTURE X(11)    VALUE 'B9999027053'.
       02   FILLER         PICTURE X(11)    VALUE 'R0050060015'.
       02   FILLER         PICTURE X(11)    VALUE 'R0100058032'.
       02   FILLER         PICTURE X(11)    VALUE 'R0200051032'.
       02   FILLER         PICTURE X(11)    VALUE 'R0500047032'.
       02   FILLER         PICTURE X(11)    VALUE 'R9999040032'.
01     RATE-TABLE1        REDEFINES RATE-TABLE.
       02   RATE-ITEM     OCCURS 9 TIMES   INDEXED BY T-INDEX.
            03   TSC         PICTURE X.
            03   TQUANTITY   PICTURE 9999.
            03   RATE        PICTURE V999.
            03   PERCENT     PICTURE V999.
```

```
01     FLAGS.
       02   MORE-DATA-FLAG     PICTURE XXX    VALUE 'YES'.
            88  MORE-DATA                     VALUE 'YES'.
            88  NO-MORE-DATA                  VALUE 'NO'.

PROCEDURE DIVISION.
MAIN-ROUTINE.
    OPEN    INPUT     READNGFILE
            OUTPUT    BILLFILE.
    READ READNGFILE
        AT END MOVE 'NO' TO MORE-DATA-FLAG.
    PERFORM READ-FILE
        UNTIL NO-MORE-DATA.
    PERFORM SEARCH-TABLE.
    CLOSE   READNGFILE
            BILLFILE.
    IF BADFILE-OPEN,
    THEN
        CLOSE BADFILE
    ELSE
        NEXT SENTENCE.
    STOP RUN.

READ-FILE.
    IF END-METER IS LESS THAN BEGIN-METER
    THEN
        COMPUTE QUANTITY = (1000 - BEGIN-METER) +
            END-METER
        PERFORM PROCESS-RECORD
    ELSE
        IF END-METER IS EQUAL TO BEGIN-METER
        THEN
            PERFORM ZCODE
        ELSE
            SUBTRACT BEGIN-METER FROM END-METER GIVING
                QUANTITY
            PERFORM PROCESS-RECORD.
    READ READNGFILE
        AT END MOVE 'NO' TO MORE-DATA-FLAG.

ZCODE.
    IF BADFILE-NOT-OPEN
    THEN
        PERFORM OPEN-BAD
    ELSE
        NEXT SENTENCE.
    MOVE 'Z' TO ECODE.
    PERFORM CREATE-BADRECORD.

PROCESS-RECORD.
    MOVE RNO TO BNO.
    MOVE RSC TO BSC.
    SET T-INDEX TO 1.

SEARCH-TABLE.
    IF TSC (T-INDEX) = RSC
        AND TQUANTITY (T-INDEX) = QUANTITY
    THEN
        PERFORM FOUND
    ELSE
        IF TSC (T-INDEX) = RSC
            AND TQUANTITY (T-INDEX) > QUANTITY
        THEN
            PERFORM FOUND
        ELSE
            IF BADFILE-NOT-OPEN
            THEN
                PERFORM OPEN-BAD
            ELSE
```

Figure 9.34 Continued.

```
               NEXT SENTENCE.
          IF RSC  =  'B' OR 'R'
          THEN
               MOVE 'T' TO ECODE
               PERFORM CREATE-BADRECORD
          ELSE
               MOVE 'S' TO ECODE
               PERFORM CREATE-BADRECORD.

CREATE-BADRECORD.
     MOVE READRECORD TO BADATA.
     WRITE BADRECORD.
     PERFORM READ-FILE.

OPEN-BAD.
     OPEN OUTPUT     BADFILE.
     MOVE 1 TO OPEN-CODE.

FOUND.
     MULTIPLY QUANTITY BY RATE (T-INDEX) GIVING GROSS
          ROUNDED.
     COMPUTE NET  =  GROSS  *  PERCENT (T-INDEX).
     WRITE BILLRECORD.
     PERFORM READ-FILE.
```

SUBSCRIPT PROBLEM

INPUT

Field	Card Columns	
Customer Number	2–5	
Item Number	6–10	
Item Cost	21–24	XX.XX
Department	30	
Customer Name	32–50	

CALCULATIONS TO BE PERFORMED

Calculate the total bill for each customer by applying the appropriate discount for the department, that is, multiply the item cost by the discount factor for the department to arrive at the charge price.

OUTPUT

Print reports as follows:

1. D I S C O U N T T A B L E

Department   Discount

2. C U S T O M E R R E P O R T

CUST. NO.   CUST. NAME   DEPT.   ITEM NO.   ITEM COST   DISCT. PER.   DISCT. AMT.   CHARGE

SUBSCRIPT PROBLEM

MAIN-ROUTINE

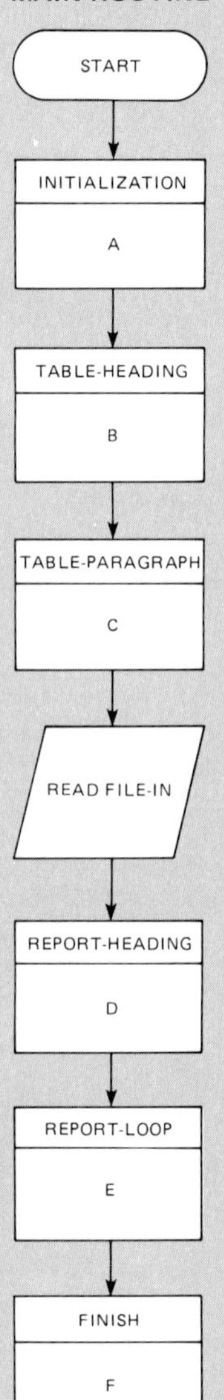

PROCESS-ROUTINES

INITIALIZATION

TABLE-HEADING

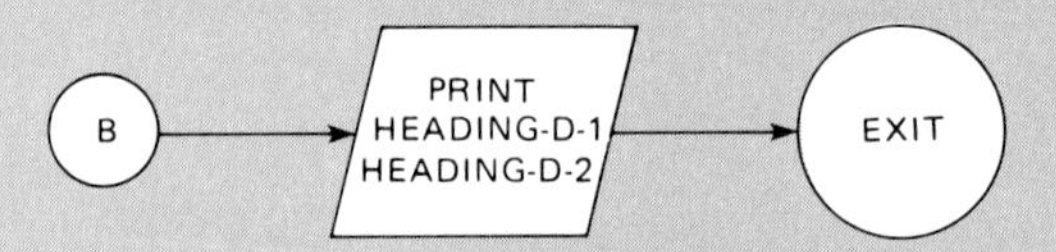

TABLE-PARAGRAPH

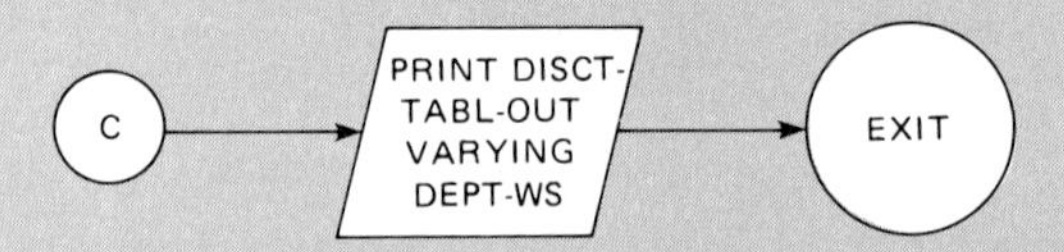

REPORT-HEADING

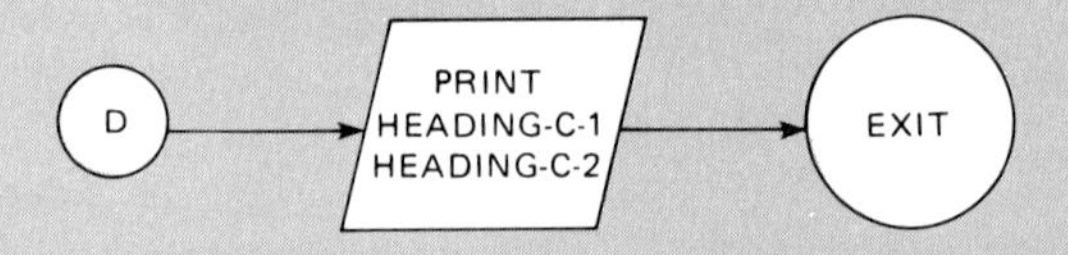

(PROCESS-ROUTINES, CONTINUED)

REPORT-LOOP

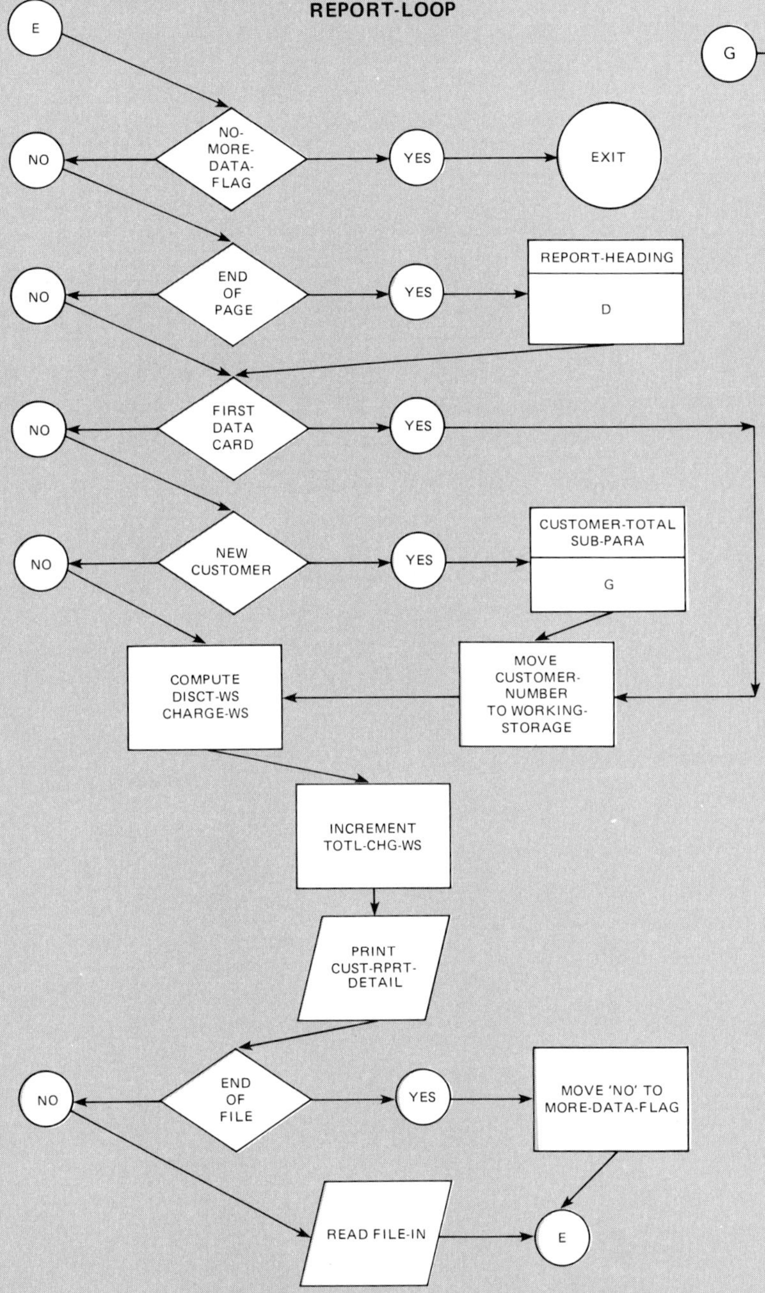

E

NO-MORE-DATA-FLAG

NO

YES

EXIT

END OF PAGE

NO

YES

REPORT-HEADING

D

FIRST DATA CARD

NO

YES

NEW CUSTOMER

NO

YES

CUSTOMER-TOTAL SUB-PARA

G

COMPUTE DISCT-WS CHARGE-WS

MOVE CUSTOMER-NUMBER TO WORKING-STORAGE

INCREMENT TOTL-CHG-WS

PRINT CUST-RPRT-DETAIL

END OF FILE

NO

YES

MOVE 'NO' TO MORE-DATA-FLAG

READ FILE-IN

E

CUSTOMER-TOTAL-SUB-PARA

G

PRINT CUST-RPRT-TOTAL

EXIT

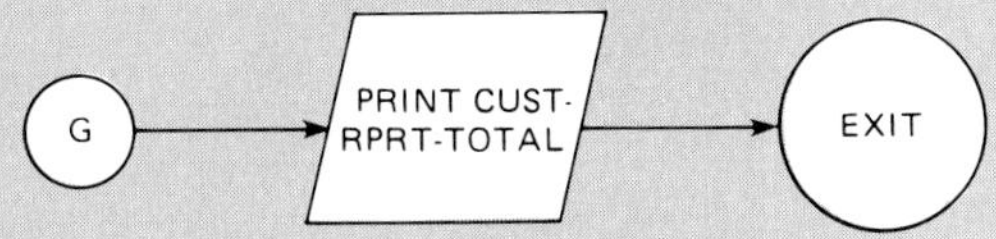

FINISH

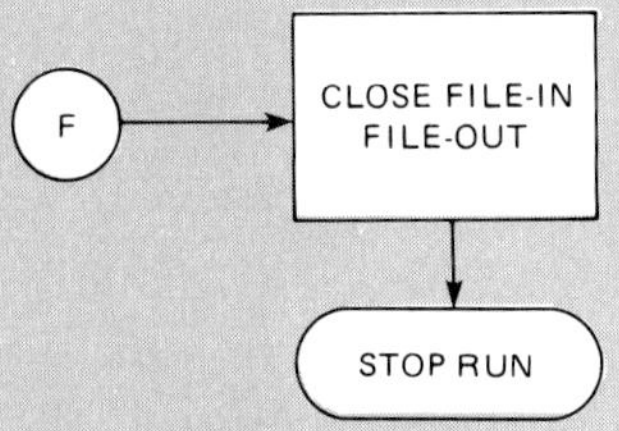

F

CLOSE FILE-IN FILE-OUT

STOP RUN

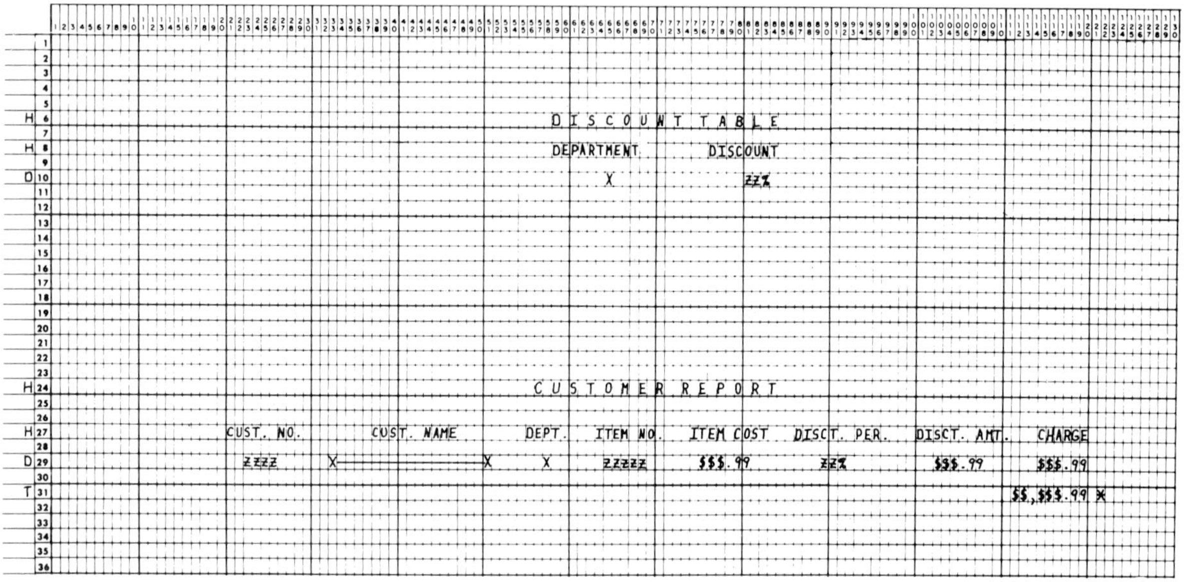

```
H  6                                          D I S C O U N T   T A B L E
H  8                                          DEPARTMENT        DISCOUNT
D 10                                              X                ZZZ

H 24                                        C U S T O M E R   R E P O R T

H 27           CUST. NO.        CUST. NAME        DEPT.   ITEM NO.   ITEM COST   DISCT. PER.   DISCT. AMT.   CHARGE
D 29             ZZZZ      X-----------------X      X      ZZZZZ      $$$.99         ZZZ          $$$.99      $$$.99
T 31                                                                                                      $$,$$$.99 *
```

```
PP 5734-CB2 V4 RELEASE 1.3 01AUG75         IBM OS AMERICAN NATIONAL STANDARD COBOL        DATE AUG  4,1977

     1

00001          IDENTIFICATION DIVISION.                                    1
00002
00003          PROGRAM-ID.                                                 2
00004              SUBSCRIPT PROBLEM.                                      3
00005          AUTHOR.                                                     4
00006              DOROTHY R HENNESSEY.                                    5
00007          INSTALLATION.                                               6
00008              WEST LOS ANGELES COLLEGE.                               7
00009          DATE-WRITTEN.                                               8
00010              16 JULY 1977.                                           9
00011          DATE-COMPILED. AUG  4,1977.                                 10
00012          ENVIRONMENT DIVISION.                                       15
00013
00014          CONFIGURATION SECTION.                                      17
00015          SOURCE-COMPUTER.                                            18
00016              IBM-360.                                                19
00017          OBJECT-COMPUTER.                                            20
00018              IBM-360.                                                21
00019          SPECIAL-NAMES. C01 IS SKIP-TO-ONE.                          22
00020          INPUT-OUTPUT SECTION.                                       23
00021          FILE-CONTROL.                                               24
00022              SELECT FILE-IN       ASSIGN TO UR-2540R-S-CARDIN.       25
00023              SELECT FILE-OUT      ASSIGN TO UR-1403-S-PRINT.         26
00024
00025
```

2

```
00027          DATA DIVISION.                                                 28
00028
00029          FILE SECTION.                                                  30
00030          FD  FILE-IN                                                    31
00031              RECORDING MODE IS F                                        32
00032              LABEL RECORDS ARE OMITTED                                  33
00033              DATA RECORD IS CARDIN.                                     34
00034          01  CARDIN.                                                    35
00035              03  FILLER              PICTURE X.                         36
00036              03  CUSTOMER-NUMBER     PICTURE 9(4).                      37
00037              03  ITEM-NUMBER         PICTURE 9(5).                      38
00038              03  FILLER              PICTURE X(10).                     39
00039              03  ITEM-COST           PICTURE 99V99.                     40
00040              03  FILLER              PICTURE X(5).                      41
00041              03  DEPARTMENT          PICTURE 9.                         42
00042              03  FILLER              PICTURE X.                         43
00043              03  CUSTOMER-NAME       PICTURE X(19).                     44
00044              03  FILLER              PICTURE X(30).                     45
00045          FD  FILE-OUT                                                   46
00046              RECORDING MODE IS F                                        47
00047              LABEL RECORDS ARE OMITTED                                  48
00048              DATA RECORD IS PRINT.                                      49
00049          01  PRINT               PICTURE X(133).                        50
00050          WORKING-STORAGE SECTION.                                       51
00051          77  DEPT-WS                 PICTURE 99      VALUE 01.          52
00052          77  CUST-NO-WS              PICTURE 9(4)    VALUE ZEROES.      53
00053          77  DISCT-WS                PICTURE 99V99.                     54
00054          77  CHARGE-WS               PICTURE 99V99.                     55
00055          77  TOTL-CHG-WS             PICTURE 9999V99 VALUE ZEROES.      56
00056          77  LINE-COUNT              PICTURE 99      VALUE ZEROES.      56.1
00057          01  FLAGS.                                                     57.1
00058              03  MORE-DATA-FLAG      PICTURE XXX     VALUE 'YES'.
00059                  88  MORE-DATA                       VALUE 'YES'.
00060                  88  NO-MORE-DATA                    VALUE 'NO '.
00061          01  EMPTY                   PICTURE X(133)  VALUE SPACES.      57.5
```

3

```
00063          01  FILLER.                                                    58
00064              03  DISCT-TAB.                                             59
00065                  05  FILLER          PICTURE V99     VALUE .05.         60
00066                  05  FILLER          PICTURE V99     VALUE .07.         61
00067                  05  FILLER          PICTURE V99     VALUE .10.         62
00068                  05  FILLER          PICTURE V99     VALUE .15.         63
00069                  05  FILLER          PICTURE V99     VALUE .06.         64
00070                  05  FILLER          PICTURE V99     VALUE .22.         65
00071                  05  FILLER          PICTURE V99     VALUE .12.         66
00072                  05  FILLER          PICTURE V99     VALUE .09.         67
00073                  05  FILLER          PICTURE V99     VALUE .20.         68
00074              03  DISCT-TABLE REDEFINES DISCT-TAB.                       69
00075                  05  DISCT           PICTURE V99     OCCURS 9 TIMES.    70
00076          01  HEADING-D-1.                                               71
00077              03  FILLER  PIC X(53)   VALUE SPACES.                      72
00078              03  FILLER  PIC X(27)   VALUE 'D I S C O U N T   T A B L E'. 73
00079              03  FILLER  PIC X(53)   VALUE SPACES.                      74
00080          01  HEADING-D-2.                                               75
00081              03  FILLER  PIC X(53)   VALUE SPACES.                      76
00082              03  FILLER  PIC X(27)   VALUE 'DEPARTMENT          DISCOUNT'. 77
00083              03  FILLER  PIC X(53)   VALUE SPACES.                      78
00084          01  HEADING-C-1.                                               79
00085              03  FILLER  PIC X(52)   VALUE SPACES.                      80
00086              03  FILLER  PIC X(18)   VALUE 'C U S T O M E R  '.         81
00087              03  FILLER  PIC X(11)   VALUE 'R E P O R T'.               82
00088              03  FILLER  PIC X(52)   VALUE SPACES.                      83
```

```
  4

00090      01  HEADING-C-2.                                                       84
00091          03  FILLER   PIC X(12)   VALUE SPACES.                             85
00092          03  FILLER   PIC X(9)    VALUE 'CUST. NO.'.                        86
00093          03  FILLER   PIC X(6)    VALUE SPACES.                             87
00094          03  FILLER   PIC X(10)   VALUE 'CUST. NAME'.                       88
00095          03  FILLER   PIC X(6)    VALUE SPACES.                             89
00096          03  FILLER   PIC X(5)    VALUE 'DEPT.'.                            90
00097          03  FILLER   PIC X(4)    VALUE SPACES.                             91
00098          03  FILLER   PIC X(8)    VALUE 'ITEM NO.'.                         92
00099          03  FILLER   PIC X(4)    VALUE SPACES.                             93
00100          03  FILLER   PIC X(9)    VALUE 'ITEM COST'.                        94
00101          03  FILLER   PIC X(4)    VALUE SPACES.                             95
00102          03  FILLER   PIC X(8)    VALUE 'DISCT. %'.                         96
00103          03  FILLER   PIC X(4)    VALUE SPACES.                             97
00104          03  FILLER   PIC X(11)   VALUE 'DISCT. AMT.'.                      98
00105          03  FILLER   PIC X(4)    VALUE SPACES.                             99
00106          03  FILLER   PIC X(6)    VALUE 'CHARGE'.                          100
00107          03  FILLER   PIC X(13)   VALUE SPACES.                            101
00108      01  DISCT-TABL-DETAIL.                                                102
00109          03  FILLER               PICTURE X(57)    VALUE SPACES.           103
00110          03  DEPT-TAB-OUT         PICTURE ZZ.                              104
00111          03  FILLER               PICTURE X(15)    VALUE SPACES.           105
00112          03  DISCT-TABL-OUT       PICTURE VZZ.                             106
00113          03  FILLER               PICTURE X        VALUE '%'.              107
00114          03  FILLER               PICTURE X(56)    VALUE SPACES.           108
```

```
  5

00116      01  CUST-RPRT-DETAIL.                                                 109
00117          03  FILLER               PICTURE X(14)    VALUE SPACES.           110
00118          03  CUSTOMER-NUMBER      PICTURE Z(4).                            111
00119          03  FILLER               PICTURE X(4)     VALUE SPACES.           112
00120          03  CUSTOMER-NAME        PICTURE X(19).                           113
00121          03  FILLER               PICTURE X(4)     VALUE SPACES.           114
00122          03  DEPARTMENT           PICTURE 9.                               115
00123          03  FILLER               PICTURE X(7)     VALUE SPACES.           116
00124          03  ITEM-NUMBER          PICTURE Z(5).                            117
00125          03  FILLER               PICTURE X(7)     VALUE SPACES.           118
00126          03  ITEM-COST            PICTURE $$$.99.                          119
00127          03  FILLER               PICTURE X(9)     VALUE SPACES.           120
00128          03  DISCT-OUT            PICTURE VZZ.                             121
00129          03  FILLER               PICTURE X        VALUE '%'.              122
00130          03  FILLER               PICTURE X(11)    VALUE SPACES.           123
00131          03  DISCT-AMT-OUT        PICTURE $$$.99.                          124
00132          03  FILLER               PICTURE X(7)     VALUE SPACES.           125
00133          03  CHARGE-OUT           PICTURE $$$.99.                          126
00134          03  FILLER               PICTURE X(19)    VALUE SPACES.           127
00135      01  CUST-RPRT-TOTAL.                                                  128
00136          03  FILLER               PICTURE X(104)   VALUE SPACES.           129
00137          03  TOTAL-CHARGE         PICTURE $$,$$9.99.                       130
00138          03  FILLER               PICTURE XX       VALUE ' *'.             131
00139          03  FILLER               PICTURE X(17)    VALUE SPACES.           132
```

```
     6

00141          PROCEDURE DIVISION.                                              134
00142
00143          MAIN-ROUTINE.                                                    136
00144              PERFORM INITIALIZATION.                                      137
00145              PERFORM TABLE-HEADING.                                       138
00146              PERFORM TABLE-PARAGRAPH                                      139
00147                  VARYING DEPT-WS FROM 1 BY 1                              140
00148                  UNTIL DEPT-WS IS > 9.                                    141
00149              READ FILE-IN
00150                  AT END MOVE 'NO ' TO MORE-DATA-FLAG.
00151              PERFORM REPORT-HEADING.
00152              PERFORM REPORT-LOOP
00153                  THRU REPORT-LOOP-EXIT
00154                  UNTIL NO-MORE-DATA.
00155              PERFORM FINISH.                                              144
00156
00157          PROCESS-ROUTINES.                                                146
00158
00159          INITIALIZATION.                                                  148
00160              OPEN INPUT  FILE-IN
00161                   OUTPUT FILE-OUT.
00162
00163          TABLE-HEADING.                                                   151
00164              WRITE PRINT FROM HEADING-D-1                                 152
00165                  AFTER ADVANCING SKIP-TO-ONE LINES.                       153
00166              WRITE PRINT FROM HEADING-D-2                                 154
00167                  AFTER ADVANCING 3 LINES.                                 155
00168
00169          TABLE-PARAGRAPH.                                                 156
00170              MOVE DEPT-WS TO DEPT-TAB-OUT.                                157
00171              MOVE DISCT (DEPT-WS) TO DISCT-TABL-OUT.                      158
00172              WRITE PRINT FROM DISCT-TABL-DETAIL                           159
00173                  AFTER ADVANCING 2 LINES.                                 160
```

```
     7

00175          REPORT-HEADING.
00176              WRITE PRINT FROM HEADING-C-1                                 162
00177                  AFTER ADVANCING SKIP-TO-ONE LINES.                       163
00178              WRITE PRINT FROM HEADING-C-2                                 164
00179                  AFTER ADVANCING 3 LINES.                                 165
00180              WRITE PRINT FROM EMPTY                                       166
00181                  AFTER ADVANCING 1 LINES.                                 167
00182              ADD 5 TO LINE-COUNT.                                         168
00183
00184          REPORT-LOOP.
00185              IF LINE-COUNT IS > 37
00186              THEN
00187                  MOVE ZEROES TO LINE-CCUNT
00188                  PERFORM REPORT-HEADING
00189              ELSE
00190                  NEXT SENTENCE.
00191
00192              IF CUST-NO-WS IS = 0
00193              THEN
00194                  MOVE CUSTOMER-NUMBER OF CARDIN TO CUST-NO-WS
00195              ELSE
00196                  NEXT SENTENCE.
00197
00198              IF CUSTOMER-NUMBER OF CARDIN IS > CUST-NO-WS
00199              THEN
00200                  PERFORM CUSTOMER-TOTAL-SUB-PARA
00201              ELSE
00202                  NEXT SENTENCE.
```

8

```
00204           COMPUTE DISCT-WS ROUNDED = ITEM-COST OF CARDIN
00205               * DISCT (DEPARTMENT OF CARDIN)
00206           COMPUTE CHARGE-WS = ITEM-COST OF CARDIN
00207               - DISCT-WS.
00208           ADD CHARGE-WS TO TOTL-CHG-WS.                                  179
00209           MOVE CORRESPONDING CARDIN TO CUST-RPRT-DETAIL.                 180
00210           MOVE DISCT (DEPARTMENT OF CARDIN) TO DISCT-OUT.                181
00211           MOVE DISCT-WS TO DISCT-AMT-OUT.                                182
00212           MOVE CHARGE-WS TO CHARGE-OUT.
00213           WRITE PRINT FROM CUST-RPRT-DETAIL                              183
00214               AFTER ADVANCING 1 LINES.                                   184
00215           ADD 1 TO LINE-COUNT.                                           185
00216           READ FILE-IN                                                   172
00217               AT END MOVE 'NO ' TO MORE-DATA-FLAG.                       173
00218       REPORT-LOOP-EXIT.  EXIT.
00219
00220       CUSTOMER-TOTAL-SUB-PARA.                                           186
00221           MOVE TOTL-CHG-WS TO TOTAL-CHARGE.                              187
00222           WRITE PRINT FROM CUST-RPRT-TOTAL                               188
00223               AFTER ADVANCING 1 LINES.                                   189
00224           WRITE PRINT FROM EMPTY                                         190
00225               AFTER ADVANCING 1 LINES.                                   191
00226           ADD 2 TO LINE-COUNT.                                           191.1
00227           MOVE CUSTOMER-NUMBER OF CARDIN TO CUST-NO-WS.                  192
00228           MOVE ZEROES TO TOTL-CHG-WS.                                    193
00229       FINISH.                                                            194
00230           PERFORM CUSTOMER-TOTAL-SUB-PARA.                               195
00231           CLOSE FILE-IN                                                  196
00232                 FILE-OUT.                                                197
00233           STOP RUN.                                                      198
```

D I S C O U N T   T A B L E

DEPARTMENT	DISCOUNT
1	05%
2	07%
3	10%
4	15%
5	06%
6	22%
7	12%
8	09%
9	20%

C U S T O M E R   R E P O R T

CUST. NO.	CUST. NAME	DEPT.	ITEM NO.	ITEM COST	DISCT. %	DISCT. AMT.	CHARGE
152	J. LANGDON	8	87653	$24.75	09%	$2.23	$22.52
152	J. LANGDON	6	64025	$9.45	22%	$2.08	$7.37
152	J. LANGDON	4	41915	$13.70	15%	$2.06	$11.64
152	J. LANGDON	1	17410	$2.51	05%	$.13	$2.38
							$43.91 *
2468	L. MORRISEY	1	18520	$3.75	05%	$.19	$3.56
2468	L. MORRISEY	2	20012	$4.20	07%	$.29	$3.91
2468	L. MORRISEY	3	31572	$10.15	10%	$1.02	$9.13
2468	L. MORRISEY	4	48792	$37.50	15%	$5.63	$31.87
2468	L. MORRISEY	5	50407	$15.15	06%	$.91	$14.24
2468	L. MORRISEY	6	61575	$20.10	22%	$4.42	$15.68
2468	L. MORRISEY	7	79204	$51.70	12%	$6.20	$45.50
2468	L. MORRISEY	8	85075	$37.84	09%	$3.41	$34.43
2468	L. MORRISEY	9	98476	$87.94	20%	$17.59	$70.35
							$228.67 *
3451	M. JACKSON	3	37847	$27.90	10%	$2.79	$25.11
3451	M. JACKSON	5	58492	$68.50	06%	$4.11	$64.39
3451	M. JACKSON	6	60010	$20.40	22%	$4.49	$15.91
3451	M. JACKSON	8	85260	$78.52	09%	$7.07	$71.45
3451	M. JACKSON	9	90520	$27.52	20%	$5.50	$22.02
							$198.88 *
4512	S. LEVITT	2	24680	$30.50	07%	$2.14	$28.36
4512	S. LEVITT	5	56784	$52.53	06%	$3.15	$49.38
4512	S. LEVITT	6	60410	$12.15	22%	$2.67	$9.48
4512	S. LEVITT	7	78952	$89.25	12%	$10.71	$78.54
4512	S. LEVITT	8	85278	$49.75	09%	$4.48	$45.27
4512	S. LEVITT	8	87492	$64.25	09%	$5.78	$58.47
4512	S. LEVITT	9	97204	$84.75	20%	$16.95	$67.80
							$337.30 *
5417	K. CONKLIN	1	13579	$35.72	05%	$1.79	$33.93

C U S T O M E R   R E P O R T

CUST. NO.	CUST. NAME	DEPT.	ITEM NO.	ITEM COST	DISCT. %	DISCT. AMT.	CHARGE
5417	K. CONKLIN	2	24615	$18.75	07%	$1.31	$17.44
5417	K. CONKLIN	3	34928	$37.45	10%	$3.75	$33.70
5417	K. CONKLIN	4	48527	$87.50	15%	$13.13	$74.37
5417	K. CONKLIN	5	50150	$18.95	06%	$1.14	$17.81
5417	K. CONKLIN	5	54652	$38.92	06%	$2.34	$36.58
5417	K. CONKLIN	5	59765	$98.95	06%	$5.94	$93.01
5417	K. CONKLIN	7	71572	$18.95	12%	$2.27	$16.68
5417	K. CONKLIN	8	85175	$80.10	09%	$7.21	$72.89
5417	K. CONKLIN	9	90275	$4.60	20%	$.92	$3.68
5417	K. CONKLIN	9	91572	$18.57	20%	$3.71	$14.86
5417	K. CONKLIN	9	97576	$84.95	20%	$16.99	$67.96
							$482.91 *
6213	Z. HAMPTON	1	15792	$64.25	05%	$3.21	$61.04
6213	Z. HAMPTON	1	19975	$98.75	05%	$4.94	$93.81
6213	Z. HAMPTON	3	34576	$51.15	10%	$5.12	$46.03
6213	Z. HAMPTON	4	49512	$85.20	15%	$12.78	$72.42
							$273.30 *
7545	M. LARSON	1	14676	$38.45	05%	$1.92	$36.53
7545	M. LARSON	1	18592	$82.51	05%	$4.13	$78.38
7545	M. LARSON	1	19994	$98.98	05%	$4.95	$94.03
7545	M. LARSON	2	21214	$15.15	07%	$1.06	$14.09
7545	M. LARSON	3	37515	$82.12	10%	$8.21	$73.91
7545	M. LARSON	3	38592	$96.15	10%	$9.62	$86.53
7545	M. LARSON	4	48485	$87.14	15%	$13.07	$74.07
7545	M. LARSON	5	52762	$37.92	06%	$2.28	$35.64
7545	M. LARSON	5	57684	$80.15	06%	$4.81	$75.34
7545	M. LARSON	7	79015	$96.25	12%	$11.55	$84.70
7545	M. LARSON	8	80123	$5.60	09%	$.50	$5.10
7545	M. LARSON	8	82462	$20.15	09%	$1.81	$18.34
7545	M. LARSON	9	91520	$18.15	20%	$3.63	$14.52
7545	M. LARSON	9	93715	$40.15	20%	$8.03	$32.12
							$723.30 *

Exercises

Write your answer in the space provided (answer may be one or more words).

1. COBOL provides a Table Handling module that has a capability for ______________ tables of ______________ data and to access an item ______________ to its ______________ in the table.
2. A table consists of a number of ______________ data that are of ______________ format.
3. The individual data items in a table are called ______________.
4. Table elements are accessed by either ______________ or ______________ techniques.
5. A subscript is an ______________ whose ______________ identifies a particular ______________ in the table.
6. An index is a computer ______________ or ______________ whose ______________ corresponds to an ______________ number.
7. ______________ is one of the values contained in the table.
8. ______________ is any number used to locate one of the values within the table.
9. The Table Handling feature enables the COBOL program to process ______________ or ______________ of ______________ data.
10. ______________ levels of subscripting are permitted in a COBOL source program.
11. Through the use of a ______________ or ______________, a data item may be accessed ______________ to its position in a file rather than its individual name.
12. The Table Handling feature also has the language facility for specifying how many ______________ an item can be ______________.
13. The feature also provides additional facilities for specifying ______________ or ______________ keys, and permitting ______________ of a table for an item satisfying a specified ______________.
14. Tables of contiguous data items are defined in COBOL by including the ______________ clause in their ______________ entries which specifies that an item is to be ______________ as many ______________ as stated.
15. Whenever the user refers to a table element, the reference must indicate which ______________ of the ______________ is intended.
16. The subscript can be represented either by a ______________ that is an ______________ or by a ______________ that is defined elsewhere.
17. The subscript is enclosed in ______________ and written ______________ following the ______________ of the ______________.
18. A table reference must include as many ______________ as there are ______________ in the table whose ______________ is being defined.
19. When more than one subscript is required, they are written in order of successively ______________ inclusive dimensions of the ______________ organization.
20. It is permissible to mix ______________ and ______________ in subscripts.
21. Subscripts may be qualified in which case, ______________ or ______________ follows the subscripting ______________.
22. The name and description of the table element applies to each ______________ and ______________ of the table element.
23. The ______________ option allows both direct and relative referencing of data.
24. An index is assigned to a given table through the use of the ______________ clause in the definition of the table.
25. An index-name must be initialized by a ______________ statement before it can be used as a table reference.
26. Direct indexing is specified by using an ______________ in the form of a subscript.
27. Relative indexing is specified when the ______________ space of the ______________ is followed by a ______________ group of items.
28. Subscripting and ______________ must not be used together in a single reference.
29. An ______________ clause eliminates the need for separate entries for repeated data and indicates the number of times an identical form item is repeated.

30. When the ________________ clause is used, the data-name which is the subject of the entry must be either subscripted or indexed whenever it is referred to.
31. A ________________ consists of all occurrences of one level of a table.
32. The ________________ option of the OCCURS clause is used when the end of the occurrences cannot be determined.
33. The KEY option is used in conjunction with the ________________ option in the execution of a ________________ statement.
34. The ________________ clause permits the programmer to specify index data items.
35. In the Procedure Division, the ________________ and ________________ statements may be used to facilitate table handling elements.
36. The SEARCH statement is used to search a table from a ________________ that satisfies the specific condition and to adjust the value of the associated ________________ to the occurrence number corresponding to the table element.
37. Upon execution of a SEARCH statement, a serial search takes place starting with the ________________ setting.
38. The search through a table file continues until one of the ________________ is satisfied.
39. If the end of the search is reached without the condition being satisfied, the ________________ statement is executed.
40. The SET statement must be used when initializing an ________________ value before the execution ________________ statement.

Answers

1. DEFINING, CONTIGUOUS, RELATIVE, POSITION
2. CONTIGUOUS, LIKE
3. TABLE ELEMENTS
4. INDEXING, SUBSCRIPTING
5. INTEGER, VALUE, ELEMENT
6. STORAGE POSITION, REGISTER, CONTENT, OCCURRENCE
7. TABLE VALUE
8. SEARCH ARGUMENT
9. LISTS, TABLES, REPEATED
10. THREE
11. SUBSCRIPTING, INDEX-NAME, RELATIVE
12. TIMES, REPEATED
13. ASCENDING, DESCENDING, SEARCHING, CONDITION
14. OCCURS, DATA DESCRIPTION, REPEATED, TIMES
15. OCCURRENCE, ELEMENT
16. LITERAL, INTEGER, DATA-NAME
17. PARENTHESIS, IMMEDIATELY, NAME, TABLE ELEMENT
18. SUBSCRIPTS, DIMENSIONS, ELEMENT
19. LESS, DATA
20. LITERAL, DATA-NAMES
21. OF, IN, DATA-NAME
22. REPETITION, OCCURRENCE
23. INDEXED BY
24. INDEXED BY
25. SET
26. INDEX-NAME
27. TERMINAL, DATA-NAME, PARENTHESIZED
28. INDEXING
29. OCCURS
30. OCCURS
31. TABLE ELEMENT
32. DEPENDING ON
33. INDEXED BY, SEARCH ALL
34. USAGE IS INDEX
35. SEARCH, SET
36. TABLE ELEMENT, INDEX-ITEM
37. CURRENT INDEX
38. CONDITIONS
39. AT END
40. INDEX-NAME, SEARCH

Questions for Review

1. Briefly describe the following standard terms of table handling: table, table element, subscripting, indexing, table value, search argument, table lookup.
2. What are the principal functions of the Table Handling feature?
3. How are tables defined and how is each table element referenced?
4. What are subscripts?
5. Explain the use of indexing. How is it more effective than subscripting?
6. What is meant by direct indexing? Relative indexing?
7. What is the function of the OCCURS clause? Explain the use of the DEPENDING ON, KEY, and INDEXED BY options.
8. What are some of the important rules governing the use of index-names?
9. What is the function of the USAGE IS INDEX clause?
10. What is the purpose of the SEARCH statement? Explain the procedures of the SEARCH statement in locating a data item in a file serially and locating an item in a nonserial manner.
11. What is the purpose of the SET statement and how is it used in table handling operations?

Problems

1. *Match each item with its proper description.*

	Item		Description
______	1. OCCURS	A.	Any number used to locate one of the values within the table.
______	2. SET	B.	A number of contiguous data items that are of like format.
______	3. Subscript	C.	A computer storage position or register whose content corresponds to an occurrence number.
______	4. Index-Name	D.	Table element that satisfies the specified condition.
______	5. Table	E.	Scanning a table until an item associated with search argument is found.
______	6. SEARCH	F.	An integer whose value identifies a particular element in a table.
______	7. Search Argument	G.	Indicates the number of times a series of items with an identical form is repeated.
______	8. Index	H.	Name given to the table associated with a table of elements.
______	9. Table Elements	I.	Establishes reference points for table handling operations
______	10. Table Lookup	J.	Individual data items within a table.

2. *A typical tax table*

Dependents	Rate
1	.175
2	.164
3	.159
4	.153
5	.147
6	.141
7	.121

In the Working-Storage Section,

a. Set up table.
b. Redefine the table describing it as one value entry repeated a certain number of times.
c. Write the procedural statement using subscripts to multiply GROSS by the fourth element in the table to arrive at TAXAMT.

3. *In the following insurance table, (the entries for ages are 1 to 65), (Premium XXX.XX)*

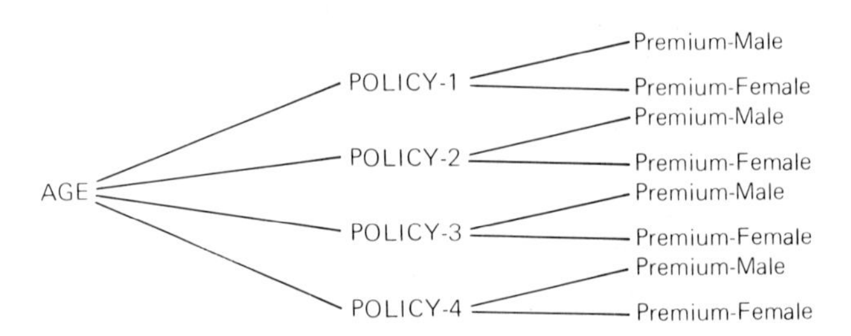

It is necessary to reference the premium for a female, age 64 for policy 2. Write the necessary data and Procedure Divisions entries to accomplish above.

4. *A deck of sales cards (in random sequence) is read and the total from each card is accumulated in the appropriate counter depending upon the value (01–50) punched in the STATE-CODE field of the card. At the end of the file, the totals of each of the 50 cards is to be printed along with the state code.*

Write the necessary Data and Procedure Division entries to accomplish the following:

a. Set up a table for 50 items. Each item is to contain eight positions to represent a sales total counter for each of the 50 states.
b. Print a report indicating the state number and total.

FLOWCHART

Input File
State sales cards in random sequence.

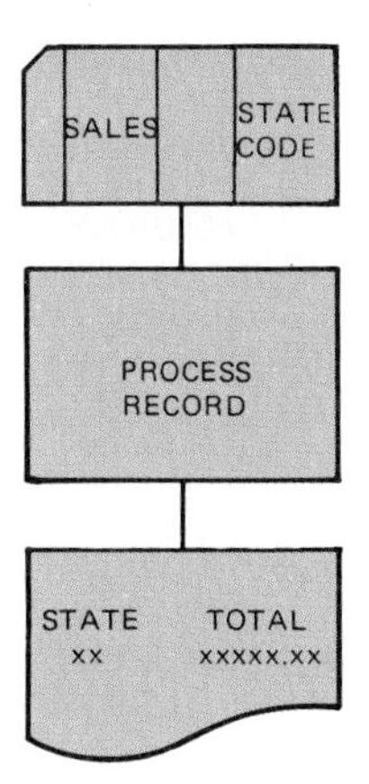

Output Record

5. 02 A OCCURS 5 TIMES.
   03 B OCCURS 4 times.
   04 C PICTURE 99.

It is necessary to reference C in the 4th B in the 5th A. Which of the following entries are correct? (There is more than one correct answer.)

a. C IN B IN A (5, 4)
b. C (5, 4) IN B IN A
c. C IN B (5, 4)
d. C IN A (5, 4)
e. C (4) IN A (5)

6. *A common method used for encoding a date punched in a card is as follows:*

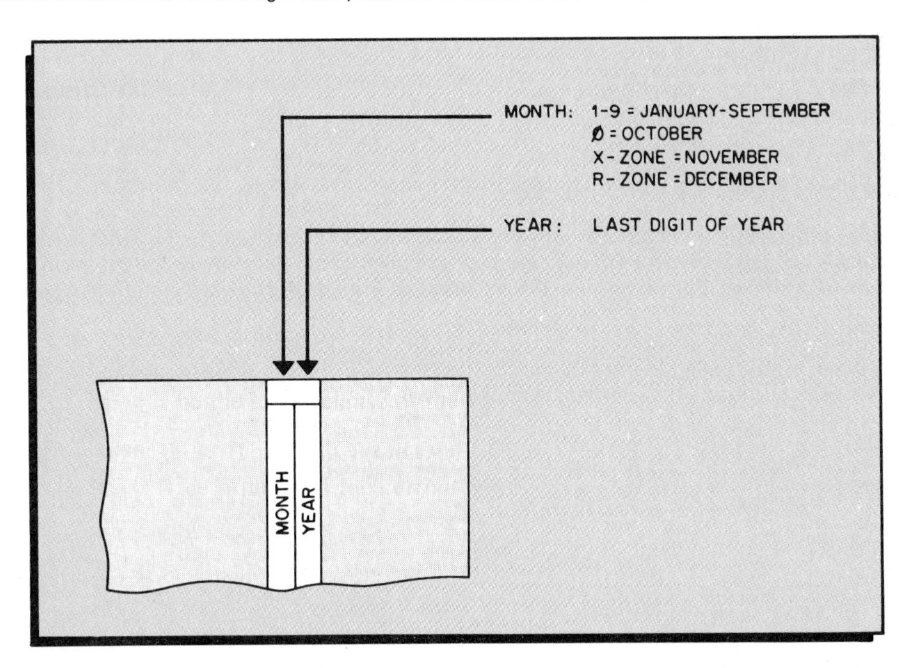

Write the necessary Data and Procedure Division entries to translate this code into a readable format for printing purposes: For example,

22 = FEBRUARY 1972

7. *Assume you wish to print a header line as follows:*

```
01 HEADING-1.
    02  H1               PICTURE X(8) VALUE IS 'ITEM-NO.'.
    02  FILLER           PICTURE X(5).
    02  DESC             PICTURE X(8).
    02  FILLER           PICTURE X(5).
    02  H2               PICTURE X(6) VALUE IS 'TOTAL.'.
    02  FILLER           PICTURE X(5).
    02  DATE.
        03  MONTH        PICTURE X(9).
        03  FILLER       PICTURE X(3).
        03  DAY          PICTURE 99.
        03  FILLER       PICTURE XX.
        03  YEAR         PICTURE 99.
```

In another work area, you have the following record:

```
01 WORK-AREA.
    02  ITEM-NO          PICTURE 99.
    02  MONTH-CODE       PICTURE 99.
    02  DAY-CODE         PICTURE 99.
    02  YEAR-CODE        PICTURE 99.
```

Write the necessary procedural entries to fill in the header-line as follows:

1. ITEM-NO can vary from 01 through 10. Depending upon its contents, DESC is set to one of the following:

01—PENCIL #2	04—PAINTSET	07—STAPLES	10—MISC.
02—PENCIL #3	05—BNDPAPERS	08—CARBONS	
03—PENCIL #4	06—STAPLERS	09—ERASERS	

2. MONTH is to be set depending upon the contents of MONTH-CODE:

01—JANUARY	04—APRIL	07—JULY	10—OCTOBER
02—FEBRUARY	05—MAY	08—AUGUST	11—NOVEMBER
03—MARCH	06—JUNE	09—SEPTEMBER	12—DECEMBER

3. DAY-CODE and YEAR-CODE are to be inserted into DAY and YEAR.

8. *In an insurance premium run, the monthly rates are determined by the risk class. Here, we have a situation where there is a wide and irregular gap between the argument assigned to one table value and the argument assigned to the table value which follows. This means that the risk class code could not be used directly as a subscript.*

Write the necessary Data and Procedure Division entries to

1. Set up the table in the Working-Storage Section and
2. Step-by-step search to locate the appropriate premium rate.

Risk Class	Premium Rate
210	17.50
273	15.50
370	12.30
420	11.95
465	14.60
481	15.25
900	19.45
950	20.01
988	18.10
1030	8.55
1245	14.03
1366	19.99
1505	20.33
1666	12.22
1899	10.00

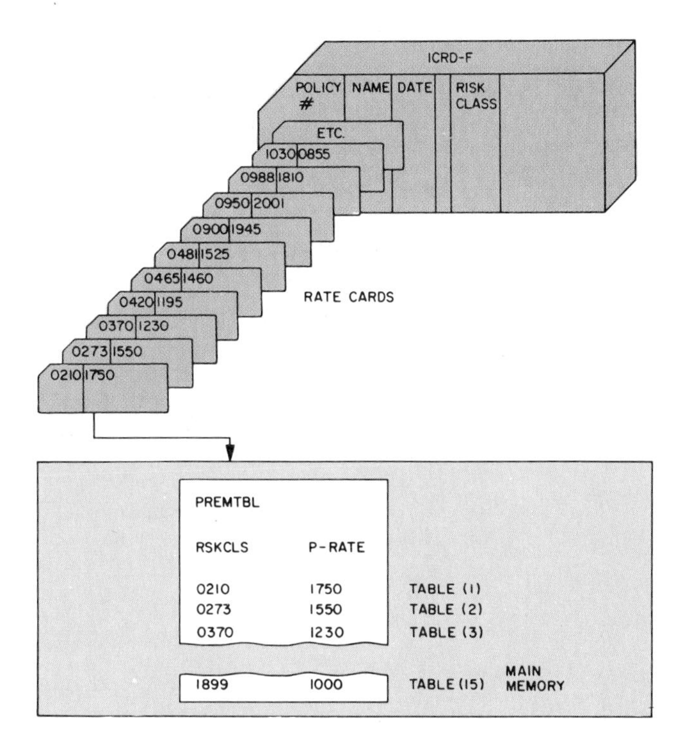

9. *Write the necessary entries for the Data and Procedure Divisions for the following:*

   1. Define the input and output areas with data description entries for each of the values shown in the employee master record diagram.
   2. Code the Working-Storage Section with the following:
      a. A variable to be used in the computation of the new pay rate.
      b. A table to allow the percent factor of one plus the percent increase shown in the diagram to be processed as a table.
      c. A variable to be used as a subscript.
      d. A work area to store the variable to which a percent factor of one plus the percent increase may be transmitted from the card reader.
   3. The Procedure Division entries to accept the table values that have been punched into cards as follows: 103, 103, 103, 105, 105, 108, 108, 108, 110 and 110; then update the employee master records.

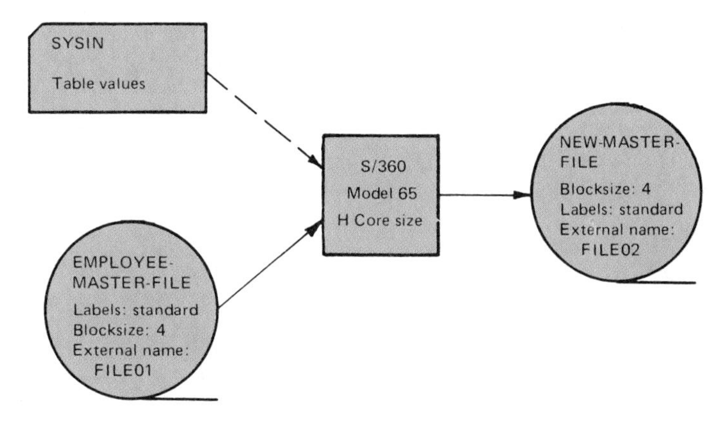

Records in EMPLOYEE-MASTER-FILE (EMPLOYEE-MASTER-RECORD)
and NEW-MASTER-FILE (NEW-MASTER-RECORD)

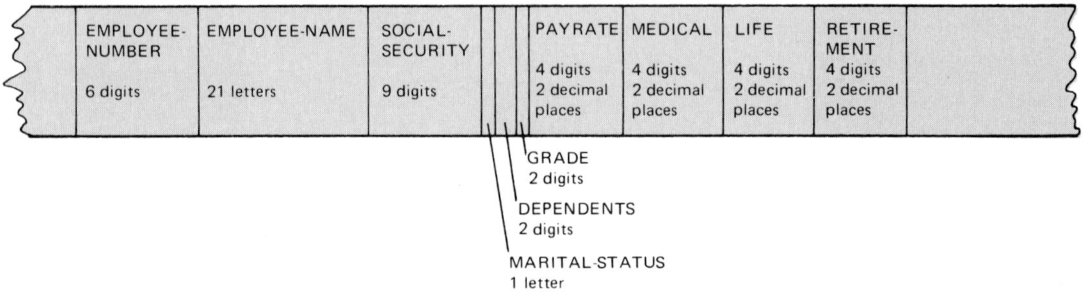

EMPLOYEE-NUMBER	EMPLOYEE-NAME	SOCIAL-SECURITY	MARITAL-STATUS	DEPENDENTS	GRADE	PAYRATE	MEDICAL	LIFE	RETIREMENT
6 digits	21 letters	9 digits	1 letter	2 digits	2 digits	4 digits 2 decimal places	4 digits 2 decimal places	4 digits 2 decimal places	4 digits 2 decimal places

```
IDENTIFICATION DIVISION.
PROGRAM-ID. EMPLOYEE-PAYRATE-UPDATE.

ENVIRONMENT DIVISION.
CONFIGURATION SECTION.
SOURCE-COMPUTER. IBM-360-H65.
OBJECT-COMPUTER. IBM-360-H65.
INPUT-OUTPUT SECTION.
FILE-CONTROL.
    SELECT EMPLOYEE-MASTER-FILE
        ASSIGN TO UT-2400-S-FILE01.
    SELECT NEW-MASTER-FILE
        ASSIGN TO UT-2400-S-FILE02.
```

Employees of a manufacturing company have been assigned a grade of 1 through 10 according to the skills and knowledge required by their jobs. Recently the company has approved pay rate increases for all employees. The increases vary with employee grade as shown in the table to the right. A program is to be written to update the pay rate portion of each employee master record. The system flowchart, a diagram of the employee master record, and the first two divisions of the program are shown.

GRADE	INCREASE
1	3%
2	3%
3	3%
4	5%
5	5%
6	8%
7	8%
8	8%
9	10%
10	10%

10. *Given the following information:*

```
01  TRANS-RECORD.
    02  CITY-ITEM   OCCURS 2 TIMES.
        03  STORE-ITEM   OCCURS 4 TIMES.
            04  DEPT-NO   PICTURE X(5).
            04  DAILY-SALES   PICTURE 9999V99   OCCURS 3 TIMES.
77  DAY     PICTURE 99.
77  DEPT    PICTURE 99.
77  STORE   PICTURE 99.
```

REQUIRED:

a. Write the procedural statements to add DAILY-SALES to TOTAL-SALES for each item using STORE, DEPT and DAY as subscripts.
b. Using the INDEXED BY option, write the necessary Data and Procedure Division entries to add DAILY-SALES to TOTAL-SALES for each item using STORE-INDEX, DEPT-INDEX and DAY-INDEX as indexes.

11. *Given the following information:*

```
02      PRODUCT-NO               PICTURE 9999.
02      WAREHOUSE-DATA           OCCURS 3 TIMES
                                 ASCENDING KEY IS QTY-ON-HAND
                                 INDEXED BY WAREHOUSE-ITEM.
        03    LOCATION           PICTURE 99.
        03    BIN-NO             PICTURE 999.
        03    QTY-ON-HAND        PICTURE 9(5).
```

Write the necessary procedural statements to search the above to locate QTY-ON-HAND equal to 100. If the item is found, the program is to branch to FOUND-ITEM where the quantity will be moved to an output area (OUTPUT-FIELD). If the item is not found, the program is to branch to NOT-IN-TABLE-ROUTINE.

12. *Orders for merchandise are received and punched into cards as follows:*

ORDER-ITEM

Field	**Card Columns**	
Code	1	
Stock Number	2–7	
Date	8–13	
Customer Number	14–19	
Customer Name	20–35	
Quantity	36–39	
Price	40–44	XXX.XX
Blank	45–80	

Set up a table for DISCOUNT-TABLE which can be used for items subject to discount. The stock number and discount rate are punched into cards and are treated as values of the DISCOUNT-TABLE as follows:

DISCOUNT-TABLE

DISCOUNT-ITEM (1)	.87
DISCOUNT-ITEM (2)	.79
DISCOUNT-ITEM (3)	.85
.	.
.	.
.	.
.	.
DISCOUNT-ITEM (15)	.69

The total number of discount items for each month will be punched into a card to determine the size of the DISCOUNT-TABLE for that month. The maximum number of discount items for one month is 15.

a. In the Working-Storage Section, write the entries to define the variables,
   TABLE-MAXIMUM—which is to contain the number of discount items.
   TABLE-VALUE—to which a table value will be transmitted prior to being moved to a specific table element.
   DISCOUNT-TABLE—whose size is determined by the TABLE-MAXIMUM.
   DISCOUNT—to be used as a subscript in referring to table elements.
b. Write the necessary Procedure Division entries to accept the TABLE-MAXIMUM and the DISCOUNT-TABLE.

13. *Using the information in problem 12,*

a. Code the record description for DISCOUNT-TABLE so that a SEARCH statement can be used for the text of discount items. (Specify an index for the table. Use D as an index.)
b. Using the SET statement at one, search the DISCOUNT-TABLE until you find a stock number equal to a discount item number, multiply the quantity of the equal stock number by the price and the discount percentage to arrive at a charge price. If an equal stock number is not found in the search, multiply the quantity by the price to arrive at the charge price.

   Write the necessary procedural entries to accomplish the above.

10 Report Writer Feature

INTRODUCTION

A report represents a pictorial organization of data. To present a report, the physical aspects of the report format must be differentiated from the conceptual characteristics of the data to be included in the report. In defining the physical aspects of the report format, consideration must be given to the width and length of the report medium, to the individual page structure, to the type of hardware device on which the report is finally to be written. Structure controls are established to insure that the report format is maintained.

To define the conceptual characteristics of the data, i.e., the logical organization of the report itself, the concept of level structure is used. Each report may be divided into respective report groups, which, in turn are subdivided into a sequence of items. Level structure permits the programmer to refer to an entire report-name, a major report group, a minor report group, an elementary item within a report group, etc.

To create the report, the approach taken is to define the types of report groups that must be considered in presenting data in a formal manner. Types may be defined as heading groups, footing groups, control groups, or detail print groups. A report group describes a set of data that is to be considered as an individual unit, irrespective of its physical format structure. The unit may be the representation of a data record, a set of constant report headings, or a series of variable control totals. The description of the report group is a separate entity. The report group may extend over several actual lines of a page and may have a descriptive heading above it, which sometimes is necessary in order to produce the desired output report format.

The Report Writer Feature provides the facility for producing reports by specifying the physical appearance of a report rather than requiring specification of the detailed procedures necessary to produce that report. The programmer can specify the format of the printed report in the Data Division, thereby minimizing the amount of Procedure Division coding he would have to write to create the report.

A hierarchy is used in defining that logical organization of the report. Each report is divided into report groups, which in turn are divided into sequence of items. Such a hierarchical structure permits explicit reference to a report group with implicit reference to other levels in the hierarchy. A report group contains one or more items to be presented on one or more lines.

The specification for the format of the printed output together with any necessary control totals and control headings can be written into the program. The detailed report group items are the basic elements of this report. The necessary data for the detail group items can be supplied from sources outside of the report or by the summation of data items within the report. This summation is the process of adding either the individual data items or other control totals.

Additional information in the form of control headings and control totals can be printed with the detail group items. Control totals and control headings occur automatically when the machine senses a control break. When the value of a specified item used for control purposes changes, a control break occurs.

Report headings at the beginning of the report, as well as totals at the end of the report, can be printed as the report is being prepared. Individual page headings at the top of each page, and totals at the bottom of each sheet, may also be printed concurrently. As the program is being executed, line and page counters are incremented automatically and are used by the program to print the various headings, as well as to control the skipping and spacing of the printed items. Data is added, and the totals are printed automatically.

The Report Writer option can print the report as the information is being processed and can also put the data into intermediate storage where it may be used for subsequent off-line printing in a specified format.

A printed report consists of the information reported in the format in which it is printed. Several reports may be printed from the same program. A special single-character identification is necessary to specify each individual report.

At program execution time the report is put in the specified format, the data to be accumulated is added, the necessary totals are printed, counters are incremented and reset, and each line and page is printed. Thus the programmer need not concern himself with any of the details of the operations.

In the Data Division, the programmer provides the necessary data-names and describes the formats of the reports he wishes to produce. In the Procedure Division he writes the necessary statements that provide the desired reports.

DATA DIVISION

The Report Writer Feature allows the programmer to describe his report pictorially in the Data Division, thereby minimizing the amount of Procedure Division coding necessary. The programmer must write in the File Section of the Data Division a description of all names and formats of the reports he wishes to produce. In addition, a complete file and record description of the input data must also be written in this section. A Report Section must be added at the end of the Data Division to define the format of each finished report. A report may be written in two files at the same time.

File Description

The file description entry furnishes the necessary information concerning the identification, physical structure, and record-names pertaining to the file

```
FD  file-name

     [BLOCK CONTAINS Clause]
     [RECORD CONTAINS Clause]
     [RECORDING MODE Clause]
     LABEL RECORDS Clause
     [VALUE OF Clause]
     [DATA RECORDS Clause]
     REPORT Clause.
```

Figure 10.1 Format—File Description Entry.

(fig. 10.1). A detailed discussion of the clauses, with the exception of the Report clause, will be found in the Data Division, chap. 6.

Report Clause

A Report clause is required in the FD entry to list the name of the report to be produced (fig. 10.2). The name or names of each report to be produced appears in this clause. The sequence of the names is not important, and these reports may be of different sizes, formats, etc. The Report Name(s) must be the same name as appears in the Report Section, since the Report clause references the description entries with their associated file description entry (fig. 10.3).

```
{REPORT IS  }
{REPORTS ARE}   report-name-1  [report-name-2] . . .
```

Figure 10.2 Format—Report Clause.

```
ENVIRONMENT DIVISION.
     SELECT FILE-1     ASSIGN UR-1403-S-PRTOUT.
     SELECT FILE-2     ASSIGN UT-2400-S-SYSUT
     .
     .
     .
DATA DIVISION.
FD   FILE-1     RECORDING MODE F
                RECORD CONTAINS 121 CHARACTERS
                REPORT IS REPORT-A.
FD   FILE-2     RECORDING MODE V
                RECORD CONTAINS 101 CHARACTERS
                REPORT IS REPORT-A.
```

For each GENERATE in the Procedure Division, the records for REPORT-A will be written on FILE-1 and FILE-2, respectively. The records on FILE-2 will not contain columns 102 through 121 of the corresponding records on FILE-1.

Figure 10.3 Example—Report Clause.

Report Section

The Report Section must begin on a separate line by itself with the Report Section header. The specification for each report is written here. The physical layout of the report, as well as the user's logical organization, are stipulated in this section (fig. 10.4). These entries should include:

1. The maximum number of lines to be printed on each page.
2. The format and contents of the headings, and when and where they are to appear.
3. The source and format of the data and where it is to appear in a report.
4. The data items that are to act as control factors during the presentation of the report.

5. The format of the totals to be printed and accumulated, and when and where they are to appear.

Structure

The Report Section consists of two types of entries for each report: one describes the physical aspects of the report format, while the other describes the conceptual characteristics of the items that make up the report and their relation to the report format. These entries are:

1. Report Description Entry-RD.
2. Report Group Description Entries.

Report Description Entry—RD

The Report Section must contain at least one report description entry (fig. 10.5). This entry contains information pertaining to the overall format of a report named in the File Section, and is uniquely identified by the level indicator RD. The characteristics of the report page are provided by describing the number of physical lines per page and the limits for presenting the specified headings,

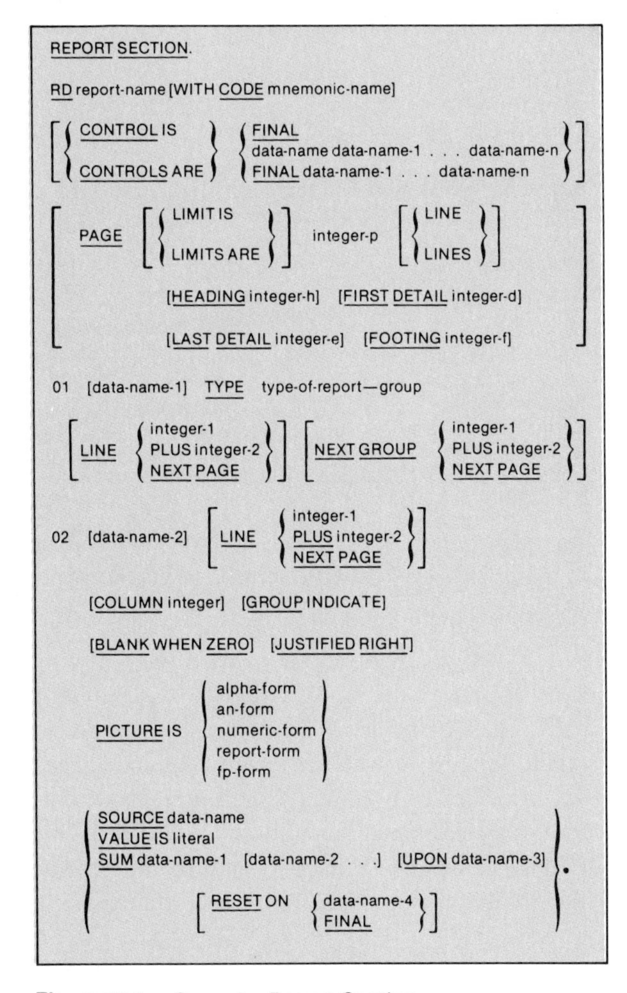

Figure 10.4 Format—Report Section.

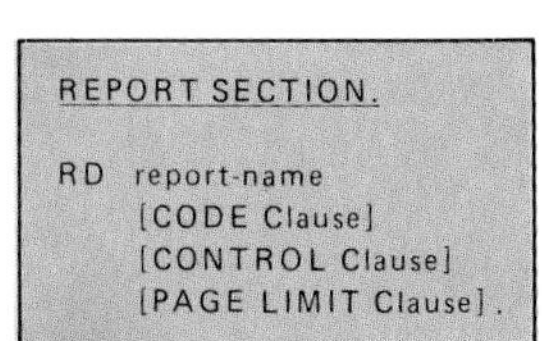

REPORT SECTION.
RD report-name
[CODE Clause]
[CONTROL Clause]
[PAGE LIMIT Clause].

Figure 10.5
Format—Report Description Entry.

footings, and details within a page structure. RD is the reserved word for the level indicator, and each report named in a FD entry in the File Section must be defined by an RD entry.

Report-Name

Report-Name is the unique name of the report and must be specified in the Report clause of the file description entry for the file in which the report is to be written (fig. 10.6).

```
{ REPORT IS   }
{ REPORTS ARE }  report-name ...
```

Figure 10.6
Format—Report Clause.

Code Clause

The CODE clause is used to specify an identifying character added at the beginning of each line produced (fig. 10.7). This is an identifying one-character code that is added to the beginning of each line when there is more than one

```
WITH CODE mnemonic-name
```

Figure 10.7 Format—Code Clause.

report to be written for the file. The identifying character is appended to the beginning of the line preceding the carriage-control line-spacing character (fig. 10.8). This clause must not be specified if the report is printed on-line.

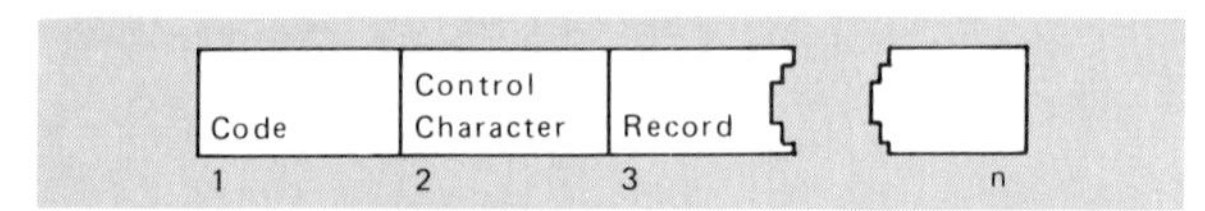

Figure 10.8
Format of a Report Record with the Code Clause Specified.

When the programmer wishes to write a report from a file, he needs merely to read a record, check the first character for the desired code, and have it printed if the desired code is found. The record should be printed starting from the third character.

Mnemonic-Name. This name must be associated with a single character in the SPECIAL-NAMES paragraph in the Environment Division (fig. 10.9).

Control Clause

The CONTROL clause is used to specify the different levels of controls to be applied to the report (fig. 10.10). *A control is a data item that is tested each time a detail group is generated.* Controls govern the basic format of the report. When a control break occurs, special actions will be taken before the next line of the report is printed. Controls are listed in hierarchical order, proceeding from the most important down to the least important. Thus, by specifying HEADING and FOOTING controls, the programmer is able to instruct the Report Writer to produce the report in the desired sequence and format.

Rules Governing the Use of the Control Clause

1. A control is a data item whose value is tested each time a detail group item is to be printed.

```
ENVIRONMENT DIVISION.
.
.
.
SPECIAL-NAMES.    'A' IS CHR-A.
                  'B' IS CHR-B.
.
.
.
DATA DIVISION.
FILE SECTION.
FD  RPT-OUT- FILE
    RECORDS CONTAIN 122 CHARACTERS
    LABEL RECORDS ARE STANDARD
    REPORTS ARE REP-FILE-A REP-FILE-B.
.
.
.
REPORT SECTION.
RD  REP-FILE-A       CODE CHR-A. . .
.
.
.
RD  REP-FILE-B       CODE CHR-B. . .
.
.
.
```

Figure 10.9 How to Create and Print a Report with a Code of A.

```
{CONTROL IS    }  {FINAL                                      }
{              }  {identifier-1 [identifier-2] . . .          }
{CONTROLS ARE  }  {FINAL identifier-1 [identifier-2]          }
```

Figure 10.10 Format—Control Clause.

```
RD. . .CONTROLS ARE YEAR MONTH WEEK DAY
```

Figure 10.11 Example—Control Clause.

2. If the test indicates a change in the value of the data item, a control break occurs, and special action is taken before the detail line is printed.
3. The special action to be taken depends upon what the programmer has stipulated. When controls are tested, the highest control level specified is tested first, then the second-highest level, etc. When a control break is indicated for a higher level, an implied lower control break occurs as well. A control heading or control totals or neither may be defined for each control break by the programmer. (See figure 10.11.)
4. The control footing and headings that are defined are printed prior to printing the original referenced data. They are printed in the following sequence: lowest-level control footing, next-higher-level footing, etc. up to and including the control footing for the level at which the control break occurred; then the control heading for that level, then the next-lower control heading, etc. down to and including the minor control heading; then the detail line is printed.

   If in the course of printing control headings and footing, an end-of-page condition is detected, the current page is ejected, and a new page is begun. If the associated report groups are given, a page footing and/or a page heading is also printed.
5. The levels of control are indicated by the order in which they are written. The identifiers specify the hierarchy of controls. Identifier-1 is the major control, Identifier-2 is the intermediate control, Identifier-3 is the minor control, etc.

6. FINAL is the exception to the rule that controls the data items. It is the highest control level possible. A control break by FINAL occurs at the beginning and at the end of the report only. All implied lower-level control breaks are taken at the same time as FINAL.
7. All level totals are printed in the sequence in which they are written with the exception of the FINAL total.
8. The CONTROL clause is required when CONTROL FOOTING or CONTROL HEADING is specified.

Page Limit Clause

The PAGE LIMIT clause is used to describe the physical format of a page of the report. The clause specifies the specific line control to be maintained within the logical presentation of a page. If there is no PAGE LIMIT clause specified, the PAGE-COUNTER and LINE-COUNTER special registers are not generated, and line control is not performed. The PAGE LIMIT clause is required when the page format is to be controlled by the Report Writer.

The fixed data-names, PAGE-COUNTER and LINE-COUNTER, are numeric counters automatically generated by the Report Writer based on the presence of specific entries, and do not require any data description clauses. The description of these two counters is included here for the purpose of explaining their resultant effect on the overall report format.

PAGE-COUNTER

A numeric counter that may be used as a source data item in order to present the page number on a report. The maximum size of a PAGE-COUNTER is based on the size specified in the PICTURE clause associated with the elementary item whose SOURCE is PAGE-COUNTER. This counter may be referred to in any Procedure Division statement. *A PAGE-COUNTER is generated for a report by the Report Writer only if the PAGE LIMIT clause is specified.*

Rules Governing the Use of the Page-Counter Clause

1. The numeric counter is generated automatically by the Report Writer to be used as a SOURCE item in order to automatically present consecutive page numbers.
2. One counter is supplied for each report described in the Report Section. The size of the counter is based on the size specified in the PICTURE clause associated with the elementary SOURCE data item description.
3. If more than one counter is given as a SOURCE data item within a given report, the number of numeric characters indicated by the PICTURE clause must be identical. The size must indicate sufficient numeric character positions to prevent overflow.
4. If more than one report description entry exists in the Report Section, the user must qualify PAGE-COUNTER by the Report Name. The PAGE-COUNTER may be referred to by Data Division clauses and Procedure Division statements.
5. The counter is automatically set to 1 initially by the Report Writer (INITI-

ATE statement); if a starting value for the PAGE-COUNTER is to be other than 1, the programmer may change the contents of the counter by a Procedure Division statement after an INITIATE statement has been executed.

6. The counter is automatically incremented by 1 each time a page break is recognized by the Report Writer, after the production of any PAGE FOOTING report group but before the production of any PAGE HEADING report group.

LINE-COUNTER

A numeric counter used by the Report Writer to determine when a PAGE HEADING and/or a PAGE FOOTING report group is to be printed. If a PAGE LIMIT clause is written in the report description entry, a LINE-COUNTER is supplied for that report. The maximum value of the counter is based on the number of lines per page as specified in the PAGE LIMIT clause.

Rules Governing the Use of the Line-Counter Clause

1. One line counter is supplied for each report with a PAGE LIMIT clause written in the report description entry.
2. If more than one report description entry exists in the Report Section, the user must qualify LINE-COUNTER by the report name. LINE-COUNTER may be referred to in Data Division clauses or by Procedure Division statements.
3. Changing the LINE-COUNTER by Procedure Division statements may cause the page format control to become unpredictable in the Report Writer.
4. The counter is automatically tested and incremented by the Report Writer based on control specification in the PAGE LIMIT clause and values specified in the LINE NUMBER and NEXT GROUP clauses.
5. The counter is automatically set to zero initially by the Report Writer (INITIATE statement); likewise, the counter is automatically set to zero when PAGE LIMIT Integer-1 LINES entry is exceeded during the execution.
6. If a relative LINE NUMBER indication or relative NEXT GROUP indication exceeds the LAST DETAIL PAGE LIMIT specification during object time, that is, a page break, the counter is set to zero. No additional setting based on the relative LINE NUMBER indication or NEXT GROUP indication that forced the page break takes place.
7. If an absolute LINE NUMBER indication or an absolute NEXT GROUP indication is equal to, or less than, the contents of the counter during object time, the LINE-COUNTER is set to the absolute LINE NUMBER indication following the implicit generation of any specified report groups.
8. The value of the counter during any Procedure Division test statement represents the number of the last line used by the previous report group or represents the number of the last line skipped to by the previous NEXT GROUP specification.

9. The Report Writer LINE-COUNTER control prohibits the printing of successive report lines or report groups on the same line of the same page.

The format of the PAGE LIMIT clause, shown in figure 10.12, is explained as follows:

LIMIT(S). LIMIT IS and LIMITS ARE are optional words and need not be included in the clause.

Integer-1. The Integer-1 LINE(S) clause is required to specify the depth of the report page; the depth of the report page may or may not be equal to the physical perforated continuous form often associated in a report with the page length. The size of the fixed data-name LINE-COUNTER is the maximum numeric size based on Integer-1 LINE(S) required for counter to prevent overflow.

HEADING Integer-2. The first line number of the first heading print group. No print group will start preceding Integer-2. Integer-2 is the first line upon which anything is printed.

FIRST DETAIL Integer-3. The first line number of the first normal print group, that is body; no DETAIL print group will start before Integer-3.

LAST DETAIL Integer-4. The last line number of the last normal print group, that is body; no DETAIL print group will extend beyond Integer-4.

FOOTING Integer-5. The last line number of the last CONTROL FOOTING print group is specified by Integer-5. No CONTROL FOOTING print group will extend beyond Integer-5. PAGE FOOTING print groups will follow Integer-5. (See figure 10.13.)

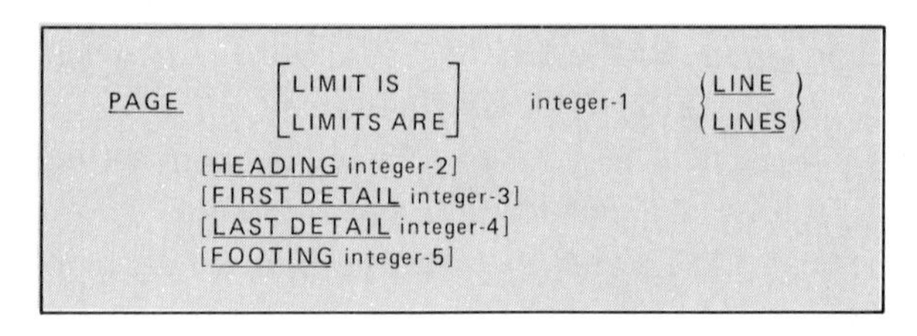

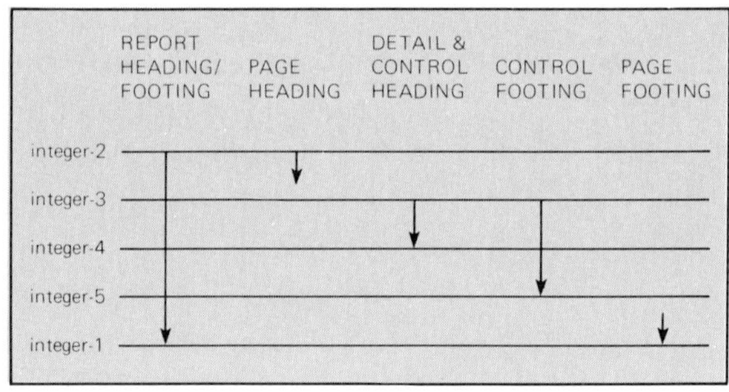

Figure 10.12 Format—Page Limit Clause.

Figure 10.13 Page Format When the Page Limit Clause is Specified.

Note: The following implicit control is assumed for omitted specifications:

1. If HEADING Integer-2 is omitted, Integer-2 is considered to be equivalent to the value 1, that is, LINE NUMBER one.
2. If FIRST DETAIL Integer-3 is omitted, Integer-3 is considered to be equivalent to the value of Integer-2.
3. If LAST DETAIL Integer-4 is omitted, Integer-4 is considered to be equivalent to the value of Integer-5.
4. If FOOTING Integer-5 is omitted, Integer-5 is considered to be equivalent to the value of Integer-4. If both LAST DETAIL Integer-4 and FOOTING

Integer-5 are omitted, Integer-4 and Integer-5 are both considered to be equivalent to the value of Integer-1.

Report Group Description Entry

A report group may be a set of data made up of several print lines with many data items, or it may consist of one print line with one data item (figs. 10.14, 10.15). Report groups may exist within report groups—all or each

```
01   [data-name-1]
     [LINE Clause]
     [NEXT GROUP Clause]
     TYPE Clause
     [USAGE Clause].
```

Figure 10.14
Format—Report Group Description Entry—Format 1.

```
level-number [data-name-1]
     [BLANK WHEN ZERO Clause]
     [COLUMN Clause]
     [GROUP Clause]
     [JUSTIFIED Clause]
     [LINE Clause]
     [NEXT GROUP Clause]
     PICTURE Clause
     [RESET Clause]
     {SOURCE}
     {SUM   }   Clause
     {VALUE }
     TYPE Clause
     [USAGE Clause].
```

Figure 10.15
Format—Report Group Description Entry—Format 2.

capable of a reference by a GENERATE or a USE statement in the Procedure Division. A description of a set of data becomes a report group by the presence of a level number and a TYPE description. The level number gives the depth of the group and the TYPE clause describes the purpose of the report group presentation. If report groups exist within report groups, all must have the same TYPE descriptions. Including a data-name with the entry permits the group to be referred to by a GENERATE or a USE statement in the Procedure Division. At object program time, report groups are created as a result of the GENERATE statement.

This entry defines the characteristics for a report group, whether a series of lines, one line, or an elementary item. The placement of an item in relation to the entire report group, the hierarchy of a particular group within a report group, the format descriptions of all items, and any control factors associated with the group—all are defined in the entry. The system of level numbers is employed here to indicate elementary items and group items within the range of 01–49.

Pictorially to the programmer, a report group is a line, or a series of lines, initially consisting of all SPACES; its length is determined by the compiler based on the environmental specifications. Within the framework of a report, the order of report groups specified is not significant. Within the framework of a report group, the programmer describes the presented elements consecutively from left to right, and then from top to bottom. The description of a report group is analogous to the data record, consists of a set of entries defining the characteristics of the included elements. However, in the report group, SPACES are associated except where a specified entry is indicated for presentation, whereas in the data record, each character position must be defined.

The Report Group Description entry specifies the characteristics of a particular report group and of the individual data-names within a report group. A report group may be comprised of one or more report groups. Each report group is described by a hierarchy of entries similar to the description of the data record. There are three types of report groups: heading, detail, and footing.

A report group is considered to be one unit of the report consisting of a line or a series of lines that are printed (or not printed) under certain conditions.

Rules Governing the Use of the Report Group Description Entry

1. Except for the data-name clause (which, when present, must immediately follow the level number) the clauses may be written in any order.
2. In order for a report group to be referred to by a Procedure Division statement, it must have a data-name.
3. If a COLUMN clause is present in the data description of an item, the data description must also contain a PICTURE clause in addition to one of the clauses, SOURCE, SUM, or VALUE.
4. In Format 2, the level numbers may be any number from 1 through 49.
5. If Format 1 is used to indicate a report group, a report group must contain a report group entry (level 01), and it must be the first entry. A report group extends from this entry either to the next group level 01 or to the end of the next group description.
6. Format 2 is used to indicate an elementary item or a group item within a report group. If a report group is an elementary item, Format 2 may include the TYPE or NEXT GROUP clause to specify the report group and elementary item in the same entry.
7. When the LINE clause is specified in Format 1, the entries for the first report line within the report group are presented on the specified line. When the LINE clause is specified in Format 2, sequential entries with the same level number in the report group are implicitly presented on the same line. A LINE NUMBER at a subordinate level must not contradict a LINE NUMBER at a group level.
8. The NEXT GROUP clause, when specified, refers to the spacing at object time between the last line of this report group and the first line of the next report group.

Data-Name-1

Report group names are required in the following instances:

1. When the data-name represents a DETAIL report group referred to by a GENERATE statement in the Procedure Division.
2. When the data-name represents a HEADING or FOOTING report group referred to by a USE statement in the Procedure Division.
3. When a reference is made in the Report Section to a DETAIL report group by a SUM UPON clause.

Line Clause

The LINE clause indicates the absolute or relative line number of this entry in reference to the page or previous entry (fig.10.16).

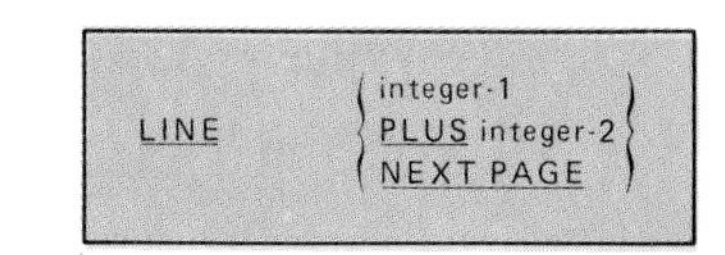

Figure 10.16 Format—Line Clause.

Rules Governing the Use of the Line Clause

1. Integer-1 and Integer-2 must be positive integers. Integer-1 must be within the range specified in the PAGE LIMIT(S) clause in the report description entry.
2. The LINE clause must be given for each report line of a report group. For the first line of a report group, the LINE clause must be given either at the report group level or prior to or for the first elementary item in the line. For report lines other than the first in a report group, it must be given prior to the first elementary item in the line.
3. When a LINE clause is encountered, subsequent entries following the entry with the LINE clause are implicitly presented on the same line until either another LINE clause or the end of the report group is encountered.
4. Absolute LINE NUMBER entries must be indicated in ascending order, and an absolute LINE NUMBER cannot be preceded by a relative LINE NUMBER.

Integer-1 indicates an absolute line number which sets the LINE-COUNTER to this value for printing the item in this and following entries within the report group until a different value for the LINE-COUNTER is specified. It indicates the fixed line of the page in which the line is to be printed.

Integer-2 indicates a relative line number that specifies the number of lines to be skipped before printing the line. The line is relative to the previous line printed or skipped. The LINE-COUNTER is incremented by the value of Integer-2 and is used for printing the items in this and following entries until a different value for LINE-COUNTER is specified.

NEXT PAGE phrase may be used to indicate an automatic skip to the next page before presenting the first line of the next report group. Appropriate TYPE PAGE FOOTINGS, TYPE PAGE HEADINGS will be produced as specified. This LINE clause may appear only in a report group entry or may be the LINE clause of the first line of a report group.

Next Group Clause

The NEXT GROUP clause is used to define the line control following the printing of the current report group being processed (fig. 10.17).

Figure 10.17 Format—Next Group Clause.

Rules Governing the Use of the Next Group Clause

1. The same rules apply for Integer-1, Integer-2, and NEXT PAGE options as enumerated previously for the LINE clause.

2. The NEXT GROUP clause must appear only at the 01 level which defines the report group. When specified for a CONTROL FOOTING/HEADING report group, the NEXT GROUP clause results in automatic line spacing only when a control break occurs on the level for which that control is specified.

Integer-1 indicates an absolute line number which sets the LINE-COUNTER to the value after printing the last line of the report group.

Integer-2 indicates a relative number which increments the LINE-COUNTER by the Integer-2 value after printing the last line of the report group.

NEXT PAGE phrase may be used to indicate an automatic skip to the next page before printing the first line of the next group. (See figure 10.18.)

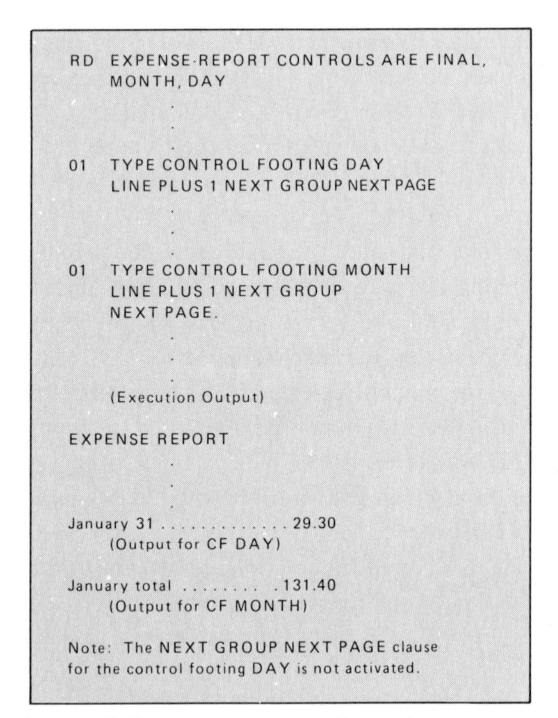

```
RD  EXPENSE-REPORT CONTROLS ARE FINAL,
    MONTH, DAY
        .
        .
01  TYPE CONTROL FOOTING DAY
    LINE PLUS 1 NEXT GROUP NEXT PAGE
        .
        .
        .
01  TYPE CONTROL FOOTING MONTH
    LINE PLUS 1 NEXT GROUP
    NEXT PAGE.
        .
        .
        .
    (Execution Output)

EXPENSE REPORT
        .
        .
        .
January 31 . . . . . . . . . . . . . 29.30
    (Output for CF DAY)

January total . . . . . . . . . 131.40
    (Output for CF MONTH)

Note: The NEXT GROUP NEXT PAGE clause
for the control footing DAY is not activated.
```

Figure 10.18 Example—Next Group Clause.

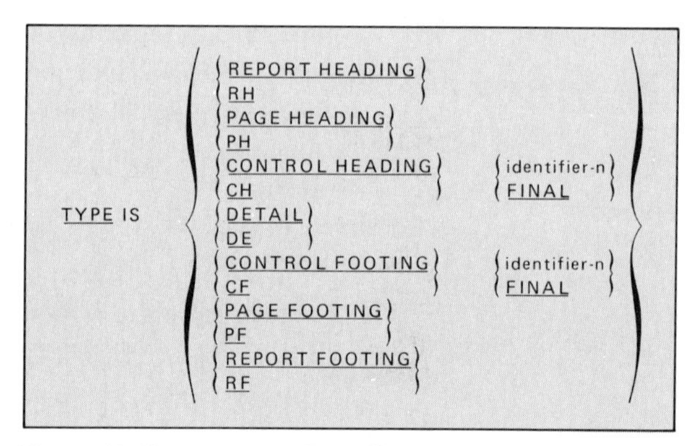

Figure 10.19 Format—Type Clause.

Type Clause

The TYPE clause specifies the particular type of report group that is described in this entry and indicates the time at which the report group is to be generated (fig. 10.19). Abbreviations may be used in the TYPE clause.

Rules Governing the Use of the Type Clause

1. The level number 01 identifies a particular report group to be generated as output and the TYPE clause indicates the time for the generation of this report group.
2. If the report group is described as other than TYPE DETAIL, its generation is an automatic Report Writer function.

3. If the report group is described with the TYPE DETAIL clause, the Procedure Division statement—GENERATE data-name—directs the Report Writer to produce the named report group.
4. Nothing precedes a REPORT HEADING entry, and nothing follows a REPORT FOOTING entry within a report.
5. A FINAL type control break may be designated only once for CONTROL HEADING or CONTROL FOOTING entries within a particular report group.
6. CONTROL HEADING report groups appear with the current values of any indicated SOURCE data items before the DETAIL report groups of the CONTROL group are produced.
7. CONTROL FOOTING report groups appear with the previous value of any indicated CONTROL SOURCE data items just after the DETAIL report groups of the CONTROL groups have been produced.
8. The USE procedure specified for a CONTROL FOOTING report group refers to:
   a. Source data items specified in the CONTROL clause affect the previous value of the data item.
   b. Source data items not specified in the CONTROL clause affect the current value of the item.

   These report groups appear whenever a control break is noted.
9. LINE NUMBER determines the absolute or relative position of the CONTROL report group exclusive of the other HEADING and FOOTING report group. (See figure 10.20.)

```
REPORT HEADING      (one occurrence only)
PAGE HEADING
      .
      .
      .
CONTROL HEADING
DETAIL
CONTROL FOOTING
      .
      .
      .
PAGE FOOTING
REPORT FOOTING      (one occurrence only)
```

Figure 10.20
Heading or Footing Report Groups Sequence.

Report Heading (RH). The REPORT HEADING entry indicates a report group that is produced only once, at the beginning of the report, during the execution of the first GENERATE statement. Only one REPORT HEADING is permitted in a report. SOURCE clauses used in REPORT HEADING report group items refer to the value of data items at the time the first GENERATE statement is executed.

Page Heading (PH). The PAGE HEADING entry indicates a report group that is automatically produced at the beginning of each page of the report. Only one page heading is permitted for each report. The page heading is printed on the first page after the REPORT HEADING is specified.

Control Heading (CH). The CONTROL HEADING entry indicates a report group that is printed at the beginning of a control group for a designated identifier, or, in the case of FINAL, is produced once before the first control at the initiation of a report during the execution of the first GENERATE statement. There can be only one report group of this type for each identifier and for the FINAL specified in a report. In order to produce any CONTROL HEADING report groups, a control break must occur. SOURCE clauses used in TYPE CONTROL HEADING FINAL report groups refer to the value of the items at the time the first GENERATE statement is executed. (See figure 10.21.)

```
Final Control Heading    (one occurrence only)
Major Control Heading
        .
        .
        .
Minor Control Heading
```

Figure 10.21 Control Heading Report Groups Sequence.

Identifier-n as well as FINAL, must be one of the identifiers described in the CONTROL clause in the report description entry.

Detail (DE). The DETAIL entry indicates a report group that is produced for each GENERATE statement in the Procedure Division. (The data-name specified in the 01 level is referred to by the GENERATE statement. This name must be unique.) There is no limit to the number of DETAIL report groups that may be included in a report.

Control Footing (CF). The CONTROL FOOTING entry indicates a report group that is produced at the end of a control group for a designated identifier or one that is produced once at the termination of a report ending in a FINAL control group. There can be only one report group for each identifier and for the FINAL entry specified in a report. In order to produce any CONTROL FOOTING report group, a control break must occur. SOURCE clauses used in TYPE CONTROL FOOTING FINAL report groups refer to the values of the items at the time the TERMINATE statement is executed.

Page Footing (PF). The PAGE FOOTING entry indicates a report group that is automatically produced at the bottom of each page of the report. There can be only one report group of this type in a report.

Report Footing (RF). The REPORT FOOTING entry indicates a report group that is produced only once at the termination of a report. There can be only one report group of this type in a report. SOURCE clauses used in TYPE REPORT FOOTING report groups refer to the value of the items at the time the TERMINATE statement is executed. (See figure 10.22.)

```
Minor Control Footing
        .
        .
        .
Major Control Footing
Final Control Footing    (one occurrence only)
```

Figure 10.22 Control Footing Report Groups Sequence.

Usage Clause

DISPLAY is the only option that may be specified for elementary or group items in a report group description entry. (See "USAGE" clause in Data Division, chap. 6.)

Column Clause

The COLUMN clause indicates the absolute column number in the printed page of the high-order (leftmost) character of an elementary item (fig. 10.23). Integer-1 must be a positive integer. The clause can only be given at the elementary level within a report group.

COLUMN NUMBER IS integer-1

Figure 10.23 Format—Column Clause.

The COLUMN clause indicates that the *leftmost* character of the elementary item is placed in the position specified by the integer. If the column number is not indicated, the elementary item, though included in the description of the report group, is suppressed when the report group is produced at object time.

Group Indicate Clause

The GROUP INDICATE clause specifies that the elementary item is to be produced only on the first occurrence of the item after any CONTROL or PAGE break. The clause must be specified only at the elementary level within a DETAIL report group. (See figure 10.24.)

GROUP INDICATE

Figure 10.24 Format—Group Indicate Clause.

The elementary item is not only group-indicated on the first DETAIL report group containing the item after a control break, but is also indicated on the first DETAIL report group containing the item on a new page, even though a control break did not occur. (See figure 10.25.)

```
REPORT SECTION.
    .
    .
    .
01  DETAIL-LINE TYPE IS DETAIL LINE
    NUMBER IS PLUS 1.
    05  COLUMN IS 2 GROUP INDICATE
        PICTURE IS A(9) SOURCE IS
        MONTHNAME OF RECORD-AREA (MONTH)
    .
    .
    .
              (Execution Output)
    .
    .
    .
JANUARY     15      A00 . . .
                    A02 . . .

PURCHASES AND COST . . .

JANUARY     21      A03 . . .
                    A03 . . .
```

GROUP INDICATE items are printed after page and control breaks.

Figure 10.25 Sample Showing Group Indicate Clause and Resultant Execution Output.

Justified Clause

The same rules are applicable to the use of the JUSTIFIED clause in a report group description as discussed in the Data Division. (See "Justified" clause in Data Division, chap. 6.)

Picture Clause

The same rules are applicable to the use of the PICTURE clause in a report group description as discussed in the Data Division. (See "Picture" clause in Data Division, chap. 6.)

Reset Clause

The RESET clause indicates the CONTROL identifier that causes the SUM counter in the elementary item entry to be reset to zero on a control break (fig. 10.26).

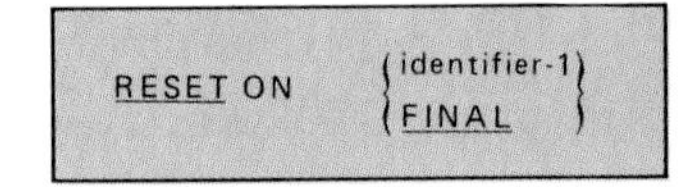

Figure 10.26 Format—Reset Clause.

After presentation of the TYPE CONTROL FOOTING report group, the counters associated with the report are reset automatically to zero unless an explicit RESET clause is given specifying resetting based on a higher-level control than the associated control for the report group.

The RESET clause may be used for programming totaling of identifiers where subtotals of identifiers may be desired without automatic resetting upon printing the report group.

The RESET clause may only be used in conjunction with a SUM clause at the elementary level.

Identifier-1 must be one of the identifiers described in the CONTROL clause in the report description entry. Identifier-1 must be a higher-level CONTROL identifier than the CONTROL identifier associated with the CONTROL FOOTING report group in which the SUM and RESET clause appear.

Blank When Zero Clause

The same rules are applicable to the use of the BLANK WHEN ZERO clause in a report group description as described in the Data Division. (See "blank when zero" clause in Data Division, chap. 6.)

Source Clause

The SOURCE clause indicates a data item to be used as the source for this report item (fig. 10.27). The item is presented according to the PICTURE and COLUMN clauses in this elementary item entry.

SOURCE IS identifier-1

Figure 10.27 Format—Source Clause.

The SOURCE clause has two functions:

1. To specify a data item that is to be printed.
2. To specify a data item that is to be summed in a CONTROL FOOTING report group.

Sum Clause

The SUM clause is used to cause automatic summation of data and may appear only in an elementary item entry of a CONTROL FOOTING report group (fig. 10.28).

Figure 10.28
Format—Sum Clause.

```
SUM   identifier-2 identifier-3 . . . [UPON data-name]
```

Rules Governing the Use of the Sum Clause

1. A SUM clause may appear only in a TYPE CONTROL FOOTING report group.
2. If a SUM counter is referred to by a Procedure Division statement or Report Section entry, a data-name must be specified with the SUM clause entry. The data-name then represents the summation counter automatically generated by the Report Writer to total the operands immediately following the SUM. If a summation counter is never referred to, the counter need not be named explicitly by a data-name entry. A SUM counter is only algebraically incremented just before the presentation of the TYPE DETAIL report group in which the item being summed appears as a source item.
3. Whether the SUM clause names the summation counter or not, the PICTURE clause must be specified for each SUM counter. Editing characters or editing clauses may be included in the description of a SUM counter. Editing of a SUM counter only occurs upon presentation of that SUM counter. At all other times, the SUM counter is treated as a numeric data item. The SUM counter must be large enough to accommodate the summed quantity without truncating of integral digits.
4. An operand of a SUM clause must be an elementary numeric data item that appears in the File, Working-Storage Sections, or data-name that is the name of a SUM counter.
5. Each item being summed—that is, Identifier-2, Identifier-3, etc., must appear as a SOURCE item in a TYPE DETAIL report group, or be the name of a SUM counter in a TYPE CONTROL FOOTING report group at an equal or lower position in the control hierarchy. Although the items must be explicitly written in a TYPE DETAIL report group, they may actually be suppressed at presentation time. In this manner, direct association without ambiguity can be made from the current data available by a GENERATE statement to the data items to be presented within the Report Section.
6. If higher-level report groups are indicated in the control hierarchy, counter updating (commonly called "rolling counter" forward) procedures take place prior to the reset operation.
7. The summation of data items defined as SUM counter in the TYPE CONTROL FOOTING report groups is accomplished explicitly or implicitly with the Report Writer automatically handling the updating function. If a sum control of a data item is not desired for presentation at a higher level, the lower-level SUM specification may be omitted. In this case the same results are obtained as if the lower-level SUM were specified.
8. The *UPON data-name* is required to obtain selective summation for a particular data item which is named as a SOURCE item in two or more TYPE DETAIL report groups. Identifier-2 and Identifier-3 must be source data

items in data-names. Data-name must be the name of a TYPE DETAIL report group. If the UPON data-name option is not used, Identifier-2 and Identifier-3, etc. respectively, are added to the SUM counter at each execution of a GENERATE statement. This statement generates a TYPE DETAIL report group that contains the SUM operands at the elementary level. (See figure 10.29.)

RD . . . CONTROLS ARE YEAR MONTH WEEK DAY

Method 1:

```
01  TYPE CONTROL FOOTING YEAR.
    05 SUM COST.
01  TYPE CONTROL FOOTING MONTH.
    05 SUM COST.
01  TYPE CONTROL FOOTING WEEK.
    05 SUM COST.
01  TYPE CONTROL FOOTING DAY.
    05 SUM COST.
```

Method 2:

```
01  TYPE CONTROL FOOTING YEAR.
    05 SUM A.
01  TYPE CONTROL FOOTING MONTH.
    05 A SUM B.
01  TYPE CONTROL FOOTING WEEK.
    05 B SUM C.
01  TYPE CONTROL FOOTING DAY.
    05 C SUM COST.
```

Method 2 will execute faster. One addition will be performed for each day, one more for each week, and one for each month. In Method 1, four additions will be performed for each day.

Figure 10.29 Example—Sum Clause.

Value Clause

The VALUE clause causes the report data item to assign the specified value each time its report group is presented only if the elementary item entry does not contain a GROUP INDICATE clause (fig. 10.30). If the GROUP INDICATE clause is present, and a given object time condition exists, the item will assume the specified value. (See "Group Indicate" rules.)

```
VALUE IS literal-1
```

Figure 10.30 Format—Value Clause.

PROCEDURE DIVISION

In a program that utilizes the Report Writer feature, records are read and data is manipulated by programmer instructions in the Procedure Division prior to entering the report phase (fig. 10.31). A report is produced by the execution

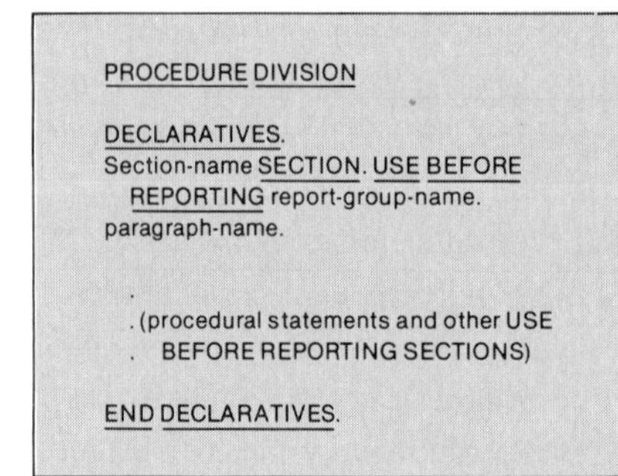

```
PROCEDURE DIVISION

DECLARATIVES.
Section-name SECTION. USE BEFORE
   REPORTING report-group-name.
paragraph-name.

   .
   . (procedural statements and other USE
   .    BEFORE REPORTING SECTIONS)

END DECLARATIVES.
```

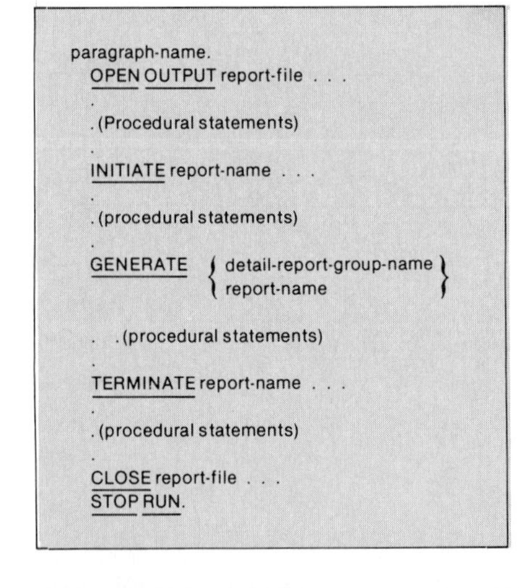

```
paragraph-name.
   OPEN OUTPUT report-file . . .
   .
   . (Procedural statements)
   .
   INITIATE report-name . . .
   .
   . (procedural statements)
   .
   GENERATE  { detail-report-group-name }
             { report-name              }
   .
   . (procedural statements)
   .
   TERMINATE report-name . . .
   .
   . (procedural statements)
   .
   CLOSE report-file . . .
   STOP RUN.
```

Figure 10.31
Format—Procedure Division.

Figure 10.31 Continued.

The INITIATE, GENERATE, and TERMINATE statements are required entries in the Procedure Division for the production of a report (or reports). The Declarative Section statement USE BEFORE REPORTING allows the programmer to gain control during the generation of any report group.

1. Procedural statements in USE BEFORE REPORTING must conform to the restrictions stipulated under DECLARATIVES and USE Declarative described in the Procedure Division section of this text. USE BEFORE REPORTING procedures permit manipulation or alteration of report-group-elements immediately prior to printing. Refer to Guides to Use of the Report Writer described later in this section.
2. All other features of ANS COBOL as contained in this text may be used in a program containing the Report Writer feature. Complete freedom in processing data, etc., is allowed before, during, and after producing a report.

of the INITIATE, GENERATE, and TERMINATE statements in the Procedure Division. The INITIATE statement initializes all counters associated with the Report Writer, the GENERATE statement is used each time a detailed portion of the report is to be produced, and the TERMINATE statement is used to end the report. The Report Writer feature allows additional manipulation of data by means of a USE BEFORE REPORTING declarative in the Declaratives Section of the Procedure Division.

Initiate Statement

The INITIATE statement begins the processing of a report (fig. 10.32).

Figure 10.32
Format—Initiate Statement.

```
INITIATE report-name-1 [report-name-2] . . .
```

Rules Governing the Use of Initiate Statement

1. Each report name must be defined by a report description entry in the Report Section of the Data Division.
2. The INITIATE statement resets all data-name entries that contain SUM clauses associated with the report. The Report Writer controls for all the TYPE report groups that are associated with the report are set up in their respective order.
3. The PAGE-COUNTER, if specified, is set to 1 prior to or during the execution of the INITIATE statement. If a different starting value for the PAGE-COUNTER other than 1 is desired, the programmer may reset this counter with a statement in the Procedure Division following the INITIATE statement.
4. The LINE-COUNTER, if specified, is set to zero prior to or during the execution of the INITIATE statement.
5. The INITIATE statement does not open the file with which the report is associated. An OPEN statement for the file must be specified. The INITIATE statement performs Report Writer functions for individually described programs analogous to the input-output functions that the OPEN statement performs for individually described files.
6. A second INITIATE statement for a particular report-name may not be executed unless a TERMINATE statement has been executed for that report-name subsequent to the first INITIATE statement.

Generate Statement

The GENERATE statement is used to produce a report (fig. 10.33). At process time it links the Procedure Division to the Report Writer described in the Report Section of the Data Division.

GENERATE identifier

Figure 10.33 Format—Generate Statement.

Rules Governing the Use of the Generate Statement

1. Identifier represents a TYPE DETAIL report group or an RD (report description) entry.
2. If the identifier is the name of a TYPE DETAIL report group, the GENERATE statement performs all the automatic operations of the Report Writer and produces an actual output detail report group in the output. This is called *detail reporting*.
3. If the identifier is the name of a RD entry, the GENERATE statement does all the automatic operations of the Report Writer and updates the footing report group(s) within a particular report group without producing an actual detail report group associated with the report. In this case, all SUM counters associated with the report descriptions are algebraically incremented each time a GENERATE statement is executed. This is called *summary reporting*. If more than one TYPE DETAIL group is specified, all SUM counters are algebraically incremented each time a GENERATE statement is executed.
4. The GENERATE statement, implicit in both detail and summary reporting, produces the following automatic operations (if defined):
   a. Steps and tests the LINE-COUNTER and/or PAGE-COUNTER to produce appropriate PAGE-FOOTING and/or PAGE HEADING report groups.
   b. Recognizes any specified control breaks to produce appropriate CONTROL FOOTING and/or CONTROL HEADING report groups.
   c. Accumulates into the SUM counters all specified IDENTIFIERS. Resets the SUM counters on an associated control break. Performs an updating procedure between control break levels for each set of SUM counters.
   d. Executes any specified routines defined by a USE statement before generation of the associated report groups.
5. During the execution of the first GENERATE statement, the following report groups associated with the report, if specified, are produced in the following order:
   a. REPORT HEADING report group.
   b. PAGE HEADING report group.
   c. All CONTROL HEADING report groups in this order: FINAL, major to minor.
   d. The DETAIL report group, if specified, in the GENERATE statement.
6. If a control break is recognized at the time of the execution of a GENER-

ATE statement (other than the first that is executed for the report), all CONTROL FOOTING report groups specified for the report are produced from the minor group up to and including the report group specified for the identifier which cause the control break. Then the CONTROL HEADING report group(s) specified for the report, from the report group specified for the identifier that causes the control break, down to the minor report group, are produced in that order. The DETAIL report group specified in the GENERATE statement is then produced.

7. Data is moved to the data item in the report group description entry of the Report Section and is edited under control of the Report Writer according to the same rules for movement and editing as described for the MOVE statement. (See "move" statement in the Procedure Division.)

Terminate Statement

The TERMINATE statement is used to terminate the processing of a report (fig. 10.34).

Figure 10.34
Format—Terminate Statement.

```
TERMINATE report-name-1 [report-name-2] . . .
```

Rules Governing the Use of the Terminate Statement

1. Each report-name given in a TERMINATE statement must be defined by an RD entry in the Data Division.
2. The TERMINATE statement produces all the control footings associated with this report, as if a control break had just occurred at the highest level, and completes the Report Writer functions for the named report. The TERMINATE statement also produces the last page footings and report footing report groups associated with this report.
3. Appropriate PAGE HEADING or FOOTING report groups are prepared in their respective order for the report description.
4. A second TERMINATE statement for a particular file may not be executed unless a second INITIATE statement has been executed for the report-name. If a TERMINATE statement has been executed for a report, a GENERATE statement for that report must not be executed unless an intervening INITIATE statement for that report is executed.
5. The TERMINATE statement does not close the file with which the report is associated. A CLOSE statement for the file must be given by the user. The TERMINATE statement performs Report Writer functions for individually described report programs analogous to the input-output functions that the CLOSE statement performs for individual described files.
6. SOURCE clauses used in TYPE CONTROL FOOTING FINAL or TYPE REPORT FOOTING report groups refer to the values of the items during the execution of the TERMINATE statement.

Use Statement

A USE BEFORE REPORTING declarative statement may be written in the Declaratives Section of the Procedure Division if the programmer wishes to alter or manipulate the data before it is presented in the report (fig. 10.35).

Figure 10.35 Format—Use Before Reporting Declarative.

```
USE BEFORE REPORTING data-name.
```

The USE statement specifies Procedure Division statements that are executed just before a report group named in the Report Section of the Data Division is produced.

Rules Governing the Use of a Use Statement

1. A USE statement, when present, must immediately follow a section header in the Declaratives portion of the Procedure Division, and must be followed by a period and a space. The remainder of the section must consist of one or more procedural paragraphs that define the procedures to be processed.
2. Data-name represents a report group named in the Report Section of the Data Division. The data-name must not be used in more than one USE statement. The data-name must be qualified by the report-name if not unique.
3. No Report Writer statement (INITIATE, GENERATE or TERMINATE) may be written in a procedural paragraph or a paragraph following the USE sentence in the Declaratives Section.
4. The USE statement itself is never executed; rather, it defines the conditions calling for the execution of the USE procedures.
5. The designated procedures are executed by the Report Writer just before the named report is produced, regardless of page or control breaks associated with report groups. The report group may be any type except DETAIL.
6. Within a USE procedure, there must not be any reference to any nondeclarative procedure. Conversely, in the nondeclarative portion, there must be no reference to procedure-names that appear in the Declaratives portion, except that PERFORM statements may refer to USE declaratives or to procedures associated with USE declaratives.

(See figure 10.36.)

(Note: When the programmer wishes to suppress the printing of the specified report group, the statement:

```
MOVE 1 TO PRINT-SWITCH
```

is used in the USE BEFORE REPORTING declarative section. When this statement is encountered, only the specified report group is not printed; the statement must be written for each report group whose printing is to be suppressed.)

The USE BEFORE REPORTING procedures must not change the contents of control data items or alter subscripts or indexes used in referencing the controls.

GUIDES TO THE USE OF THE REPORT WRITER

The Report Writer is a valuable feature of the COBOL language, making possible the automatic generation of a report, or reports, in either detail or summary form. The data processing potential of the Report Writer is greatly enhanced in that it may be combined with any other COBOL language element,

The following coding example illustrates the use and operation of the use before reporting declarative section. Part 1 of the example shows the definition of two CONTROL FOOTING report groups in the Report Section of the Data Division. Part 2 of the example shows a method of using the USE BEFORE REPORTING declarative. In this example, the USE BEFORE REPORTING declarative section in Part 2 will be executed before MINOR control footing (describing in Part 1) is produced.

Part 1

```
01  MINOR TYPE CF C-1 LINE PLUS 2.
  02  A SUM P PICTURE $$$.99
    BLANK WHEN ZERO COLUMN 3.
  02  B SUM Q PICTURE $$$99.99CR COLUMN 10.
  02  SOURCE E PICTURE **,***.** COLUMN 25.
01  TYPE CF C-2 LINE PLUS 1.
  02  SUM A PICTURE $$$$$.99
    BLANK WHEN ZERO COLUMN 5.
```

P, Q, and E can be defined in any of the File, Working- Storage, or Linkage Sections. C-1 and C-2 are control data items in the CONTROLS clause.

Part 2

```
DECLARATIVES.
RW SECTION. USE BEFORE REPORTING MINOR.
  MINOR-PARA.
  IF A = 0 AND B = 0 MOVE 1 TO PRINT-SWITCH.
END DECLARATIVES.
```

When a control break occurs for C-1, the following operations are performed:

1. The RW SECTION is executed.
2. The PRINT-SWITCH is tested. If it is 1, the PRINT-SWITCH is reset to zero, and steps 3 and 4 are bypassed.
3. The print line is constructed under control of the Report Writer as follows:
   a. moving the correct character to the first part of the record (to cause a double-space);
   b. moving the sum counters to the print line (edited); and
   c. moving E to the print line (edited).
4. The line is printed and 2 is added to LINE-COUNTER.
5. The sum counter A is added to the sum counter defined in the second report group.
6. The sum counters in the first report group are set to zero.

Figure 10.36 Example—Use Before Reporting Declarative.

e.g., SORT-verb, ENTER-verb, INPUT-OUTPUT-verbs, computation, data-manipulation and other procedural statements. However, the actual processing of the report is automatic. The construction of print-lines, "crossfooting" and "rolling" of sum counters, control break-testing and printing, heading and footing generation, etc., require no programming effort once the report has been defined in the Report Section.

A report definition specifies the page and line format, and describes the source data to be printed and the numerical data to be summed. The page and line formats are defined in the Report Section with the clauses CONTROL, PAGE, TYPE, LINE, NEXT GROUP, COLUMN, and GROUP INDICATE. How these clauses should be defined for a particular report is best determined by reference to a print spacing chart. The source data which is to appear on each print-line is referred to by a SOURCE clause that causes it to be moved to the print buffer as the report-group is generated. The numerical data to be summed is referred to by a SUM clause that causes it to be added to a counter and printed, with or without editing, as defined in the associated PICTURE clause. (Figure 10.37-10.38)

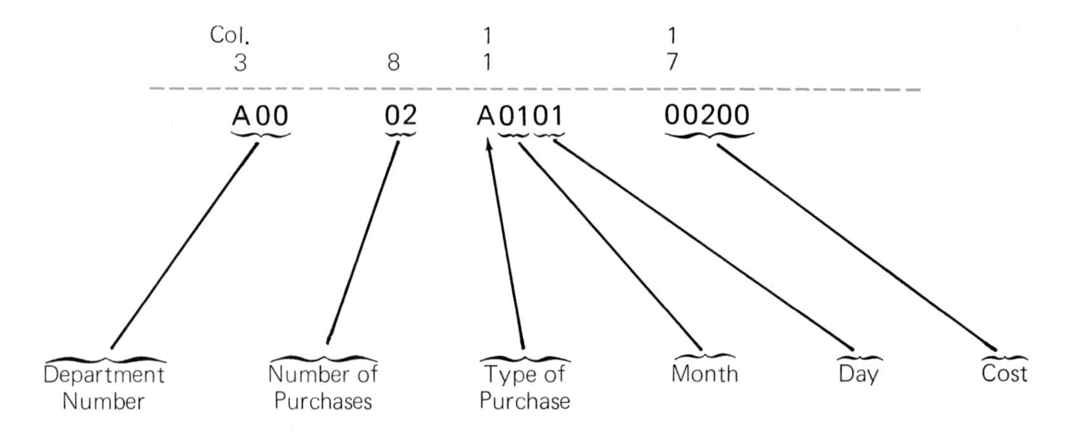


```
000000 IDENTIFICATION DIVISION.
000010 PROGRAM-ID. ACME.
000020 INSTALLATION. ACME ACCOUNTING DEPARTMENT.
000030 REMARKS. THE REPORT WAS PRODUCED BY THE REPORT WRITER FEATURE.
000040 ENVIRONMENT DIVISION.
000050 CONFIGURATION SECTION.
000060 SOURCE-COMPUTER. IBM-370.
000070 OBJECT-COMPUTER. IBM-370.
000080 INPUT-OUTPUT SECTION.
000090 FILE-CONTROL.
000100     SELECT INFILE ASSIGN TO SYS015-UT-2400-S.
000110     SELECT REPORT-FILE ASSIGN TO SYS001-UT-2400-S.
000120 DATA DIVISION.
000130 FILE SECTION.
000140 FD  INFILE LABEL RECORDS ARE OMITTED DATA RECORD IS INPUT-RECORD
000150     RECORDING MODE F.
000180 01  INPUT-RECORD.
000190     02  FILLER          PICTURE AA.
000200     02  DEPT            PICTURE XXX.
000210     02  FILLER          PICTURE AA.
000220     02  NO-PURCHASES    PICTURE 99.
000230     02  FILLER          PICTURE A.
000240     02  TYPE-PURCHASE   PICTURE A.
000250     02  MONTH           PICTURE 99.
000260     02  DAY             PICTURE 99.
000270     02  FILLER          PICTURE A.
000280     02  COST            PICTURE 999V99.
000281     02  FILLER PICTURE X(59).
000290 FD  REPORT-FILE, REPORT IS EXPENSE-REPORT
000300     LABEL RECORDS ARE STANDARD.
000310 WORKING-STORAGE SECTION.
000320 77  SAVED-MONTH PICTURE 99 VALUE 1.
000330 77  SAVED-DAY   PICTURE 99 VALUE 0.
000340 77  CONTINUED   PICTURE X(11) VALUE SPACE.
000380 01  FILLER.
000390     02  RECORD-MONTH.
000400         03  FILLER PICTURE A(9) VALUE IS 'JANUARY  '.
000410         03  FILLER PICTURE A(9) VALUE IS 'FEBRUARY '.
000420         03  FILLER PICTURE A(9) VALUE IS 'MARCH    '.
000430         03  FILLER PICTURE A(9) VALUE IS 'APRIL    '.
000440         03  FILLER PICTURE A(9) VALUE IS 'MAY      '.
000450         03  FILLER PICTURE A(9) VALUE IS 'JUNE     '.
```

Figure 10.37 Example—COBOL Program Report Writer Feature.

Figure 10.37 Continued.

```
000460          03  FILLER PICTURE A(9) VALUE IS 'JULY     '.
000470          03  FILLER PICTURE A(9) VALUE IS 'AUGUST   '.
C00480          03  FILLER PICTURE A(9) VALUE IS 'SEPTEMBER'.
CC0490          03  FILLER PICTURE A(9) VALUE IS 'OCTOBER  '.
000500          03  FILLER PICTURE A(9) VALUE IS 'NOVEMBER '.
000510          03  FILLER PICTURE A(9) VALUE IS 'DECEMBER '.
000520       02  RECORD-AREA REDEFINES RECORD-MONTH OCCURS 12 TIMES.
000530          03  MONTHNAME PICTURE A(9).
000531 01  FLAGS.
000532     05  MORE-DATA-FLAG  PICTURE XXX    VALUE 'YES'.
000533         88  MORE-DATA                  VALUE 'YES'.
000534         88  NO-MORE-DATA               VALUE 'NO '.
CC0540 REPORT SECTION.
0C0550 RD  EXPENSE-REPORT CONTROLS ARE FINAL, MONTH, DAY
000560       PAGE 59 LINES HEADING 1 FIRST DETAIL 9 LAST DETAIL 48
CC0570          FOOTING 52.
000580 01  TYPE REPORT HEADING.
000590     02  LINE 1 COLUMN 27 PICTURE A(26) VALUE IS
000600     'ACME MANUFACTURING COMPANY'.
000610     02  LINE 3 COLUMN 26 PICTURE A(29) VALUE IS
000620     'QUARTERLY EXPENDITURES REPORT'.
000630 01  PAGE-HEAD TYPE PAGE HEADING LINE 5.
000640     02  COLUMN 30 PICTURE A(9) SOURCE MONTHNAME OF
C00650     RECORD-AREA (MONTH).
0C0660     02  COLUMN 39 PICTURE A(12) VALUE IS 'EXPENDITURES'.
C00661     02  COLUMN 52 PICTURE X(11) SOURCE CONTINUED.
000670     02  LINE 7 COLUMN 2 PICTURE X(35) VALUE IS
0CC680     'MONTH      CAY     DEPT  NO-PURCHASES'.
000690     02  COLUMN 40 PICTURE X(33) VALUE IS
0C0700      'TYPE         COST  CUMULATIVE-COST'.
000710 01  DETAIL-LINE TYPE DETAIL LINE PLUS 1.
000720     02  COLUMN 2 GROUP INDICATE PICTURE A(9)  SOURCE MONTHNAME
0C0730     OF RECORD-AREA (MONTH).
0C0740     02  COLUMN 13 GROUP INDICATE PICTURE 99 SOURCE DAY.
000750     02  COLUMN 19 PICTURE XXX SOURCE DEPT.
C00760     02  COLUMN 31 PICTURE Z9 SOURCE NO-PURCHASES.
000770     02 COLUMN 42 PICTURE A SOURCE TYPE-PURCHASE.
000780     02  COLUMN 50 PICTURE ZZ9.99 SOURCE COST.
000790 01  TYPE CONTROL FOOTING DAY LINE PLUS 2.
000800     02  CCLUMN 2 PICTURE X(22) VALUE IS 'PURCHASES AND COST FOR'.
000810     02  COLUMN 24 PICTURE Z9 SOURCE SAVED-MONTH.
00C820     02  COLUMN 26 PICTURE X VALUE IS '-'.
00C830     02  COLUMN 27 PICTURE 99 SOURCE SAVED-DAY.
000840     02  CCLUMN 30 PICTURE ZZ9 SUM NO-PURCHASES.
000850     02  MIN COLUMN 49 PICTURE $$$9.99 SUM COST.
000860     02  COLUMN 65 PICTURE $$$$9.99 SUM COST RESET ON FINAL.
000870     02  LINE PLUS 1 COLUMN 2 PICTURE X(70) VALUE ALL '*'.
000880 01  TYPE CONTROL FOOTING MONTH LINE PLUS 1 NEXT GROUP NEXT PAGE.
000890     02  COLUMN 16 PICTURE A(14) VALUE IS 'TOTAL COST FOR'.
000900     02  COLUMN 31 PICTURE A(9) SOURCE MONTHNAME OF RECORD-AREA
0C0910         (SAVED-MONTH).
0C0920     02  COLUMN 40 PICTURE AAA VALUE 'WAS'.
0CC930     02  INT COLUMN 49 PICTURE $$$9.99 SUM MIN.
0C0940 01  TYPE CCNTRCL FOOTING FINAL LINE PLUS 1.
0C0950     02  COLUMN 16 PICTURE A(26) VALUE IS
CCC960     'TOTAL COST FOR QUARTER WAS'.
0C0970     02  COLUMN 48 PICTURE $$$$9.99 SUM INT.
000980 01  TYPE PAGE FOOTING LINE 55.
001000     02  LINE 57 COLUMN 59 PICTURE X(12) VALUE IS, 'REPORT-PAGE-'.
001010     02  COLUMN 71 PICTURE 99 SOURCE PAGE-COUNTER.
001020 01  TYPE REPORT FOOTING.
001030     02  LINE PLUS 1 COLUMN 32 PICTURE A(13) VALUE IS
001040     'END OF REPORT'.
```

Figure 10.37 Continued.

```
001060  *PROCEDURE DIVISION.
001070   DECLARATIVES.
001080   PAGE-HEAD-RTN SECTION.
001090           USE BEFORE REPORTING PAGE-HEAD.
001100       IF NO-MORE-DATA
001110           MOVE 1 TO PRINT-SWITCH
001120       ELSE
001130           IF MONTH EQUAL SAVED-MONTH
001140                   MOVE '(CONTINUED)' TO CONTINUED
001150           ELSE
001160                   MOVE SPACES TO CONTINUED
001170                   MOVE MONTH TO SAVED-MONTH.
001180   END DECLARATIVES.
001190   MAIN-ROUTINE.
001200       OPEN   INPUT  INFILE
001210              OUTPUT REPORT-FILE.
001220       INITIATE EXPENSE-REPORT.
001230       READ INFILE
001240           AT END MOVE 'NO ' TO MORE-DATA-FLAG.
001250       PERFORM WRITE-REPORT
001260           UNTIL NO-MORE-DATA.
001270       PERFORM DONE.
001280   WRITE-REPORT.
001290       GENERATE DETAIL-LINE.
001300       MOVE DAY TO SAVED-DAY.
001310       READ INFILE
001320           AT END MOVE 'NO ' TO MORE-DATA-FLAG.
001330   DONE.
001340       TERMINATE EXPENSE-REPORT.
001350       CLOSE INFILE
001360             REPORT-FILE.
001370       STOP RUN.
```

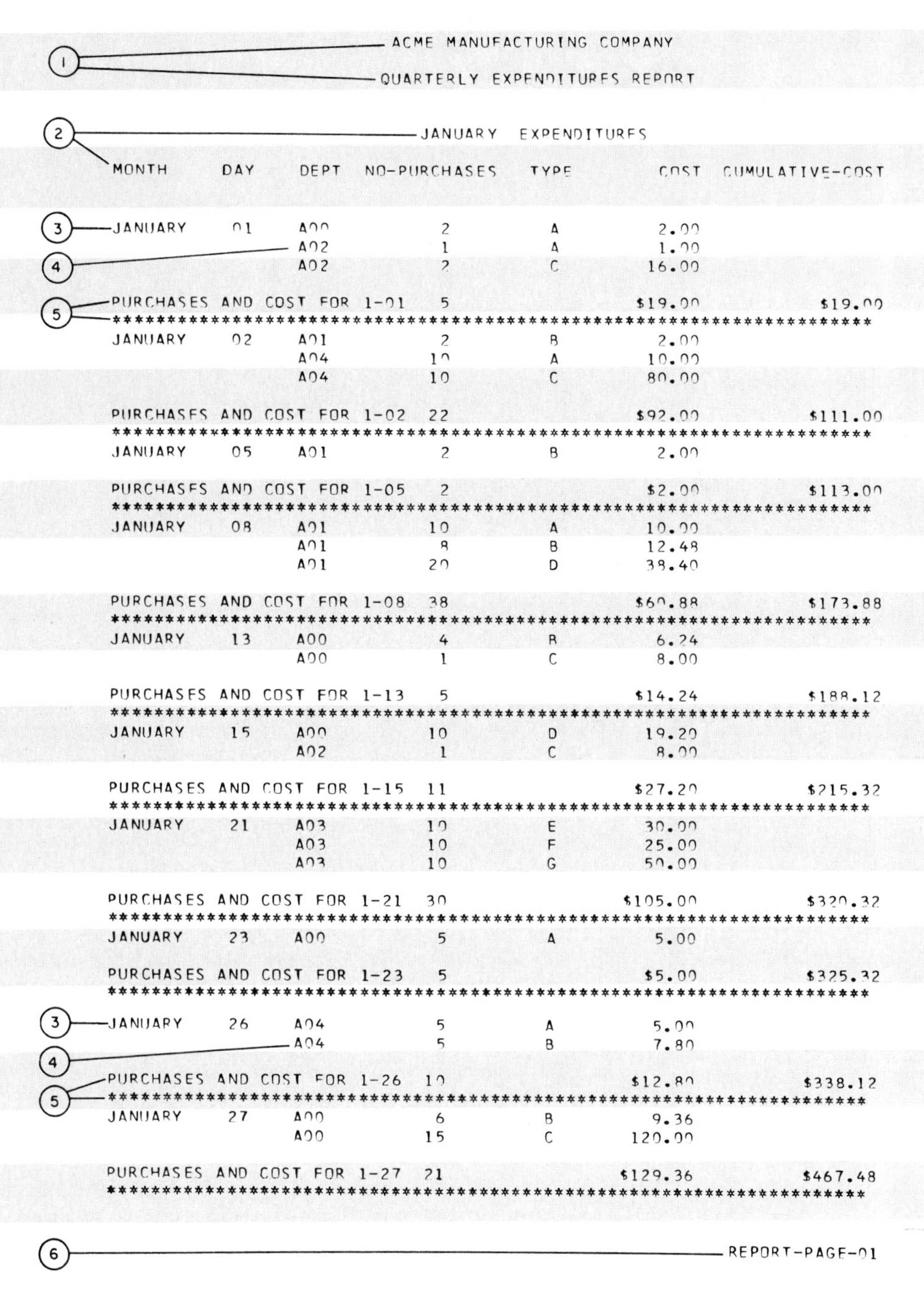

```
                              ACME MANUFACTURING COMPANY
                             QUARTERLY EXPENDITURES REPORT

                              JANUARY  EXPENDITURES
MONTH       DAY     DEPT  NO-PURCHASES   TYPE        COST  CUMULATIVE-COST

JANUARY      01     A00         2         A          2.00
                    A02         1         A          1.00
                    A02         2         C         16.00

PURCHASES AND COST FOR 1-01     5                  $19.00            $19.00
***************************************************************************
JANUARY      02     A01         2         B          2.00
                    A04        10         A         10.00
                    A04        10         C         80.00

PURCHASES AND COST FOR 1-02    22                  $92.00           $111.00
***************************************************************************
JANUARY      05     A01         2         B          2.00

PURCHASES AND COST FOR 1-05     2                   $2.00           $113.00
***************************************************************************
JANUARY      08     A01        10         A         10.00
                    A01         8         B         12.48
                    A01        20         D         38.40

PURCHASES AND COST FOR 1-08    38                  $60.88           $173.88
***************************************************************************
JANUARY      13     A00         4         B          6.24
                    A00         1         C          8.00

PURCHASES AND COST FOR 1-13     5                  $14.24           $188.12
***************************************************************************
JANUARY      15     A00        10         D         19.20
                    A02         1         C          8.00

PURCHASES AND COST FOR 1-15    11                  $27.20           $215.32
***************************************************************************
JANUARY      21     A03        10         E         30.00
                    A03        10         F         25.00
                    A03        10         G         50.00

PURCHASES AND COST FOR 1-21    30                 $105.00           $320.32
***************************************************************************
JANUARY      23     A00         5         A          5.00

PURCHASES AND COST FOR 1-23     5                   $5.00           $325.32
***************************************************************************

JANUARY      26     A04         5         A          5.00
                    A04         5         B          7.80
PURCHASES AND COST FOR 1-26    10                  $12.80           $338.12
***************************************************************************
JANUARY      27     A00         6         B          9.36
                    A00        15         C        120.00

PURCHASES AND COST FOR 1-27    21                 $129.36           $467.48
***************************************************************************

                                                             REPORT-PAGE-01
```


Figure 10.38 Example—Report Produced by the Report Writer Feature.

Figure 10.38 Continued.

```
(2)───────────────────────────JANUARY  EXPENDITURES (CONTINUED)
   MONTH        DAY     DEPT  NO-PURCHASES    TYPE           COST  CUMULATIVE-COST

   JANUARY       30     A00           2         B            3.12
                        A02          10         A           10.00
                        A02           1         C            8.00
                        A04          15         B           23.40
                        A04          10         C           80.00

   PURCHASES AND COST FOR 1-30   38                       $124.52          $592.00
   *******************************************************************************
   JANUARY       31     A00           1         A            1.00
                        A04           6         A            6.00

   PURCHASES AND COST FOR 1-31    7                         $7.00          $599.00
   *******************************************************************************
(7)──────────────TOTAL COST FOR JANUARY  WAS    $599.00

~~~~~~~~~~~~~~~~~~~~~~~~~~~~~~~~~~~~~~~~~~~~~~~~~~~~~~~~~~~~~~~~~~~~~~~~~~~~~~~~~~

(8)──── TOTAL COST FOR QUARTER WAS    $2010.02

(6)───────────────────────────────────────────────────────────────REPORT-PAGE-06

(9)───────────────────────────────END OF REPORT
~~~~~~~~~~~~~~~~~~~~~~~~~~~~~~~~~~~~~~~~~~~~~~~~~~~~~~~~~~~~~~~~~~~~~~~~~~~~~~~~~~
```

Key Relating Report to Report Writer Source Program

(1) is the report heading resulting from source lines 00580–00620.

(2) is the page heading resulting from source lines 00630–00700.

(3) is the detail line resulting from source lines 00710–00780 (note that since it is the first detail line after a control break, the fields defined with 'group indicate', lines 00720–00740, appear).

(4) is a detail line resulting from the same source lines as 3. In this case, however, the fields described as 'group indicate' do not appear (since the control break did not immediately precede the detail line).

(5) is the control footing (for DAY) resulting from source lines 00790–00870.

(6) is the page footing resulting from source lines 00980–01010.

(7) is the control footing (for MONTH) resulting from source lines 00880–00930.

(8) is the control footing (for FINAL) resulting from source lines 00940–00970.

(9) is the report footing resulting from source lines 01020–01040.

Lines 01070–01104 of the example illustrate a use of 'USE BEFORE REPORTING'. The effect of the source is that each time a new page is started, a test is made to determine if the new page is being started because a change in MONTH has been recognized (the definition for the control footing for MONTH specifies 'NEXT GROUP NEXT PAGE') or because the physical limits of the page were exhausted. The calculation involved sets up a fixed ('PAGE GROUP') which is referenced by a SOURCE clause in the PAGE FOOTING description. Consequently, two page counters can be maintained: one indicating physical pages and one indicating logical pages.

```
(6)───────────────────────────────────────────────────────────────REPORT-PAGE-02
```

REPORT WRITER PROBLEM

INPUT

Field	Card Columns	
Department	1–3	
Serial Number	4–10	
Name	14–32	
Hours	38–40	XX.X
Net Pay	41–47	XXX.XX

OUTPUT

Headings

Report Headings

Description	Print Positions
LABOR LISTING	18–30

Page Headings

Line 1

Description	Print Positions
SERIAL	12–17
HOURS	44–48

Line 2

Description	Print Positions
DEPT.	5–9
NUMBER	12–17
N A M E	27–33
WORKED	43–48
NET PAY	54–60

Detail

Description	Print Positions	
Department	6–8	
Serial Number	11–17	
Name	20–38	
Hours Worked	44–47	XX.X
Net Pay	51–60	$$$,$$$.99

Control Footing MINOR—Department FINAL

Description		Print Positions	
HEADING—TOTAL FOR DEPARTMENT (Dept. No.)		25–47	
MINOR	Net Pay	50–60	$$$$,$$$.99
		61	*
FINAL	Net Pay	49–60	$$$$$,$$$.99
		61–62	**

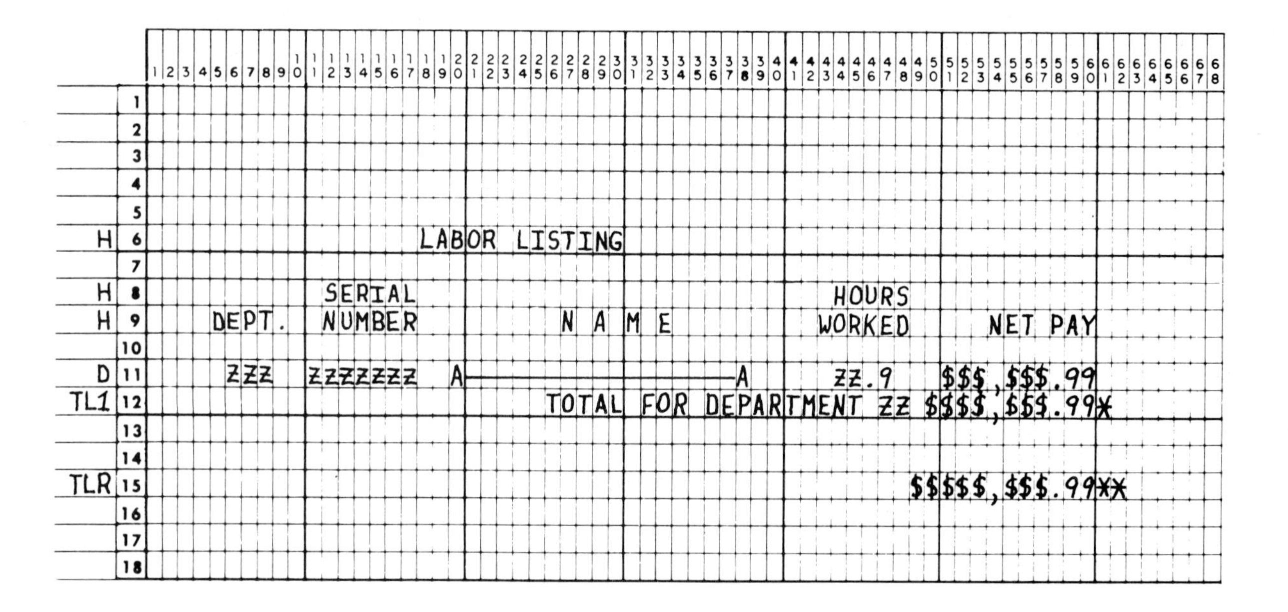

```
001010 IDENTIFICATION DIVISION.
001020 PROGRAM-ID.            REPORT-WRITER.
001030 AUTHOR.                C FEINGOLD.
001040 INSTALLATION.          DISTRICT OFFICE.
001050 DATE-WRITTEN.          JUNE 14 1977.
001060 DATE-COMPILED. 10/27/77.
001070 REMARKS.        THIS IS A PROGRAM TO PREPARE A LABOR LISTING.
002010 ENVIRONMENT DIVISION.
002020 CONFIGURATION SECTION.
002030 SOURCE COMPUTER.
002040      IBM-370-I155.
002050 OBJECT-COMPUTER.
002060      IBM-370-I155.
002070 INPUT-OUTPUT SECTION.
002080 FILE-CONTROL.
002090     SELECT FILE-IN
002100         ASSIGN TO UR-2540R-S-CARDIN.
002110     SELECT FILE-OUT
002120         ASSIGN TO UR-1403-S-PRINT.
003010 DATA DIVISION.
003020 FILE SECTION.
003030 FD  FILE-IN
003040     RECORDING MODE F
003050     LABEL RECORDS OMITTED
003060     RECORD CONTAINS 80 CHARACTERS
003070     DATA RECORD IS CARD-IN.
003080 01  CARD-IN.
003090     02  DEPT-IN         PICTURE 999.
003100     02  SERIAL-IN       PICTURE 9(7).
003110     02  FILLER          PICTURE X(3).
003120     02  NAME-IN         PICTURE A(19).
003130     02  FILLER          PICTURE X(5).
003140     02  HOURS-IN        PICTURE 99V9.
003150     02  NET-IN          PICTURE 9(5)V99.
003160     02  FILLER          PICTURE X(33).
003170 FD  FILE-OUT
003180     LABEL RECORDS OMITTED
003190     REPORT IS LABOR-LIST.
003200 WORKING-STORAGE SECTION.
003210 01  FLAGS.
003220     05  MORE-DATA-FLAG  PICTURE XXX    VALUE 'YES'.
003230         88  MORE-DATA                  VALUE 'YES'.
003240         88  NO-MORE-DATA               VALUE 'NO '.
004010 REPORT SECTION.
004020 RD  LABOR-LIST
004030     CONTROLS ARE FINAL, DEPT-IN
004040     PAGE               52 LINES
004050     HEADING            1
004060     FIRST DETAIL       6
004070     LAST DETAIL        46
004080     FOOTING            52.
004090 01  TYPE REPORT HEADING.
004100     02  LINE 1  COLUMN 18 PICTURE X(13)
004110         VALUE IS 'LABOR LISTING'.
004120 01  TYPE PAGE HEADING LINE 3.
004130     02  COLUMN 12  PICTURE X(6)
004140         VALUE IS 'SERIAL'.
004150     02  COLUMN 44  PICTURE X(5)
004160         VALUE IS 'HOURS'.
004170     02  LINE 4  COLUMN 5  PICTURE X(5)
004180         VALUE IS 'DEPT.'.
004190     02  COLUMN 12  PICTURE X(6)
004200         VALUE IS 'NUMBER'.
004210     02  COLUMN 27  PICTURE X(7)
```

```
004220              VALUE IS 'N A M E'.
004230          02  COLUMN 43  PICTURE X(16)
004240              VALUE IS 'WORKED'.
005010          02  COLUMN 54  PICTURE X(7)
005020              VALUE IS  'NET PAY'.
005030 01   DETAIL-LINE  TYPE DETAIL LINE PLUS 1.
005040          02  COLUMN 6  GROUP INDICATE  PICTURE ZZZ SOURCE DEPT-IN.
005050          02  COLUMN 11  PICTURE Z(7)  SOURCE SERIAL-IN.
005060          02  COLUMN 20  PICTURE A(19) SOURCE NAME-IN.
005070          02  COLUMN 44  PICTURE ZZ.9  SOURCE HOURS-IN.
005080          02 COLUMN 51   PICTURE $$$,$$$.99   SOURCE NET-IN.
005090 01   TYPE CONTROL FOOTING DEPT-IN LINE PLUS 1
005095      NEXT GROUP PLUS 1.
005100          02  COLUMN 25  PICTURE X(20)
005110              VALUE IS 'TOTAL FOR DEPARTMENT'.
005120          02  COLUMN 45  PICTURE ZZZ   SOURCE  DEPT-IN.
005130          02  MIN COLUMN 50 PICTURE $$$$,$$$.99 SUM NET-IN.
005140          02  COLUMN 61  PICTURE X  VALUE IS '*'.
005150 01   TYPE CONTROL FOOTING FINAL  LINE PLUS 3.
005160          02  COLUMN 49  PICTURE $$$$$,$$$.99 SUM MIN.
005170          02  COLUMN 61  PICTURE XX VALUE IS '**'.
006010 PROCEDURE DIVISION.
006020 MAIN-ROUTINE.
006030      OPEN INPUT  FILE-IN
006040           OUTPUT FILE-OUT.
006050      INITIATE LABOR-LIST.
006060      READ FILE-IN
006070          AT END MOVE 'NO ' TO MORE-DATA-FLAG.
006080      PERFORM WRITE-REPORT
006090           UNTIL NO-MORE-DATA.
006100      TERMINATE LABOR-LIST.
006110      CLOSE      FILE-IN
006120                 FILE-OUT.
006130      STOP RUN.
006140 WRITE-REPORT.
006150      GENERATE DETAIL-LINE.
006160      READ FILE-IN
006170          AT END MOVE 'NO ' TO MORE-DATA-FLAG.
```

```
                LABOR LISTING

       SERIAL                                 HOURS
DEPT.  NUMBER          N A M E               WORKED       NET PAY

   1  1475176   P ACKLEY                       42.1       $215.67
      2497561   A BECKER                       47.1       $462.54
      3948252   N CHAPMAN                      34.1       $276.84
      4987124   E EELS                         43.0       $424.31
      8526941   H KUBO                         31.9       $315.20
      9247617   G WITT                         51.4       $610.40
                    TOTAL FOR DEPARTMENT   1          $2,304.96*

   2  1472481   B ANCHETA                      35.2       $324.21
      2759415   D BLACK                        41.5       $410.51
      4107151   S CHOW                         48.2       $574.27
      4326415   M EISNER                       37.6       $242.50
      6929412   F SACKS                        31.8       $204.60
                    TOTAL FOR DEPARTMENT   2          $1,756.09*

   3  2468015   T BARRET                       41.8       $412.60
      3461915   M KARL                         42.7       $432.75
      4671421   Y LERNER                       31.5       $261.72
      5176234   J MCFADDEN                     48.9       $515.72
      8349512   R STERNBERG                    52.5       $672.42
      9567415   H WALKER                       41.2       $412.62
                    TOTAL FOR DEPARTMENT   3          $2,707.83*
```

```
        1   1475176   P ACKLEY                         42.1        $215.67
            2497561   A BECKER                         47.1        $462.54
            3948252   N CHAPMAN                        34.1        $276.84
            4987124   E FELS                           43.0        $424.31
            8526941   H KUBO                           31.9        $315.20
            9247617   G WITT                           51.4        $610.40
                           TOTAL FOR DEPARTMENT  1               $2,304.96*

        2   1472481   B ANCHETA                        35.2        $324.21
            2759415   D BLACK                          41.5        $410.51
            4107151   S CHOW                           48.2        $574.27
            4326415   M EISNER                         37.6        $242.50
            6929412   F SACKS                          31.8        $204.60
                           TOTAL FOR DEPARTMENT  2               $1,756.09*

        3   2468015   T BARRET                         41.8        $412.60
            3461915   M KARL                           42.7        $432.75
            4671421   Y LERNER                         31.5        $261.72


             SERIAL                                    HOURS
    DEPT.    NUMBER            N A M E                 WORKED      NET PAY

        3   5176234   J MCFADDEN                       48.9        $515.72
            8349512   P STERNBERG                      52.5        $672.42
            9567415   H WALKER                         41.2        $412.62
                           TOTAL FOR DEPARTMENT  3               $2,707.83*

        1   1475176   P ACKLEY                         42.1        $215.67
            2497561   A BECKER                         47.1        $462.54
            3948252   N CHAPMAN                        34.1        $276.84
            4987124   E FELS                           43.0        $424.31
            8526941   H KUBO                           31.9        $315.20
            9247617   G WITT                           51.4        $610.40
                           TOTAL FOR DEPARTMENT  1               $2,304.96*

        2   1472481   B ANCHETA                        35.2        $324.21
            2759415   D BLACK                          41.5        $410.51
            4107151   S CHOW                           48.2        $574.27
            4326415   M EISNER                         37.6        $242.50
            6929412   F SACKS                          31.8        $204.60
                           TOTAL FOR DEPARTMENT  2               $1,756.09*

        3   2468015   T BARRET                         41.8        $412.60
            3461915   M KARL                           42.7        $432.75
            4671421   Y LERNER                         31.5        $261.72
            5176234   J MCFADDEN                       48.9        $515.72
            8349512   R STERNBERG                      52.5        $672.42
            9567415   H WALKER                         41.2        $412.62
                           TOTAL FOR DEPARTMENT  3               $2,707.83*
                                                                 $20,306.64**
```

Exercises

Write your answer in the space provided (answer may be one or more words).

1. A report represents a ________________ organization of data.
2. In defining the physical aspects of the report format, consideration must be given to the ________________ and ________________ of the report medium, to the individual ________________ structure, to the type of ________________ device on which the ________________ is finally written.
3. To define the conceptual characteristics of the data, the concept of ________________ is used.
4. To create the report, the approach taken is to define the types of ________________ that must be considered in presenting ________________ in a ________________ manner.
5. The Report Writer feature provides the facility for producing reports by specifying the ________________ appearance of a report rather than requiring the specifications for the detailed ________________ necessary to produce the report.
6. A ________________ of levels is used in defining the logical organization of the report.
7. Each report is divided into ________________ groups which in turn are divided into a sequence of ________________.
8. The ________________ report group items are the basic element of the report.
9. Control ________________ and ________________ occur automatically when the machine senses a control break.
10. ________________ headings at the beginning of the report as well as ________________ at the end of the report can be printed as the report is being prepared.
11. ________________ and ________________ counters are incremented automatically as the report is being prepared.
12. The Report Writer feature can place data into ________________ storage where it may be used for subsequent ________________ printing in a specified format.
13. The Report Writer feature allows the programmer to describe the report ________________.
14. A ________________ section must be added at the end of the Data Division to define the format of each finished report.
15. The ________________ clause is required in every FD entry to list the name of the report to be produced.
16. The Report Section must begin on a ________________ line by itself.
17. The Report Section describes the ________________ aspects of the report format and the ________________ characteristics of the items of the report.
18. The report description entry contains information pertaining to the ________________ format of the report.
19. The ________________ must be specified in the Report clause of the file description entry.
20. The Code clause is used when ________________ than one report is written for the file.
21. A ________________ is a data item that is tested each time a detail group is to be written.
22. All level totals are printed in ________________ in which they are written with the exception of the ________________ total.
23. The ________________ clause describes the physical format of a page of the report.
24. The ________________ and ________________ are numeric counters generated automatically by the Report Writer feature.
25. The Page-Counter clause may be referred to by ________________ clauses and ________________ statements.
26. The Page-Counter clause is set to ________________ initially by the ________________ statement in the Procedure Division.
27. The Line-Counter is automatically set to ________________ initially by the Initiate statement in the Procedure Division.
28. The value of a Line-Counter during any Procedure Division test represents the number of the ________________ line used by the previous group.
29. The report group description entry specifies the ________________ of a particular report group of the individual ________________ within a report group.
30. The Line clause indicates the ________________ or ________________ line number of this entry in reference to the page or previous entry.

31. Absolute line number entries must be indicated in ________________ order and cannot be preceded by a ________________ line number.
32. The ________________ clause is used to indicate an automatic skip to the next page before the ________________ of the next group.
33. The Next Group clause is used to define the line control ________________ the printing of the current group being processed.
34. Any report group other than ________________ is automatically generated by the Report Writer feature.
35. Nothing precedes a ________________ entry and nothing follows a ________________ entry within a report.
36. A Final total control break can be designated only once for ________________ or ________________ entries within a report.
37. The ________________ entry is produced only once at the beginning of the report during the execution of the first ________________ statement.
38. The ________________ entry indicates that the report group is to be produced for each Generate statement.
39. ________________ is the only Usage option that can be specified for an elementary or group item in a report group description entry.
40. The Column clause indicates the ________________ character of the printing position.
41. The Group Indicate clause specifies that the elementary item is to be printed only on the ________________ occurrence of the item after any ________________ or ________________ break.
42. The Reset clause causes the ________________ counter in the elementary item to be reset to ________________ on a ________________.
43. The Source clause specifies a data item to be ________________ or to be ________________.
44. The Sum clause is used to cause automatic ________________ of data and may appear only in ________________ item entry of a ________________ report group.
45. A ________________ declarative may be written in the Declaratives section of the Procedure Division if the programmer wishes to ________________ or manipulate the data in any manner before it is printed on a report.
46. The ________________ statement begins processing of a report.
47. The Initiate statement does not ________________ the file with which the report is associated.
48. The Generate statement is used to ________________ a report and link the ________________ to the Report Writer described in the ________________ of the Data Division at process time.
49. The Generate statement does all the ________________ operations of the Report Writer feature.
50. The identifier of a Generate statement must be a ________________ report group to produce detailed reporting.
51. If the identifier of a Generate statement is a Report entry, a ________________ report is produced without any detailed printing of the items.
52. The Terminate statement is used to terminate the processing of a ________________ and to produce all ________________ associated with the report.
53. A Terminate statement does not ________________ the files.
54. The ________________ statement specifies Procedure Division statements that are executed just before the named report group is processed.
55. The ________________ statement, when present, must immediately follow a section header in the ________________ portion of the Procedure Division.

Answers

1. PICTORIAL
2. WIDTH, LENGTH, PAGE, HARDWARE, REPORT
3. LEVEL STRUCTURE
4. REPORT GROUPS, DATA, FORMAT
5. PHYSICAL, PROCEDURES
6. HIERARCHY
7. REPORT, ITEMS
8. DETAIL
9. TOTALS, HEADINGS
10. REPORT, TOTALS
11. LINE, PAGE
12. INTERMEDIATE, OFF-LINE
13. PICTORIALLY
14. REPORT
15. REPORT
16. SEPARATE
17. PHYSICAL, CONCEPTUAL
18. OVERALL
19. REPORT NAME

20. MORE
21. CONTROL
22. SEQUENCE, FINAL
23. PAGE LIMIT
24. PAGE-COUNTER, LINE-COUNTER
25. DATA DIVISION, PROCEDURE DIVISION
26. ONE, INITIATE
27. ZERO
28. LAST
29. CHARACTERISTICS, DATA-NAMES
30. ABSOLUTE, RELATIVE
31. ASCENDING, RELATIVE
32. NEXT PAGE, FIRST LINE
33. FOLLOWING
34. TYPE DETAIL
35. REPORT HEADING, REPORT FOOTING
36. CONTROL HEADING, CONTROL FOOTING
37. REPORT HEADING, GENERATE
38. DETAIL
39. DISPLAY
40. LEFTMOST
41. FIRST, CONTROL, PAGE
42. SUM, ZERO, CONTROL BREAK
43. PRINTED, SUMMED
44. SUMMATION, ELEMENTARY, CONTROL FOOTING
45. USE BEFORE REPORTING, ALTER
46. INITIATE
47. OPEN
48. PRODUCE, PROCEDURE DIVISION, REPORT SECTION
49. AUTOMATIC SUMMATION
50. DETAIL
51. SUMMARY
52. REPORT, CONTROL FOOTINGS
53. CLOSE
54. USE
55. USE, DECLARATIVES

Questions for Review

1. What are the important items to be considered when presenting a report?
2. What are the report groups that must be considered in presenting data in a format manner?
3. What is the purpose of the Report Writer feature?
4. Describe the principal features of the Report Writer feature.
5. What is meant by a control break?
6. What is the purpose of the Report clause?
7. What should the entries include in the Report Section?
8. Explain the structure of the Report Section.
9. What is the Code clause used for?
10. How is the Control clause used in the Report Writer feature?
11. What is the importance of the Page Limit clause?
12. Describe the operation of the Page Counter and Line Counter.
13. What is a report group? What are the three types of Report Groups?
14. Explain the use of the Line clause and how it operates.
15. What is the Type clause? Briefly describe the different types of Type clauses.
16. Explain the uses of the following clauses: Column, Group Indicate, Reset, Source.
17. What is the purpose of the Sum clause? How does it operate on data?
18. Explain the purpose and use of the Initiate, Generate, and Terminate statements.
19. What are the uses of the Use Before Reporting declarative?

Problems

1. *Match each item with its proper description.*

_____ 1. Report Section
_____ 2. Report Description Entry
_____ 3. Report Group Description Entry
_____ 4. Initiate Statement
_____ 5. Generate Statement
_____ 6. Terminate Statement
_____ 7. Use Statement

A. Begins processing of a ıeport.
B. Stops processing of a report.
C. Specification for each report.
D. Characteristics of particular report group.
E. Information pertaining to overall format of a report.
F. Statements executed before report group.
G. Produces a report.

2. *Match each clause with its proper description.*

_____ 1. Report
_____ 2. Code
_____ 3. Control
_____ 4. Page Limit
_____ 5. Line
_____ 6. Next Group
_____ 7. Type
_____ 8. Group Indicate
_____ 9. Source
_____10. Sum

A. Physical format of a page.
B. Line control following current report group.
C. First occurrence after control or page break.
D. Name of report to be produced.
E. Automatic summation of data.
F. Value to be tested.
G. More than one report from a file.
H. Time that report group is to be generated.
I. Absolute or relative number.
J. Item to be used for report item.

3. *Prepare the Report Writer entries for the following:*

INPUT

DEPT | SERIAL NUMBER | NAME | HOURS | NET PAY

OUTPUT

DEPT.	SERIAL NUMBER	N A M E		HOURS WORKED	NET PAY
1	1234560	SMITH	JW	40.1	285.15
1	1892750	JONES	RA	39.6	152.16
1	8929016	MAUS	JB	62.5	182.55
					619.86
2	0238648	GOLDMAN	H	31.7	100.25
2	0333367	WOLFE	DJ	9.5	26.60
					126.85

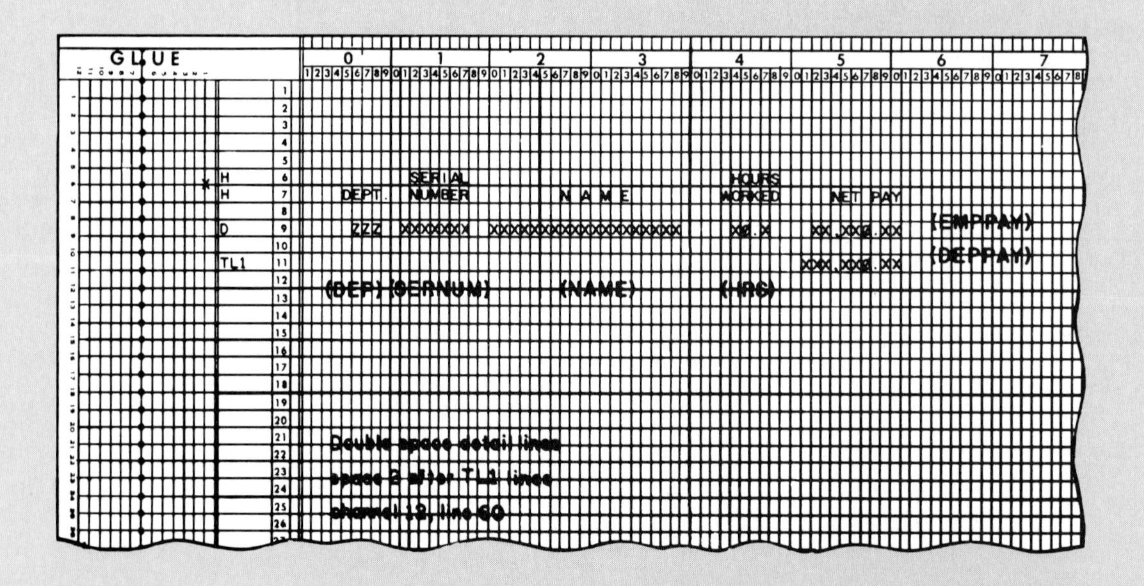

4. *Prepare the Report Writer entries for the following:*

REPORT

```
                 CUSTOMER NAME AND ADDRESS LISTING

ARTSON        H V    123 WOOD LANE        DE MONES, CAL.     1

BELEBOR       G      784 GRAND DRIVE      SEMMDALE, VA.      2

BREIGHT       H M    NEW SPRING BLVD      HEER, MD.          3

CALIPHANDER   A C    STRETCH BLVD         MITTAK, ALA.       4

DIERR         D      1 MADISON ROAD       HEAROLD, N.M.      5
```

CHART

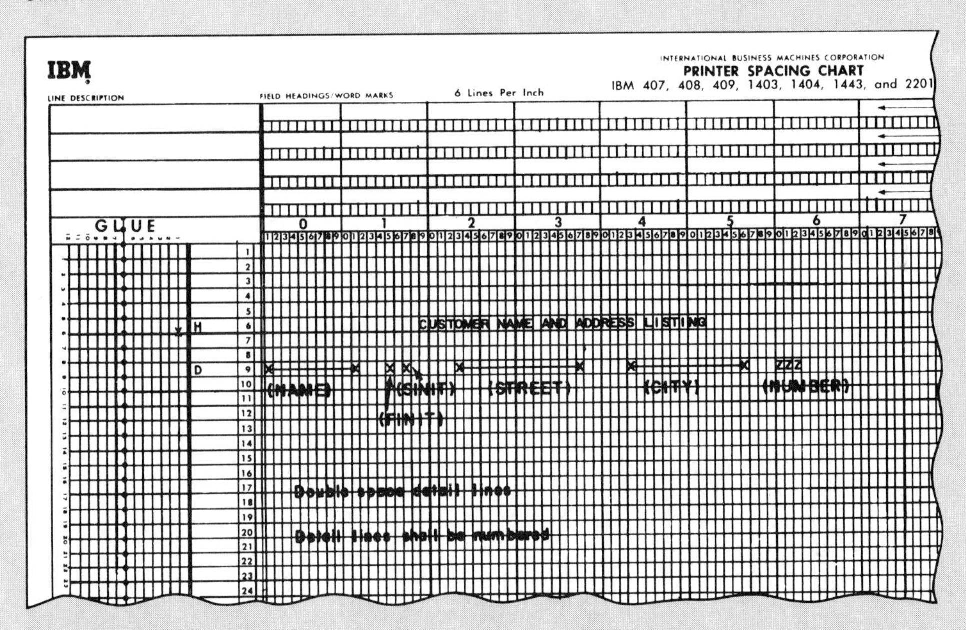

INPUT RECORD

FIRST INITIAL
SECOND INITIAL
NAME
ADDRESS
CITY / STATE

5. *Prepare the Report Writer entries for the following:*

INPUT

Field	Card Columns
Date	1–6
Location	7–9
Department	10–13
Employee Number	14–17
Employee Name	18–32 First Initial, second initial, name.
Group Insurance	33–36 XX.XX
Union Dues	37–40 XX.XX
Bonds	41–44 XX.XX
United Fund	45–48 XX.XX
Pension	49–52 XX.XX
Major Medical	53–56 XX.XX
Other	57–60 XX.XX
Total Deductions	61–65 XXX.XX

CALCULATIONS: Cards are in sequence by departments within locations.

a. Totals by departments for Group Insurance, Union Dues, Bonds, United Fund, Pension, Major Medical, Other and Total Deductions.

b. Totals by location for Group Insurance, Union Dues, Bonds, United Fund, Pension, Major Medical, Other and Total Deductions.

OUTPUT

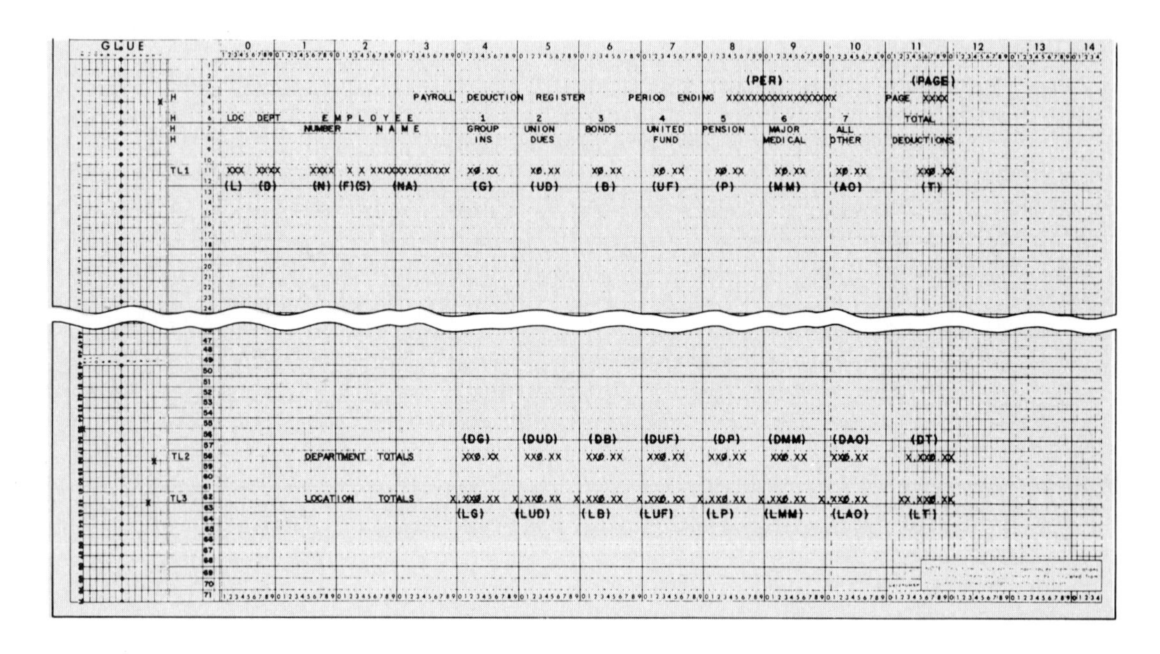

6. *Prepare the Report Writer entries for the following:*

INPUT

Field	Card Columns
Date	1–6
Customer Number	7–10
Old Balance	11–15
Purchases	16–20
Payments	21–25
New Balance	26–30

CALCULATIONS:

a. Totals by customer number for Old Balance, Purchases, Payments and New Balance.
b. Final totals for Old Balance, Purchases, Payments and New Balance.

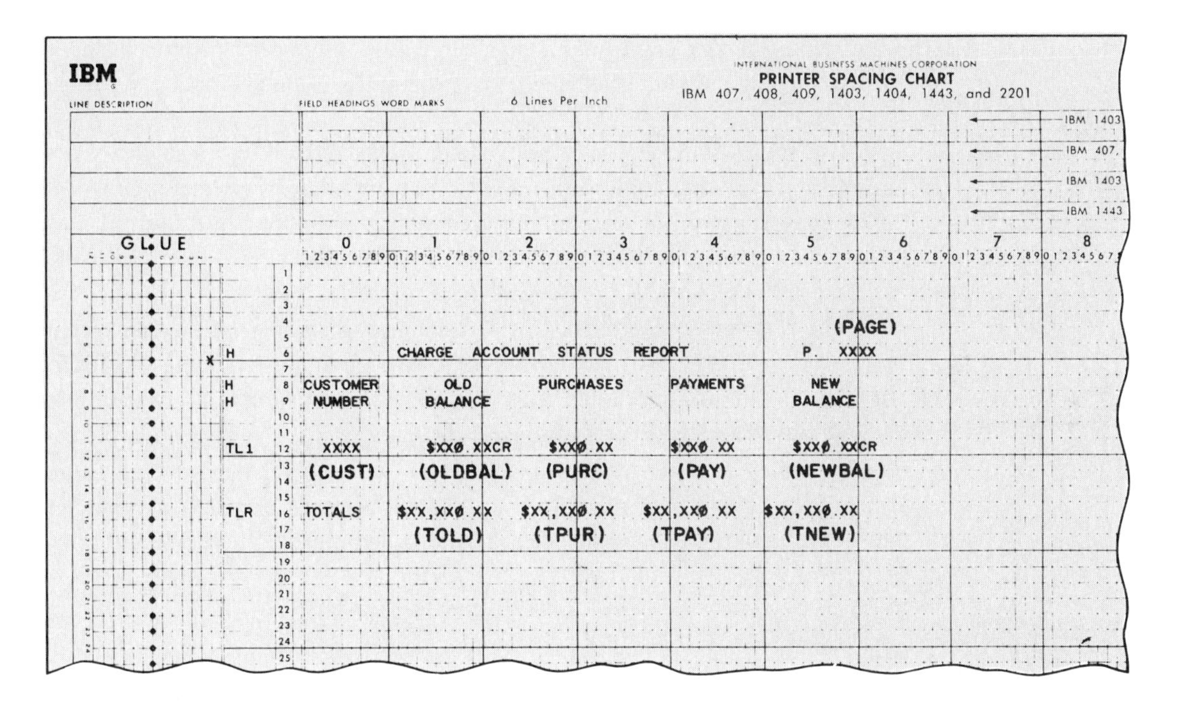

11 Sort Feature

INTRODUCTION

Because sorting constitutes a large percentage of the workload in a business data processing system, an efficient sort program is a highly necessary part of that system.

The SORT statement arranges a collection of records into a sequence determined by the programmer. Each SORT statement has its own internal working file, called a *sort file,* created and used during the sorting operation. It may be either magnetic disk or magnetic tape. The sort file itself is referred to and accessed only by the SORT statement. (See figure 11.1.)

The Sort Feature provides the COBOL programmer with a convenient access to the sorting capacity of the Sort/Merge program by including a COBOL SORT statement and specifying other necessary elements of the Sort Feature in his program. The Sort Feature provides the capability for sorting a file of records according to a set of user-specified keys within each record (see figure 11.2). Sorting operations for fixed or variable records of varying modes of data representation can be specified by the programmer. Optionally, the pro-

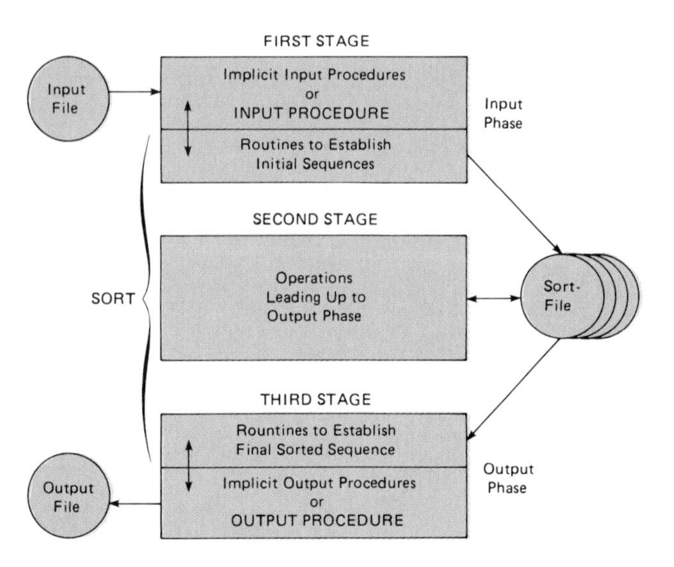

Figure 11.1 Schematic—Sort Operation.

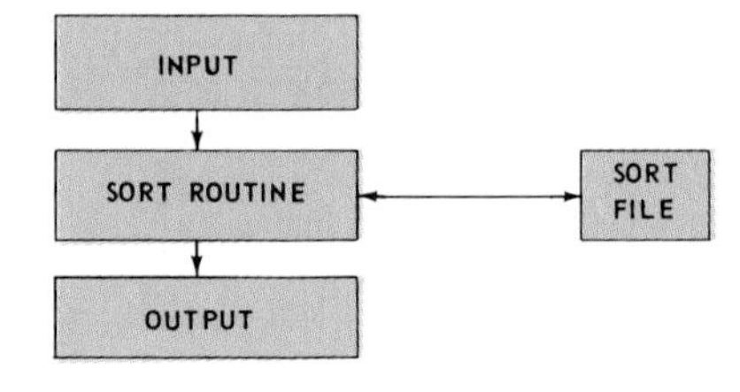

In this illustration the execution of the SORT statement performs the following functions:

1. Opens the input, sort, and output files.
2. Transcribes all of the input file to the sort file.
3. Sorts the file according to the prescribed specifications.
4. Merges the output of the sort to the output file.
5. Closes the input, sort, and output files.

Figure 11.2 Example—Sort with No Input or Output Procedure.

grammer may apply some special processing which may consist of addition, deletion, creation, alteration, editing, or other modification of individual records by input or output procedures. This special processing allows the programmer to summarize, delete, shorten, or otherwise alter the records being sorted during the initial or final phases of the sort. (See figure 11.3.)

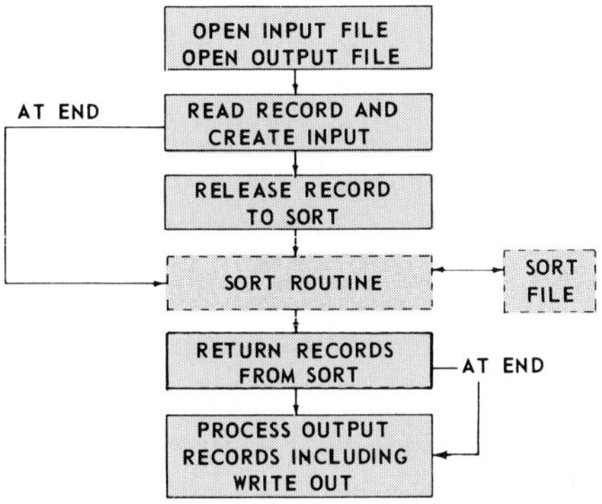

OPEN INPUT FILE
OPEN OUTPUT FILE
AT END
READ RECORD AND CREATE INPUT
RELEASE RECORD TO SORT
SORT ROUTINE
SORT FILE
RETURN RECORDS FROM SORT
AT END
PROCESS OUTPUT RECORDS INCLUDING WRITE OUT

In this operation the following functions are performed:

1. When a SORT statement is executed, control is transferred to the input procedures.
2. In the INPUT procedures, the programmer opens the input file(s) to the sort. Records are read, created, and released to the sort. When the input data is exhausted, the file is to be closed.
3. The sort is performed when all records to be sorted have been passed to the sort by the RELEASE verb and the last logical instruction of the INPUT procedure has been executed.
4. On the final merge pass of the sort, control is given to the programmer to perform output procedures in which the output file is opened, sorted records are processed, and the output file is closed.
5. The sort terminates when all records have been sorted and passed by the RETURN verb to the OUTPUT procedures and the last logical instruction of the OUTPUT procedures have been executed.

Figure 11.3 Example—Sort with Input and Ouput Procedures.

BASIC ELEMENTS OF THE SORT FEATURE

To use the Sort Feature, the COBOL programmer must provide additional information in the Environment, Data, and Procedure Divisions of the source program. The basic elements of the Sort Feature are the SORT statement in the Procedure Division and the Sort-File-Description (SD) entry, with its associated record description entries in the Data Division.

1. The programmer must name in the Environment Division, with SELECT sentences for all files to be used as input to and output from Sort-file.
2. In the Data Division, the programmer must write file description entries (FD) for all files that are to provide input and output to the sort program. In addition, a Sort-File-Description entry (SD) must be written describing the records to be sorted, including the sorting-key fields. The record description entry associated with the Sort-File-Description may be considered as redefining the records being sorted.
3. The programmer must specify the records being sorted, the sort-key names to be sorted on, and whether the sort is to be in ascending or descending sequence or a mixture of both—that is, sort-keys may be specified as ascending or descending, independent of one another and the sequence of the records will conform to the mixture specified, and whether the records are to be processed before and/or after the sort. The programmer writes a SORT statement in the Procedure Division specifying all of the aforementioned options.
4. The sort-work files are provided by the Sort/Merge program to serve as intermediate work files during the sorting process.

The SORT statement in the Procedure Division is the primary element of a source program that performs one or more sorting operations. A sorting operation is based on the sort-keys named in the SORT statement. A sort-key specifies the fields within a record on which the file is sorted. Sort-keys are defined in the record description clauses associated with the Sort-File-Description (SD) entry. The term sorting operation means not only the manipulation of the Sort Program of sort-work files on the basis of sort-keys designated by the COBOL program, but it also includes the method of making available to, and returning records from, these sort-work files. A *sort-work file* is a collection of records that is involved in the sorting operation as it exists in intermediate device(s). Records are made available to the Sort/Merge program by the USING or INPUT PROCEDURE options of the Sort Statement. Sorted records are returned from the Sort/Merge program by either the GIVING or OUTPUT PROCEDURE options of the SORT statement.

ENVIRONMENT DIVISION

Input-Output Section

File-Control Paragraph

The File-Control paragraph of the Input-Output Section names each file, identifies the file medium, and permits particular hardware assignments. SELECT entries must be written in the File-Control paragraph of the Input-Output Section for all files used within input and output procedures, and files specified in the USING or GIVING options of the SORT statement in the Procedure Division. (See figure 11.4.) The SELECT clause may be specified for the sort-file (fig. 11.5). The file-name identifies the sort-file to the compiler.

```
SELECT file-name
    ASSIGN TO [integer-1] system-name-1 [system-name-2] . . .
       OR system-name-3 [FOR MULTIPLE {REEL} ]
                                      {UNIT}
       [RESERVE {integer-2} ALTERNATE [AREA ] ].
                {NO       }           [AREAS]
```

Figure 11.4 Format—Select Clause.

```
SELECT sort-file-name
    ASSIGN TO [integer-1] system-name-1 [system-name-2] . . .
```

Figure 11.5 Format—Select Clause for Sort File.

The presence of the OR clause means the file-name emerging from the sorting operation is either on the assigned hardware units preceding the key word OR, or on the hardware units following the key word OR. At the conclusion of the sorting operation, an indication will be given as to which hardware units contain the file. The proper hardware units are addressed when this file is opened for input.

The RESERVE clause may be used in conjunction with the above format. The RESERVE clause is applicable as described in the Environment Division. (See "Reserve" clause in the Environment Division, chap. 5.)

I-O-Control Paragraph

The I-O-Control paragraph of the Input-Output Section specifies the memory areas to be shared by different files.

Same Record/Sort Area Clause. When the RECORD option is used, the clause specifies that two or more named files are to use the same memory for processing of the current logical record (fig. 11.6). All of the files may be opened at the same time; however, the logical record of only one of the files can exist in the record area at one time.

SAME {RECORD | SORT} AREA FOR file-name-1 {file-name-2}. . .

Figure 11.6
Format—Same Record/Sort Area Clause.

If the SAME SORT AREA clause is used, at least one of the file-names must represent a sort-file. Files that do not represent sort-files may also be named in this clause. This clause specifies that storage is shared as follows:

1. The SAME SORT AREA clause specifies a memory area which will be made available for use in sorting each sort-file named. Thus, any memory area allocated for the sorting of a sort-file is available for reuse in sorting any of the other sort-files.
2. In addition, storage areas assigned to files that do not represent sort-files may be allocated as needed for sorting the sort-files named in the SAME SORT AREA clause. The extent of such allocation will be specified by the implementor.
3. Files other than sort-files do not share the same storage area with each other. If the user wishes these files to share the same storage area with each other, he must also include in the program a SAME SORT AREA or SAME RECORD AREA naming these files.
4. During the execution of a SORT statement that refers to a sort-file named in this clause, any nonsort-file named in this clause must not be opened.

In the IBM System/360 or 370 computer, the function of the SORT option is to optimize the assignment of storage areas to a given SORT statement. The system handles storage assignments automatically; hence, the SORT option, if given, is treated as a comment.

DATA DIVISION

In the Data Division, the programmer must include file description entries for files to be sorted, sort-file description entries for sort-work files, and record description entries for each.

File Section

The File Section of a program which contains a sorting operation must furnish information concerning the physical structure, identification, and record-names of the records to be sorted. This is provided in the Sort-File-Description only.

Sort-File-Description Entry

The Sort-File-Description entry describes the records to be processed by the Sort Feature (fig. 11.7). The entry must appear in the File Section for every file named as the first operand of a SORT statement.

SD

A COBOL reserved word that must appear at the A margin. The level indicator SD identifies the beginning of the Sort-File-Description and must precede the file-name.

```
SD sort-file-name
   [RECORDING MODE IS mode]
   [RECORD CONTAINS [integer-1 TO]
   integer-2 CHARACTERS]
   DATA  {RECORD IS   } record-name . . .
         {RECORDS ARE }
```

Figure 11.7
Format—Sort-File-Description Entry.

Sort-File Name

This is a programmer-supplied name that is used in the Sort Feature. The SORT statement in the Procedure Division specifies this name. All rules for data-names apply, and it must not be qualified or subscripted.

Data Record and Record Contains Clauses

These clauses follow the same rules prescribed for File Section entries in the Data Division. (See "Data Record and Record Contains" entries in the Data Division, chap. 6.)

Record Description Entry

The format of the Record Description entry will vary according to the type of item being described. All rules for Record Description entries in the Data Division apply to this entry. (See "Record Description Entry" in the Data Division.)

Sort Keys

Sort-Keys are identified by data-names assigned to each field involved in the sorting operation.

Rules Governing the Use of Sort-Keys

1. The keys must be physically located in the same position and must have the same data format in every logical record of the sort-file.
2. Key items must not contain an OCCURS clause or be subordinate to entries that contain an OCCURS clause.
3. A maximum of twelve keys may be specified. The total length of all the keys must not exceed 256 bytes.
4. All keys must be at a fixed displacement from the beginning of a record; that is, they cannot be located after a variable table in a record.
5. All key fields must be located within the first 4,092 bytes of a logical record.
6. The data-names describing the keys may be qualified.

PROCEDURE DIVISION

The COBOL procedural statements that are available for use in a sort program are as follows:

- The SORT statement for performing a sorting operation on a collection of records.
- The RELEASE statement for transferring records to the initial phase of a sort operation.
- The RETURN statement for obtaining sorted records from the final phase of a sort operation.

The Procedure Division must contain a SORT statement to describe the sorting operation and, optionally, any necessary input and output procedures. The procedure-names constituting the input and output procedures are specified in the SORT statements.

The Procedure Division may contain more than one SORT statement appearing anywhere except in the Declaratives portion or in the input or output procedures associated with the SORT statement.

Sort Statement

The SORT statement provides the information necessary to execute the SORT feature (fig. 11.8).

```
SORT file-name-1 ON {DESCENDING} KEY {data-name-1}...
                    {ASCENDING }
     [ON {DESCENDING} KEY {data-name-2} ...]...
         {ASCENDING }
     {INPUT PROCEDURE IS section-name-1 [THRU section-name-2]}
     {USING file-name-2                                      }

     {OUTPUT PROCEDURE IS section-name-3 [THRU section-name-4]}
     {GIVING file-name-3                                      }
```

Figure 11.8
Format—Sort Statement.

Functions of the SORT Statement

1. Directs the sorting operation to obtain the necessary records to be sorted either from an INPUT PROCEDURE or the USING file.
2. The records are then sorted on a set of specified keys.
3. After the last sort is completed, the sorted records are made available to either an OUTPUT PROCEDURE or to the GIVING file.

(See figure 11.9.)

File-name-1

This is the name given in a Sort-File-Description entry in the Data Division that describes the records to be sorted.

```
SORT SALES-RECORDS ON ASCENDING KEY
  CUSTOMER-NUMBER, DESCENDING KEY DATE,

    USING FN-1,
    GIVING FN-2.
```

Figure 11.9
Example—Sort Statement.

Ascending or Descending

One of these clauses must be included (both may be included) to specify the sequence of records to be sorted. The sequence is applicable to all sort-keys immediately following the clause. More than one data-name may be specified for sorting after the ASCENDING or DESCENDING word. Every data-name used must have been described in a clause associated with the Sort-File-Description entry. Sort-keys are always listed from left to right in order of decreasing significance regardless of whether they are ASCENDING or DESCENDING. The sort-keys must be specified in the logical sequence in which the records are to be sorted.

When the ASCENDING clause is used, the sorted sequence is from the lowest value of the key to the highest value, according to the collating sequence for the COBOL character set.

When the DESCENDING clause is used, the sorting sequence is from the highest value of the key to the lowest, according to the collating sequence for the COBOL character set.

Data-Name

This is the data-name assigned to each sort-key; it is required in every statement.

Rules Governing the Use of Data-Names

1. More than one data-name may be specified after the ASCENDING or DESCENDING options.
2. The same data-name must not be used twice in the same SORT statement.
3. Every data-name must have been defined in a Record Description entry associated with the Sort Description entry.
4. The sort-keys are specified in the desired order of sorting.

Sort Program With Input and Output Procedures

For many sort applications, it is necessary to apply some processing to the content of a sort file. The special processing may consist of addition, deletion, creation, alteration, editing, or other modification of the individual records. The special processing may be necessary before, after, or both before and after the records are reordered by the sort. The COBOL sort allows the programmer to express these procedures in the COBOL language and to specify at which point, before or after the sort, they are to be executed. When the procedures are executed before the sort, they are called *input procedures*. When the procedures are executed after the sort, they are called *output procedures*.

A COBOL sort may contain any number of sorts; each may have its own independent special procedures. The SORT feature automatically executes these procedures at the specified point in such a way that extra passes over the sort file are not required.

Before the SORT operation is executed, the programmer specifies input procedures in which records are read and operated on. In the input procedures, the RELEASE statement creates the sort file. At the conclusion of the input procedures those records that have been output by the RELEASE statement comprise the sort file.

At the end of the input procedures, the SORT statement arranges the entire set of records in the sort file according to the keys specified in the SORT statement.

After the records have been sorted, they are available for output procedures. In these output procedures, the RETURN statement makes the next record available in sorted order from the sort file. At the end of the output procedures are the records that have been made available by the RETURN statement for further processing. This processing must include the writing of records into the output file.

Input Procedure

The presence of an INPUT PROCEDURE indicates that the programmer has written an input procedure to process the records before they are sorted. This procedure is included in the Procedure Division in one or more sections. This procedure passes one record at a time to the Sort Feature after it has completed its processing.

Section-name-1 is the name of the first or only section in the main program that contains the input procedures. This section is required if the INPUT PROCEDURE is used.

Section-name-2 is the name of the last section and is required if the procedure terminates in a section other than that in which it was started.

The INPUT PROCEDURE consists of one or more sections that are written into a source program.

Rules Governing the Use of Input Procedures

1. The INPUT PROCEDURE can include any statements needed to select, create, or modify records.
2. Control must not be passed on to an INPUT PROCEDURE unless the related SORT statement is executed, because the RELEASE statement in the INPUT PROCEDURE has no meaning unless it is controlled by a SORT statement.
3. The INPUT PROCEDURE must not include any SORT statements.
4. Any files used as sort-work files may not be opened or referred to by an INPUT PROCEDURE.
5. The INPUT PROCEDURE must build the records to be sorted one at a time. The record must have been described and assigned a data-name in the record description entry associated with the Sort Description entry.
6. The INPUT PROCEDURE must make the record available to the sorting operation after it has been processed. A RELEASE statement is used for this purpose.
7. After all the records have been released to the sorting operation, the INPUT PROCEDURE must transfer control to the last statement in the INPUT PROCEDURE to terminate the procedure.
8. The INPUT PROCEDURE must not contain any transfers of control to points outside the INPUT PROCEDURE.
9. The remainder of the Procedure Division must not contain any transfer of control to points inside the INPUT PROCEDURES (with the exception of the return of control from a Declarative section).

If an INPUT PROCEDURE is specified, control is passed to the INPUT PROCEDURE when the SORT program input phase is ready to receive the first record. The compiler inserts a return mechanism at the end of the last section of the INPUT PROCEDURE, and when control passes the last statement in the INPUT PROCEDURE, the records that have been released to file-name-1 are sorted. The RELEASE statement transfers records from the INPUT PROCEDURE to the input phase of the sort operation. (See "Release" statement.)

Using

The USING option indicates that the records to be sorted are all in one file and are to be passed to the sorting operation as one unit when the SORT statement is executed. If the option is used, all records to be sorted must be in the same files.

If this option is used, all the records in file-name-2 are transferred automatically to file-name-1. At the time of the execution of the SORT statement, file-name-2 must not be open. File-name-2 must be a standard sequential file. For the USING option, the compiler will automatically OPEN, READ, RELEASE, and CLOSE file-name-2 without the programmer specifying these functions.

Output Procedure

This procedure indicates that the programmer has written an OUTPUT PROCEDURE to process the sorted records. This procedure is included in the Procedure Division in the form of one or more sections. The procedure returns records one at a time from the Sort Feature after they have been sorted.

Section-name-3 is the name of the first or only section in the main program that contains the OUTPUT PROCEDURE. This section is required if the output procedures are used.

Section-name-4 is the name of the last section that contains output procedures and is required if the procedure terminates in a section other than the one in which it started.

The OUTPUT PROCEDURE consists of one or more sections that are written into a source program.

Rules Governing the Use of Output Procedures

1. The OUTPUT PROCEDURE can include any statement necessary to select, modify, or copy the sorted records being returned one at a time in sorted order from the sort-file.
2. Control must not be passed to the OUTPUT PROCEDURE, except when a related SORT statement is being executed, because RETURN statements in the OUTPUT PROCEDURE have no meaning unless they are controlled by the SORT statement.
3. The OUTPUT PROCEDURE must not contain any SORT statements.
4. Any files used as sort-files may not be opened or referred to by the OUTPUT PROCEDURE.
5. Records must be obtained one at a time from Sort/Merge program over the RETURN statement. Once a record is returned, the previously returned record is no longer available.
6. The OUTPUT PROCEDURE must manipulate the returned sorted record by referring to the data record that has been described and assigned a data-name in the record description entry. If the records are to be written on an output file, the programmer must provide the appropriate OPEN statement prior to the execution of the SORT statement or the OUTPUT PROCEDURE itself.
7. The OUTPUT PROCEDURE must not contain any transfer of control outside the OUTPUT PROCEDURE.
8. The remainder of the Procedure Division must not contain any transfers of control to points inside the OUTPUT PROCEDURE (with the exception of the return of control from a Declaratives section).

9. After all the records have been returned by the Sort Feature, and the OUTPUT PROCEDURE attempts to execute another RETURN statement, the AT END clause of the RETURN statement will be executed. The AT END clause should direct control of the program to the last statement of the OUTPUT PROCEDURE to terminate the OUTPUT PROCEDURE.

If an OUTPUT PROCEDURE is specified, control passes to it after file-name-1 has been placed in sequence by the SORT statement. The compiler inserts a return mechanism at the end of the last section of the OUTPUT PROCEDURE. When control passes the last statement in the OUTPUT PROCEDURE, the return mechanism provides for the termination of the sort and then passes control to the next statement after the SORT statement.

When all records are sorted, control is passed to the OUTPUT PROCEDURE. The RETURN statement in the OUTPUT PROCEDURE is a request for the next record. (See "Return" statement.)

Giving

The GIVING option indicates where the records are to be placed after sorting. If this option is used, all sorted records will be placed in one file, and if this option is used, all sorted records in file-name-1 are automatically transferred to file-name-3. At the time of execution of the SORT statement, file-name-3 must not be open. File-name-3 must name a standard sequential file. For the GIVING option, the compiler will OPEN, RETURN, WRITE, and CLOSE file-name-3 without the programmer specifying these functions.

(*Note:* The SORT statements and INPUT and OUTPUT PROCEDURES are permitted anywhere in the Procedure Division except in the Declaratives Section.)

Release Statement

The RELEASE statement transfers records from the INPUT PROCEDURE to the input phase of the sort operation (fig. 11.10).

Figure 11.10
Format—Release Statement.

```
RELEASE sort-record-name [FROM identifier]
```

Rules Governing the Use of the Release Statement

1. A RELEASE statement may only be used within the range of an INPUT PROCEDURE.
2. If the INPUT PROCEDURE option is specified, the RELEASE statement must be included within the given set of procedures.
3. Sort-record-name must be the name of a logical record in the associated sort-file-description entry and may be qualified.
4. If FROM option is used, the contents of the identifier data area are moved to the record-name, then the contents of sort-record-name are released to the sort-file. Moving takes place according to the rules of the MOVE statement with the CORRESPONDING option. The information in the record area

is no longer available, but the information in the data area associated with identifier is available.

5. After the RELEASE statement is executed, the logical record is no longer available. When control passes from the INPUT PROCEDURE, the file consists of all those records that were placed in it by the execution of the RELEASE statement.
6. Sort-record-name and identifier must not refer to the same storage area. (See figure 11.11.)

```
RELEASE INPUT-RECORD.
RELEASE RECORD-ONE.
```

Figure 11.11 Example—Release Statement.

Return Statement

The RETURN statement obtains individual records in sorted order from the final phase of the sort program (fig. 11.12).

```
RETURN sort-file-name RECORD [INTO identifier]
    AT END imperative-statement
```

Figure 11.12
Format—Return Statement.

Rules Governing the Use of the Return Statement

1. Sort-file-name must be described by a Sort File Description entry in the Data Division.
2. A RETURN statement may be used only within the range of an OUTPUT PROCEDURE associated with a SORT statement for a file-name.
3. The INTO option may be used only when the input file contains just one type of record. The storage area associated with identifier and the storage area which is the record area associated with file-name must not be the same storage area.
4. The identifier must be the names of a Working-Storage area or an output record area. Use of the INTO option has the same affect as the MOVE statement for alphanumeric items.
5. The imperative statement in the AT END phrase specifies the action to be taken when all the sorted records have been obtained for the sorting operation.
6. When a file consists of more than one type of logical record, these records automatically share the storage area. This is equivalent to saying that there exists an implicit redefinition of the area, and only the information that is present in the current record is available.
7. After the execution of the imperative statement in the AT END phrase, no RETURN statement may be executed within the OUTPUT PROCEDURE. (See figure 11.13.)

```
RETURN FILE-ONE AT END GO TO END-PROGRAM.
RETURN FILE-SORT AT END GO TO LAST.
```

Figure 11.13
Example—Return Statement.

CONTROL OF INPUT AND OUTPUT PROCEDURES

The INPUT and OUTPUT PROCEDURES function in a manner similar to option 1 of the PERFORM statement; for example, naming a section in an INPUT PROCEDURE clause causes execution of that section during the sorting operation to proceed as though that section had been the subject of a PERFORM statement. As with the PERFORM statement, the execution of the section is terminated after execution of its last statement.

Return is back to the next statement after the INPUT or OUTPUT PROCEDURES have been executed. The EXIT verb may be used as a common exit point for conditional exits from INPUT or OUTPUT PROCEDURES. If the EXIT verb is used, it must appear as the last paragraph of the INPUT or OUTPUT PROCEDURES (see "Exit" statement in Procedure Division chap. 8). (See figures 11.14, 11.15, 11.16, 11.17.)

```
IDENTIFICATION DIVISION.
PROGRAM-ID.  BASICSRT.

ENVIRONMENT DIVISION.
CONFIGURATION SECTION.
SOURCE-COMPUTER.  NCR-CENTURY-200.
OBJECT-COMPUTER.  NCR-CENTURY-200   MEMORY SIZE 64000 WORDS.
INPUT-OUTPUT SECTION.
FILE-CONTROL.
    SELECT RAWTFILE       ASSIGN TO NCR-TYPE-32.
    SELECT SWFILE         ASSIGN TO NCR-TYPE-32.
    SELECT TRANFILE       ASSIGN TO NCR-TYPE-32.

DATA DIVISION.
FILE SECTION.
FD  RAWTFILE, LABEL RECORD IS STANDARD.
01  RAWTRECORD          PICTURE X(13).
SD  SWFILE, RECORD CONTAINS 13 CHARACTERS
01  SWRECORD.
    02    TNO           PICTURE X(6).
    02    FILLER        PICTURE X(6).
    02    TCODE         PICTURE X.

FD  TRANFILE, LABEL RECORD IS STANDARD.
01  TRANRECORD          PICTURE X(13).

PROCEDURE DIVISION.
BEGIN.
    SORT SWFILE ON ASCENDING KEY TNO, TCODE
        USING RAWFILE,
        GIVING TRANFILE.
    STOP RUN.
```

Figure 11.14 Example—Sort Program—Using and Giving Options.

```
IDENTIFICATION DIVISION.
PROGRAM-ID.    CONTEST.

ENVIRONMENT DIVISION.
CONFIGURATION SECTION.
SOURCE-COMPUTER.        IBM-370-I155.
OBJECT-COMPUTER.        IBM-370-I155.
SPECIAL-NAMES.          SYSOUT IS PRINTER.
INPUT-OUTPUT SECTION.
FILE-CONTROL.
    SELECT NET-FILE-IN        ASSIGN TO UT-2400-S-INFILE.
    SELECT NET-FILE-OUT       ASSIGN TO UT-2400-S-SORTOUT.
    SELECT NET-FILE           ASSIGN TO UT-2400-S-NETFILE.

DATA DIVISION.
FILE SECTION.
SD  NET-FILE            DATA RECORD IS SALES-RECORD.
01  SALES-RECORD.
    05    EMPL-NO       PICTURE 9(6).
    05    DEPT          PICTURE 9(2).
    05    NET-SALES     PICTURE 9(7)V99.
    05    NAME-ADDR     PICTURE X(55).
FD  NET-FILE-IN         LABEL RECORDS ARE OMITTED
                        DATA RECORD IS NET-CARD-IN.
01  NET-CARD-IN.
    05    EMPL-NO-IN        PICTURE 9(6).
    05    DEPT-IN           PICTURE 9(2).
    05    NET-SALES-IN      PICTURE 9(7)V99.
    05    NAME-ADDR-IN      PICTURE X(55).
FD  NET-FILE-OUT        LABEL RECORDS ARE OMITTED
                        DATA RECORD IS NET-CARD-OUT.
01  NET-CARD-OUT.
    05    EMPL-NO-OUT       PICTURE 9(6).
    05    DEPT-OUT          PICTURE 9(2).
    05    NET-SALES-OUT     PICTURE 9(7)V99.
    05    NAME-ADDR-OUT     PICTURE X(55).
WORKING-STORAGE SECTION.
01  FLAGS.
    05    MORE-INPUT-FLAG          PICTURE XXX    VALUE 'YES'.
          88    MORE-INPUT                        VALUE 'YES'.
          88    NO-MORE-INPUT                     VALUE 'NO'.
    05    MORE-TRANS-FLAG          PICTURE XXX    VALUE 'YES'.
          88    MORE-TRANS                        VALUE 'YES'.
          88    NO-MORE-TRANS                     VALUE 'NO'.
PROCEDURE DIVISION.
ELIM-DEPT-7-9-NO-PRINTOUT.
    SORT NET-FILE
        ASCENDING KEY DEPT, DESCENDING KEY NET-SALES
        INPUT PROCEDURE SCREEN-DEPT
        GIVING NET-FILE-OUT.

SCREEN-DEPT SECTION.
INPUT-ROUTINE.
    OPEN    INPUT    NET-FILE-IN.
```

Figure 11.15 Example—Sort Program—Input Procedure and Giving Options.

Figure 11.15 Continued.

```
     READ NET-FILE-IN
          AT END MOVE 'NO' TO MORE-INPUT-FLAG.
     PERFORM PROCESS-INPUT-ROUTINE
          UNTIL NO-MORE-INPUT.
     CLOSE     NET-FILE-IN.
     GO TO SCREEN-DEPT-EXIT.

 PROCESS-INPUT-ROUTINE.
     DISPLAY EMPL-NO-IN, DEPT-IN, NET-SALES-IN,
          NAME-ADDR-IN UPON PRINTER.
     IF DEPT-IN IS EQUAL TO 7 OR 9
     THEN
          NEXT SENTENCE
     ELSE
          MOVE NET-CARD-IN TO SALES-RECORD.
          RELEASE SALES-RECORD.
 SCREEN-DEPT-EXIT.     EXIT.

CHECK-RESULTS SECTION.
CHECK-ROUTINE.
     OPEN     INPUT     NET-FILE-OUT.
     READ NET-FILE-OUT
          AT END MOVE 'NO' TO MORE-TRANS-FLAG.
     PERFORM PROCESS-CHECK-ROUTINE
          UNTIL NO-MORE-TRANS.
     CLOSE     NET-FILE-OUT.
     STOP RUN.

PROCESS-CHECK-ROUTINE.
     DISPLAY EMPL-NO-OUT, NET-SALES-OUT, NAME-ADDR-OUT
          UPON PRINTER.
     READ NET-FILE-OUT
          AT END MOVE 'NO' TO MORE-TRANS-FLAG.
```

The above program illustrates a sort based on a sales contest. The records to be sorted contain data on salesman; name and address, employee number, department number, and precalculated net sales for the contest period.

The salesman with the highest net sales in each department wins a prize, and smaller prizes are awarded for second highest sales, third highest, etc. The order of the SORT is (1) by department, the lowest numbered first (ASCENDING KEY DEPT); and (2) by net sales within each department, the highest net sales first (DESCENDING KEY NET-SALES).

The records for the employees of departments 7 and 9 are eliminated in an input procedure (SCREEN-DEPT) before sorting begins. The remaining records are then sorted, and the output is placed on another file for use in a later job step.

```
IDENTIFICATION DIVISION.
PROGRAM-ID.     SORTMFILE.

ENVIRONMENT DIVISION.
CONFIGURATION SECTION.
SOURCE-COMPUTER.   NCR-CENTURY-200.
OBJECT-COMPUTER.   NCR-CENTURY-200   MEMORY SIZE 64000 WORDS.
SPECIAL-NAMES.     SYSOUT IS PRINTER.
INPUT-OUTPUT SECTION.
FILE-CONTROL.
    SELECT RAWMFILE        ASSIGN TO NCR-TYPE-41.
    SELECT SFILE           ASSIGN TO NCR-TYPE-41-02-03-04.
    SELECT MFILE           ASSIGN TO NCR-TYPE-41.

DATA DIVISION.
FILE SECTION.
FD     RAWMFILE      LABEL RECORD IS STANDARD.
01     RAWMRECORD.
       02     FILLER             PICTURE X(26).
       02     ACCOUNT-STATUS     PICTURE X.
       02     FILLER             PICTURE X(11).

SD     SFILE         RECORD CONTAINS 38 CHARACTERS.
01     SRECORD.
       02     MNO                PICTURE X(6).
       02     FILLER             PICTURE X(32).

FD     MFILE         LABEL RECORD IS STANDARD.
01     MRECORD.
       02     MNOO               PICTURE X(6).
       02     BALANCE            PICTURE 9(5)V99.
       02     FILLER             PICTURE X(25).
WORKING-STORAGE SECTION.
01     FLAGS.
       02     MORE-INPUT-FLAG          PICTURE XXX      VALUE 'YES'.
              88     MORE-INPUT                         VALUE 'YES'.
              88     NO-MORE-INPUT                      VALUE 'NO'.
       02     MORE-OUTPUT-FLAG         PICTURE XXX      VALUE 'YES'.
              88     MORE-OUTPUT                        VALUE 'YES'.
              88     NO-MORE-OUTPUT                     VALUE 'NO'.

PROCEDURE DIVISION.
BEGIN.
       SORT SFILE
            ASCENDING KEY MNO,
            INPUT PROCEDURE CHECK-STATUS
            OUTPUT PROCEDURE CHECK-BALANCE.
       STOP RUN.

CHECK-STATUS SECTION.
INPUT-ROUTINE.
       OPEN     INPUT     RAWMFILE.
       PERFORM PROCESS-INPUT-ROUTINE
            UNTIL NO-MORE-INPUT.
       CLOSE     RAWMFILE.
       GO TO CHECK-STATUS-EXIT.

PROCESS-INPUT-ROUTINE.
       IF ACCOUNT-STATUS IS EQUAL TO ZERO,
       THEN
            NEXT SENTENCE
       ELSE
            RELEASE SRECORD FROM RAWMRECORD.
CHECK-STATUS-EXIT.     EXIT.

CHECK-BALANCE SECTION.
OUTPUT-ROUTINE.
       OPEN     OUTPUT     MFILE.
       PERFORM PROCESS-OUTPUT-ROUTINE
            UNTIL NO-MORE-OUTPUT.
       CLOSE     MFILE.
       GO TO CHECK-BALANCE-EXIT.

PROCESS-OUTPUT-ROUTINE.
       RETURN SFILE INTO MRECORD
            AT END MOVE 'NO' TO MORE-OUTPUT-FLAG.
       IF BALANCE IS EQUAL TO ZERO,
       THEN
            DISPLAY 'ACCOUNT NUMBER' MNOO 'HAS A ZERO
                 BALANCE'
                 UPON PRINTER
       ELSE
            WRITE MRECORD.
CHECK-BALANCE-EXIT.     EXIT.
```

Figure 11.16 Example—Sort Program—Input Procedure and Output Procedure Options.

```
IDENTIFICATION DIVISION.
PROGRAM-ID.     SORT2.

ENVIRONMENT DIVISION.
CONFIGURATION SECTION.
SOURCE-COMPUTER.          IBM-370-I155.
OBJECT-COMPUTER.          IBM-370I155.
INPUT-OUTPUT SECTION.
FILE-CONTROL.
    SELECT INFILE          ASSIGN TO UR-2540R-S-CARDIN.
    SELECT OUTFILE         ASSIGN TO UT-2400-S-WORK.
    SELECT SORTFILE        ASSIGN TO DA-2314-S-DSKFILE.

DATA DIVISION.
FILE SECTION.
FD  INFILE
    RECORDING MODE F
    LABEL RECORDS OMITTED
    DATA RECORD IS INREC.
01  INREC.
    02  FILLER             PICTURE X(80).

FD  OUTFILE
    RECORDING MODE F
    LABEL RECORDS STANDARD
    DATA RECORD IS OUTREC.
01  OUTREC.
    02  FILLER             PICTURE X(80).

SD  SORTFILE
    RECORDING MODE F
    DATA RECORD IS SRT-REC.
01  SRT-REC.
    02  KEY1               PICTURE X(5).
    02  KEY2               PICTURE X(12).
    02  KEY3               PICTURE X(15).
    02  FILLER             PICTURE X(20).
    02  KEY4               PICTURE X(28).

WORKING-STORAGE SECTION.
01  FLAGS.
    02  MORE-INPUT-FLAG        PICTURE XXX      VALUE 'YES'.
        88  MORE-INPUT                          VALUE 'YES'.
        88  NO-MORE-INPUT                       VALUE 'NO'.
    02  MORE-OUTPUT-FLAG       PICTURE XXX      VALUE 'YES'.
        88  MORE-OUTPUT                         VALUE 'YES'.
        88  NO-MORE-OUTPUT                      VALUE 'NO'.

PROCEDURE DIVISION.
BEGIN.
    DISPLAY ' START-EXAMPLE ' UPON CONSOLE.
    SORT SORTFILE.
        DESCENDING KEY KEY1, KEY2,
        ASCENDING KEY KEY3,
        DESCENDING KEY KEY4
        INPUT PROCEDURE INP-PROC
        OUTPUT PROCEDURE OUT-PROC.
    STOP RUN.

INP-PROC SECTION.
INPUT-ROUTINE.
    OPEN    INPUT    INFILE.
    PERFORM PROCESS-INPUT-ROUTINE
        UNTIL NO-MORE-INPUT.
    CLOSE INFILE.
    DISPLAY ' END OF INPUT ' UPON CONSOLE.
    GO TO INP-PROC-EXIT.

PROCESS-INPUT-ROUTINE.
    READ INFILE
        AT END MOVE 'NO' TO MORE-INPUT-FLAG.
    MOVE INREC TO SRT-REC.
    RELEASE SRT-REC.
INP-PROC-EXIT.    EXIT.

OUT-PROC SECTION.
OUTPUT-ROUTINE.
    OPEN    OUTPUT    OUTFILE.
    PERFORM PROCESS-OUTPUT-ROUTINE
        UNTIL NO-MORE-OUTPUT.
    CLOSE    OUTFILE.
    DISPLAY ' END OF OUTPUT '.
    GO TO OUT-PROC-EXIT.

PROCESS-OUTPUT-ROUTINE.
    RETURN SORTFILE
        AT END MOVE 'NO' TO MORE-OUTPUT-FLAG.
    MOVE SRT-REC TO OUTREC.
    WRITE OUTREC.
OUT-PROC-EXIT.    EXIT.
```

Figure 11.17 Example—Sort Program—Input Procedure and Output Procedure Options.

Exercises

Write your answer in the space provided (answer may be one or more words).

1. The Sort statement arranges a collection of ______________ into a ______________ determined by the ______________.
2. Each Sort statement has its own internal ______________, called a ______________, ______________ and ______________ during the sorting operation.
3. The Sort feature provides the COBOL programmer a convenient access to the ______________ of the ______________ program.
4. A file of records are sorted according to a set of user-supplied ______________ within each record.
5. A programmer may ______________, ______________, ______________ or otherwise alter the records during the ______________ or ______________ phases of the sort.
6. The basic element of the Sort feature are the ______________ statement in the Procedure Division and the ______________ entry in the Data Division.
7. In the Environment Division, the programmer must name with ______________ sentences for all files to be used as input to or output from Sort-file.

8. In the Data Division, ________________ entries must be written for all files that provide input to the Sort procedure.
9. The programmer must specify the ________________ being sorted, the ________________ names to be sorted, whether the sort is to be in ________________ or ________________ sequence or mixture of both and whether the records are to be processed ________________ and/or ________________ the sort.
10. A sorting operation is based on the ________________ named in the ________________ statement.
11. A ________________ is a collection of records that is involved in the sorting operation as it exists in intermediate devices.
12. Records are made available to the Sort/Merge program by the ________________ or ________________ option of the Sort statement.
13. Sorted records returned from the Sort/Merge program may be written with the ________________ or ________________ options of the Sort statement.
14. Select entries in the Environment Division must be written for files used with ________________ and ________________ procedures and files specified in the ________________ or ________________ options of the Sort statement in the Procedure Division.
15. When two or more files are to use the same area for processing of the current logical record, the ________________ clause in the I-O-Control paragraph is used.
16. All sort keys must be physically located in the ________________ position and have the ________________ data formats in every logical record.
17. A maximum of ________________ sort keys may be specified.
18. The Sort statement provides the necessary information to ________________ the Sort feature.
19. Either ________________ or ________________ clause or both must be included to specify the sequence of records to be sorted.
20. A ________________ must be assigned to each sort key.
21. The sort keys are specified in the ________________ order of sorting.
22. The Input Procedure indicates a procedure to process records ________________ they are sorted.
23. The Input Procedure passes ________________ record at a time to the Sort feature.
24. The Input Procedure can include any statements to ________________, ________________, or ________________ records.
25. An Input Procedure must not include any ________________ statements.
26. The Using option indicates that all records are in ________________ file.
27. For the Using option, the compiler will automatically ________________, ________________, ________________, and ________________ the file without the programmer having to specify these operations.
28. The Output Procedure is used to process the ________________ records.
29. The ________________ option indicates where records are to be placed after sorting.
30. The Sort statements and Input and Output Procedures are permitted anywhere in the Procedure Division except in the ________________.
31. The Release statement transfers records from the ________________ to the ________________ phase of the sort operation.
32. The Release statement must be included and may only be used within the range of ________________.
33. The Return statement causes individual records to be obtained from the sorting operation after they have been ________________ and indicate what ________________ are to be taken with each sorted record.
34. A Return statement must be included and may only be used within the range of ________________.
35. When a file contains more than one type of logical record, these records automatically share the ________________ storage area.
36. The Input and Output Procedure functions in the same manner as the ________________ statement.
37. An ________________ verb may be used as a common exit point for the Input or Output Procedures.

Answers

1. RECORDS, SEQUENCE, PROGRAMMER
2. WORKING FILE, SORT-FILE, CREATED, USED
3. SORTING CAPACITY, SORT/MERGE
4. KEYS
5. SUMMARIZE, DELETE, SHORTEN, INITIAL, FINAL
6. SORT, SORT-FILE-DESCRIPTION
7. SELECT
8. FILE DESCRIPTION
9. RECORD, SORT KEY, ASCENDING, DESCENDING, BEFORE, AFTER
10. SORT KEYS, SORT
11. SORT WORK FILE
12. USING, INPUT PROCEDURE
13. GIVING, OUTPUT PROCEDURE
14. INPUT, OUTPUT, USING, GIVING
15. SAME RECORD
16. SAME, SAME
17. 12
18. EXECUTE
19. ASCENDING, DESCENDING
20. DATA-NAME
21. DESIRED
22. BEFORE
23. ONE
24. SELECT, CREATE, MODIFY
25. SORT
26. ONE
27. OPEN, READ, RELEASE, CLOSE
28. SORTED
29. GIVING
30. DECLARATIVES SECTION
31. INPUT PROCEDURE, INPUT
32. INPUT PROCEDURE
33. SORTED, ACTIONS
34. OUTPUT PROCEDURE
35. SAME
36. PERFORM
37. EXIT

Questions for Review

1. Briefly describe the operation of the Sort statement.
2. What is the purpose of the Sort feature?
3. What are the basic elements of the Sort feature?
4. What is the purpose of the Sort Description entry and how it differs from a File Description entry?
5. What are sort keys and how are they used in the Sort feature?
6. What is a Sort statement and what are its main functions?
7. When is the Using option used?
8. What is an Input Procedure and how does it operate with the Sort feature?
9. What is the function of the Release statement?
10. What is an Output Procedure and how does it operate with the Sort feature?
11. What is the function of the Return statement?
12. In what manner are Input and Output Procedures and Perform statements alike?

Problems

1. *Match the item with its proper description.*

____ 1. Using
____ 2. Sort Keys
____ 3. Output Procedure
____ 4. Return Statement
____ 5. Select Entry
____ 6. Input Procedure
____ 7. Giving
____ 8. Release Statement
____ 9. Sort Description Entry
____ 10. Sort Statement

A. Process records before sorting.
B. Execute the Sort feature.
C. Specify files in Sort statement.
D. Causes records to be transferred to sort.
E. Describes the records to be sorted.
F. Where records are to be placed after sort.
G. Causes records to be obtained from sort.
H. Process sorted records.
I. Data-names assigned to fields in sort.
J. Records in one file.

2. *Given the following information:*

TAPE FORMAT:

IRG	Dept. No.	Clock No.	Name	SS#	YTD gross	YTD W/T	YTD fica	IRG
	2 pos.	3 pos.	20 pos.	9 pos.	7 pos.	6 pos.	5 pos.	

52 character logical record, fixed length
10 logical records/block

Write a program to

a. Input the tape record.
b. Sort the record on Social Security Number sequence.
c. Output the tape record in the same format as the input record.

3. *Given the following information:*

CARD FORMAT

Dept. No.	Clock No.		Name	Soc. Sec. No.		Current gross	Current W/T	Current fica	code
1-2	3-5	6	7 26	27-35		62-68	69-74	75-79	80

TAPE FORMAT

IRG	Dept. No.	Clock No.	Name	Soc. Sec. No.	Gross	W/T	Fica	IRG
	2 pos.	3 pos.	20 pos.	9 pos.	7 pos.	6 pos.	5 pos.	

52 characters per logical record, fixed length
5 records per block

Write the program to accomplish the following:

a. Input the card record.
b. Sort record to Department Number and Clock Number sequence.
c. Write card record on tape output.

4. *Given the following information:*

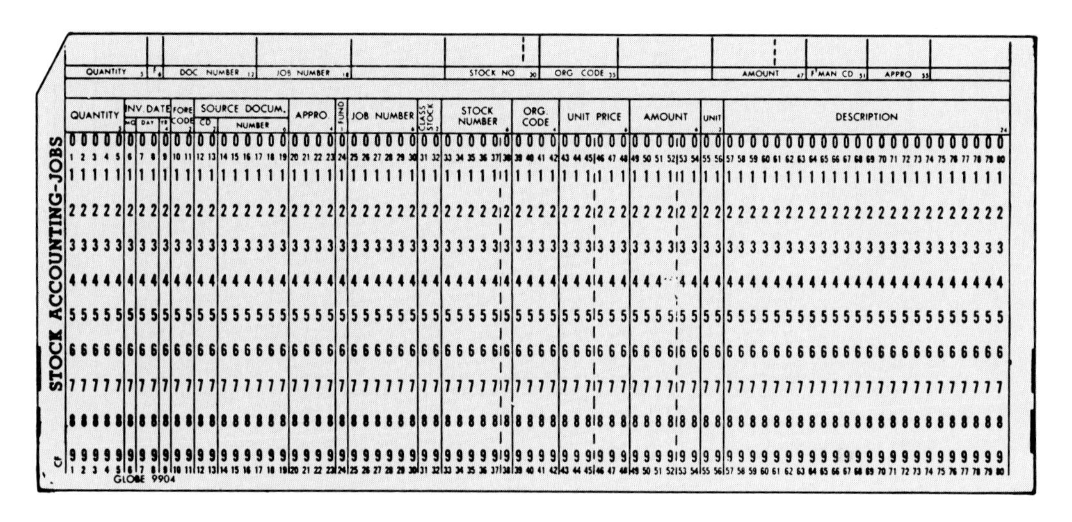

Write the necessary program to accomplish the following:

a. Read in the input records.
b. Multiply quantity by unit price giving amount.
c. Sorting the records into stock number sequence.
d. Out put a report on the printer as follows:

Record Positions	Field	Print Positions	
31—32	Class Stock	3—4	
33—38	Stock Number	6—11	
55—56	Unit	13—14	
57—80	Description	16—39	
1—5	Quantity	41—46	XX,XXX
43—48	Unit Price	48—55	$XXX.XXX
49—54	Amount	58—66	$X,XXX.XX

Final total of all amounts.

5. *Using the following information, write a program to accomplish the following:*

a. Read the input record.
b. Calculate the net pay by multiplying hours by the rate.
c. Sort records into Department and Serial Number sequence.
d. Punch card using the same format as the input record: including net pay.
e. Write report per format using the Report Writer Feature.

INPUT

DEPT | SERIAL NUMBER | NAME | HOURS | NET PAY

REPORT

DEPT.	SERIAL NUMBER	N A M E		HOURS WORKED	NET PAY
1	1234560	SMITH	JW	40.1	285.15
1	1892750	JONES	RA	39.6	152.16
1	8929016	MAUS	JB	62.5	182.55
					619.86
2	0238648	GOLDMAN	H	31.7	100.25
2	0333367	WOLFE	DJ	9.5	26.60
					126.85

CHART

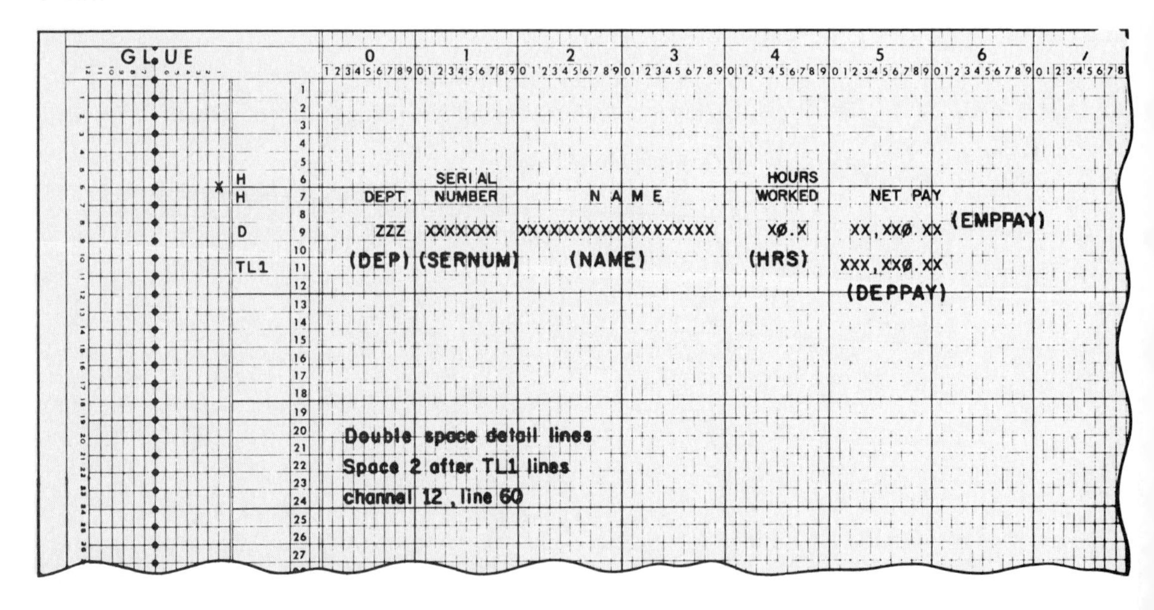

12 Direct Access (Mass Storage Devices)

INTRODUCTION

"Inline" processing denotes the ability of the data processing system to process data as soon as it becomes available. This implies that the input data does not have to be sorted in any manner, manipulated, or edited before it can be entered into a system, whether the input consists of transactions of a single application or of many applications.

Direct-access mass storage devices have made inline processing feasible for many applications (fig. 12.1). While sorting transactions are still advantageous before certain processing runs, in most instances the necessity for presorting has been eliminated. The ability to process data inline provides solutions to problems that heretofore were thought impractical.

Direct-access mass storage enables the user to maintain current records of diversified applications and to process nonsequential and intermixed data for multiple application area. The term *direct-access* implies access at random by multiple users of data (files, programs, subroutines, programming aids) involving mass storage devices. These storage devices differ in physical appearance, capacity, and speed, but functionally they are similar in terms of data recording, checking, and programming. The direct-access devices used for mass memory storage are disk storage, drum, and data cells. The disk storage is the most popular of the mass storage devices in use today.

TERMINOLOGY

Direct access terminology and concepts are a prerequisite to an understanding of programming using direct access devices (fig. 12.2).

Direct-Access Storage Device (DASD)

A direct-access storage device (DASD) is one in which each physical record has a discrete location and a unique address. Thus records can be stored on a DASD in such a way that the location of any one record can be determined without extensive searching. Records can be accessed directly as well as serially.

File

The term file *can mean a physical unit (a DASD, for instance), or an organized collection of related information.* In this text, the latter definition applies. An inventory, for example, contains all of the data concerning a particular inventory. It may occupy several physical units, or part of one physical unit. The term *data set* is often used instead of *file* to describe an organized collection of related information.

Figure 19-1. IBM 2311 Disk Storage Drive.

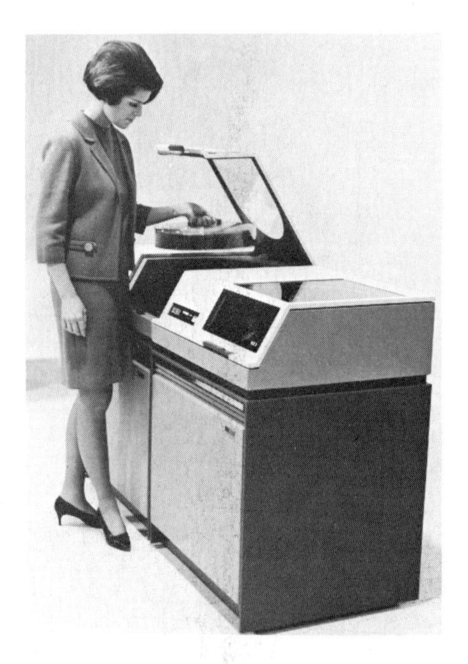

Figure 19-2. Removable Disk.

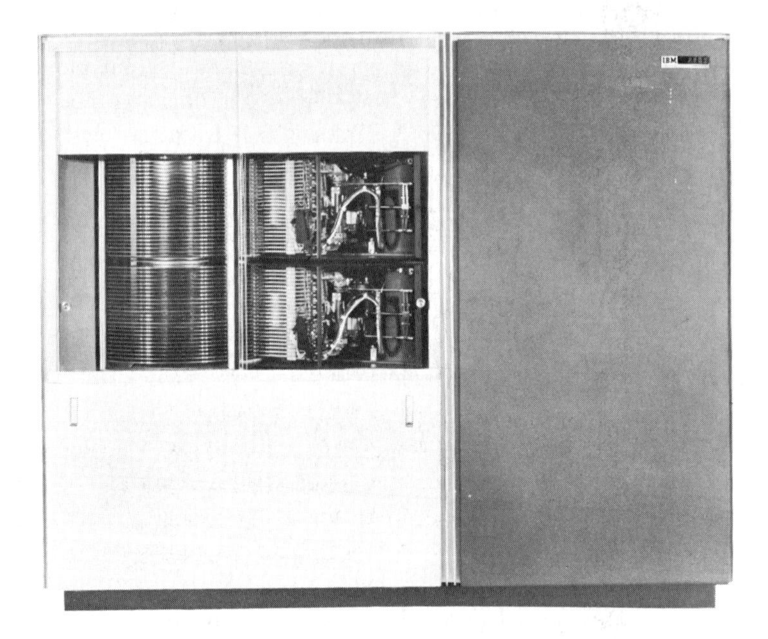

Figure 19-3. IBM 2302 Disk Storage—Non-Removable Disks.

Figure 12.1 Disk Storage Devices.

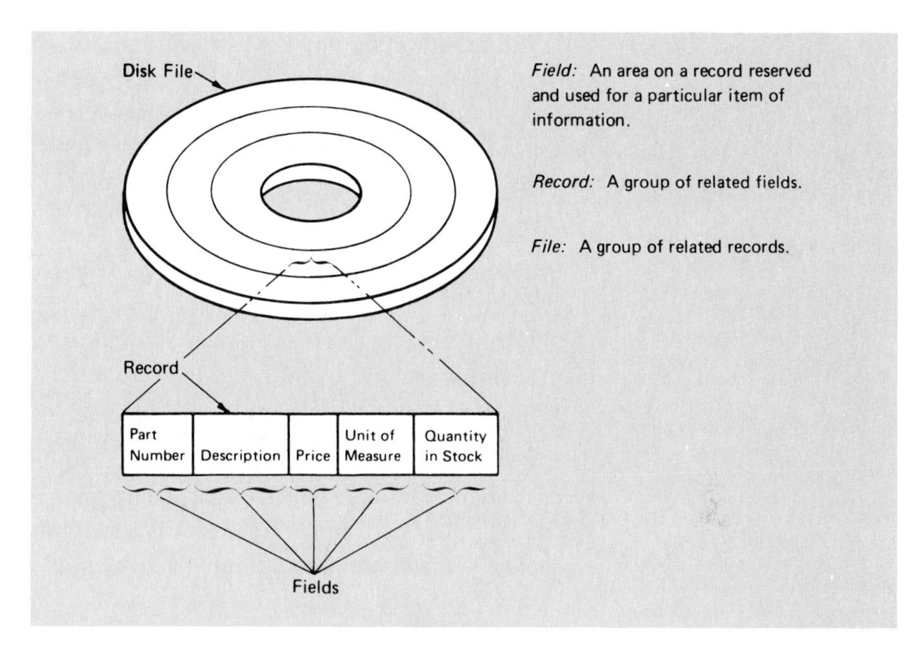
Disk File
Field: An area on a record reserved and used for a particular item of information.
Record: A group of related fields.
File: A group of related records.
Record
Part Number
Description
Price
Unit of Measure
Quantity in Stock
Fields

Figure 12.2 File, Record, Field Example.

Record The term *record* can also mean a physical unit or a logical unit. *A logical record may be defined as a collection of data related to a common identifier.* An inventory file, for example, would contain a record (logical record) for each part number in the inventory. A physical record consists of one or more logical records. The term *block* is equivalent to the term *physical record*. On a DASD, certain "nondata" information required by the control unit of the device is recorded in the same record area as the physical record. This nondata information and the physical record may be referred to as a whole with the term *data record*.

Key Each logical record contains a control field or key that uniquely identifies it. The key of the inventory record, for example, would probably be the part number.

Data sets may be organized in several ways. Some of the more popular data organizations are:

Sequential Records are placed in physical rather than logical sequence. Given one record, the location of the next record is determined by its physical position in the data set. Sequential organization is used for all magnetic tape devices, and may be selected for direct-access devices. Punched tape, punched cards, and printed output are sequentially organized.

Indexed Sequential Records are arranged in sequence according to a key that is part of every record, on the tracks of a direct-access volume. An index or set of indexes maintained by the system gives the location of certain principal records. This permits direct as well as sequential access to a record.

Direct The record within the data set, which must be on a direct-access volume, may be organized in any manner that the programmer chooses. All space attached to the data set is available for data records. No space is required for indexes. The programmer specifies addresses by which records are stored and retrieved directly.

Partitioned Independent groups of sequentially organized records, called *members,* are on direct-access storage. Each member has a simple name stored in a directory that is part of the data set and contains the location of the member's starting point. Partitioned data sets are generally used to store programs, such as compilers, subroutines, etc. As a result, they are often referred to as libraries.

Volume *Volume* is a generic term used to refer to a standard unit of auxiliary storage. A volume may be a reel of magnetic tape, a disk pack, or a drum. Direct-access volumes are used to store executable programs, including the operating system itself. Direct-access storage is also used for data and for temporary working storage. One direct-access volume may be used for many different data sets, and space on it may be relocated and reused.

PROCESSING TECHNIQUES

Sequential Processing

In sequential processing, input transactions are grouped together and sorted into a predetermined sequence and processed against a master file. When information is recorded in magnetic tape or punched cards, the most efficient method of processing is sequential. Direct-access storage devices are also efficient sequential processors, especially when the percentage of activity against the master file is high. (Refer to figure 12.3.)

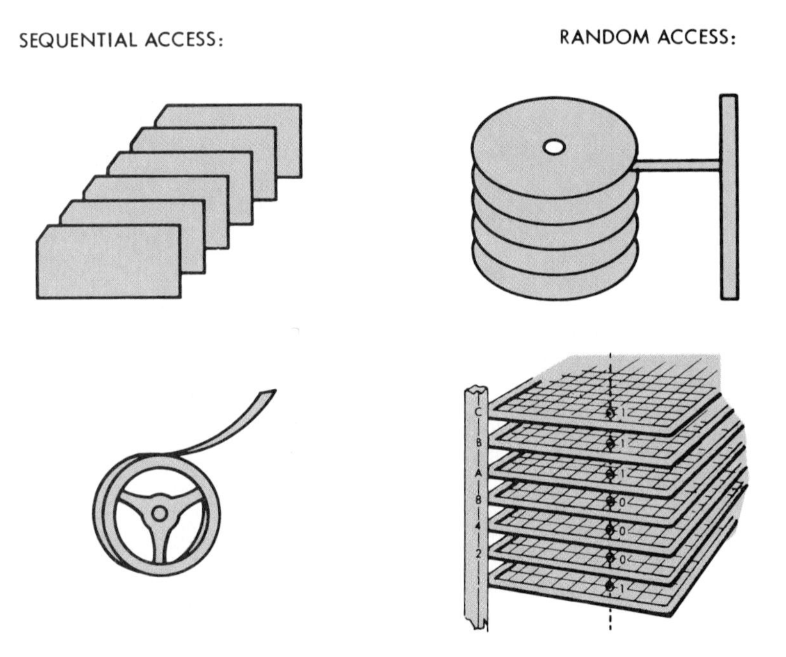

Figure 12.3 Processing Techniques.

Random Processing

The processing of detail transactions against a master file regardless of the sequence of the input documents is called *random processing.* With direct-access storage devices it can be very efficient, especially if the files are organized in such a manner that each record can be located quickly. It is possible to process input transactions against more than one file in a single run. This saves time in both setup and sorting, and minimizes the control problems, since the transactions are handled less frequently.

The use of mass storage devices makes it possible to select the processing technique that will suit the application best. Thus, some applications may be processed sequentially, while those in which the time required to sort, or the delay with the batching process is a material factor, can be processed randomly. Real savings in overall processing time for a job can be made by combining runs in which the same input data affects several files. The detail items can be processed sequentially against a primary file and randomly against the secondary file, all in one run. This is the basis of "inline" processing.

DATA FILE ORGANIZATIONS

Data file organizations refers to the physical arrangement of data records within a file. To give the programmer maximum flexibility in reading and writing data sets from mass storage devices, the following methods of data organization are most commonly used for disk operations. Sequential (Standard), Indexed Sequential, and Direct (Random).

Sequential Organization

In a sequential file, records are organized solely on the basis of their successive physical locations in the file (fig. 12.4). The records are written one after the other—track by track, cylinder by cylinder—at successively higher locations. The records are usually, but not necessarily, in sequence according to their keys (control numbers). The records are usually read or updated in the same sequence as that in which they appear.

Figure 12.4
Sequentially Organized Data Set.

Updating a sequential file that is located on a mass storage device is more efficient than updating a file located on a magnetic tape or punched cards. After the file is opened, the record can be read, updated, and written back in the same location in the file without creating a new file. Thus the file can be used for both input and output activities without the necessity for opening and closing files between operations.

A file on a mass storage device may have the same standard sequential data file organization as any of the unit record or magnetic tape files. The mass storage file, however, may be differently organized so that any record may be accessed merely by specifying the "key" or unique field that tells the system where the desired record is located. This differs from standard sequential organization in that the desired records can be accessed at random without accessing all previous records.

Nonsequential processing of a sequential file is, at best, very inefficient. If it is done infrequently, the time required to locate the records may not matter. There are several ways to program nonsequential processing, with significant differences in the time required. The slowest way is to read the records sequentially until the desired one is located. On the average, half of the files would have to be read. A sequential search takes less time if the records are formatted with keys. The search is done in Search Key High or Equal at the speed of one revolution per track. When the search condition is satisfied, the corresponding record is read.

Additions and deletions require a complete rewrite of a sequential file. Therefore, sequential organization is used in direct-access storage devices primarily for tables and intermediate storage rather than for master files. Its use is recommended for master files, only, if there is a high percentage of activity and if virtually all processing is sequential.

Creating a Standard Sequential Disk File

ENVIRONMENT DIVISION. In order to specify that a file will be on a mass storage device, certain entries must be specified in the SELECT clause. (See figure 12.5.)

```
       SELECT DISK-SEQ
           ASSIGN TO UT-2311-S-SEQUEND.

FD     DISK-SEQ
       LABEL RECORDS ARE STANDARD
       BLOCK CONTAINS 4 RECORDS.
```

Figure 12.5
Example—Environment and Data Division Entries for a Sequential Disk File.

Select Clause. A file-name (programmer supplied) specified for the file. The SELECT clause for a sequential disk file should contain the following.

Assign Clause. The file assigned to a particular mass storage device.

Class Indicator (UT). A two-character field that specifies the device class. A class indicator for a file on a mass storage device may be either DA (Direct Access) or UT (Utility). Since sequential disk files are used in the same manner as magnetic tape files, the class indicator ordinarily used for sequential disk files would be UT.

Device Number (2311). A four- or five-character field used to specify a particular device within a device class. This device number specifies that the file is to be located on a model 2311 disk unit.

Organization (S). A one-character field that indicates the file organization. The organization symbol *S* indicates that the file will have sequential organization.

Name. A one- to eight-character field specifying the external name by which the file is known to the system. If the name is not specified, the symbolic name (SYSnnn) is used as the external name. This name (SYSnnn) may be used as the first name in the system name, in which case this would represent the symbolic unit to which the file is assigned.

DATA DIVISION. The Data Division entries required for sequential disk files are as follows.

FD—File-name of disk file.

Block Contains—This clause is required if records are blocked.

Label Records Are Standard—All files in direct-access devices must have standard labels.

PROCEDURE DIVISION. To create a sequential file, you must use an output file and a WRITE statement.

Write Statement. The INVALID option of the WRITE statement must be used when a sequential disk file is being created (fig. 12.6). The statement following the reserved word INVALID is activated when an attempt is made to write beyond the limit of space reserved for the file. An appropriate action to be specified in an INVALID option might be a branch to a routine that will display a message on the console to indicate that the area is filled. The amount of space reserved for a file is usually specified on job-control cards according to the operating system in a particular installation. (See figures 12.7, 12.8.)

```
WRITE record-name [FROM identifier-1]
    INVALID imperative-statement
```

```
WRITE DISKFACT
    INVALID MOVE 'YES' TO FULL-DISK-FLAG.

WRITE DISKFACT FROM INRECORD
    INVALID MOVE 'YES' TO FULL-DISK-FLAG.
```

Figure 12.6
Format—Write Statement with Invalid Option Examples—Write Statement with Invalid Option.

```
ENVIRONMENT DIVISION.


FILE-CONTROL.
    SELECT DISKFILE         ASSIGN TO UT-2311-S-OUTDISK.
    SELECT TAPEFILE         ASSIGN TO UT-2400-S-INTAPE.

PROCEDURE DIVISION.
MAIN-ROUTINE.
    OPEN  INPUT      TAPEFILE
          OUTPUT     DISKFILE.
    READ TAPEFILE INTO DISKRECORD
        AT END MOVE 'NO' TO MORE-DATA-FLAG.
    PERFORM PROCESS-ROUTINE
        UNTIL NO-MORE-DATA.
    PERFORM TERMINATION-ROUTINE.

PROCESS-ROUTINE.
    WRITE DISK-RECORD
        INVALID DISPLAY 'DISK IS FULL' UPON CONSOLE
            PERFORM TERMINATION-ROUTINE.
    READ TAPEFILE INTO DISKRECORD
        AT END MOVE 'NO' TO MORE-DATA-FLAG.

TERMINATION-ROUTINE.
    CLOSE  TAPEFILE
           DISKFILE.
    STOP RUN.
```

Figure 12.7 Program for Creating a Sequential Disk File from a Tape File.

Data from a disk called QUANFILE is to be used to create data records to be written onto one of two disk files called CALFILE and TFILE. The following is the record layout of each record in the input file QUANFILE.

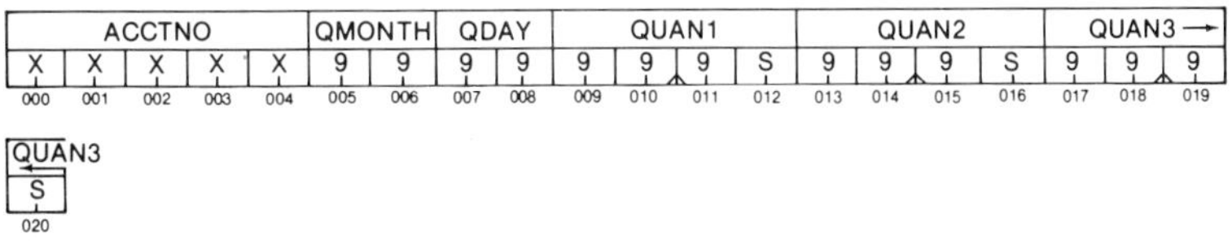

ACCTNO					QMONTH		QDAY		QUAN1				QUAN2				QUAN3 →			
X	X	X	X	X	9	9	9	9	9	9	9	S	9	9	9	S	9	9	9	S
000	001	002	003	004	005	006	007	008	009	010	011	012	013	014	015	016	017	018	019	020

Figure 12.8 Example—Create Sequential Disk Files.

Figure 12.8 Continued.

The contents of the data fields within the record of QUANFILE are as follows:

ACCTNO: Account number	QUANT1: Quantity 1
QMONTH: Month	QUANT2: Quantity 2
QDAY: Day	QUANT3: Quantity 3

The record layout of each record in the output file, CALFILE, is as follows:

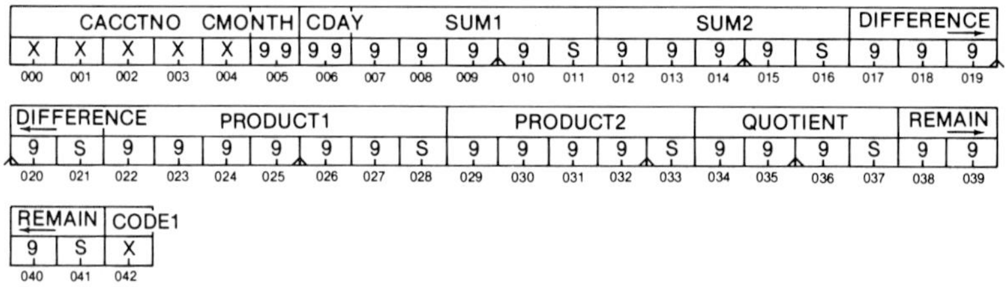

CACCTNO | CMONTH | CDAY | SUM1 | SUM2 | DIFFERENCE

X	X	X	X	X	9 9	9 9	9	9	9	9	S	9	9	9	9	S	9	9	9
000	001	002	003	004	005	006	007	008	009	010	011	012	013	014	015	016	017	018	019

DIFFERENCE | PRODUCT1 | PRODUCT2 | QUOTIENT | REMAIN

9	S	9	9	9	9	9	9	S	9	9	9	9	S	9	9	9	S	9	9
020	021	022	023	024	025	026	027	028	029	030	031	032	033	034	035	036	037	038	039

REMAIN | CODE1

9	S	X
040	041	042

The contents of the data fields within the record of CALFILE are as follows:

CACCTNO: Account number
CMONTH: Month
CDAY: Day
SUM1: Sum 1
SUM2: Sum 2
DIFFENCE: Difference
PRODUCT1: Product 1
PRODUCT2: Product 2
QUOTIENT: Quotient
REMAIN: Remainder
CODE1: Code

If at least one of the three fields, quantity 1, quantity 2, and quantity 3, in the QUANFILE record is greater than 10 or equal to 10, then one record is created and written onto CALFILE. The following formulas determine the contents of the fields in the record of CALFILE.

Sum 1 = quantity 1 + quantity 2

Sum 2 = quantity 3 + 15

Difference = quantity 3 − quantity 2

Product 1 = quantity 3 × quantity 2

Product 2 = quantity 3 × quantity 2 (rounded to nearest whole number)

Quotient = $\frac{\text{quantity 1 + quantity 3}}{\text{quantity 2}}$ with the remainder placed in REMAIN.

Code = A if quantity is less than 50

Code = X if quantity 2 is greater than 50

Code = E if quantity is equal to 50.

If the three fields, quantity 1, quantity 2, and quantity 3, in the QUANFILE record are all less than 10, then one record is created and written onto TFILE. Each record in the output file TFILE has the same structure as each record in the output file CALFILE.

```
IDENTIFICATION DIVISION.
PROGRAM-ID. CALPROGRAM.
AUTHOR. C. FEINGOLD.
DATE-WRITTEN. JUNE 15, 1977.
DATE-COMPILED.
REMARKS.    CREATE TWO DISK FILES SEQUENTIALLY.

ENVIRONMENT DIVISION.
CONFIGURATION SECTION.
SOURCE-COMPUTER. NCR-CENTURY-200.
OBJECT-COMPUTER. NCR-CENTURY-200  MEMORY SIZE 32000 WORDS.
INPUT-OUTPUT SECTION.
FILE-CONTROL.
    SELECT QUANFILE     ASSIGN TO NCR-TYPE-32.
    SELECT CALFILE      ASSIGN TO NCR-TYPE-32.
    SELECT TFILE        ASSIGN TO NCR-TYPE-32.

DATA DIVISION.
FILE SECTION.
FD  QUANFILE            BLOCK CONTAINS 24 RECORDS
                        RECORD CONTAINS 21 CHARACTERS
                        LABEL RECORD IS STANDARD
                        DATA RECORD IS QUANRECORD.
01  QUANRECORD.
    02  ACCTNO              PICTURE X(5).
    02  QMONTH              PICTURE 99.
    02  QDAY                PICTURE 99.
    02  QUAN1               PICTURE S99V9.
    02  QUAN2               PICTURE S99V9.
    02  QUAN3               PICTURE S99V9.
FD  CALFILE                 BLOCK CONTAINS 11 RECORDS
                            RECORD CONTAINS 43 CHARACTERS
                            LABEL RECORD IS STANDARD
                            DATA RECORD IS CALRECORD.
01  CALRECORD.
    02  CACCTNO             PICTURE X(5).
    02  CMONTH              PICTURE 99.
    02  CDAY                PICTURE 99.
    02  SUM1                PICTURE S999V9.
    02  SUM2                PICTURE S999V9.
    02  DIFFERENCE          PICTURE S999V9.
    02  PRODUCT1            PICTURE S9999V99.
    02  PRODUCT2            PICTURE S9999.
    02  QUOTIENT            PICTURE S99V9.
    02  REMAIN              PICTURE S999.
    02  CODE1               PICTURE X.
```

Figure 12.8 Continued.

```
FD   TFILE                    BLOCK CONTAINS 11 RECORDS
                              RECORD CONTAINS 43 CHARACTERS
                              LABEL RECORD IS STANDARD
                              DATA RECORD IS TRECORD.
01   TRECORD                  PICTURE X(43).

WORKING-STORAGE SECTION.
77   WSUM                   PICTURE S999V9.
01   FLAGS.
     02   MORE-DATA-FLAG          PICTURE XXX       VALUE 'YES'.
          88  MORE-DATA                             VALUE 'YES'.
          88  NO-MORE-DATA                          VALUE 'NO'.

PROCEDURE DIVISION.
MAIN-ROUTINE.
     OPEN    INPUT     QUANFILE
             OUTPUT    CALFILE
                       TFILE.
     READ QUANFILE
          AT END MOVE 'NO' TO MORE-DATA-FLAG.
     PERFORM PROCESS-ROUTINE
          UNTIL NO-MORE-DATA.
     CLOSE  QUANFILE
            CALFILE
            TFILE.

PROCESS-ROUTINE.
     MOVE ACCTNO TO CACCTNO.
     MOVE QMONTH TO CMONTH.
     MOVE QDAY TO CDAY.
     ADD QUAN1, QUAN2 GIVING SUM1.
     ADD QUAN3, 15 GIVING SUM2.
     SUBTRACT QUAN2 FROM QUAN3 GIVING DIFFERENCE.
     MULTIPLY QUAN3 BY QUAN2 GIVING PRODUCT1.
     MULTIPLY QUAN3 BY QUAN2 GIVING PRODUCT2 ROUNDED.
     ADD QUAN1, QUAN3, GIVING WSUM.
     DIVIDE WSUM BY QUAN2 GIVING QUOTIENT REMAINDER REMAIN.
     IF QUAN2 IS LESS THAN 50
     THEN
          MOVE 'A' TO CODE1
          NEXT SENTENCE
     ELSE
          IF QUAN2 IS GREATER THAN 50
          THEN
               MOVE 'X' TO CODE1
          ELSE
               MOVE 'E' TO CODE1.
     IF QUAN1 IS LESS THAN 10
          AND QUAN2 IS LESS THAN 10
          AND QUAN3 IS LESS THAN 10
     THEN
          WRITE TRECORD FROM CALRECORD
     ELSE
          WRITE CALRECORD.
     READ QUANFILE
          AT END MOVE 'NO' TO MORE-DATA-FLAG.
```

Updating a Standard Sequential Disk File

After a sequential disk file has been created, it may be maintained by updating the records and placing them back on the file. This can be accomplished only if the file has been opened with the I-O option. The I-O option must be used if the sequential file is used for both input and output operations.

The I-O option may be specified only for files that are stored on direct-access devices. The INVALID option of the WRITE statement is unnecessary when the sequential file has been opened as I-O because there is no possibility of overfilling the file, since each record is replaced in the same position from which it was accessed. (See figure 12.9.)

Indexed Sequential Organization

An indexed sequential organization file is a sequential file with indexes that permit rapid access to individual records as well as rapid sequential processing (fig. 12.10). The indexes are created and written by the system as the file is created or organized. A key provided by the user precedes each block of data and is used to provide the index. An index sequential file is similar to a sequential file; however, by referring to the indexes maintained with the file, it is possible to quickly locate individual records for random processing. Moreover, a separate area can be set aside for additions, making it unnecessary to rewrite the entire file, a process that would be required for sequential processing. Although the records are not maintained in key sequence, the indexes are referred to in order to retrieve the added records in key sequence, thus making rapid sequential processing possible. (See figure 12.11.)

The programming system has control over the location of the individual records in this method of organization. The user need do very little input or output programming; the programming system does most of it since the characteristics of the file are known.

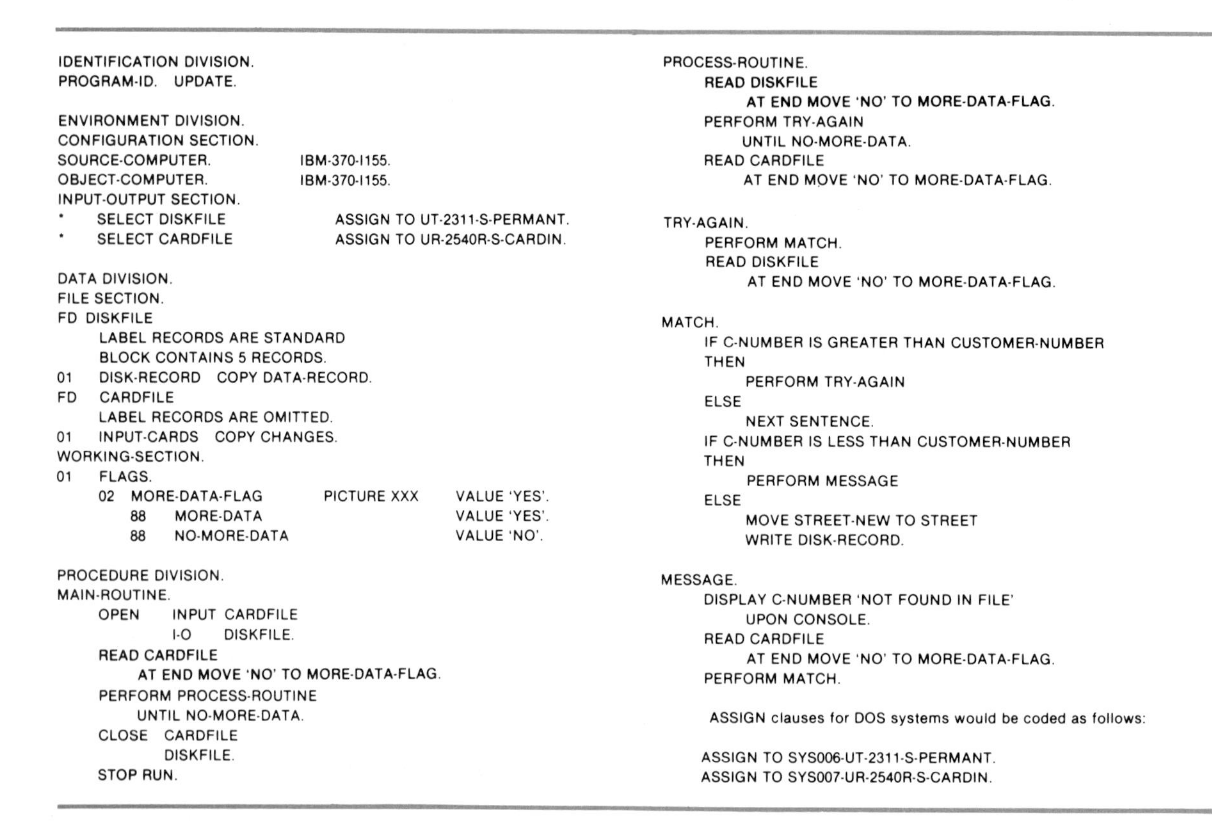

```
IDENTIFICATION DIVISION.
PROGRAM-ID.  UPDATE.

ENVIRONMENT DIVISION.
CONFIGURATION SECTION.
SOURCE-COMPUTER.          IBM-370-I155.
OBJECT-COMPUTER.          IBM-370-I155.
INPUT-OUTPUT SECTION.
*    SELECT DISKFILE           ASSIGN TO UT-2311-S-PERMANT.
*    SELECT CARDFILE           ASSIGN TO UR-2540R-S-CARDIN.

DATA DIVISION.
FILE SECTION.
FD DISKFILE
     LABEL RECORDS ARE STANDARD
     BLOCK CONTAINS 5 RECORDS.
01   DISK-RECORD  COPY DATA-RECORD.
FD   CARDFILE
     LABEL RECORDS ARE OMITTED.
01   INPUT-CARDS  COPY CHANGES.
WORKING-SECTION.
01   FLAGS.
     02  MORE-DATA-FLAG        PICTURE XXX     VALUE 'YES'.
         88   MORE-DATA                        VALUE 'YES'.
         88   NO-MORE-DATA                     VALUE 'NO'.

PROCEDURE DIVISION.
MAIN-ROUTINE.
     OPEN     INPUT  CARDFILE
              I-O    DISKFILE.
     READ CARDFILE
          AT END MOVE 'NO' TO MORE-DATA-FLAG.
     PERFORM PROCESS-ROUTINE
          UNTIL NO-MORE-DATA.
     CLOSE  CARDFILE
            DISKFILE.
     STOP RUN.

PROCESS-ROUTINE.
     READ DISKFILE
          AT END MOVE 'NO' TO MORE-DATA-FLAG.
     PERFORM TRY-AGAIN
          UNTIL NO-MORE-DATA.
     READ CARDFILE
          AT END MOVE 'NO' TO MORE-DATA-FLAG.

TRY-AGAIN.
     PERFORM MATCH.
     READ DISKFILE
          AT END MOVE 'NO' TO MORE-DATA-FLAG.

MATCH.
     IF C-NUMBER IS GREATER THAN CUSTOMER-NUMBER
     THEN
          PERFORM TRY-AGAIN
     ELSE
          NEXT SENTENCE.
     IF C-NUMBER IS LESS THAN CUSTOMER-NUMBER
     THEN
          PERFORM MESSAGE
     ELSE
          MOVE STREET-NEW TO STREET
          WRITE DISK-RECORD.

MESSAGE.
     DISPLAY C-NUMBER 'NOT FOUND IN FILE'
          UPON CONSOLE.
     READ CARDFILE
          AT END MOVE 'NO' TO MORE-DATA-FLAG.
     PERFORM MATCH.
```

ASSIGN clauses for DOS systems would be coded as follows:

```
ASSIGN TO SYS006-UT-2311-S-PERMANT.
ASSIGN TO SYS007-UR-2540R-S-CARDIN.
```

Figure 12.9 Program to Update a Sequential Disk File.

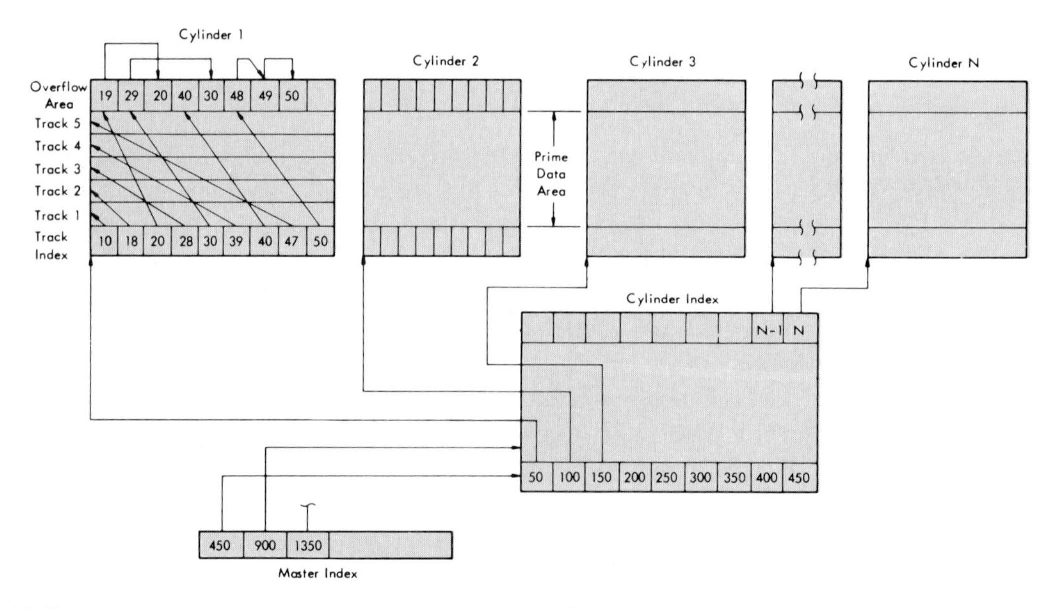

Figure 12.10 Index Structure for an Indexed Sequential Data Set.

Indexed sequential organization gives the programmer greater flexibility in the operations he can perform on the data file. He is provided with the facilities for reading or writing records in a manner similar to that for sequential organization. He can also read or write individual records whose keys may be in any order, and can add logical records with new keys. The system locates the proper position in the data file for the new record and makes all the necessary adjustments to the indexes.

The indexed sequential file must be stored on a direct-access device. Just as with standard sequential files, the indexed sequential files must be created sequentially. Identify-data in the control field of the record, called *keys,* must be in ascending sequence in succeeding records. As the records are written into the file, the system creates indexes based on the key, or control field, in each record to make possible quick location of any record in the file. Thus any record in an indexed sequential file may be accessed by specifying the appropriate key.

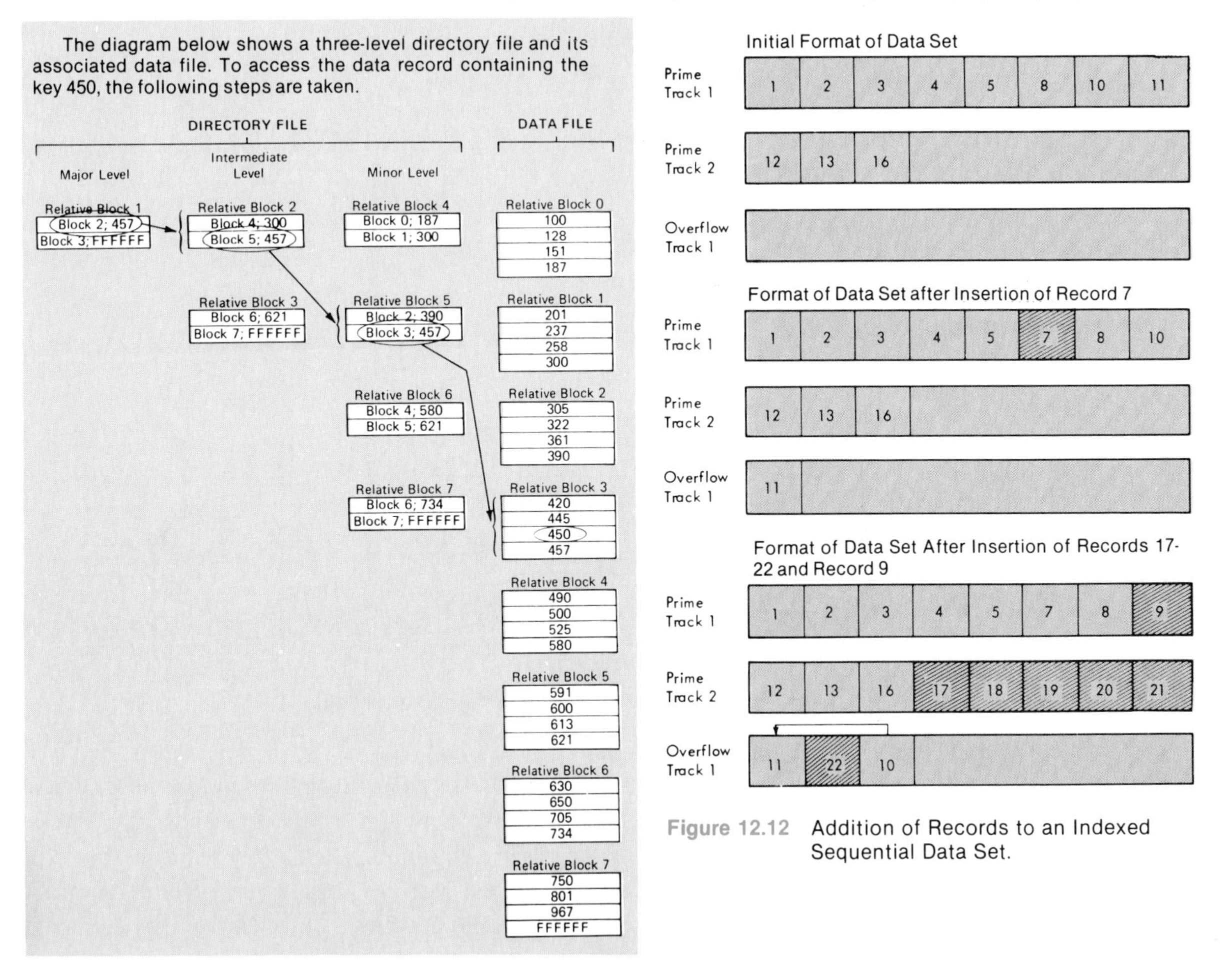

Figure 12.11 Example—Retrieving Records in an Indexed Sequential File.

Figure 12.12 Addition of Records to an Indexed Sequential Data Set.

In addition to quick access of any record, an advantage of indexed sequential file is that records may be added to any part of the file after it has been created, and the system will keep all records in logical sequence, although some records may technically be in a special "overflow" area. In accessing the file sequentially, the system will access records in logical sequence by key rather than in physical sequence by position on the device. (See figure 12.12.)

An indexed sequential file is a sequential file with indexes that permit rapid access to individual records as well as rapid sequential processing. An indexed sequential file has three distinct areas; a prime area, indexes, and an overflow area. (See figure 12.13.)

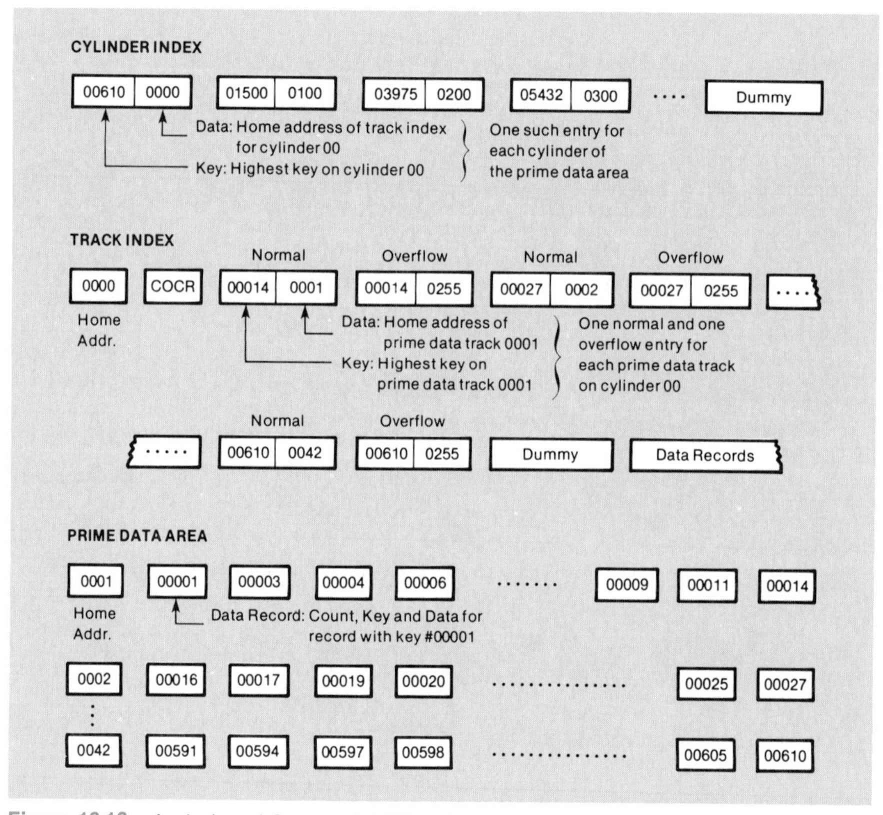

Figure 12.13 An Indexed Sequential File with No Additions.

Prime Area

The prime area is the area in which records are written when the file is created or subsequently reorganized. Additions to the file may also be written in the prime area. The prime area may span multiple volumes and consist of several noncontiguous areas. The records in the prime area are in key sequence.

Prime records must be formatted with keys. They may be blocked or unblocked. If blocked, each logical record contains its key, and the key area contains the key of the highest record in the block.

Indexes

There are two or more indexes of different levels. They are created and written by the operating system when the file is created or reorganized.

Track Index

This is the lowest level of index and is always present. Its entries point to data records. There is one track address for each cylinder in the prime area. It is always written on the first track(s) of the cylinder that it indexes. There is a pair of entries for each prime data track in the cylinder containing the home address of the prime track and the key of the highest record in the track (normal entry), and the overflow area. The last entry of each track index is a dummy entry indicating the end of the index. The rest of the index track contains prime records if there is enough room for them.

Cylinder Index

This is a higher level of indexes and is always present. Its entries point to track indexes. There is one cylinder index for the file. It may reside on a different type of DASD than the rest of the file. It consists of one entry for each cylinder in the prime area, followed by a dummy entry. The entries are formatted in the same fashion as the track index entries. The key area contains the key of the highest record in the cylinder to which the entry points. The data area contains the Home Address of the track index for that cylinder.

Overflow Area

A certain number of whole tracks, as specified by the user, are reserved in each cylinder for overflow data from prime tracks in that cylinder. (See figures 12.14, 12.15.)

Creating an Indexed Sequential Disk File

ENVIRONMENT DIVISION. In order to specify that a file will be on a mass storage device, certain entries must be specified in the SELECT clause.

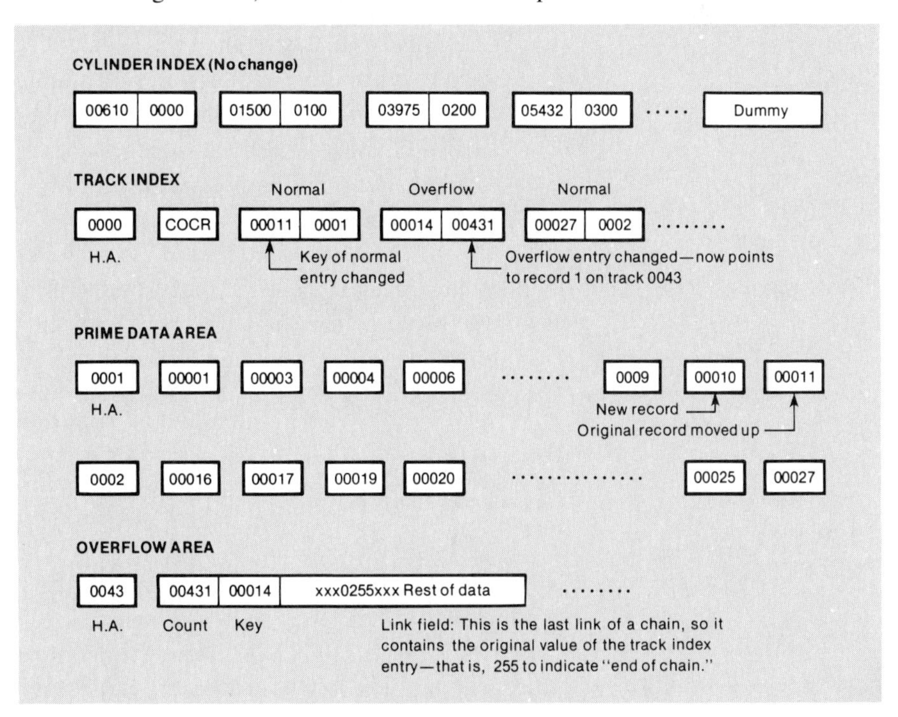

Figure 12.14 An Indexed Sequential File After the First Addition to a Prine Track.

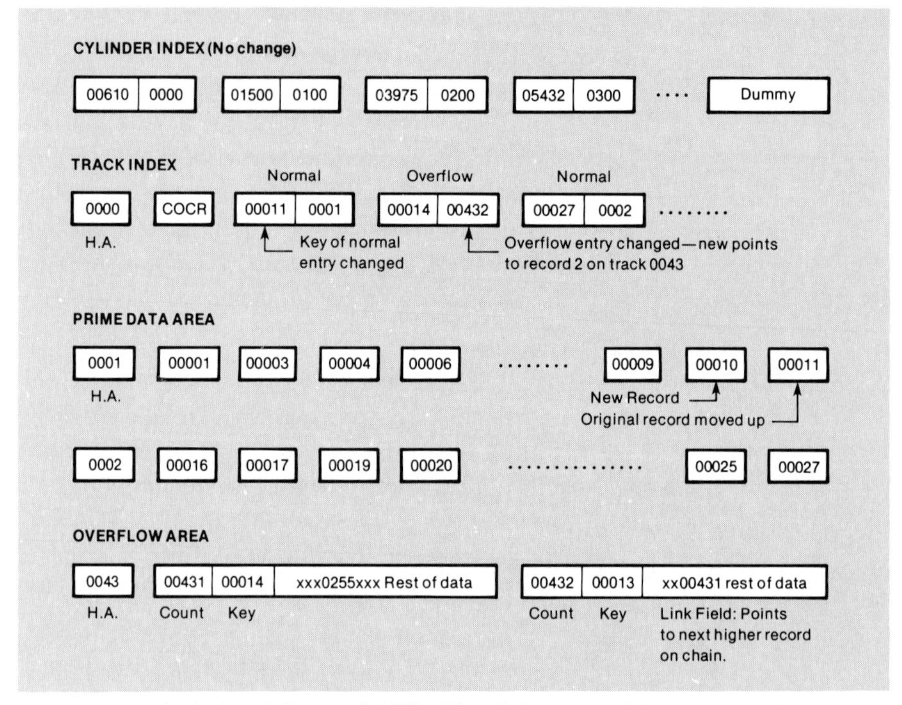

An indexed Sequential File After Subsequent Additions to a Track.

1. *Select Clause.* A file-name (programmer supplied) for the file. The SELECT clause for an indexed sequential disk file should contain the following:
   a. *Class Indicator (DA).* The class indicator must be direct-access. All indexed storage files must be stored on a direct-access device.
   b. *Device Number (2311).* This device number specifies that the file is to be located on a model 2311 disk unit.
   c. *Organization (I).* This organization symbol indicates that the file has an indexed sequential organization.
   d. *Name.* The name identifies the external name by which the file is known to the system. If the name is not specified, the symbolic name (SYSnnn) is used as the actual external name. This name (SYSnnn) may be used as the first name in system-name, in which case this would represent the symbolic unit to which the file is assigned. (See figure 12.16.)

Figure 12.16
Example—Select Clause for an Index Sequential File.

```
SELECT PAYROLL-REGISTER
       ASSIGN TO DA-2311-I-PAYFILE
       RECORD KEY IS SOCIAL-SECURITY-NO.
```

2. *Access Mode Clause.* The ACCESS MODE clause defines the manner in which the records or a file are to be accessed. If the clause is not specified, ACCESS IS SEQUENTIAL is assumed and records are placed or obtained sequentially. The next logical record is made available from the file when the READ statement is executed, or the next logical record is placed into the file

when a WRITE statement is executed. ACCESS IS SEQUENTIAL may be applied to files assigned to magnetic tape, unit record, or direct-access devices.

When the ACCESS IS RANDOM option is used, storage and retrieval are based on an ACTUAL KEY or NOMINAL KEY associated with each record. When the RANDOM option is used, the file must be assigned to a direct-access device.

In creating an indexed sequential disk file, the clause ACCESS IS SEQUENTIAL is optional.

3. *Record Key Clause.* The RECORD KEY clause is used to access an indexed sequential file. It specifies the elementary variable within the file record that identifies the record and is required for all records stored in indexed sequential files. (See figures 12.17.)

Any unique elementary variable in the record associated with the indexed sequential file can be specified as the RECORD KEY.

DATA DIVISION. The Data Division entries required for indexed sequential disk files are as follows.

FD—Name of disk file.

Block Contains—This clause required of all records that are blocked.

Label Records Are Standard—All files on direct-access devices must have standard labels.

PROCEDURE DIVISION. To create an indexed sequential file, you must use an output file and the INVALID KEY option in any WRITE statement.

Write Statement. When the WRITE statement is being used to refer to an indexed sequential file, the optional word KEY must be used (fig. 12.18).

The INVALID KEY option is activated if the key field of the record being written does not contain a greater value than the key field of the record just written; that is, a key out of sequence or a duplicate key may generate an INVALID KEY option. An imperative statement after the word INVALID KEY directs the program to an error procedure.

```
01  WAGE-RECORD.
    02  EMPLOYEE-NUMBER        PICTURE 9(5).
    02  HOURS.
        03  REGULAR            PICTURE 9(3).
        03  OVERTIME           PICTURE 9(3).
    02  WAGES.
        03  REGULAR            PICTURE 9(3).
        03  OVERTIME           PICTURE 9(3).
    02  FILLER                 PICTURE X(61).

SELECT WAGE-FILE
    ASSIGN TO DA-2311-I-WAGESOUT
    RECORD KEY IS EMPLOYEE-NUMBER.
```

Figure 12.17 Example—Record Key Clause.

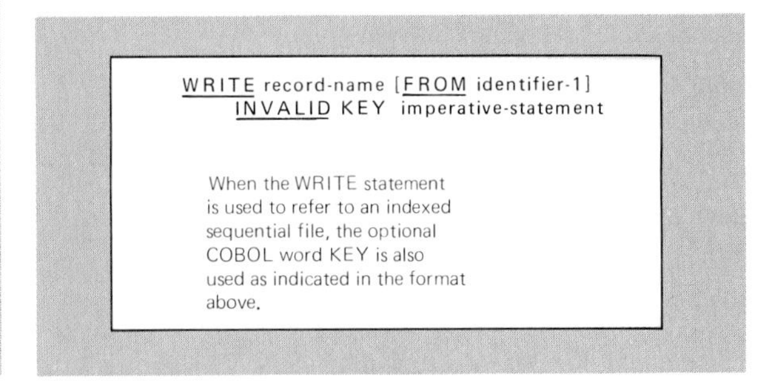

WRITE record-name [FROM identifier-1]
    INVALID KEY imperative-statement

When the WRITE statement is used to refer to an indexed sequential file, the optional COBOL word KEY is also used as indicated in the format above.

Figure 12.18 Format—Write Statement.

When the WRITE statement with the INVALID KEY option is executed, the key of the record is checked for correct sequence before the record is written. If the INVALID KEY option is activated in an attempt to write a record into an indexed sequential file, that record is not placed into the indexed sequential file. Subsequent records, however, may still be placed into the file. The record with the INVALID KEY may be placed into a sequential file, if desired, for subsequent individual checking of the key. The imperative statement following the words INVALID KEY could direct the program to the proper routine to accomplish this. (See figures 12.19, 12.20, 12.21.)

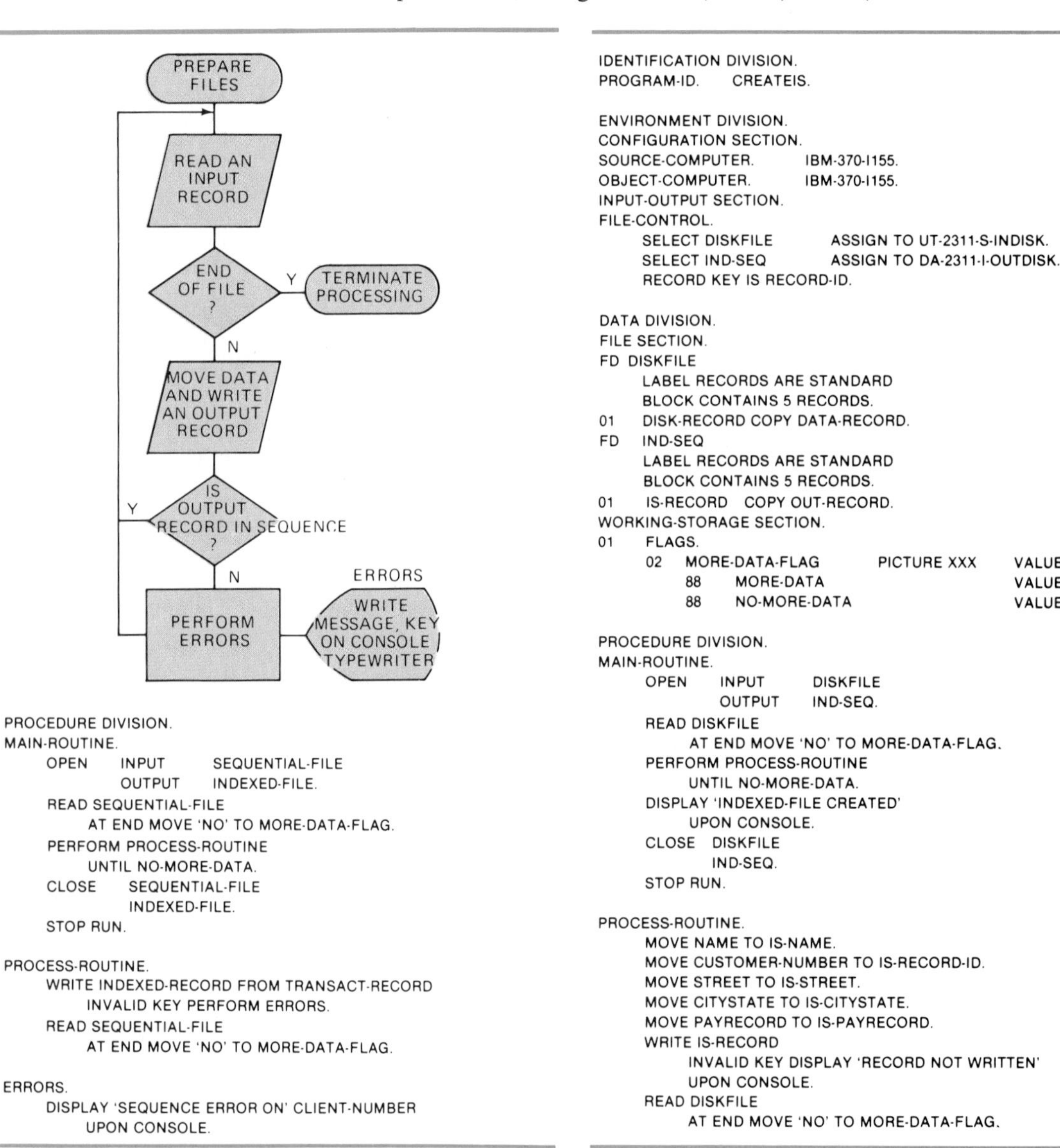


```
PROCEDURE DIVISION.
MAIN-ROUTINE.
    OPEN    INPUT     SEQUENTIAL-FILE
            OUTPUT    INDEXED-FILE.
    READ SEQUENTIAL-FILE
        AT END MOVE 'NO' TO MORE-DATA-FLAG.
    PERFORM PROCESS-ROUTINE
        UNTIL NO-MORE-DATA.
    CLOSE    SEQUENTIAL-FILE
             INDEXED-FILE.
    STOP RUN.

PROCESS-ROUTINE.
    WRITE INDEXED-RECORD FROM TRANSACT-RECORD
        INVALID KEY PERFORM ERRORS.
    READ SEQUENTIAL-FILE
        AT END MOVE 'NO' TO MORE-DATA-FLAG.

ERRORS.
    DISPLAY 'SEQUENCE ERROR ON' CLIENT-NUMBER
        UPON CONSOLE.
```

Figure 12.19 Example—Creation of Indexed Sequential File.

```
IDENTIFICATION DIVISION.
PROGRAM-ID.    CREATEIS.

ENVIRONMENT DIVISION.
CONFIGURATION SECTION.
SOURCE-COMPUTER.        IBM-370-I155.
OBJECT-COMPUTER.        IBM-370-I155.
INPUT-OUTPUT SECTION.
FILE-CONTROL.
    SELECT DISKFILE         ASSIGN TO UT-2311-S-INDISK.
    SELECT IND-SEQ          ASSIGN TO DA-2311-I-OUTDISK.
    RECORD KEY IS RECORD-ID.

DATA DIVISION.
FILE SECTION.
FD  DISKFILE
    LABEL RECORDS ARE STANDARD
    BLOCK CONTAINS 5 RECORDS.
01  DISK-RECORD COPY DATA-RECORD.
FD  IND-SEQ
    LABEL RECORDS ARE STANDARD
    BLOCK CONTAINS 5 RECORDS.
01  IS-RECORD   COPY OUT-RECORD.
WORKING-STORAGE SECTION.
01  FLAGS.
    02  MORE-DATA-FLAG          PICTURE XXX     VALUE 'YES'.
        88   MORE-DATA                          VALUE 'YES'.
        88   NO-MORE-DATA                       VALUE 'NO'.

PROCEDURE DIVISION.
MAIN-ROUTINE.
    OPEN    INPUT       DISKFILE
            OUTPUT      IND-SEQ.
    READ DISKFILE
        AT END MOVE 'NO' TO MORE-DATA-FLAG.
    PERFORM PROCESS-ROUTINE
        UNTIL NO-MORE-DATA.
    DISPLAY 'INDEXED-FILE CREATED'
        UPON CONSOLE.
    CLOSE   DISKFILE
            IND-SEQ.
    STOP RUN.

PROCESS-ROUTINE.
    MOVE NAME TO IS-NAME.
    MOVE CUSTOMER-NUMBER TO IS-RECORD-ID.
    MOVE STREET TO IS-STREET.
    MOVE CITYSTATE TO IS-CITYSTATE.
    MOVE PAYRECORD TO IS-PAYRECORD.
    WRITE IS-RECORD
        INVALID KEY DISPLAY 'RECORD NOT WRITTEN'
        UPON CONSOLE.
    READ DISKFILE
        AT END MOVE 'NO' TO MORE-DATA-FLAG.
```

Figure 12.20 Program to Create an Indexed Sequential File.

This program creates an indexed sequential file. These records are presented in ascending sequence by RECORD KEY. The APPLY clause builds the master index.

```
IDENTIFICATION DIVISION.
PROGRAM-ID.        CREATIS.
REMARKS. ILLUSTRATE CREATION OF INDEXED SEQUENTIAL FILE.

ENVIRONMENT DIVISION.
CONFIGURATION SECTION.
SOURCE-COMPUTER.    IBM-370-I155.
OBJECT-COMPUTER.    IBM-370-I155.
INPUT-OUTPUT SECTION.
FILE-CONTROL.
    SELECT IS-FILE       ASSIGN TO SYS015-DA-2311-I-MASTER
                         ACCESS IS SEQUENTIAL
                         RECORD KEY IS REC-ID.
    SELECT CARD-FILE     ASSIGN TO SYS007-UR-2540R-S
                         RESERVE 1 ALTERNATE AREA.

I-O-CONTROL.
    APPLY MASTER-INDEX TO 2311 ON IS-FILE.

DATA DIVISION.
FILE SECTION.
FD IS-FILE
    BLOCK CONTAINS 5 RECORDS
    RECORD MODE IS F
    LABEL RECORDS ARE STANDARD
    DATA RECORD IS DISK.
01  DISK.
    05  DELETE-CODE              PICTURE X.
    05  REC-ID                   PICTURE 9(10).
    05  DISK-FLD1                PICTURE X(10).
    05  DISK-NAME                PICTURE X(20).
    05  DISK-BAL                 PICTURE 99999V99.
    05  FILLER                   PICTURE X(52).
FD  CARD-FILE
    RECORDING MODE IS F
    LABEL RECORDS ARE OMITTED
    DATA RECORD IS CARDS.
01  CARDS.
    05  KEY-ID                   PICTURE 9(10).
    05  CD-NAME                  PICTURE X(20).
    05  CD-BAL                   PICTURE 99999V99.
    05  FILLER                   PICTURE X(43).
WORKING-STORAGE SECTION.
01  FLAGS.
    05  MORE-DATA-FLAG           PICTURE XXX     VALUE 'YES'.
        88  MORE-DATA                            VALUE 'YES'.
        88  NO-MORE-DATA                         VALUE 'NO'.

PROCEDURE DIVISION.
MAIN-ROUTINE.
    OPEN    INPUT     CARD-FILE
            OUTPUT    IS-FILE.
    READ CARD-FILE
        AT END MOVE 'NO' TO MORE-DATA-FLAG.
    PERFORM PROCESS-ROUTINE
        UNTIL NO-MORE-DATA.
    DISPLAY 'END OF JOB'
        UPON CONSOLE.
    CLOSE   CARD-FILE
            IS-FILE.
    STOP RUN.

PROCESS-ROUTINE.
    MOVE LOW-VALUE TO DELETE-CODE.
    MOVE KEY-ID TO REC-ID.
    MOVE CD-NAME TO DISK-NAME.
    MOVE CD-BAL TO DISK-BAL.
    WRITE DISK
        INVALID KEY PERFORM ERR.
    READ CARD-FILE
        AT END MOVE 'NO' TO MORE-DATA-FLAG.

ERR.
    DISPLAY 'DUPLICATE OR SEQ-ERR'
        UPON CONSOLE.
```

Figure 12.21 Program to Create an Indexed Sequential File.

```
FILE-CONTROL.
    SELECT STUDENT-FILE     ASSIGN TO DA-2311-I-PERM
                            RECORD KEY IS STUDENT-NUMBER.
    SELECT PRINTOUT         ASSIGN TO UR-1403-S-PRINT.

PROCEDURE DIVISION.
MAIN-ROUTINE.
    OPEN    INPUT     STUDENT-FILE
            OUTPUT    PRINTOUT.
    READ STUDENT-FILE
        AT END MOVE 'NO' TO MORE-DATA-FLAG.
    PERFORM PROCESS-ROUTINE
        UNTIL NO-MORE-DATA.
    CLOSE   STUDENT-FILE
            PRINTOUT.
    STOP RUN.

PROCESS-ROUTINE.
    MOVE STUDENT-NUMBER TO S-NUMBER.
    MOVE CORRESPONDING STUDENT-DATA TO PRINTDATA.
    WRITE PRINTDATA.
    READ STUDENT-FILE
        AT END MOVE 'NO' TO MORE-DATA-FLAG.
```

Figure 12.22 Accessing Records Sequentially in an Indexed Sequential File.

Accessing an Indexed Sequential File Sequentially

It is frequently necessary to access records sequentially from an indexed sequential file. These records may be accessed sequentially, record by record, until the file is closed. This program would be similar to a program for accessing a standard sequential file, with the exception of the File-Control paragraph entries describing the file as indexed sequential. (See figure 12.22.)

Records in an indexed sequential file may be accessed at a record that is not the first record in a file, continuing until the file is closed. This is not possible in a standard sequential file, but can be accomplished in an indexed sequential file by specifying the desired beginning key, and positioning the file at that desired key prior to the accessing of the record.

ENVIRONMENT DIVISION. The Environment Division entries for an indexed sequential file from which all records will be accessed sequentially are identical to the entries required when an indexed sequential file is to be created.

DATA DIVISION. The Data Division entries for an indexed sequential file from which all records will be accessed sequentially are similar to entries previously described for standard and indexed sequential programs.

PROCEDURE DIVISION. The Procedure Division coding is similar to standard sequential accessing of records.

Whenever sequential access of records from an indexed sequential file will begin at some other record than the beginning of the file, the NOMINAL KEY clause must be specified in addition to the RECORD KEY clause in the File-Control paragraph of the Environment Division.

Nominal Key Clause. A NOMINAL KEY clause is used with indexed sequential files and specifies any elementary variable described in the Working-Storage Section of the program. (See figure 12.23.)

```
DATA DIVISION.
FILE SECTION.
FD  INDEXED-FILE
    LABEL RECORDS ARE STANDARD
    BLOCK CONTAINS 5 RECORDS.
01  IF-RECORD.
    02  FILLER                              PICTURE X.
    02  RECORD-ID                           PICTURE 9(10).
    02  FIELD-1                             PICTURE X(20).
    02  FIELD-2                             PICTURE X(10).
    02  FILLER                              PICTURE X(62).
WORKING-STORAGE SECTION.
77  RECORD-NUMBER                           PICTURE 9(10).
```

```
    RECORD KEY IS RECORD-ID
    NOMINAL KEY IS RECORD-NUMBER
```

Figure 12.23
Example—Record Key and Nominal Key Clauses.

In order to access sequentially the records in an indexed sequential file beginning at some record other than the first record of a file, the value of the NOMINAL KEY variable must be set equal to the RECORD KEY of the record at which the sequential access is to begin. Since the NOMINAL KEY variable must be given a value before accessing of records begins, a value must be set in Working-Storage through the execution of a MOVE, ACCEPT, or VALUE clause.

After the NOMINAL KEY variable has been set to the value of the RECORD KEY variable of the record at which accessing of records is to begin, the indexed sequential file must be positioned so that a READ statement will access the desired record. A START statement is used to position the file. The absence of START statement causes the searching of a file to start at the first record.

Start Statement. The START statement initiates the processing of a segment of a sequentially indexed sequential file at a specified key (fig. 12.24).

```
START file-name  INVALID KEY imperative-statement
```

```
NOMINAL KEY IS PATIENT-NUMBER

77  PATIENT-NUMBER             PICTURE X(7).

    MOVE 'A208787' TO PATIENT-NUMBER.
    START PATIENT-FILE
        INVALID KEY PERFORM END-ROUTINE.
```

Figure 12.24
Format—Start Statement and to Position File before Accessing an Indexed Sequential File Sequentially.

File-name is the name of the indexed sequential file to be processed. The value must be stored in the data-name specified in the NOMINAL KEY clause before executing the START statement.

INVALID KEY option is executed when the value of the NOMINAL KEY at the time the START statement is executed is not equal to some RECORD KEY variable in the file.

A START statement must be executed after the OPEN statement but before the first READ statement. Processing will continue sequentially until a START statement or a CLOSE statement, or until the end of file is reached.

If processing is to begin at the first record, a START statement is unnecessary before the first READ statement.

The START statement will be used only to position the file, and the READ statement will access the subsequent records in the file. (See figure 12.25.)

Accessing an Indexed Sequential File Randomly

ENVIRONMENT DIVISION. The SELECT, ASSIGN, RECORD KEY, and NOMINAL KEY clauses are used in the same manner as accessing an indexed sequential file sequentially.

ACCESS IS RANDOM option is required for random access of records from a file. Storage and retrieval are based on the ACTUAL KEY or NOMINAL KEY associated with each record. The RECORD KEY must be specified in the File-Control paragraph when sequential access is to begin at some record

Figure 12.25
Example—Program Accessing Records Sequentially from an Indexed Sequential File Starting at any Point.

A large department store wishes to send a special brochure to selected sample of customers. The manager has decided to make a preliminary listing of some customers. The first fifty customers in the file will be checked and then fifty customers will be checked starting at a key that the operator will key in at the appropriate time. Of these 100 customers, only the addresses of those who have had charge accounts for longer time than three years will be listed.

Figure 12.25 Continued.

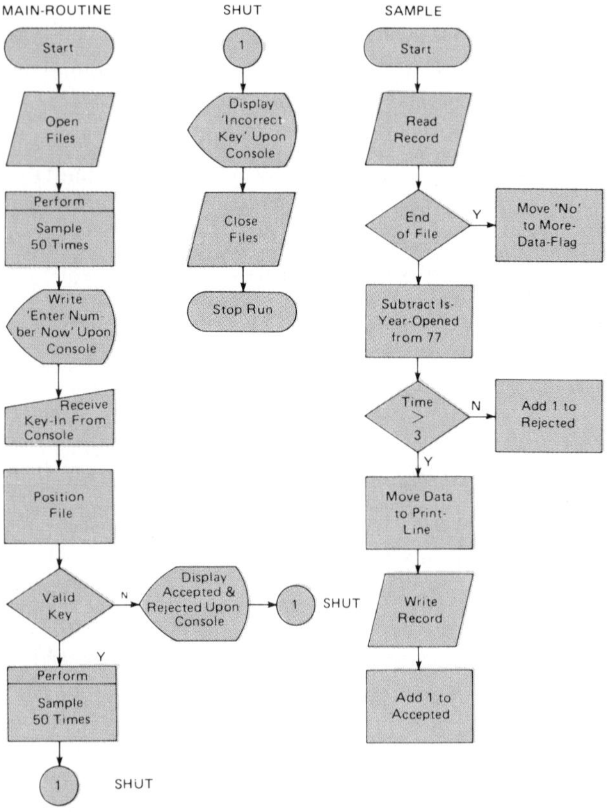

MAIN-ROUTINE

Start
Open Files
Perform Sample 50 Times
Write 'Enter Number Now' Upon Console
Receive Key-In From Console
Position File
Valid Key — N → Display Accepted & Rejected Upon Console → 1 SHUT
Y
Perform Sample 50 Times
1 SHUT

SHUT

1
Display 'Incorrect Key' Upon Console
Close Files
Stop Run

SAMPLE

Start
Read Record
End of File — Y → Move 'No' to More-Data-Flag
Subtract Is-Year-Opened from 77
Time > 3 — N → Add 1 to Rejected
Y
Move Data to Print-Line
Write Record
Add 1 to Accepted

```
IDENTIFICATION DIVISION.
PROGRAM-ID.      SEQ-ACCESS.

ENVIRONMENT DIVISION.
CONFIGURATION SECTION.
SOURCE-COMPUTER.        IBM-370-I155.
OBJECT-COMPUTER.        IBM-370-I155.
INPUT-OUTPUT SECTION.
FILE-CONTROL.
     SELECT IND-SEQ            ASSIGN TO DA-2311-I-ISFILE
                               RECORD KEY IS RECORD-ID
                               NOMINAL KEY IS KEY-IN.
     SELECT PRINT-FILE         ASSIGN TO UR-1403-S-OUTFILE.
DATA DIVISION.
FILE SECTION.
FD  IND-SEQ
     LABEL RECORDS ARE STANDARD
     BLOCK CONTAINS 5 RECORDS.
01   DISK-RECORD.
     02   IS-PERSONAL.
          03   IS-NAME              PICTURE X(21).
          03   RECORD-ID            PICTURE X(6).
          03   IS-ADDRESS.
               04   IS-STREET       PICTURE X(15).
               04   IS-CITYSTATE       PICTURE X(15).
     02   IS-PAYRECORD.
          03   IS-YEAR-OPENED       PICTURE 99.
          03   IS-MAXIMUM-CREDIT    PICTURE 9999V99   USAGE COMP-3.
          03   IS-MAXIMUM-BILL      PICTURE 9999V99   USAGE COMP-3.
          03   IS-BALANCE-DUE       PICTURE 9999V99   USAGE COMP-3.
          03   IS-PAYCODE           PICTURE 9.
               88   BAD               VALUE 1.
               88   POOR              VALUE 2.
               88   SLOW              VALUE 3.
               88   AVERAGE           VALUE 4.
               88   GOOD              VALUE 5.
               88   EXCELLENT         VALUE 6.
               88   NONE              VALUE 7.
FD  PRINT-FILE
     LABEL RECORDS ARE OMITTED.
01   OUTGO                          PICTURE X(133).
WORKING-STORAGE SECTION.
77   TIME-WS                        PICTURE 99.
77   KEY-IN                         PICTURE X(6).
77   ACCEPTED     PICTURE 999       USAGE COMP-3   VALUE ZEROS.
77   REJECTED     PICTURE 999       USAGE COMP-3   VALUE ZEROS.
01   PRINT-LINE.
     02   FILLER            PICTURE X(9)        VALUE SPACES.
     02   FILLER            PICTURE XX          VALUE '19'.
     02   YEARS             PICTURE 99.
     02   FILLER            PICTURE X(10)       VALUE SPACES.
     02   IDENT             PICTURE X(6).
     02   FILLER            PICTURE X(10)       VALUE SPACES.
     02   CUSTOMER          PICTURE X(20).
     02   FILLER            PICTURE X(15)       VALUE SPACES.
     02   O-STREET          PICTURE X(15).
     02   FILLER            PICTURE X(10)       VALUE SPACES.
     02   O-CITYSTATE       PICTURE X(15).
     02   FILLER            PICTURE X(18)       VALUE SPACES.
01   FLAGS.
     02   MORE-DATA-FLAG            PICTURE XXX     VALUE 'YES'.
          88   MORE-DATA                            VALUE 'YES'.
          88   NO-MORE-DATA                         VALUE 'NO'.-

PROCEDURE DIVISION.
MAIN-ROUTINE.
     OPEN     INPUT      IND-SEQ
              OUTPUT     PRINT-FILE.
     PERFORM SAMPLE 50 TIMES.
     DISPLAY 'ENTER NUMBER NOW'
          UPON CONSOLE.
     ACCEPT KEY-IN FROM CONSOLE.
     START IND-SEQ
          INVALID KEY   DISPLAY 'INCORRECT KEY' UPON CONSOLE
               PERFORM SHUT.
     PERFORM SAMPLE 50 TIMES.
     PERFORM SHUT.

SAMPLE.
     READ IND-SEQ
          AT END MOVE 'NO' TO MORE-DATA-FLAG.
     SUBTRACT IS-YEAR-OPENED FROM 77 GIVING TIME-WS.
     IF TIME-WS IS GREATER THAN 3
     THEN
          MOVE IS-YEAR-OPENED TO YEARS
          MOVE RECORD-ID TO IDENT
          MOVE IS-NAME TO CUSTOMER
          MOVE IS-STREET TO O-STREET
          MOVE IS-CITYSTATE
          WRITE OUTGO FROM PRINT-LINE
          ADD 1 TO ACCEPTED
     ELSE
          ADD 1 TO REJECTED.

SHUT
     DISPLAY 'ACCEPTED      'ACCEPTED 'REJECTED    'REJECTED
          UPON CONSOLE.
     CLOSE   IND-SEQ
               PRINT-FILE.
     STOP RUN.
```

other than the first record of the file or when the ACCESS IS RANDOM option is specified.

When the records are accessed randomly, the NOMINAL KEY field must be set to the value of the RECORD KEY field of the desired record before a READ statement is executed for the file. The NOMINAL KEY in Working-Storage will contain the key of the desired record.

Any READ statement that refers to a randomly accessed indexed sequential file must include the INVALID KEY option (fig. 12.26). The end-of-file record

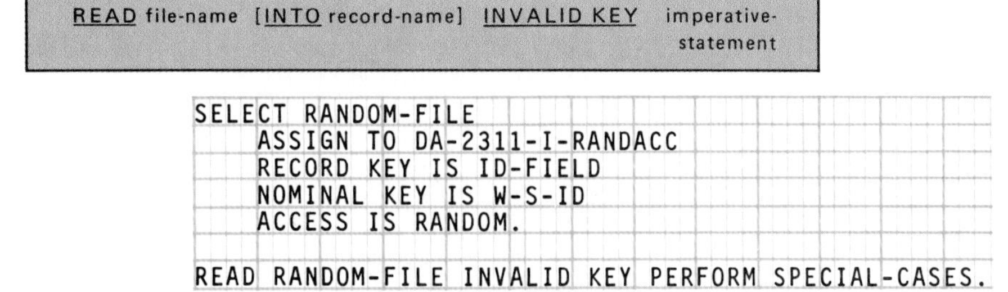

READ file-name [INTO record-name] INVALID KEY imperative-statement
SELECT RANDOM-FILE
ASSIGN TO DA-2311-I-RANDACC
RECORD KEY IS ID-FIELD
NOMINAL KEY IS W-S-ID
ACCESS IS RANDOM.
READ RANDOM-FILE INVALID KEY PERFORM SPECIAL-CASES.

Figure 12.26 Format—Read Statement and Necessary Entries to Access a Record Randomly from an Indexed Sequential File.

is not checked when a file is accessed randomly. The INVALID KEY option of a READ statement is activated when no record with a control field equal to the value of the NOMINAL KEY variable can be found in the file. (See figure 12.27.)

The random accessing of a file having the indexed sequential type of organization is illustrated by the COBOL solution of a program having an input file called TRFILE on disk. The records of TRFILE have been sorted by customer number. The following is the record layout of each record in TRFILE.

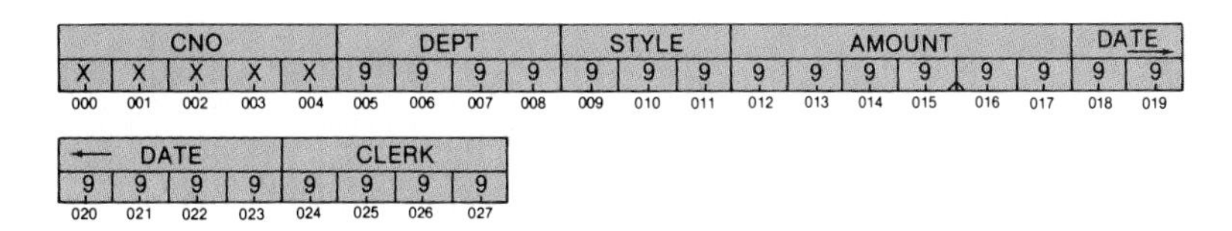

The contents of the fields within the TRFILE record are as follows:

CNO: Customer number
DEPT: Department number
STYLE: Style number
AMOUNT: Amount of purchase
DATE: Date of transaction
CLERK: Clerk number

A disk file called DESFILE has its records sorted by style number. Each record of DESFILE contains descriptive information concerning a

Figure 12-27 Program—Accessing an Indexed Sequential File Randomly.

Figure 12.27 Continued.

specified style number. The following is the record layout of each record in DESFILE.

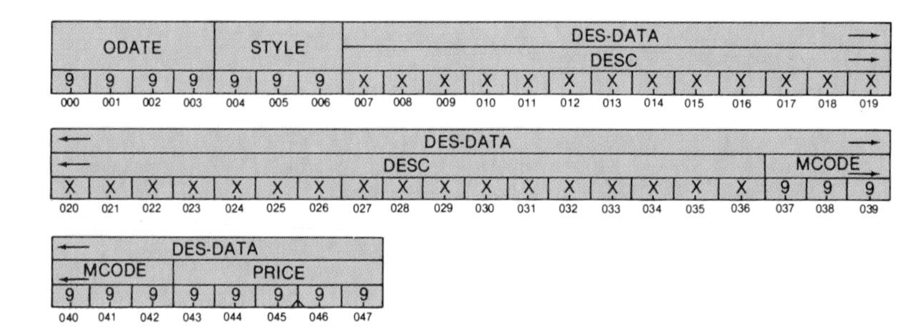

The contents of the fields within the DESFILE record are as follows:

ODATE: Original order date
STYLE: Style number
DESC: Description
MCODE: Manufacturer code
PRICE: Unit price

Using the style number from the TRFILE record, the DESFILE record having the same style number is to be located. Fields from the DESFILE record are to be combined with fields from the TRFILE record to create a record for output to a disk file called TR2FILE. The following is the record layout of each record written onto TR2FILE.

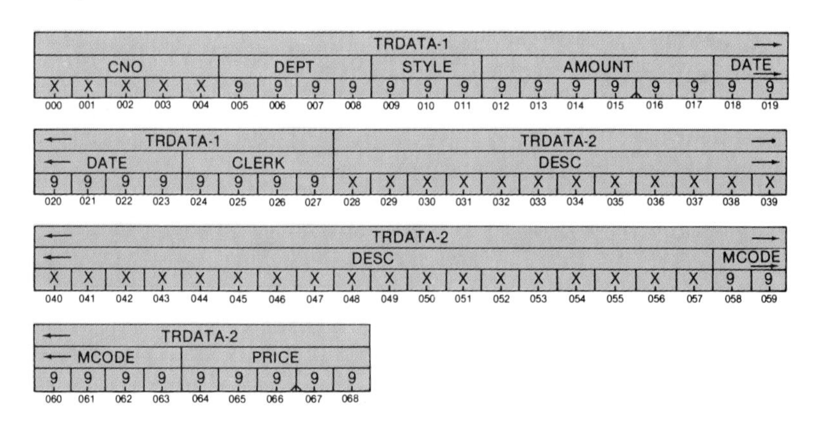

The contents of the fields within TR2FILE record are as follows:

CNO: Customer number
DEPT: Department number
STYLE: Style number
AMOUNT: Amount of purchase
CLERK: Clerk number
DESC: Description
MCODE: Manufacturer code
PRICE: Unit price

Figure 12.27 Continued.

If a style number equal to the TRFILE style number cannot be found in the description file DESFILE, then the TRFILE record is to be written onto an error file called EFILE. The following is the record layout of each record written onto EFILE.

CNO					DEPT				STYLE			AMOUNT						DATE →	
X	X	X	X	X	9	9	9	9	9	9	9	9	9	9	9	9	9	9	9
000	001	002	003	004	005	006	007	008	009	010	011	012	013	014	015	016	017	018	019

← DATE				CLERK			
9	9	9	9	9	9	9	9
020	021	022	023	024	025	026	027

The contents of the fields within the EFILE record are as follows.

CNO: Customer number
DEPT: Department number
STYLE: Style number.
AMOUNT: Amount of purchase
DATE: Date of transaction
CLERK: Clerk number

```
IDENTIFICATION DIVISION.
PROGRAM-ID.  ISPROGRM.

ENVIRONMENT DIVISION.
CONFIGURATION SECTION.
SOURCE-COMPUTER.   NCR-CENTURY-200.
OBJECT-COMPUTER.   NCR-CENTURY-200   MEMORY SIZE 32000 WORDS.
INPUT-OUTPUT SECTION.
FILE-CONTROL.
    SELECT DYSTEM          ASSIGN TO NCR-TYPE-32
                           ACCESS MODE IS RANDOM
                           NOMINAL KEY IS STYLE.
    SELECT TRFILE          ASSIGN TO NCR-TYPE-32.
    SELECT TR2FILE         ASSIGN TO NCR-TYPE-32.
    SELECT EFILE           ASSIGN TO NCR-TYPE-32.

DATA DIVISION.
FILE SECTION.
FD TRFILE                BLOCK CONTAINS 18 RECORDS
                         RECORD CONTAINS 28 CHARACTERS
                         LABEL RECORD IS STANDARD
                         DATA RECORD IS TRECORD.
01    TRECORD.
      02  FILLER           PICTURE X(9).
      02  STYLE            PICTURE 999.
      02  FILLER           PICTURE X(16).
FD    DYSTEM               BLOCK CONTAINS 10 RECORDS
                           RECORD CONTAINS 48 CHARACTERS
                           LABEL RECORD IS STANDARD
                           DATA RECORD IS DESRECORD.
01    DESRECORD.
      02  FILLER           PICTURE XXXX.
      02  DSTYLE           PICTURE 999.
      02  DES-DATA         PICTURE X(41).
FD    TR2FILE              BLOCK CONTAINS 7 RECORDS
                           RECORD CONTAINS 69 CHARACTERS
                           LABEL RECORD IS STANDARD
                           DATA RECORD IS TR2RECORD.
01    TR2RECORD.
      02  TRDATA-1         PICTURE X(28).
      02  TRDATA-2         PICTURE X(41).
FD    EFILE                BLOCK CONTAINS 18 RECORDS
                           RECORD CONTAINS 28 CHARACTERS
                           LABEL RECORD IS STANDARD
                           DATA RECORD IS ERECORD.
01    ERECORD              PICTURE X(28).

WORKING-STORAGE SECTION.
01    FLAGS.
      02  MORE-DATA-FLAG          PICTURE XXX     VALUE 'YES'.
          88  MORE-DATA                           VALUE 'YES'.
          88  NO-MORE-DATA                        VALUE 'NO'.

PROCEDURE DIVISION.
MAIN-ROUTINE.
      OPEN   INPUT   DYSTEM
                     TRFILE
             OUTPUT  TR2FILE
                     EFILE.
      READ TRFILE
          AT END MOVE 'NO' TO MORE-DATA-FLAG.
      PERFORM PROCESS-ROUTINE
           UNTIL NO-MORE-DATA.
      CLOSE   DYSTEM
              TRFILE
              TR2FILE
              EFILE.
      STOP RUN.

PROCESS-ROUTINE.
      READ DYSTEM
          INVALID KEY PERFORM ERROR-ROUTINE.
      MOVE DES-DATA TO TRDATA-2.
      MOVE TRECORD TO TRDATA-1.
      WRITE TR2 RECORD.
      READ TRFILE
          AT END MOVE 'NO' TO MORE-DATA-FLAG.

ERROR-ROUTINE.
      MOVE TRECORD TO ERECORD.
      WRITE ERECORD.
      PERFORM PROCESS-ROUTINE.
```

Adding Records to an Indexed Sequential File Randomly

When adding records to an indexed sequential file, it is not necessary to recreate an indexed sequential file as it was with standard sequential files. Any record with a RECORD KEY that is not currently in the indexed sequential file may be added to it.

ENVIRONMENT DIVISION. SELECT, ASSIGN, RECORD KEY, NOMINAL KEY, and ACCESS IS RANDOM clauses similar to those clauses used in indexed sequential file accessing randomly are required.

DATA DIVISION. Similar entries for indexed sequential files accessed randomly.

PROCEDURE DIVISION. When records are to be added to an indexed sequential file, the file must have random access and be opened as I-O file. The INVALID KEY option of the WRITE statement is activated if the NOMINAL KEY field associated with the record duplicates the RECORD KEY field of a record already in the file. (See figure 12.28.)

```
    SELECT STUDENT-MASTER        ASSIGN TO DA-2311-I-PERM
                                 RECORD KEY IS STUDENT-NUM
                                 NOMINAL KEY IS KEY-NUMBER
                                 ACCESS IS RANDOM.
    SELECT UPDATE-DATA           ASSIGN TO UR-2540R-S-CARDS.

PROCEDURE DIVISION.
MAINROUTINE.
    OPEN    INPUT UPDATE-DATA
            I-O     STUDENT-MASTER.
    READ UPDATE-DATA
        AT END MOVE 'NO' TO MORE-DATA-FLAG.
    PERFORM PROCESS-ROUTINE
        UNTIL NO-MORE-DATA.
    CLOSE   UPDATE-DATA
            STUDENT-MASTER.
    STOP RUN.

PROCESS-ROUTINE.
    MOVE CARD-NUMBER TO KEY-NUMBER.
    WRITE STUDENT-DATA FROM TRANSFER
        INVALID KEY PERFORM BAD-KEY.
    READ UPDATE-DATA
        AT END MOVE 'NO' TO MORE-DATA-FLAG.

BAD-KEY.
    DISPLAY KEY-NUMBER'   BAD KEY'
        UPON CONSOLE.
    PERFORM PROCESS-ROUTINE.
```

Figure 12.28 Example—Adding Records Randomly to an Indexed Sequential File.

Updating and Replacing Records in an Indexed Sequential File Randomly

ENVIRONMENT DIVISION. Same entries as *adding* records to an indexed sequential file randomly.

DATA DIVISION. Same entries as *adding* records to an indexed sequential file randomly.

PROCEDURE DIVISION. Every *READ* statement that refers to a randomly accessed indexed sequential file must have the INVALID KEY option. The INVALID KEY option of a READ statement refers to a randomly accessed indexed sequential file opened as I-O is activated under the same circumstances as the INVALID KEY option of a READ statement that refers to randomly accessed files opened as INPUT. The INVALID KEY option is activated when no RECORD KEY equal to the current value of the NOMINAL KEY variable can be located in the file.

Before a READ statement can be executed for a randomly accessed indexed sequential file, the NOMINAL KEY variable must be set to the desired value, and the file must be opened as I-O or INPUT.

After a READ statement is executed for an indexed sequential file opened as I-O, the accessed record may be updated and placed back in the same position in the file. An updated record is placed back into an indexed sequential file with a REWRITE statement. The next input or output statement for an indexed sequential file opened as I-O after a READ statement may be a statement to place the record back into the file with a REWRITE statement. The key fields should not be altered between a READ and a REWRITE statement.

The READ statement and its associated REWRITE statement may be separated by any number of statements provided that they are not separated by any other input or output statements that refer to the indexed sequential file.

Rewrite Statement. The function of the REWRITE statement is to place a logical record on a direct-access device with a specified record, if the contents of the associated ACTUAL KEY or NOMINAL KEY are found to be valid (fig. 12.29).

```
REWRITE record-name  [FROM identifier]
    INVALID KEY imperative-statement
```

NOTE: The INVALID KEY option must be used for the IBM DOS 360/370 compilers. The INVALID KEY option is optional for the IBM OS 360/370 compiler.

Figure 12.29 Format—Rewrite Statement.

The READ statement for a file must be executed before a REWRITE statement for the file can be executed. A REWRITE statement can be executed only for direct or indexed sequential files opened as I-O. If the ACCESS IS RANDOM option is specified for the file, the ACTUAL or NOMINAL KEY must be set to the desired value prior to the execution of the REWRITE statement. The record name-is the name of the logical record in the Data Division.

The record-name must be associated with an indexed sequential file that was opened as I-O. The record that is placed back into the file is the last record accessed by the READ statement referring to that file.

If the INVALID KEY option of a READ statement is activated, no record has been accessed. The next input or output statement for an indexed sequential file after the INVALID KEY of a READ statement is activated could be another READ statement to access a different record. (See figures 12.30, 12.31, 12.32.)

Direct (Random) Organization

A file organized in a direct (random) manner is characterized by some predictable relationship between the key of the record and the address of that record in a direct-access storage device. The relationship is established by the user and permits the rapid access to any record of the file if the file is carefully organized. The records will probably be distributed nonsequentially throughout the file. If so, processing the record in key sequence requires a preliminary sort or the use of a finder file.

When a request to store or retrieve a record is made, an address relative to the beginning of the file or an actual address (i.e., device, cylinder, track, record position) must be furnished. This address can be specified as being the address of the desired record or as a starting point within the file where the search for the record is to begin. When a record search is specified, the programmer must also furnish the key (i.e., part number, customer number, etc.)

The Toluca Community College has a problem with students who move frequently and needs a program to update addresses in its indexed sequential master file.

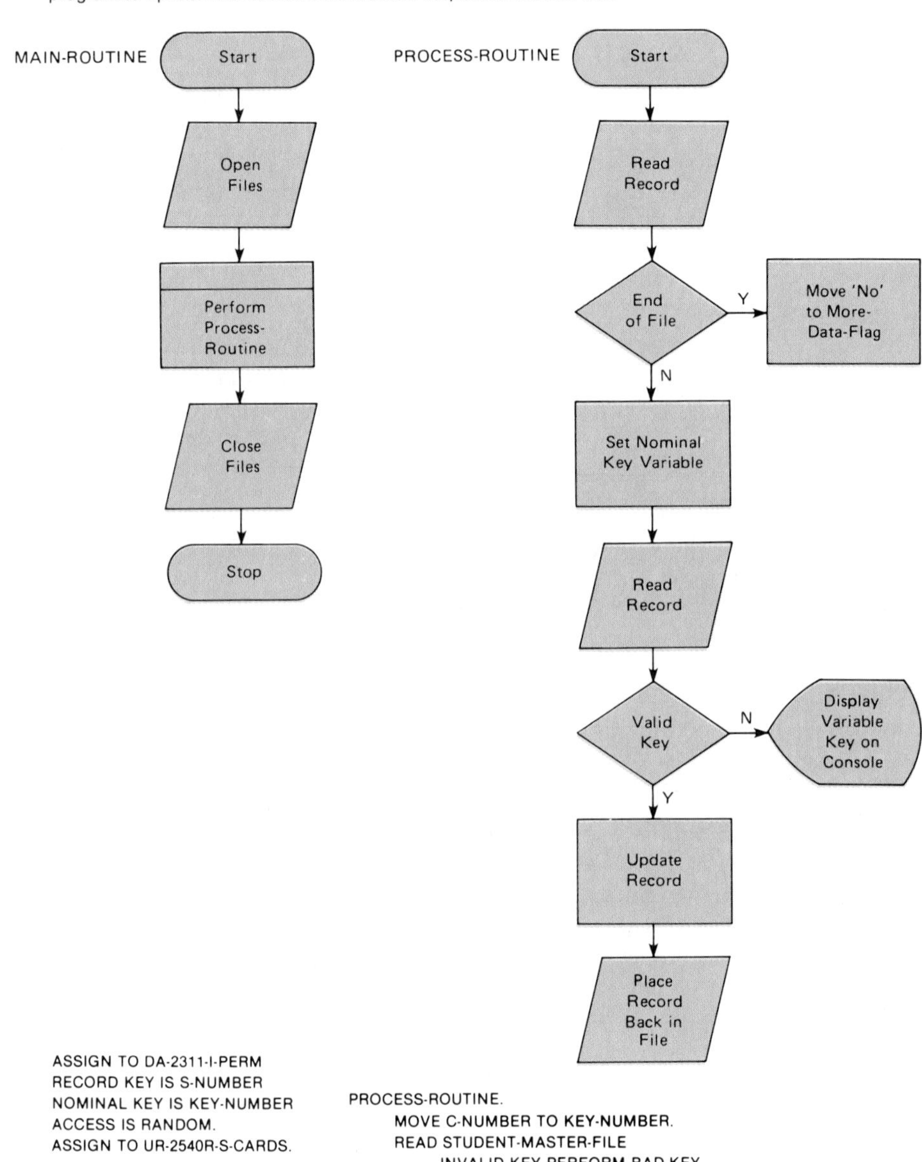


```
SELECT STUDENT-MASTER-FILE     ASSIGN TO DA-2311-I-PERM
                               RECORD KEY IS S-NUMBER
                               NOMINAL KEY IS KEY-NUMBER
                               ACCESS IS RANDOM.
SELECT UPDATE-DATA-FILE        ASSIGN TO UR-2540R-S-CARDS.

PROCEDURE DIVISION.
MAIN-ROUTINE.
    OPEN    INPUT  UPDATE-DATA
            I-O    STUDENT-MASTER-FILE.
    READ UPDATE-DATA-FILE
        AT END MOVE 'NO' TO MORE-DATA-FLAG.
    PERFORM PROCESS-ROUTINE
        UNTIL NO-MORE-DATA.
    CLOSE   UPDATE-DATA-FILE
            STUDENT-MASTER-FILE.
    STOP RUN.

PROCESS-ROUTINE.
    MOVE C-NUMBER TO KEY-NUMBER.
    READ STUDENT-MASTER-FILE
        INVALID KEY PERFORM BAD-KEY.
    MOVE C-STREET TO STREET.
    MOVE C-CITY TO CITY.
    MOVE C-STATE TO STATE.
    REWRITE STUDENT-DATA-RECORD.
    READ UPDATE-DATA-FILE
        AT END MOVE 'NO' TO MORE-DATA-FLAG.

BAD-KEY.
    DISPLAY KEY-NUMBER 'NOT IN FILE'
        UPON CONSOLE.
    PERFORM PROCESS-ROUTINE.
```

Figure 12.30 Program to Add and Change Records Randomly in an Indexed Sequential File.

The random accessing of a master file having an indexed sequential type of organization is illustrated by the COBOL solution of a program having a transaction file called TFILE on disk. The records of TFILE have not been sorted. One or more transaction records may have the same customer number. The following is the layout of each record in TFILE.

TNO						TAMOUNT						TCODE
X	X	X	X	X	X	9	9	9	9	9	9	9
000	001	002	003	004	005	006	007	008	009	010	011	012

The contents of the data fields within the TFILE records are as follows:

TNO: Customer number
TAMOUNT: Transaction amount
TCODE: Transaction code.

The type of transaction record is determined by the content of the transaction code field as follows:

1 indicates an insertion transaction record.
2 indicates a deposit transaction record.
3 indicates an interest transaction record.
4 indicates a withdrawal transaction record.
5 indicates a deletion transaction record.

A disk file record called MPILE is a master file whose records have been sorted by ascending customer number. Each master record has a unique customer number. The structure of each record in this master file is shown below.

MNO						BALANCE							LTCODE	LTAMOUNT					
X	X	X	X	X	X	9	9	9	9	9	9	9	9	9	9	9	9	9	9
000	001	002	003	004	005	006	007	008	009	010	011	012	013	014	015	016	017	018	019

LTDATE						MDATA											
9	9	9	9	9	9	X	X	X	X	X	X	X	X	X	X	X	X
020	021	022	023	024	025	026	027	028	029	030	031	032	033	034	035	036	037

The contents of the data fields within the MFILE record are as follows.

MNO: Customer number
BALANCE: Balance
LTCODE: Last transaction code
LTAMOUNT: Last transaction amount
LTDATE: Last transaction date in the order day, month, year
MDATA: Miscellaneous data

The *master file* called MFILE must be *reorganized as a data file of an indexed sequential system called MSYSTEM.* Using the customer number from the transaction record the master record having the same customer number is to be located. A master record is to be updated by the transaction record having the same customer number. The updating actions to be taken are determined by the transaction code.

An *insertion transaction* is processed by inserting a new record into the master file. The inserted record is created by placing the transaction customer nuer into the master customer number field. The transaction amount is placed into both the current balance and the last transaction amount fields of the master record. The transaction code is placed into the last transaction code field. The data is placed into the last transaction date field. Zero characters are placed into the remaining fields of the master record.

A *deposit or interest transaction* record updates the master record by adding the amount of the transaction to the current balance in the master record. The transuction code is placed into the last transaction code field. The transaction amount is placed into the last transaction amount field. The date is placed into the last transaction date field.

A *withdrawal transaction* record updates the master record by subtracting the amount of the transaction from the current balance in the master record. The transaction code is placed into the last transaction code field. The transaction amount is placed into the last transaction amount field. The date is placed into the last transaction date field.

A*deletion transaction* is processed by deleting the record from the master file.

A record is created and output to an error file called ERRORFILE whenever one of the six error conditions occur.

The following is the record layout of each record in ERRORFILE.

ENO						EAMOUNT							ETCODE	ECODE
X	X	X	X	X	X	9	9	9	9	9	9	9	9	X
000	001	002	003	004	005	006	007	008	009	010	011	012	013	014

The contents of the data fields within ERRORFILE record are as follows:

ENO: Customer number
EAMOUNT: Transaction amount or balance if master record is being deleted
ETCODE: Transaction code
ECODE: Error code

The value of the error code is determined by the type of error condition. The error code for each of the six conditions are as follows.

1. The error code is equal to the letter D when a master record has been deleted.
2. The error code is equal to the letter E when a deletion record cannot update the master file.
3. The error code is equal to the letter I when a master record already exists for the customer number in an insertion record.
4. The error code is equal to the letter J when an insertion record cannot update the master file.

Figure 12.31 Program to Create, Add, Delete, and Update Records Randomly in an Indexed Sequential File.

Figure 12.31 Continued.

5. The error code is equal to the letter M when a master record does not exist for the customer number in a deposit, interest, withdrawal, or deletion transaction record.
6. The error code is equal to the letter C when a transaction record does not have a transaction code equal to 1, 2, 3, 4, or 5.

```
IDENTIFICATION DIVISION.
PROGRAM-ID.  ISUPDATE.

ENVIRONMENT DIVISION.
CONFIGURATION SECTION.
SOURCE-COMPUTER.    NCR-TYPE-200.
OBJECT-COMPUTER.    NCR-TYPE-200     MEMORY SIZE 32000 WORDS.
INPUT-OUTPUT SECTION.
FILE-CONTROL.
    SELECT MSYSTEM              ASSIGN TO NCR-TYPE-32
                                ACCESS MODE IS RANDOM
                                NOMINAL KEY IS TNO.
    SELECT TFILE                ASSIGN TO NCR-TYPE-32.
    SELECT ERRORFILE            ASSIGN TO NCR-TYPE-32.

DATA DIVISION.
FILE SECTION.
FD  TFILE                   BLOCK CONTAINS 39 RECORDS
                            RECORD CONTAINS 13 CHARACTERS
                            LABEL RECORD IS STANDARD
                            DATA RECORD IS TRECORD.
01    TRECORD.
      02   TNO              PICTURE X(6).
      02   TAMOUNT          PICTURE 9999V99.
      02   TCODE            PICTURE 9.
FD    MSYSTEM               BLOCK CONTAINS 13 RECORDS
                            RECORD CONTAINS 38 CHARACTERS
                            LABEL RECORD IS STANDARD
                            DATA RECORD IS MRECORD.
01    MRECORD.
      02   MNO              PICTURE X(6).
      02   BALANCE          PICTURE 9(5)V99.
      02   LTCODE           PICTURE 9.
      02   LTAMOUNT         PICTURE 9999V99.
      02   LTDATE           PICTURE X(6).
      02   FILLER           PICTURE X(12)
FD  ERRORFILE               BLOCK CONTAINS 34 RECORDS
                            RECORD CONTAINS 15 CHARACTERS
                            LABEL RECORD IS STANDARD
                            DATA RECORD IS ERECORD.
01  ERECORD.
    02  ENO                 PICTURE X(6).
    02  EAMOUNT             PICTURE 9(5)V99.
    02  ETCODE              PICTURE 9.
    02  ECODE               PICTURE X.

WORKING-STORAGE SECTION.
01  WRECORD.
    02  WNO                 PICTURE 9(6).
    02  WBALANCE            PICTURE 9(5)V99.
    02  WLTCODE             PICTURE 9.
    02  WLTAMOUNT           PICTURE 9999V99.
    02  WLTDATE             PICTURE X(6).
    02  FILLER              PICTURE X(12)           VALUE ZEROS.
01  FLAGS.
    02  MORE-DATA-FLAG      PICTURE XXX             VALUE 'YES'.
        88  MORE-DATA                               VALUE 'YES'.
        88  NO-MORE-DATA                            VALUE 'NO'.

PROCEDURE DIVISION.
MAIN-ROUTINE.
    OPEN    INPUT    TFILE
            I-O      MSYSTEM
            OUTPUT   ERRORFILE.
    READ TFILE
        AT END MOVE 'NO' TO MORE-DATA-FLAG.
    PERFORM PROCESS-ROUTINE THRU PROCESS-ROUTINE-EXIT
        UNTIL NO-MORE-DATA.
    CLOSE    TFILE
             MSYSTEM
             ERRORFILE.
    STOP RUN.

PROCESS-ROUTINE.
    GO TO INSERTION, UPDATE, UPDATE, UPDATE, DELETION
        DEPENDING ON TCODE.
    PERFORM TRANSERR.
    READ TFILE
        AT END MOVE 'NO' TO MORE-DATA-FLAG.
    GO TO PROCESS-ROUTINE-EXIT.

INSERTION.
    MOVE TNO TO WNO.
    MOVE TAMOUNT TO WBALANCE, WLTAMOUNT.
    MOVE VDATE TO LTDATE.
    MOVE TCODE TO WLTCODE.
    WRITE MRECORD FROM WRECORD
        INVALID KEY PERFORM CHECK-FLAG.
    GO TO PROCESS-ROUTINE-EXIT.

UPDATE.
    READ MSYSTEM
        INVALID KEY PERFORM BADTRAN, GO TO PROCESS-
ROUTINE-EXIT.
    IF TCODE  =  2 OR 3,
    THEN
        ADD TAMOUNT TO BALANCE
    ELSE
        SUBTRACT TAMOUNT FROM BALANCE.
    MOVE TCODE TO LTCODE.
    MOVE TAMOUNT TO LTAMOUNT.
    MOVE VDATE TO LTDATE.
    REWRITE MRECORD.
    GO TO PROCESS-ROUTINE-EXIT.

DELETION.
    MOVE 'D' TO ECODE.
    MOVE BALANCE TO EAMOUNT.
    ENTER NEAT/3
        ISDELRMSYSTEM, TNO, BADDEL
        ENTER COBOL
        GO TO PROCESS-ROUTINE-EXIT.

PROCESS-ROUTINE-EXIT.
    EXIT.

TRANSERR.
    MOVE 'C' TO ECODE.
    PERFORM E1 THRU E2.

CHECK-FLAG.
    IF MSYSTEM $ISCONTFLG  =  49,
    THEN
        MOVE 'I' TO ECODE
    ELSE
        MOVE 'J' TO ECODE.
    PERFORM E1 THRU E2.

BADTRAN.
    MOVE 'M' TO ECODE.
    PERFORM E1 THRU E2.

BADDEL.
    MOVE 'E' TO ECODE.
    MOVE ZERO TO EAMOUNT.
    PERFORM E2.

E1.
    MOVE TAMOUNT TO EAMOUNT.

E2.
    MOVE TNO TO ENO.
    MOVE TCODE TO ETCODE.
    WRITE ERECORD.
```

This program randomly updates an existing indexed sequential file. The READ IS-FILE statement causes a search of indexes for an equal compare between the NOMINAL KEY obtained from the input record and the RECORD KEY of the I-O file. If an equal compare occurs, the record is updated, and the details of this update are printed. If a matching record is not found, the invalid key branch is taken.

```
IDENTIFICATION DIVISION.
PROGRAM-ID.        RANDOMIS.
REMARKS.           ILLUSTRATE RANDOM RETRIEVAL FROM IS-FILE.

ENVIRONMENT DIVISION.
CONFIGURATION SECTION.
SOURCE-COMPUTER.        IBM-370-I155.
OBJECT-COMPUTER.        IBM-370-I155.
INPUT-OUTPUT SECTION.
FILE-CONTROL.
    SELECT IS-FILE             ASSIGN TO DA-2311-I-MASTER
                               ACCESS IS RANDOM
                               NOMINAL KEY IS KEY-ID
                               RECORD KEY IS REC-ID.
    SELECT CARD-FILE           ASSIGN TO UR-1442R-S-INFILE
                               RESERVE 10 ALTERNATE AREAS.
    SELECT PRINT-FILE          ASSIGN TO UT-2400-S-PROUT
                               RESERVE NO ALTERNATE AREAS.
I-O-CONTROL.
    RERUN ON UT-2400-S-CKPT EVERY 10000 RECORDS OF IS-FILE.

DATA DIVISION.
FILE SECTION.
FD  IS-FILE
    BLOCK CONTAINS 5 RECORDS
    RECORD CONTAINS 100 CHARACTERS
    LABEL RECORDS ARE STANDARD
    RECORDING MODE IS F
    DATA RECORD IS DISK.
01  DISK.
    05  DELETE-CODE            PICTURE X.
    05  REC-ID                 PICTURE 9(10).
    05  DISK-FLD1              PICTURE X(10).
    05  DISK-NAME              PICTURE X(20).
    05  DISK-BAL               PICTURE 99999V99.
    05  FILLER                 PICTURE X(52).
FD  CARD-FILE
    RECORDING MODE IS F
    LABEL RECORDS ARE OMITTED
    DATA RECORD IS CARDS.
01  CARDS.
    05  KEY-IDA                PICTURE 9(10).
    05  CD-NAME                PICTURE X(20).
    05  CD-AMT                 PICTURE 99999V99.
    05  FILLER                 PICTURE X(43).
FD  PRINT-FILE
    RECORDING MODE IS F
    LABEL RECORDS ARE STANDARD
    DATA RECORD IS PRINTER.
01  PRINTER.
    05  FORMSC                 PICTURE X.
    05  PRINT-ID               PICTURE X(10).
    05  FILLER                 PICTURE X(10).
    05  PRINT-NAME             PICTURE X(20).
    05  FILLER                 PICTURE X(10).
    05  PRINT-BAL              PICTURE $ZZZ,999.99.
    05  FILLER                 PICTURE X(10).
    05  PRINT-AMT              PICTURE $ZZZ,ZZZ.99.
    05  FILLER                 PICTURE X(10).
    05  PRINT-NEW-BAL          PICTURE $ZZZ,ZZZ.99.
WORKING-STORAGE SECTION.
77  KEY-ID                     PICTURE 9(10).
01  FLAGS.
    05  MORE-DATA-FLAG         PICTURE XXX          VALUE 'YES'.
        88  MORE-DATA                               VALUE 'YES'.
        88  NO-MORE-DATA                            VALUE 'NO'.

PROCEDURE DIVISION.
MAIN-ROUTINE.
    OPEN    INPUT    CARD-FILE
            OUTPUT   PRINT-FILE
            I-O      IS-FILE.
    READ CARD-FILE
        AT END MOVE 'NO' TO MORE-DATA-FLAG.
    PERFORM PROCESS-ROUTINE
        UNTIL NO-MOREDATA.
    CLOSE    CARD-FILE
             PRINT-FILE
             IS-FILE.
    DISPLAY 'END OF JOB'
        UPON CONSOLE.
    STOP RUN.

PROCESS-ROUTINE.
    MOVE SPACES TO PRINTER.
    MOVE KEY-IDA TO KEY-ID.
    READ IS-FILE
        INVALID KEY PERFORM NO-RECORD.
    MOVE REC-ID TO PRINT-ID.
    MOVE DISK-NAME TO PRINT-NAME.
    MOVE DISK-BAL TO PRINT-BAL.
    MOVE CD-AMT TO PRINT-AMT.
    ADD CD-AMT TO DISK-BAL.
    MOVE DISK-BAL TO PRINT-NEW-BAL.
    REWRITE DISK
        INVALID KEY PERFORM NO-RECORD.
    WRITE PRINTER AFTER ADVANCING 2 LINES.
    READ CARD-FILE
        AT END MOVE 'NO' TO MORE-DATA-FLAG.

NO-RECORD.
    DISPLAY 'NO RECORD FOUND'
        UPON CONSOLE.
    DISPLAY KEY-ID
        UPON CONSOLE.
    PERFORM PROCESS-ROUTINE.
```

Figure 12.32 Program for Random Retrieval and Updating of an Indexed Sequential File.

that is associated with the record. With direct addressing, every possible key in the file converts to a unique address, thus making it possible to locate any record in the file with one search and one read.

The user has complete freedom in deciding where records are to be located in a direct organized file. When creating or making additions to the file, the user may specify the location for a record key by supplying the track address and identifier, or simply the track address, while letting the system find the location for the record. The record is written in the first available location on the track specified. If the specified track is full, the system continues to search successive tracks until a location is found.

Direct organization is generally used for files whose characteristics do not permit the use of sequential or indexed sequential files, or for situations in which the time required to locate individual records must be kept at a minimum. This method has considerable flexibility, but it has a serious disadvantage in that the programming system must provide the routines to read a file of direct organization. The user is largely responsible for the logic and programming requirements to locate records since he establishes the relationship between the key of the record and the addresses in the direct-access storage device.

In a direct organization, the records will probably be distributed nonsequentially throughout the file. If so, processing the records in key sequence requires a preliminary sort or the use of a finder file.

With direct addressing, every possible key in the file converts to a unique address. This makes it possible to locate any record in the file with one seek and one record.

ENVIRONMENT DIVISION. The SELECT, ASSIGN, and ACCESS clauses are written in the same manner as *indexed sequential files accessed randomly,* with the exception of *NOMINAL KEY*. When direct data organization is used, the position of the logical records in a file is controlled by the user through the specifications of an ACTUAL KEY defined in the Environment Division. The ACTUAL KEY has two components. The first is the track identifier, which identifies the relative or actual track at which a record is to be placed or at which the search for a record is to begin. The second component is a record identifier, which serves as a unique logical identifier for a specific record on the track. Files with direct organization must be assigned to direct-access devices.

DATA DIVISION. Same entries as *indexed sequential files accessed randomly.*

Procedure Division

Read Statement. The INVALID KEY option must be specified for files in the random-access mode. The imperative statements following INVALID KEY is executed when the contents of the ACTUAL KEY field are invalid. Only the track specified in ACTUAL KEY is searched for records being read.

The contents of the ACTUAL KEY must be set to the desired value before the READ statement. The READ statement implicitly performs the functions of the SEEK statement, unless a SEEK statement for the file has been executed prior to the READ statement.

```
SEEK file-name RECORD
```

Figure 12.33
Format—Seek Statement.

Seek Statement. The SEEK statement serves only as documentation, and is meant to initiate the accessing of a mass storage data record for subsequent reading or writing (fig. 12.33). The file-name must be defined by a file description entry in the Data Division.

A SEEK statement pertains only to direct files in the random-access mode and may be executed prior to the execution of a READ or WRITE statement.

The SEEK statement uses the contents of the data-name in the ACTUAL KEY clause for the location of the record to be accessed. If the key is invalid, when the next READ or WRITE statement for the associated file is executed, control will be passed to the imperative statement following the INVALID KEY option.

Write Statement. The INVALID KEY phrase must be specified for a file that resides on a direct-access device. The INVALID KEY option is executed when the file is opened as I-O or OUTPUT and where the track address specified in the ACTUAL KEY is outside the limits of the file, or if a direct file is opened as I-O, or when a record is not found, or where the track number is outside the limits of the file.

For randomly accessed files the WRITE statement performs the functions of the SEEK statement, unless the SEEK statement for this record is executed prior to the WRITE statement. A WRITE statement executed for a direct file in the random-access mode assumes the meaning of a REWRITE statement when the file is opened, as I-O and the WRITE statement is the next output operation following a READ for a record with the same key. (See figures 12.34, 12.35.)

This program creates a file with direct organization through the use of an ACTUAL KEY. The ACTUAL KEY consists of a relative track address and a unique record identifier. In the program, a field in the input record (CD-ITEM-CODE) is converted to a track address (TRACK-ID) through the use of a simple remainder randomizing technique. This technique consists of dividing the value in the field of the input record (CD-ITEM-CODE) by 19, and using the resulting remainder (TRACK-ID) as the relative track address.

```
IDENTIFICATION DIVISION.
PROGRAM-ID.    CREATEDF.
REMARKS.  ILLSTRATE CREATION OF A DIRECT FILE.

ENVIRONMENT DIVISION.
CONFIGURATION SECTION.
SOURCE-COMPUTER.     IBM-370-I155.
OBJECT-COMPUTER.     IBM-370-I155.
INPUT-OUTPUT SECTION.
FILE-CONTROL.
    SELECT DA-FILE         ASSIGN TO DA-2311-D-MASTER
                           ACCESS IS RANDOM
                           ACTUAL KEY IS FILEKEY.
    SELECT CARD-FILE       ASSIGN TO UR-1442R-S-INFILE
                           RESERVE 3 ALTERNATE AREAS.

DATA DIVISION.
FILE SECTION.
FD  DA-FILE
    DATA RECORD IS DISK
    LABEL RECORDS ARE STANDARD.
01  DISK.
    05  DISK-ITEM-CODE        PICTURE X(3).
    05  DISK-ITEM-NAME        PICTURE X(29).
    05  DISK-STOCK-ON-HAND    PICTURE S9(6)      USAGE COMP SYNC.
    05  DISK-UNIT-PRICE       PICTURE S999V99    USAGE COMP SYNC.
    05  DISK-STOCK-VALUE      PICTURE S9(9)V99   USAGE COMP SYNC.
    05  DISK-ORDER-POINT      PICTURE S9(3)      USAGE COMP SYNC.
FD  CARD-FILE
    LABEL RECORDS ARE OMITTED
    DATA RECORD IS CARDS.
01  CARDS.
    05  CD-ITEM-CODE          PICTURE X(3).
    05  CD-ITEM-NAME          PICTURE X(29).
    05  CD-STOCK-ON-HAND      PICTURE S9(6).
    05  CD-UNIT-PRICE         PICTURE S999V99.
    05  CD-STOCK-VALUE        PICTURE S9(9)V99.
    05  CD-ORDER-POINT        PICTURE S9(3).
    05  FILLER                PICTURE X(23).
WORKING-STORAGE SECTION.
77  SAVE                    PICTURE S9(5)    USAGE COMP SYNC RIGHT.
77  QUOTIENT                PICTURE S9(4)    USAGE COMP SYNC RIGHT.
77  PRODUCT                 PICTURE S9(4)    USAGE COMP SYNC RIGHT.
01  FILEKEY.
    05  TRACK-ID            PICTURE S9(5)    USAGE COMP SYNC RIGHT.
    05  RECORD-ID           PICTURE X(29).
01  FLAGS.
    05  MORE-DATA-FLAG  PICTURE XXX          VALUE 'YES'.
        88  MORE-DATA                        VALUE 'YES'.
        88  NO-MORE-DATA                     VALUE 'NO'.

PROCEDURE DIVISION.
MAIN-ROUTINE.
    OPEN     INPUT     CARD-FILE
             OUTPUT    DA-FILE.
    READ CARD-FILE
        AT END MOVE 'NO' TO MORE-DATA-FLAG.
    PERFORM PROCESS-ROUTINE
        UNTIL NO-MORE-DATA.
    CLOSE      CARD-FILE
               DA-FILE.
    DISPLAY 'END OF JOB'
        UPON CONSOLE.
    STOP RUN.

PROCESS-ROUTINE.
    MOVE CD-ITEM-CODE TO SAVE.
    DIVIDE 19 INTO SAVE GIVING QUOTIENT
        REMAINDER TRACK-ID.
    MOVE CD-ITEM-NAME TO RECORD-ID.
    MOVE CD-ITEM-CODE TO DISK-ITEM-CODE.
```

Figure 12.34 Program for Creating Direct File.

Figure 12.34 Continued.

```
    MOVE CD-ITEM-NAME TO DISK-ITEM-NAME.
    MOVE CD-STOCK-ON-HAND TO DISK-STOCK-ON-HAND.
    MOVE CD-UNIT-PRICE TO DISK-UNIT-PRICE.
    MOVE CD-STOCK-VALUE TO DISK-STOCK-VALUE.
    MOVE CD-ORDER-POINT TO DISK-ORDER-POINT.
    WRITE DISK
        INVALID KEY PERFORM ERROR-ROUTINE.
    READ CARD-FILE
        AT END MOVE 'NO' TO MORE-DATA-FLAG.
    ERROR-ROUTINE.
        DISPLAY 'UNABLE TO WRITE RECORD'
            UPON CONSOLE.
        DISPLAY TRACK-ID
            UPON CONSOLE.
        PERFORM PROCESS-ROUTINE.
```

```
IDENTIFICATION DIVISION.
PROGRAM-ID.        UPDATING.
REMARKS.   THIS IS A SIMPLIFIED UPDATE PROGRAM.
ENVIRONMENT DIVISION.
CONFIGURATION SECTION.
SOURCE-COMPUTER.        IBM-370-I155.
OBJECT-COMPUTER.        IBM-370-I155.
INPUT-OUTPUT SECTION.
FILE-CONTROL.
    SELECT MASTER-FILE          ASSIGN TO DA-2311-D-MASTER
                                ACCESS MODE IS RANDOM
                                ACTUAL KEY IS FILEKEY.
    SELECT DETAIL-FILE          ASSIGN TO UT-2400-S-INFILE
                                ACCESS IS SEQUENTIAL.
    SELECT ACTION-FILE          ASSIGN TO UT-2400-S-OUTFILE.
DATA DIVISION.
FILE SECTION.
FD  MASTER-FILE
    LABEL RECORDS ARE STANDARD
    DATA RECORD IS MASTER-RECORD.
01  MASTER-RECORD.
    05  ITEM-CODE           PICTURE X(3).
    05  ITEM-NAME           PICTURE X(29).
    05  STOCK-ON-HAND       PICTURE S9(6)        USAGE COMP SYNC.
    05  UNIT-PRICE          PICTURE S999V99      USAGE COMP SYNC.
    05  STOCK-VALUE         PICTURE S9(9)V99     USAGE COMP SYNC.
    05  ORDER-POINT         PICTURE S9(3)        USAGE COMP SYNC.
FD  DETAIL-FILE
    LABEL RECORDS ARE OMITTED
    DATA RECORD IS DETAIL-RECORD.
01  DETAIL-RECORD.
    05  ITEM-CODE           PICTURE X(3).
    05  ITEM-NAME           PICTURE X(29).
    05  RECEIPTS            PICTURE S9(3)        USAGE COMP SYNC.
    05  SHIPMENTS           PICTURE S9(3)        USAGE COMP SYNC.
FD  ACTION-FILE
    LABEL RECORDS ARE OMITTED
    DATA RECORD IS ACTION-RECORD.
01  ACTION-RECORD.
    05  ITEM-CODE           PICTURE X(3).
    05  ITEM-NAME           PICTURE X(29).
    05  STOCK-ON-HAND       PICTURE S9(6)        USAGE COMP SYNC.
    05  UNIT-PRICE          PICTURE S999V99      USAGE COMP SYNC.
    05  ORDER-POINT         PICTURE S9(3)        USAGE COMP SYNC.
WORKING-STORAGE SECTION.
77  SAVE                    PICTURE S9(10)       USAGE COMP SYNC.
77  QUOTIENT                PICTURE S999         USAGE COMP SYNC.
01  FILEKEY.
    05  TRACK-ID            PICTURE S9(5)        USAGE COMP SYNC.
    05  RECORD-ID           PICTURE X(29).
01  ERROR-MESSAGE.
    05  ERROR-MESSAGE-1       PICTURE X(20).
    05  ERROR-MESSAGE-2       PICTURE X(36).
    05  ERROR-MESSAGE-3       PICTURE X(46).
01  FLAGS.
    05  MORE-DATA-FLAG        PICTURE XXX        VALUE 'YES'.
        88  MORE-DATA                            VALUE 'YES'.
        88  NO-MORE-DATA                         VALUE 'NO'.
PROCEDURE DIVISION.
MAIN-ROUTINE.
    OPEN     INPUT      DETAIL-F-FILE
             I-O        MASTERILE
             OUTPUT     ACTION-FILE.
    READ DETAIL-FILE
        AT END MOVE 'NO' TO MORE-DATA-FLAG.
    PERFORM PROCESS-ROUTINE
        UNTIL NO-MORE-DATA.
    CLOSE      DETAIL-FILE
               MASTER-FILE
               ACTION-FILE.
    STOP RUN.

PROCESS-ROUTINE.
    MOVE ITEM-CODE IN DETAIL-RECORD TO SAVE.
    DIVIDE 19 INTO SAVE GIVING QUOTIENT
        REMAINDER TRACK-ID.
    MOVE ITEM-NAME IN DETAIL-RECORD TO RECORD-ID.
    READ MASTER-FILE
        INVALID KEY PERFORM INPUT-ERROR, PERFORM ERROR-
            WRITE.
    COMPUTE STOCK-ON-HAND IN MASTER-RECORD = STOCK-ON-
        HAND IN MASTER-RECORD + RECEIPTS - SHIPMENTS.
    IF STOCK-ON-HAND IN MASTER-RECORD IS LESS THAN ZERO
    THEN
        PERFORM DATA-ERROR
        PERFORM ERROR-WRITE
    ELSE
        MULTIPLY STOCK-ON-HAND IN MASTER-RECORD BY UNIT-
            PRICE IN MASTER-RECORD GIVING STOCK-VALUE IN
            MASTER-RECORD.
    IF-STOCK-ON-HAND IN MASTER-RECORD IS NOT GREATER THAN
            ORDER-POINT IN MASTER-RECORD
    THEN
        PERFORM REORDER-ROUTINE
    ELSE
        NEXT SENTENCE.
    WRITE MASTER-RECORD
        INVALID KEY PERFORM OUTPUT-ERROR, PERFORM ERROR-
            WRITE.
    READ DETAIL-FILE
        AT END MOVE 'NO' TO MORE-DATA-FLAG.

INPUT-ERROR.
    MOVE ' KEY ERROR ON INPUT ' TO ERROR-MESSAGE-1.
    MOVE SPACES TO ERROR-MESSAGE-3.

DATA-ERROR.
    MOVE 'DATA ERROR ON INPUT ' TO ERROR-MESSAGE-1.
    MOVE MASTER-RECORD TO ERROR-MESSAGE-3.

OUTPUT-ERROR.
    MOVE 'KEY ERROR ON OUTPUT ' TO ERROR-MESSAGE-1.
    MOVE SPACES TO ERROR-MESSAGE-3.

ERROR-WRITE.
    MOVE DETAIL-RECORD TO ERROR-MESSAGE-2.
    DISPLAY ERROR-MESSAGE
        UPON CONSOLE.
    PERFORM PROCESS-ROUTINE.

REORDER-ROUTINE.
    MOVE CORRESPONDING MASTER-RECORD TO ACTION-RECORD.
    WRITE ACTION-RECORD.
```

Figure 12.35 Complete Updating Program—Direct File.

Relative Organization

Relative file organization permits accessing of records of a mass storage device in either a random or a sequential manner. Each record in a relative file is uniquely identified by an integer value greater than zero that specifies the record's logical ordinal position in the file.

Relative organization does not use an index or record key to identify each record in a file. The relative file consists of records which are identified by relative record numbers. The file may be thought of as composed of a serial string of areas, each capable of holding a logical record. Each of these areas is denominated by a relative record number. Records are stored and retrieved based on this number. For example, the tenth record is the one addressed by relative record number 10 and is in the tenth record area, whether or not records have been written in the first through ninth record areas.

Access Modes

In the *sequential* access mode, the sequence in which records are accessed is the ascending order of the relative record numbers of all records which currently exist within the file.

In the *random* access mode, the sequence in which records are accessed is controlled by the programmer. The desired record is accessed by placing its relative record number in a relative key data item.

The relative file organization can be best used when the record identification key can be used as a record number or can be easily converted to a record number.

Environment Division

The SELECT, ASSIGN, and ACCESS clauses are written in the same manner as indexed sequential files, with the exception of the RELATIVE KEY.

The RELATIVE KEY clause applies only to files with relative organization. The relative key data item contains the logical ordinal position of the record in the file. The first logical record has a relative record number of 1, and subsequent logical records have relative record numbers of 2, 3, 4, etc. The format of the RELATIVE KEY clause is

```
RELATIVE KEY IS data-name-1
```

Data-name-1 must not be in a record description entry for that file. The value contained must be an unsigned integer in the range of 1 . . . . . .

If ACCESS IS SEQUENTIAL, this clause is optional; however, if specified, data-name-1 will contain the current relative record number.

When the relative file is opened in the output mode, the file may be created by one of the following:

1. If the ACCESS MODE IS SEQUENTIAL, the WRITE statement causes a record to be released by the system. The first record will have a relative record number of 1 and subsequent records released will have relative numbers of 2, 3, 4, etc. If the RELATIVE KEY data item has been specified in the SELECT clause for the associated file, the relative record number of the

record just released will be placed in the RELATIVE KEY data item during execution of the WRITE statement .

2. If the ACCESS MODE IS RANDOM, prior to the execution of the WRITE statement, the value of the RELATIVE KEY data item must be initialized in the program, while the relative record number to be associated with the record is then released by the WRITE statement.

When a relative file is opened in the I-O mode and the ACCESS MODE IS RANDOM, records are to be inserted in the associated file. The value of the RELATIVE KEY data item must be initialized by the program, with the record to be associated with relative record number in the record area. Execution of the WRITE statement then causes the contents of the record area to be released.

Data Division

Same entries as indexed sequential except that the BLOCK CONTAINS clause is optional for relative I-O, and if present will be ignored.

Procedure Division

The same entries may be used as for indexed sequential organization. (See figure 12.36.)

```
IDENTIFICATION DIVISION.
PROGRAM-ID.  REL-IO.
*       1.  TO ILLUSTRATE CREATION OF AN RELATIVE FILE IN
            RANDOM MODE.
*       2.  TO ILLUSTRATE SEQUENTIAL READ OF A RELATIVE FILE.

ENVIRONMENT DIVISION.
CONFIGURATION SECTION.
SOURCE-COMPUTER.        XEROX-530.
OBJECT-COMPUTER.        XEROX-530.
INPUT-OUTPUT SECTION.
FILE-CONTROL.
    SELECT CARD-FILE            ASSIGN TO READER
                                RESERVE 2 AREAS.
    SELECT RELATIVE-CR          ASSIGN TO SYS030
                                ORGANIZATION IS RELATIVE
                                ACCESS MODE IS RANDOM
                                RELATIVE KEY IS WORK-KEY.
    SELECT RELATIVE-RD          ASSIGN TO SYS030
                                ORGANIZATION IS RELATIVE
                                ACCESS MODE IS SEQUENTIAL.
    SELECT PRINT-FILE           ASSIGN TO PRINTER
                                RESERVE 2 AREAS.

I-O-CONTROL.
    SAME AREA FOR RELATIVE-CR, RELATIVE-RD.

DATA DIVISION.
FILE SECTION.
FD  CARD-FILE                   LABEL RECORDS OMITTED
                                DATA RECORD IS CARD-RECORD.

01  CARD-RECORD.
    02  CARD-KEY.
        03  CARD-REGION         PICTURE 9.
        03  CARD-ACCOUNT-NO     PICTURE 999.
    02  CARD-SALESMAN           PICTURE X(30).
    02  CARD-LAST-YEAR          PICTURE 9(7)V99.
    02  CARD-TYPE               PICTURE XX.
    02  FILLER                  PICTURE X(35).
FD  RELATIVE-CR                 LABEL RECORDS OMITTED
                                RECORD CONTAINS 60 CHARACTERS
                                DATA RECORD IS REL-CR-RECORD.

01  REL-CR-RECORD.
    02  REL-CR-SALESMAN         PICTURE X(30).
    02  REL-CR-TYPE             PICTURE XX.
    02  REL-CR-LAST-YEAR        PICTURE 9(7)V99.
    02  FILLER                  PICTURE X(19).
FD  RELATIVE-RD                 LABEL RECORDS OMITTED
                                RECORD CONTAINS 60 CHARACTERS
                                DATA RECORD IS REL-RD-RECORD.

01  REL-RD-RECORD.
    02  REL-RD-IMAGE            PICTURE X(60).
FD  PRINT-FILE                  LABEL RECORDS OMITTED
                                DATA RECORD IS PRINT-RECORD.

01  PRINT-RECORD.
    02  PRINT-IMAGE             PICTURE X(133).
WORKING-STORAGE SECTION.
77  WORK-KEY                    PICTURE 9(4).
01  FLAGS.
    02  MORE-INPUT-FLAG             PICTURE XXX     VALUE 'YES'.
        88  MORE-INPUT                              VALUE 'YES'.
        88  NO-MORE-INPUT                           VALUE 'NO'.
    02  MORE-TRANS-FLAG             PICTURE XXX     VALUE 'YES'.
        88  MORE-TRANS                              VALUE 'YES'.
        88  NO-MORE-TRANS                           VALUE 'NO'.

PROCEDURE DIVISION.
MAIN-ROUTINE.
    OPEN    INPUT   CARD-FILE
            OUTPUT  RELATIVE-CR.
    READ CARD-FILE
        AT END MOVE 'NO' TO MORE-INPUT-FLAG.
    PERFORM CREATE-RELATIVE
        UNTIL NO-MORE-INPUT.
```

Figure 12.36 Program to Create and Access a Relative File Sequentially.

Figure 12.36 Continued.

```
    CLOSE     CARD-FILE
              RELATIVE-CR.
    OPEN    INPUT    RELATIVE-RD
            OUTPUT   PRINT-FILE.
    READ RELATIVE-RD
        AT END MOVE 'NO' TO MORE-TRANS-FLAG.
    PERFORM READ-RELATIVE
        UNTIL NO-MORE-TRANS.
    CLOSE     RELATIVE-RD
              PRINT-FILE.
    STOP RUN.

CREATE-RELATIVE.
    MOVE CARD-KEY TO WORK-KEY.
    MOVE CARD-SALESMAN TO REL-CR-SALESMAN.
    MOVE CARD-LAST-YEAR TO REL-CR-LAST-YEAR.
    MOVE CARD-TYPE TO REL-CR-TYPE.
    WRITE REL-CR-RECORD
        INVALID KEY    PERFORM WRITE-ERROR.
    READ CARD-FILE
        AT END MOVE 'NO' TO MORE-INPUT-FLAG.

WRITE-ERROR.
    DISPLAY ' INVALID KEY ' CARD-KEY
        UPON CONSOLE.
    CLOSE     CARD-FILE
              RELATIVE-CR.
    STOP RUN.

READ-RELATIVE.
    WRITE PRINT-RECORD FROM REL-RD-RECORD.
    READ RELATIVE-RD
        AT END MOVE 'NO' TO MORE-TRANS-FLAG.
```

Exercises

Write your answer in the space provided (answer may be one or more words).

1. ________ processing denotes the ability of the system to process data as soon as it becomes available.
2. Mass direct access storage enables the user to maintain ________ records of diversified applications and also to process ________ data for multiple application areas.
3. The direct access storage devices used for mass memory storage are ________, ________, and ________.
4. A direct access storage device is one in which each ________ has a ________ location and a ________ address.
5. A file is a ________ or an ________ of related information.
6. A logical record is a ________ of data related to a ________.
7. A key is a ________ that uniquely identifies a ________.
8. Sequential is an organization where records are placed in ________ rather than ________ sequence.
9. In an indexed sequential organization, records are arranged in ________ according to a ________ that is part of every ________ with the ________ maintained by the ________.
10. In a direct organization, records are arranged in a manner organized by the ________.
11. In a partitioned organization, there are ________ groups of ________ organized records.
12. Volume is a ________ term used to refer to a ________ unit or ________ storage.
13. In sequential processing, input transactions are grouped together and sorted into a ________ sequence.
14. Mass storage devices are efficient sequential processors when the percentage of activity against the master file is ________.
15. Random processing is the processing of all transactions against a master file regardless of the ________ of the documents.
16. The data file organization refers to the ________ of data records within a file.
17. In sequential organization, the data records are organized solely in their ________ physical locations.
18. After a sequential file is opened, the record can be ________, ________, and ________ in the same location without ________ a new file.
19. A sequential file can be used for both input and output activities without the necessity for ________ and ________ files between operations.
20. In creating a standard sequential file, the file must be assigned to a particular ________ device.
21. To create a sequential file, the procedural statements must use an ________ file, and a ________ statement.
22. The ________ option of the Write statement must be used when a sequential disk file is being created.

23. The statement following ______________ is activated when an attempt is made to write beyond the space reserved for the file.
24. After a sequential disk file has been created, it may be maintained by ______________ the records and placing them back in the file.
25. The ______________ option of the Write statement may be used if the sequential file is to be used for both input and output operations.
26. An indexed sequential file is a ______________ file with ______________ that permit rapid access to records.
27. A ______________ provided by the user precedes each ______________ of data and is used to provide the index.
28. In an indexed sequential file, the ______________ has control over the location of the individual records.
29. The ______________ in an indexed sequential file are created and written by the system as the file is created or organized.
30. In an indexed sequential file, the system will access records sequentially by ______________ written rather than by the physical sequence in the file.
31. The prime area of an indexed sequential file is the area in which ______________ are written when the file is ______________ or subsequently ______________.
32. The track index is the ______________ level of index and points to ______________.
33. The cylinder index is the ______________ level of index and points to ______________.
34. An overflow area contains a certain number of whole ______________ as specified by the ______________ and are reserved in each ______________ for ______________ data from ______________ tracks in that ______________.
35. The Access Mode clause defines the manner in which the ______________ of a file are to be ______________.
36. If the Access Mode clause is not specified ______________ is assumed.
37. When the Access Is Random clause is used, the storage and retrieval of records are based on an ______________ or ______________ key associated with each record.
38. The ______________ clause is used to access an indexed sequential file.
39. Records in an ascending sequential file may be accessed at a record that is not the ______________ record in a file and continued until the file is ______________.
40. A Nominal Key is used with indexed sequential file to specify an elementary ______________ defined in the ______________ of the program.
41. A Start statement is used to ______________ a file.
42. The Start statement ______________ the processing of a segment of a ______________ indexed sequential file at a specified ______________.
43. A Start statement must be executed after the ______________ statement but before the ______________ statement.
44. When accessing an indexed sequential file randomly, the ______________ field must be set to the ______________ field of the desired record before a ______________ statement is executed for the file.
45. When adding records to an indexed sequential file, it is not necessary to ______________ a file.
46. An updated record is placed back in a disk file with a ______________ statement.
47. A Read statement for a file must be executed before a ______________ statement for the file can be executed.
48. A direct organized file is characterized by some predictable relationship between the ______________ of the record and the ______________ of that record in a direct access storage device.
49. The relationship in a direct organized file is established by the ______________ and permits access to records in a file.
50. Direct file organization is usually for files where the time required to ______________ individual records must be kept at a ______________.
51. The Seek statement serves only as ______________.
52. The Seek statement pertains only to a ______________ file in the ______________ mode.

53. A Seek statement uses the contents of the data in the ______________ clause for the ______________ of the record to be accessed.
54. Relative organization permits accessing records of a mass storage device in either a ______________ or ______________ manner.
55. Relative organization does not use an ______________ or ______________ to identify each ______________ in the file.
56. The relative file consists of ______________ which are identified by ______________ numbers.
57. The relative key data item contains the logical ______________ position of the record in the file.
58. The value of the relative key data item must be initialized by the ______________, with the record to be associated with the relative ______________ in the record area.

Answers

1. INLINE
2. CURRENT, NONSEQUENTIAL
3. DISK STORAGE, DRUM, DATA CELLS
4. PHYSICAL RECORD, DISCRETE, UNIQUE
5. PHYSICAL UNIT, ORGANIZED COLLECTION
6. COLLECTION, COMMON IDENTIFIER
7. CONTROL FIELD, RECORD
8. PHYSICAL, LOGICAL
9. SEQUENCE, KEY, RECORD, INDEX, SYSTEM
10. PROGRAMMER
11. INDEPENDENT, SEQUENTIALLY
12. GENERIC, STANDARD, AUXILIARY
13. PREDETERMINED
14. HIGH
15. SEQUENCE
16. PHYSICAL ARRANGEMENT
17. SUCCESSIVE
18. READ, UPDATED, WRITTEN BACK, CREATING
19. OPENING, CLOSING
20. MASS STORAGE
21. OUTPUT, WRITE
22. INVALID
23. INVALID
24. UPDATING
25. I-O
26. SEQUENTIAL, INDEXES
27. KEY, BLOCK
28. PROGRAMMING SYSTEM
29. INDEXES
30. KEYS
31. RECORDS, CREATED, REORGANIZED
32. LOWEST, DATA RECORDS
33. HIGHER, TRACK INDEXES
34. TRACKS, USER, CYLINDER, OVERFLOW, PRIME, CYLINDER
35. RECORDS, ACCESSED
36. ACCESS IS SEQUENTIAL
37. ACTUAL, NOMINAL
38. RECORD KEY
39. FIRST, CLOSED
40. VARIABLE, WORKING-STORAGE SECTION
41. POSITION
42. INITIATES, SEQUENTIALLY, KEY
43. OPEN, READ
44. NOMINAL KEY, RECORD KEY, READ
45. RELOCATE
46. REWRITE
47. REWRITE
48. KEY, ADDRESS
49. USER
50. LOCATE, MINIMUM
51. DOCUMENTATION
52. DIRECT, RANDOM ACCESS
53. ACTUAL KEY, LOCATION
54. RANDOM, SEQUENTIAL
55. INDEX, RECORD KEY, RECORD
56. RECORDS, RELATIVE RECORD
57. ORDINAL
58. PROGRAM, RECORD NUMBER

Questions for Review

1. Explain the "inline" processing technique and how it is used with mass storage devices.
2. Briefly define the following terms: direct access storage device, file, record, key, sequential, indexed sequential, direct, partitioned, volume.
3. What is sequential processing and how does it differ from random processing?
4. How is a file organized in a sequential organization manner?
5. What are the necessary entries in the Environment, Data, and Procedure divisions to create a sequential disk file?
6. How is a standard sequential file updated?
7. Explain the operations of a standard sequential organized file.
8. What is an indexed sequential file organization?
9. What are the important advantages of an indexed sequential file?
10. Briefly describe the three distinct areas of an indexed sequential file.
11. What are the necessary entries in the Environment, Data and Procedure divisions to create an indexed sequential file?
12. How are the records accessed in an indexed sequential file?
13. What is the purpose of the Start statement and how is it used in processing records in an indexed sequential file?
14. How are records added to an indexed sequential file randomly?
15. What is the function of the Rewrite statement and how is it used with mass storage devices?
16. What is direct organization?
17. Explain the operation of a direct organization file.

18. What are the necessary entries in the Environment, Data and Procedure divisions to create a direct organization file?
19. What is the Seek statement used for?
20. What is relative organization?
21. Explain the operation of a relative organized file.
22. When is the relative file organization best used?

Problems

1. *Match each item with its proper description.*

Item	Description
______ 1. Prime Area	A. Process of transactions against a master file regardless of sequence.
______ 2. Inline	B. Access at random by multiple users of data involving mass storage devices.
______ 3. Sequential Processing	C. Records are written when file is created or subsequently reorganized.
______ 4. Key	D. Process data as soon as available.
______ 5. Indexes	E. Input transactions grouped and sorted into a predetermined sequence against a master file.
______ 6. Direct Access	F. A control field in a logical record that uniquely identifies it.
______ 7. Overflow Area	G. Created and written by the operating system when indexed sequential file is created or reorganized.
______ 8. Direct Access Storage Device	H. A term used to refer to a standard unit or auxiliary storage.
______ 9. Random Processing	I. Each physical record has a discrete location and a unique address.
______10. Volume	J. Certain number of whole tracks reserved for each cylinder for data that will not fit in prime areas.

2. *Match each item with its proper description.*

Item	Description
______ 1. START	A. Initiates accessing of a mass storage record for subsequent reading or writing based on actual key.
______ 2. NOMINAL KEY	B. Place logical record on a direct access storage device with a specified record if contents of key are found to be valid.
______ 3. ACCESS MODE	C. Activated when no record with a control field equal to key is found.
______ 4. SEEK	D. Initiates processing of a segment of a sequentially organized file at a specified key.
______ 5. RELATIVE KEY	E. Beyond limit of space reserved for file.
______ 6. RECORD KEY	F. Initiates processing sequentially indexed sequential files at a specified record.
______ 7. INVALID KEY	G. Used in direct organization to access a record.
______ 8. INVALID	H. Defines the manner in which records are to be accessed.
______ 9. REWRITE	I. Contains logical ordinal position of the record in the file.
______10. ACTUAL KEY	J. Used to access records in an indexed sequential file.

3. *Match each data file organization with its proper description.*

Item	Description
______ 1. Direct	A. Independent groups of sequentially organized records.
______ 2. Relative	B. Sequential file with indexes that permit rapid access to records.
______ 3. Sequential	C. Organized on basis of some predictable relationship between key of record and address in storage.
______ 4. Partitioned	D. Records are organized solely on the basis of their successive physical location in the file.
______ 5. Indexed Sequential	E. Each record uniquely identified by an integer value greater than zero which specifies record's logical ordinal position in the file.

4. *The Johnson Corporation is converting its master tape file for sales to a master sequential disk file with the same organization.*

   The file-name for the tape is SALES-TAPE and the record-name is TAPE-RECORD. The record lengths are fixed at 100 characters and recorded in blocks of five with standard labels.

   The disk file name is SALES-DISK and the record-name is DISK-RECORD with the same organization as the tape file.

   Write the program to create the disk file. Assume following device numbers.

   (In all subsequent problems, if the device numbers are not mentioned, the following will be assumed.)

Computer	— IBM 370 Model H155.
Tape Unit	— Model 2400.
Disk Unit	— Model 2311.
Printer	— Model 1403.
Card Reader	— Model 2540.
Card Punch	— Model 2540.

5. *The Acme Manufacturing Company wishes to create a sequential disk file from a set of cards.*

   a. Write the necessary entries for the Environment and Data Divisions based on the following information:

      Computer to be used — IBM 370 Model H155.
      Input — File-name FILE-IN Record-name CARD-IN.
      Output — File-name DISK-SEQ Record-name DISK-IN
      Record Size 80 characters.
      All records on the disk are in blocks of five and are of a fixed length.
      Output Device — Disk Model 2311.

   b. Write the necessary procedural statements to create the disk file by transferring the input data to the output file. Assume both records are of the same size. The program should branch to a routine called FULL-DISK when the disk is full and display a message on the console to that effect.

6. *The Bryan Tool Corporation is converting its present master tape file to a disk with the same organization. You are asked to write a program based on the following:*

   a. Write the necessary entries for the Environment and Data Division based on the following:

      Computer to be used — IBM 370 Model H155.
      Input — File-name TAPE-FILE Record-name TAPE-IN
      Record Size 100 characters.
      All records on the tape are in blocks of five and are of a fixed length. All label records are standard.
      Output — File-name DISK-SEQ Record-name DISK-IN
      Record Size 100 characters.

   b. Write the necessary procedural statements to create the disk file by transferring the input data to the output file. The program should branch to a routine called FULLDISK when the disk is full and display a message on the console to that effect.

7. *Write a program to update the Bryan Tool Corporation sequential disk file (Problem 6) for new addresses. The following information is provided:*

   Disk Record Record name RECORD-IN has the following format.

Field	**Record Positions**
Date	1–6
Customer Number	7–12
Customer Name	13–27
Address—Street	28–42
—City and State	43–57
Balance	58–64
Credit Limit	65–71
Unused in this program	72–100

Changes Card File-name CHANGES Record-name CHANGE-CARD

Field	Record Positions
Date	1–6
Customer Number	7–12
Street	13–27
City and State	28–42
Unused	43–80

If a customer number is not in the disk file, the program should branch to a routine called NOT-IN-FILE where the appropriate message together with the customer number will be displayed on console.

8. *Using the same information in Problem 6, the Bryan Tool Corporation wishes to create an indexed sequential file from its present master tape file. The Record Key will be the customer number.*

   If the record is not written for any reason, the program should branch to ERROR-ROUTINE where the proper message will be displayed.

   When the program is complete, display INDEXED FILE FOR CUSTOMER ACCOUNTS CREATED on console.

9. *An indexed file contains the following record:*

DISK-RECORD

Field	Record Position	
Balance	1–6	
Date	7–12	
Cumulative Disbursements	13–20	
Cumulative Receipts	21–28	
Minimum Balance	29–36	
Class of Stock	37–40	
Stock Number	41–47	
Unit Price	48–53	XXX.XXX
Amount	54–61	XXXXXX.XX
Unit	62–65	
Description	66–95	
Unused	96–100	

Write the necessary entries for

a. The File-Control paragraph. The file name is DISK-FILE and the record name is DISK-RECORD. The disk device is a model 2311. The elementary variable stock-number contains the identifier for each record.
b. The File Section entry. The records in the file are blocked in groups of five.
c. The procedural statements to access all records sequentially and list them on a model 1403 printer in the same format as the input record leaving two spaces between each field. Appropriate headings for each field should be printed and totals of the balance and amount should be accumulated and printed at end of report.

10. *Using the same record as Problem 9, the outside auditors wish to sample check our inventory as follows:*

    A listing of the accounts in the same print format as Problem 9 for the following groups of accounts.
    Five accounts starting with stock-number 1456845, 5152467, and 8759415.

Write the necessary entries for

a. The File-Control paragraph for above.
b. The procedural statements to access the specified records and print them.

11. *Using the disk record described in Problem 9, you wish to add records to the indexed sequential disk file without having to recreate the file. Write the necessary entries for the File-Control paragraph and Procedure Division to accomplish this.*

12. *Using the same record as Problem 9, you wish to access randomly to print selected critical inventory items.*

The finder cards will contain the following information:

File-name	FINDER-FILE	Record-name	FINDER-CARD

Field	Card Columns
Date	1-6
Stock Number	7-13

The card reader is model 2540.

Write the necessary entries for

a. The File-Control paragraph.
b. The procedural statements to access the records and print the selected records in the same format as Problem 9. If record is not found, display a message on the console to that effect and proceed to the next record.

13. *Using the same record as Problem 9, write a program to update the records in the disk randomly.*

The format for the transaction card is as follows: (The card will be read on a model 2540 card reader).

File-name	TRANSACTION-FILE	Record-name	TRANSACTION-CARD

Field	Card Columns
Date	1-6
Stock Number	7-13
Transaction Code	14
Receipts—1	
Disbursements—2	
Quantity	15-19
Amount	20-26
Unused	27-80

Write the necessary entries for the

a. File-Control paragraph.
b. The procedural entries for the following:

1. *Receipts Card.*
   Add Quantity to Balance.
   Add Quantity to Cumulative Receipts.
   Add Amount to Amount in Disk Record.
2. *Disbursement Card.*
   Subtract Quantity from Balance.
   Add Quantity to Cumulative Disbursements.
   Subtract Amount from Amount in Disk Record.
3. If Stock Number not found in Disk Record, branch to ERROR-ROUTINE and display appropriate message and read another transaction record.
4. Compute new Unit Price by dividing Balance into Amount.
5. Print TRANSACTION REGISTER as follows: Same print format as Problem 9.
   List the old balance, then the transaction record and finally the new balance.
6. Write new record back on disk file.

13 Declaratives and Linkage Sections

DECLARATIVES SECTION

The Declaratives Section is written in the Procedure Division to specify any special circumstance under which a procedure is to be executed in the object program. Although the COBOL compiler provides error recovery routines in the case of input/output errors, the programmer may wish to specify additional procedures of his own to supplement those supplied by the compiler. The system automatically handles checking and creation of labels in tape files, but the programmer may wish to use his own tape table-handling procedures. The Report Writer feature may also use declarative sections.

Since these procedures can be executed only at a time when an error occurs in the reading or writing of records, or when the labels of a file are to be processed, or before a report group is to be produced, they cannot appear in the regular sequence of procedural statements. The procedures are invoked nonsynchronously; that is, they are not executed as part of the sequential coding written by the programmer, but rather when a condition occurs which cannot normally be tested by the program.

The Declaratives Section is written as a subdivision at the beginning of the Procedure Division prior to the execution of the first procedure. A group of declarative procedures constitutes a declaratives section. Although the declaratives sections are located at the beginning of the Procedure Division, execution of the object program actually starts with the first procedure following the termination of the declaratives section.

The declaratives section subdivision of the Procedure Division must begin with the key word DECLARATIVES at the A margin followed by a period and a space. The declaratives section is terminated by the key words END DECLARATIVES followed by a period and a space (fig. 13.1). Both DECLARATIVES and END DECLARATIVES must appear on a line by themselves with no other coding permissible on the same line. Every declaratives

```
PROCEDURE DIVISION.
DECLARATIVES.
{section-name SECTION.  USE-sentence.
{paragraph-name.      sentence . . .   .} . . . } . .
END DECLARATIVES.
```

Figure 13.1
Format—Declaratives Section.

section is terminated by the occurrence of another declaratives section or the words END DECLARATIVES.

A declaratives section consists of

1. A section-name followed by the key word SECTION written at the A margin.
2. A USE statement must follow the section header on the same line after an intervening space or spaces, and terminated by a period.

Declaratives are instructions to the COBOL compiler and the run-time system to perform special operations during object programs execution before or after other standard operations. All declaratives sections must be grouped at the beginning of the Procedure Division preceded by the key word DECLARATIVES and followed by the key words END DECLARATIVES. On the line following the declaratives line, a section-name must be specified followed by a USE statement. The section continues until another section-name is executed or until the END DECLARATIVES line is used.

Use Statement

The USE statement specifies procedures to be performed for each type of declarative that is added to the standard procedures provided by the compiler. The USE statement is itself never executed; rather, it defines the conditions calling for the execution of the USE procedures. The remainder of the section must consist of one or more procedural paragraphs that specify the procedures to be performed.

Within a USE procedure, there must be no reference to any nondeclarative procedure. Conversely, in the nondeclaratives portion, there must be no reference to procedure names that appear in the declaratives section, except that the PERFORM statement may refer to a USE declarative handling label and error procedures associated with a USE statement.

The following types of procedures are associated with the USE statement.

1. Input/output label-handling procedures.
2. Input/output error-handling procedures.
3. Report-writing procedures.

A USE statement must identify the conditions under which each section is executed. Procedures specified in a USE statement must be self-contained. One USE section (that is, one declaratives section) may be accessed from another USE section via the PERFORM statement, but it is illegal to branch outside the declaratives area, or into the declaratives area from the outside.

Label-Processing Declaratives

The statements in these declaratives are used to handle user-created header labels (see fig. 13.2). Nonstandard labels can be specified only for tape files. These procedures are in addition to the standard procedures provided by the input/output system.

The declarative statement USE BEFORE STANDARD BEGINNING LABEL PROCEDURE transfers control to the declaratives section *before* the OPEN statement is executed. The procedures in this declaratives section cannot include input/output statements, such as, OPEN, CLOSE, READ or WRITE,

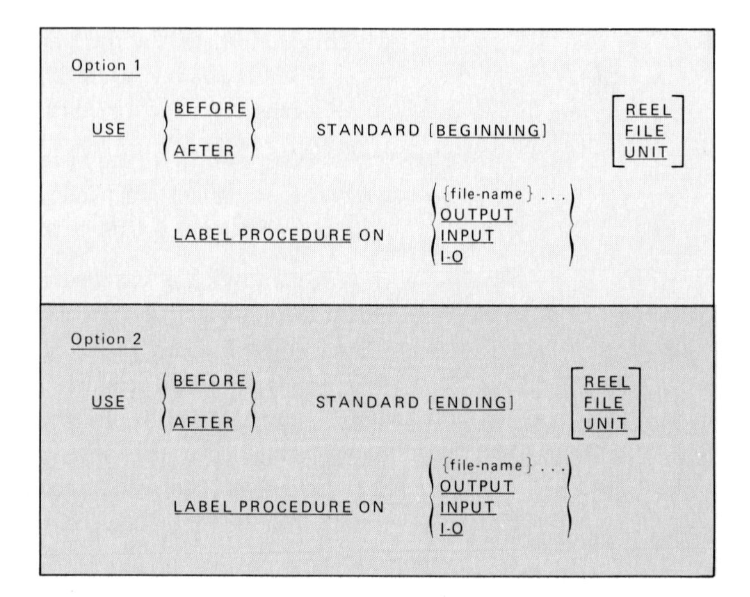

Figure 13.2 Format—Label Processing Declaratives.

for the file which caused the transfer. It is permissible to include input/output statements for other files except one whose usage is modified by OPEN statements. Such a file cannot have input/output during the declarative procedures. The ACCEPT and DISPLAY statements cannot be included in the declarative procedures.

The declarative statement USE AFTER STANDARD BEGINNING LABEL PROCEDURE transfers control to the declaratives section during the OPEN statement's execution. The declarative procedures can include READ or WRITE statements for the file that caused execution of the declarative; OPEN and CLOSE statements on this file are not allowed. It is *not* permissible to include input/output statements for other files. The ACCEPT and DISPLAY statements can be included in the declarative procedures.

The declarative statement USE BEFORE STANDARD ENDING LABEL PROCEDURE transfers control to the declaratives section before the CLOSE statement is executed. The declarative procedures may contain input/output statements for any file and the ACCEPT and DISPLAY statements.

The declarative statement USE AFTER STANDARD ENDING LABEL PROCEDURE transfers control to the declaratives section during the execution of the CLOSE statement. The declarative procedure can include READ or WRITE statements for the file but not OPEN and CLOSE statements. It is *not* permissible to include input/output statements for other files. The ACCEPT and DISPLAY statements can be included in the declarative procedures.

Rules Governing the Use of Label-Processing Declaratives

1. An OPEN statement in the Procedure Division causes the execution of the USE statement associated with the BEGINNING label.
2. A CLOSE statement in the Procedure Division causes the execution of the USE statement associated with the ENDING label.

3. The word BEGINNING refers to the user's header labels and the word ENDING refers to any trailer labels. If neither is specified, both header and trailer labels are processed.
4. The labels must be listed as data-names in the LABEL RECORDS clause in the file description entry for the file and must be described as level 01 data item subordinate to the file entry.
5. If neither UNIT, REEL, nor FILE is included, the designated procedures are executed for both REEL or UNIT (whichever is appropriate) and FILE labels. The REEL option is not applicable to direct-access (mass storage) files. The UNIT option is not applicable to random-access mode since only FILE labels are processed in this mode.
6. The same file-name can appear in a different specific arrangement of a format. However, appearance of a file-name in a USE statement must not cause the execution or the simultaneous request for execution of more than one USE declarative. No file may request a sort-file.
7. The file-name option must not be used with a LABEL RECORDS ARE OMITTED clause.

   The designated procedures of a USE statement are executed as follows. The OUTPUT, INPUT, or I-O options are specified, when

   a. The OUTPUT is specified only for files opened as OUTPUT.
   b. The INPUT is specified only for files opened as INPUT.
   c. The I-O is specified only for files opened as I-O.

If the OUTPUT, INPUT, or I-O option is specified, the USE procedure does not apply respectively to OUTPUT, INPUT, or I-O files that are described with the LABEL RECORDS ARE OMITTED clause.

Within the procedures of a USE declarative in which the USE statement specifies an option other than file-name-1 option, reference to common label items need not be qualified by a file-name. A common label is an elementary data item that appears in every label record of the program but does not appear in any data record of the program. Such items must have the same name, description, and relative position in every label record.

The exit from the declaratives section is inserted following the last statement of the section. All logical program paths within the section must lead to the exit point. (See figure 13.3.)

Input-Output Error-Processing Declaratives

These declaratives are used to specify procedures to be followed if an input/output error occurs during the processing (fig. 13.4). This option provides the users with input/output correction procedures in addition to those specified by the compiler.

When a USE statement is present, it must immediately follow a section header in the declaratives section, and must be followed by a period and a space. The remainder of the section must consist of one or more procedural programs that define the procedures to be used. USE is not an executable statement; rather,

The following program creates a file with user labels. To create the labels, the program contains a DECLARATIVES section, with USE procedures for creating both header and trailer labels.

The program illustrates the following items:

1. For the file requiring the creation of user labels, the LABEL RECORDS clause uses the data-name option.
2. THE USE AFTER BEGINNING/ENDING LABEL option is specified to create user labels.
3. The program creates two user header labels, utilizing the special exit GO TO MORE LABELS to create the second label.
4. The information to be inserted in the user labels comes from input file records. Therefore, records containing the information must be read and stored before the output file is opened, and the header label procedures are invoked.

```
IDENTIFICATION DIVISION.
PROGRAM-ID.          LABELPGM.

ENVIRONMENT DIVISION.
CONFIGURATION SECTION.
SOURCE-COMPUTER.          IBM-370-I155.
OBJECT-COMPUTER.          IBM-370-I155.
INPUT-OUTPUT SECTION.
FILE-CONTROL.
       SELECT NO-LBL              ASSIGN TO UT-2400-S-INFILE.
       SELECT USER                ASSIGN TO UT-2400-S-USRFILE.

DATA DIVISION.
FILE SECTION.
FD  NO-LBL
       RECORD CONTAINS 80 CHARACTERS
       LABEL RECORD IS OMITTED.
01     IN-REC.
       05  TYPEN               PICTURE X(4).
       05  DEPT-ID             PICTURE X(11).
       05  BIL-PERIOD          PICTURE X(5).
       05  NAME                PICTURE X(20).
       05  AMOUNT              PICTURE 9(6).
       05  FILLER              PICTURE X(15).
       05  SECUR-CODE          PICTURE XX.
       05  FILLER              PICTURE 9.
       05  ACCT-NUM            PICTURE 9(10).
       05  FILLER              PICTURE 9(6)
01     IN-LBL-HIST             REDEFINES IN-REC.
       05  FILLER              PICTURE X(4).
       05  FILE-HISTORY        PICTURE X(76).
FD     USER
       RECORD CONTAINS 80 CHARACTERS
       BLOCK CONTAINS 5 RECORDS
       LABEL RECORDS ARE USR-LBL USR-LBL-HST.
01     USR-LBL.
       05  USR-HDR             PICTURE X(4).
       05  DEPT-ID             PICTURE X(11).
       05  USR-REC-CNT         PICTURE 9(8)     COMP-3.
       05  BIL-PERIOD          PICTURE X(5).
       05  FILLER              PICTURE X(50).
       05  SECUR-CODE          PICTURE XX.
01     USR-LBL-HIST            REDEFINES USR-LBL.
       05  FILLER              PICTURE X(4).
       05  LBL-HISTORY         PICTURE X(76).
01     USR-REC.                PICTURE X(4).
       05  TYPEN               PICTURE X(5).
       05  FILLER              PICTURE X(20).
       05  NAME                PICTURE X(4).
       05  FILLER              PICTURE 9(10).
       05  ACCT-NUM            PICTURE 9(6)     COMP-3.
       05  AMOUNT              PICTURE X(23).
       05  FILLER              PICTURE 9(8).
       05  U-SEQ-NUMB
WORKING-STORAGE SECTION.
77     U-REC-NUMB              PICTURE 9(8)     VALUE ZEROS.
77     SAV-DEPT-ID             PICTURE X(11).
77     LBL-SWITCH              PICTURE 9        VALUE ZERO.
01     STOR-REC.
       05  DEPT-ID             PICTURE X(11).
       05  BIL-PERIOD          PICTURE X(5).
       05  SECUR-CODE          PICTURE XX.
01     FLAGS.
       05  MORE-DATA-FLAG              PICTURE XXX    VALUE 'YES'.
           88  MORE-DATA                              VALUE 'YES'.
           88  NO-MORE-DATA                           VALUE 'NO'.

PROCEDURE DIVISION.
DECLARATIVES.
USR-HDR-LBL SECTION.     USE AFTER BEGINNING FILE LABEL
       PROCEDURE ON USER.
PAR-1.
     IF LBL-SWITCH = 0
     THEN
          MOVE SPACES TO USR-LBL
          MOVE ZEROS TO USR-REC-CNT
          MOVE 'UHL1' TO SUR-HDR
          MOVE CORRESPONDING STOR-REC TO USR-LBL
          ADD 1 TO LBL-SWITCH GO TO MORE-LABELS
     ELSE
          MOVE 'UHL2' TO USR-HDR
          MOVE FILE-HISTORY TO LBL-HISTORY.
USR-TRLR-LBL SECTION.    USE AFTER ENDING FILE LABEL
       PROCEDURE ON USER.
     MOVE SPACES TO USR-LBL.
     MOVE 'UTL1' TO USR-HDR.
     MOVE SAV-DEPT-ID TO DEPT-ID IN USR-LBL.
     MOVE U-REC-NUMB TO USR-REC-CNT.
END DECLARATIVES.


MAIN-ROUTINE.
     OPEN     INPUT     NO-LBL.
     PERFORM PROCESS-ROUTINE
          UNTIL NO-MORE-DATA.
     CLOSE     NO-LBL
               USER.
     STOP RUN.

PROCESS-ROUTINE.
     READ NO-LBL
          AT END MOVE 'NO' TO MORE-DATA-FLAG.
     IF USER-SWITCH = 1
     THEN
          NEXT SENTENCE
     ELSE
          ADD 1 TO USER-SWITCH
          MOVE CORRESPONDING IN-REC TO STOR-REC
          MOVE DEPT-ID OF IN-REC TO SAV-DEPT-ID
          OPEN OUTPUT USER
          PERFORM PROCESS-ROUTINE.
     MOVE SPACES TO USR-REC.
     ADD 1 TO U-REC-NUMB.
     MOVE CORRESPONDING IN-REC TO USR-REC.
     MOVE U-REC-NUMB TO U-SEQ-NUMB.
     WRITE USR-REC.
```

Figure 13.3 Sample Label Declarative Program.

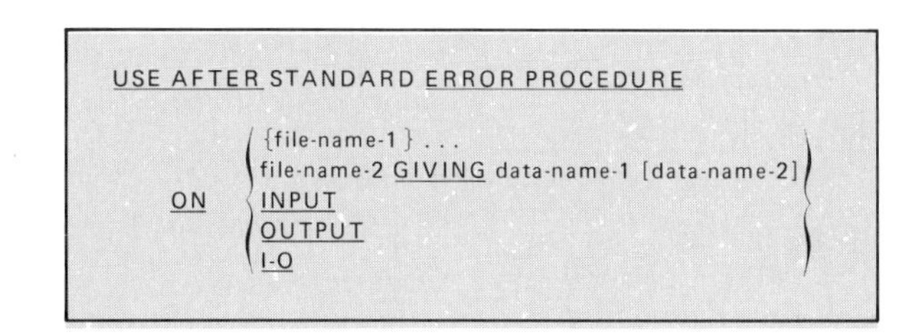

Figure 13.4
Format—Input-Output Processing Declarative.

it defines the conditions under which the associated procedure, the declaratives section itself, is to be executed.

Rules Governing the Use of Input-Output Error-Processing Declaratives

1. The error-handling procedures are activated when an input/output error occurs during the execution of a READ, WRITE, REWRITE, or START statement.
2. Automatic system error routines are executed before user-specified procedures.
3. When the file-name is used, error-handling procedures are executed for input/output errors occurring for that file(s) only.
4. A file-name must not be referenced, implicitly or explicitly, by more than one USE statement.
5. User handling procedures are executed for invalid key conditions if the INVALID KEY option is not specified in the statement causing the condition.
6. Within the error procedures, the allowable executable statements depend on the organization and the access specified for the file in error.
7. The user error-handling procedures are executed when the INPUT, OUTPUT, or I-O option is as follows:

   a. If the INPUT option is specified and input/output error occurs, only the files opened as INPUT will be affected.
   b. If the OUTPUT option is specified, and input/output error occurs, only the files opened as OUTPUT will be affected.
   c. When the I-O option is specified, and input/output error occurs, only the files opened as I-O will be affected.

An exit form of this type of declarative can be affected by executing the last statement in the section (normal return) or by means of a GO TO statement. This is the normal return from an error declarative to the statement following the input/output statement that causes the error.

To gain access to error information, the GIVING option must be used. Error-processing statements must be coded in the COBOL program or in a subprogram to analyze the contents of data-name-1, or data-name-1 and data-name-2.

When an uncorrectable input/output error occurs, the declarative is entered. Data-name-1 will contain information indicating the condition. Data-name-2 is an area large enough to hold the largest physical block that exists on or can be written on file-name, and will contain the block in error when the error occurs during the READ operation.

Continued Processing of a File

The continued processing of a file is permitted under the following conditions.

1. An error-processing procedure exists in the declaratives section.
2. The detection of the error results in an automatic transfer to the error-processing procedure which permits the programmer to examine the error condition before it enters the process.
3. At the conclusion of the processing of the error, it is the programmer's responsibility to update the parameters normally returned by the Input/Output Control System.

Report Writer Declaratives

The USE BEFORE REPORTING sentence specifies Procedure Division statements that are to be executed just before a report group named in the Report Section of the Data Division is produced (fig. 13.5).

Figure 13.5
Format—Report Writer Declarative.

```
USE BEFORE REPORTING data-name.
```

Rules Governing the Use of Report Writer Declaratives

1. Data-name represents a report group named in the Report Section of the Data Division and must not appear in more than one USE statement. Data-name must be qualified if not unique.
2. No Report Writer statement (INITIATE, GENERATE, or TERMINATE) may be written in any procedural paragraph(s) following the USE statement in the declaratives.
3. The designated procedures are executed by the Report Writer just before the named report is produced, regardless of page or control breaks associated with the report group. The report group may be of any type except DETAIL.
4. There must not be any reference to any nondeclarative procedure. Conversely, in the nondeclarative portion, there must be no reference of procedure-names that appear in the declarative portion, except that PERFORM statements may refer to a USE declarative or to procedures associated with the USE declarative.

(*Note:* When the user wishes to suppress the printing of a specified report group, the statement MOVE 1 TO PRINT-SWITCH is used in the USE BEFORE REPORTING declarative section. When this statement is encountered, only the specified report group is not printed. The statement must be written for each report group whose printing is to be suppressed.)

The use of PRINT-SWITCH to suppress printing of a report group implies that:

```
PROCEDURE DIVISION.
DECLARATIVES.
TRANFILE-LABEL SECTION.
       USE AFTER STANDARD BEGINNING FILE LABEL PROCEDURE ON
             TRANFILE.
LABEL-ROUTINE.
       MOVE......................
       IF...........................
MASTER-LABEL SECTION.
       USE AFTER STANDARD BEGINNING FILE LABEL PROCEDURE ON
             MASTERFILE.
M-LABEL.
       IF..........................
FILE-ERROR SECTION.
       USE AFTER STANDARD ERROR PROCEDURE-ON INPUT.
ERROR-ROUTINE.
       MOVE.......................
END DECLARATIVES.
BEGIN.
       OPEN INPUT TRANFILE, MASTERFILE.
```

Figure 13.6
Examples—Declaratives.

1. Nothing is to be printed.
2. The LINE-COUNTER is not changed.
3. The function of the NEXT GROUP clause, if one appears in the report group, is nullified.

(See figure 13.6.)

LINKAGE SECTION

The Linkage Section describes data that is common between programs that communicate with each other within a single run unit (fig. 13.7). Program interaction requires that both programs have access to the same data items.

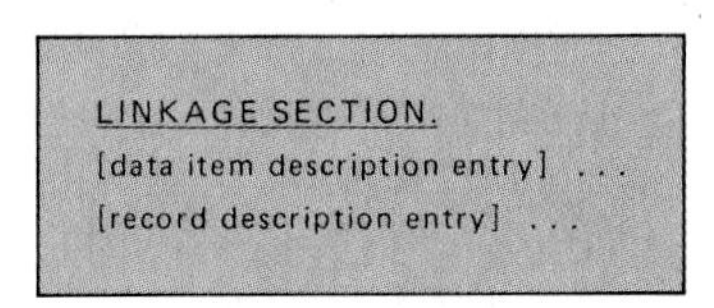
LINKAGE SECTION.
[data item description entry] . . .
[record description entry] . . .

Figure 13.7 Format—Linkage Section.

The Linkage Section is used for describing data that is available through the calling program that is to be referred to in both the *calling* and *called* program. A program that refers to another program is a *calling* program. A program that is referred to is a *called* program. (See figure 13.8.)

The Linkage Section is written in the Data Division to describe data from another program. The data item description entries and record description entries in the Linkage Section provide names and descriptions, but storage within

1. A is considered a calling program by B
2. B is considered a called program by A
3. B is considered a calling program by C
4. C is considered a called program by B

A: Calling program of B
B: Called program of A; Calling program of C
C: Called program of B

Figure 13.8 Example—Calling and Called Program.

the program is not reserved, inasmuch as the data area exists elsewhere. Any data description clause may be used to describe items in the Linkage Section, with one exception: the VALUE clause may not be specified for other than level 88 items.

Concept of Interprogram Communication

Complex data processing systems are frequently solved by the use of separately compiled but logically coordinated programs, which at execution time form logical and physical subdivisions of a single run unit. This approach lends itself to dividing a large problem into smaller, more manageable portions that can be programmed and debugged independently. At execution time, control is transferred from program to program by the use of CALL, ENTRY, GOBACK, and EXIT PROGRAM statements.

In COBOL terminology, a program is either a source program or an object program, depending on content; a source program is a syntactically correct set of COBOL statements, and an object program is the set of instructions, constants, and other machine-oriented data resulting from the operation of a compiler on a source program. A run unit is the total machine language necessary to solve a data processing problem; it includes one or more object programs as defined above, and it may include machine language from sources other than a COBOL compiler.

Subprogram Linkage Statements

Subprogram linkage statements are special statements permitting communication between object programs. These statements are CALL, ENTRY, GOBACK, and EXIT PROGRAM.

Calling and Called Programs

Transfer of Control

The CALL statement provides the means whereby control can be passed from one program to another within a run unit. A program that is activated by a CALL statement may itself contain CALL statements. However, results are unpredictable in those cases wherein circularity of control is initiated; that is, where program A calls program B, then program B calls program A or another program that calls program A.

When control is passed to a called program, execution proceeds in the normal way from the procedure statement to the procedure statement beginning with the first nondeclarative statement. If control reaches a STOP RUN statement, this signals the logical end of the run unit. If control reaches a GOBACK or EXIT PROGRAM statement, this signals the logical end of the called program only, and control then reverts to the point immediately following the CALL statement in the calling program. Stated briefly, the GOBACK or EXIT PROGRAM statement terminates only the program in which it occurs, whereas the STOP RUN statement terminates the entire run unit.

If the called program is not COBOL, then the termination of the run unit or the return to the calling program must be programmed in accordance with the language of the called program.

A COBOL program can refer to and pass control to other COBOL programs, or to programs written in other languages. A program in another language can refer to and pass control to a COBOL program. Control is returned from the called program to the first instruction following the calling sequence in the calling program.

A called program can also be a calling program, that is, a called program can in turn call another program.

Interprogram Data Storage

Program interaction requires that both programs have access to the same data items. In the calling program the common data items are described along with all other data items in the File Section or Working-Storage Section. At execution time, memory is allocated for the entire Data Division. In the called program, common data items are described in the Linkage Section. At execution time, memory space is not allocated for this section. Communication between the called program and the common data items stored in the calling program is affected through USING clauses contained in both programs. The USING clause in the calling program is contained in the CALL statement and the operands are a list of common data-names described in the Data Division. The USING clause in the called program follows the Procedure Division header or Entry statement, and the operands are a list of common data-names described in the Linkage Section. The data-names specified by the USING clause of the CALL statement indicate those data items available to a calling program that may be referred to by using the called program. The sequence of appearance of the data-names in the USING clause of the CALL statement and the USING clause in the Procedure Division and Entry statement is significant. Corresponding data-names refer to a single set of data that is available to the calling program. The correspondence is positional, and not by name. While the called program is being executed, every reference to an operand whose data-name appears in the called program's USING clause is treated as if it were a reference to the corresponding operand in the USING clause of the active CALL statement.

Thus, the interprogram communication feature allows a program to communicate with one or more other programs. This communication is provided by (1) the ability to transfer control from one program to another within a run unit, and (2) the capability of both programs to have access to the same data.

Specifying Linkage

Whenever a program calls another program, linkage must be established between the two. The calling program must state the entry point of the called program and must specify any arguments to be passed. The called program must have an entry point and must be able to accept the arguments. Further, the called program must establish the linkage for the return of control to the calling program.

A calling COBOL program must contain the following statement at the point where another program is to be called.

Call Statement

The CALL statement permits the communication between the COBOL object program and one or more subprograms or other language subprograms (fig. 13.9).

```
CALL literal [USING identifier-1 [identifier-2] ... ]
```

Figure 13.9
Format—Call Statement.

Rules Governing the Use of the Call Statement

1. The CALL statement appears in the calling program. It may not appear in the called program.
2. *Literal* is a nonnumeric literal and is the name of the program being called.
   a. Literal must conform to rules for the formation of a program-name.
   b. The first eight characters of the literal are used to make correspondence between the calling and called programs.
   c. If the called program is to be entered at the beginning of the Procedure Division, the literal must specify the program-name in the PROGRAM-ID paragraph of the called program.
3. If there is a USING clause in the CALL statement that invoked it, the called program must have a USING clause as part of its Procedure Division header.
4. When the called program is to be entered at entry points other than the beginning of the Procedure Division, these alternate entry points are identified by an ENTRY statement and a USING option corresponding to the USING option of the invoking CALL statement. In the case of a CALL statement with a corresponding ENTRY, literal must be a name other than the program-name but must follow the same rules as those for the formation of a program-name.
5. The *identifier* specified in the USING option of the CALL statement indicates those data items available to a calling program that may be referred to in the called program. When the called subprogram is a COBOL program, each of the USING options of the calling program must be identified as a data item in the File Section, Working-Storage Section, or Linkage Section. If the called subprogram is in a language other than COBOL, the operands may either be a file-name or a procedure-name.
6. Names in the USING lists (that of CALL in the main program and that of the Procedure Division header or the ENTRY statement in the subprogram) are paired on one-for-one correspondence, even though there is no necessary relationship between the actual names for the paired items; but the data-names must be equivalent.
7. The USING option is used only if there is a USING option in the called entry point either at the beginning of the Procedure Division of the called program, or included in an ENTRY statement of the called program. The number of operands in the USING option of the CALL statement should be the same as the number of operands in the USING option of the Procedure Division header, or an ENTRY statement.

Linkage in a Called Program

A called program must contain two sets of statements.

A. One of the following statements must appear at the point where the program is entered. If the called program is entered as the first instruction in the Procedure Division, the program is called, and arguments are passed by using the Procedure Division header with the USING option.

If the entry point of the called program is not the first statement of the Procedure Division, then the ENTRY statement is used.

Entry Statement

The ENTRY statement establishes an entry point in a COBOL subprogram (fig. 13.10).

Rules Governing the Use of Entry Statement

1. Control is transferred to the ENTRY point by a CALL statement in an invoking program.
2. Literal must not be the name of the called program but is formed according to the same rules followed for program-name. Literal must not be the name of any other entry point or specified in the CALL statement that invoked it.
3. A called program, once invoked, is entered at that ENTRY statement literal.

(See figure 13.11.)

```
ENTRY literal    [USING identifier-1    [identifier-2] . . . ]
```

Figure 13.10 Format—Using Option—Calling and Called Programs.

Format 1 (Within a Calling Program)
CALL literal-1 [USING identifier-1 [identifier-2] . . .]

Format 2 (Within a Called Program)
Option 1 ENTRY literal-1 [USING identifier-1 [identifier-2] . . .]
Option 2 PROCEDURE DIVISION [USING identifier-1 [identifier-2] . . .].

Figure 13.11 Format—Using Options—Calling and Called Programs.

B. Either of the following statements must be inserted where control is to be returned to calling program: GOBACK and EXIT PROGRAM.

Both the GOBACK and EXIT PROGRAM statements cause the restoration of the necessary registers and the return of control to the point in the calling program immediately following the calling sequence.

Goback Statement

The GOBACK statement marks the logical end of a called program (fig. 13.12).

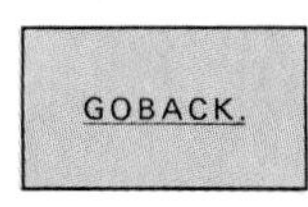

Figure 13.12 Format—Goback Statement.

Rules Governing the Use of the Goback Statement

1. A GOBACK statement must appear as the only statement or as the last of a series of imperative statements in a sentence.

2. If control reaches a GOBACK statement while operating under the control of a CALL statement, control returns to the point in the calling program immediately following the CALL statement.
3. If control reaches a GOBACK statement and there is no CALL statement active, there will be an abnormal termination of the job.

Exit Program Statement

This form of the EXIT statement marks the logical end of a called program (fig. 13.13). The statement is used in the same manner and performs as other EXIT statements.

```
paragraph-name.    EXIT PROGRAM.
```

Figure 13.13
Format—Exit Program Statement.

Rules Governing the Use of the Exit Program Statement

1. The statement must be preceded by a paragraph-name and must be the only statement in a paragraph.
2. If the control reaches an EXIT PROGRAM statement while operation is under the control of a CALL statement, control returns to the point in the calling program immediately following the CALL statement.
3. If control reaches an EXIT PROGRAM statement and no CALL statement is active, control passes through the exit point to the first sentence of the next paragraph.

OPERATION OF CALLING AND CALLED PROGRAMS

The execution of a CALL statement causes control to pass to the called program. The first time a called program is entered, its state is that of fresh copy of the program. Each subsequent time a called program is entered, the state is as it was upon the last exit from that program. The reinitiation of items in the called program is the responsibility of the programmer.

When a called program has a USING option in its Procedure Division header and linkage was affected by a CALL statement where literal is the name of the called program, execution of the called program begins with the first instruction in the Procedure Division after the Declaratives Section.

When linkage to a called subprogram is affected by a CALL statement when literal is the name of an entry point specified in the ENTRY statement of the called program, that execution of the called program begins with the first statement following the ENTRY statement.

When the USING option is present, the object program operates as though each occurrence of Identifier-1, Identifier-2, etc. in the Procedure Division had been replaced by the corresponding identifier for the USING option of the CALL statement of the calling program; that is, corresponding identifiers refer to a single set of data which is available to the calling program. The correspondence is positional, not nominal.

When control reaches the GOBACK or EXIT PROGRAM statement in the called program, control returns to the point in the calling program immediately following the CALL statement. (See figures 13.14, 13.15, 13.16, 13.17, 13.18.)

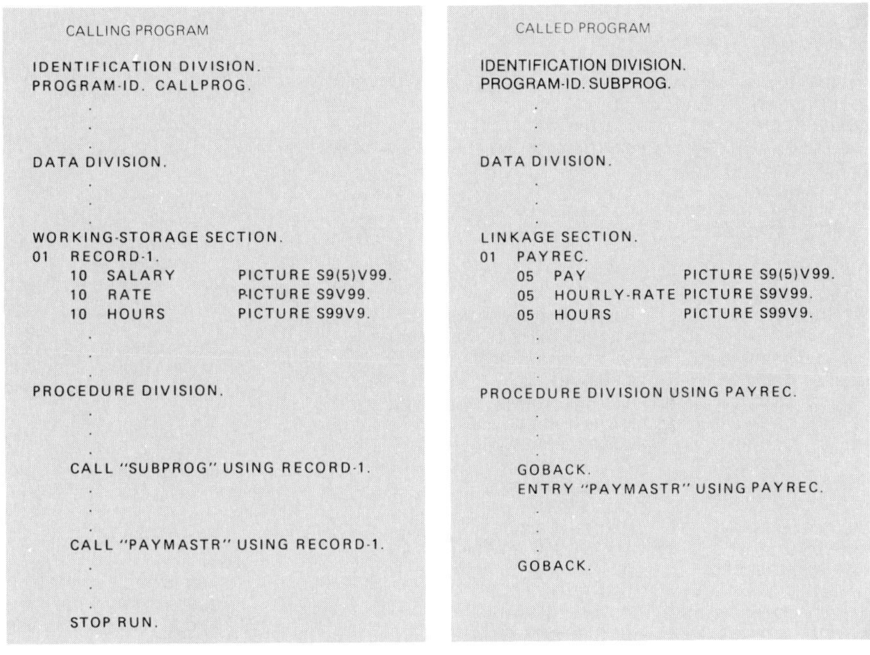

```
CALLING PROGRAM

IDENTIFICATION DIVISION.
PROGRAM-ID. CALLPROG.
    .
    .
    .
DATA DIVISION.
    .
    .
    .
WORKING-STORAGE SECTION.
01  RECORD-1.
    10  SALARY          PICTURE S9(5)V99.
    10  RATE            PICTURE S9V99.
    10  HOURS           PICTURE S99V9.
    .
    .
    .
PROCEDURE DIVISION.
    .
    .
    .
    CALL "SUBPROG" USING RECORD-1.
    .
    .
    .
    CALL "PAYMASTR" USING RECORD-1.
    .
    .
    .
    STOP RUN.
```

```
CALLED PROGRAM

IDENTIFICATION DIVISION.
PROGRAM-ID. SUBPROG.
    .
    .
    .
DATA DIVISION.
    .
    .
    .
LINKAGE SECTION.
01  PAYREC.
    05  PAY             PICTURE S9(5)V99.
    05  HOURLY-RATE     PICTURE S9V99.
    05  HOURS           PICTURE S99V9.
    .
    .
    .
PROCEDURE DIVISION USING PAYREC.
    .
    .
    .
    GOBACK.
    ENTRY "PAYMASTR" USING PAYREC.
    .
    .
    .
    GOBACK.
```

Figure 13.14 Example—Calling and Called Program.

```
CALLING PROGRAM

IDENTIFICATION DIVISION.
PROGRAM-ID.   LK-IF-MOVE.

ENVIRONMENT DIVISION.
CONFIGURATION SECTION.
SOURCE-COMPUTER.        XEROX-530.
OBJECT-COMPUTER.        XEROX-530.

DATA DIVISION.
FILE SECTION.
WORKING-STORAGE SECTION.
01  R1.
    02  RG.
        03  RG1         PICTURE X(8).
        03  RG2         PICTURE X(8).
        03  RG3         PICTURE X(7).
    02  RN.
        03  RN1         PICTURE 9(2)   COMP.
        03  RN2         PICTURE 9(6)   COMP.
        03  RN3         PICTURE 9(9)V99   COMP.
        03  RN4         PICTURE S9(3)   COMP-3.
        03  RN5         PICTURE 9(4).
        03  RN6         PICTURE XBXXBBXX.
        03  RN7         PICTURE $$99.99.

PROCEDURE DIVISION.
START-HERE.
    CALL 'LD-IF-MOVE' USING R1.
    STOP RUN.
```

```
CALLED PROGRAM

IDENTIFICATION DIVISION.
PROGRAM-ID.   LD-IF-MOVE.

ENVIRONMENT DIVISION.
CONFIGURATION SECTION.
SOURCE-COMPUTER.        XEROX-530.
OBJECT-COMPUTER.        XEROX-530.

DATA DIVISION.
LINKAGE SECTION.
01  L1.
    02  LG.
        03  LG1         PICTURE X(8).
        03  LG2         PICTURE X(8).
        03  LG3         PICTURE X(7).
    02  LN.
        03  LN1         PICTURE 9(2)   COMP.
        03  LN2         PICTURE 9(6)   COMP.
        03  LN3         PICTURE 9(9)V99   COMP.
        03  LN4         PICTURE . S9(3)   COMP-3.
        03  LN5         PICTURE 9(4).
        03  LN6         PICTURE XBXXBBXX.
        03  LN7         PICTURE $$99.99.

PROCEDURE DIVISION.
START-HERE.
    MOVE 789.56 TO LN7.
    DISPLAY LN7.
LINK-EXIT.
    EXIT PROGRAM.
```

Figure 13.15 Example—Calling and Called Program.

CALLING PROGRAM

```
IDENTIFICATION DIVISION.
PROGRAM-ID.  SUMMARY1.

ENVIRONMENT DIVISION.
CONFIGURATION SECTION.
SOURCE-COMPUTER.    NCR-CENTURY-200.
OBJECT-COMPUTER.    NCR-CENTURY-200   MEMORY SIZE 64000 WORDS.
INPUT-OUTPUT SECTION.
FILE-CONTROL.
    SELECT CHARGEFILE          ASSIGN TO NCR-TYPE-32.
SELECT TOTALFILE               ASSIGN TO NCR-TYPE-32.

DATA DIVISION.
FILE SECTION.
FD CHARGEFILE               LABEL RECORD IS STANDARD
                            DATA RECORD IS CHANGERECORD.
01  CHARGERECORD.
    02  CNO                 PICTURE X(5).
    02  CSTORE-A            PICTURE S999V99.
    02  CSTORE-B            PICTURE S999V99.
    02  CSTORE-C            PICTURE S999V99.
    02  PAYMENT             PICTURE 9999V99.
    02  BALANCE             PICTURE S9999V99.
    02  RATE                PICTURE 99V99.
FD TOTALFILE                LABEL RECORD IS STANDARD
                            DATA RECORD IS TOTALRECORD.
01  TOTALRECORD.
    02  TNO                 PICTURE X(5).
    02  TOTAL-CHARGES        PICTURE S9999V99.
    02  TOTAL-BALANCE        PICTURE S9999V99.
    02  TOTAL-BAL-DUE        PICTURE S9999V99.
WORKING-STORAGE SECTION.
77  WORK-TOTAL              PICTURE S9999V99.
77  WORK-BALANCE            PICTURE 9999V99.
01  FLAGS.
    05  MORE-DATA-FLAG              PICTURE XXX         VALUE 'YES'.
        88  MORE-DATA                                   VALUE 'YES'.
        88  NO-MORE-DATA                                VALUE 'NO'.

PROCEDURE DIVISION.
MAIN-ROUTINE.
    OPEN     INPUT         CHARGEFILE
             OUTPUT        TOTALFILE.
    READ CHARGEFILE
        AT END MOVE 'NO' TO MORE-DATA-FLAG.
    PERFORM PROCESS-ROUTINE
        UNTIL NO-MORE-DATA.
    CLOSE     CHARGEFILE
              TOTALFILE.
    STOP RUN.

PROCESS-ROUTINE.
    MOVE CNO TO TNO.


    CALL 'SUMMARY2' USING TOTALRECORD, WORK-BALANCE,
        CHARGERECORD.

    WRITE TOTALRECORD.
    READ CHARGEFILE
        AT END MOVE 'NO' TO MORE-DATA-FLAG.
```

CALLED PROGRAM

```
IDENTIFICATION DIVISION.
PROGRAM-ID.  SUMMARY2.

ENVIRONMENT DIVISION.
CONFIGURATION SECTION.
SOURCE-COMPUTER.  NCR-CENTURY-200.
OBJECT-COMPUTER.  NCR-CENTURY-200   MEMORY SIZE 64000 WORDS.

DATA DIVISION.
LINKAGE SECTION.
77  W-BALANCE               PICTURE 9999V99.
01  CHARGERECORD.
    02  FILLER              PICTURE X(5).
    02  STORE-A             PICTURE S999V99.
    02  STORE-B             PICTURE S999V99.
    02  STORE-C             PICTURE S999V99.
    02  PAYMENT             PICTURE 9999V99.
    02  FILLER              PICTURE X(11).
01  TOTAL.
    02  FILLER              PICTURE X(5).
    02  TOTAL-CHARGES       PICTURE S9999V99.
    02  TOTAL-BALANCE       PICTURE S9999V99.
    02  FILLER              PICTURE X(7).

PROCEDURE DIVISION USING TOTAL, W-BALANCE, CHARGERECORD.
BEGIN.
    ADD STORE-A, STORE-B, STORE-C GIVING TOTAL-CHARGES.
    SUBTRACT PAYMENT FROM W-BALANCE.
    ADD TOTAL-CHARGES, W-BALANCE GIVING TOTAL-BALANCE.
END-MODULE.
    EXIT PROGRAM.
```

Figure 13.16 Example—Calling and Called Program.

The following is an illustration of a run unit consisting of a main program with overlays arranged into two groups; therefore memory contains the main program, one overlay from group 1, and one overlay from group 2.

MAIN PROGRAM

PROGRAM1
(CALL PROGRAM4)
(CALL PROGRAM7)
(CALL PROGRAM2)

PROGRAM2
(CALL PROGRAM6)
(CALL PROGRAM3)

PROGRAM3
(CALL PROGRAM8)
(STOP RUN)

GROUP 1

PROGRAM4
(CALL PROGRAM5)
(EXIT PROGRAM)

PROGRAM6
(CALL PROGRAM5)
(EXIT PROGRAM)

PROGRAM8
(EXIT PROGRAM)

GROUP 2

PROGRAM5
(EXIT PROGRAM)

PROGRAM7
(CALL PROGRAM8)
(EXIT PROGRAM)

The calling sequence of the programs would be as follows.

Execute PROGRAM1
CALL PROGRAM4
    Execute PROGRAM4
    CALL PROGRAM5
        Execute PROGRAM5
        EXIT PROGRAM from PROGRAM5 to PROGRAM4
    Continue execution of PROGRAM4
    EXIT PROGRAM from PROGRAM4 to PROGRAM1
Continue execution of PROGRAM1
CALL PROGRAM7
    Execute PROGRAM7
    CALL PROGRAM8
        Execute PROGRAM8
        EXIT PROGRAM from PROGRAM8 to PROGRAM7
    Continue execution of PROGRAM7
    EXIT PROGRAM from PROGRAM7 to PROGRAM1
Continue execution of PROGRAM1
CALL PROGRAM2
    Execute PROGRAM2
    CALL PROGRAM6
        Execute PROGRAM6
        CALL PROGRAM5
            Execute PROGRAM5
            EXIT PROGRAM from PROGRAM5 to PROGRAM6
        Continue execution of PROGRAM6
        EXIT PROGRAM from PROGRAM6 to PROGRAM2
    Continue execution of PROGRAM2
    CALL PROGRAM3
        Execute PROGRAM3
        CALL PROGRAM8
            Execute PROGRAM8
            EXIT PROGRAM from PROGRAM8 to PROGRAM3
        Continue execution of PROGRAM3
        STOP RUN terminates execute of this run unit

Figure 13.17 Example—Overlay Programs.

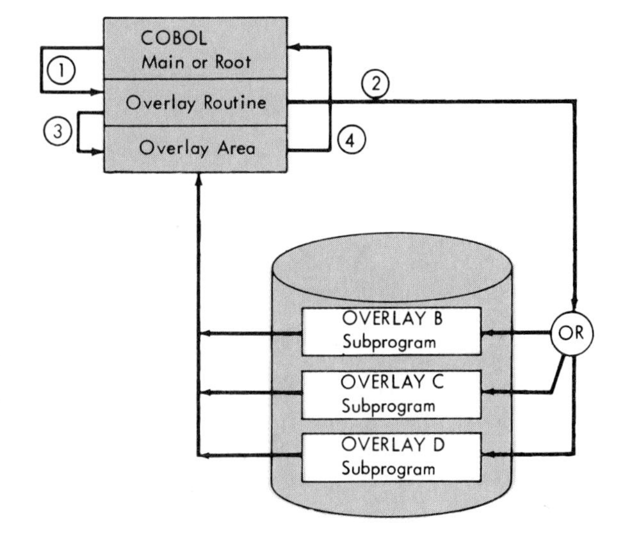

Flow Diagram of Overlay Logic

1. The main program calls the overlay routine.
2. The overlay routine fetches the particular COBOL subprogram and places it in the overlay area.
3. The overlay routine transfers control to the first instruction of the called program.
4. The called program returns to the COBOL calling program (*not* to the assembler language overlay routine).

Figure 13.18 Examples—Overlay Programs.

Figure 13.18 Continued.

COBOL Program Main (Root or Main Program)

```
IDENTIFICATION DIVISION.
PROGRAM-ID. MAINLINE.
.
.
.
ENVIRONMENT DIVISION.
.
.
.
DATA DIVISION.
.
.
.
WORKING-STORAGE SECTION.
77  PROCESS-LABEL PICTURE IS X(8) VALUE IS "OVERLAYB".
77  PARAM-1 PICTURE IS X.
77  PARAM-2 PICTURE IS XX.
77  COMPUTE-TAX PICTURE IS X(8) VALUE IS "OVERLAYC".

01  NAMET.
    02  EMPLY-NUMB PICTURE IS 9(5).
    02  SALARY PICTURE IS 9(4)V99.
    02  RATE PICTURE IS 9(3)V99.
    02  HOURS-REG PICTURE IS 9(3)V99.
    02  HOURS-OT PICTURE IS 9(2)V99.
01  COMPUTE-SALARY PICTURE IS X(8) VALUE IS "OVERLAYD".
01  NAMES.
    02  RATES PICTURE IS 9(6).
    02  HOURS PICTURE IS 9(3)V99.
    02  SALARYX PICTURE IS 9(2)V99.
.
.
.
PROCEDURE DIVISION.
.
.
.
    CALL "OVRLAY" USING PROCESS-LABEL' PARAM-1, PARAM-2.
.
.
.
    CALL "OVRLAY" USING COMPUTE-TAX, NAMET.
.
.
.
    CALL "OVRLAY" USING COMPUTE-SALARY, NAMES.
.
.
.
```

COBOL Subprogram B

```
IDENTIFICATION DIVISION.
PROGRAM-ID. OVERLAY1.
.
.
.
ENVIRONMENT DIVISION.
.
.
.
DATA DIVISION.
.
.
LINKAGE SECTION.

01  PARAM-10 PICTURE X.
01  PARAM-20 PICTURE XX.
.
.
.
PROCEDURE DIVISION.
PARA-NAME. ENTRY "OVRLAY1" USING PARAM-10, PARAM-20.
.
.
.
            GOBACK.
```

COBOL Subprogram C

```
IDENTIFICATION DIVISION.
PROGRAM-ID. OVERLAY2.
.
.
.
ENVIRONMENT DIVISION.
.
.
.
DATA DIVISION.
.
.
LINKAGE SECTION.

01  NAMEX.
    02  EMPLY-NUMBX PICTURE IS 9(5).
    02  SALARYX PICTURE IS 9(4) V99.
    02  RATEX PICTURE IS 9(3)V99.
    02  HOURS-REGX   PICTURE IS 9(3)V99.
    02  HOURS-OTX PICTURE IS 9(2)V99.

PROCEDURE DIVISION.
PARA-NAME. ENTRY "OVRLAY2" USING NAMEX.
.
.
.
            GOBACK.
```

Exercises

Write your answer in the space provided (answer may be one or more words).

1. The Declaratives Section is written in the ________________ Division to specify any special circumstance under which a ________________ is to be executed in the ________________ program.
2. The programmer may wish to supply additional ________________ of his own to supplement those supplied by the ________________.
3. The declaratives cannot appear in the ________________ sequence of procedural statements.
4. A declarative procedure is invoked ________________, not part of the ________________ coding of the program but when a condition arises that cannot normally be ________________ by the program.

5. The Declaratives Section is written at the ________ of the Procedure Division prior to the execution of the ________ procedure.
6. The Declaratives Section must begin with the key word ________ at the ________ margin followed by a ________ and a ________.
7. The Declaratives Section is terminated by the key words ________.
8. A Use statement specifies the ________ to be performed for each ________ of declarative.
9. A Use statement is never itself ________ but defines conditions calling for the ________ of the ________ procedure.
10. The type of procedures associated with the Use statement are ________, ________, and ________.
11. Label Processing declaratives are used to handle ________ header labels.
12. An Open statement in the Procedure Division causes the execution of the Use statement, associated with the ________ labels which refers to the users ________ labels.
13. A Close statement in the Procedure Division causes the execution of the Use statement associated with the ________ labels which refers to the users ________ labels.
14. The labels in a Label Processing declarative must be listed as ________ in the ________ clause for file description entry.
15. Error Processing declaratives are used to specify procedures if an ________ error occurs during the processing and provides input/output ________ procedures in addition to those specified by the ________.
16. The error-handling procedures are activated when an error occurs during the execution of a ________, ________, ________ or ________ statement.
17. Automatic ________ routines are executed before ________ procedures.
18. The continued processing of a file is permitted if an ________ procedure exists in the ________ section.
19. A Report Writer declarative specifies Procedure Division statements that are to be executed before a ________ name in the ________ Section of the Data Division is produced.
20. When the user wishes to suppress the printing of a specified report group, he writes the statement ________ in the ________ declaratives section.
21. The Linkage Section is written in the ________ Division to describe ________ from ________ program.
22. The entries in the Linkage Section provide name and descriptions but ________ storage areas are reserved.
23. Complex data processing systems are frequently solved by the use of ________ compiled but logically ________ programs.
24. A run unit is the total ________ language necessary to solve a data processing problem.
25. The Call statement provides the means whereby ________ can be passed from ________ program to ________ within a ________.
26. ________ linkage statements permit communication between object programs.
27. When control is passed to a called program, execution proceeds with the first ________ statement.
28. The Goback or Exit Program statement terminates only the program in which ________ occurs and the Stop Run statement terminates the entire ________.
29. A COBOL program can ________ and ________ to other COBOL programs or to programs written in ________ languages.
30. A program that refers to another program is called a ________ program.
31. A program that is referred to is a ________ program.
32. Control is returned from the ________ program to the first instruction following the ________ sequence in the ________ program.
33. Program interaction requires that both programs have access to the ________ data items.

34. In the calling program, the common data items are described in the ______ or ______ sections.
35. In the called program, common data items are described in the ______ Section.
36. Whenever the program calls another program, ______ must be established between the two programs.
37. The calling program must state the ______ point of the called program and must specify the ______ to be passed.
38. The called program must have an ______ point and must be able to accept the ______.
39. The called program must establish the ______ for return of ______ to the ______ program.
40. The Call statement permits communication between the ______ program and one or more ______ or other ______.
41. The Call statement appears in the ______ program and may not appear in the ______ program.
42. A called program must contain a statement where the program is to be ______.
43. The Entry statement establishes an entry in a COBOL ______ of a ______ program.
44. If the called program is entered as the first instruction in the Procedure Division, the ______ option is used.
45. If the entry point of a called program is other than the first statement of the Procedure Division, the ______ statement is used.
46. Either the ______ or ______ statement must be used to return control to the calling program.
47. The Goback statement marks the ______ end of a called program.
48. The Exit Program statement is used in the same manner as other ______ statements.

Answers

1. PROCEDURE, PROCEDURE, OBJECT
2. PROCEDURES, COMPILER
3. REGULAR
4. NONSYNCHRONOUSLY, SEQUENTIAL, TESTED
5. BEGINNING, FIRST
6. DECLARATIVES, A, PERIOD, SPACE
7. END DECLARATIVES
8. PROCEDURES, TYPE
9. EXECUTED, EXECUTION, USE
10. LABEL-HANDLING, ERROR-HANDLING, REPORT WRITING
11. USER CREATED
12. BEGINNING, HEADER
13. ENDING, TRAILER
14. DATA-NAMES, LABEL RECORDS
15. INPUT/OUTPUT, CORRECTION, COMPILER
16. READ, WRITE, REWRITE, START
17. SYSTEM ERRORS, USER SPECIFIED
18. ERROR-PROCESSING, DECLARATIVES
19. REPORT GROUP, REPORT
20. MOVE 1 TO PRINT-SWITCH, USE BEFORE REPORTING
21. DATA, DATA, ANOTHER
22. NO
23. SEPARATELY, COORDINATED
24. MACHINE
25. CONTROL, ONE, ANOTHER, RUN UNIT
26. SUBPROGRAM
27. NON-DECLARATIVE
28. IT, RUN UNIT
29. REFER TO, PASS CONTROL, OTHER
30. CALLING
31. CALLED
32. CALLED, CALLING, CALLING
33. SAME
34. FILE, WORKING-STORAGE
35. LINKAGE
36. LINKAGE
37. ENTRY, ARGUMENTS
38. ENTRY, ARGUMENTS
39. LINKAGE, CONTROL, CALLING
40. OBJECT, SUBPROGRAMS, LINKAGE SUBPROGRAMS
41. CALLING, CALLED
42. ENTERED
43. SUBPROGRAM, CALLED
44. USING
45. ENTRY
46. GOBACK, EXIT PROGRAM
47. LOGICAL
48. EXIT

Questions for Review

1. Why is a Declaratives Section written? Give the steps necessary to write the section.
2. What is the primary function of the Use statement? List the types of procedures associated with the Use statement.
3. What are Label Processing declaratives? What are their purposes and uses?

4. What are the Error Processing declaratives? What are their purposes and uses?
5. Under what conditions is the continued processing of a file containing errors permitted?
6. When is the Use Before Reporting declarative used? What is its purpose and how is it used?
7. How can the printing of a report group in the Report Writer be suppressed?
8. What is the function of the Linkage Section?
9. Briefly describe the concept of interprogram communication.
10. What are Subprogram Linkage statements? Give examples.
11. Differentiate between a calling and called program.
12. Briefly describe the interprogram data storage.
13. How is linkage specified in calling and called programs?
14. What is the function of the Call statement and how is it used in a calling program?
15. What is the function of the Entry statement and how is it used in a called program?
16. What is the function of the Goback and Exit Program statements and when is each used in a called program?
17. Explain the operation of a calling and called program.

Problems

1. *Match each item with its proper description.*

______ 1. Declaratives Section	A.	Correction procedures.
______ 2. Label-Processing Declarative	B.	Execution of statements just before the report group named.
______ 3. Error-Processing Declarative	C.	Handle user created labels.
______ 4. Report Writer Declarative	D.	Special circumstance under which a procedure is to be executed.
______ 5. Linkage Section	E.	Describes data from another program.

2. *Match each item with its proper description.*

______ 1. Called Program	A.	Permits communication between object program and subprogram.
______ 2. Calling Program	B.	Entry point in a subprogram.
______ 3. Call Statement	C.	A program that refers to another program.
______ 4. Entry Statement	D.	Logical end of a called program.
______ 5. Goback Statement	E.	Operates in same manner as Exit statement.
______ 6. Exit Program Statement	F.	A program that is referred to.

3. *The following is the Procedure Division of a program to create a Direct file organization.*

```
PROCEDURE DIVISION.
DECLARATIVES.
ERROR-PROCEDURE SECTION. USE AFTER STANDARD ERROR PROCEDURE
    ON D-FILE GIVING ERROR-COND.
ERROR-ROUTINE.
    EXHIBIT NAMED ERROR-COND.
    IF ERROR = 1
    THEN
        PERFORM SYNONYM-ROUTINE
    ELSE
        DISPLAY 'OTHER STANDARD ERROR    REC-ID
        PERFORM EOJ.
SYNONYM-ROUTINE.
    IF CC = 84
        AND HD = 9
    THEN
        DISPLAY 'OVERFLOW AREA FULL'
        PERFORM EOJ
    ELSE
        NEXT SENTENCE.
    IF CC = 84
```

```
    THEN
        ADD 1 TO HD
        PERFORM ADJUST-HD
    ELSE
        NEXT SENTENCE.
    IF HH = 9
    THEN
        PERFORM END-CYLINDER
    ELSE
        ADD 1 TO HH
        PERFORM WRITES.
    END-CYLINDER.
        MOVE 84 TO CC.
        MOVE HD TO HH.
        PERFORM WRITES.
    ADJUST-HD.
        MOVE HD TO HH.
        PERFORM WRITES.
    END DECLARATIVES.
    FILE-CREATION SECTION.
    INIT.
        OPEN    INPUT     C-FILE
                OUTPUT    D-FILE.
        PERFORM READS
            UNTIL NO-MORE-DATA.
        PERFORM EOJ.
READS.
    READ C-FILE
        AT END MOVE'NO' TO MORE-DATA-FLAG.
    MOVE CORRESPONDING C-REC TO D-REC.
    MOVE PART-NUM OF C-REC TO REC-ID
        SAVE.
    DIVIDE SAVE BY 829 GIVING QUOTIENT
        REMAINDER TRACK-1.
    ADD 10 TO TRACK-1.
    MOVE CYL TO CC.
    MOVE HEAD TO HH.
    EXHIBIT NAMED TRACK-ID, C-REC, CC, HH.
    WRITE D-REC
        INVALID KEY PERFORM INVALID-KEY.
WRITES.
    EXHIBIT NAMED TRACK-ID, C-REC, CC, HH.
    WRITE D-REC
        INVALID KEY PERFORM INVALID-KEY.
    PERFORM READS.
INVALID-KEY.
    DISPLAY 'INVALID KEY  ' REC-ID.
    PERFORM EOJ.
EOJ.
    CLOSE    C-FILE
             D-FILE.
    STOP RUN.
```

Explain the function of the Error-Processing declarative.

4.

```
PROCEDURE DIVISION.
DECLARATIVES.
RW-1 SECTION. USE BEFORE REPORTING PAGE-HED.
RW-2.
    IF CBL-CTR IS LESS THAN 1
        OR-EQUAL TO 1
    THEN
        MOVE 'BEGIN' TO BEGIN-FIELD-FOR-PH
        MOVE SPACES TO CONTINUED-FIELD-FOR-PH
    ELSE
        MOVE SPACES TO BEGIN-FIELD-FOR-PH
        MOVE 'CONTINUED' TO CONTINUED-FIELD-FOR-PH.
RW-3 SECTION. USE BEFORE REPORTING PAGE-FOOT.
RW-4.
    IF CBL-CTR IS GREATER THAN 1
    THEN
        MOVE MONTHNAME (MONTH) TO MONTH-FOR-PF
        MOVE 'CONTINUED ON NEXT PAGE' TO TEXT-FOR-PF
    ELSE
        MOVE SPACES TO MONTH-FOR-PF
        MOVE SPACES TO TEXT-FOR-PF.
END DECLARATIVES.
```

In the above declarative procedural statements, what is the main purpose of each. Give examples to illustrate your point.

5. *Write the Report Writer declarative to suppress printing for a footing group (FOOT-LINE) for job numbers 15–20.*

6. *Write a Label Processing declarative to process the user header labels on an input file, FILE-IN. The condition to be tested is as follows:*

If the LAB-ID is greater than ALLOWABLE-DATE, stop the run and display the message "File-In Is Incorrect; Label Error, Mount Proper Reel" on the console and wait for a reply. Branch back to recheck the labels after correction.

If the LAB-ID is not greater than ALLOWABLE-DATE, display "Label Is OK" upon console and proceed to the procedural statements to execute the program.

7. *Given the following information:*

```
IDENTIFICATION DIVISION.
PROGRAM-ID.  INVENTORY.
    .
    .
    .
DATA DIVISION.
    .
    .
    .
WORKING-STORAGE SECTION.
01  TRANS-RECORD.
    02  QUANTITY            PICTURE 9(5).
    02  UNIT-PRICE          PICTURE 999V99.
    02  AMOUNT              PICTURE 9(5)V99.
    .
    .
    .
PROCEDURE DIVISION.
    .
    .
    .
    CALL 'INVENTORYSUB' USING TRANS-RECORD.
    .
    .
    .
    CALL 'INVUPDATE' USING TRANS-RECORD.
    .
    .
IDENTIFICATION DIVISION.
PROGRAM-ID.  INVENTORYSUB.
    .
    .
    .
DATA DIVISION.
    .
    .
LINKAGE SECTION.
01  INVREC.
    02  UNITS               PICTURE 9(5).
    02  PRICE               PICTURE 999V99.
    02  INV-BALANCE         PICTURE 9(5)V99.
    .
    .
PROCEDURE DIVISION USING INVREC.
    .
    .
    GOBACK.
    ENTRY 'INVUPDATE' USING INVREC.
    .
    .
    GOBACK.
```

In the above programs, explain the processing routines between the calling and called programs. How are the different values handled in each program by what data name? What happens with each execution of the Call statement?

14 Additional and Optional Features

ENVIRONMENT DIVISION

Configuration Section—Special-Names Paragraph

The SPECIAL-NAMES paragraph enables the programmer to assign mnemonic-names or condition-names to the option switches and the system tags used in the program (fig. 14.1). This paragraph also allows the programmer to assign a currency sign symbol other than $ and to exchange the function of the comma and period in editing currency amounts. If present, it must immediately follow the OBJECT-COMPUTER paragraph. (These function-names apply to IBM 360/370 computers only.)

```
SPECIAL-NAMES.
        [function-name-1 IS mnemonic-name] . .
        [function-name-2 [IS mnemonic-name]
            {ON STATUS IS condition-name-1
            {OFF STATUS IS condition-name-2
                        [OFF STATUS IS condition-name-2]}
                        [ON STATUS IS condition-name-1] }   ] . . .
        [CURRENCY SIGN IS literal]   [DECIMAL-POINT IS COMMA].
```

Figure 14.1 Format—Special-Names Paragraph.

Rules Governing the Use of Special-Names Paragraph

1. In ACCEPT and DISPLAY statements, associated mnemonic-names may be used to identify the function-names identified with the input or output device. The function-name may be chosen only from the following list.

SYSIN
SYSOUT
SYSPUNCH
CONSOLE

2. When the mnemonic-name option is used in a WRITE statement with the ADVANCING option, the mnemonic-name must be defined as a function-name in the SPECIAL-NAMES paragraph. It is used to skip to channel 1–12 and to suppress spacing if desired. It may also be used for pocket selection for a card punch file.
(See figures 14.2, 14.3.)

Function-Name	Action Taken
CSP	Suppress spacing
C01-C12	Skip to channel 1 to 12 respectively
S01-S02	Pocket selection

Figure 14.2 Function Names and Their Meanings.

```
SPECIAL-NAMES.
    C01   IS SKIP-TO-1.

PRT-HEADING.
    WRITE PRINTOUT FROM HDG-1
         AFTER ADVANCING SKIP-TO-1 LINES.
```

Figure 14.3 Example—Function Name and Mnemonic Name.

3. Mnemonic-names with a single character enclosed in quotation marks are used in the CODE clause in the report description entry of the Report Writer Feature to identify output where more than one type of output is desired from a single input. (See "Report Writer." Chapter 10)
4. Mnemonic-names may not be used in a source program except in the verb format which permits their usage.
5. The literal that appears in the CURRENCY SIGN IS clause is used in the PICTURE clause to represent the currency symbol. The literal is limited to a single character, must be a nonnumeric, and must not be any of the following:

   a. Digits 0–9.
   b. Alphabetic characters A–Z and space.
   c. Special characters * – , . ; ( ) + " or '

   If the CURRENCY SIGN clause is not used, only the $ can be used as the currency symbol in the PICTURE clause. (See figure 14.4.)

(See figures 14.5, 14.6.)

In the following example, the character R is used as the currency symbol in the PICTURE clause.

```
SPECIAL-NAMES.
    CURRENCY SIGN IS 'R'.
```

If the clause CURRENCY SIGN IS literal is not present, then only the symbol $ can be used as the currency symbol in the PICTURE clause.

Figure 14.4 Example—Currency Sign.

The OPEN SWITCH is a function-name of this computer that enables the programmer to test the status of any one of the eight hardware option switches. The switch can set to an on or off status by the computer operator. The status of the switch can then be interrogated by testing the condition-name assigned to the switch. For example, the on status of switch 3 could be assigned a condition-name as follows:

```
SPECIAL-NAMES.
    OPTION SWITCH-3 ON STATUS IS PAPER-TAPE-INPUT.
```

The on status of option switch 3 can be tested by this conditional statement:

```
IF PAPER-TAPE-INPUT
THEN
     PERFORM PT-ROUTINE
ELSE
     PERFORM PC-ROUTINE.
```

This conditional statement goes to procedure with the procedure-name PT-ROUTINE if option switch 3 is on; it goes to PC-ROUTINE if option switch is off.

Figure 14.5 Example—Option Switch.

6. The clause DECIMAL-POINT IS COMMA means that the function of the comma and the period are exchanged in the PICTURE character string and in numeric literals. When the clause is used, the user must use a comma to represent a decimal point when required in numeric literals or in a PICTURE clause. The period is used for all functions ordinarily served by the comma.

Suppose the programmer wants to perform one set of calculations if switch SW1 is on, and another set of calculations if SW1 is off. The paragraph entry might be,

```
SPECIAL-NAMES.
    SW1 IS BIT-ONE, ON STATUS IS BIT-ONE-ON, OFF
    STATUS IS BIT-ONE-OFF.
```

where SW1 is function-name, BIT-ONE is a mnemonic-name, and BIT-ONE-ON and BIT-ONE-OFF are condition-names. Either or both of the condition-names can then be used in Procedure Division conditional test statements. One such statement might be,

```
    IF BIT-ONE-ON, PERFORM PAR-3.
```

In this case, if switch SW1 is in an on status, control will be transferred to the paragraph named PAR-3. If switch SW1 is in an off status, control will fall through to the next statement after this one.

Figure 14.6
Example—Option Switch.

Input-Output Section—File-Control Paragraph

Reserve Clause

The RESERVE clause is used to reserve additional input or output areas in addition to the buffers allocated by the compiler (fig. 14.7). The RESERVE clause allows the programmer to document the number of input or output or input/output areas in storage allocated to a specific file by the compiler. In American National Standard (ANS) COBOL, each file is assigned one storage area. The RESERVE clause indicates that storage areas are to be reserved for the file in addition to the usual single area. For example, RESERVE 1 ALTERNATE AREA indicates that two storage areas are allocated to a file.

The clause RESERVE NO ALTERNATE AREA indicates that no additional areas are to be reserved for a file.

Rules Governing the Use of the Reserve Clause

1. The clause specifies the number of buffers represented by the integer to be reserved for standard sequential or an indexed sequential file that is accessed sequentially in addition to the one buffer which is reserved automatically.
2. The clause must not be specified for direct files. If specified, the clause is ignored, and only one buffer is reserved.
3. If NO is written, no additional areas are assigned besides the minimum of one.

   (See figure 14.8.)

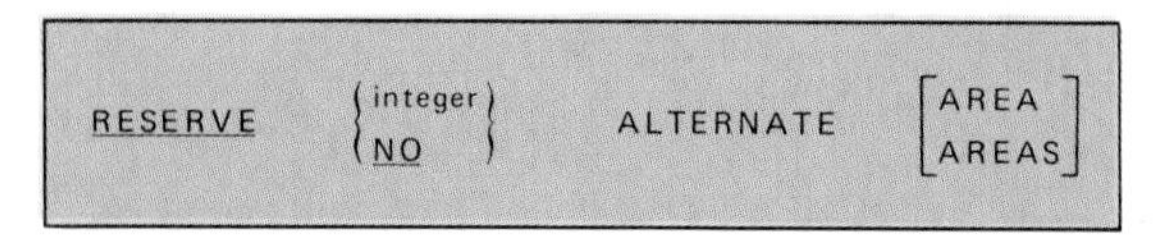

Figure 14.7 Format—Reserve Clause.

```
INPUT-OUTPUT SECTION.
FILE-CONTROL.
    SELECT CARD-FILE        ASSIGN TO READER
                            RESERVE 2 AREAS.
    SELECT PRINT-FILE       ASSIGN TO PRINTER
                            RESERVE 2 AREAS.
```

Two additional areas will be reserved for the input CARD-FILE and the output PRINT-FILE in addition to the buffers allocated by the compiler.

Figure 14.8 Example—Reserve Clause.

File-Limit Clause

The FILE-LIMIT clause is used to specify the logical beginning and logical end of a file on a mass storage (direct-access) device (fig. 14.9). The logical beginning of a mass storage file is the address specified in the first operand of the clause, and the logical end is the address specified in the last operand of the clause.

Because file boundaries are determined by the operating system job-control card, no allocation is made by this clause, and the entire clause is treated as comments, serving only as documentation for the program. (See figure 14.10.)

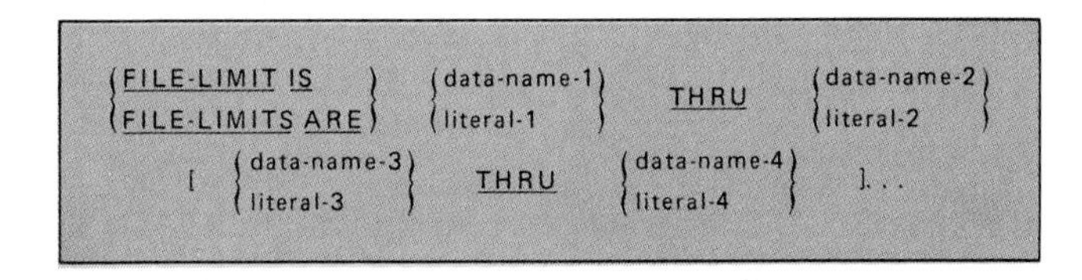

Figure 14.9 Format—File-Limit Clause.

In the following example, the limits of the file ABCFILE are modified at execution time by changing the contents of the data item SEGMENT-END. SEGMENT-START contains zero; therefore, if SEGMENT-END contains 50, the size of the file would be 50 sectors.

```
SELECT ABCFILE     ASSIGN TO NCR-TYPE-32
    FILE-LIMIT IS SEGMENT-START THRU SEGMENT-END.

WORKING-STORAGE SECTION.
77    SEGMENT-START          PICTURE 9999     VALUE IS ZERO.
77    SEGMENT-END            PICTURE 9999.
```

Figure 14.10 Example—File-Limit Clause.

Track-Area Clause

The TRACK-AREA clause is used to specify a required area where records are to be added to a random-access device with indexed sequential organization (fig. 14.11). The size of the track must be at least the size of an entire track plus one logical record.

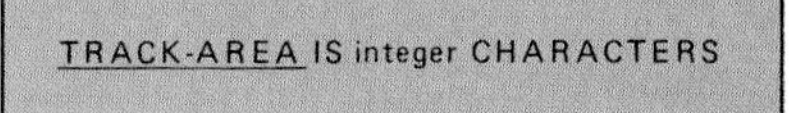

Figure 14.11 Format—Track-Area Clause.

Rules Governing the Use of the Track-Area Clause

1. The clause is required when variable-length records are added to the file. Efficiency in adding a record can be improved when the TRACK-AREA clause is specified.
2. If an integer is specified, an area equal to the number specified is obtained by the system when the file is opened. When the file is closed, the assigned area is released to the system.
3. If a record is added to an indexed sequential file and the TRACK-AREA clause is not used for the file, the contents of the NOMINAL KEY field are unpredictable after a WRITE statement is executed.

Input-Output Section—I-O-Control Paragraph

(See figure 14.12.)

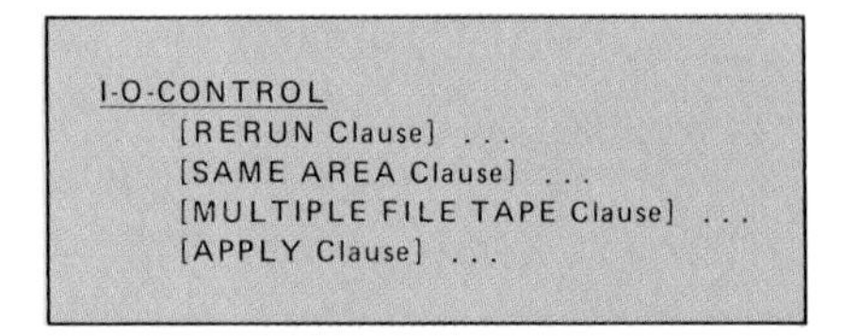

```
I-O-CONTROL
      [RERUN Clause] ...
      [SAME AREA Clause] ...
      [MULTIPLE FILE TAPE Clause] ...
      [APPLY Clause] ...
```

Figure 14.12
Format—Environment Division—I-O-Control Paragraph.

Rerun Clause

When object programs process large volumes of data, it is desirable to indicate points at which the program may be restarted in case its running is interrupted by a power failure or some other unexpected event.

The RERUN clause is used to establish any rerun or restart procedures (fig. 14.13). This clause specifies that checkpoint areas are to be written on the actual device at the time of the checkpoint, and that they can be read back into core storage to restart the program from that point. A checkpoint record is the recording of the status of the problem program and main storage resources at desired intervals. The presence of this clause specifies that checkpoint records are to be taken. Checkpoint records are read sequentially and must be assigned to tape or mass storage devices. (See figure 14.14.)

```
RERUN ON system-name
          EVERY integer RECORDS OF file-name
```

Figure 14.13
Format—Rerun Clause.

- To write single checkpoint records using tape:

```
//CHECKPT   DD   DSNAME=CHECK1,                    X
//               VOLUME=SER=ND003,                 X
//               UNIT=2400,DISP=(NEW,KEEP),        X
//               LABEL=(,NL)
                 .
                 .
                 .
        ENVIRONMENT DIVISION.
                 .
                 .
                 .
            RERUN ON UT-2400-S-CHECKPT EVERY
            5000 RECORDS OF ACCT-FILE.
```

- To write single checkpoint records using disk (note that more than one data set may share the same external-name):

```
//CHEK      DD   DSNAME=CHECK2,                    X
//               VOLUME=(PRIVATE,RETAIN,           X
//                  SER=DB030,                     X
//               UNIT=2314,DISP=(NEW,KEEP),        X
//               SPACE=(TRK,300)
```

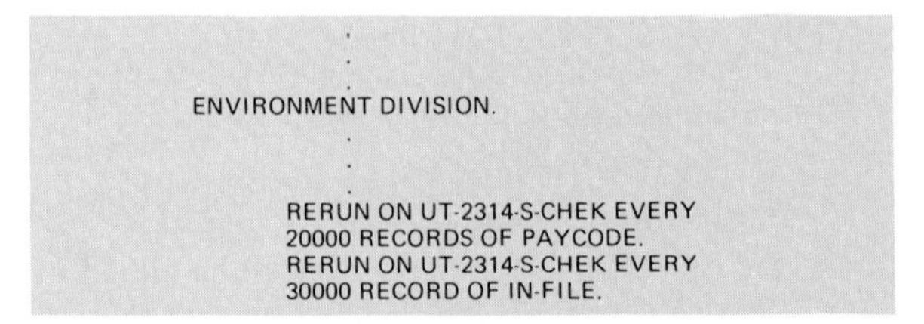

```
                 .
                 .
                 .
        ENVIRONMENT DIVISION.
                 .
                 .
                 .
            RERUN ON UT-2314-S-CHEK EVERY
            20000 RECORDS OF PAYCODE.
            RERUN ON UT-2314-S-CHEK EVERY
            30000 RECORD OF IN-FILE.
```

- To write multiple contiguous checkpoint records (on tape):

```
//CHEKPT    DD   DSNAME=CHECK3,                    X
//               VOLUME=SER=111111,                X
//               UNIT=2400,DISP=(MOD,PASS),        X
//               LABEL=(,NL)
                 .
                 .
                 .
        ENVIRONMENT DIVISION.
                 .
                 .
                 .
            RERUN ON UT-2400-S-CHEKPT EVERY
            10000 RECORDS OF PAY-FILE.
```

Figure 14.14 Examples—Rerun and Checkpoint Procedures—IBM.

System-name specifies the external medium for the checkpoint file, the file upon which the checkpoint records are written. It must not be the same as the system-name used in the File-Control ASSIGN clause, but it follows the same rules of formation. System-name must specify a tape or mass storage device.

File-name represents the file for which checkpoint records are to be written. It must be described with a file description entry in the Data Division.

Same Area Clause

The SAME AREA clause specifies that two or more files will occupy the same area during processing (fig. 14.15). The shared area consists of all storage

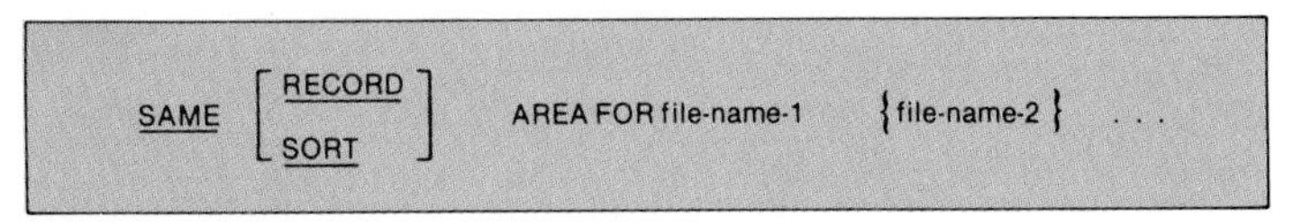

Figure 14.15 Format—Same Area Clause.

areas (including alternate areas) assigned for the file specified. Only one of these files may be opened at a time. A file in use must be closed before any other file named in the clause may be opened. More than one SAME AREA clause may be written in a program; however, a file-name must not appear in more than one of these clauses.

Multiple File Tape Clause

The MULTIPLE FILE TAPE clause documents the sharing of the same physical reel of magnetic tape by two or more files (fig. 14.16). Regardless of the number of files in a single reel, only those used in the object program need be specified. (See figure 14.17.)

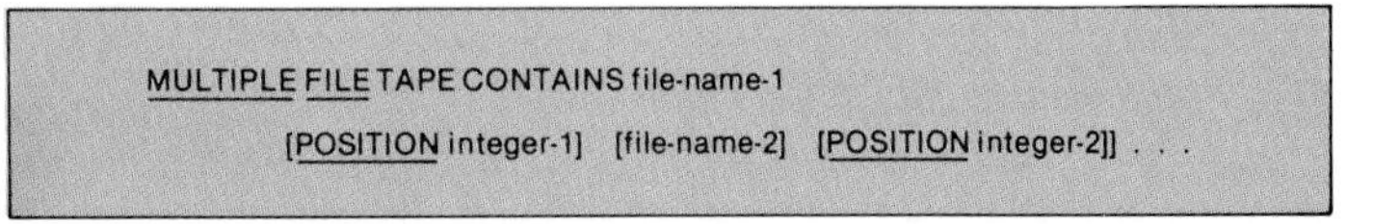

Figure 14.16 Format—Multiple File Tape Clause.

In the example below, a reel may contain ABCFILE, GOODFILE, BESTFILE, and XYZFILE in the order listed. If only ABCFILE and BESTFILE will be used in a program, the MULTIPLE FILE TAPE clause would have the following forms:

```
I-O-CONTROL.
    MULTIPLE FILE TAPE CONTAINS ABCFILE POSITION 1, BESTFILE
      POSITION 3.
```

If all four files were used in a program, the MULTIPLE FILE TAPE clause would have this format:

```
I-O-CONTROL.
    MULTIPLE FILE TAPE CONTAINS ABCFILE, GOODFILE, BESTFILE,
      XYZFILE.
```

Figure 14.17 Example—Multiple File Tape Clause.

Apply Write-Only

This option is used to make optimum use of buffers and device space when creating a file, the recording mode of which is V (fig. 14.18). Normally a buffer

```
APPLY WRITE-ONLY ON file-name-1 [file-name-2] . . .
```

Figure 14.18 Format—Apply Write-Only Clause.

is truncated when there is not enough space remaining to accommodate the maximum size record. The use of this option will cause the buffer to be truncated only when the next record does not fit in the unused remainder of the buffer. This option is meaningful only when the file is opened as OUTPUT.

The files named in this option must be standard sequential.

Every WRITE statement associated with this option must use WRITE RECORD NAME FROM identifier option.

Apply Core-Index on Clause

This option may be specified only for an indexed sequential file whose access mode is random (fig. 14.19). It is used to specify the highest-level index to be processed in core. The area will be obtained at open time and released at close time.

```
APPLY CORE-INDEX TO data-name ON file-name-1 [file-name-2] . . .
```

Figure 14.19 Format—Apply Core-Index Clause.

Apply Record-Overflow on Clause

If the record-overflow feature is available for mass storage devices being specified, the amount of unused space on a volume may be reduced by specifying this option for files on that volume (fig. 14.20). If the option is used, a block that does not fit on this track is partially written on that track and continued on the next available track.

```
APPLY RECORD-OVERFLOW ON file-name-1 [file-name-2] . . .
```

Figure 14.20 Format—Apply Record-Overflow Clause.

This option may be specified only for a standard sequential file (with F, U, or V recording mode records) assigned to a mass storage device, or a direct file with fixed-length records. (See figure 14.21.)

```
ENVIRONMENT DIVISION.
CONFIGURATION SECTION.
SOURCE-COMPUTER.  NCR-CENTURY-200.
OBJECT-COMPUTER.  NCR-CENTURY-200   MEMORY SIZE 32000 CHARACTERS.
SPECIAL-NAMES.
    SWITCH-5 ON STATUS IS PROCESS-FLIGHTFILE,
    OVERHAUL-FILE IS OVFILE.
INPUT-OUTPUT SECTION.
FILE-CONTROL.
    SELECT TRANFILE             ASSIGN TO NCR-TYPE-41.
    SELECT FLIGHTFILE           ASSIGN TO NCR-TYPE-00.
    SELECT AIRFILE              ASSIGN TO NCR-TYPE-41.
    SELECT NEWAIRFILE           ASSIGN TO NCR-TYPE-41.
    SELECT OVERHAUL-FILE        ASSIGN TO NCR-TYPE-25.
I-O-CONTROL.
    RERUN EVERY END OF REEL OF NEW NEWAIRFILE.
```

Figure 14.21 Sample—Environment Division Entries.

DATA DIVISION

Renames Clause

The RENAMES clause permits alternate, possible overlapping, groupings of elementary items. The RENAMES data description entry consists of level number 66, the data-name, and the RENAMES clause. One or more RENAMES clauses may be written for a record (fig. 14.22). All RENAMES entries associated with a given logical record must immediately follow the last data description entry of that record.

```
66 data-name-1 RENAMES data-name-2 [THRU data-name-3]
```

Figure 14.22 Format—Renames Clause.

Rules Governing the Use of the Renames Clause

1. Level number 66 must be used with RENAMES clause.
2. All logical entries associated with a given logical record must immediately follow its last data description entry.
3. Data-name-2 and data-name-3 must be the name of elementary items or groupings of elementary items in the associated logical record, and cannot be the same data-name.

4. Data-name-3 cannot be subordinate to data-name-2.
5. A level 66 cannot rename another level 66 entry nor can it rename a level 77, 88, or 01 entry.
6. Data-name-1 cannot be used as a qualifier and can be qualified only by level 01 or FD entries.
7. Data-name-2 and data-name-3 may be qualified.
8. An OCCURS clause may not appear in data-name-2, data-name-3, or any item subordinate to it.
9. Data-name-2 must precede data-name-3 in the record description entry. (See figures 14.23, 14.24, 14.25.)

```
01  CORRECTED-RECORD.
    05  GROUP-A.
        10  FIELD-1A.
            15  ITEM-1A   PICTURE XX.
            15  ITEM-2A   PICTURE XXX.
            15  ITEM-3A   PICTURE XX.
        10  FIELD-2A.
            15  ITEM-4A   PICTURE XX.
            15  ITEM-5A   PICTURE XX.
            15  ITEM-6A   PICTURE XX.
    05  GROUP-B REDEFINES GROUP-A.
        10  FIELD-1B.
            15  ITEM-1B   PICTURE XXXX.
            15  ITEM-2B   PICTURE XXX.
        10  FIELD-2B.
            15  ITEM-3B   PICTURE XXX.
            15  ITEM-4B   PICTURE XXX.
66  NEW-REC RENAMES ITEM-2A THRU ITEM-3B.

01  OUT-REC.
    05  FIELD-X.
        10  SUMMARY-GROUPX.
            15  FILE-1 PICTURE X.
            15  FILE-2 PICTURE X.
            15  FILE-3 PICTURE X.
    05  FIELD-Y.
        10  SUMMARY-GROUPY.
            15  FILE-1 PICTURE X.
            15  FILE-2 PICTURE X.
            15  FILE-3 PICTURE X.
    05  FIELD-Z.
        10  SUMMARY-GROUPZ.
            15  FILE-1 PICTURE X.
            15  FILE-2 PICTURE X.
            15  FILE-3 PICTURE X.
66  SUM-X RENAMES FIELD-X.
66  SUM-XY RENAMES FIELD-X THRU FIELD-Y.
66  SUM-XYZ RENAMES FIELD-X THRU FIELD-Z.
```

Figure 14.23 Example—Renames Clause.

```
01  MASTERRECORD
    02  ACCOUNT-NUMBER          PICTURE X(6).
    02  CHARGES-1               PICTURE 999V99.
    02  CHARGES-2               PICTURE 999V99.
    02  CHARGES-3               PICTURE 999V99.
    02  TOTAL-CHARGES           PICTURE 9999V99.
    02  TOTAL-PAYMENT           PICTURE 9999V99.
    02  PAYMENTS-1              PICTURE 9999V99.
    02  PAYMENTS-2              PICTURE 9999V99.
    66  CHARGE-INFO   RENAMES CHARGES-1 THRU TOTAL-
        CHARGES.
    66  TOTAL-INFO   RENAMES TOTAL-CHARGES THRU TOTAL-
        PAYMENTS.
    66  PAYMENT-INFO RENAMES TOTAL-PAYMENTS THRU
        PAYMENTS-2.
    66  PAST-DUE     RENAMES PAYMENTS-2.
```

Figure 14.24 Example—Renames Clause.

Consider the following data description:

```
01  PAY-MASTER.
    04  EMP-REC.
        08  EMP-NO              PICTURE 9(6).
        08  EMP-NAME            PICTURE X(25).
        08  DEPT                PICTURE 9(4).
        08  MAIL-STOP           PICTURE X(4).
        08  EXTENSION           PICTURE 9(4).
    06  PAY-REC.
        10  PAY-RATE            PICTURE 9V999.
        10  GROSS               PICTURE 9(4).
        10  GROSS-YTD           PICTURE 9(5).
    66  TELEPHONE-DIRECTORY  RENAMES
        EMP-NAME THRU EXTENSION.
```

By using the RENAMES clause, the programmer has created a new "pseudo record," which appears below:

```
    07  TELEPHONE-DIRECTORY.
        08  EMP-NAME            PICTURE X(25).
        08  DEPT                PICTURE 9(4).
        08  MAIL-STOP           PICTURE X(4).
        08  EXTENSION           PICTURE 9(4).
```

Figure 14.25 Example—Renames Clause.

PROCEDURE DIVISION

Transform Statement

This statement is used to alter characters according to a transformation rule (figure 14.26). For example, it may be necessary to change the characters in an item to a different collating sequence.

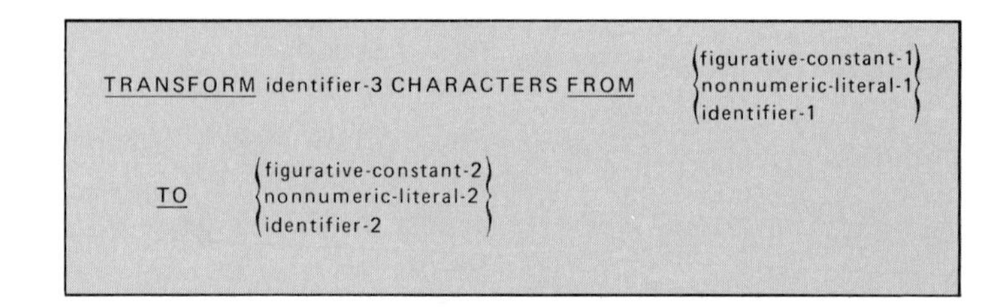

Figure 14.26 Format—Transform Statement.

Rules Governing the Use of the Transform Statement

1. Identifier-3 must be an elementary alphabetic, alphanumeric, numeric edited item, or a group item.
2. The combination of the FROM and TO options determine what the transformation rule is to be.
3. Nonnumeric literals require enclosing quotation marks.
4. Identifier-1 and Identifier-2 must be elementary alphabetic or alphanumeric items or any fixed-length group items not to exceed 255 characters in length.
5. A character may not be repeated in a nonnumeric-literal-1 or in the area defined by Identifier-1. If a character is repeated, the results will be unpredictable.
6. The allowable figurative constants are ZERO, ZEROS, ZEROES, SPACE, SPACES, QUOTE, QUOTES, HIGH-VALUE, HIGH-VALUES, LOW-VALUE, and LOW-VALUES.
7. When either Identifier-1 or Identifier-2 appears as an operand of the specific transformation, the user can change the transformation rule at object time.

(See figures 14.27, 14.28.)

Identifier-3 (Before)	FROM	TO	Identifier-3 (After)
1b7bbABC	SPACE	QUOTE	1"7" "ABC
1b7bbABC	"17CB"	"QRST"	QbRbbATS
1b7bbABC	b17ABC	CBA71b	BCACC71b
1234WXY89	98YXW4321	ABCKEFGHI	IHGFEDCBA

Figure 14.27 Examples—Data Transformation.

Operands	Transformation Rule
FROM figurative-constant-1 TO figurative-constant-2	All characters in the data item represented by identifier-3 equal to the single character figurative-constant-1 are replaced by the single character figurative-constant-2.
FROM figurative-constant-1 TO nonnumeric-literal-2	All characters in the data item represented by identifier-3 equal to the single character figurative-constant-1 are replaced by the single character nonnumeric-literal-2.
FROM figurative-constant-1 TO identifier-2	All characters in the data item represented by identifier-3 equal to the single character figurative-constant-1 are replaced by the single character represented by identifier-2.
FROM nonnumeric-literal-1 TO figurative-constant-2	All characters in the data item represented by identifier-3 that are equal to any character in nonnumeric-literal-1 are replaced by the single character figurative-constant-2.
FROM nonnumeric-literal-1 TO nonnumeric-literal-2	Nonnumeric-literal-1 and nonnumeric-literal-2 must be equal in length or nonnumeric-literal-2 must be a single character. If the nonnumeric-literals are equal in length, any character in the data item represented by identifier-3 equal to a character in nonnumeric-literal-1 is replaced by the character in the corresponding position of nonnumeric-literal-2. If the length of nonnumeric-literal-2 is one, all characters in the data item represented by identifier-3 that are equal to any character appearing in nonnumeric-literal-1 are replaced by the single character given in nonnumeric-literal-2.
FROM nonnumeric-literal-1 TO identifier-2	Nonnumeric-literal-1 and the data item represented by identifier-2 must be equal in length or identifier-2 must represent a single character item.

Figure 14.28 Transform Statement Rules.

Figure 14.28 Continued.

Operands	Transformation Rule
	If nonnumeric-literal-1 and identifier-2 are equal in length, any character represented by identifier-3 equal to a character in nonnumeric-literal-1 is replaced by the character in the corresponding position of the item represented by identifier-2. If the length of the data item represented by identifier-2 is one, all characters represented by identifier-3 that are equal to any character appearing in nonnumeric literal-1 are replaced by the single character represented by identifier-2.
FROM identifier-1 TO figurative-constant-2	All characters represented by identifier-3 that are equal to any character in the data item represented by identifier-1 are replaced by the single character figurative-constant-2.
FROM identifier-1 TO nonnumeric-literal-2	The data item represented by identifier-1 and nonnumeric-literal-2 must be of equal length or nonnumeric-literal-2 must be one character. If identifier-1 and nonnumeric-literal-2 are equal in length, any character in identifier-3 equal to a character in identifier-1 is replaced by the character in the corresponding position of nonnumeric-literal-2. If the length of nonnumeric-literal-3 is one, all characters represented by identifier-3 that are equal to any character represented by identifier-1 are replaced by the single character given in nonnumeric-literal-2.
FROM identifier-1 TO identifier-2	Any character in the data item represented by identifier-3 equal to a character in the data item represented by identifier-1 is replaced by the character in the corresponding position of the data item represented by identifier-2. Identifier-1 and identifier-2 can be one or more characters, but must be equal in length.

SOURCE PROGRAM LIBRARY FACILITY

The writing of the data definitions for a master file that is accessed by many COBOL programs is a frequently-encountered data processing problem. In order to avoid the repetitious writing of such data definitions, the data definitions are placed into the direct access file as recompilation master. Then when a source program needs the definition of the master file, just one COPY statement is written in place of the many data description entries.

The library module provides a capability for specifying text that is to be copied from a library. The COBOL library contains text that is available to a source program at compile time. Prewritten source program entries can be included in a source program at compile time. Thus an installation can use standard file descriptions, record descriptions, or procedures without recoding them. The effect of the compilation of the library text is the same as if the text were actually written as part of the source program.

The COBOL library contains text that is available to a source program at compile time. These entries and procedures are contained in these user-created libraries; they are included in a source program by means of a COPY statement.

COBOL library text is placed in the COBOL library as a function independent of the COBOL program and according to implementor-defined techniques. An entry in the COBOL library may contain source program texts for the Environment, Data, or Procedure Divisions, or any combination of the three.

Copy Statement

The COPY statement permits the inclusion of prewritten Environment, Data, and Procedure Division entries at compile time in the source program (figs. 14.29, 14.30).

Library-name is the name of a member of a partitioned data set contained in the user's library; it identifies the library subroutine to the control program. Library-name must follow the rules for the formation of a program-name. The first eight characters are used to identify the name.

Rules Governing the Use of Copy Statement

1. The COPY statement may appear as follows:
   a. In any of the paragraphs of the Environment Division.
   b. In any of the level indicators entries or any 01 level entry in the Data Division.
   c. In a Section or a paragraph in the Procedure Division.
2. No other statement may appear in the same entry as the COPY statement.
3. The library text is copied from the library and the result of the compilation is the same as if the text were actually a part of the source program. The words COPY *Library-name* are replaced by the information by library-name. This information comprises the sentences or clauses which are necessary to complete the paragraph, sentence, or entry containing the COPY statement.
4. The copying process is terminated by the end of the library text itself.

```
Option 1 (within the Configuration Section):

    SOURCE-COMPUTER.  COPY statement.
    OBJECT-COMPUTER.  COPY statement.
    SPECIAL-NAMES.  COPY statement.

Option 2 (within the Input-Output Section):

    FILE-CONTROL.  COPY statement.
    I-O-CONTROL.  COPY statement.

Option 3 (within the FILE-CONTROL Paragraph):

    SELECT file-name COPY statement.

Option 4 (within the File Section):

    FD  file-name COPY statement.
    SD  sort-file-name COPY statement.

Option 5 (within the Report Section):

    FD  report-name COPY statement.
    RD  report-name [WITH CODE mnemonic-name] COPY statement.

Option 6 (within a File or Sort description entry, or within the
Working-Storage Section or the Linkage Section):

    01  data-name COPY statement.

Option 7 (with a Report Group):

    01  [data-name] COPY statement.

Option 8 (within the Working-Storage Section or the Linkage Section):

    77  data-name COPY statement.

Option 9 (within the Procedure Division):

    section-name SECTION [priority-number].  COPY statement.
    paragraph-name.  COPY statement.
```

Figure 14.29 Format—Copy Statement.

```
COPY library-name [SUPPRESS]
                                   {word-2      }
        [REPLACING word-1  BY      {literal-1   }
                                   {identifier-1}
                           {word-4      }
        [word-3  BY        {literal-2   }   ] . . . ] .
                           {identifier-2}
```

Figure 14.30 Format—Copy Statement—Replacing Option.

5. The text in the library must not contain a COPY statement.
6. If the REPLACING option is used, each word specified in the format is replaced by the stipulated word, identifier, or literal that is associated with it in the format. The use of the REPLACING option does not alter the material as it appears in the library.

(See figures 14.31, 14.32, 14.33, 14.34.)

For example, suppose the library entry named WORKREC consists of the following Data Division record

```
02 AAA.
   03 HOURS-WORKED          PICTURE 99.
   03 SICK-LEAVE            PICTURE 99.
   03 HOLIDAY-PAY           PICTURE 99.
```

The user may copy this record into his source program with the first of the following two statements:

```
01 WORK-RECORD  COPY  WORKREC.
01 NEXT-RECORD.
```

After the record has been copied, the compiled source program will appear as if it had been written as:

```
01 WORK-RECORD.
   02 AAA.
      03 HOURS-WORKED       PICTURE 99.
      03 SICK-LEAVE         PICTURE 99.
      03 HOLIDAYPAY         PICTURE 99.
01 NEXT-RECORD.
```

Figure 14.31 Example—Copy Statement.

If the programmer wishes to retrieve the member, CFILEA, catalogued previously, he writes the statement:

```
FD  FILEA COPY CFILEA
```

The compiler translates this instruction to read:

```
FD  FILEA           BLOCK CONTAINS 20 RECORDS
                    RECORD CONTAINS 120 CHARACTERS
                    LABEL RECORDS ARE STANDARD
                    DATA RECORD IS FILE-OUT.
```

Note that CFILEA itself does not appear in the statement. CFILEA is a name identifying the entries. It acts as a header record but is not itself retrieved. The compiler source listing, however, will print out the COPY statement as the programmer wrote it.

Figure 14.33 Example—Copy Statement.

If the library entry PAYLIB consists of the following Data Division record:

```
01  A.
    05    B    PICTURE S99.
    05    C    PICTURE S9(5)V99.
    05    D    PICTURE S9999 OCCURS 0 TO 52 TIMES
               DEPENDING ON B OF A.
```

the programmer can use the COPY statement in the Data Division of his program as follows:

```
01  PAYROLL COPY PAYLIB.
```

In this program, the library entry is then copied and appears in the source listing as follows:

```
01  PAYROLL.
    05    B    PICTURE S99.
    05    C    PICTURE S9(5)V99.
    05    D    PICTURE S9999 OCCURS 0 TO 52 TIMES
               DEPENDING ON B OF A.
```

Note that the data-name A has not been changed in the DEPENDING ON option. To change some (or all) of the names within the library entry to names he wishes to reference in his program, the programmer can use the REPLACING option:

```
01  PAYROLL COPY PAYLIB REPLACING A BY PAYROLL
    B BY PAY-CODE C BY GROSSPAY.
```

In this program the library entry is copied and appears in the source listing as follows:

```
01  PAYROLL.
    05  PAY-CODE    PICTURE S99.
    05  GROSSPAY    PICTURE S9(5)V99.
    05  D    PICTURE S9999 OCCURS 0 TO 52 TIMES
             DEPENDING ON PAY-CODE OF PAYROLL.
```

The entry as it appears in the library remains unchanged.

Figure 14.32 Example—Copy Statement.

The following examples shows the effect of a COPY statement which copies the complete contents of CLIBOVER01.

```
    MOVE BAL TO NEW-BAL.
OVER-DRAFT.      COPY CLIBOVER01.
GOOD-BAL.
    MOVE NEW-BAL TO OUT-BAL.
```

Recompilation master having the name CLIBOVER01:

```
    ADD AMT TO WORK.
    MOVE WORK TO NEW-WORK.
    WRITE OVER-RECORD.
```

Source statements after the COPY operation:

```
    MOVE BAL TO NEW-BAL.
OVER-DRAFT.
    ADD AMT TO WORK.
    MOVE WORK TO NEW-WORK.
    WRITE OVER-RECORD.
GOOD-BAL.
    MOVE NEW-BAL TO OUT-BAL.
```

Figure 14.34 Example—Copy Statement.

Exercises

Write your answer in the space provided (answer may be one or more words).

1. The Special-Names paragraph is used to equate special ________________ names with ________________ mnemonic-names.
2. The Special-Names paragraph is written only if the user specifies a ________________ -name in the ________________ Division.
3. When the ________________ option is used in a Write statement with the Advancing option, the ________________ name must be defined as a ________________ name in the Special-Names paragraph.
4. Mnemonic-names with a single character enclosed in quote marks is used in the ________________ clause of the report description entry.
5. If the Currency Sign clause is not used, only the ________________ can be used as the currency symbol in the Picture clause.
6. If the function of commas and periods are to be exchanged in the Picture clause or in a numeric literal, the clause ________________ is used.
7. The Reserve clause is used to reserve additional ________________ or ________________ areas in addition to the buffer allocated by the ________________.
8. The Reserve clause must not be specified for ________________ files.
9. If no additional input/output areas aside from the minimum of one are needed, the ________________ option is used.
10. The File-Limit clause is used for the logical ________________ and logical ________________ of a file on a ________________ device.
11. The Multiple-File Tape clause indicates that two or more tape files share the same ________________ reel of tape.
12. The Track-Area clause may be used when records are to be added to an ________________ file in the ________________ access mode.
13. The Apply Write-Only On clause is used to make optimum use of ________________ and ________________ space when creating a file whose recording mode is ________________.
14. The Apply Core-Index On option is used with ________________ files whose access mode is ________________ and is used to specify the highest level ________________ to be processed in core storage.
15. The Apply Record-Overflow On clause is used to reduce the amount of ________________ space in disk storage.
16. The Rerun clause is used to check the status for a ________________ of a problem program at ________________ intervals.
17. The Renames clause permits possible ________________ and ________________ of elementary data.
18. Level number ________________ must be used with the Renames clause.
19. The Transform statement is used to ________________ characters according to a ________________ rule.
20. The library module contains ________________ that is available to a ________________ program at ________________ time.
21. ________________ source items can be included at ________________ time.
22. The Copy statement permits the inclusion of prewritten ________________, ________________ and Procedure Division entries at ________________ time in a ________________ program.
23. The words Copy ________________ are replaced by the prewritten library entries.
24. The text in a library must not contain a ________________ statement.
25. If the Replacing option is used each word is replaced by the stipulated ________________, ________________ or ________________ which is associated with it in the format.

Answers

1. FUNCTION, USER SPECIFIED
2. MNEMONIC, PROCEDURE
3. MNEMONIC-NAME, MNEMONIC, FUNCTION
4. CODE
5. $
6. DECIMAL POINT IS COMMA
7. INPUT, OUTPUT, COMPILER
8. DIRECT
9. NO
10. BEGINNING, END, MASS STORAGE
11. PHYSICAL
12. INDEXED, RANDOM
13. BUFFER, DEVICE, V
14. INDEXED, RANDOM, INDEX
15. UNUSED
16. RECORD, DESIRED
17. OVERLAPPING, GROUPINGS
18. 66
19. ALTER, TRANSFORMATION
20. TEXT, SOURCE, COMPILE
21. PREWRITTEN, COMPILE
22. ENVIRONMENT, DATA, COMPILE, SOURCE
23. LIBRARY-NAME
24. LIBRARY
25. WORD, IDENTIFIER, LITERAL

Questions for Review

1. What is the purpose of the Special-Names paragraph?
2. How is the Special-Names paragraph used with the Advancing option of the Write statement? Give an example.
3. What are the main uses of the Reserve clause?
4. How is the File-Limit clause used with mass storage devices?
5. What are the main purposes of the Multiple-File Tape clause?
6. Give the main uses of the following Apply clauses: Apply Write-Only On, Apply Core-Index On, Apply Record-Overflow On.
7. What is the main function of the Rerun clause? Explain the operation of the Rerun clause.
8. What is the purpose of the Renames clause?
9. How is the Transform statement used? Give an example.
10. What purposes does the Copy statement serve? Give examples of its use.
11. What are libraries and how are they used to COBOL programs?
12. How are standardized library texts included in the users program?

Problems

1. *Match each item with its proper description.*

   ______ 1. Special-Names — A. User specified name.
   ______ 2. Mnemonic Name — B. Additional input/output areas.
   ______ 3. Function Name — C. Equate function name with user specified mnemonic name.
   ______ 4. Reserve — D. Two files sharing the same physical reel of tape.
   ______ 5. File-Limit — E. Name of device or action taken.
   ______ 6. Multiple-File Tape — F. Logical beginning and ending of a mass storage file.

2. *Match each item with its proper description.*

   ______ 1. Track-Area — A. Optimum use of buffer and device space.
   ______ 2. Apply Write-Only — B. Reduce amount of unused space on a volume.
   ______ 3. Apply Core-Index On — C. Checkpoint record.
   ______ 4. Apply Record-Overflow On — D. Inclusion of prewritten library entries.
   ______ 5. Rerun — E. Records added to indexed file in random access mode.
   ______ 6. Renames — F. Highest level index in core storage.
   ______ 7. Copy — G. Overlapping and grouping of elementary data items.

3. *You are using an IBM 370 Model H155 computer, write the necessary entries to accomplish the following:*

   a. The Configuration Section of the Environment Division entries including the Special-Names paragraph to define the mnemonic name FIRST-LINE as the first line of the form.
   b. The procedural Write statement with the After Advancing option to write the output record RECORD-OUT after skipping to the first line of the next form after reaching the last line of the form.

4. *A file called CUSTOMER-FILE with a record name of CUSTOMER-RECORD is to be outputted on a printer in the same format as the input record.*

   Using the following hardware devices:

Device	Model Number
Card Reader	2540
Printer	1403

   Write the necessary entries to accomplish the following:

   a. Configuration Section including the Special-Names paragraph.
   b. File-Control paragraph including the Reserve clause.
   c. The procedural entries to move the input record to the output including end of page procedures.

5.

```
01  CORRECTED-RECORD.
    05  GROUP-A.
        10  FIELD-1A.
            15  ITEM-1A  PICTURE XX.
            15  ITEM-2A  PICTURE XXX.
            15  ITEM-3A  PICTURE XX.
        10  FIELD-2A.
            15  ITEM-4A  PICTURE XX.
            15  ITEM-5A  PICTURE XX.
            15  ITEM-6A  PICTURE XX.
    05  GROUP-B REDEFINES GROUP-A.
        10  FIELD-1B.
            15  ITEM-1B  PICTURE XXXX.
            15  ITEM-2B  PICTURE XXX.
        10  FIELD-2B.
            15  ITEM-3B  PICTURE XXX.
            15  ITEM-4B  PICTURE XXX.
```

   In the above record it is necessary to group items ITEM-3A to ITEM-4B for future processing.

   Write the necessary entry to accomplish the above using the Renames clause.

6. *Since we are using the same Configuration Section in all of our programs, write the Configuration Section using the following library names:*

Paragraph	Library Name
Source-Computer	S-COMPUTER
Object-Computer	O-COMPUTER
Special-Names	S-NAMES

7. *In many of our programs, we are using the same headings and detail line formats. We have decided to put these formats in our library as follows:*

Library Name	Function
SALES-HEAD	Name of report.
COLUMN-HEAD	Column headings of report.
DETAIL-RECORD	Detail line of the report.

   Write the necessary entries for

   Working-Storage items for headings and detail using the COPY clause for the report heading (HDG-1), column headings (HDG-2), and detail line (DETAIL-LINE).

8. *Most of our programs contain the same routine for printing headings on the first page and each overflow. We have decided to put these procedures in our library. Assume that the library name of the routine is HEADING-ROUTINE, write the Procedure Division entry using the COPY clause for the paragraph called PRT-HDG.*

9.

	Identifier-3 (Before)	From	To	Identifier-3 (After)
1.	149b07b9965	SPACE	'—'	________
2.	JONES	'SNOEJ'	'HIMTS'	________
3.	2b49b7512	'b729514'	'7586931'	________
4.	12345658	'5386142'	'OSLBACN'	________

In the above, fill in identifier-3 after the data transformation.

Note: b denotes blank character.

15 COBOL Programming Techniques

INTRODUCTION

The writing of a computer program is a difficult and tedious task. Many times a program rapidly solves a particular problem without any regard for the efficiency of the program itself. In the writing of COBOL programs, this problem is more serious than in other programming languages. As mentioned earlier, COBOL does not produce a program as efficient as one written in the basic language of the particular computer. Thus it is imperative that the programmer adopt techniques that will increase the efficiency of the COBOL program.

Prior to the writing of the program, the problem should be properly defined, and the appropriate flowcharts, source document formats, and output formats should be available to the programmer.

PROGRAMMING STANDARDS

COBOL has extensive features that provide the capability for a source program to be highly self-documenting. These features make it possible for a new programmer to assume maintenance of a COBOL program with little or no prior training by the originating programmer. Yet, if an established set of standards is not placed in use, each programmer will establish his own. Nonstandard methods of programming, even in COBOL, greatly complicate program maintenance for a newly-assigned programmer.

One of the major objectives in the original design of COBOL was to have a COBOL program that would be essentially self-documenting. Features of the language tending to promote this objective are the compartmentalizing of the program into divisions, the requirement of an Identification Division, many of the options in the Environment Division, and particularly the use of a subset of the English language in the Procedure Division.

However, in the haste of getting a program "on the air," a programmer may choose a way of writing his program which makes the later debugging and maintenance of that program unnecessarily difficult. Generally speaking, the same program could have been written without any further expenditure of energy in a somewhat different manner which would aid rather than hinder future maintenance. The key to easier debugging and maintenance is the establishment, understanding, and enforcement of a small set of programming standards.

Source Program Appearance

In order to improve the "English" readability of a COBOL source program, many things can be done to present an organized appearance. For example, COBOL completely ignores input cards that are entirely blank. Blank cards can thus be interspersed within a deck to set off paragraphs, sections, etc. A card punched with all asterisks (*) in columns 7 through 72 can be used to divide portions of the source program into unmistakable divisions.

Source Program Conventions

Potentially difficult-to-read statements should be simplified. Improved readability and program logic understanding can result if the following suggestions are followed:

Use Indentation to Indicate Subordination

Those rules of COBOL which require certain items to begin in Area A do not dictate that these items necessarily begin in column 8. Such items might appear beginning in column 8, or in column 9, 10, or 11. Furthermore, language elements which must appear in Area B may begin in column 12 or any column thereafter. Therefore, the programmer is free to do the following:

1. Begin any division header in column 8, preceded by four blank lines, and followed by two blank lines.
2. Begin any section header in column 9, followed by one blank line.
3. Begin in column 10, preceded by one blank line, any of the following:
   a. a paragraph name in the Identification, Environment, or Procedure Division, or
   b. an FD, 01, or 77 level entry in the Data Division, with not more than one of these items in a single card.
4. In the Data Division, begin any entry which is immediately subordinate to another entry two spaces to the right of that entry.
5. In the Procedure Division, whenever a statement is conditioned by an "IF," indent the conditioned statement to at least column 16, in order to show its dependency on the condition expressed in the "IF."
6. In the Procedure Division, start each new sentence in column 12. Start each subsequent line of the same sentence in column 16. (Avoid compound sentences.)

Start Each New Statement on a New Line

This suggestion is a requirement for division headers, section headers, and Data Division entries. It is not required, but it is suggested for the following items:

- Paragraph-names—place each paragraph-name on a line by itself. Any later rewrite or rearrangement of the contents of the paragraph need not disturb the line containing the paragraph-name.
- Procedure Division Verbs—write each statement beginning with a verb on a line by itself. Any change made to that statement requires modification of only that source program part.
- Conditional statements—write an IF condition on a line by itself, indenting all statements conditioned by it.

Make Each Nonconditioned Procedure Division Statement a Separate Sentence

It is not possible to make each Procedure Division statement a separate sentence where a series of Procedure Division statements are conditioned by an IF. In this case, of course, only the last such statement is followed by a period. Reading is made easier by the fact that any statements not followed by a period are conditioned statements.

Indent When Using GO TO with the DEPENDING ON Option

When using the DEPENDING ON option, indent each paragraph-name from the GO TO, putting each on a separate line. As an example:

```
12

GO TO

     PXXXX-ERROR-RTE
     PXXXX-VALID-TRX-RTE
     PXXXX-EXCEPTION-PRLS
     DEPENDING ON PROCESS-SWITCH.
```

Do Not Split Words or Numeric Literals Between Lines

A nonnumeric literal may be as much as 120 characters in length. It might, therefore, be split between two or maybe even three lines. On the other hand, a COBOL key word, a programmer-defined name, or a numeric literal need never be split between two lines. If a word cannot fit on a given line, begin it on the next line. Although the capability of splitting words and numeric literals exists in COBOL, the use of this option by the programmer tends to make the program difficult to read and maintain.

Use Blank Lines Freely to Improve Readability

A blank line may appear anywhere in a COBOL program, except between two lines across which a word or literal is split. In other words, a blank may appear anywhere except immediately preceding a continuation line. The use of blank lines makes any search for the major routines within a program much easier.

Use Section Names to Improve Understanding of Program Organization

The use of section identification is recommended. By giving two or more successive paragraphs a section-name, those paragraphs can be executed by a PERFORM statement that need not have a THRU clause. Section identification in a large source module also provides clearer organization of COBOL routines.

The first cards of each section should contain comments. If columns 7–9 and 70–72 of the source cards of containing comments are punched with asterisks, such cards are not only treated as comments due to the asterisk in column 7, but they also stand out clearly on the source listings, thus highlighting the comments punched between columns 11 through 68. Alternately, a NOTE paragraph can be used, if desired, for prose readability.

Section-names should be descriptive of the routine they identify. Consider the following example:

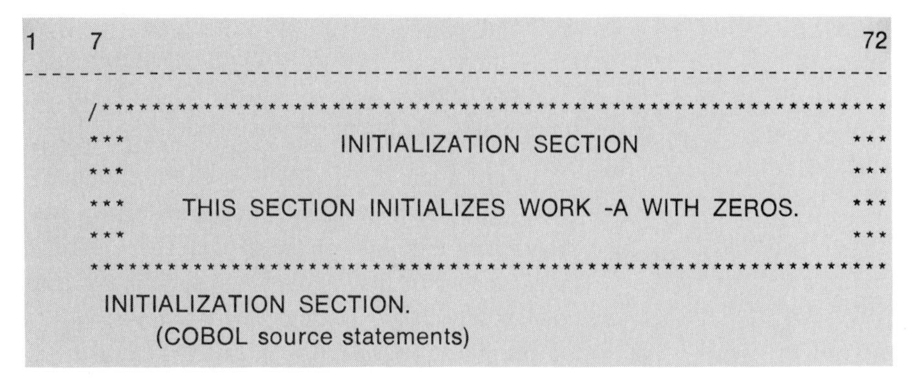

```
1     7                                                                    72
-------------------------------------------------------------------------------
      /*****************************************************************
      ***                 INITIALIZATION SECTION                       ***
      ***                                                              ***
      ***     THIS SECTION INITIALIZES WORK -A WITH ZEROS.             ***
      ***                                                              ***
      ******************************************************************
       INITIALIZATION SECTION.
           (COBOL source statements)
```

Use Comments Freely for Explanation

When a program is being written, by no means all that is in the programmer's mind concerning the problem being solved actually goes into the code. In order to facilitate understanding of the logic of a program, it may be necessary to include many comments along with pure program text. For example, blocks of comments should precede major sections of the program. Comments cards identified by an * symbol in column 7 can provide such text comprehension. Such a line appears on the compilation listing, but serves no purpose other than comments.

Comments cards identified by a / symbol in column 7 provide for page ejection in the source listing prior to printing the comment. This type of comments should be used to clearly identify major sections of the program.

Note: Many compilers do not support (1) to achieve ejection on a source program listing. To achieve ejection on these computers, the word eject on a card before the line that is to appear at the top of the new page is required.

While COBOL is designed to be as self-documenting as possible, often the logic used by a programmer may not be apparent to someone else reading the program. In such a case (and such cases are far more prevalent than programmers realize), comment lines are invaluable. For added readability, it is recommended that whenever a comment line is to be included, three comment lines be included. The first and third lines are blank, except for the asterisk in column 7, and the desired comments. Optionally, the blank lines need not contain an asterisk in column 7.

The writing of the COBOL program can be simplified and the efficiency increased if the following programming techniques are applied. All four divisions of the COBOL program must be completed before the source program can be compiled and executed.

IDENTIFICATION DIVISION

This is the simplest division of the four. It contains the information that identifies the program and is intended to provide information to the reader of the program. The name of the program must be stated, and other information about the program may optionally be mentioned.

Entries That Should Appear in the Identification Division

1. The name of the division-Division Header.
2. The name of the program.
3. The name of the programmer.

4. When the program was written.
5. Remarks that will explain the data processing job from which the program was written.
6. Any other optional information.

Only items (1) and (2) are necessary for the proper execution of the program. Additional information is desirable for adequate documentation of the program. The reader of the program would be interested in the REMARKS paragraph where the intent of the program is mentioned. The pertinent information contained therein should provide the reader with a better understanding of the program. Therefore, it is essential that the REMARKS paragraph be as thorough as possible.

The standards shown in figure 15.1 are suggested for uniformity within an installation: (See also figure 15.2.)

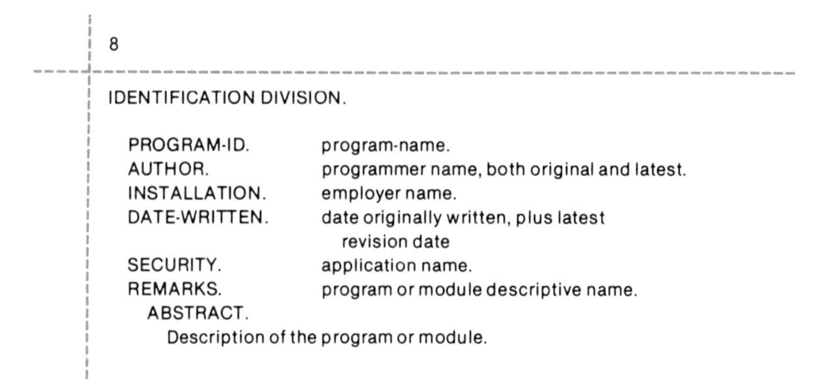

```
  8
IDENTIFICATION DIVISION.

   PROGRAM-ID.          program-name.
   AUTHOR.              programmer name, both original and latest.
   INSTALLATION.        employer name.
   DATE-WRITTEN.        date originally written, plus latest
                           revision date
   SECURITY.            application name.
   REMARKS.             program or module descriptive name.
     ABSTRACT.
       Description of the program or module.
```

Figure 15.1 Identification Division Standards.

```
  8
IDENTIFICATION DIVISION.

   PROGRAM-ID.          PROGA.
   AUTHOR.              JON QUEUE CODER.
   INSTALLATION.        ABC ENTERPRISES.
   DATE-WRITTEN.        OCTOBER 12, 1978.
   SECURITY.            MONTHLY PAYROLL.
   REMARKS.             CALCULATE GROSS TO NET PAY.
*    ABSTRACT.
*      PURPOSE AND METHOD OF ACCOMPLISHMENT—THIS PROGRAM
*           CALCULATES PAYROLL DEDUCTION TOTALS, CROSSFOOTING TO
*           OBTAIN NET PAY.
*      RETURN CONDITIONS—ANY CHANGE IN PERCENTAGES.
*      LANGUAGE USED—SYSTEM/3 COBOL.
*      SOURCE MODULE ATTRIBUTES—REUSABLE INDEFINITELY.
*      RESTRICTIONS—USE FOR SALARIED PERSONNEL ONLY.
*      REASON FOR CHANGE—NEW.
```

Figure 15.2 Sample Identification Division.

ENVIRONMENT DIVISION

The Environment Division is the only division in the COBOL program that is machine oriented. The division contains information about the equipment to be used when the object program is compiled and executed. Most importantly, it links the devices of the computer system and the data files to be processed.

In this division of a COBOL program, somewhat more flexibility is required from program to program than is required for the Identification Division. And yet, since the hardware does not change between programs, and there are certain common usages of input/output devices for a given user, the coding can be simplified, and reading of source program listings can be expedited, when standard entries are devised.

Advantages of Using the COBOL Source Library for Environment Division Coding

With the exception of the File-Control paragraph, the entire contents of the Environment Division can be standardized for all programs. For this reason it is strongly suggested that once such standard text has been devised, it be cataloged into the COBOL source language library. File-Control entries for each file to be processed can then be hand-coded, or these too can be included from separate source modules cataloged for each standard file.

Entries That Should Appear in the Environment Division

1. The name of the division-Division Header.
2. Configuration Section—Source and Object Computer Paragraphs.
3. Input-Output Section—File-Control and I-O-Control Paragraphs.

The external device names in the File-Control paragraph should be checked with the system programmers, as the names will vary with each data processing unit.

All files mentioned in the File-Control paragraph should be properly defined in the Data Division, and opened and closed in the Procedure Division.

Any special input/output techniques should be defined in the I-O-Control paragraph. (See figure 15.3.)

```
ENVIRONMENT DIVISION.
CONFIGURATION SECTION.
SOURCE-COMPUTER paragraph
OBJECT-COMPUTER paragraph
[SPECIAL-NAMES paragraph]
[INPUT-OUTPUT SECTION.
FILE-CONTROL paragraph
[I-O-CONTROL paragraph]]
```

Figure 15.3
Structure—Environment Division.

DATA DIVISION

The Data Division describes the information to be processed by the object program. Each file mentioned in the Environment Division must be described in the Data Division. In addition, each data item within these files must be described. All data items that comprise the Working-Storage Section, such as constants and work areas, must also be described.

The Data Division should be written using copies of the source document formats and output formats as guides for writing the file description entries and the record description entries.

Unlike the Procedure Division, which is substantially different for each program, the Data Division offers opportunities for work-saving standardization. Most computer installations operate with a fixed number of permanent files. The Data Division entries for each such file can be cataloged into the COBOL source language library on disk, and included in each program requiring use of such file(s) via the COPY statement. By this means, data-names assigned to each field can be standardized in all programs. Updates to file or record descriptions can, in many cases, be accomplished by merely recataloging the revised source module, and recompiling all programs making use of this module.

Include Common Data Definitions with COPY Statements

The use of the Copy statement in data definition is a very powerful COBOL feature. The COPY statement permits the programmer to include in the source program at compile time prewritten Data Division or Environment Division entries from the COBOL source language library. There are many advantages in this approach:

1. Standardization of data definitions is achieved—every programmer uses identical file definitions with common data-names and descriptions.
2. Programming effort is minimized because only file description needs to be written, keypunched, and debugged only once, and is then available to all programmers.
3. Program maintenance becomes considerably easier. Because of a common understanding of the data definitions, programmer can pick up and more quickly modify programs written by others. In addition, a modification of the data definition itself can be made once and recataloged. Then all affected modules can be recompiled, minimizing direct programming effort.
4. The library member becomes a final authority on file contents. Consider including comments for each field (group and elementary) containing a complete description of the usage, coding values or valid ranges, variance and inter-field relationships, etc. References may be made to other documents when further clarification is needed.

Use Meaningful Names

One feature of COBOL is its capability for using programmer-defined names of as long as thirty characters. While a name this long would seldom be used, the freedom to have names longer than five or eight or ten characters provides the programmer with the opportunity of defining names that are meaningful with regard to the items referenced by those names. Failure to take advantage of this capability may prevent a reader of the program from understanding the data involved. Experience shows that most successful naming conventions result in names of eight to twelve characters. Obviously this can vary depending upon the nature of the item to be named, but the general principle still applies.

All programs of any complexity eventually require maintenance. This task usually falls to the original programmer, if he is still available. While at the time of the original writing all programs may be clear, after the passage of time usually little is recalled. As a result, a programmer should consider the use of self-documenting names and frequent comment lines describing what is being accomplished. It is this very technique that enables one to perform successful maintenance with a minimum of effort.

In reading programs written by experienced, competent programmers, the repetitious use of the hyphen symbol in COBOL names is apparent. Not only does the hyphen enable construction of compound names, but also ensures against the inadvertent use of a COBOL reserved word.

One technique that reduces differences in program code produced by more than one programmer is the establishment of a list of standard abbreviations. For example, once standardized within an installation, the abbreviation "MAST" for "MASTER" is clear to everyone.

Entries That Should Appear in the Data Division

1. The name of the division-Division Header.
2. File Section-file description entries and record description entries.
3. Working-Storage Section record description entries for constants and work areas.
4. Linkage Section-record description entries used for subprograms.
5. Report Section—report description and report group description entries.

(See figure 15.4.)

```
DATA DIVISION.
FILE SECTION.
{file description entry
{record description entry}...}...
WORKING-STORAGE SECTION.
[data item description entry]...
[record description entry]...
LINKAGE SECTION.
[data item description entry]...
[record description entry]...
REPORT SECTION.
{report description entry
{report group description entry}...}...
```

Figure 15.4
Structure—Data Division.

File Section

The File Section specifies the characteristics of the file.

File Description Entries

FD File-name is required and must agree with the name specified in the Environment Division.

Recording Mode Clause is optional but should be included if records are fixed, variable or undefined; otherwise, the compiler will generate an algorithm that does not always give V.

Block Contains clause must be included when the records are blocked (e.g., when records are blocked on a tape). If the records vary in size, the character option should be used instead of records to specify the total number of characters in each block. When there is only one record per block, the clause may be omitted.

Record Contains clause is used when variable size records are used. The clause should specify the length of the shortest and longest record in the file. This clause may be omitted since the compiler determines the record size from the record description entries.

A good programming practice is to include this clause in every FD entry for the following reasons:

1. The compiler will check the agreement of the record count in the record description entry with the RECORD CONTAINS clause in the FD entry. This will assure that no data fields were erroneously omitted in the record description entry. Otherwise, the compiler will assume the count in the record description entries as being the correct record length.
2. It provides the programmer and reader of the program with the size of the record without the necessity of counting all field lengths stated in the record description entries.

Label Records clause must be included in every record description entry even if the files are located in cards where the OMITTED option is used.

Data Records clause is optional, and each record-name should be written in sequence as it appears in the record description entries. The record description entries must appear immediately after the file description entries.

Record Description Entries

Level Numbers are required for each entry. Each level is a given number, always beginning with 01 for the data record itself. Each succeeding level is given a larger number to indicate a further breakdown of the data item. These larger numbers need not be in sequence. This may provide the programmer with more flexibility in assigning numbers. All level numbers may be written at the A margin, although only the 01 and 77 level numbers are required to be written at the A margin. However, indentation should be used, as this improves the readability of the program.

Data-names may be unique or otherwise qualified. The highest qualifier must be a unique name. In the File Section, the highest qualifier is the file-name; thus it is possible for two records to have the same name.

Names of independent items in the Working-Storage Section must be unique since they cannot be qualified. The highest qualifier in the Working-Storage Section is the record-name.

A recommended programming procedure for writing Data Division entries for output files for the printer is to describe the formats of the output file in the Working-Storage section and to use the WRITE verb with the FROM option referring to the Working-Storage item. This technique provides the programmer with

1. The ability to define heading with appropriate VALUE clauses (forbidden in the File Section) in the Working-Storage Section. Areas in the output can be blanked where necessary.
2. The ability to use one record description entry to define the output for both headings and detail lines.

Picture clause tells the number of characters to be stored and what type they will be. Picture clauses are only found in the descriptions of elementary items. Picture and usage clauses must be compatible. For example, an alphabetic item cannot have a usage clause of computational.

Value clause is used to assign initial values. The Value clause is not permitted in the File section, except level-88 (Condition-name) entries. The Value clause must agree with its picture. For example,

77 Discount Picture SV99, Value + .02.

77 Total-Identification, Picture A(15), Value "Pay This Amount".

Usage clause is allowed at both the group and elementary levels. If the clause is omitted, the items usage is assumed to be display.

If a data item field is to be used in a series of arithmetic operations, it would be advisable to define an area in Working-Storage Section in the computational mode. The data item would be moved from the input area to the Working-Storage area. After all the computations are completed, the computed item would be returned to the output area in the display mode.

This technique would save processing time of the compiler in changing the item from display to computational mode and back to display mode for each arithmetic operation.

Suffixes

To make it easier for programmers to locate items in a program listing, especially during the debugging stage, it is good practice to attach a suffix to a data-name to indicate where the item is to be found in the program. For example, if a QUANTITY item appears in the input and output records as well as in a Working-Storage area, the data-names may be assigned as follows:

Input item — QUANTITY-IN
Output item — QUANTITY-OUT
Working-Storage item — QUANTITY-WS

The same principle may be applied to all items that appear in these areas in order to make it simpler for the reader to know which fields are logically part of the same record or area.

Move Corresponding

The MOVE CORRESPONDING statement can save time in writing many MOVE statements of identical items, usually from an input to an output record. However, this technique involves qualification whenever an identical item is involved in a Procedure Division statement. To eliminate excessive qualifying, a REDEFINES or RENAMES statement may be used with the corresponding items. For example,

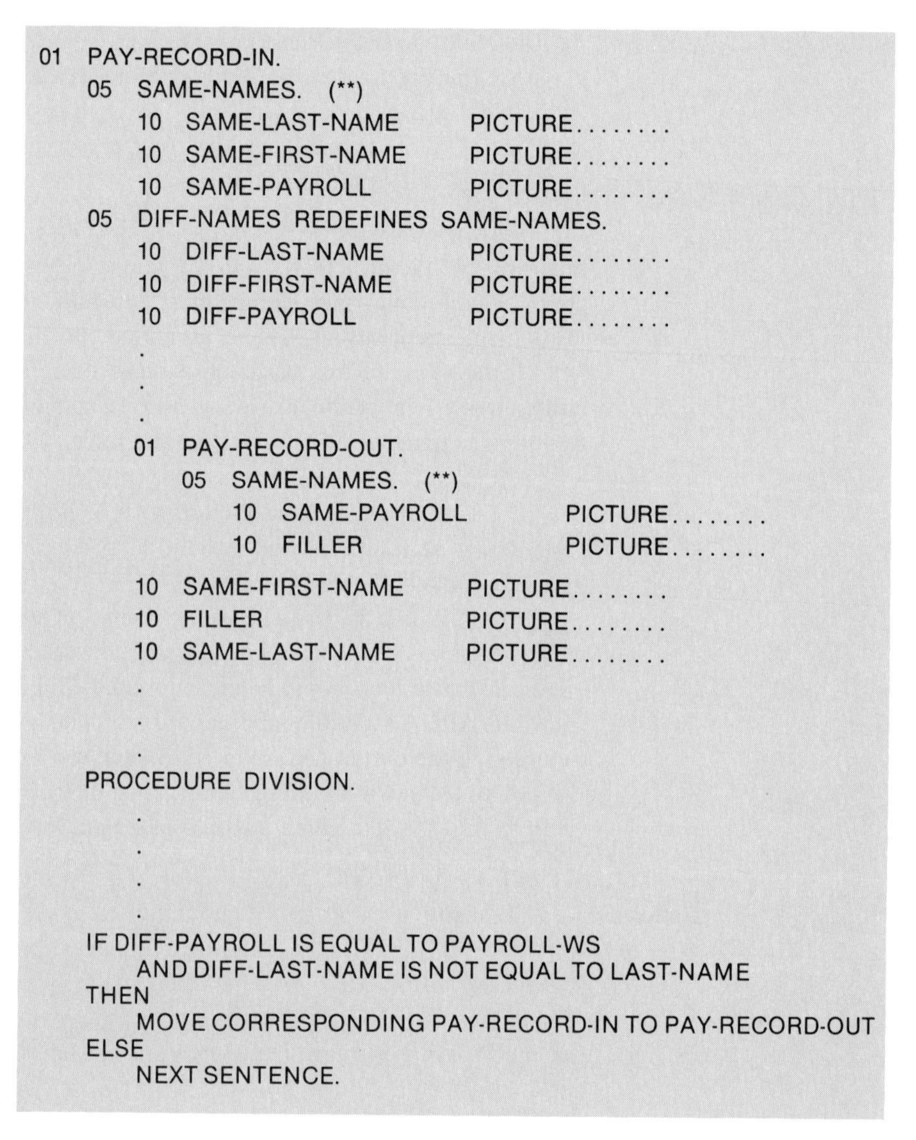

```
01  PAY-RECORD-IN.
    05  SAME-NAMES.  (**)
        10  SAME-LAST-NAME       PICTURE........
        10  SAME-FIRST-NAME      PICTURE........
        10  SAME-PAYROLL         PICTURE........
    05  DIFF-NAMES REDEFINES SAME-NAMES.
        10  DIFF-LAST-NAME       PICTURE........
        10  DIFF-FIRST-NAME      PICTURE........
        10  DIFF-PAYROLL         PICTURE........
        .
        .
        .
        01  PAY-RECORD-OUT.
            05  SAME-NAMES.  (**)
                10  SAME-PAYROLL         PICTURE........
                10  FILLER               PICTURE........
        10  SAME-FIRST-NAME      PICTURE........
        10  FILLER               PICTURE........
        10  SAME-LAST-NAME       PICTURE........
        .
        .
        .
    PROCEDURE DIVISION.
        .
        .
        .
        .
    IF DIFF-PAYROLL IS EQUAL TO PAYROLL-WS
        AND DIFF-LAST-NAME IS NOT EQUAL TO LAST-NAME
    THEN
        MOVE CORRESPONDING PAY-RECORD-IN TO PAY-RECORD-OUT
    ELSE
        NEXT SENTENCE.
```

Figure 14.29 Format—Copy Statement.

(NOTE: Fields marked with a double asterisk (**) in the foregoing listing must have exactly the same names for their subordinate fields in order to be considered corresponding. The same names must not be the redefining ones, or they will not be regarded as corresponding.)

Level Numbers

The programmer should use widely incremented level numbers (i.e., 01, 05, 10, 15, etc., instead of 01, 02, 03, 04, etc.) in order to allow room for future insertions of group levels. For readability, indent level numbers. Use level-88 numbers for codes. Then, if the codes must be changed, the Procedure Division coding for these tests need not be changed.

INCREASING EFFICIENCY OF DATA DIVISION ENTRIES

Conserving Core Storage

When writing Data Division statements, the COBOL programmer need not concern himself with data problems such as decimal alignment and mixed format: the compiler generates extra instructions to perform the necessary adjustments. Entries in the Data Division can significantly affect the amount of core storage required by a program.

Decimal Alignment

Procedure Division operations are most efficient when the decimal positions of the data items involved are aligned. If they are not, the compiler generates instructions to align the decimal positions before any operations involving the data items can be executed.

In a typical source program, the frequency of the most common verbs written in the Procedure Division of a COBOL program averaged over a number of programs is:

```
        Moves — 50%
        Go to — 20%
        If    — 15%

Miscellaneous (Arithmetic, Calculations, Input/Output, Perform) 15%.

An example of a pair of fields:
        77 A  Picture 99V9       Computational-3.  (Sending Field).
        77 B  Picture 999V99     Computational-3.  (Receiving Field).
MOVE A TO B.
```

Because the receiving field is one decimal position larger than the sending field, decimal alignment must be performed. Each time the move is executed, 2,250 bytes of storage are used. Adding one additional decimal position in the data sending or receiving field is small in cost compared to the savings possible in the Procedure Division.

Unequal-Length Fields

An intermediate operation may be required when handling fields of unequal length. For example, zeros may have to be inserted in numeric fields and blanks in the alphabetic or alphanumeric fields in order to pad out to the proper length. The compiler will have to generate instructions to perform these insertions.

To avoid these operations, the number of digits should be equal. Any increase in data fields is more than compensated for by the savings in the generated object program. For example,

```
SENDFIELD              Picture S999
RECFIELD               Picture S99999
Change SENDFIELD to    Picture S99999
```

Mixed Data Formats

When fields are used together in move, arithmetic, or relational statements, they should be in the same format whenever possible. Conversions require additional storage and longer execution time. Operations involving mixed data formats require one of the items to be converted to a matching data format before the operation is executed.

For maximum efficiency, avoid mixed data formats or use a one-time conversion; that is, move the data to a work area, thus converting it to the matching format. By referencing the work area in procedural statements, the data is converted only once instead of for each operation.

The following examples show what must logically be done before indicated operations can be performed when working with mixed data fields.

Display to Computational-3

To execute a move. No additional code is required (if proper alignment exists) because one instruction can both move and convert the data.

To execute a compare. Before a Compare is executed, Display data must be converted to Computational-3 format.

To perform arithmetic calculations. Before arithmetics are performed, Display data is converted to Computational-3 format.

Computational-3 to Display

To execute a move. Before a Move is executed, Computational-3 data is converted to Display data format.

To execute a compare. Before a compare is executed, Display data is converted to Computational-3 data format.

To perform arithmetic calculations. Before arithmetic calculations are performed, Display data is converted to Computational-3 data format. The result is generated in a Computational-3 work area, which is then converted and moved to Display result field.

Display to Display

To perform arithmetic calculations. Before arithmetic calculations are performed, Display data is converted to Computational-3 format. The result is generated in Computational-3 work area, which is then converted and moved to the Display result field.

Sign Control

The absence or presence of a plus or minus sign in the description of an arithmetic field can affect the efficiency of a program. For numeric fields specified, the compiler attempts to insure that a positive is present so that the values are treated as absolute.

The use of unsigned numeric fields increases the possibility of error (an unintentional negative sign could cause invalid results) and requires additional generated code to control the sign. The use of unsigned fields should be limited to fields treated as absolute values.

For example, if data is defined as

```
A Picture 999
B Picture S999
C Picture S999,
```

the following moves are made: MOVE B TO A. MOVE B TO C. Moving B to A causes four more bytes to be used than moving B to C because A is an absolute value.

Conditional Statements

Computing arithmetic values separately and then comparing them may produce more accurate results than including arithmetic statements in conditional statements. The final result of an expression included in a conditional statement is *limited to an accuracy of six decimal places*. The following example shows how separating computations from conditional statements can improve the accuracy.

The data is defined as

```
77 A   Picture  S9V9999          Computational-3
77 B   Picture  S9V9999          Computational-3
77 C   Picture  S999V9(8)        Computational-3
```

and the following conditional statement is written.

```
If A * B = C, PERFORM EQUAL-X.
```

the final result will be 99V9(6). Although the receiving field for the final result (C) specifies eight decimal positions, the final result actually obtained in the example specifies six decimal places. For increased accuracy, define the final result field as desired, perform the computation, and then make the desired comparison as follows:

```
77 X   Picture  S999V9(8)              Computational-3
COMPUTE X = A * B.
IF X = C, PERFORM EQUAL-X.
```

Summary-Basic Principles of Effective Coding

1. Match decimal places in related fields (decimal point alignment).
2. Match integer places in related fields (unequal length fields).
3. Do not mix usage of data formats (mixed data formats).
4. Include an S (sign) in all numeric pictures (sign control).
5. Keep arithmetic expressions out of conditionals (conditional statements).

Reusing Data Areas

The main storage area can be used more efficiently by writing different data descriptions for the same data area. For example, the coding that follows shows how the same area can be used as a work area for records of several input files that are not processed concurrently.

```
Working-Storage Section.
      01  Work-Area-File-1        (Largest record description for File-1.)
      01  Work-Area-File-2        Redefines Work-Area-File-1
                                  (Largest record description for File-2.)
```

Alternate Groupings and Descriptions

Program data can be described more efficiently by providing alternate groupings or data descriptions for the same data. As an example of alternate groupings, suppose a program makes references to both a field and its subfields where it could be more efficient to describe the subfields with different usages.

This can be done with the Redefines clause, as follows:

```
01  Payroll-Record.
    02  Employee-Record      Picture  X(28).
    02  Employee-Field Redefines Employee-Record.
```

```
        03  Name           Picture  X(24).
        03  Number         Picture  S9(4)     Computational-3.
    02  Date-Record        Picture  X(10).
```

An example of different data descriptions specified for the same data, the following example illustrates how a table can be utilized.

```
    02  Value-A.
        03 A1    Picture  S9(9)    Computational-3    Value is Zeroes.
        03 A2    Picture  S9(9)    Computational-3    Value is 1.
    02  Table-A Redefines Value-A    Picture S9(9)    Computational-3
                                                      Occurs 2 times.
```

Data Formats in Computer

The following examples illustrate how various COBOL data formats appear in the System 360/370 computer in Extended Binary Coded-Decimal Interchange Code (EBCDIC) format.

Numeric Display (External Decimal)

The value of an item is −1234.

```
    Picture 9(4) Display            F1 F2 F3 F4
                                             Byte
    Picture S9(4). Display          F1 F2 F3 D4
```

Hexadecimal F is treated arithmetically as a plus in the low-order byte. The hexadecimal D represents a negative sign.

```
    COMPUTATIONAL-3  (INTERNAL DECIMAL)
```

The value of an item is +1234.

```
    Picture 9(4)     Computational-3     01 23 4F
                                               Byte
    Picture S9(4)    Computational-3     01 23 4C
```

Hexadecimal F is treated arithmetically as plus.
Hexadecimal C represents a positive sign.

```
    COMPUTATIONAL  (BINARY)
```

The value of an item is 1234.

```
    Picture S9(4)    Computational    0 000 0000 0000 0100 1101 0010
                                      -                   ---------
                                      Sign                     Byte
```

An 0 bit in the sign position means the number is positive. Negative numbers appear in the 2s complement form with a 1 in the sign position.

Redundant Coding

To avoid redundant coding of usage designations, use computational at the group level (this does not affect the object program).

For example,

Instead of

```
     02  Fuller.
         03 A  Computational-3  Picture 99V9.
         03 B  Computational-3  Picture 99V9.
         03 C  Computational-3  Picture 99V9.
```

write

```
     02  Fuller     Computational-3.
         03 A  Picture 99V9.
         03 B  Picture 99V9.
         03 C  Picture 99V9.
```

WORKING-STORAGE SECTION

Group Related Data Items Together

Group similar fields together under their own group level, such as SWITCHES, SUBSCRIPTS, ACCUMULATORS, COUNTERS, etc. Literals that are used often and might change in the future should be defined in the Working-Storage Section. Rather than searching through an entire module to modify a frequently-used literal, only one modification should then be necessary in working storage. When this is done, however, consider using a name other than the spelled-out value of the literal. It may be extremely upsetting to find, after a long debugging session, that the item in working storage named FIVE has a value of 4. Conversely, literals that are used infrequently and that are not likely to change, should be used directly in Procedure Division statements, thereby increasing readability and causing more efficient code to be generated.

PROCEDURE DIVISION

The Procedure Division should be written using the Data Division as a guide to file, data-names, constants, and work areas used. The programmer actually writes the COBOL program using the program flowchart as a guide for the procedure entries.

The Procedure Division specifies the action, such as input/output, data movement, and arithmetic operations that are required to process the data. A series of English-like statements are written in the sequence of the program flowchart.

Entries That Appear In the Procedure Division

1. The name of the division-Division Header.
2. Any optional sections.
3. A series of procedural paragraphs specifying the actions to be performed.

INCREASING THE EFFICIENCY OF PROCEDURE DIVISION ENTRIES

A program can be made more efficient in the Procedure Division with some of the techniques described below. (See also figure 15.5.)

```
PROCEDURE DIVISION [USING identifier-1 [identifier-21]...].
[[DECLARATIVES.
{section-name SECTION. USE Sentence.
{paragraph-name.  {sentence}...}...}...
END DECLARATIVES.]
{section-name SECTION [priority].]
{paragraph-name.  {sentence}...}...}...
```

Figure 15.5
Structure—
Procedure Division.

Modular Programming

Modular programming, as its name implies, means the arrangement of a program into separate modules of logic.

Modularizing involves organizing the Procedure Division into at least three functional levels: a main line routine, processing subroutines, and input/output subroutines. When the Procedure Division is modularized, programs are easier to maintain and document. In addition, modularizing makes it simple to break down a large program, resulting in a more efficient program if the Segmentation Feature is used.

Advantages of modular programming are many; disadvantages are few. For example, a module that is written and tested but once can be used in more than one program, by cataloging it in the source language library from which it can be retrieved by the COBOL COPY statement. Examples of such common modules are: program initialization, error routines, determination of control breaks, calculation routines, and end-of-job routines.

Maintenance is easier for programs written using the modular approach. A common module needs to be recompiled just once. The savings in terms of quicker and easier recompilations can be particularly significant at the testing stage, or in an installation with programs subject to a high degree of modification.

Modular programming can result in a big savings in main storage usage because of the ability to overlay, and thus provide room for maintenance changes as they occur in major programs. One-time or frequently-used routines written in a modular fashion can be changed to overlays very quickly and easily by assigning priority numbers to their section-names.

A common modular approach might include one section that would make decisions, and direct control among the various sections. As an example, every file maintenance program reads details and masters, and performs different logic for master high, low, or equal conditions. PERFORM or GO TO statements can then be executed for the processing sections required for each condition.

In summary, advantages of modular programming include the following:

1. Use of prewritten, common modules
2. Better organized logic
3. Modular testing
4. Easier maintenance
5. Ability to overlay

Main-Line Routine

The main-line routine should be short and simple, and should contain all of the major logical decisions of the program. This routine controls the order in which second-level routines are executed. All second-level routines should be invoked from the main-line routine by PERFORM or CALL statements.

Processing Routines

Processing routines should be subdivided into as many functional levels as necessary, depending on the complexity of the program. These routines should have a single entry point as the first statement of the routine and a single exit, which should be the EXIT statement.

Make All Conditional Expressions as Simple as Logically Possible

COBOL provides the facility for complex conditional expressions. It is normally not necessary to use this facility, however, and may lead to confusion on the part of less experienced programmers trying to read the program. There is an appropriate time for very complex conditional expressions, and that is when the only alternative to the complex expression is a series of individual simple IF conditions spread all over the program. In this case, the difficulty of jumping from one IF to another is greater than the difficulty in comprehending the complex expression. In all other cases, however, including the case where a number of different branches are to be made, the use of simple IF clauses is preferable.

Use of GO TO with the DEPENDING ON Option Instead of a Series of IF Statements

When the possible values of a control variable are the numeric values 1, 2, 3, etc., the GO TO statement with the DEPENDING ON option is more efficient in regard to storage space and execution time than a long series of relational expressions and IF statements. (See figure 15.6.)

```
    GO TO HAIRCUT, SHAMPOO, STYLIST, PERM-WAVE
        DEPENDING ON TYPE-OPERATOR.
    PERFORM TYPE-ERROR-ROUTINE

SUBROUTINES.
    NOTE THE FOLLOWING ARE SUBROUTINES CALLED UPON BY
    THE MAIN PROGRAM.

HAIRCUT.
    MULTIPLY CUSTOMERS BY 8.00 GIVING GROSS.
    PERFORM PRINTOUT.
SHAMPOO.
    MULTIPLY CUSTOMERS BY 10.00 GIVING GROSS.
    PERFORM PRINTOUT.
STYLIST.
    MULTIPLY CUSTOMERS BY 15.00 GIVING GROSS.
    PERFORM PRINTOUT.
PERM-WAVE.
    MULTIPLY CUSTOMERS BY 16.00 GIVING GROSS.
    PERFORM PRINTOUT.
```

Figure 15.6
Example—Go To Statement With Depending On Option.

Replace Blanks in Numeric Fields With Zeros

On certain COBOL compilers, should an input field read from a punched card and defined as numeric contain all blanks, a data exception occurs if an arithmetic operation is attempted.

A method of avoiding the situation is to perform a numeric class test on all numeric fields from a card, for example,

```
IF FIELD-NAME IS NOT NUMERIC PERFORM CARD-ERROR-ROUTINE.
```

Open, READ, CLOSE, and WRITE Statements

The OPEN and READ statement must reference a file assigned in the Environment Division and described in the Data Division. The CLOSE statement must be written for each file opened. The WRITE statement references a record-name.

Sequence of Divisions Important

Since the last statement in the Procedure Division denotes the end of the source program, it is imperative that all divisions remain in the proper sequence. If the sequence is disturbed, many diagnostic errors will be generated unnecessarily.

Intermediate Results

The compiler treats arithmetic statements as a succession of operations and sets up intermediate result fields to contain the results of these operations. The compiler can process complicated statements, but not always with the same efficiency of storage utilization as the source program. Because truncation may occur during compilation, unexpected intermediate results may occur.

Binary Data

If an operation involving binary operands requires an intermediate result greater than 18 digits, the compiler converts the operands to internal decimal before performing the operation. If the result field is binary, the result will be converted from internal decimal to binary.

If an intermediate result will not be greater than nine digits, the operation is performed most efficiently as binary data fields.

COBOL Library Subroutines

If a decimal multiplication operation requires an intermediate result greater than 30 digits, a COBOL library subroutine is used to perform the multiplication. The result of this multiplication is truncated to 30 digits.

A COBOL library subroutine is used to perform division if (1) the divisor is equal to or greater than 15 digits, (2) the length of the divisor plus the length of the dividend is greater than 16 bytes, or (3) the scaled dividend is greater than 30 digits (a scaled dividend is a number that has been multiplied by a power of 10 in order to obtain the desired number of decimal places in the quotient).

Intermediate Result Greater than 30 Digits

When the number of digits in a decimal intermediate result field is greater than 30, the field is truncated to 30. A warning message will be generated at compilation time, but the program flow will not be interrupted at execution time. This truncation may cause the result to be invalid.

On Size Error

The ON SIZE ERROR option applies only to the final tabulated results, not to intermediate result fields.

A method of avoiding unexpected intermediate results is to make critical computations by assigning maximum (or minimum) values to all fields and analyzing the results by testing the critical computations for results expected.

Because of concealed intermediate results, the final result is not always obvious.

The necessity for computing the worst case (or best case) results can be eliminated by keeping statements simple. This can be accomplished by splitting up the statement and controlling the intermediate results to be sure unexpected final results are not obtained.

For example,

```
COMPUTE B = (A + 3) / C + 27.600.
```

First define adequate intermediate result fields, i.e.,

```
02  INTERMEDIATE-RESULT-A     PICTURE  S9(6)V999.
02  INTERMEDIATE-RESULT-B     PICTURE  S9(6)V999.
```

Then split up the expression as follows:

```
ADD A, 3 GIVING INTERMEDIATE-RESULT-A.
```

Then write:

```
DIVIDE C INTO INTERMEDIATE-RESULT-A GIVING INTERMEDIATE-
RESULT-B.
```

Then compute the final results by writing:

```
ADD INTERMEDIATE-RESULT-B, 27.600 GIVING B.
```

Arithmetic Fields

Initialize arithmetic fields before using them in computation. Failure to do so may result in invalid results, or the job might terminate abnormally.

Comparison Fields

Numeric comparisons are usually done in Computational-3 format; therefore, Computational-3 is the most efficient data format.

Because the compiler inserts slack bytes which can contain meaningless data, group comparisons should not be attempted when slack bytes are within the group unless the programmer knows the contents of the slack bytes.

Open and Close Statements

Each opening or closing of a file requires the use of main storage that is directly proportional to the number of files being opened. Opening or closing more than one file with the same statement is faster than using a separate statement for each file. Separate statements, however, require less storage area.

For example,

one statement OPEN INPUT FILE-A, FILE-B, FILE-C. rather than

```
OPEN INPUT FILE-A.
OPEN INPUT FILE-B.
OPEN INPUT FILE-C.
```

Accept Verb

The Accept verb does not provide for the recognition of the last card being read from the card reader. When COBOL detects /* card, it drops through to the next statement. Because no indication of this is given by COBOL, the end-of-file detection requires special treatment. Thus the programmer must provide his own end card (some card other than /*) which he can test to detect an end-of-file condition.

Paragraph-Names

Paragraph-names use storage when the PERFORM verb is used in the program. Use of paragraph-names for comments requires more storage than the use of a Note or a blank card. Use Note and/or a blank card for identifying inline procedures where paragraph-names are not required.

For example, avoid writing the following:

```
MOVE A TO B.
PERFORM JOES-ROUTINE.


JOES-ROUTINE. COMPUTE A = D + C * F.
```

Recommended Coding

```
MOVE A TO B.
PERFORM ROUTINE.
NOTE JOES-ROUTINE.
ROUTINE. COMPUTE A = D + C * F.
```

COMPUTE Statement

The use of the COMPUTE statement generates more efficient coding than does the use of individual arithmetic statements because the compiler can keep track of internal work areas and does not have to store the results of intermediate calculations. It is the user's responsibility, however, to insure that the data is defined with the level of significance required in the answer.

IF Statement

Nested and computed IF statements should be avoided, as the logic is difficult to debug. Performing an IF operation for an item greater than 256 bytes in length requires the generation of more instructions than are required for that of an IF operation of an item of 256 bytes or less.

MOVE Statement

Performing a MOVE operation for an item greater than 256 bytes in length requires the generation of more instructions than is required for that of a MOVE statement for an item of 256 bytes or less.

When a MOVE statement with CORRESPONDING option is executed, data items are considered CORRESPONDING only if their respective data-names are the same, including all implied qualifications, up to but not including the data-names used in the MOVE statement itself.

NOTE Statement

An asterisk (*) in column 7 should be used in place of the NOTE statement because there is the possibility that when NOTE is the first sentence in a paragraph, it will inadvertently cause the whole paragraph to be treated as part of the NOTE.

PERFORM Verb

PERFORM is a useful verb if the programmer adheres to the following rules.

1. Always execute at the last statement of a series of routines being operated on by a PERFORM statement. When branching out of the routine, make sure control will eventually return to the last statement of the routine. This statement should be an EXIT statement. Although no code is generated, the EXIT statement allows a programmer to immediately recognize the extent of a series of routines within the range of a PERFORM statement.
2. Always either PERFORM routine-name THRU routine name-exit, or PERFORM a section name. A PERFORM paragraph-name can cause trouble for the programmer trying to maintain the program. For example, if a paragraph must be broken into two paragraphs, the programmer must examine every statement to determine whether or not this paragraph is within the range of the PERFORM statement. Then all statements referencing the paragraph-name must be changed to PERFORM THRU statements.

Read Into and Write From Options

Use READ INTO and WRITE FROM to do all the processing in the Working-Storage Section. This is suggested for two reasons.

1. Debugging is much simpler. Working-Storage areas are easier to locate in a dump and so are buffer areas. And, if files are locked, it is much easier to determine which record in a block was being processed when the abnormal termination occurred.
2. Trying to access a record area after the AT END condition has occurred (for example, AT END MOVE HIGH-VALUE TO INPUT-RECORD) can cause problems if the record area is only in the File Section.

(*Note:* The programmer should be aware that additional time is used to execute the move operations involved in each READ INTO or WRITE FROM instruction.)

HIPO

A precise definition of user requirements is essential to the development of a correct data processing system. Ideally the user, perhaps assisted by data processing personnel, prepares a specification package. This describes what the system is and what it is to accomplish. Requirements for input, output, stored data, logical processing, control, testing, performance, and documentation are detailed. These completed specifications are accepted by data processing personnel, which now assures project responsibility. A system analyst transforms the specifications into system design and programming specifications. The latter are delivered to programmers for implementation of coding. This is a sound approach to specification development but there are serious problems inherent therein, such as:

1. Often the user has neither the time nor the experienced personnel to produce adequate business specifications.
2. In seemingly complete specifications, omissions may not be discovered until programming has been finished, causing costly modifications.
3. Misunderstanding specification detail becomes more likely with each level of interface between the user and the programmer. If these misunderstandings are not corrected early in the development cycle, built-in programming errors are the consequence.
4. Users may not gain many of the possible benefits of a proposed system. If they do not understand data processing, users may not request valuable output which could readily be obtained at little or no additional expense.

A better means of deriving statements of user requirements is needed, one which ensures that the statements are complete and correct. HIPO fulfills this need.

What is HIPO?

HIPO is an acronym for Hierarchy plus Input, Process, and Output. It is a method of graphically describing a software entity such as a system or program as an arrangement of functions to be performed. HIPO is an approach to functional specification and documentation of programs. Each function is designed using a HIPO diagram, in which inputs and outputs are listed and the processing that is to be carried out is specified. A visual table of contents diagram points to the HIPO diagrams in the package and thereby shows the functions and subfunctions to be carried out by the various parts of the program, and the relationships between them. At the detailed design level, it also shows the hierarchy of segments.

Although primarily developed for design and documentation purposes, HIPO has evolved as part of the improved programming technology of structured programming. The operation of HIPO is as follows:

1. The hierarchy portion of HIPO involves a tree-like structure similar to an organization chart. It is composed of functions or actions. Each function on the hierachy is represented as a box and can be described within that box as a verb (action) and an object (data affected). The verb-object format thus names as well as defines the functions. (See figure 15.7.)
2. The top box on the hierarchy describes the entire piece of software in terms of a single function. Each level below is a subset of the function above it. This

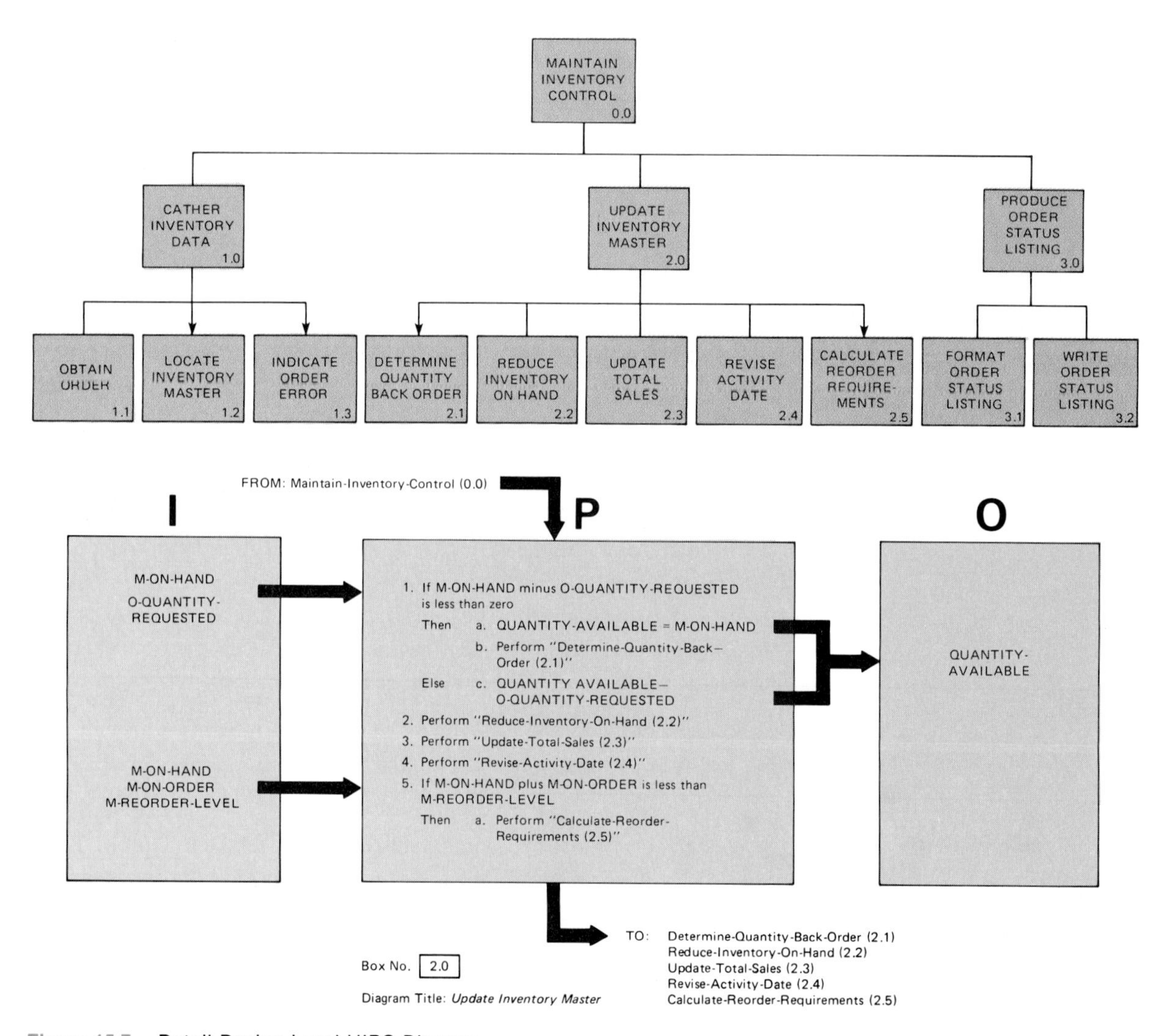

Figure 15.7 Detail Design Level HIPO Diagram.

hierarchy of functions is created by a technique known as *functional decomposition,* whereby a function is exploded into increasingly lower levels of detail until all subfunctions have been defined. Determining the main function of the software, decomposing it into a hierarchy of subfunctions, and naming the subfunctions is not a trivial exercise. It requires a great deal of insight, creativity, and experience on the part of the designer.

Every box on the hierarchy has a corresponding IPO (input, process, and output) diagram. The IPO provides a visual description of what takes place within each box. In producing IPO diagrams, any conditions for the execution of the subfunctions are evaluated; thus the performance of the subfunctions is controlled. (See figure 15.8.)

Top level functions on a hierarchy contain the control logic. They determine when and in what order lower level functions are to be invoked. They consist primarily of CALLs, PERFORMSs, DOWHILEs, DOUNTILs, and IF-

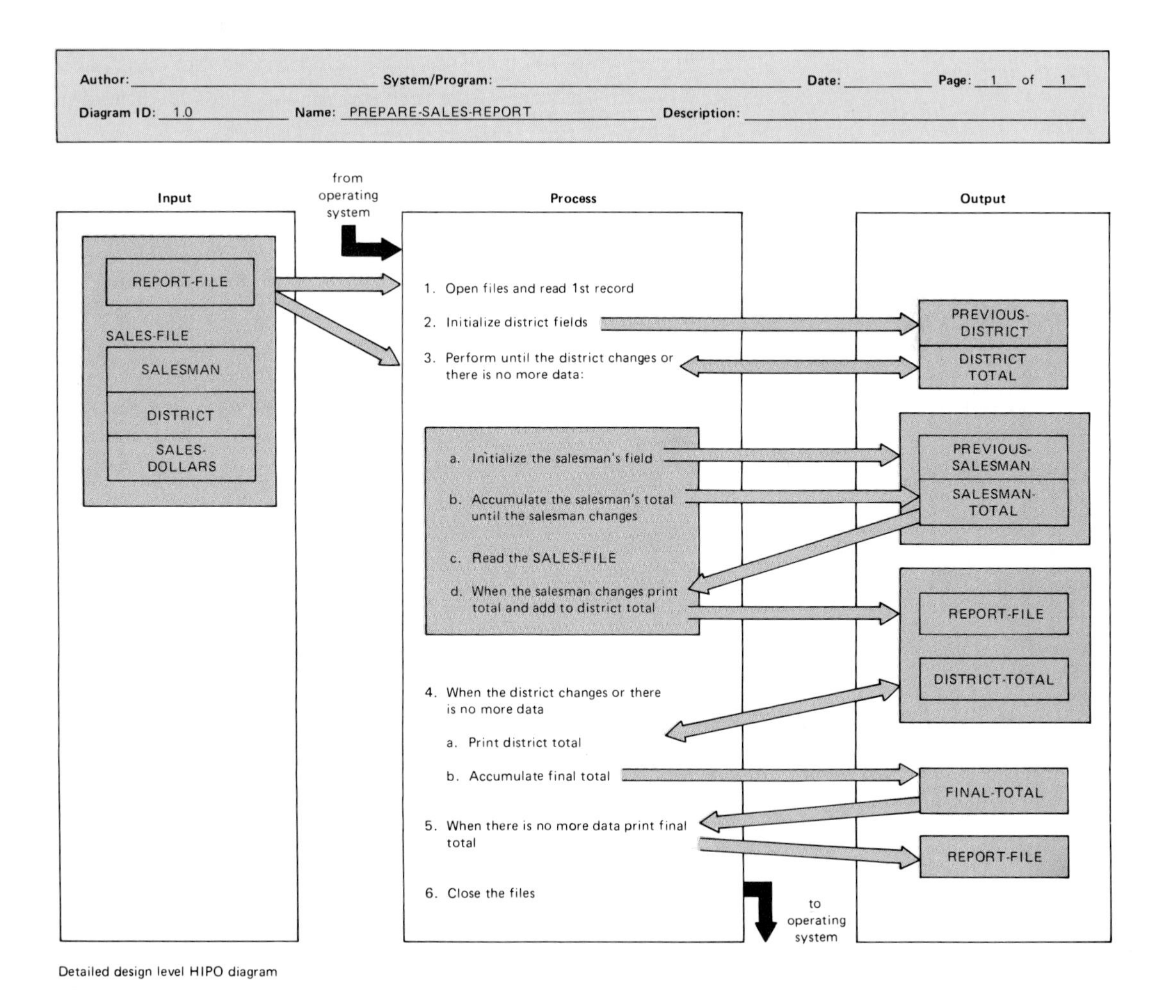

Figure 15.8 Detail Design Level HIPO Diagram.

THENELSE statements. Lower level functions are the workers; here sequential coding statements are found to predominate.

Benefits

The major benefits of using HIPO in specification development are:

1. A description of system requirements represented in HIPO form and mutually accepted by the user area and data processing is more complete, more accurate, more transparent, and more concise than standard business specifications.
2. Since requirements and design are being developed simultaneously, a saving of time and money is realized. There is no need to develop programming specifications from business specifications. With minor modifications, requirement IPOs can be refined into design IPOs, implementation IPOs, and finally documentation IPOs.
3. When the specification-design phase is completed, a more exact estimate of remaining resource requirements is possible. Detailed HIPOs represent a measurable checkpoint in the development cycle of a system. Manpower and computer costs for implementation and testing can be reassessed with a greater degree of accuracy than they can for the use of traditional methods of development.
4. Users become acquainted with hierarchical design, IPOs, and the concept of "function" during specification development. This familiarity helps them to understand their system more readily.
5. The capability of developing specifications through HIPO allows an application area in need of a data processing system to request that system, even though the resources to deliver written business specifications might be lacking. Not only can a quality system that meets all requirements be delivered to the user, but data processing can introduce the user to system enhancements available at little or no additional cost.

Conclusion

The Improved Programming Technologies are in a state of evolution. HIPO, as a technique, has evolved from use solely as a documentation tool to use as a design aid as well. In this latest advance, HIPO has successfully fulfilled two of the greatest needs in the data processing industry: by providing a means of precisely defining user requirements; and by expanding the essential lines of communication between user and data processing.

Structured programming refers primarily to the coding phase rather than the design phase of the program development cycle. HIPO is one good way to approach the design task, and one that is complementary to structured programming. (See figure 15.9.)

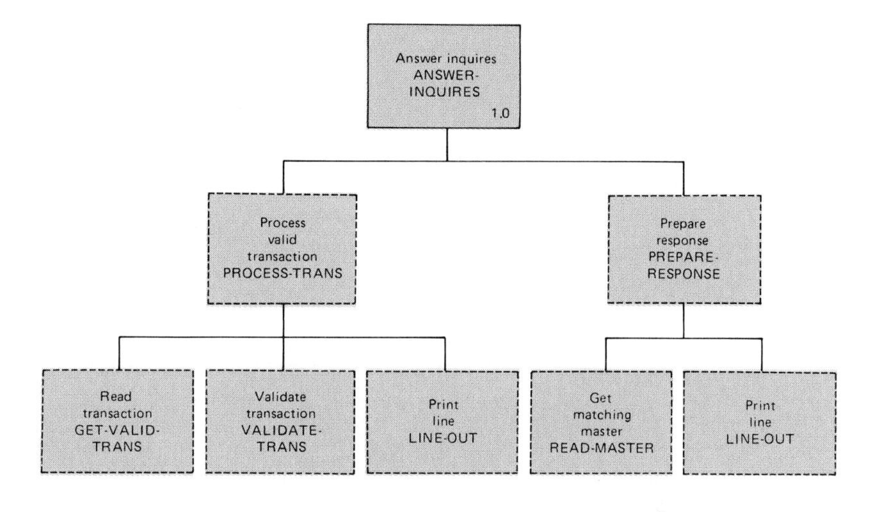

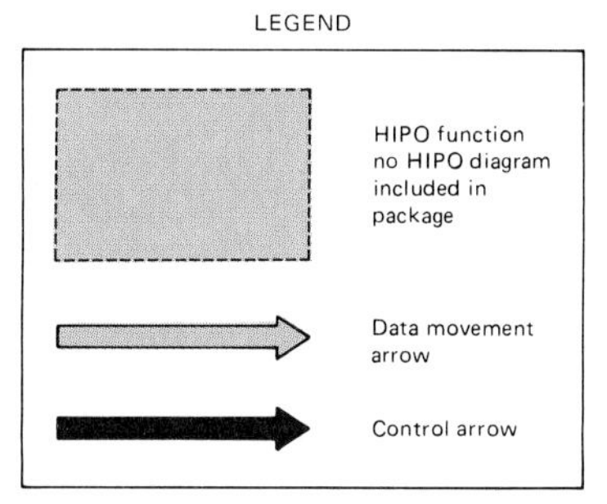

Detailed design level visual table of contents for the inquiry response application

Figure 15.9 HIPO Diagram for figure 7.26.

Figure 15.9 Continued.

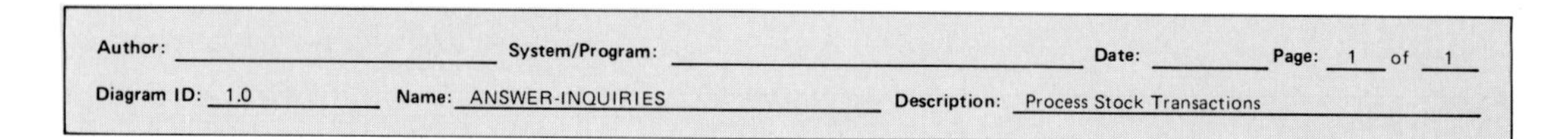

Author: ______ System/Program: ______ Date: ______ Page: 1 of 1

Diagram ID: 1.0 Name: ANSWER-INQUIRIES Description: Process Stock Transactions

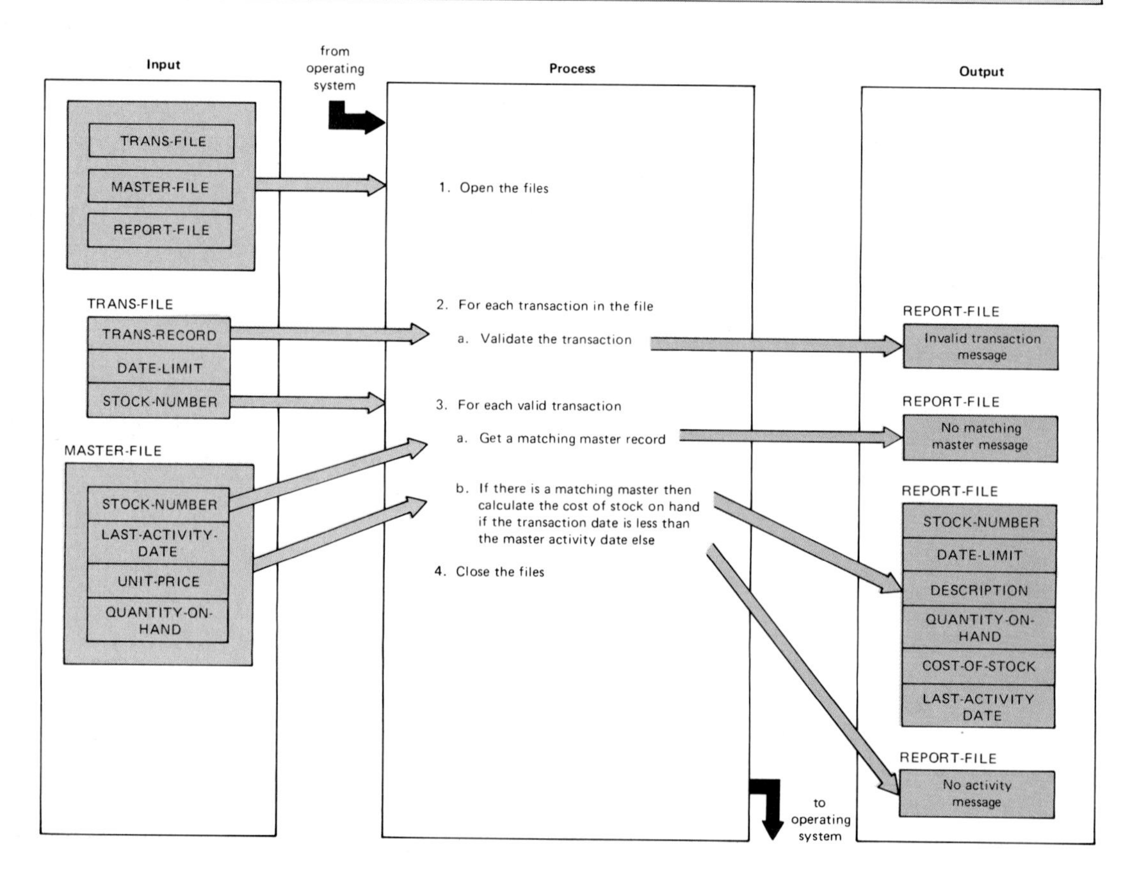

Extended Description

Notes	Paragraph	Ref.
2.	PROC-TRANS	
	GET-VALID-TRANS	
2a. All fields numeric,		
date less than 70001	VALIDATE-TRANSACTION	
Print	LINE-OUT	
3.	PREPARE-RESPONSE	
3a.	READ-MASTER	
Print	LINE-OUT	

Extended Description

Notes	Paragraph	Ref.

A HIPO diagram for the inquiry response application

CONVENTIONAL-LINKAGE FOR COBOL PROGRAMMING

The following are *only* suggestions, inasmuch as no conventional linkage for structured COBOL programming has as yet been developed.

Conventional-Linkage combines both program design and a program structure. As a program design, Conventional-Linkage is top-down in nature. The concept requires that high-level coding be developed first (and possibly tested if physically modular—perhaps tested if logically modular and large). Then in the pattern of top-down design, each successively lower level is developed and coded. As a top-down program design, Conventional-Linkage requires that interfaces between COBOL sections and units of coding be predefined.

As a program structure (structured programming) proper use of Conventional-Linkage requires a rigid placement of program functions, and also the concepts of a single entry and exit point in a unit of code. For example, the code never branches out of set boundaries.

Conventional-Linkage helps to accomplish the above in the following manner:

1. Functions are almost always located in the same section, so that one is not concerned with where these operations will be performed. This allows the programmer to devote his time to the logic of the application.
2. Sections are independent in that the programmer is not concerned with what is happening in other sections.
3. Basic interfaces are predefined, as, for example,
   a. how data is passed from section to section, and
   b. how sections communicate (switches, keys, etc.).

Conventional-Linkage consists of seven COBOL sections:

1. CONVENTIONAL-LINKAGE SECTION.
   Contains the control logic for the program.
2. INITIALIZATION SECTION.
   Executes housekeeping functions and runs through one processing cycle to get the program started.
3. READ-RECORD SECTION.
   Reads a record from appropriate input file designated by FILE-PRIORITY-SELECT SECTION.
4. FILE-PRIORITY-SELECT SECTION.
   Selects next record to be processed based on lowest keys and highest priority.
5. CONTROL-BREAK-TEST SECTION.
   Determines whether a control break has occurred and takes the appropriate actions.
6. DETAIL-PROCESSING SECTION.
   Processes records, even to the extent of moving a record or parts of a record to a work area.
7. SUBROUTINES SECTION.

All units of coding which may be used in more than one place or in different sections including the termination routines.

Conventional-Linkage

The control coding for the execution of a program using all sections which are performed as sections appears in the following manner:

```
CONVENTIONAL-LINKAGE SECTION.
     PERFORM INITIALIZATION.
     PERFORM DRIVER-SECTION.
DRIVER-SECTION.
     PERFORM READ-RECORD.
     PERFORM FILE-PRIORITY-SELECT.
     PERFORM CONTROL-BREAK-TEST.
     PERFORM DETAIL-PROCESSING
```

Initialization Section

1. *Housekeeping.*
   a. Edit control data such as date, run code and so on. Control data may come from a card or any other source.
   b. Clear working storage fields that cannot be initialized with VALUE clauses.
   c. Open files.
2. *Prime Program.*
   a. Perform READ-RECORD once for each file.
   b. Perform FILE-PRIORITY-SELECT once.
   c. Move CONTROL-BREAK-HOLD to FIELD-CONTROL.

      CONTROL-BREAK-HOLD and FIELD-CONTROL are group items that contain keys tested in the CONTROL-BREAK-TEST SECTION.
   d. Perform DETAIL-PROCESSING.

Read-Record Section

The file to be read in is controlled by a switch (FILE-POINT) set in the FILE-PRIORITY-SELECT SECTION.

1. Read appropriate file (for end of file condition, see 6).
2. Sequence-check the file.
3. Move record key to field used by FILE-PRIORITY-SELECT. This may involve some modification or rearrangement of the key, and this field should have a name suggesting both file and function, such as FPS-SELECT-KEY.
4. Increment memory record count.
5. Edit if appropriate. This edit should be limited to testing of record type.
6. End of file. Each AT END imperative does the following:
   a. Turns on the appropriate end-of-file switch.
   b. Moves high-values to appropriate FPS-KEY.
   c. Performs TERMINATION-ROUTINE.
   d. Branches to READ-RECORD-EXIT.

File-Priority-Select Section

1. Selects record of lowest key. The FPS-SELECT-KEYs passed by READ-RECORD SECTION are used for comparison. When there are equal FPS-SELECT-KEYs, the program logic must determine which record has priority.
2. Moves the lowest FPS-SELECT-KEY (or priority FPS-SELECT-KEY) to CONTROL-BREAK-HOLD. CONTROL-BREAK-HOLD is never up-

dated anywhere else in the program. CONTROL-BREAK-HOLD is used by the CONTROL-BREAK-TEST SECTION.

3. Set a switch (FILE-POINT). The values of this switch will:
   a. Tell READ-RECORD SECTION which file will be read next.
   b. Tell DETAIL-PROCESSING SECTION which record from which file is to be processed next.
   c. Be available to CONTROL-BREAK-TEST SECTION if needed.

(Note: After INITIALIZATION SECTION has been performed, FILE-POINT will never be updated anywhere other than in the FILE-PRIORITY-SELECT SECTION.)

Control-Break-Test Section

1. This section determines whether a control break has occurred, and at what level. A control break is the result of an unequal compare between CONTROL-BREAK-HOLD and FIELD-CONTROL, which are control keys containing one or more segments. If the control key has only one segment, e.g., INVOICE-NO, no further compares are necessary. If the control key is made of multiple segments, such as WAREHOUSE-CENTER-NO, PRODUCER-NO, and INVOICE-NO, and an unequal comparison occurs, segments are tested individually starting with the highest level, e.g., WAREHOUSE-CENTER-NO, to discover the highest level of change.
2. In addition to testing for control breaks, the section contains two types of break functions, they are,
   a. EOF-BREAK—wraps up the last group processed. This may include:
      1) producing records,
      2) balancing,
      3) producing total records,
      4) checking results of processing and taking appropriate action,
      5) rolling counters,
      6) and anything else dictated by the logic of the program.
   b. GROUP-BREAK—does whatever is necessary to allow the next group to be processed. This may include,
      1) clearing counters and indexes,
      2) clearing work areas,
      3) resetting switches, and
      4) updating FIELD-CONTROL by moving in CONTROL-BREAK-HOLD.
3. EOF-BREAKs are performed from low-level to high-level. All EOF-BREAKs are completed before any GROUP-BREAKs are performed.
4. GROUP-BREAKs are performed from high-level to low-level.
5. The example of CONTROL-BREAK-TEST coding shown below is based on the following WORKING-STORAGE SECTION definitions of the control fields:

```
01  CONTROL-BREAK-HOLD.
    05  CB-WAREHOUSE-CENTER-NO       PICTURE X.
    05  CB-PRODUCER-NO               PICTURE X(6).
    05  CB-INVOICE-NO                PICTURE X(8).
01  FIELD-CONTROL.
    05  FC-WAREHOUSE-CENTER-NO       PICTURE X.
    05  FC-PRODUCER-NO               PICTURE X(6).
    05  FC-INVOICE-NO                PICTURE X(8).
FIELD-BREAK SECTION.
    IF CONTROL-BREAK = FIELD-CONTROL
    THEN
        PERFORM CONTROL-BREAK-TEST-EXIT
    ELSE
        NEXT SENTENCE.
TEST-FOR-PC-BREAK.
    IF CB-WAREHOUSE-CENTER-NO = FC-WAREHOUSE-CENTER-NO
    THEN
        NEXT SENTENCE
    ELSE
        PERFORM INVOICE-EOF-BREAK THRU
            INVOICE-GROUP-BREAK-EXIT
        PERFORM CONTROL-BREAK-TEST-EXIT.
TEST-FOR-PRODUCER-BREAK.
    IF CB-PRODUCER-NO = FC-PRODUCER-NO
    THEN
        NEXT SENTENCE
    ELSE
        PERFORM INVOICE-EOF-BREAK THRU
            PRODUCER-EOF-BREAK-EXIT
        PERFORM PRODUCER-GROUP-BREAK THRU
            INVOICE-GROUP-BREAK-EXIT
        PERFORM FIELD-BREAK-EXIT.
INVOICE-BREAK.
        PERFORM INVOICE-EOF-BREAK THRU
            INVOICE-EOF-BREAK-EXIT
        PERFORM INVOICE-GROUP-BREAK THRU
            INVOICE-GROUP-BREAK-EXIT
        PERFORM FIELD-BREAK-EXIT.
INVOICE-EOF-BREAK.
        •
        •
INVOICE-EOF-BREAK-EXIT.
        EXIT.
PRODUCER-EOF-BREAK.
        •
        •
PRODUCER-EOF-BREAK-EXIT.
        EXIT.
```

```
WAREHOUSE-CENTER-EOF-BREAK.
          .
          .
WAREHOUSE-CENTER-EOF-BREAK-EXIT.
          EXIT.
WAREHOUSE-CENTER-GROUP-BREAK.
          .
          .
WAREHOUSE-CENTER-GROUP-BREAK-EXIT.
          EXIT.
PRODUCER-GROUP-BREAK.
          .
          .
PRODUCER-GROUP-BREAK-EXIT.
          EXIT.
INVOICE-GROUP-BREAK.
          .
          .
INVOICE-GROUP-BREAK-EXIT.
          EXIT.
          .
          .
CONTROL-BREAK-TEST-EXIT.
          EXIT.
FIELD-BREAK-EXIT.
          EXIT.
```

(Note: Normally, it is recommended in structured programming that the THRU option of the PERFORM verb *not* be used. However, there are situations wherein the use of the THRU option will save coding time, as in the above example, and in CASE program structures.)

Detail-Processing Section

1. Uses FILE-POINT to select processing routine to be applied. This assumes different routines for different record types.
2. Processing routines may do some or all of the following:
   a. Set switches (for use in CONTROL-BREAK-TEST, etc.),
   b. Establish tables,
   c. Add to counters,
   d. Produce output, and
   e. Any other data manipulation required.

Subroutines Section

1. The routines residing in this section are generally those units of coding that are used in more than one place in a section or in
   a. Write routines,
   b. Headings,
   c. Conversions,
   d. Accumulations (formulas),
   e. Other arithmetic or mathematical functions,

f. Balancing,
g. Other, and
h. TERMINATION-ROUTINE.

2. TERMINATION-ROUTINE is performed from READ-RECORD SECTION only.
   a. Tests for end-of-job condition. If not end-of-job, branches to exit.
   b. At end-of-job, wraps up the program by performing all levels of EOF-BREAKS.
   c. Closes files.
   d. Writes messages.
   e. Displays counts and balances.

CONCLUSION

Some of the programming techniques that will aid the programmer in preparing efficient programs have been explained and illustrated. Additional programming techniques will be found in the reference manuals of the computer manufacturers. It is hoped that as the programmer becomes more proficient in COBOL programming, he will develop his own programming techniques.

Exercises

Write your answer in the space provided (answer may be one or more words).

1. COBOL does not produce a program as ______ as one written in the ______ of a particular computer.
2. Prior to the writing of the program, the program should be properly ______ and the appropriate ______, ______ formats and ______ formats be available to the programmer.
3. COBOL has extensive features providing the capability for a source program to be highly ______.
4. The key to debugging and maintenance is the ______, ______, and ______ of a small set of ______ standards.
5. ______ can be inserted to set off paragraphs and sections.
6. Start each new ______ on a ______ line.
7. Make each ______ Procedure Division statement a ______ sentence.
8. Do not split ______ or ______ between lines.
9. Use ______ name to improve understanding of program organization.
10. Use ______ freely for explanations.
11. The Identification Division is the ______ of the four and provides information that ______ the program and provides ______ to the reader relative to the program.
12. The required entries in the Identification Division are the name of the ______ and ______.
13. The Remarks paragraph provides the reader with information regarding the ______ of the program.
14. The Environment Division is the only COBOL division that is ______ oriented.
15. The Environment Division contains information about the ______ to be used, the computer upon which the program will be ______ and ______ and links the ______ to the computer system and the ______ to be processed.
16. All files mentioned in the File-Control paragraph should be properly ______ in the Data Division and ______ and ______ in the Procedure Division.
17. All device numbers used in the File-Control paragraph should be ______.
18. The Data Division describes information to be ______ by the ______ program.
19. Each file mentioned in the ______ Division should be ______ in the Data Division.

20. The Copy statement permits the programmer to include in the __________ program at __________ time prewritten __________ and __________ entries from the COBOL source language __________.
21. A programmer should use __________ names and frequent __________ describing what is being accomplished.
22. The __________ enables construction of compound names.
23. The Data Division should be written using copies of the __________ formats and __________ formats as guides before writing the __________ entries and __________ entries.
24. The File Section contains __________ entries and __________ entries.
25. The Working-Storage Section is used for __________ and __________.
26. The Linkage Section contains record description entries for __________.
27. The Report Section contains entries related to the __________ feature.
28. The Record Contains clause should be included in the FD entry because the compiler will check the agreement of the __________ entry with the __________ clause of the record to be processed without the necessity of __________ all field positions.
29. When variable records are used, the Record Contains clause should specify the length of the __________ and __________ record in the file.
30. Record description entries must immediately follow __________ entries.
31. The data record must begin with a __________ level number.
32. The highest qualifier of a data-name permitted in the File Section is a __________.
33. Value clauses, outside of condition-name entries, are forbidden in the __________ Section.
34. If a data item is to be used in a series of arithmetic operations, it would be advisable to define an area in the __________ Section with the __________ mode.
35. When writing Data Division statements, the programmer is not concerned with __________ alignment and __________ formats as the __________ generates __________ instructions to perform the necessary __________.
36. Procedure Division operations are most efficient when the decimal positions of the data involved are __________.
37. The most used verb in the Procedure Division is the __________ verb.
38. When fields are used together in move, arithmetic or relational statements, they should be in the same __________ whenever possible.
39. The absence or presence of a __________ or __________ sign in the description of an arithmetic field can affect the __________ of a program.
40. Computing arithmetic values __________ and then comparing them may produce more __________ results than including __________ statements in __________ statements as the final result of a conditional statement is limited to an accuracy of __________ decimal places.
41. The main storage area can be used more efficiently by writing different __________ for the __________ data area.
42. Program data can be described more efficiently by providing __________ groupings or __________ descriptions for the same data.
43. In the IBM 360/370 computer, the hexadecimal __________ is treated arithmetically as a plus sign in the __________ order byte.
44. The usage designation of __________ at the __________ level can avoid redundant coding.
45. The Procedure Division should be written using the Data Division as a guide to __________, __________, __________, and __________.
46. __________ flowchart acts as a guide for procedure entries and a series of __________ statements are written in the __________ specified in the flowchart.
47. Modular programming is the arrangement of a __________ into separate __________ of __________.

48. The main line routine should be ________________ and ________________ and should contain all the major ________________ divisions of the ________________.
49. Processing should be subdivided into as many ________________ levels as necessary.
50. A numeric ________________ test should be performed on all ________________ fields from a card.
51. The ________________ and ________________ statement must reference a file assigned in the ________________ Division.
52. The ________________ statement must be written for each file opened.
53. The Procedure Division can be written as a series of ________________ programs which will permit the programmer to write ________________ in any ________________ he wishes.
54. The ________________ statement in the Procedure Division signals the end of the source program.
55. The computer cannot process complicated statements with the same efficiency of ________________ utilization as the ________________ program.
56. If an intermediate result will not be greater than nine digits, the operation is performed more efficiently as ________________ data file.
57. If a decimal multiplication requires an intermediate result greater than 30 digits, a COBOL ________________ subroutine is used for the multiplication and the result is truncated to ________________ digits.
58. The On Size Error option applies only to the ________________ tabulated result.
59. A good practice is to ________________ arithmetic fields before using them in a computation.
60. Numeric comparisons are more efficient when the data is in the ________________ mode.
61. ________________ and ________________ more than one file with one statement is faster than any separate statement for each file.
62. The ________________ verb does not provide for the recognition of the ________________ card being read from the card reader.
63. Use of paragraph-names for comments requires more storage than the use of a ________________ statement or a ________________ card.
64. The use of the Compute statement generates more efficient coding than does the use of individual ________________ statements.
65. ________________ and ________________ IF statements should be avoided as the logic is difficult to debug.
66. Performing a move operation for an item greater than ________________ bytes in length requires the generation of more instructions than that required for one with less.
67. An ________________ should be used in place of a Note statement if it is the first statement of a paragraph.
68. A Perform statement can be used if the ________________ of the flow of data is returned to the ________________ statement of the routine referred.
69. A Read Into and Write option can make debugging simpler because ________________ and ________________ areas are easier to locate in a dump.
70. A precise definition of ________________ requirements is essential to the development of a correct ________________.
71. HIPO is an acronym for ________________.
72. HIPO is a method of ________________ describing a ________________ entity.
73. In ________________ each function is exploded into increasingly ________________ levels of detail until all ________________ have been defined.
74. Every box in a HIPO hierarchy has a corresponding ________________ diagram.
75. Top level functions in a hierarchy contain the ________________.
76. HIPO is one good way to approach the ________________ task and one that is complementary to ________________.
77. Conventional-Linkage combines both ________________ and ________________.
78. As a program structure, proper use of Conventional-Linkage requires a rigid placement of ________________ functions as well as the concept of a ________________ and ________________ in a unit of code.

Answers

1. EFFICIENT, BASIC LANGUAGE
2. DEFINED, FLOWCHARTS, SOURCE DOCUMENT, OUTPUT
3. SELF-DOCUMENTING
4. ESTABLISHMENT, UNDERSTANDING, ENFORCEMENT, PROGRAMMING
5. BLANK CARDS
6. STATEMENT, NEW
7. NONCONDITIONED, SEPARATE
8. WORDS, NUMERIC LITERALS
9. SECTION
10. COMMENTS
11. SIMPLEST, IDENTIFIES, INFORMATION
12. DIVISION, PROGRAM
13. INTENT
14. MACHINE
15. EQUIPMENT, COMPILED, EXECUTED, DEVICES, DATA FILES
16. DEFINED, OPENED, CLOSED
17. VERIFIED
18. PROCESSED, OBJECT
19. ENVIRONMENT, DESCRIBED
20. SOURCE, COMPILE, DATA DIVISION, ENVIRONMENT DIVISION, LIBRARY
21. SELF-DOCUMENTING, COMMENT LINES
22. HYPHEN
23. SOURCE DOCUMENT, OUTPUT, FILE DESCRIPTION, RECORD DESCRIPTION
24. FILE DESCRIPTION, RECORD DESCRIPTION
25. CONSTANTS, WORK AREAS
26. SUBPROGRAMS
27. REPORT WRITER
28. RECORD DESCRIPTION, RECORD CONTAINS, COUNTING
29. SHORTEST, LONGEST
30. FILE DESCRIPTION
31. 01
32. FILE NAME
33. FILE
34. WORKING-STORAGE, COMPUTATIONAL-3
35. DECIMAL, MIXED, COMPILER, EXTRA, ADJUSTMENT
36. ALIGNED
37. MOVE
38. FORMAT
39. PLUS, MINUS, EFFICIENCY
40. SEPARATELY, ACCURATE, ARITHMETIC, COMPUTATIONAL, SIX
41. DATA DESCRIPTION, SAME
42. ALTERNATE, DATA
43. F, LOW
44. COMPUTATIONAL, GROUP
45. FILE, DATA-NAMES, CONSTANTS, WORK AREAS USED
46. PROGRAM, ENGLISH-LIKE, SEQUENCE
47. PROGRAM, MODULES, LOGIC
48. SHORT, SIMPLE, LOGICAL, PROGRAM
49. FUNCTIONAL
50. CLASS, NUMERIC
51. OPEN, READ, ENVIRONMENT
52. CLOSE
53. SUBROUTINE, PROCEDURES, SEQUENCE
54. LAST
55. STORAGE, SOURCE
56. BINARY
57. LIBRARY, 30
58. FINAL
59. INITIALIZE
60. COMPUTATIONAL-3
61. OPENING, CLOSING
62. ACCEPT, LAST
63. NOTE, BLANK
64. ARITHMETIC
65. NESTED, COMPUTER
66. 256
67. ASTERISK
68. CONTROL, LAST
69. WORKING-STORAGE, BUFFER
70. USER, DATA PROCESSING SYSTEM
71. HIERARCHY-PLUS-INPUT-PROCESS-OUTPUT
72. GRAPHICALLY, SOFTWARE
73. FUNCTIONAL DECOMPOSITION, LOWER, SUBFUNCTIONS
74. IPO
75. CONTROL LOGIC
76. DESIGN, STRUCTURED PROGRAMMING
77. PROGRAM DESIGN, PROGRAM STRUCTURE
78. PROGRAMMING, SINGLE ENTRY, EXIT POINT

Questions for Review

1. Why is it so important to write an efficient COBOL program?
2. What are some suggestions for improving readability and program logic understanding?
3. What additional information should be included in the Identification Division?
4. What precautions should be taken in writing the Environment Division?
5. How can the use of the Copy statement improve COBOL programs? What are some of its main advantages?
6. Why should the Records Contains clause be included in every file description entry in the Data Division?
7. Why is it desirable to write Data Division entries for output files in the Working-Storage Section?
8. When should the computational mode be used with data items?
9. How could storage be conserved in decimal alignment?
10. How does the compiler align fields of unequal length?
11. What problem is caused by operations involving data items of mixed data format? How can this be overcome?
12. Explain the operation of sign control in the efficiency of a program.
13. Why is it possible to have an inaccurate answer as a result of an arithmetic statement in a conditional statement?
14. List the basic principles of effective coding.
15. How may alternate groupings provide a more efficient program?
16. How may the Procedure Division be written more efficiently?

17. What is the danger of intermediate results in arithmetic statements?
18. What is the problem involved in using an Accept statement?
19. Why is it more efficient to use a Compute statement rather than a series of arithmetic statements?
20. Why should nested and computed IF statements be avoided?
21. Why is Perform a useful verb?
22. Why is the Read Into and Write From options suggested for input/output operations?
23. What does a specification package usually contain?
24. What is HIPO and how does it satisfy the requirements for specification package?
25. What is functional decomposition?
26. What are the major benefits of HIPO?
27. What is Conventional-Linkage and what is its main advantages?
28. What are the seven sections of Conventional-Linkage and what is the main function of each?

Problems

1. *Match each item with its proper description.*

______	1. HIPO	A.	Applies only to the final tabulated results.
______	2. Mixed Data Format	B.	Indicate where item is to be found in program.
______	3. ON SIZE ERROR	C.	Machine oriented.
______	4. COPY	D.	Require one item to be converted to matched format.
______	5. Environment Division	E.	Limited to 30 digits.
______	6. Suffix	F.	Arrange program into separate modules of logic.
______	7. Conventional-Linkage	G.	Include prewritten Data and Environment Division entries.
______	8. Modular Program	H.	Method of graphically describing a software entity as an arrangement of functions to be performed.
______	9. Alternate Descriptions	I.	Redefine fields.
______	10. Intermediate Results	J.	Combines both program design and program structure.

2. *Write the minimum number of entries required for the following information for the Identification and Environment Divisions.*

   a. The program is to be written to process data on a IBM 370 computer model H155.
   b. The input/output devices to be used are the model 2540 card reader and a 2400 tape unit.

3. *In the following Compute statement, the accuracy of the result (X) will be affected by the number of integers and decimal digits returned in the various intermediate results.*

   Rewrite the statement into a series of Compute statements to assure that this does not occur. Assume all fields will not exceed eight digits.

```
COMPUTE X = A + (B / C) + ((D ** E) * F) - G.
```

4. *A program references a field and its subfields, each with different usages. The fields are as follows:*

Field	Record Positions
Name	1–24
Number	25–28

   The number will be involved in numerous calculations. The entire field by itself will be used for display purposes.

   Write the record description entries for the above so that both the name and number fields can be referenced as well as each subfield with different Usage clauses.

5. *Set up Working-Storage areas for WORK-AREA-FILE1 and WORK-AREA-FILE2 so that the same area can be used as a work area for records of several input files that can not be processed concurrently.*

6. *A program contains the following instructions:*

```
77  FLD-A     PICTURE  S9(5)V9999.
77  FLD-B     PICTURE  S99V99.
 .
 .
 .
PROCEDURE  DIVISION.
 .
 .
 .
     ADD  FLD-A  TO  FLD-B.
```

What Picture clause must be changed to make the program correct and more efficient?

Write the correct Picture clause for the item to be changed.

7. *Which of the following Perform statements is incorrect?*

A. x PERFORM a THRU m

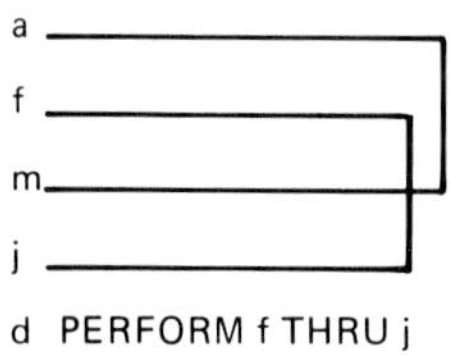

```
a
f
m
j
d PERFORM f THRU j
```

B. x PERFORM a THRU m

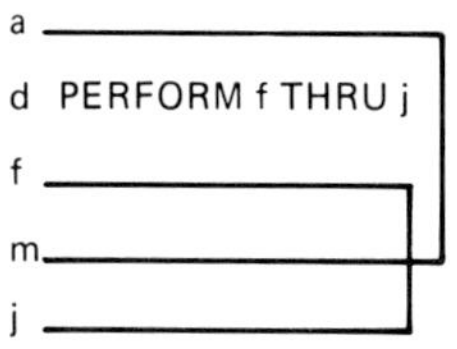

```
a
d PERFORM f THRU j
f
m
j
```

C. x PERFORM a THRU m

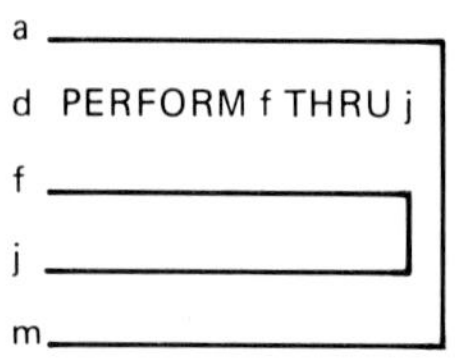

```
a
d PERFORM f THRU j
f
j
m
```

D. x PERFORM a THRU m

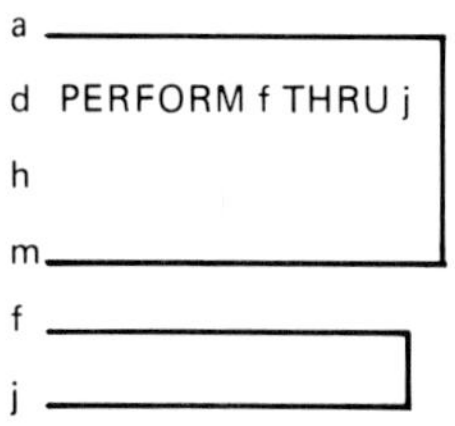

```
a
d PERFORM f THRU j
h
m
f
j
```

Appendix A Debugging COBOL Programs

COMPILER DIAGNOSTICS

Diagnostic messages are generated by the compiler and listed on the systems printer when errors are found in the source program (fig. A.1). A complete listing of diagnostic messages will be found in the programmer's guide reference manual for that particular computer.

Debugging Diagnostic Messages

1. Approach each diagnostic message in sequence as it appears in the compilation source listing. It is possible to get compound diagnostic messages as frequently as an earlier diagnostic message indicates the reason for a later diagnostic message. For example, a missing quotation mark for a nonnumeric literal could involve the inclusion of some clauses not intended for that particular literal. This could cause an apparently valid clause to be diagnosed as invalid because it is not complete, or because it is in conflict with something that preceded it.
2. Check for missing or superfluous punctuation or errors of this type.
3. Frequently, a seemingly meaningless message is clarified when the invalid syntax or format of the clause or statement in question is referenced.

Diagnostic Messages

The diagnostic messages associated with the compilation are always listed. The format of the diagnostic message for IBM System 360/370 computers is:

1. *Compiler Generated Card Number*. This is the number of a line in the source program related to the error.
2. *Message Identifier*. Message identification for the system.
3. *The Severity Level*. There are four severity levels as follows:

(W) WARNING. This level indicates that an error was made in the source program. However, it is not serious enough to interfere with the execution of the program.

(C) CONDITIONAL. This level indicates that an error was made, but the compiler usually makes a corrective assumption. The statement containing the error is retained. Execution can be attempted.

(E) ERROR. This level indicates that a serious error was made. Usually the compiler makes no corrective assumption. The statement or operand containing the error is dropped. Compilation is completed, but execution of the program should not be attempted.

```
     5


CARD   ERROR MESSAGE

       ILA1100I-W    1 SEQUENCE ERROR IN SOURCE PROGRAM.
7      ILA1095I-W    WORD 'SECTION' OR 'DIVISION' MISSING. ASSUMED PRESENT.
12     ILA1132I-E    INVALID SYSTEM-NAME. SKIPPING TO NEXT CLAUSE.
13     ILA1132I-E    INVALID SYSTEM-NAME. SKIPPING TO NEXT CLAUSE.
19     ILA1056I-E    FILE-NAME NOT DEFINED IN A SELECT. DESCRIPTION IGNORED.
35     ILA1056I-E    FILE-NAME NOT DEFINED IN A SELECT. DESCRIPTION IGNORED.
46     ILA1077I-C    ALPHANUMERIC LIT CONTINUES IN A-MARGIN. ASSUME B-MARGIN.
50     ILA1077I-C    ALPHANUMERIC LIT CONTINUES IN A-MARGIN. ASSUME B-MARGIN.
51     ILA1077I-C    ALPHANUMERIC LIT CONTINUES IN A-MARGIN. ASSUME B-MARGIN.
55     ILA1077I-C    ALPHANUMERIC LIT CONTINUES IN A-MARGIN. ASSUME B-MARGIN.
56     ILA1077I-C    ALPHANUMERIC LIT CONTINUES IN A-MARGIN. ASSUME B-MARGIN.
56     ILA1076I-C    ALPHANUMERIC LIT EXCEEDS 120 CHARACTERS. TRUNCATED TO 120.
79     ILA1037I-E    * INVALID IN DATA DESCRIPTION. SKIPPING TO NEXT CLAUSE.
79     ILA2039I-C    PICTURE CONFIGURATION ILLEGAL. PICTURE CHANGED TO 9 UNLESS USAGE IS 'DISPLAY-ST',
                     THEN L(6)BDZ9BDZ9.
83     ILA1037I-E    ** INVALID IN DATA DESCRIPTION. SKIPPING TO NEXT CLAUSE.
83     ILA2039I-C    PICTURE CONFIGURATION ILLEGAL. PICTURE CHANGED TO 9 UNLESS USAGE IS 'DISPLAY-ST',
                     THEN L(6)BDZ9BDZ9.
85     ILA3001I-E    FILE-IN NOT DEFINED. DELETING TILL LEGAL ELEMENT FOUND.
85     ILA3001I-E    FILE-OUT NOT DEFINED. DELETING TILL LEGAL ELEMENT FOUND.
85     ILA4002I-E    OPEN STATEMENT INCOMPLETE. STATEMENT DISCARDED.
86     ILA4050I-E    SYNTAX REQUIRES RECORD-NAME . FOUND DNM=1-337 . STATEMENT DISCARDED.
86     ILA3001I-E    HDG-1 NOT DEFINED.
86     ILA3001I-E    0 NOT DEFINED.
87     ILA4050I-E    SYNTAX REQUIRES RECORD-NAME . FOUND DNM=1-337 . STATEMENT DISCARDED.
88     ILA4050I-E    SYNTAX REQUIRES RECORD-NAME . FOUND DNM=1-337 . STATEMENT DISCARDED.
89     ILA3001I-E    FILE-IN NOT DEFINED. STATEMENT DISCARDED.
94     ILA3001I-E    TOT-DED-WS NOT DEFINED. SUBSTITUTING TALLY .
105    ILA4091I-E    SYNTAX REQUIRES OPERAND. FOUND END OF PAGE . TEST DISCARDED.
106    ILA4050I-E    SYNTAX REQUIRES RECORD-NAME . FOUND DNM=1-337 . STATEMENT DISCARDED.
108    ILA5011I-W    HIGH ORDER TRUNCATION MIGHT OCCUR.
110    ILA4050I-E    SYNTAX REQUIRES RECORD-NAME . FOUND DNM=1-337 . STATEMENT DISCARDED.
114    ILA1077I-C    ALPHANUMERIC LIT CONTINUES IN A-MARGIN. ASSUME B-MARGIN.
113    ILA4003I-E    EXPECTING NEW STATEMENT. FOUND TO . DELETING TILL NEXT VERB OR PROCEDURE NAME.
116    ILA5011I-W    HIGH ORDER TRUNCATION MIGHT OCCUR.
117    ILA4050I-E    SYNTAX REQUIRES RECORD-NAME . FOUND DNM=1-337 . STATEMENT DISCARDED.
117    ILA3001I-E    FINAL-TOTAL NOT DEFINED.
```

Figure A.1 Diagnostic Messages.

(D) DISASTER. This level indicates that a serious error was made. Compilation is not completed and results are unpredictable.

4. *Message Text.* The text identifies the condition that causes the error and indicates the actions taken by the compiler.

Execution Output

The output generated by the program execution (in addition to data written on output files) may include

1. Data displayed on the console or on the printer.
2. Messages to the operator.
3. System informative messages.
4. System diagnostic messages.
5. A system dump.

A dump and system diagnostic messages are generated automatically during the program execution only if the program contains errors that cause the abnormal termination of the program.

Operator Messages

The COBOL phase may issue operator messages. In the message, XX denotes a system-iterated 2-character numeric file that is used to identify the program issuing the message. (See figure A.2.)

C110A STOP literal

Explanation: The programmer has issued a STOP literal statement in the American National Standard COBOL source program.

System Action: Awaits operator response.

Programmer Response: Not applicable.

Operator Response: Operator should respond with end-of-block, or with any character in order to proceed with the program.

C111A AWAITING REPLY

Explanation: This message is issued in connection with the American National Standard COBOL ACCEPT statement.

System Action: Awaits operator response.

Programmer Response: Not applicable.

Operator Response: The operator should reply as specified by the programmer.

Figure A.2 Object Time Messages—Console.

STOP Statement

The following message is generated by the STOP statement with the *literal* option:

```
XX  C110A  STOP  'literal'
```

This message is issued at the programmer's discretion to indicate possible alternative action to be taken by the operator.

The operator responds according to the instructions given both by the message and on the job request form supplied by the programmer. If the job is to be resumed, the programmer presses the end-of-block key on the console.

ACCEPT Statement

The following message is generated by an ACCEPT statement with the FROM CONSOLE option:

```
XX  C111A  "AWAITING REPLY"
```

This message is issued by the object program when operator intervention is required.

The operator responds by entering the reply and by pressing the end-of-block key on the console. (The contents of the text field should be supplied by the programmer on the job request form.)

System Output

Informative and diagnostic messages may appear in the listing during the execution of the object program.

Each of these messages contains an identification code in the first column of the message to indicate the portion of the operating system that generated the message.

Dump

If a serious error occurs during the execution of the problem program, the programmer can request a printout of storage through the use of the DUMP option in the job-control cards. The job would be abnormally terminated, any remaining steps bypassed, and a program phase dump is generated. The programmer can use the dump to checkout his program. In cases where a serious error occurs in other than the problem program (for example, in the control program), a dump is not produced. (*Note:* the program phase dump can be suppressed if the NODUMP option of the job-control card statement has been specified.)

How to Use a Dump

When a job is abnormally terminated due to a serious error in the problem program, a message is written on the system output device which indicates the following:

1. Type of interrupt (for example, program check).
2. The hexadecimal (IBM) address of the instruction that caused the interrupt.
3. Condition code.

The hexadecimal address of the instruction that caused the dump is subtracted from the load address of the module (which can be obtained from the map of main storage generated by the Linkage Editor) to obtain the relative instruction address as shown in the Procedure Division map. If the interrupt occurred within the COBOL program, the programmer can use the error address to locate the specific statement which caused a dump to be generated. Examination of the statement and fields associated with it may produce information as to the specific nature of the error.

Figure A.3 illustrates a sample dump caused by a data exception. Invalid data (for example, data that did not correspond to its usage) was placed in the numeric field B as a result of redefinition. Letters identify the text corresponding to the letter in the program listing.

```
// JOB DTACHK                                                        05.00.19
// OPTION NODECK,LINK,LIST,LISTX,SYM,ERRS
// EXEC FCOBOL




 CBL QUOTE,SEQ
00001    000010 IDENTIFICATION DIVISION.
00002    000020 PROGRAM-ID. TESTRUN.
00003    000030     AUTHOR. PROGRAMMER NAME.
00004    000040     INSTALLATION. NEW YORK PROGRAMMING CENTER.
00005    000050     DATE-WRITTEN.  FEBRUARY 4, 1971
00006    000060 DATE-COMPILED. 04/24/71
00007    000070    REMARKS. THIS PROGRAM HAS BEEN WRITTEN AS A SAMPLE PROGRAM FOR
00008    000080     COBOL USERS. IT CREATES AN OUTPUT FILE AND READS IT BACK AS
00009    000090     INPUT.
00010    000100
00011    000110 ENVIRONMENT DIVISION.
00012    000120 CONFIGURATION SECTION.
00013    000130 SOURCE-COMPUTER. IBM-360-H50.
```

Figure A.3 Sample Dump Resulting from Abnormal Termination.

Figure A.3 Continued

```
00014   000140 OBJECT-COMPUTER. IBM-360-H50.
00015   000150 INPUT-OUTPUT SECTION.
00016   000160 FILE-CONTROL.
00017   000170     SELECT FILE-1 ASSIGN TO SYS008-UT-2400-S.
00018   000180     SELECT FILE-2 ASSIGN TO SYS008-UT-2400-S.
00019   000190
00020   000200 DATA DIVISION.                                        \
00021   000210 FILE SECTION.                                          |
00022   000220 FD  FILE-1                                             |
00023   000230     LABEL RECORDS ARE OMITTED                          |
00024   000240     BLOCK CONTAINS 5 RECORDS                           |
00025   000250     RECORDING MODE IS F                                |
00026   000255     RECORD CONTAINS 20 CHARACTERS                      |
00027   000260     DATA RECORD IS RECORD-1.                           |
00028   000270 01  RECORD-1.                                          |
00029   000280     05 FIELD-A PIC X(20).                              > (E)
00030   000290 FD  FILE-2                                             |
00031   000300     LABEL RECORDS ARE OMITTED                          |
00032   000310     BLOCK CONTAINS 5 RECORDS                           |
00033   000320     RECORD CONTAINS 20 CHARACTERS                      |
00034   000330     RECORDING MODE IS F                                |
00035   000340     DATA RECORD IS RECORD-2.                           |
00036   000350 01  RECORD-2.                                          |
00037   000360     05 FIELD-A PIC X(20).                             /

00038   000370 WORKING-STORAGE SECTION.
00039   000380 01  FILLER.
00040   000390     02 COUNT PIC S99 COMP SYNC.
00041   000400     02 ALPHABET PIC X(26) VALUE IS "ABCDEFGHIJKLMNOPQRSTUVWXYZ".
00042   000410     02 ALPHA REDEFINES ALPHABET PIC X OCCURS 26 TIMES.
00043   000420     02 NUMBR PIC S99 COMP SYNC.
00044   000430     02 DEPENDENTS PIC X(26) VALUE "01234012340123401234012340".
00045   000440     02 DEPEND REDEFINES DEPENDENTS PIC X OCCURS 26 TIMES.
00046   000450 01  WORK-RECORD.
00047   000460     05 NAME-FIELD PIC X.
00048   000470     05 FILLER PIC X.
00049   000480     05 RECORD-NO PIC 9999.
00050   000490     05 FILLER PIC X VALUE IS SPACE.
00051   000500     05 LOCATION PIC AAA VALUE IS "NYC".
00052   000510     05 FILLER PIC X VALUE IS SPACE.
00053   000520     05 NO-OF-DEPENDENTS PIC XX.
00054   000530     05 FILLER PIC X(7) VALUE IS SPACES.
00055   000534 01  RECORDA.
00056   000535     02 A PICTURE S9(4) VALUE 1234.
00057   000536     02 B REDEFINES A PICTURE S9(7) COMPUTATIONAL-3.
00058   000540
00059   000550 PROCEDURE DIVISION.
00060   000560 BEGIN. READY TRACE.
00061   000570     NOTE THAT THE FOLLOWING OPENS THE OUTPUT FILE TO BE CREATED
00062   000580     AND INITIALIZES COUNTERS.
00063   000590 STEP-1. OPEN OUTPUT FILE-1. MOVE ZERO TO COUNT, NUMBR.
00064   000600     NOTE THAT THE FOLLOWING CREATES INTERNALLY THE RECORDS TO BE
00065   000610     CONTAINED IN THE FILE, WRITES THEM ON TAPE, AND DISPLAYS
00066   000620     THEM ON THE CONSOLE.
00067   000630 STEP-2. ADD 1 TO COUNT, NUMBR. MOVE ALPHA (COUNT) TO
00068   000640     NAME-FIELD.
00069   000645         COMPUTE B = B + 1. <--(D)
00070   000650     MOVE DEPEND (COUNT) TO NO-OF-DEPENDENTS.
00071   000660     MOVE NUMBR TO RECORD-NO.
00072   000670 STEP-3. DISPLAY WORK-RECORD UPON CONSOLE. WRITE RECORD-1 FROM
00073   000680     WORK-RECORD.
00074   000690 STEP-4. PERFORM STEP-2 THRU STEP-3 UNTIL COUNT IS EQUAL TO 26.
00075   000700     NOTE THAT THE FOLLOWING CLOSES THE OUTPUT FILE AND REOPENS
00076   000710     IT AS INPUT.
00077   000720 STEP-5. CLOSE FILE-1. OPEN INPUT FILE-2.
00078   000730     NOTE THAT THE FOLLOWING READS BACK THE FILE AND SINGLES
00079   000740     OUT EMPLOYEES WITH NO DEPENDENTS.
00080   000750 STEP-6. READ FILE-2 RECORD INTO WORK-RECORD AT END GO TO STEP-8.
00081   000760 STEP-7. IF NO-OF-DEPENDENTS IS EQUAL TO "0" MOVE "Z" TO
00082   000770     NO-OF-DEPENDENTS. EXHIBIT NAMED WORK-RECORD. GO TO STEP-6.
00083   000780 STEP-8. CLOSE FILE-2.
00084   000790     STOP RUN.
```

Figure A.3 Continued

INTRNL NAME	LVL	SOURCE NAME	BASE	DISPL	INTRNL NAME	DEFINITION	USAGE	R	O	Q	M
DNM=1-148	FD	FILE-1	DTF=01		DNM=1-148		DTFMT				F
DNM=1-178	01	RECORD-1	BL=1	000	DNM=1-178	DS OCL20	GROUP				
DNM=1-199	02	FIELD-A	BL=1	000	DNM=1-199	DS 20C	DISP				
DNM=1-216	FD	FILE-2	DTF=02		DNM=1-216		DTFMT				F
DNM=1-246	01	RECORD-2	BL=2	000	DNM=1-246	DS OCL20	GROUP				
DNM=1-267	02	FIELD-A	BL=2	000	DNM=1-267	DS 20C	DISP				
DNM=1-287	01	FILLER	BL=3	000	DNM=1-287	DS OCL56	GROUP				
DNM=1-306	02	COUNT	BL=3	000	DNM=1-306	DS 1H	COMP				
DNM=1-321	02	ALPHABET	BL=3	002	DNM=1-321	DS 26C	DISP				
DNM=1-339	02	ALPHA	BL=3	002	DNM=1-339	DS 1C	DISP	R	O		
DNM=1-357	02	NUMBR	BL=3	01C	DNM=1-357	DS 1H	COMP				
DNM=1-372	02	DEPENDENTS	BL=3	01E	DNM=1-372	DS 26C	DISP				
DNM=1-392	02	DEPEND	BL=3	01E	DNM=1-392	DS 1C	DISP	R	O		
DNM=1-408	01	WORK-RECORD	BL=3	038	DNM=1-408	DS OCL20	GROUP				
DNM=1-432	02	NAME-FIELD	BL=3	038	DNM=1-432	DS 1C	DISP				
DNM=1-452	02	FILLER	BL=3	039	DNM=1-452	DS 1C	DISP				
DNM=1-471	02	RECORD-NO	BL=3	03A	DNM=1-471	DS 4C	DISP-NM				
DNM=1-490	02	FILLER.	BL=3	03E	DNM=1-490	DS 1C	DISP				
DNM=2-000	02	LOCATION	BL=3	03F	DNM=2-000	DS 3C	DISP				
DNM=2-018	02	FILLER	BL=3	042	DNM=2-018	DS 1C	DISP				
DNM=2-037	02	NO-OF-DEPENDENTS	BL=3	043	DNM=2-037	DS 2C	DISP				
DNM=2-063	02	FILLER	BL=3	045	DNM=2-063	DS 7C	DISP				
DNM=2-082	01	RECORDA	BL=3	050	DNM=2-082	DS OCL4	GROUP				
DNM=2-102	02	A	BL=3	050	DNM=2-102	DS 4C	DISP-NM				
DNM=2-113	02	B ← (J)	BL=3	050	DNM=2-113	DS 4P	COMP-3	R			

MEMORY MAP

TGT	003E8
SAVE AREA	003E8
SWITCH	00430
TALLY	00434
SORT SAVE	00438
ENTRY-SAVE	0043C
SORT CORE SIZE	00440
NSTD-REELS	00444
SORT RET	00446
WORKING CELLS	00448
SORT FILE SIZE	00578
SORT MODE SIZE	0057C
PGT-VN TBL	00580
TGT-VN TBL	00584
SORTAB ADDRESS	00588
LENGTH OF VN TBL	0058C
LNGTH OF SORTAB	0058E
PGM ID	00590
A(INIT1)	00598
UPSI SWITCHES	0059C
OVERFLOW CELLS	005A4
BL CELLS ← (N)	005A4
DTFADR CELLS	005B0 ← (F)
TEMP STORAGE	005B8
TEMP STORAGE-2	005C0
TEMP STORAGE-3	005C0
TEMP STORAGE-4	005C0
BLL CELLS	005C0
VLC CELLS	005C4
SBL CELLS	005C4
INDEX CELLS	005C4
SUBADR CELLS	005C4
ONCTL CELLS	005CC
PFMCTL CELLS	005CC
PFMSAV CELLS	005CC
VN CELLS	005D0
SAVE AREA =2	005D4
XSASW CELLS	005D4
XSA CELLS	005D4
PARAM CELLS	005D4
RPTSAV AREA	005D8
CHECKPT CTR	005D8
IOPTR CELLS	005D8

Figure A.3 Continued

```
                    REGISTER ASSIGNMENT

                     REG 6     BL =3  <--(K)
                     REG 7     BL =1
                     REG 8     BL =2


 67        0006FC  41 40 6 002                  LA    4,002(0,6)               DNM=1-339
           000700  48 20 6 000                  LH    2,000(0,6)               DNM=1-306
           000704  4C 20 C 03A                  MH    2,03A(0,12)              LIT+2
           000708  1A 42                        AR    4,2
           00070A  58 40 C 038                  S     4,038(0,12)              LIT+0
           00070E  50 40 D 1DC                  ST    4,1DC(0,13)              SBS=1
           000712  58 E0 D 1DC                  L     14,1DC(0,13)             SBS=1
           000716  D2 00 6 038 E 000            MVC   038(1,6),000(14)         DNM=1-432     DNM=1-339
 69        00071C  FA 30 6 050 C 03C     (C)--> AP    050(4,6),03C(1,12)       DNM=2-113     LIT+4
 70        000722  41 40 6 01E                  LA    4,01E(0,6)               DNM=1-392
           000726  48 20 6 000                  LH    2,000(0,6)               DNM=1-306
           00072A  4C 20 C 03A                  MH    2,03A(0,12)              LIT+2
           00072E  1A 42                        AR    4,2
           000730  5B 40 C 038                  S     4,038(0,12)              LIT+0
           000734  50 40 D 1E0                  ST    4,1E0(0,13)              SBS=2
           000738  58 E0 D 1E0                  L     14,1E0(0,13)             SBS=2
           00073C  D2 00 6 043 E 000            MVC   043(1,6),000(14)         DNM=2-37      DNM=1-392
           000742  92 40 6 044                  MVI   044(6),X'40'             DNM=2-37+1

 // EXEC LNKEDT


     PHASE    XFR-AD  LOCORE  HICORE  DSK-AD   ESD TYPE    LABEL      LOADED  REL-FR

 PHASE***     0032A0  0032A0  004ADB  53 01 2  CSECT       TESTRUN    0032A0  0032A0  <--(B)

                                               CSECT       IJFFBZZN   003C50  003C50
                                               *  ENTRY    IJFFZZZN   003C50
                                               *  ENTRY    IJFFBZZZ   003C50
                                               *  ENTRY    IJFFZZZZ   003C50

                                               CSECT       ILBDSAE0   0049F0  0049F0
                                                  ENTRY    ILBDSAE1   004A06

                                               CSECT       ILBDMNS0   0049E8  0049E8

                                               CSECT       ILBDDSP0   0041B8  0041B8
                                               *  ENTRY    ILBDDSP1   004708
                                               *  ENTRY    ILBDDSP2   0047A0
                                               *  ENTRY    ILBDDSP3   004958

                                               CSECT       ILBDIML0   004990  004990

                                               CSECT       IJJCPD1    003FC0  003FC0
                                                  ENTRY    IJJCPD1N   003FC0
                                               *  ENTRY    IJJCPD3    003FC0


 // ASSGN SYS008,X'182'
 // EXEC


 0S03I PROGRAM CHECK INTERRUPTION - HEX LOCATION 0039BC - CONDITION CODE 0 - DATA EXCEPTION
 0S00I JOB DTACHK    CANCELED
```

Figure A.3 Continued

```
        DTACHK                                                          (L)

GR 0-7  00003850 00003960 00000001 00000001   0000338A 50003C12 00003388 00003550
GR 8-F  000035B8 00003BE2 000032A0 000032A0   00003880 00003688 0000338A 000041B8
FP REG  00000000 00000000 00000000 00000000   00000000 00000000 00000000 00000000
COMREG  BG ADDR IS 000208


000000  00000000 00000000 00000000 00000000   00000000 00000208 FF050000 00000000
000020  FF050007 40002E06 FF150007 C00039C2   5B5BC2C5 D6D1F440 FF05000E 80002E00
000040  00002F28 08000000 00002F18 00000000   FCBF1CB3 015005E8 00040000 0F0014BA
000060  00040000 00000336 00040000 0000147A   00000000 00000BBC 00040000 000002D4
000080  00000000 00000000 00000000 00000003   00050003 06B006B0 06B041BB 00734570
0000A0  0146940F B47B41A0 C0544570 0B8418A8   41900156 4180B2CE 47F000DA 06B006B0
0000C0  06B006B0 06B006B0 06B041BB 001741BB   00504570 01464180 01569640 A0019120
0000E0  A00C4710 00EA9260 A00195E2 A0024780   0DC695C1 A0024780 0DC69561 A0024780
000100  010E9104 A0004780 010E9203 008F9281   A0004BA0 0262487A C00049A0 027641AA
000120  C0440778 94F9703B D7017058 70589283   A0009680 A0014400 04080788 947FA001
000140  4570B218 07F842B0 00E748B0 02C847F0   BC704570 BC70D205 BEEEBEF5 DC05BEEE
000160  C0441BAA DD06BEEE 000C43A1 000742A0   023741AA C0444400 A0045890 A0044220
000180  A0009140 A0014710 BB64D207 01F09008   68009058 68209060 68409068 68609070
0001A0  4BA00262 41AAC000 D2010016 A0009898   90108200 01F04400 A0045890 A0049818
0001C0  9030989D 01F08200 00389284 C0A4D207   01F0BF50 9890BF58 820001F0 9680A000
0001E0  41100030 47F0B166 96030039 82000038   FF050007 40002E06 00001000 00002000
000200  00003000 80001048 F0F461F2 F461F7F1   32A03000 00000000 00000000 00000000
000220  C4E3C1C3 C8D24040 0007AFFF 00004ADB   00004ADB 00000010 0007FFFF F875ECD1
000240  A8A07CD0 00C62171 21782269 226A0000   25102514 25183CF0 F4F2F4F7 F1F1F1F4
000260  00002044 0000000C 22E21E4E 1EF41F04   1F140020 214C0010 5B5BC2D6 00130001
000280  01001F98 20000000 00000000 02080000   00000294 00000000 000025AC 00000044
0002A0  00001F2C 00000000 00000000 00000000   00000000 00000000 00000000 00000000
0002C0  0000289C 00003228 100020CE 00001DC8   00002A9C 923801C9 909D01F0 4190086C
0002E0  48A00236 4AA00262 9180A000 47100306   58B0A004 9018B030 48B002C8 41CBB000
000300  41DCB000 07F99601 A00048B0 02C841CB   B00041DC B00095FF A00F0789 90E0BF6C
000320  48E001C8 D207BF50 E00094FD BF51D213   BF5801F0 07F9909D 01F09220 01C94590
000340  02E04190 01B69500 00234780 03F49526   002347B0 00BE4860 00221A66 487002CA
000360  48667000 07F6181F 1B664121 000F4570   BCAC4860 BE5C1B22 43201007 4130001F
000380  1B234740 03904130 00151B23 47B00392   1A234220 04D94320 100747F0 04584720
0003A0  00CA4230 04D94820 02364322 C0031A23   950B1007 47F00454 472000C8 96801002
0003C0  960C1004 07F91858 41430002 43540000   41455000 1A444A40 BE5495FF 40004770
0003E0  03CC4284 000007F9 95FF04B1 07891B00   5000BF74 95FF04B1 4780B238 48600236
000400  95600237 47800366 D502A005 02814770   04204111 00004910 BE3A47B0 04201B66
000420  4121000F 4570BCAC D5021009 023947B0   00C61B33 43301007 95011006 4770039E
000440  92FF04D9 D200044D A00D4123 000BD500   1007A00E 47B000CA D4031002 C0B04182
000460  20004870 02544338 70004930 BE3847B0   03B88930 00034A30 024891F0 30044780
000480  B60ED501 0022BB2C 4780B8E4 D501BE4C   BE5A4720 04AE4930 BE744770 04AE950F
0004A0  00234780 04AE9110 30064780 B2384180   00014148 80001A44 4A40BE54 18584A50
0004C0  0274D200 04B14000 50104000 92FF4000   42205000 4260500C 92035018 91F03004
0004E0  4780B62E 47F004EC 5880BF8C 44000CB8   9560C09C 47700566 D20202CD 10095860
000500  02CC9507 60004770 0566D202 02CD6001   587002CC 1B444340 30054C40 BE485A40
000520  02D04144 0000D503 70014000 47800566   9120100C 47100560 91051002 47700560
000540  91406004 47800560 94BF6004 91101002   47800558 96401002 96141002 9601100C
000560  D2034000 700195FF 30024770 03C64280   30029198 30060779 43203000 4322C09D
000580  48603000 95003000 47800592 9F006000   07694060 05E29550 30044780 080C9504
0005A0  50184780 08409101 100C4710 05D0940F   06FF91F0 500C4780 05D0D300 C09C3004
0005C0  95035018 47D00634 9560C09C 4780067C   D2020049 1009940F 0703D300 0048500C
0005E0  9C00000E 477005F2 4032C0B4 96803006   07F94730 0BC69106 00454770 0E9C913F
000600  00454770 060C91AF 00440789 D201003A   05E29550 30044770 0620D202 00491009
000620  58600048 4A60BDDC 50600040 4032C0B4   47F0089A 95015018 4720065C 9560C09C
000640  47700654 45700B84 9120800F 47100094   47F0065C 91203006 47100094 9560C09C
000660  477005D0 45700B84 4B800262 4878C000   91407038 471005D0 96F006FF 95003003
```

Figure A.3 Continued

```
            DTACHK

0032E0   00005218 00000208 00000000 00000000    00000000 00000000 00000000 00000000
003300   00000000 --SAME--
003320   00000000 58C0F0C6 58E0C000 58D0F0CA    9500E000 4770F0A2 9610D048 92FFE000
003340   47F0F0AC 98CEF03A 90ECD00C 185D989F    F0BA9110 D0480719 07FF0700 00003BE2
003360   000032A0 000032A0 00003880 00003688    000038EC 00003BC8 C3D6C2C6 F0F0F0F1
003380   E3C5E2E3 D9E4D540 0001C1C2 C3C4C5C6    C7C8C9D1 D2D3D4D5 D6D7D8D9 E2E3E4E5
0033A0   E6E7E8E9 0001F0F1 F2F3F4F0 F1F2F3F4    F0F1F2F3 F4F0F1F2 F3F4F0F1 F2F3F4F0
0033C0   C1000000 000040D5 E8C34000 00404040    40404040 00000000 F1F2F3C4 00004C40
0033E0   01010014 00000000 00000000 00000000    0E000000 04000000 00009200 00000108
003400   00003430 00000000 10003C50 1160E2E8    E2F0F0F8 40400162 10000000 04000000
003420   00000000 86BCF018 41E0E001 58201044    010034E8 20000064 00003550 00003550
003440   00000014 000035B3 00640063 00000000    00000000 000049F0 01010014 00000000
003460   00000000 00000000 00000000 04000000    00008200 00000108 000034A8 00000000
003480   10003C50 1168E2E8 E2F0F0F8 40400272    00000000 20000000 00000000 86BCF018
0034A0   41E0E001 58201044 020035B8 00000064    00003620 00000000 00000014 00000000
0034C0   00640063 00000000 00004A06 000049F0    00000000 00000000 00000000 00000000
0034E0   00000000 00000000 00004770 30129261    10004110 100107F3 D20467CE 6017D201
003500   67D56274 C6C3D6C2 D6D3F8F0 F8F0F1F0    F1F2F1F1 F2F0F2F2 F2F1F3F0 F4F0F5F0
003520   F5F1F6F0 F6F1F7F0 0100DDA8 10006670    20006148 40005DC8 70004C40 41110004
003540   41110004 41110004 58110000 58F10010    45EF0018 41105342 07FB0000 000032B0
003560   000035A4 000035F8 00003DB4 000039B0    00003944 00004096 00003D0A 000032B0
003580   000062B8 00004478 00004C94 00005704    00005A4C 00005B68 0000373E 000035E4
0035A0   000036A6 060C40FF C4B2DE09 D2106276    D207601C D212F363 603B6276 96F06041
0035C0   4110601C 5840C65C 41200008 05301B24    47403018 95401000 47703012 92611000
0035E0   41101001 00000000 6276C494 58F0C340    077F9240 6820D206 00004218 000042B0
003600   00004348 000043E0 000062B8 00004478    00004510 000045F8 D500627C 00000000
003620   000001FF 00003800 00003982 00003968    00003F2C 00003BAA 00003BAA 00003C40
003640   000037DA 00003BAA 00003E6C 00003D60    00004090 00003DBE 00003B20 00003BC6
003660   00003BC6 00000203 02030001 04050104    00000203 00000105 00000404 00000104
003680   04040202 01030000 00202020 20210000    1C404040 40404000 00200000 00006148
0036A0   00180014 0F0F0000 000C1C0C 00000000    58F0C010 000036F4 10000006 0C000822
0036C0   00000000 00040D00 01E40267 00000003    7000004B 00000000 00000000 000038EC
0036E0   00000000 00000000 000033F8 00003550    000032A0 000033F8 50003C12 02AA1000
003700   00100C00 09EE0000 FFFFD201 6030C49A    4810C4A6 06104C10 C48C5010 D24C4810
003720   C4A60610 4C10C48C 5010D264 414062AE    5A40D24C F871D208 4000D205 00000000
003740   50D05362 41D053F6 5430536A 98675366    18809506 800041E0 568E58F0 52BA078F
003760   43680000 8C600004 89600002 8870001B    58B6536E 91508000 477054D0 91A08000
003780   47E054D0 00003958 00003550 01005366    70003934 000041B8 00003850 00003960
0037A0   00003550 000032A0 000033F8 50003C12    00003388 00003550 000035B8 00003BE2
0037C0   000032A0 000032A0 00003880 00003960    000041B8 00003850 00004708 00003550
0037E0   00015540 00003958 58F10010 45EF0008    180747F0 568ED703 532E532E 47F0568E
003800   49A053E2 58C053E6 078C91FF 53D14780    566845B0 55F445B0 00000000 00000000
003820   E000590C 478056AC 91FF53D0 47105686    41B05686 47F055F4 000032A0 91FF53D1
003840   47105598 00003550 000035B8 00003388    000033F8 00003470 00000000 0000001C
003860   00000000 0000338A 42F90000 88F00008    00003A3E 17671776 1767D201 60045366
003880   000049E8 000041B8 00004990 00003950    00003A3E 00003AE0 00003B2C 00003B88
0038A0   00003A5E 00003A72 00003B26 00003B58    00003A3E 504088AE 00000001 1C00001A
0038C0   5B5BC2D6 D7C5D540 5B5BC2C3 D3D6E2C5    5B5BC2C6 C3D4E4D3 F0E90000 C0000000
0038E0   E6D6D9D2 60D9C5C3 D6D9C420 58F0C004    051F0001 4004F6F0 404040AA 9640D048
003900   58F0C004 051F0001 4004F6F3 40404010    4110C040 5800D1C8 184005F0 5000F008
003920   4500F00C 000033F8 0A024100 D1C858F0    C00805EF 5810D1C8 96101020 5020D1BC
003940   5870D1BC D2016000 C038D201 601CC038    58F0C004 051F0001 4004F6F7 404040F1
003960   4830C03A 4A306000 4E30D1D0 D705D1D0    D1D0940F D1D64F30 D1D04030 60004830
003980   C03A4A30 601C4E30 D1D0D705 D1D0D1D0    940FD1D6 4F30D1D0 4030601C 41406002
0039A0   48206000 4C20C03A 1A425B40 C0385040    D1DC58E0 D1DCD200 6038E000 FA306050
0039C0   C03C4140 601E4820 60004C20 C03A1A42    5B40C038 5040D1E0 58E0D1E0 D2006043
0039E0   E0009240 60444830 601C4E30 D1D0F331    603AD1D6 96F0603D 58F0C004 051F0001
003A00   4004F7F2 4040404F 58F0C004 051F0002    00000014 0D0001C4 0038FFFF D2137000
```

Figure A-3 is a sample dump which was caused by a data exception. Invalid data (i.e., data which did not correspond to its usage) was placed in the numeric field B as a result of redefinition. The following notes illustrate the method of finding the specific statement in the program which caused the dump. Letters identifying the text correspond to letter in the program listing.

(A) The program interrupt occurred at HEX LOCATION 0039BC. This is indicated in the SYSLST message printed just before the dump.

Figure A.3 Continued

Ⓑ The linkage editor map indicates that the program was loaded into address 0032A0. This is determined by examining the load point of the control section TESTRUN. TESTRUN is the name assigned to the program module by the source coding:
PROGRAM-ID. TESTRUN.

Ⓒ The specific instruction which caused the dump is located by subtracting the load address from the interrupt address (i.e., subtracting 32A0 from 39BC). The result, 71C, is the relative interrupt address and can be found in the object code listing. In this case the instruction in question is an AP (add decimal).

Ⓓ The left-hand column of the object code listing gives the compiler-generated card number associated with the instruction. It is card 69. As seen in the source listing, card 69 contains the COMPUTE statement.

Ⓔ The DTF for FILE-1 procedes the DTF for FILE-2.

Ⓕ DTFADR CELLS begin at relative location 5B0.

Ⓖ Since the relocation factor is 32A0, the DTRADR CELLS begin at location 3850 in the dump.

Ⓗ The DTF for FILE-1 begins at location 33F8, and the DTF for FILE-2 begins at location 3470.

Since the problem program in Figure A-3 interrupted because of a data exception, the programmer should locate the contents of field B at the time of the interruption. This can be done as follows?

Ⓙ Locate data-name B in the glossary. It appears under the column headed SOURCE-NAME. Source-Name B has been assigned to base locator 3 (i.e., BL = 3) with a displacement of 050. The sum of the value of base locator 3 and the displacement value 50 is the address of data-name B.

Ⓚ The Register Assignment table lists the registers assigned to each base locator. Register 6 has been assigned to BL = 3.

Ⓛ The contents of the 16 general registers at the time of the interrupt are displayed at the beginning of the dump. Register 6 contains the address 00003388.

Ⓜ The location of data-name B can now be determined by adding the contents of register 6 and the displacement value 50. The result, 33D8, is the address of the leftmost byte of the 4-byte field B.

Note: Field B contains F1F2F3C4. This is external decimal representation and does not correspond to the USAGE COMPUTATIONAL-3 defined in the source listing.

Ⓝ The location assigned to a given data-name may also be found by using the BL CELLS pointer in the TGT Memory Map. Figure 10 indicates that the BL cells begin at location 3844 (add 5A4 to the load point address, 32A0, of the object module). The first four bytes are the first BL cell, the second four bytes

Figure A.3 Continued

are the second BL cell, etc. Note that the third BL cell contains the value 3388. This is the same value as that contained in register 6.

Note: Some program errors may destroy the contents of the general registers or the BL cells. In such cases, alternate methods of locating the DTF's are useful.

Errors That Can Cause the Dump

A dump may be caused by one of many different types of errors. Several of these errors may occur at the COBOL language level while others can occur at job-control levels.

The following are examples of COBOL language errors that can cause a dump.

1. *A* GO TO *statement with no procedure-name following it.* This statement may have been improperly initialized with an ALTER statement, and the execution of this statement will cause an invalid branch to be taken with unpredictable results.
2. *Moves of arithmetic calculations that have not been properly initialized.* For example, neglecting to initialize the object of an OCCURS clause with the DEPENDING ON option, referencing data fields prior to the first READ statement may cause a program interrupt or dump.
3. Invalid data placed in a numeric field as a result of redefinition.
4. Input/output errors that are nonrecoverable.
5. An input file contains invalid data, such as blanks or partially blank numeric fields or data incorrectly specified by its data description.

The compiler does not generate a test to check the sign position for a valid configuration before the item is used as operand. The programmer must test for valid data by means of the class test and by using either the EXAMINE or TRANSFORM statement to convert it to valid data.

For example, if the high-order positions of a numeric data field contains blanks and is to be involved in a calculation requiring a numeric PICTURE, the blank positions could be transformed to zeros through the use of the TRANSFORM or EXAMINE verbs, thus creating a valid numeric field.

Locating Data in a Dump

The location assigned to a given data-name may be found by using the BL number and displacement given for that entry in the glossary and then locating the appropriate BL in the TGT. The hexadecimal sum of the glossary displacement and the contents of the cell should give the relative address of the desired area. This can be converted to an absolute address.

A programmer using the COBOL compiler has several methods available to him for testing his programs, debugging them, and revising them for increased efficiency in operation.

The COBOL debugging language can be used by itself or in conjunction with other COBOL statements. A dump can also be used for program checkout.

THE DEBUG LANGUAGE

The COBOL debugging language is designed to aid the COBOL programmer in producing an error-free program in the shortest possible time. The sections that follow discuss the use of the debug language and other methods of program checkout.

The three debug language statements are TRACE, ON, and EXHIBIT. Any one of these statements can be used as often as necessary. They can be interspersed throughout a COBOL source program, or they can be in a packet in the input stream to the compiler.

Program checkout may not be desired after testing is completed. A debug packet can be removed after testing. This allows elimination of the extra object program coding generated for the debug statements.

The output produced by the TRACE and EXHIBIT statements is listed on the system logical output device.

The following discussions describe ways to use the debug language.

Following the Flow of Control

The READY TRACE statement causes each section and paragraph-name (or number) to be listed on the system output unit when control passes to that point. The output appears as a list of unqualified procedure names.

To reduce the number of names that are generated and the time taken to generate them, a trace can be stopped with a RESET TRACE statement. The READY TRACE/RESET TRACE combination is helpful in examining a particular area of the program. The READY TRACE statement can be coded so that the trace begins before control passes to that area. The RESET TRACE statement can be coded so that the trace stops when the program has passed the area. The two TRACE statements can be used together where the flow of control is difficult to determine, e.g., with a series of PERFORM statements or with nested conditionals.

Trace Statement

The format of the TRACE statement is

```
READY }
RESET }   TRACE
```

After a READY TRACE statement is executed, a message is written each time execution of a paragraph or section begins. The READY TRACE statement is placed where the trace is to begin.

The execution of a RESET TRACE statement terminates the functions of a previous READY TRACE statement. The RESET TRACE statement is placed in the location where the trace is to terminate.

Compile-Time Debugging Packet

Debugging statements for a given paragraph or section in a program may be grouped together into a debugging packet. These statements will be compiled with the source language program, and will be executed at object time. Each packet refers to a specified paragraph-name or section-name in the Procedure Division.

Each compile-time debug packet is headed by the control card DEBUG. The general form of this card is

Card Column	Card Column
1	8
DEBUG	location

where the parameters are described as follows:

Location

Location is the COBOL section-name or paragraph-name (qualified, if necessary) indicating the point in the program at which the packet is to be executed. Effectively, the statements in the packet are executed as if they were physically placed in the source program following the section-name or paragraph-name, but preceding the text associated with the name. The same *location* must not be used in more than one DEBUG control card. *Location* may not be a paragraph-name within the DEBUG packet itself.

(*Note:* Location can start anywhere within Margin A.)

A debug packet may consist of any procedural statements conforming to the requirements of COBOL. A GO TO, PERFORM, or ALTER statement in a debug packet may refer to a procedure-name in any debug packet or in the main body of the Procedure Division.

Another way to control the amount of tracing so that it is done conditionally is to use the ON statement with the TRACE statement. When the COBOL compiler encounters an ON statement, it sets up a mechanism, such as a counter which is incremented during execution whenever control passes through the ON statement. For example, if an error occurs when a specific record is processed, the ON statement can be used to isolate the problem record. The statement should be placed where control passes only once for each record that is read. When the contents of the counter equal the number of the record (as specified in the ON statement), a trace can be taken on that record.

ON (Count-Conditional Statement)

The ON statement is a conditional statement. It specifies when the statements it contains are to be executed (fig. A.4). ELSE (or OTHERWISE) NEXT SENTENCE may be omitted if it immediately precedes the period for the sentence. All integers contained in the statement must be positive.

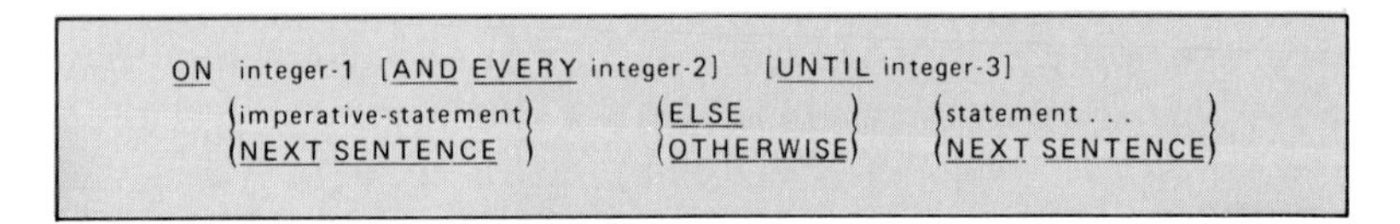

ON integer-1 [AND EVERY integer-2] [UNTIL integer-3]
{imperative-statement} {NEXT SENTENCE}
{ELSE} {OTHERWISE}
{statement . . .} {NEXT SENTENCE}

Figure A.4 Format On Statement.

The count-condition (Integer-1 AND EVERY Integer-2 UNTIL Integer-3) is evaluated as follows:

Each ON statement has a compiler-generated counter associated with it. The counter is initialized in the object program with a value of zero.

Each time the path of program flow reaches the ON statement, the counter is advanced by 1. Where *m* is any positive integer, if the value of the counter is equal to Integer-1 + (m * Integer-2), but is less than Integer-3 (if specified), then the imperative statements (or NEXT SENTENCE) are executed. Otherwise, the statements after ELSE (or NEXT SENTENCE) are executed. If the ELSE option does not appear, the next sentence is executed.

If Integer-2 is not given, but Integer-3 is given, it is assumed that Integer-2 has a value of 1. If Integer-3 is not given, no upper limit is assumed for it.

If neither Integer-2 nor Integer-3 is specified, the imperative statements are executed only once.

Examples: ON 2 AND EVERY 2 UNTIL 10 DISPLAY A ELSE DISPLAY B.

On the second, fourth, sixth, and eighth times, A is displayed. B is displayed at all other times.

ON 3 DISPLAY A.

On the third time through the count-conditional statement, A is displayed. No action is taken at any other time.

The following example shows a way in which the 200th record could be selected for a TRACE statement.

```
Col.
1         Area A
-------------------------------------------------------------------
          RD-REC.
            .
            .
            .
DEBUG     RD-REC
          PARA-NM-1.      ON 200 READY TRACE.
                          ON 201 RESET TRACE.
```

If the TRACE statement were used without the ON statement, every record would be traced.

An example of a common program error is failing to break a loop or unintentionally creating a loop in the program. If many iterations of the loop are required before it can be determined that there is a program error, the ON statement can be used to initiate a trace only after the expected number of iterations has been completed. (*Note:* If an error occurs in an ON statement, the diagnostic may refer to the previous statement number.)

Displaying Data Values During Execution

A programmer can display the value of a data item during program execution by using the EXHIBIT statement (see fig. A.5). The three forms of this statement display: (1) the names and values of the data-names listed in the EXHIBIT statement (EXHIBIT NAMED) whenever the statement is

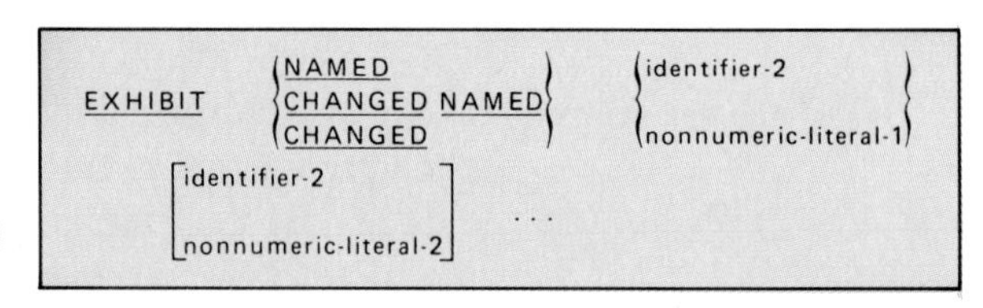

Figure A.5 Format Exhibit Statement.

encountered during execution; (2) the values of the data-names listed in this statement only if the value has changed since the last execution (EXHIBIT CHANGED); and (3) the data-names listed in the statement and the value of the data-names only if the values have changed since the previous execution (EXHIBIT CHANGED NAMED).

Exhibit Statement

The execution of an EXHIBIT NAMED statement causes a formatted display of the data-names (or nonnumeric literals) listed in the statement (see

Data values can be used to check the accuracy of the program. For example, using EXHIBIT NAMED, the programmer can display specified fields from records, compute the calculations himself, and compare his calculations with the output from his program. The coding for a payroll problem might be:

```
Col.
1          Area A
           .
           .
           .
           GROSS-PAY-CALC.
               COMPUTE GROSS-PAY =
               RATE-PER-HOUR * (HRSWKD
               + 1.5 * OVERTIMEHRS).
           NET-PAY-CALC.
           .
           .
DEBUG      NET-PAY-CALC
           SAMPLE-1. ON 10 AND
               EVERY 10 EXHIBIT NAMED
               RATE-PER-HOUR, HRSWKD,
               OVERTIMEHRS, GROSS-PAY.
```

This coding will cause the values of the four fields to be listed for every tenth data record before net pay calculatioms are made. The output could appear as:

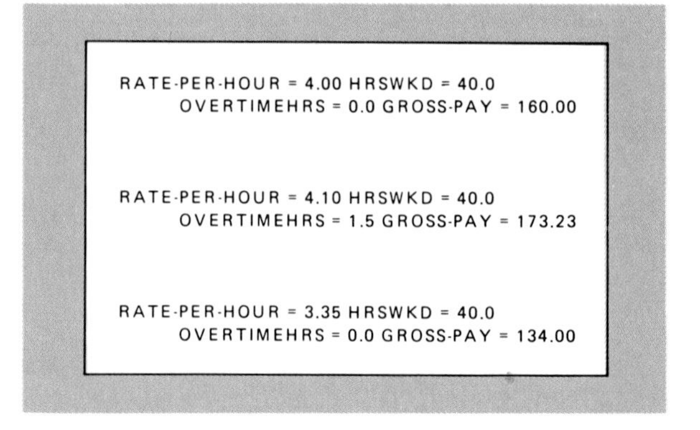

RATE-PER-HOUR = 4.00 HRSWKD = 40.0
OVERTIMEHRS = 0.0 GROSS-PAY = 160.00
RATE-PER-HOUR = 4.10 HRSWKD = 40.0
OVERTIMEHRS = 1.5 GROSS-PAY = 173.23
RATE-PER-HOUR = 3.35 HRSWKD = 40.0
OVERTIMEHRS = 0.0 GROSS-PAY = 134.00

Note: Decimal points are included in this example for clarity, but actual printouts depend on the data description in the program.

Figure A.6 Examples—Exhibit Named Statement.

fig. A.6). The format of the output for each data-name listed in the NAMED or CHANGED NAMED form of an EXHIBIT statement is:

original data-name (including qualifiers, if written)
blank
equal sign
blank
value of data-name
blank

Literals listed in the statement are preceded by a blank, when displayed.

The CHANGED form of the EXHIBIT statement provides for a display of items when they change value, compared to the value at the previous time the

```
EXHIBIT CHANGED STATE CITY RATE
```

The output from the EXHIBIT statement with the CHANGED option could appear as:

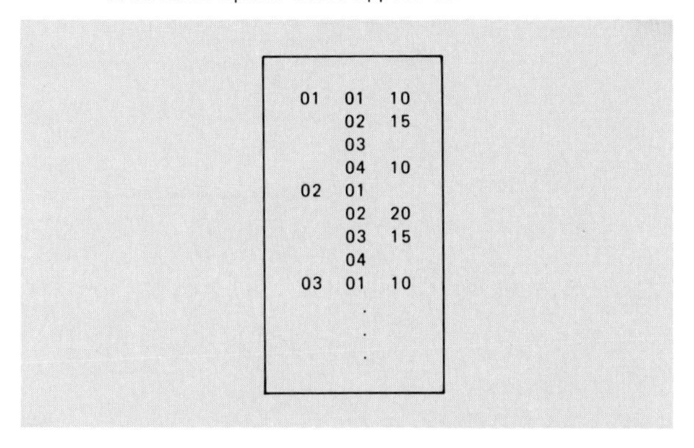

```
01  01  10
    02  15
    03
    04  10
02  01
    02  20
    03  15
    04
03  01  10
    .
    .
    .
```

Figure A.7 Example—Exhibit-Changed Statement.

The first column contains the code for a state, the second column contains the code for a city, and the third column contains the code for the postage rate. The value of a data-name is listed only if it has changed since the previous execution. For example, since the postage rate to city 02 and city 03 in state 01 are the same, the rate is not printed for city 03.

EXHIBIT CHANGED statement was executed (see fig. A.7). The initial time such a statement is executed, all values are considered changed; they are displayed and saved for purposes of comparison.

Note that if two distinct EXHIBIT CHANGED data-name statements appear in a program, changes in *data-name* are associated with the two separate statements. Depending on the path of program flow, the values of *data-name* saved for comparison may differ for the two statements.

If the list of operands in an EXHIBIT CHANGED statement includes literals, they are printed as remarks and are preceded by a blank. A check of any unusual conditions can be made by using various combinations of COBOL statements in the debug packet. For example:

```
IF OVERTIMEHRS GREATER THAN 2.0 EXHIBIT NAMED PAYRCDHRS.
```

In connection with the previous example, this statement could cause the entire pay record to be displayed whenever an unusual condition (overtime exceeding two hours) is encountered.

The EXHIBIT CHANGED statement also can be used to monitor conditions that do not occur at regular intervals. The values of data-names are listed only if the value has changed since the last execution of the statement. For example, suppose the program calculates postage rates to various cities. The flow of the program might be:

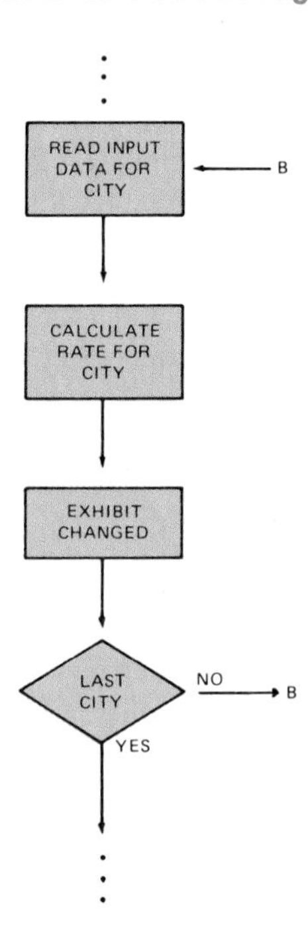

The EXHIBIT CHANGED NAMED statement lists the data-name and the value of that data-name if the value has changed. For example, the program might calculate the cost of various methods of shipping to different cities. After the calculations are made, the figure A.8 statement could be in the program. Note that a data-name and its value are listed only if the value has changed since the previous execution.

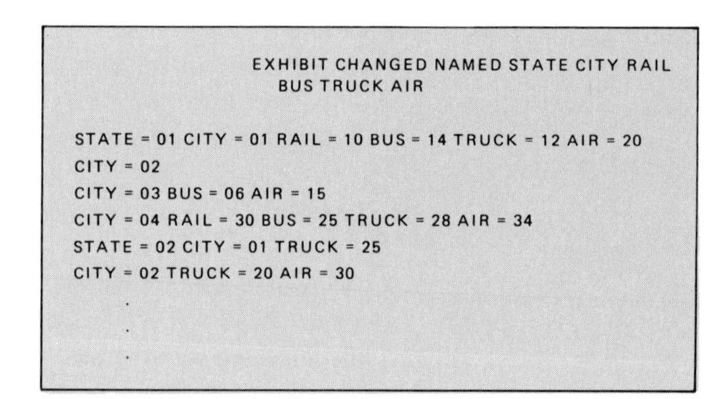

```
EXHIBIT CHANGED NAMED STATE CITY RAIL
   BUS TRUCK AIR

STATE = 01 CITY = 01 RAIL = 10 BUS = 14 TRUCK = 12 AIR = 20
CITY = 02
CITY = 03 BUS = 06 AIR = 15
CITY = 04 RAIL = 30 BUS = 25 TRUCK = 28 AIR = 34
STATE = 02 CITY = 01 TRUCK = 25
CITY = 02 TRUCK = 20 AIR = 30
   .
   .
   .
```

Figure A.8
Example—Exhibit Changed Named Statement.

Testing a Program Selectively

A debug packet allows the programmer to select a portion of the program for testing. The packet can include test data and can specify operations the programmer wants to be performed. When the testing is completed, the packet can be removed. The flow of control can be selectively altered by the inclusion of debug packets, as illustrated in the example of selective testing of B (see fig. A.9).

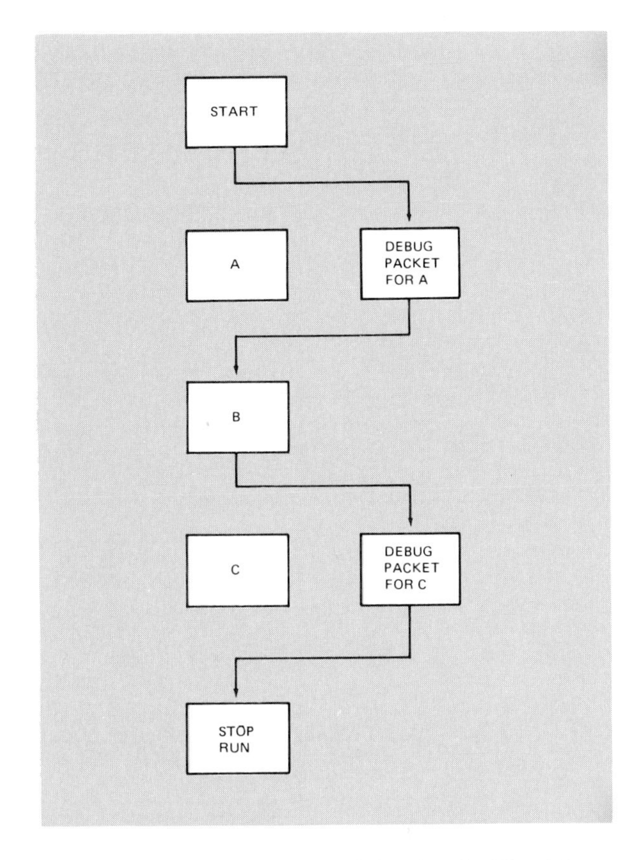

Figure A.9
Example—
Selective Testing.

In this program, A creates data, B processes it, and C prints it. The debug packet for A simulates test data. It is first in the program to be executed. In the packet, the last statement is GO TO B, which permits A to be bypassed. After B is executed with the test data, control passes to the debug packet for C, which contains a GO TO statement that transfers control to the end of the program, bypassing C.

TESTING CHANGES AND ADDITIONS TO PROGRAMS

If a program runs correctly, but changes or additions can make it more efficient, a debug packet can be used to test changes without modifying the original source program.

If the changes to be incorporated are in the middle of a paragraph, the entire paragraph, with the changes included, must be written in the debug packet. The last statement in the packet should be a GO TO statement that transfers control to the next procedure to be executed.

There are usually several ways to perform an operation. Alternative methods can be tested by putting them in debug packets.

The source program library facility can be used for program checkout by placing a source program in a library (see "Libraries"). Changes or additions to the program can be tested by using the BASIS card and any number of INSERT and DELETE cards (see fig. A.10). Such changes or additions remain in effect only for the duration of the run.

A debug packet can also be used in conjunction with the BASIS card to debug a program or to test deletions or additions to it. The debug packet is inserted in the input stream immediately following the BASIS card and any INSERT or DELETE cards.

```
000730              IF ANNUAL-PAY GREATER THAN 9000  GO TO PAY-WRITE.
000735              IF ANNUAL-PAY GREATER THAN 9000 - BASE-PAY GO TO LAST-FICA.
000740  FICA-PAYR.  COMPUTE FICA-PAY = BASE-PAY * .052
000745                      MOVE FICA-PAY TO OUTPUT-FICA.
000750  PAY-WRITE.          MOVE BASE-PAY TO OUTPUT-BASE.
000755                      ADD BASE- PAY TO ANNUAL-PAY.
  .                 .
  .                 .
  .                 .
000850              STOP RUN.
```

Sample Coding to Calculate FICA

```
// JOB PGM2
// OPTION LOG,DECK,LIST,LISTX,ERRS
// EXEC FCOBOL
 CBL QUOTE
BASIS PAYROLL
DELETE 000730,  000735
              IF OCCUPATION-CODE = "DR" PERFORM PAY-INCREASE THRU EX1.
INSERT  000850
        PAY-INCREASE.   MULTIPLY 1.05 BY BASE-PAY.
        EX1.            EXIT.
```

Altering a Program from the Source Statement Library Using INSERT and DELETE Cards

```
              IF OCCUPATION-CODE = "DR" PERFORM PAY-INCREASE THRU EX1.
000740  FICA-PAYR.  COMPUTE FICA-PAY = BASE-PAY * .052
000745            MOVE FICA-PAY TO OUTPUT-FICA.
000750  PAY-WRITE.  MOVE BASE-PAY TO OUTPUT-BASE.
000755            ADD BASE-PAY TO ANNUAL-PAY.
.                 .
.                 .
.                 .
000850            STOP RUN.
        PAY-INCREASE. MULTIPLY 1.05 BY BASE-PAY.
        EX1.          EXIT.
```

Effect of INSERT and DELETE Cards

Figure A.10 Example—Insert and Delete Cards.

Appendix B Job-Control Language

INTRODUCTION

Job-Control Cards (JCL) establish the communication link between the COBOL programmer and the control system of the computer. The control system consists of a number of processing programs and a control program. The processing programs will include the COBOL compiler, service programs, as well as any user-written programs.

The control program supervises the execution and loading of the processing programs; controls the location, storage, and retrieval of data; and schedules the jobs for continuous processing of problem programs.

EXECUTING A COBOL PROGRAM

The basic operations to be performed to execute a COBOL program are:

A. *Compilation.* The process of translating a COBOL source program into a series of instructions comprehensible to the computer. In computer terminology, the input (source program) to the compiler is called the *source module.* The output (compiled source program) from the compiler is called the *object module.*

B. *Linkage Editing.* The Linkage Editor is a source program that prepares object modules for execution. It can also be used to combine two or more separately compiled object modules into a format suitable for execution as a single program. The executed output of the Linkage Editor is called a *load module.* The Linkage Editor may also combine previously edited load modules with or without one or more object modules to form one load module.

C. *Loading.* The loader is a service program that processes COBOL object and load modules, resolves any references to subprograms, and executes the loaded module. All these functions are specified in one step.

D. *Execution.* Actual execution is under the supervision of the control program, which obtains a load module, loads it into main storage, and initiates execution of the machine language instructions contained in the load module.

(See figure B.1.)

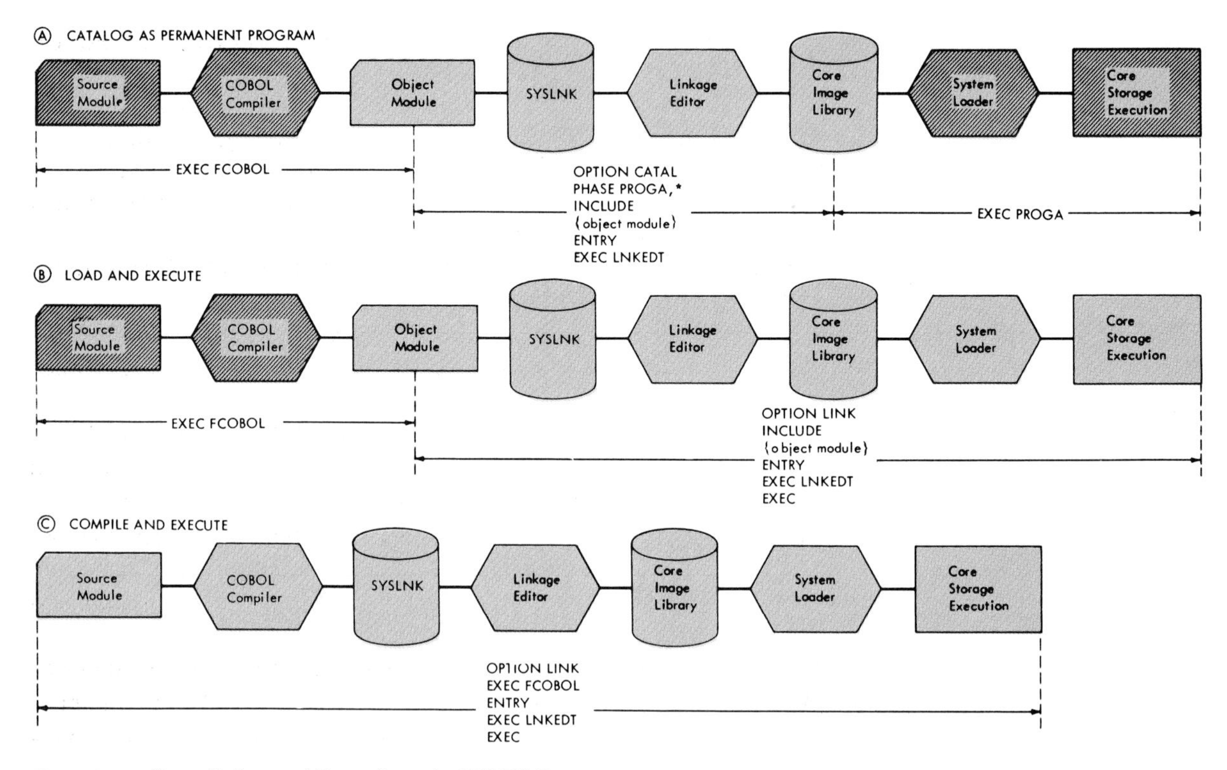

Figure B.1 Compilation and Execution of a COBOL Program.

The JOB statement identifies the beginning of a job and the job to be performed. It may be also used by the installation's accounting routines.

The EXEC statement describes the job step and calls for execution.

Sample DOS JCL Statements

All JCL cards start in card column 1. Slashes in column 1 and 2 (//) identify the cards to the system as JCL, and not program or data cards. If commas are indicated in certain JCL statements, they must be included.

```
 1.  // JOB EXAMPLE
 2.  // OPTION LINK
 3.  // EXEC FCOBOL
 4.     Source Program Cards
 5.  /*
 6.  // EXEC LNKEDT
 7.  // EXEC
 8.     Data Cards
 9.  /*
10.  /&
```

Card 1. The JOB statement identifies the program to the system. There must be at least one space before and after JOB. Name is limited to eight characters or less and need not be related to the PROGRAM-ID.

Card 2. The OPTION LINK card may vary from installation to installation depending upon the needs and requirements of the programmer. For example, if the program is unexecutable, a dump of storage may be requested.

Card 3. The EXEC FCOBOL card is calling for the execution of the COBOL compiler to translate the source program into the object module.

Card 4. The punched source program cards would be inserted here.

Card 5. The delimiter (/*) signals the end of the source program and separates the source program from subsequent JCL statements.

Card 6. The execution of the EXEC LNKEDT statement will prepare the object module for execution.

Card 7. The EXEC statement controls the execution of the machine language instructions contained in the load module upon the data deck which follows.

Card 8. The data deck containing the source data to be processed.

Card 9. Another delimiter (/*) to signal the end of the data deck and separate the data from any subsequent JCL statements.

Card 10. The delimiter (/&) to signal the end of the program and to separate the executed program from any subsequent programs. An important note to remember is that the omission of this delimiter may cause two programs to be executed together.

It is beyond the scope of this text to discuss the various combinations of JCL cards permissible. There are many options available to the programmer, and they should be carefully studied in the programmer's guide reference manual for the particular computer. The foregoing is a typical example of the use and purpose of JCL cards. Following is a brief explanation of the format, use, and purpose of IBM DOS JCL cards.

JOB CONTROL LANGUAGE (DOS) (IBM)

Job control statements prepare the system for the execution of COBOL programs.

(See figure B.2.)

```
// JOB PROG1
.
.
.
// EXEC FCOBOL
   {source deck - main program}
/*
.
.
.
// EXEC FCOBOL
   {source deck - first subprogram}
/*
.
.
.
// EXEC FCOBOL
   {source deck - second subprogram}
/*
.
.
.
// EXEC LNKEDT
.
.
.
// EXEC
```

Figure B.2
Sample Structure of Job Deck for Compiling, Link Editing and Executing a Main Program and Two Subprograms.

Job Control Statements

Job control statements are designed for an eighty-column card format. The statements are punched into cards in essentially free form, but certain rules must be observed, as follows:

1. *Name.* Two slashes (//) identify the statement as a job-control statement. These punches must be in columns one and two of the card followed immediately by at least one blank.

   Exception: The end-of-job statement delimeter contains /& in columns one and two; the end-of-data statement delimeter contains /* in columns one and two, and the comments statement contains an * in column one and a blank in column two.
2. *Operation.* This identifies the operation to be performed. It can be up to eight characters long with at least one blank following its last character.
3. *Operand.* This may be a blank or may contain one or more entries separated by commas. The last term must be followed by a blank, unless its last character is in column 71.
4. *Comments.* Optional programmer comments must be separated from the operand by at least one space.

All JCL cards are read by the systems input device specified by the symbolic name SYSRDR.

Comment Statements

Comment statements (statements preceded by an asterisk in column one and followed by a blank) may be placed anywhere in the job deck. They may contain any character and are usually for communication with the operator; accordingly, they are written on the console printer keyboard as well as on the printer.

JOB DEFINITION

A *JOB* is a specified unit of work to be performed under the control of the operating system. A typical job might be the processing of a COBOL program—compiling the source program, editing the module to form a phase, and then executing the phase.

A *JOBSTEP* is exactly what the name implies—one step in the processing of the job. Thus, in the JOB mentioned above, one job step is the compilation of the source statements, another is the line editing of the module, and the other is the execution of the phase. A compilation requires the execution of the COBOL compiler, an editing process implies the execution of the Linkage Editor, and, finally, the execution phase is the execution of the problem program itself.

COMPILATION OF JOB STEPS

The compilation of a COBOL program may necessitate more than one job step (more than one execution of the COBOL compiler). In some cases, a COBOL program consists of a main program and one or more subprograms. To compile such a program, a separate job step must be specified for the main program and for each of the subprograms. Thus the COBOL compiler is executed once for the main program and once for each subprogram. Each execution produces a module. The separate modules can be then combined into one phase by a single job step—the execution of the Linkage Editor.

There are four job-control statements that are used for job definition: the JOB statement, the EXEC statement, end-of-data statement (/*), and the end-of-job statement (/&). These are optional job-control statements that may be used to specify specific JCL functions.

The *JOB* statement defines the start of a job. One JOB statement is required for every job; it must be the first statement in the job deck. The programmer must name his job on the JOB statement.

The *EXEC* statement requests the execution of a program. Therefore, one EXEC statement is required for each job step within a job stream. The EXEC statement identifies the program that is to be executed (for example, the COBOL compiler, the Linkage Editor).

The end-of-data statement, also referred to as the slash asterisk (/*) statement, defines the end of a programmers input data. The slash asterisk statement immediately follows the input data. For example, COBOL source statements would be placed immediately after the EXEC statement for the COBOL compiler; a /* statement would follow the last COBOL source statement. If input data is kept separate, the /* immediately follows each set of input data.

The end-of-job statement, also referred to as slash ampersand (/&) statement, defines the end of the job. A /& statement must appear as the last statement in the job deck. If this statement is omitted, the preceding job would be combined with this job. (See figure B.3.)

COMPILATION

Compilation is the execution of the COBOL compiler. The programmer requests compilation by placing in the job deck an EXEC statement specifying the name of the COBOL compiler. Input to the compiler is a set of COBOL source statements consisting of either a main program or a subprogram.

Output from the COBOL compiler is dependent upon the options specified. This output may include a listing of source statements exactly as they appear in the input deck. Separate data and/or Procedure Division maps, a symbolic cross-reference list, and diagnostic messages can also be produced. The format of the compiler output is described and illustrated in the "Debugging" section.

The programmer can override any of the compiler options specified when the system was generated or can include some not specified by specifying the OPTION control statement in the compiler job step.

EDITING

Editing is the execution of the Linkage Editor. The programmer requests editing by placing in the job deck an EXEC statement that contains the name LNKEDT, the name of the Linkage Editor.

Output from the Linkage Editor consists of one or more phases. A phase may be an entire program or it may be part of an overlay structure (multiple phases).

Statement	Function
// ASSGN	Input/output assignments.
// CLOSE	Closes a logical unit assigned to magnetic tape.
// DATE	Provides a date for the Communication Region.
// DLAB	Disk file label information.
// DLBL	Disk file label information.
// EXEC	Execute program.
// EXTENT	Disk file extent.
// JOB	Beginning of control information for a job.
// LBLTYP	Reserves storage for label information.
// LISTIO	Lists input/output assignments.
//MTC	Controls operations on magnetic tape.
// OPTION	Specifies one or more job control options.
// PAUSE	Creates a pause for operator intervention.
// RESET	Resets input/output assignments to standard assignments.
// RSTRT	Restarts a checkpointed program.
// TLBL	Tape label information.
// TPLAB	Tape label information.
// UPSI	Sets user-program switches.
// VOL	Disk/tape label information.
// XTENT	Disk file extent.
/*	End-of-data-file or end-of-job-step.
/&	End-of-job.
*	Comments.

Figure B.3 Job Control Statement.

A phase produced by the Linkage Editor can be executed immediately after it is produced (that is, in the job step immediately following the Linkage Editor), or it can be executed later, either in a subsequent job step of the same job or a subsequent job step.

PHASE EXECUTION

Phase execution is the execution of the problem program—for example, the program written by the COBOL programmer. If the program is an overlay structure (multiple phases), the execution job step actually involves the execution of all phases in the program.

The programmer requests the execution of a phase by placing in the job deck an EXEC statement that specifies the name of the phase. However, if the phase to be executed was produced in the immediately preceding job step, it is not necessary to specify its name in the EXEC statement.

SEQUENCE OF JOB-CONTROL STATEMENTS

The job deck for a specific job always begins with a JOB statement and ends with a /& (end-of-job) statement. A specific job consists of one or more job steps. The beginning of a job step is indicated by the appearance of an EXEC statement. When an EXEC statement is encountered, it initiates the execution of the job step, which includes all preceding control statements up to but not including a previous EXEC statement.

ASSGN Statements

The ASSGN control statement assigns a logical input/output unit to a physical device. An ASSGN control statement must be present in the job deck for each data file assigned to an external storage device in the COBOL program where these assignments differ from those established at system generation time. Data files are assigned to programmer logical units in COBOL by means of the source language ASSGN clause. The ASSGN control statement may also be used to change a system standard assignment for the duration of the job. Device assignments made by the ASSGN statement are considered temporary until another ASSGN statement appears.

JOB CONTROL LANGUAGE (OS) (IBM)

The job control language statements for an operating system (OS) are similar to those for a DOS system. The functions and rules for punching these statements are the same. There are some additional statements required in an OS system.

The types of job control statements used to compile, linkage edit, and execute programs in an OS environment are:

Statement	Function
JOB	Indicates the beginning of a new job and describes that job.
EXEC	Indicates a job step and describes that job step; indicates the load module or catalogued procedure to be executed.
DD	Describes data sets, and controls device and volume assignment.
delimiter	Separates data sets in the input stream from control statements; it must follow each data set that appears in the input stream, e.g., after a COBOL source module punched deck.
comment	Contains miscellaneous remarks and notes written by the programmer; it may appear anywhere in the job stream after the JOB statement.

The general format of control statements are as indicated in figure B.4.

The nine job control language statements used to describe a job to the system are:

1. Job (JOB) statement.
2. Execute (EXEC) statement.
3 Data definition (DD) statement.
4. Delimiter statement.

Statement	Columns 1	2	3	4 Fields
Job	/	/	name	JOB operand[1] comments[1]
Execute	/	/	name[1]	EXEC operand comments[1]
Data Definition	/	/	name[1]	DD operand comments[1]
Procedure	/	/	name[1]	PROC operand comments[1]
Command	/	/		operation(command) operand comments[1]
Delimiter	/	*		comments[1]
Null	/	/		
Comment	/	/	*	comments
Pend	/	/	name[1]	PEND

[1]Optional.

Figure B.4 General Format of Job Control Statement.

5. Null statement.
6. Procedure (PROC) statement.
7. Procedure end (PEND) statement.
8. Comment statement.
9. Command statement.

A job control statement consists of one or more 80-byte records. Most jobs are submitted to the operating system for execution in the form of 80-column punched cards or as card images off direct access devices. The operating system is able to distinguish a job control statement from data included in the input stream. In columns 1 and 2 of all the statements except the delimiter statement, a // is coded. For the delimiter statement, a /* is coded in columns 1 and 2 and this notifies the operating system that the statement is a delimiter statement (see fig. B.5). For a comment statement, a //* is coded in columns 1, 2, and 3 respectively.

```
//JOB1       JOB
//STEP1      EXEC  PGM=IKFCBL00,PARM=DECK
//SYSUT1     DD    DSNAME=&&UT1,UNIT=SYSDA,SPACE=(TRK,(40))
//SYSUT2     DD    DSNAME=&&UT2,UNIT=SYSSQ,SPACE=(TRK,(40))
//SYSUT3     DD    DSNAME=&&UT3,UNIT=SYSSQ,SPACE=(TRK,(40))
//SYSUT4     DD    DSNAME=&&UT4,UNIT=SYSSQ,SPACE=(TRK,(40))
//SYSPRINT   DD    SYSOUT=A
//SYSPUNCH   DD    SYSOUT=B
//SYSIN      DD    *
     (source deck)
/*
```

Figure B.5 Job Control Procedure.

Parameters coded on these JCL statements help the job scheduler to regulate the execution of jobs and job steps, retrieve and dispose of data, allocate I/O resources, and communicate with the operator.

JOB Statement

The job statement indicates to the system at what point a job begins (fig. B.6). The name of the job is coded and is used to identify messages to the operator and to identify the program output. Additional information, such as accounting information, conditions for early termination of a job, job priority, and maximum amount of time, can be specified in parameters on the card (fig. B.7).

EXEC Statement

The EXEC statement marks the beginning of a job step and the end of the preceding step (fig. B.8). The program to be executed is identified or the

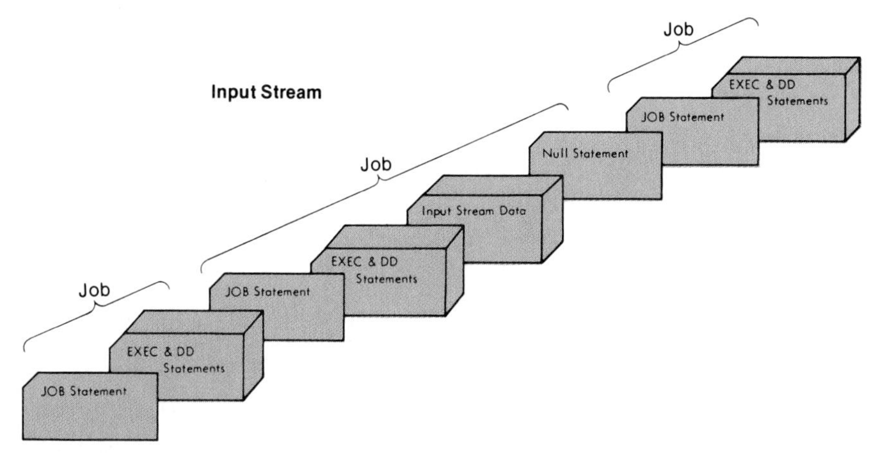

Figure B.6 Defining Job Boundaries.

```
|Name        |Operation|                          Operand
|            |         |
|            |         |           Positional Parameters
|            |         |
|//jobname   |JOB      |[([account-number] [,accounting-information])¹ ² ³]
|            |         |
|            |         |[,programmer-name]⁴ ⁵
|            |         |
|            |         |           Keyword Parameters
|            |         |
|            |         |[MSGLEVEL=(x,y)]⁶
|            |         |[TIME=(minutes,seconds)]
|            |         |[CLASS=jobclass]
|            |         |[COND=((code,operator) [,(code,operator)]...⁷)⁸]
|            |         |[PRTY=job priority]
|            |         |[MSGCLASS=classname]
|            |         |[REGION=(nnnnnxK[,nnnnnyK])]
|            |         |[ROLL=(x,y)]
|            |         |[TYPRUN=HOLD]
|            |         |[RD=request]
|            |         |           (*                       )
|            |         |[RESTART=( {stepname                 }  [,checkid])]
|            |         |           (stepname.procstepname   )

¹If the information specified (account-number and/or accounting-information) contains
 blanks, parentheses, or equal signs, the information must be delimited by single
 quotation marks instead of parentheses.
²If only account-number is specified, the delimiting parentheses may be omitted.
³The maximum number of characters allowed between the delimiting quotation marks is
 142.
⁴If programmer-name contains any special characters other than the period, it must be
 enclosed within single quotation marks.
⁵The maximum number of characters allowed for programmer-name is 20.
⁶x = 0, 1, or 2 is the JCL message.
 y = 0 or 1 is the allocation message level.
 Note that the value 1 may be used in place of (1,1).
⁷The maximum number of repetitions allowed is 7.
⁸If only one test is specified, the outer pair of parentheses may be omitted.
```

Figure B.7 Job Statement.

catalogued procedure or in-stream procedure is called. A catalogued procedure is a set of job control language statements that has been assigned a name and placed in a partitioned data set known as the procedure library.

The EXEC statement may also be used to provide job step accounting information, to give conditions for bypassing or executing a job step, etc.

DD Statement

A DD statement identifies a data set and describes its attributes (fig. B.9). There must be a DD statement for each data set used or created in a job step.

```
Name              Oper-   Operand
                  ation
                                        Positional Parameters

//[stepname]1     EXEC    { PGM=progname                        }
                          { PGM=*.stepname.ddname               }
                          { PROC=procname                       }
                          { procname                            }
                          { PGM=*.stepname.procstep.ddname      }

                                        Keyword Parameters

                          [{ACCT2          }                              3 4 5]
                          [{ACCT.procstep  } = (accounting-information)        ]

                          [{COND2          }                                                       6 7]
                          [{COND.procstep  } = ((code,operator[,stepname[.procstep]])...)             ]

                          [{PARM2          }                        3 8 9]
                          [{PARM.procstep  } = (option[,option]...)      ]

                          [{TIME           }                      ]
                          [{TIME.procstep  } = (minutes,seconds)  ]

                          [{REGION         }                      ]
                          [{REGION.procstep} = nnnnnxK[,nnnnnyK]  ]

                          [{ROLL           }         ]
                          [{ROLL.procstep  } = (x,y) ]

                          [{RD             }           ]
                          [{RD.procstep    } = request ]

                          [{DPRTY          }                     ]
                          [{DPRTY.procstep } = (value 1, value 2)]

1Stepname is required when information from this control statement is referred to in a
 later job step.
2If this format is selected, it may be repeated in the EXEC statement once for each
 step in the cataloged procedure.
3If the information specified contains any special characters except hyphens, it must
 be delimited by single quotation marks instead of parentheses.
4If accounting-information contains any special characters except hyphens, it must be
 delimited by single quotation marks.
5The maximum number of characters allowed between the delimiting quotation marks or
 parentheses is 142.
6The maximum number of repetitions allowed is 7.
7If only one test is specified, the outer pair of parentheses may be omitted.
8If the only special character contained in the value is a comma, the value may be
 enclosed in quotation marks.
9The maximum number of characters allowed between the delimiting quotation marks or
 parentheses is 100.
```

Figure B.8 Exec Statement.

The DD statements are placed after the EXEC statement for the step. The DD statement provides such information as the name of the data set, the name of the volume on which it resides, the type of I/O device that holds the data set, the format of the records in the data set, whether the data set is old or new, the size of newly created data sets, and the method that will be used to create or access a data set. The name of the DD statement provides a symbolic link between the data set (on data file) named in the program and the actual name and location of the corresponding data set. This symbolic link allows one to relate the data set in the program to different data sets on different occasions.

Delimiter and Null Statements

The delimiter statement (or /* statement) and the null statement (or // statement) are markers in an input stream. The delimiter statement is used to separate data placed in the input stream from any JCL statement that may follow the data. The null statement can be used to mark the end of the JCL statements and data for a job.

```
Name                          | Operation   | Operand
// {ddname          } 1       | DD          | (see below and next page)
   {procstep.ddname }         |             |

Operand2

        Positional Parameters

  [*    ] 3
  [DATA ]
  [DUMMY]

        Keyword Parameters 4 5

[DDNAME=ddname]
[                 {dsname                          }] 11
[                 {dsname(element)                 }]
[ {DSNAME}        {*.ddname                        }]
[ {DSN   }   =    {*.stepname.ddname               }]
[                 {*.stepname.procstep.ddname      }]
[                 {&&name                          }]
[                 {&&name(element)                 }]

[        [dsname                     ]                          ] 6
[ DCB=(  [*.ddname                   ]  [,subparameter-list])   ]
[        [*.stepname.ddname          ]                          ]
[        [*.stepname.procstep.ddname ]                          ]

  [SEP=(subparameter list)7] 10
  [AFF=ddname              ]

          Positional Subparameters    Keyword Subparameters

  [UNIT=(name[,[n/P][,DEFER]][,SEP=(list of up to 8 ddnames)])8] 10 12
  [UNIT=(AFF=ddname)                                           ]
                                  Positional Subparameters

          {TRK                   }
SPACE=(   {CYL                   }    ,(primary-quantity[,secondary-quantity],
          {average-record-length }

                                                          [,MXLG  ]
                 [directory- or index-quantity])[,RLSE]   [,ALX   ]   [, ROUND])
                                                          [,CONTIG]

SPACE=(ABSTR,(quantity,beginning-address[,directory- or index-quantity]))

              {CYL                   }
SPLIT=(n,     {                      }  ,(primary-quantity[,secondary-quantity]))
              {average-record-length }

              {TRK                   }
SUBALLOC=(    {CYL                   }  ,(primary-quantity[,secondary-quantity]
              {average-record-length }

                                          {ddname                    }
              [,directory-quantity]),     {stepname.ddname           } )
                                          {stepname.procstep.ddname  }

              Positional Subparameters

  {VOLUME}
  {      }=([PRIVATE],[RETAIN],[volume sequence number],[volume count])
  {VOL   }

              Keyword Subparameters

  [,SER=(volume-serial-number[volume-serial-number]9...)]
  [                                                      ]
  [          {dsname                      }              ]
  [,REF=     {*.ddname                    }              ]
  [          {*.stepname.ddname           }              ]
  [          {*.stepname.procstep.ddname  }              ]

  [                                        {NL }                                 ]
  [LABEL=([data-set-sequence-number],      {SL }  [,EXPDT=yyddd]  [,PASSWORD])   ]
  [                                        {NSL}  [,RETPD=xxxx ]                 ]
  [                                        {SUL}                                 ]
```

Figure B.9 The DD Statement 1 of 2.

Figure B.9 Continued

```
Operand2 (cont.)

 ┌           ┌ NEW ┐  ┌,DELETE  ┐ ┌,DELETE  ┐ ┐
 │           │ OLD │  │,KEEP    │ │,KEEP    │ │
 │ DISP=(    │ SHR │  │,PASS    │ │,CATLG   │ │ )
 │           └ MOD ┘  │,CATLG   │ └,UNCATLG ┘ │
 │                    └,UNCATLG ┘             │
 │ SYSOUT=classname                           │
 └ SYSOUT=(x[,program-name][,form-no.])       ┘

1The name field must be blank when concatenating data sets.
2All parameters are optional to allow a programmer flexibility in the use of the DD
 statement; however, a DD statement with a blank operand field is meaningless.
3If the positional parameter is specified, keyword parameters other than DCB cannot be
 specified.
4If subparameter-list consists of only one subparameter and no leading comma
 (indicating the omission of a positional subparameter) is required, the delimiting
 parentheses may be omitted.
5If subparameter-list is omitted, the entire parameter must be omitted.
6See "User-Defined Files" for the applicable subparameters.
7See the publication IBM System/360 Operating System: Job Control Language Reference.
8If only name is specified, the delimiting parentheses may be omitted.
9If only one volume-serial-number is specified, the delimiting parentheses may be
 omitted.
10The SEP and AFF parameters should not be confused with the SEP and AFF subparameters
 of the UNIT parameter.
11The value specified may contain special characters if the value is enclosed in
 apostrophes.  If the only special character used is the hyphen, the value need not be
 enclosed in apostrophes.  If DSNAME is a qualified name, it may contain periods
 without being enclosed in apostrophes.
12The unit address may contain a slash, and the unit type number may contain a hyphen,
 without being enclosed in apostrophes, e.g., UNIT=293/5,UNIT=2400-2.
```

PROC and PEND Statements

The PROC statement may appear as the first JCL statement in a catalogued or in-stream procedure. For catalogued procedures or in-stream procedures, the PROC statement is used to assign default values to parameters defined in a procedure (fig. B.10). An in-stream procedure is a set of job control language statements that appear in the input stream. The PROC statement is used to mark the beginning of an in-stream procedure. The PEND statement is used to mark the end of an in-stream procedure.

Figure B.10
Catalog Procedure.

In the figure at right, JOB2 is the name of the job, STEPA is the name of the single job step. The EXEC statement calls the cataloged procedure containing STEP1 to execute the job step (PROC = CATPROC).

```
//JOB2          JOB
//STEPA         EXEC    PROC=CATPROC
//STEP1.SYSIN   DD      *
    (source deck)
/*
```

Comment Statement

The comment statement can be inserted either before or after any JCL statement that follows the JOB statement, and can contain any information that would be helpful to one interested in the program.

Command Statement

The command statement is used to enter commands through the input stream. Commands can activate and deactivate system input and output units, request printouts and displays, and perform a number of other operator functions.

JOB CONTROL FIELDS

The name contains from one through eight alphanumeric characters, the first of which must be alphabetic. The name begins in card column 3. It is followed by one or more blanks. The name is used as follows:

Name Field

1. To identify the control statement to the operating system

2. To enable other control statements in the job to refer to information contained in the named statement
3. To relate DD statements to files named in a COBOL source program

Operation Field

The operation field is preceded and followed by one or more blanks. It may contain one of the following operation codes:

JOB
EXEC
DD
PROC
PEND

If the statement is a delimiter statement, there is no operation field and comments may start after one blank.

Operand Field

The operand field is preceded and followed by one or more blanks and may continue through column 71 and onto one or more continuation cards. It contains the parameters and subparameters that give required and optional information to the operating system. Parameters and subparameters are separated by commas. A blank in the operand field causes the system to treat the remaining data on the card as a comment. There are two types of parameters: positional and keyword.

Positional Parameters

Positional parameters are the first parameters in the operand field, and they must appear in the specified sequence. If a positional parameter is omitted and other positional parameters follow, the omission must be indicated by a comma. If other positional parameters do not follow, no comma is needed.

Keyword Parameters

A keyword parameter may be placed anywhere in the operand field following the positional parameters. A keyword parameter consists of a keyword, followed by an equal sign, followed by a single value or a list of parameters. If there is a subparameter list, it must be enclosed in parentheses or single quotation marks; the subparameters in the list must be separated by commas. Keyword parameters may appear in any sequence.

Comments Field

Optional comments must be separated from the last parameter (or the /* in a delimiter statement) by one or more blanks and may appear in the remaining columns up to and including column 71. An optional comment may be continued onto one or more continuation cards. Comments can contain blanks. (See figure B.11.)

Figure B.11 Example of Job Control Statement.

JOB CONTROL STATEMENTS

Card No.	Function
1	JOB card. Provides the following information to the system: A. Name of job—IW016003 B. Accounting Information—(0160,33,,,,198), the necessary information relative to the job being executed with the required positional parameters commas. C. Name of programmer—HENESYDR D. Job Priority—CLASS = I
2	EXEC card. Identifies program to be executed and the PROC statement is used to reference the catalogued procedure "COBOL."
3	DD card. Identifies data set with an asterisk (*) as following.
4	DD card. Calls for a dump of core storage in case of abnormal termination of the program during execution.
5	DD card. Identifies output is to printed for data set.
6	DD card. Identifies data set with an asterisk (*) as following.
7	Null card. Marks the end of the JCL statements and data for job.

The delimiter statement cards (/*) will appear as markers behind the input stream after card 3 and card 6.

All statements without the characters // or /* in columns 1 and 2 are generated by the system.

```
①  //IW016003 JOB (0160,33,,,,198),HENESYDR,CLASS=I              JOB  456
②  //STEP1 EXEC PROC=COBOL                                        2
        XXCOB      EXEC PGM=IKFCBL00,REGION=192K,                          00000010
        XX         PARM='LOA,NOCLI,NODMA,NOPMA,SUP,NOXRE,CSY',TIME=2       00000020
   ***                                                           00000030
   ***                                                           00000040
   ***      THIS PROCEDURE PROVIDED FOR THE USE OF LACCD STUDENTS.   00000050
   ***      IF MORE INFOMRATION IS NEEDED, CONTACT:              00000060
   ***                                                           00000070
   ***        SOFTWARE GROUP - LACCD DATA PROCESSING DIV.        00000080
   ***        2140 W. OLYMPIC BLVD, LOS ANGELES, CA 90006        00000090
   ***                                                           00000100
        XXSTEPLIB   DD DSNAME=SYS2.COBV4LIB,DISP=SHR                        00000110
        XXSYSLIB    DD DSN=INST.COPYLIB,DISP=SHR                            00000120
        XXSYSPRINT  DD SYSOUT=A                                             00000130
        XXSYSPUNCH  DD DUMMY                                                00000140
        XXSYSUT1    DD UNIT=SYSDA,SPACE=(460,(700,100))                     00000150
        XXSYSUT2    DD UNIT=SYSDA,SPACE=(460,(700,100))                     00000160
        XXSYSUT3    DD UNIT=SYSDA,SPACE=(460,(700,100))                     00000170
        XXSYSUT4    DD UNIT=SYSDA,SPACE=(460,(700,100))                     00000180
        XXSYSLIN    DD DSNAME=&LOADSET,DISP=(MOD,PASS),                     00000190
        XX          UNIT=SYSDA,SPACE=(80,(720,100))                         00000200
③  //SYSIN DD *                                                  3
    ALLOC. FOR IW016003 COB       STEP1
    155   ALLOCATED TO STEPLIB
    155   ALLOCATED TO SYSLIB
    C34   ALLOCATED TO SYSPRINT
    151   ALLOCATED TO SYSUT1
    150   ALLOCATED TO SYSUT2
    15A   ALLOCATED TO SYSUT3
    151   ALLOCATED TO SYSUT4
    150   ALLOCATED TO SYSLIN
    D11   ALLOCATED TO SYSIN
    - STEP WAS EXECUTED - COND CODE 0000
      SYS2.COBV4LIB                              KEPT
      VOL SER NOS= SPOOL1.
      INST.COPYLIB                               KEPT
      VOL SER NOS= SPOOL1.
      SYS77202.T223946.RV000.IW016003.R0004920   DELETED
```

```
          VOL SER NOS= CC3306.
          SYS77202.T223946.RV000.IW016003.R0004921        DELETED
          VOL SER NOS= CC3300.
          SYS77202.T223946.RV000.IW016003.R0004922        DELETED
          VOL SER NOS= CC3304.
          SYS77202.T223946.RV000.IW016003.R0004923        DELETED
          VOL SER NOS= CC3306.
          SYS77202.T223946.RV000.IW016003.LOADSET         PASSED
          VOL SER NOS= CC3300.

        STEP /COB       / START 77202.2239
        STEP /COB       / STOP  77202.2240 CPU    0MIN 03.52SEC STOR VIRT 100K
              XXGO          EXEC PGM=LOADER,                                              00000210
              XX            PARM='NOMAP,NOPRINT,NOXREF,LET',                              00000220
              XX            COND=(5,LT,COB),REGION=192K,TIME=1                            00000230
              XXSYSLIN DD DSNAME=*.COB.SYSLIN,DISP=(OLD,DELETE)                          00000240
              XXSYSPRINT  DD SYSOUT=A                                                    00000250
              XXSYSOUT    DD SYSOUT=A                FOR "DISPLAY" VERBS                 00000260
              XXSYSLOUT   DD SYSOUT=A                                                    00000270
              XXSYSLIB    DD DSNAME=SYS2.COBV4SUB,DISP=SHR                               00000280
              XXSORTLIB   DD DSNAME=SYS1.SORTLIB,DISP=SHR                                00000290
       ***                                                                        00000300
       ***  UPDATED 11-14-76 TO REFLECT CHANGES IN STEPLIB AND/OR SYSLIB          00000310
       ***  UPDATED 11-19-76 TO ADD AN "INSTRUCTIONAL" COPYLIB                    00000320
       ***  UPDATED 02-17-77 TO ADD A SORTLIB DD CARD                             00000330
       ***  UPDATED 04-17-77 TO ADD A "SYSOUT" DD CARD                            00000340
       ***  UPDATED 04-26-77 TO INCLUDE A TIME LIMIT ON STEP "COB"                00000341
       ***  HOWARD DEAN                                                           00000350
(4)    //GO.SYSUDUMP DD SYSOUT=A                                                  178
(5)    //GO.PRINT DD SYSOUT=A                                                     180
(6)    //GO.CARDIN DD *                                                           181
(7)    //
        ALLOC. FOR IW016003 GO        STEP1
        150    ALLOCATED TO SYSLIN
        D34    ALLOCATED TO SYSPRINT
        C35    ALLOCATED TO SYSOUT
        D36    ALLOCATED TO SYSLOUT
        155    ALLOCATED TO SYSLIB
        155    ALLOCATED TO SORTLIB
        D37    ALLOCATED TO SYSUDUMP
        C38    ALLOCATED TO PRINT
        C11    ALLOCATED TO CARDIN
IEC020I 001-5,IW016003,GO,CARDIN,D11
IEC020I GET OR READ ISSUED AFTER END-OF-FILE
IEW1991 ERROR - USER PROGRAM HAS ABNORMALLY TERMINATED
COMPLETION CODE - SYSTEM=001  USER=0000
          SYS77202.T223946.RV000.IW016003.LOADSET         DELETED
          VOL SER NOS= CC3300.
          SYS2.COBV4SUB                                   KEPT
          VOL SER NOS= SPOOL1.
          SYS1.SORTLIB                                    KEPT
          VOL SER NOS= SPOOL1.
        STEP /GO        / START 77202.2240
        STEP /GO        / STOP  77202.2240 CPU    0MIN 02.41SEC STOR VIRT 144K
         JOB /IW016003/ START 77202.2239
         JOB /IW016003/ STOP  77202.2240 CPU    0MIN 05.93SEC
```

Appendix C Segmentation Feature

INTRODUCTION

The Segmentation Feature allows the problem programmer to communicate with the compiler to specify object program overlay requirements, but permits segmentation of procedures only. Segmentation will allow the programmer to divide the Procedure Division of the source program into sections. Through the use of a system of priority numbers, certain sections are designated as permanently resident in core storage and other sections as overlayable fixed segments and/or independent segments. Thus a large program can be executed in a defined area of core storage by limiting the number of segments in the program that are permanently resident in core.

Although COBOL segmentation deals only with segmentation of procedures, the Environment Division must be considered in determining the segmentation requirements for object programs.

ORGANIZATION

Program Segments

Although it is not mandatory, the Procedure Division for a source program is usually written as a consecutive group of sections, each of which is composed of a series of closely related operations that are designed to collectively perform a particular function. However, when segmentation is used, *the entire Procedure Division must be in sections.* In addition, each section must be classified as belonging either to the fixed portion or to one of the independent segments of the object program. Segmentation in no way affects the need for qualification of procedure-names to insure uniqueness. (See figures C.1, C.2.)

Fixed Portion

The fixed portion is defined as that part of the object program that is logically treated as if it were always in storage. This portion of the program is composed of two types of storage segments: permanent segments and overlayable fixed segments.

A permanent segment is a segment in the fixed portion which cannot be overlaid by any other part of the program.

An *overlayable segment* is a segment in the fixed portion which, although logically treated as if it were always in storage, can be overlaid (if necessary) by another segment to optimize storage utilization. However, such a segment, if called for by the program, is always available in the state it was in when last used.

```
section-name SECTION [priority-number] .
```

Figure C.1 Format Program Segment

```
IDENTIFICATION DIVISION
PROGRAM-ID. SAVECORE.
.
.
ENVIRONMENT DIVISION.
OBJECT-COMPUTER. IBM-360-H50
        SEGMENT-LIMIT IS 15.
.
.
DATA DIVISION.
.
.
PROCEDURE DIVISION.
SECTION-1    SECTION 8.
.
.
SECTION-2    SECTION 8.
.
.
SECTION-3    SECTION 16.
.
.
SECTION-4    SECTION 8.
.
.
SECTION-5    SECTION 50.
.
.
SECTION-6    SECTION 16.
.
.
SECTION-7    SECTION 50.
.
.
```

Figure C.2 Example—Program Segment

Depending on the availability of storage, the number of permanent segments in the fixed portion can be varied through the use of a special facility called SEGMENT-LIMIT. (See discussion of SEGMENT-LIMIT later in section.) (See figure C.3.)

Independent Segments

An independent segment is defined as that part of the object program which can overlay, and can be overlaid by another independent segment. An independent segment is effectively in its initial state each time the segment is made available to the program.

Segment Classification

Segments which are to be segmented are classified, using a system of priority numbers and the following criteria.

1. *Logic Requirements.* Sections that must be available for reference at all times, or which are referred to very frequently, are normally classified as belonging to one of the permanent segments; sections that are less frequently used are normally classified as belonging either to one of the overlayable fixed segments or to one of the independent segments, depending on logic requirements.

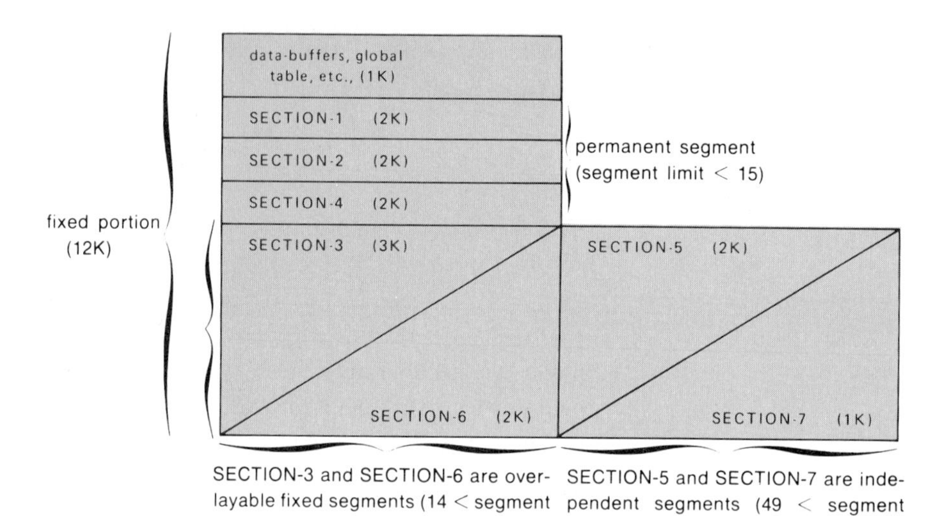

Figure C.3 Example—Storage Layout for Segment Program.

2. *Frequency of Use.* Generally, the more frequently a section is referred to, the lower its priority-number; the less frequently it is referred to, the higher its priority-number.
3. *Relationship to Other Sections.* Sections which frequently communicate with one another should be given equal priority-numbers. All sections with the same priority-number constitute a single program segment.

Segmentation Control

The logical sequence of a program is the same as the physical sequence except for specific transfers of control. A reordering of the object module will be necessary if a given segment has its sections scattered throughout the source program. However, the compiler will provide transfers to maintain the logic flow of the source program. The compiler will also insert instructions necessary to load and/or initialize a segment when necessary. Control may be transferred within a source program to any paragraph in a section; that is, it is not mandatory to transfer control to the beginning of a section.

Structure of Program Segments

Priority-Numbers. Section classification is accomplished by means of a system of priority-numbers. The priority-number is included in the section header.

Rules Governing the Use of Priority-Numbers

1. The priority-number must be an integer ranging in value from 0 through 99.
2. If the priority-number is omitted from the section header, the priority is assumed 0.
3. All sections which have the same priority-number must be together in the source program, and they constitute a program segment with that priority.
4. Segments with priority-number 0 through 49 belong to the fixed portion of the object program.

5. Segments with priority-numbers 50 through 99 are independent segments.
6. Sections in the declaratives portion of the Procedure Division must *not* contain priority-numbers in their section headers. They are treated as permanent segments with a priority-number of 0.

Segment-Limit

Ideally, all program segments having priority-numbers ranging from 0 through 49 are treated as permanent segments. However, when insufficient storage is available to contain all permanent segments plus the largest overlayable segment, it becomes necessary to decrease the number of permanent segments. The SEGMENT-LIMIT feature provides the user with a means by which he can reduce the number of permanent segments in his program, while these permanent segments still retain the logical properties of fixed portion segments (priority numbers 0 through 49).

Rules Governing the Use of the Segment-Limit Clause

1. The SEGMENT-LIMIT clause is coded in the OBJECT-COMPUTER paragraph in the Environment Division.
2. The priority-number must be an integer that ranges in value from 1 through 49.
3. When the SEGMENT-LIMIT clause is specified, only those segments having priority-numbers from 0 up to, but not including, the priority number designated as the segment limit are considered as permanent segments of the object program.

(See figure C.4.)

[SEGMENT-LIMIT IS priority-number]

Figure C.4 Format Segment-Limit Clause.

4. Those segments having priority numbers from the segment limit through 49 are considered as overlayable fixed segments.
5. When the SEGMENT-LIMIT clause is omitted, all segments having priority numbers from 0 through 49 are considered to be permanent segments of the object program.

(See figure C.5.)

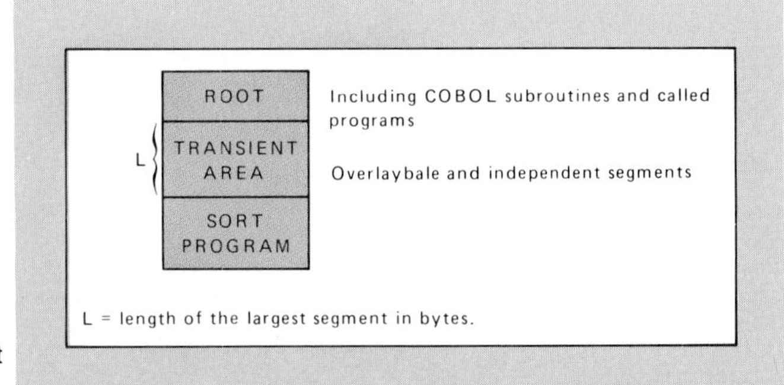

Figure C.5 Location of Sort Program in Segment of Structure.

Restrictions on Program Flow

When segmentation is used, the following restrictions are placed on the ALTER and PERFORM statements.

Alter Statement

1. A GO TO statement is a section whose priority-number is 50 or greater must not be referred to by an ALTER statement in a section with a different priority-number.
2. A GO TO statement in a section whose priority-number is lower than 50 may be referred to by an ALTER statement in any section, even if the GO TO statement to which the ALTER refers is in a segment of the program that has not been called for execution.

Perform Statement

1. A PERFORM statement that appears in a section whose priority-number is lower than SEGMENT-LIMIT can have within its range only the following:
   a. Sections with priority-numbers lower than 50.
   b. Sections wholly contained in a single segment whose priority-number is higher than 49.
2. A PERFORM statement that appears in a section whose priority-number is equal to or higher than the SEGMENT-LIMIT can have within its range only the following:
   a. Sections with the same priority-number as the section containing the PERFORM statement.
   b. Sections with priority-numbers that are lower than the SEGMENT-LIMIT.

Appendix D COBOL Differences-ANSI 1968 and ANSI 1974 COBOL

While much of the ANSI 1968 COBOL was included in the ANSI 1974 version, some of the material was changed, some deleted, and some new material was added. All of this is thoroughly detailed in publication X23. 1974 of the American National Standards Institute.

The major changes are listed below according to modules.

NUCLEUS

1. The REMARKS paragraph and the NOTE statement have been deleted in favor of a generalized comment facility. An asterisk (*) in character position 7 (continuation column) now identifies any line as a comment line. A further refinement has been added. A slash (/) in character position 7 causes the line to be treated as a comment and causes a page ejection.
2. The EXAMINE statement has been deleted in favor of the more general and powerful INSPECT statement. The INSPECT statement provides the facility to count (Format 1), replace (Format 2), or count and replace (Format 3) occurrences of single characters or groups of characters in a data item.
3. Level 77 items need no longer precede Level 01 items in the Working-Storage Section.
4. The punctuation rules with regard to spaces have been relaxed. For example, spaces may now optionally precede the comma, period, or semicolon, and may optionally precede or follow a left parenthesis.
5. Two contiguous quotation marks may be used within a nonnumeric literal to represent a single occurrence of the character quotation mark.
6. A SIGN clause has been added that permits the specification of the position that the sign is to occupy in a signed numeric item (either leading or trailing) and/or that it is to occupy a separate character position.
7. The ACCEPT statement has been expanded to provide access to internal DATE, DAY, and TIME.
8. GIVING identifier series has been added to the arithmetic statements; identifier series has been added to the COMPUTE statement; and INTO identifier series has been added to the DIVIDE statement.
9. The STRING statement has been added. This statement provides for the

juxtapositioning within a single data item of the partial or complete contents of two or more data items. A companion statement, the UNSTRING statement, has also been added. This statement causes contiguous data within a single data item to be separated and placed in multiple receiving fields.

10. Certain ambiguities in abbreviated combined conditions with regard to NOT and the use of parentheses have been eliminated. Where any portion of an abbreviated combined condition is enclosed in parentheses, all subjects and operators required for the expansion of that portion must be included within the same set of parentheses.
11. The PROGRAM COLLATING SEQUENCE clause has been added to permit specification of the collating sequence used in nonnumeric comparisons. Native, ASCII, implementor-defined and user-defined collating sequences may be specified. This makes possible the processing of ASCII files without changing source program logic.

TABLE HANDLING

1. The left parenthesis enclosing subscripts need not be preceded by a space. Commas are not required between subscripts or indices. Literals and index-names may be mixed in a table reference.
2. A data description entry that contains an OCCURS DEPENDING ON clause may be followed, within that record description, only by data description entries that are subordinate to it. Thus, the "fixed" portion of a record must entirely precede any "variable" portion. The effect of the OCCURS DEPENDING ON clause was clarified to state explicitly that internal operations involving tables described with this clause reference only the portion of the table that is "active" (i.e., the actual size as defined by the current value of the operand of the DEPENDING ON phrase is used).
3. An index may be set up or down by a negative value.
4. The subject of the condition in the WHEN phrase of the SEARCH ALL statement must be a data item named in the KEY phrase of the referenced table; the object of this condition may not be such a data item. ANSI 1968 specified that either the subject or the object could be a data item named in the KEY phrase.

SEQUENTIAL I-O

1. The FILE-LIMITS clause, the MULTIPLE REEL/UNIT clause, and the integer implementor-name phrase of the file control entry were deleted because it was felt that these functions could be handled better outside of the COBOL program.
2. The SEEK statement was deleted because it was felt to be redundant (it is implied by the READ, WRITE, etc.) and ineffective.
3. OPEN REVERSED statement now positions a file at its end. OPEN EXTEND statement was added to permit the addition of records at the end of an existing sequential file.
4. USE AFTER STANDARD ERROR was changed to read USE AFTER STANDARD ERROR/EXCEPTION; the function was expanded to per-

mit invocation of the associated on both error (e.g., boundary violation) or exception (i.e., AT END) conditions.

5. The AT END phrase of the READ statement was made optional; it must appear, however, if no applicable USE procedure appears.
6. The INVALID KEY phrase of the WRITE statement was deleted since there is no user-defined key for sequential files. Error and/or exception conditions can be monitored through appropriate USE statements.
7. The FILE STATUS clause was added to permit the system to convey information to the program concerning the status of I/O operations. Codes for "error," AT END, etc., have been defined.
8. The REWRITE statement has been added to permit the explicit updating of records on a sequential file.
9. The LINAGE clause was added to permit programmer definitions of logical page size and of the size of the top and bottom margins on the logical page.
10. The PAGE phrase was added to the WRITE statement to permit presentation of a line before or after advance to the top of the next logical page.
11. The facility to define, initialize, and access user-defined labels has been deleted.
12. The CODE-SET clause has been added to provide for conversion of sequential nonmass storage files encoded in ASCII or implementor-specified codes from/to the native character code.

The Random Access module of the ANSI 1968 COBOL has been replaced by two new modules, the Relative I-O and Indexed I-O modules.

RELATIVE I-O

Among the major features of the Relative I-O module are:

1. An ORGANIZATION IS RELATIVE clause.
2. A RELATIVE KEY clause.
3. An ACCESS MODE clause which specifies random, sequential, or dynamic access. Dynamic access permits the file to be accessed both randomly and sequentially.
4. FILE STATUS and USE AFTER STANDARD ERROR/EXCEPTION clauses as outlined in the Sequential I-O module. Here also the USE procedure may be used in place of the AT END and INVALID KEY phrases of the READ, WRITE, etc.
5. In addition to OPEN, CLOSE, READ, and WRITE, the DELETE, REWRITE, and START verbs are provided. READ NEXT statement provides for the intermixing sequential with random accesses of the file (when access mode is dynamic). START statement provides the facility to position the file such that the next sequential READ statement will reference a specified record.

INDEXED I-O

Among the major features of the Indexed I-O module are:

1. An ORGANIZATION IS INDEXED clause.
2. An ACCESS MODE clause with characteristics similar to that of the Relative I-O module.
3. FILE STATUS and USE procedures, as in the Relative I-O module.
4. The RECORD KEY specifies the data item that serves as the unique identifier for each record. The data item is known as the prime record key. The ALTERNATE KEY clause specifies additional (alternate) keys for the file. All insertion, updating, or deletion of records is done on the basis of the prime record key. Retrieval, however, may be on the basis of either prime or alternate record keys, thus providing more than one access path through the file.
5. As in the Relative I-O module, the new verbs DELETE, START, and REWRITE are available. READ NEXT and READ . . . KEY IS . . . are also available; the latter provides the means of specifying the key upon which retrieval is to be based (prime or alternate). The START statement also provides the means of specifying whether the prime area or alternate key is to be used for positioning the file.

SORT-MERGE

The major change to the Sort module of the previous standard has been the addition of a MERGE statement to permit the combination of two or more identically ordered files. The MERGE statement parallels the SORT statement in format, except that no input procedure is provided. The COLLATING SEQUENCE phrase has been added to permit the overriding of the programming collating sequence when executing a SORT or MERGE statement.

REPORT WRITER

The Report Writer module was completely rewritten in order to remove existing ambiguities and to provide a stronger and more useful facility. Care was taken in the rewrite not to imply that reports had to be presented on a printer (rather than on a type of graphic device).

SEGMENTATION

1. There is no logical difference between fixed and fixed overlayable segments. (ANSI 1968 COBOL placed certain restrictions on the range of PERFORM's involving fixed overlayable segments.)
2. A PERFORM statement in a nonindependent segment may have only one of the following within its range: (1) nonindependent segments, or (2) sections wholly contained in a single independent segment, except that (2) reads "Sections wholly contained in the same independent segment." Where a SORT or MERGE statement appears in a segmented program, then any associated input/output procedures are subject to the same constraints that apply to the range of a PERFORM (e.g., where the SORT is in a nonindependent segment, the associated input/output procedures must be either wholly contained in nonindependent segments or wholly contained in a single independent segment).

LIBRARY

The major changes introduced are:

1. The COPY statement may appear anywhere in the program that a COBOL word or separator may appear (ANSI 1968 COBOL permitted the COPY statement to appear only in certain specified places).
2. More than one library can be available.
3. All occurrences of a given literal, identifier, word or group of words in the library text can be replaced. (ANSI 1968 COBOL did not permit replacement of groups of words.)
4. The matching and replacement process has been significantly clarified.

DEBUG

The new Debug module provides a means by which the programmer can specify a debugging algorithm, including the conditions under which data items or procedures are to be monitored during program execution. The major features of this module are:

1. A USE FOR DEBUGGING statement which permits full or selective procedure and data-name monitoring; control is passed to the procedure when the specified condition arises. Associated with the execution of each debugging section (i.e., the declarative procedure associated with the USE FOR DEBUGGING statement) is the special register DEBUG-ITEM. This is updated by the system each time a debugging section is executed with such information as the name (with occurrence numbers if it should be the name of a table element) that caused the execution, the line number upon which the name appears, etc. The USE FOR DEBUGGING statements and their associated declarative procedures are treated as comment lines if the WITH DEBUGGING MODE clause does not appear in the program. An object time switch is also provided, outside the COBOL program, through which the USE FOR DEBUGGING procedure can be "turned off" without the need to recompile the program.
2. Debugging Lines. Any line with a "D" in the continuation area is a debugging line and will be compiled and executed only if the WITH DEBUGGING MODE clause appears in the program. Where this compile item switch does not appear in the program, these lines are treated as comment lines. The setting of the object time switch has no effect on the execution of the debugging lines. Through the debugging line facility, the programmer has at his disposal the full power of the COBOL language for debugging purposes.

INTER-PROGRAM COMMUNICATION

The new Inter-Program Communication module provides a facility by which a program can communicate with one or more other programs. This communication is made possible by: (a) enabling control to be transferred from one program to another within a run unit, and (b) enabling both programs to have access to the same data items. The major features of this module are:

1. The CALL statement causes control to be transferred from one object program to another. The CALL statement can be "static" (i.e., the name

of the called program is known at compile time) or dynamic (i.e., the name of the called program is not known until program execution time). The USING phrase of the CALL statement names the data to be shared with the called program: a USING phrase in the Procedure Division header of the called program specifies the name by which this shared data is to be known in the called program. The ON OVERFLOW phrase of the CALL statement will cause control to be transferred to an associated imperative statement if there is not enough memory available at execution time to permit the loading of the called program.

2. The CANCEL statement releases the areas occupied by called programs that are no longer required to be in memory.
3. The EXIT PROGRAM statement marks the logical end of a called program and causes control to be returned to the calling program (i.e., the program in which the CALL statement appears).
4. The Linkage Section appears in a program that is to operate under the control of a CALL statement. It is used in the called program to describe data that is to be made available from the calling program through the CALL USING facility described above.

COMMUNICATION

The new Communication module provides the ability to access, process, and create messages or portions thereof. It provides the ability to communicate through a Message Control System with local and remote communication devices. The major features of this module are:

1. The communication description entry (CD) specifies the interface area between the Message Control System (MCS) and a COBOL program. The CD specifies the input message queue structure, the symbolic names of destination for output messages, and such things as message data, message time, and text length.
2. The ENABLE and DISABLE statements notify the MCS to permit or inhibit the transfer of data between specified output queues and destinations for output, or between sources and input queues for input.
3. The RECEIVE statement makes available to the COBOL program a message, a portion thereof, and pertinent information about the message, from a queue maintained by the Message Control System.
4. The SEND statement causes a message or portion of a message to be released to one or more output queues maintained by the MCS.
5. The ACCEPT MESSAGE COUNT statement causes the number of messages in a queue to be made available.
6. The FOR INITIAL INPUT clause of the CD entry permits the MCS to schedule a program for execution upon receipt of a message for that program.

SUBSTANTIVE CHANGES THAT ARE INCLUDED IN THE ANSI 1974 COBOL

The following is a summary of the substantive changes that are included in the ANSI 1974 COBOL. The codes used are as follows:

Module Affected

The mnemonic names that are used in these codes are the following:

Mnemonic Name	Meaning
NUC	Nucleus
TBL	Table Handling
SEQ	Sequential I-O
REL	Relative I-O
INX	Indexed I-O
SRT	Sort-Merge
RPW	Report Writer
SEG	Segmentation
LIB	Library
DEB	Debug
IPC	Inter-Program Communication
COM	Communication

Remarks

The code reflected under the remarks column is as follows:

1. Indicates the change will not impact existing programs. For example, a new verb or an additional capability for an old verb.
2. Indicates the change could impact existing programs and some reprogramming may be needed. For example, where the semantics or syntax of an existing verb are changed.
3. Indicates that the change impacts an area that was implementor-defined in the original standard. As such it may or may not affect existing programs.

Additions to the reserved word list that will impact existing programs are not included in the list.

Language elements associated with the Report Writer modules are not assigned codes because the report writer specifications were completely rewritten, and comparison with the previous standard is therefore not meaningful.

	Substantive Change	Module Affected	Remarks
1.	Space may immediately precede or may immediately follow a parenthesis (except in a PICTURE character-string).	NUC	(1) Relaxes punctuation rules.
2.	Period, comma, or semicolon may be preceded by a space.	NUC TBL	(1) Relaxes punctuation rules.
3.	Semicolon and comma are interchangeable.	NUC	(1)
4.	An asterisk (*) in the continuation area (seventh character position) causes the line to be treated as a comment by the compiler. The comment line may appear in any division.	NUC	(1) New feature; replaces the NOTE statement and REMARKS paragraph.
5.	A stroke (slash, '/' virgule) in the continuation area (seventh character position) of a line causes page ejection of the compilation listing. (The line is treated as comment.)	NUC	(1)
6.	A phrase or clause (as well as sentence or entry) may be continued by starting subsequent lines in area B.	NUC	(1)
7.	Two contiguous quotation marks may be used to represent a single quotation mark character in a nonnumeric literal.	NUC	(1) New feature.
8.	Last line in a program may be a comment line.	NUC	(1)
9.	Mnemonic-name must have at least one alphabetic character.	NUC	(3) ANSI 1968 had no such restriction.
10.	Number of qualifiers permitted is implementor-defined, but must be at least five.	NUC	(2) ANSI 1968 specified no such lower limit.
11.	Complete set of qualifiers for a name may not be the same as the partial list of qualifiers for another name.	NUC	(2)
12.	REMARKS paragraph is deleted.	NUC	(2) Function was replaced by the comment line.
13.	Continuation of Identification Division comment-entries must not have a hyphen in the continuation indicator area.	NUC	(2)
14.	PROGRAM COLLATING SEQUENCE clause specifies that the collating sequence associated with alphabet-name is used in nonnumeric comparisons.	NUC	(1) New feature.
15.	SPECIAL-NAMES paragraph: 'L', '/', and '=' may not be specified in the CURRENCY SIGN clause.	NUC	(2) This restriction did not exist in ANSI 1968.
16.	Alphabet-name clause relates a user-defined name to a specified collating sequence or character code set (ANSI, native, or implementor-specifier).	NUC	(1) New feature.
17.	Alphabet-name clause: the literal phrase specifies a user-defined collating sequence.	NUC	(1) New feature.
18.	Condition-name may be given the status of an implementor-defined switch. Switches are implementor-defined and may be either software or hardware switches.	NUC	(1) ANSI 1968 specified hardware switches only.

	Substantive Change	Module Affected	Remarks
19.	All items which are immediately subordinate to group item must have the same level number.	NUC	(2)
20.	Level 77 items need not precede level 01 items in the Working-Storage Section.	NUC	(1) New feature.
21.	Level numbers 02–49 may appear anywhere to the right of margin A. (Margin A is defined as being between character positions 7 and 8.)	NUC	(1)
22.	Object of a REDEFINES clause can be subordinate to an item described with an OCCURS clause, but must not be referred to in the REDEFINES clause with a subscript or an index.	NUC	(1) New feature.
23.	REDEFINES: No entry with lower level-number can appear between the redefined and redefining items.	NUC	(2) ANSI 1968 had no such restriction.
24.	Multiple redefinition of same storage area permitted.	NUC	(3)
25.	An asterisk used as a zero suppression symbol in a PICTURE clause and the BLANK WHEN ZERO clause may not appear in the same entry.	NUC	(2)
26.	Alphabetic PICTURE character-string may contain the character B.	NUC	(1) New feature.
27.	The number of digit positions that can be described by a numeric PICTURE character-string cannot exceed 18.	NUC	(2) ANSI 1968 had no such rule.
28.	Stroke (/) permitted as an editing character.	NUC	(1) New feature.
29.	PICTURE character-string is limited to 30 characters.	NUC	(3) ANSI 1968 defines limit as 30 symbols where one symbol could have been two characters.
30.	SIGN clause allows the specification of the sign position.	NUC	(1) New feature.
31.	A signed numeric literal cannot be used in a VALUE clause unless it is associated with a signed PICTURE character-string.	NUC	(2)
32.	If the item is numeric edited, the literal in the VALUE clause must be nonnumeric.	NUC	(2)
33.	In the Procedure Division, a section may contain zero or more paragraphs and a paragraph may contain zero or more sentences.	NUC	(1) New feature.
34.	The unary + is permitted in arithmetic expressions.	NUC	(1) New feature.
35.	The TO is not required in the EQUAL TO of a relation condition.	NUC	(1) ANSI 1968 required the word TO.
36.	In relation and sign conditions, arithmetic expressions must contain at least one reference to a variable.	NUC	(2)
37.	Comparison of nonumeric operands; if one of the operands is described as numeric, it is treated as though it were moved to an alphanumeric item of the same size and the con-	NUC	(3)

	Substantive Change	Module Affected	Remarks
	tents of this alphanumeric item were then compared to the nonnumeric operand.		
38.	Abbreviated combined relation condition: When a portion is enclosed in parentheses, all subjects and operators required for the expansion of that portion must be included within the same set of parentheses.	NUC	(2) No such restriction appeared in ANSI 1968.
39.	Abbreviated combined relation condition: If NOT is immediately followed by a relational operator, it is interpreted as part of the relational operator.	NUC	(2) In ANSI 1968, NOT was a logical operator in such cases.
40.	Class condition: The numeric test cannot be used with a group item composed of elementary items described as signed.	NUC	(3)
41.	In an arithmetic operation, the composite of operands must not contain more than 18 decimal digits.	NUC	(2) ANSI 1968 specified limits only for ADD and SUBTRACT.
42.	ACCEPT identifier FROM DATE/DAY/TIME allows the programmer to access the date, day, and time.	NUC	(1) New feature.
43.	ADD statement: the GIVING identifier series.	NUC	(1) New feature.
44.	COMPUTE statement: the identifier series.	NUC	(1) New feature.
45.	DISPLAY statement: If the operand is a numeric literal, it must be an unsigned integer.	NUC	(2)
46.	DIVIDE statement: the INTO identifier series and the GIVING identifier series.	NUC	(2)
47.	DIVIDE statement: the remainder item can be numeric edited.	NUC	(1) New feature.
48.	GO TO statement; the word TO is not required.	NUC	(1) ANSI 1968 requires the word TO.
49.	EXAMINE statement and the special register TALLY were deleted.	NUC	(2) Function was replaced by the INSPECT statement.
50.	INSPECT statement provides ability to count or replace occurrences of single characters or groups of characters.	NUC	(1) New feature.
51.	MOVE statement: A scaled integer item (i.e., the rightmost character of the PICTURE character is a P) may be moved to an alphanumeric or an alphanumeric edited item.	NUC	(1) New feature.
52.	MULTIPLY statement: the BY identifier series and the GIVING identifier series.	NUC	(1) New feature.
53.	Perform statement: Format 4 (PERFORM . . . VARYING, not using index-names) identifiers need not be described as integers.	NUC	(1) New feature.
54.	PERFORM statement: Changing the FROM variable during execution can affect the number of times the procedures are executed in a Format 4 PERFORM if more than one AFTER phrase is specified.	NUC	(2)

	Substantive Change	Module Affected	Remarks
55.	PERFORM statement: There is no logical difference to the user between fixed and fixed overlayable segments.	NUC	(1) ANSI 1968 did not permit fixed overlayable segments to be treated the same as a fixed segment.
56.	A PERFORM statement in a non-independent segment can have in its range only one of the following: a) Nonindependent segment (fixed/fixed overlayable) b) Section and/or paragraphs wholly contained in a single independent segment.	NUC SEG	(3)
57.	A PERFORM statement in an independent segment can have in its range only one of the following: a) Nonindependent segments (fixed/fixed overlayable). b) Sections and/or paragraphs wholly contained in the same independent segment as that PERFORM statement.	NUC SEG	(3)
58.	PERFORM statement: Control is passed only once for each execution of a Format 2 PERFORM statement. (i.e., an independent segment referred to by such a PERFORM is made available in its initial state only once for each execution of that PERFORM statement.)	NUC SEG	(3)
59.	STOP statement: If the operand is numeric literal, it must be an unsigned integer.	NUC	(2)
60.	STRING statement provides for the juxtaposition of the partial or complete contents of two or more data items into a single data item.	NUC	(1) New feature.
61.	STRING statement: Delimiter identifiers need not be fixed length items.	NUC	(1)
62.	SUBTRACT statement: the GIVING identifier series.	NUC	(1) New feature.
63.	UNSTRING statement permits contiguous data in sending field to be separated and placed into multiple receiving fields.	NUC	(1) New feature.
64.	Commas are not required between subscripts or index-names.	TBL	(1)
65.	Literal subscripts may be mixed with index-names when referencing a table item.	TBL	(1) New feature.
66.	The DEPENDING phrase is now required in the Format 2 of the OCCURS clause.	TBL	(2) ANSI 1968 has no restriction.
67.	Integer-1 cannot be zero in the Format 2 of the OCCURS clause.	TBL	(2)
68.	A data description entry with an OCCURS DEPENDING clause may be followed within that record, only by entries subordinate to it. (i.e., only the last part of the record may have a variable number of occurrences.)	TBL	(2) This rule did not appear in the ANSI 1968.
69.	When a group item, having subordinate to it an entry that specifies Format 2 of the	TBL	(2)

	Substantive Change	Module Affected	Remarks
	OCCURS clause, is referenced, only part of the table area that is defined by the value of the operand of the DEPENDING phrase will be used in the operation. (i.e., the actual size of a variable length item is used, not the maximum size.)		
70.	If SYNCHRONIZED is specified for an item containing an OCCURS clause, any implicit FILLER generated for items in the same table are generated for each occurrence of those items.	TBL	(3)
71.	The results of SEARCH ALL operation are predictable only when the data in the table is ordered as described by the ASCENDING/ DESCENDING KEY clause associated with identifier-1.	TBL	(3)
72.	The subject of the condition in the WHEN phrase of the SEARCH ALL statement must be a data item named in the KEY phrase of the table; the object of this condition may not be a data item named in the KEY phrase.	TBL	(2) ANSI 1968 specified that either the subject or object could be a data item named in the KEY phrase.
73.	SEARCH . . VARYING identifier-2: If identifier-2 is an index data item, it is incremented as the associated index is incremented.	TBL	(3) In ANSI 1968, the data item is incremented by the same amount as occurrence number, i.e., by one.
74.	In Format 2 of the SET statement, the literal may be negative.	TBL	(1) New feature.
75.	File control entry: The ASSIGN TO implementor-name-1 OR implementor-name-n clause for the GIVING file of SORT statement was deleted.	SRT	(2)
76.	MERGE statement.	SRT	(1) New feature.
77.	RELEASE . . FROM identifier is placed in Level 1 of Sort-Merge Module.	SRT	(1) Was a level 2 feature.
78.	RETURN . . INTO identifier is placed in Level-1 of Sort-Merge module.	SRT	(1) Was a level 2 feature.
79.	SORT statement: the USING file-name series.	SRT	(1) ANSI 1968 allowed only one file name.
80.	SORT statement: semicolon deleted from format.	SRT	(2)
81.	SORT statement: COLLATING SEQUENCE phrase provides the ability to override the program collating sequence.	SRT	(1) New feature.
82.	No more than one file-name from a multiple file reel can appear in a SORT statement.	SRT	(2)
83.	Where a SORT or MERGE statement appears in a segmented program, then any associated input/output procedures are subject to the same constraints that apply to the range of a PERFORM.	SRT SEG	(2) No such restriction in ANSI 1968.

	Substantive Change	Module Affected	Remarks
84.	Segment-numbers are permitted in declaratives.	SEG	(1)
85.	PAGE-COUNTER and LINE-COUNTER are described as unsigned integers that must handle values from 0 to 999999.	RPW	
86.	The value in LINE-COUNTER must not be changed by the user.	RPW	
87.	LINE-COUNTER, PAGE-COUNTER, and sum counter must not be used as subscripts in the Report Section.	RPW	
88.	PAGE-COUNTER is always generated.	RPW	
89.	PAGE-COUNTER does not need to be qualified in the Report Section.	RPW	
90.	LINE-COUNTER is always generated.	RPW	
91.	LINE-COUNTER does not need to be qualified in the Report Section.	RPW	
92.	The words LINE and LINES are optional in the PAGE clause.	RPW	
93.	The DATA RECORDS clause and the REPORT clause are mutually exclusive.	RPW	
94.	A report may not be sent to more than one file.	RPW	
95.	RESET is no longer a clause; it is a phrase under the SUM clause.	RPW	
96.	Multiple SUM clauses may be specified in an item; multiple UPON phrases may be specified.	RPW	
97.	Up to three hierarchical levels are permitted in a report group description.	RPW	
98.	A report group level 01 entry cannot be elementary.	RPW	
99.	An entry that contains a LINE NUMBER clause must not have a subordinate entry that also contains a LINE NUMBER clause.	RPW	
100.	An entry that contains a COLUMN NUMBER clause but no LINE NUMBER clause must be subordinate to an entry that contains a LINE NUMBER clause.	RPW	
101.	An entry that contains a VALUE clause must also have a COLUMN NUMBER clause.	RPW	
102.	In the CODE clause, a mnemonic-name has been replaced by a literal (a two-character nonnumeric literal placed in the first two character positions of the logical record).	RPW	
103.	If the CODE clause is specified for any report in a file, it must be specified for all reports in the same file.	RPW	
104.	Control data items may not be subscripted or indexed.	RPW	
105.	Each data-name in the CONTROL clause must identify a different data item.	RPW	
106.	The GROUP INDICATE clause may only appear in a DETAIL report group entry that defines a printable item (contains a COLUMN and PICTURE clause).	RPW	

	Substantive Change	Module Affected	Remarks
107.	Line clause integers must not exceed three significant digits in length.	RPW	
108.	The NEXT PAGE phrase of the LINE clause is no longer legal in RH, PH, and PF groups.	RPW	
109.	A relative LINE NUMBER clause can no longer be the first LINE NUMBER clause in a PAGE FOOTING group.	RPW	
110.	A NEXT GROUP clause without a LINE clause is no longer legal.	RPW	
111.	Integer-2 in the NEXT GROUP clause must not exceed three significant digits in length.	RPW	
112.	If the PAGE clause is omitted, only a relative NEXT GROUP clause may be specified.	RPW	
113.	The NEXT PAGE phrase of the NEXT GROUP clause must not be specified in a PAGE FOOTING report group.	RPW	
114.	The NEXT GROUP clause must not be specified in a REPORT FOOTING report group.	RPW	
115.	The phrases of the PAGE clause may be written in any order.	RPW	
116.	In the PAGE clause, the maximum size of the integer is three significant digits.	RPW	
117.	It is no longer possible to sum upon an item in another report.	RPW	
118.	Source-sum correlation is not required. (Operands of a SUM clause need not be operands of a SOURCE clause in DETAIL groups.)	RPW	
119.	TYPE clause data-names may not be subscripted or indexed.	RPW	
120.	PAGE HEADING and PAGE FOOTING report groups may be specified only if a PAGE clause is specified in the corresponding report group description entry.	RPW	
121.	In CONTROL FOOTING, PAGE HEADING, PAGE FOOTING, and REPORT FOOTING report groups, SOURCE clauses and USE statements may not reference: a) Group data items containing control data items. b) Data items subordinate to a control data item. c) A redefinition or renaming of any part of a control data item. In PAGE HEADING and PAGE FOOTING report groups, SOURCE clauses and USE statements must not reference control data-name.	RPW	
122.	In summary reporting, only one detail group is allowed.	RPW	
123.	The description of a report must include at least one body group.	RPW	
124.	Report files must be opened with either the OPEN INPUT or OPEN EXTEND statement.	RPW	

	Substantive Change	Module Affected	Remarks
125.	A file described with a REPORT clause cannot be referenced by any input-output statement except the OPEN or CLOSE statement.	RPW	
126.	The SUPPRESS statement.	RPW	
127.	If no GENERATE statements have been executed for a report during the interval between the execution of an INITIATE statement and a TERMINATE statement for that report, the TERMINATE statement does not cause the Report Writer Control System to perform any of the related processing.	RPW	
128.	A USE procedure may refer to a DETAIL group.	RPW	
129.	FILE STATUS clause: data-name is updated by the system at the completion of each input-output operation.	SEQ REL INX	(1) New feature.
130.	ACCESS MODE IS DYNAMIC clause: provides ability to access a file sequentially or randomly in the same program.	REL INX	(1) New feature.
131.	ALTERNATE RECORD KEY clause: allows specification of multiple keys, any of which can be used to access an indexed file.	INX	(1) New feature.
132.	ACTUAL KEY clause deleted.		(2)
133.	RELATIVE KEY clause added for relative organization.	REL	(1) New feature.
134.	RECORD KEY clause added for indexed organization.	INX	(1) New feature.
135.	FILE-LIMITS clause deleted.		(2)
136.	PROCESSING MODE clause deleted.		(2)
137.	FILE-CONTROL paragraph: except for the ASSIGN clause, the order of clauses following file-name is optional.	SEQ REL INX	(1)
138.	ORGANIZATION IS RELATIVE clause.	REL	(2) New feature.
139.	ORGANIZATION IS SEQUENTIAL clause.	SEQ	(2) New feature.
140.	ORGANIZATION IS INDEXED clause.	INX	(2) New feature.
141.	MULTIPLE REEL/UNIT clause deleted.		(2)
142.	RESERVE . . ALTERNATIVE AREAS deleted.		(2)
143.	RESERVE integer AREAS allows the user to specify the exact number of areas to be used.	SEQ REL INX	(1) New feature.
144.	The file description entry for file-name must be equivalent to that used when this file was created.	SEQ REL INX	(3) No such rule in ANSI 1968.
145.	The data-name option of the LABEL RECORDS clause was deleted.	SEQ REL INX	(2) ANSI 1968 provided for user-defined label records.
146.	Data-name in the VALUE OF clause must be an implementor-name.	SEQ	(2) ANSI 1968 provided for user-defined field in label records.
147.	LINAGE clause permits programmer definition of logical page size.	SEQ	(1) New feature.

	Substantive Change	Module Affected	Remarks
148.	CLOSE . . . FOR REMOVAL statement.	SEQ	(1) New feature.
149.	DELETE statement.	REL INX	(1) New feature.
150.	OPEN REVERSED positions file at its end.	SEQ	(2)
151.	OPEN INPUT or OPEN I-O makes a record available to the program.	SEQ REL INX	(1) New feature.
152.	OPEN EXTEND statement: adds records to an existing file.	SEQ	(1) New feature.
153.	The OPEN and CLOSE statements with the NO REWIND phrase apply to all devices that claim support for this function.	SEQ	(1) ANSI 1968 restricted the application of this phrase.
154.	The OPEN REVERSED statement applies to all devices that claim support for this function.	SEQ	(1) ANSI 1968 restricted the application of this phrase.
155.	READ statement: AT END phrase required only if no applicable USE AFTER ERROR/EXCEPTION procedure specified.	SEQ REL INX	(1) New feature.
156.	READ statement: INVALID KEY phrase required only if no applicable USE AFTER ERROR/EXCEPTION procedure specified.	REL INX	(1) New feature.
157.	READ statement: INTO phrase placed in Level 1.	SEQ REL INX	(1) Level 2 feature in ANSI 1968.
158.	READ . . . NEXT statement: use to retrieve the next logical record from a file when the access mode is dynamic.	REL	(1) New feature.
159.	REWRITE statement.	SEQ REL INX	(1) New feature.
160.	SEEK statement was deleted.		(2)
161.	START statement: provides for logical positioning within a relative or indexed file for sequential retrieval of records.	REL	(1) New feature.
162.	USE statement: the label processing options are deleted.	SEQ REL	(2) ANSI 1968 provided for the processing of user-defined labels.
163.	USE . . . ERROR/EXCEPTION statement.	SEQ REL INX	(1) New feature.
164.	Recursive invocation of USE procedures prohibited.	SEQ REL INX	(2)
165.	WRITE statement: INVALID KEY phrase deleted.	SEQ	(2)
166.	WRITE statement: INVALID KEY phrase required only if no applicable USE AFTER ERROR/EXCEPTION procedure specified.	REL INX	(1)
167.	WRITE statement: FROM phrase placed in Level 1.	SEQ REL INX	(1) Level 1 feature in ANSI 1968.

	Substantive Change	Module Affected	Remarks
168.	WRITE statement: BEFORE/AFTER PAGE phrase provides ability to skip to top of a page.	SEQ	(1)
169.	WRITE statement: END-OF-PAGE phrase.	SEQ	(1) New feature.
170.	Debugging line: defined by a 'D' in the continuation column.	DEB	(1) New feature.
171.	WITH DEBUGGING MODE clause: a compile time switch; in addition an object time switch can be used to suppress coding at object time.	DEB	(1) New feature.
172.	USE FOR DEBUGGING statement.	DEB	(1) New feature.
173.	DEBUG-ITEM.	DEB	(1) New feature.
174.	Linkage Section.	IPC	(1) New feature.
175.	Procedure Division header: the USING phrase.	IPC	(1) New feature.
176.	CALL identifier statement.	IPC	(1) New feature.
177.	CALL identifier ON OVERFLOW statement.	IPC	(1) New feature.
178.	CANCEL statement.	IPC	(1) New feature.
179.	EXIT PROGRAM statement.	IPC	(1) New feature.
180.	COPY statement may appear anywhere a COBOL word may appear.	LIB	(1) New feature.
181.	Identifier, COBOL word, or a group of COBOL words may be replaced.	LIB	(1) New feature.
182.	Multiple libraries are permitted.	LIB	(1) New feature.
183.	Library-name is a user-defined word.	LIB	(1) New feature.
184.	Communication description entry (CD).	COM	(1) New feature.
185.	ACCEPT cd-name MESSAGE COUNT statement.	COM	(1) New feature.
186.	ENABLE statement.	COM	(1) New feature.
187.	DISABLE statement.	COM	(1) New feature.
188.	RECEIVE statement.	COM	(1) New feature.
189.	SEND statement.	COM	(1) New feature.

ELEMENTS DELETED FROM ANSI 1968 COBOL

The following elements were deleted from the ANSI 1968 standards.

REMARKS Paragraph. The REMARKS paragraph of the Identification Division was deleted and the function replaced by the asterisk (*) comment line.

EXAMINE Statement. The EXAMINE statement and the special register TALLY were deleted in favor of the new and more powerful INSPECT statement.

NOTE Statement. The NOTE statement was deleted and the function replaced by the asterisk (*) comment line.

FILE-LIMITS Clause. This clause was deleted from the file control entry because the function could be handled better outside the COBOL program.

SEEK Statement. This statement was redundant; it is implied by the READ, WRITE, etc.

MULTIPLE REEL/UNIT Clause. This clause was deleted from the file control entry because the function could be handled better outside the COBOL program.

ACTUAL KEY Clause. This clause was replaced by the RELATIVE KEY clause.

RESERVE integer ALTERNATE AREAS Clause. This clause was replaced by the RESERVE integer AREAS Clause.

OR implementor-name. This clause was deleted from the file control entry because the function could be handled better outside the COBOL program.

integer implementor-name. This clause was deleted from the file control entry because the function could be handled better outside the COBOL program.

PROCESSING MODE IS SEQUENTIAL Clause. This clause was deleted from the file control entry as not being needed in a synchronous environment.

USE . . . . LABEL Statement. An extensive revision to label processing is currently under way to remove ambiguities and provide for the processing of ANSI standard labels. This work was not completed in time for inclusion in the 1974 revision. In order not to hinder the introduction of this new facility, it was decided to define only a minimum label processing capability in the revised standard.

LABEL RECORDS IS data-name Clause. An extensive revision to label processing is currently underway to remove ambiguities and provide for the processing of ANSI standard labels. This work was not completed in time to be included in the 1974 revision. In order not to hinder the introduction of this new facility, it was decided to define only a minimum label processing capability in the revised standard.

Ten Problems

Problem 1

Shampoo Payroll Problem

In a beauty salon, operators are paid by the amount and type of work they do. The shampoo operators receive $4.00 per customer, the hair cutters receive $5.50 per customer, the hair setters receive $6.00 per customer, the stylists receive $8.00 per customer, and the permanent wave operators receive $10.00 per customer.

Given: Operator's name
Type of operator
Number of customers.

INPUT

Field	Card Columns
Name	1–25
Type	26
Customers	27–29
Blanks	30–80

FORMULA

Gross-pay = Rate × Customers

PRINTED OUTPUT

```
TYPE OF OPERATOR.
     SHAMPOO - - - - - - - 1
     HAIR CUTTERS- - - - 2
     HAIR SETTERS- - - - 3
     STYLISTS- - - - - - 4
     PERMANENT WAVE- - - 5

NAME OF OPERATOR         TYPE    NO. OF CUSTOMERS    GROSS PAY

SUSAN CALDWELL             1           100             $400.00

BETTY JANE CLANCY          1           120             $480.00

RUTH ANN CORBETT           1           150             $600.00

MARGARET CUSHING           2           100             $550.00
```

Problem 2

Department Store Problem

Given: Customer's name
Customer's address
Customer's account no.
Last month's balance
Payments made
Purchases made

on a card;

INPUT

Field	Card Columns
Customer's name	1 - 15
Customer's address	16 - 50
Account no.	51 - 55
Last balance	56 - 60
Month sales	61 - 65
Payments	66 - 70
Blanks	71 - 80

FORMULAS TO BE USED

Service charge = .015 * (last-balance – payments) (rounded)
Amount due = (last-balance – payments) + service charge + month's sales

OUTPUT

NAME OF CUSTOMER	ADDRESS OF CUSTOMER	ACCOUNT	PREVIOUS BALANCE	SALES	PAYMENT	SERVICE CHARGE	AMT DUE
DAVID ANDERSON	18745 MOBILE ST., RESEDA, CALIF.	62986	$100.00	$100.00	$20.00	$1.20	$181.20
BETTY L. BREWER	10321 LUNDY DR., INGLEWOOD, CALIF.	61477	$350.25	$50.00	$30.00	$4.80	$375.05
ARTHUR BROWN	12145 MADISON ST., L.A. 45, CALIF.	38940	$450.00	$30.00	$50.00	$6.00	$436.00
THOMAS CASSIDY	3726 HOPE AV., LYNWOOD, CALIF.	62180	$121.50	$40.00	$20.00	$1.52	$143.02
BOB CHAMBERS	3840 HOPE AV., LYNWOOD, CALIF.	58920	$320.00	$15.50	$35.00	$4.28	$304.78
JACK T. CROSS	6421 BELMAR ST., RESEDA, CALIF.	43313	$105.80	$150.85	$20.00	$1.29	$237.94
KENT B. DAVIS	11621 PENN DRIVE, ENCINO, CALIF.	84082	$320.75	$75.75	$35.00	$4.29	$365.79
SAMUAL FELLOWS	10732 LINDLEY AV., ENCINO, CALIF.	41750	$290.60	$40.37	$30.00	$3.91	$304.88
MICHAEL FISHER	6345 TAMPA AV., TARZANA, CALIF.	30040	$444.35	$65.25	$50.00	$5.92	$465.52
GLADYS BUTTONS	3701 BALBOA AV., VAN NUYS, CALIF.	67542	$375.00	$89.30	$45.00	$4.95	$424.25
PATRICK HANEY	4218 VICTORY ST., L.A. 45, CALIF.	72111	$450.10	$27.95	$55.00	$5.93	$428.98
LYNN HUBBARD	13245 VENTURA BLVD., RESEDA CALIF.	64375	$195.75	$36.45	$25.00	$2.56	$209.76
MARVIN JACOBS	13118 VENTURA BLVD., RESEDA, CALIF	62550	$225.95	$44.95	$30.00	$2.94	$243.84
LINDA JOHNSON	20715 VAN NUYS ST., ENCINO, CALIF.	58214	$300.00	$87.50	$35.00	$3.98	$356.48
HOWARD KEYES	2181 SHERMAN WAY, RESEDA, CALIF.	46615	$279.80	$101.75	$30.00	$3.75	$355.30

Problem 3

Bank Balance Problem

Given: *in an 80 column card*

Account number

Type code: either a 1 punch or a 2 punch:

1 indicates a checking account
2 indicates a savings account

Deposits
Withdrawals
Last Balance

INPUT

Field	Card Column
Account no.	1 - 5
Type code	6
Blanks	7 - 8
Deposits	9 - 16
Withdrawals	17 - 24
Last balance	25 - 32
Blanks	33 - 80

FORMULAS

New balance = Last-balance − withdrawals + deposits.
Interest = rate x (last-balance − withdrawals + deposits)
New balance = Last-balance − withdrawals + deposits + interest.

OUTPUT

ACCOUNT NO.	NEW BALANCE
61788	$7,280.23
61003	$10,500.00
58440	$102,685.50
57905	$38,650.00
60756	$37,500.00
61880	$25,590.00
59425	$50,500.50
63740	$45,000.00
65500	$10,500.00
62711	$5,250.00
60912	$10,500.00
57280	$7,500.00
59014	$20,000.00
60545	$50,500.00

Problem 4

"Africa" Payroll Problem

Each month a payroll is to be processed in the following manner:

A card file contains a master record for every employee in the company. Each record contains the employee's name, his number, the regular and overtime hours he has worked during the month, the wages he has earned so far this year, his rate of pay, and the number of dependents he has. The payroll is processed in the usual manner: computation of GROSS-PAY, F!CA, WH-TAX and NET PAY. The results are used to print checks and to create new master records. These new records contain the new YTD-GROSS and zeros in the hours field (both regular and overtime). These new records create the CARD-OUT file which will be used next month as CARD-IN. (During the month the regular hours and overtime hours are added in by another program.)

This company has a subsidiary in Africa whose employees, though U. S. citizens, are not required to pay income tax. These employees have A's in front of their numbers—other employee numbers will have spaces in these character positions.

This program will use a table of income tax exemptions according to the number of dependents which looks like this:

DEPENDENTS

0	1	2	3	4	5	6	7	8	9	10
$0	$56	$112	$168	$224	$280	$336	$392	$448	$504	$560

Subtract from GROSS-PAY to find the taxable amount.

The formulas which the program will use are:

In America:

Gross-Pay = (regular hours * rate) + (overtime hours * 1½ rate).

FICA = .0585 * gross pay (If YTD is less than $16,500).

WH-TAX = .18 * (gross-pay – sub) sub is exemption according to dependents chart.

NET PAY = gross-pay – FICA – WH-TAX.

In Africa:

Gross-Pay = (regular hours * rate) + (overtime hours * 1½ rate).

WH-TAX = 0.

FICA = .0585 * gross pay (If YTD is less than $16,500).

NET PAY = gross-pay – FICA.

Card-In

Field	Card Columns	
Employee Name	1 - 29	
Employee Number	30 - 39	
Blank	40	
Rate	41 - 43	
Dependents	44 - 45	
Regular Hours	46 - 50	(xxx.xx)
Overtime Hours	51 - 55	(xxx.xx)
YTD-Gross	56 - 62	(xxxxx.xx)
Blanks	63 - 80	

Print-Out

	Field	Print Positions
CHECK-LINE-1		
	Name	1 - 29
	Blanks	30 - 106
	Date	107 - 114
	Blanks	115 - 120
CHECK-LINE-2		
	Blanks	1 - 90
	Net-Pay	91 - 96
	Blanks	97 - 120
CHECK-LINE-3		
	Employee Name	1 - 29
	Blanks	30 - 31
	Employee Number	32 - 41
	Blanks	42 - 43
	Gross Pay	44 - 50
	Blanks	51 - 52
	Wh-Tax	53 - 57
	Blanks	58 - 59
	FICA	60 - 64
	Blanks	65 - 66
	Net-Pay	67 - 73
	Blanks	74 - 120

```
PAUL A. EVANS                                                                      20 AUG 77
                                                                          $339.99
PAUL A. EVANS              000053487     $420.00   $55.44   $24.57    $339.99


MARY DIXCN                                                                         20 AUG 77
                                                                          $434.65
MARY DIXON                A000062986     $461.66    $0.00   $27.01    $434.65
```

Problem 5 Sales and Commission Problem

Commissions are paid to salesmen based upon the number of units that are sold. The unit commission varies with the product sold and the total commission is based upon the number of units sold of each particular product.

Sales are determined by the number of units sold times the individual product selling price.

Given:

Product	Commission Rate	Selling Price
1	$ .10	$ 16.00
2	$ .20	$ 30.00
3	$ .30	$ 43.00
4	$ .40	$ 60.00
5	$ .50	$ 75.00

INPUT RECORD FILE

Field	Card Columns
Territory Number	1 - 2
Salesman Number	3 - 5
Date	6 - 11
Name	12 - 30
Units Sold	31 - 35
Product Number	36

PROBLEM

1. Prepare a table of commissions for the five different products so that the product number itself will serve as a subscript.
2. Prepare a table of selling prices for the five different products so that the product number itself will serve as a subscript.
3. Write a COBOL program that will read both the commission table and price table and then process each data card to calculate the commission for each salesman. At the same time, prepare a report of the number of units sold and the amount of sales for each product by salesman, by territory and an overall total of sales.

OUTPUT

MONTHLY SALES AND COMMISSION REPORT
JANUARY 1977 PAGE 1

TERRITORY NUMBER	SALESMAN NUMBER	DATE	NAME	PRODUCT NUMBER	UNITS SOLD	TOTAL SALES	COMMISSION
10	111	010372	JONES HENRY	1	100	$1,600.00	$160.00
10	111	011072	JONES HENRY	2	2,301	$69,030.00	$13,806.00
10	111	011772	JONES HENRY	3	60	$2,580.00	$774.00
10	111	012472	JONES HENRY	4	20,502	$1,230,120.00	$492,048.00
					22,963	$1,303,330.00	$506,788.00

Problem 6

Payroll Register Problem

Write a COBOL program that will calculate and print the Payroll Register as indicated.

INPUT RECORD FILE

Field	Card Columns
Month	1–3
Day	4–5
Year	6–7
Department	14–16
Serial	17–21
Gross Earnings	57–61
Insurance	62–65
Withholding Tax	69–72
Not Used	73–75
Miscellaneous Deductions	76–79
Code (letter E)	80

CALCULATIONS

1. FICA = Gross Earnings × .0585 (round to two decimal positions).
2. State UCI = Gross Earnings × .01 (round to two decimal positions).
3. Net Earnings = Gross Earnings − Insurance − FICA − Withholding Tax − State UCI − Miscellaneous Deductions.
4. If the Net Earnings are zero or negative, branch to an error routine.
5. The Department Earnings value is the sum of the Net Earnings for each employee.
6. Calculate totals for all columns by department as well as an overall total for the entire payroll.

PRINTED OUTPUT

Heading Line	Field	Printing Positions
1	WEEKLY PAYROLL REGISTER	40 - 62
2	WEEK ENDING	40 - 62
3	EMPLOYEE NO.	3 - 14
	GROSS	22 - 26
	WITHHOLDING	51 - 61
	STATE	65 - 69
4	DEPT.	2 - 6
	SERIAL	10 - 15
	EARNINGS	20 - 27
	INSURANCE	31 - 39

	Field	Printing Positions	
Heading Line			
	FICA	44 - 47	
	TAX	56 - 58	
	UCI	66 - 68	
	MISC. DEDNS.	74 - 85	
	NET EARNINGS	89 - 100	
Detail			
Department		3 - 5	
Serial		10 - 14	
Gross Earnings		21 - 26	XXXX.XX
Insurance		33 - 37	XXX.XX
FICA		43 - 47	XXX.XX
Withholding Tax		54 - 58	XXX.XX
State UCI		65 - 68	XX.XX
Miscellaneous Deductions		76 - 80	XXX.XX
Net Earnings		92 - 97	XXXX.XX

OUTPUT

```
                                   WEEKLY PAYROLL REGISTER
                                   WEEK ENDING SEP  3 1977                                    PAGE     1

  EMPLOYEE NO.       GROSS                                WITHHOLDING  STATE
 DEPT.   SERIAL     EARNINGS    INSURANCE     FICA            TAX        UCI      MISC. DEDNS.    NET EARNINGS

  200    10670      $202.00       $3.10     $11.82         $28.00      $2.02         $0.00          $157.06
  200    10695      $203.00       $3.10     $11.88         $28.00      $2.03         $5.00          $152.99
  200    10700      $204.00       $3.10     $11.93         $28.00      $2.04         $0.00          $158.93
  200    10703      $205.00       $3.10     $11.99         $28.00      $2.05         $0.00          $159.86
  200    10725      $207.00       $3.10     $12.11         $28.00      $2.07        $10.00          $151.72
  200    10730      $208.00       $3.10     $12.17         $28.00      $2.08         $0.00          $162.65
  200    10742      $209.00       $3.10     $12.23         $28.00      $2.09         $0.00          $163.58
  200    10800      $210.00       $3.10     $12.29         $28.00      $2.10         $0.00          $164.51
  200    10890      $211.00       $3.10     $12.34         $28.00      $2.11         $1.00          $164.45

DEPT. TOTALS:      $1859.00      $27.90    $108.76        $252.00     $18.59        $16.00         $1435.75

  300    10904      $212.00       $3.10     $12.40         $28.00      $2.12         $0.00          $166.38
  300    10905      $213.00       $3.10     $12.46         $28.00      $2.13         $0.00          $167.31
  300    10906      $214.00       $3.10     $12.52         $28.00      $2.14         $0.00          $168.24
  300    10907      $215.00       $3.10     $12.58         $28.00      $2.15         $0.00          $169.17

DEPT. TOTALS:       $854.00      $12.40     $49.96        $112.00      $8.54         $0.00          $671.10

  400    11215      $218.00       $3.10     $12.75         $28.00      $2.18         $0.00          $171.97
  400    11225      $219.00       $3.10     $12.81         $28.00      $2.19         $0.00          $172.90
  400    11240      $220.00       $3.15     $12.87         $28.50      $2.20         $0.00          $173.28
  400    11250      $221.00       $3.15     $12.93         $28.50      $2.21         $0.00          $174.21

DEPT. TOTALS:       $878.00      $12.50     $51.36        $113.00      $8.78         $0.00          $692.36

  600    12330      $225.00       $3.15     $13.16         $28.50      $2.25         $2.00          $175.94
  600    12340      $226.00       $3.15     $13.22         $28.50      $2.26         $2.00          $176.87
  600    12350      $227.00       $3.15     $13.28         $28.50      $2.27         $0.00          $179.80
  600    12366      $228.00       $3.15     $13.34         $28.50      $2.28         $0.00          $180.73
  600    12400      $229.00       $3.15     $13.40         $28.50      $2.29         $0.00          $181.66
  600    12400      $229.00       $3.15     $13.40         $28.50      $2.29         $0.00          $181.66

DEPT. TOTALS:      $1364.00      $18.90     $79.80        $171.00     $13.64         $4.00         $1076.66


T O T A L S :      $4955.00      $71.70    $289.88        $648.00     $49.55        $20.00         $3875.87
```

Problem 7

Updated Payroll Problem

Write a COBOL program to update a master tape file with a current card file. Both input files are in Social Security Number sequence. Input tape records are in blocks of ten 52-character records. The updated output tape blocks will be the same size.

The exception list shall be double spaced.

INPUT TAPE RECORD

Field	Positions
Employee Name	6 - 25
Social Security Number	26 - 34
Old Year-to-date Gross Earnings	35 - 41
Old Year-to-date Withholding Tax	42 - 47
Old Year-to-date FICA	48 - 52

INPUT CARD RECORD

Field	Card Columns
Department Number	1 - 2
Clock Number	3 - 5
Social Security Number	26 - 34
Current Gross Earnings	62 - 68
Current Withholding Tax	69 - 74
Current FICA	75 - 78
Code (Digit 1)	80

OPERATIONS TO BE PERFORMED

1. New Year-to-date Gross = Old Year-to-date Gross plus Current Gross
2. New Year-to-date Withholding Tax = Old Year-to-date Withholding Tax plus Current Withholding Tax.
3. New year-to-date FICA = Old Year-to-date FICA plus Current FICA.
4. If New Year-to-date FICA record exceeds $965.25, print Department Number, Clock Number, Employee Name, Social Security Number, New Year-to-date Gross, New Year-to-date Withholding Tax and Excess FICA amount.

OUTPUT UPDATED TAPE RECORD

Field	Positions
Department Number	1 - 2
Clock Number	3 - 5
Employee Name	6 - 25
Social Security Number	26 - 34
New Year-to-date Gross Earnings	35 - 41
New Year-to-date Withholding Tax	42 - 47
New Year-to-date FICA	48 - 52

OUTPUT PRINTED RECORD

Field	Print Positions	
Department Number	4 - 5	
Clock Number	9 - 11	
Employee Name	15 - 34	
Social Security Number	37 - 47	XXX-XX-XXXX
New Year-to-date Gross Earnings	52 - 60	XX,XXX.XX
New Year-to-date Withholding Tax	64 - 71	X,XXX.XX
Excess FICA	77 - 81	XX.XX

XX EMPLOYEES OVER $965.25.

```
                              EXCEPTION LIST
                          EMPLOYEES OVER $965.25                     PAGE 1

DEPT CLOCK      EMPLOYEE NAME        SOC SEC NO.     YTD GROSS  YTD W&H     EX FICA

 75   925    FOX,WILLIAM             130-09-5294     22,695.45 4,829.50     362.43

 37   857    PHILLIPS,ROBERT         364-20-8841     17,234.70 2,925.70      42.98

 15   375    JACKSON,KENNETH         543-01-2232     16,782.50 2,685.10      16.53

 42   902    SAWYER,DAVID            556-32-0201     19,822.82 3,290.50     194.38

 21   472    YOUNG,SAMUEL            557-16-7782     18,347.82 3,621.30     113.36

 87   524    HEPNER,ELMER            559-10-9299     20,007.25 3,972.90     205.17

 63   708    HORNE,ALBERT            578-20-1141     16,921.84 1,874.90      24.68

                    7 EMPLOYEES OVER $965.25
```

Problem 8

Sales Problem (Report Writer Feature)

REPORT

```
               D A I L Y       S A L E S      R E G I S T E R

WEEK OF 12-16-77                                           PAGE  01

       ENTRY      CUSTOMER    SALESMAN          SALE
        DAY        NUMBER      NUMBER          AMOUNT

        16         08257        071          $ 1,189.80
        16         11234        079              168.06
        16         29031        079               63.00
        16         79992        095               87.74

                         DAY 16 SALES       $  1,508.60 *

        17         02965        037          $12,716.92
        17         09002        001              842.17
        17         01179        002            7,071.12
        17         13605        001            3,092.72
        17         27654        009              217.90

                         DAY 17 SALES       $ 23,940.83 *

        18         00390        092          $    27.00
        18         05006        056              897.32
        18         12125        181              371.98

                         DAY 18 SALES       $  1,296.30 *

        19         00298        100          $ 2,020.60
        19         00106        024            1,494.73

                         DAY 19 SALES       $  3,515.33 *

        20         00256        003          $    79.53
        20         00652        008               95.18
        20         18569        090              421.15
```

INPUT RECORD FILE

Field	Card Columns
Month	1 - 2
Day	3 - 4
Year	5 - 6
Salesman Number	7 - 9
Customer Number	10 - 14
Sales Amount	51 - 57

CALCULATIONS

Compute the total sales values for each day of week.
Compute the total sales values for week.
Compute the number of sales for week.

Problem 9 Commission Problem (Report Writer Feature)

REPORT

```
                      S A L E S   C O M M I S S I O N   R E P O R T

                            FOR THE MONTH OF JANUAARY 1977

SALESMAN   CUSTOMER        INVOICE         NET AMOUNT        RATE        COMMISSION

  4490       49690           25170       $ 12,359.20          11        $ 1,359.51
  4490      115121           25460          1,250.00          10            125.00
  4490       78345           25198          8,255.12           8            660.41
  4490       72914           44483            690.70          14             96.70
                 SALESMAN 4490 TOTAL   $    22,555.02   *            $    2,241.62   *

  2513       14983           14152       $    110.20           6        $     6.61
  2513      712129           13444         10,986.00          12          1,318.32
  2513       11110           12136          9,850.40           8            788.03
  2513       11110           12136          9,850.40           8            788.03
                 SALESMAN 2513 TOTAL   $    30,797.00   *            $    2,900.99   *


     TOTAL FOR MONTH OF JANUARY  $          53,352.02   **         $      5,142.61 **
```

INPUT RECORD FILE

Card Columns	Field
1	Code (digit 5)
2 - 6	Invoice Number
13 - 19	Customer Number
35 - 42	Net Amount
43 - 46	Salesman Number
54 - 55	Commission Rate
56 - 62	Commission Amount

CALCULATIONS

Find the total sales and total commissions for each salesman. Find the total sales and total commissions for entire force for month.

Problem 10

Hospital Problem (Sort Feature)

A card is punched for the number of patients in each hospital of the United States.

A report is prepared indicating the various patient totals for cities and counties within each state of the United States. An overall total is indicated for the entire United States.

INPUT RECORD FILE

Field	Card Columns
Date	1 - 6
State	7 - 8
County	9 - 11
City	12 - 14
Hospital Number	15 - 18
Number of Patients	70 - 75

OPERATIONS REQUIRED:

1. Sort data cards in the following sequence; major-State, intermediate-County and minor-City.
2. Prepare listing per output record format.

OUTPUT RECORD FORMAT.

```
                                HOSPITAL PATIENT REPORT
                                  JANUARY 31 1977                 PAGE      1

STATE          COUNTY           CITY                     NUMBER OF PATIENTS


  AL             SEC              NBI                            2,000

                                 COUNTY TOTAL                     2,000

                                 STATE  TOTAL                     2,000
  AR             ABA              CAL                            3,716

                                 COUNTY TOTAL                     3,716

                                 STATE  TOTAL                     3,716
  CA             LA               LA                               800

                                 COUNTY TOTAL                       800
  CA             ORA              LB                               400

                                 COUNTY TOTAL                       400
  CA             SCL              SUN                            3,556

                                 COUNTY TOTAL                     3,556
  CA             SD               SD                             2,168

                                 COUNTY TOTAL                     2,168
  CA             VEN              PH                             1,284

                                 COUNTY TOTAL                     1,284

                                 STATE  TOTAL                     8,208
```

Data for Ten Problems

```
.........1.........2.........3.........4.........5.........6.........7.........8
PROBLEM 1
INPUT CARDS

SUSAN CALDWELL           1100
BETTY JANE CLANCY        1120
RUTH ANN CORBETT         1150
MARGARET CUSHING         2100
ROSEMARY DUPUIS          2075
MAURICE ERICKSON         2050
LILLIAN FELLING          3100
NANCY HAMILTON           3125
BARBARA HICKMAN          3110
JOSEPHINE HOUSTON        4050
LORETTA JOHNSON          4040
ELAINE LEONARD           4030
LORRAINE CLARK           5030
TERRY MCDONNELL          5035
MARY ANN PALMER          5040
```

```
.........1.........2.........3.........4.........5.........6.........7.........8
PROBLEM 2
INPUT CARDS

DAVID ANDERSON 18745 MOBILE ST., RESEDA, CALIF.   62986100001000002000
BETTY L. BREWER10321 LUNDY DR., INGLEWOOD, CALIF. 61477350250500003000
ARTHUR BROWN   12145 MADISON ST., L.A. 45, CALIF. 38940450000300005000
THOMAS CASSIDY 3726 HOPE AV., LYNWOOD, CALIF.     62180121500400002000
BOB CHAMBERS   3840 HOPE AV., LYNWOOD, CALIF.     58920320000155003500
JACK T. CROSS  6421 BELMAR ST., RESEDA, CALIF.    43313105801508502000
KENT B. DAVIS  11621 PENN DRIVE, ENCINO, CALIF.   84082320750757503500
SAMUAL FELLOWS 10732 LINDLEY AV., ENCINO, CALIF.  41750290600403703000
MICHAEL FISHER 6345 TAMPA AV., TARZANA, CALIF.    30040444350652505000
GLADYS BUTTONS 3701 BALBOA AV., VAN NUYS, CALIF.  67542375000893004500
PATRICK HANEY  4218 VICTORY ST., L.A. 45, CALIF.  72111450100279505500
LYNN HUBBARD   13245 VENTURA BLVD., RESEDA CALIF. 64375195750364502500
MARVIN JACOBS  13118 VENTURA BLVD., RESEDA, CALIF.62550225950449503000
LINDA JOHNSON  20715 VAN NU,S ST., ENCINO, CALIF. 58214300000875003500
HOWARD KEYES   2181 SHERMAN WAY, RESEDA, CALIF.   46615279801017503000
HAZEL J. MEYER 11702 ROSCOE WAY, LYNWOOD, CALIF.  30881380500300004000
DONALD OLSON   11650 ROSCOE WAY, LYNWOOD, CALIF.  43779277600159503000
FRANK PALMER   17521 LINDLEY AV., ENCINO, CALIF.  58090251750855003500
DOROTHY RANSOM 13155 MADISON ST., L.A. 45, CALIF. 60070355250479504000
```

```
.........1.........2.........3.........4.........5.........6.........7.........8
PRØBLEM 3
INPUT CARDS

617882  0095000000040000000175000
610031  0320000000365000001500000
584401  0500000000041310005681650
579051  0006500004052650078526500
607561  0125000000250000000500000
618801  0005200000000900002516000
594251  0010000000078900005028950
637401  0025000000001000004260000
655001  0000000000009000001140000
627111  0002750000007550000505050
609121  0012750000072250001645000
572801  0002500000100000000825000
590141  0001000000100000002990000
605451  0006750000000500004987500
```

```
.........1.........2.........3.........4.........5.........6.........7.........8
PRØBLEM 4
INPUT CARDS

PAUL A. EVANS                  000053487 30002140000000000751640
MARY DIXØN                    A000062986 27503160000005250775020
GUY T. GØØDWIN                 000020981 28705160000010500950070
JAMES F. KING                 A000053219 31204160001201010470090
MILTØN C. MØRGAN               000064378 29007160000000000645092
JØHN L. REED                   000052887 35002160000017505222564
JACK L. SHAFER                 000061758 32504160000105009600075
```

```
.........1.........2.........3.........4.........5.........6.........7.........8
PRØBLEM 5
INPUT CARDS

1020304050     1600300043006000750 0
10111010372JØNES HENRY        001001
10111011072JØNES HENRY        023012
10111011772JØNES HENRY        000603
10111012472JØNES HENRY        205024
10222010372SMITH RØBERT       000551
10222010372SMITH RØBERT       000702
10222010372SMITH RØBERT       008003
16431012472HØDGES JAMES       000105
16431011772HØDGES JAMES       006234
16431010372HØDGES JAMES       405003
24350012472JØHNSTØN HØWARD    000085
24350011072JØHNSTØN HØWARD    000423
24565012472MØNTGØMERY DAN     008204
24565011772MØNTGØMERY DAN     072551
```

```
.........1.........2.........3.........4.........5.........6.........7.........8
PRØBLEM 6
INPUT CARDS
             20010670                                  202000310   2800080000
             20010695                                  203000310   2800080050
             20010700                                  204000310   2800080000
             20010703                                  205000310   2800080000
             20010725                                  207000310   2800080100
             20010730                                  208000310   2800080000
             20010742                                  209000310   2800080000
             20010800                                  210000310   2800080000
             20010890                                  211000310   2800080010
             30010904                                  212000310   2800080000
             30010905                                  213000310   2800080000
             30010906                                  214000310   2800080000
             30010907                                  215000310   2800080000
             40011215                                  218000310   2800080000
             40011225                                  219000310   2800080000
             40011240                                  220000315   2850085000
             40011250                                  221000315   2850085000
             60012330                                  225000315   2850085020
             60012340                                  226000315   2850085020
             60012350                                  227000315   2850085000
             60012366                                  228000315   2850085000
             60012400                                  229000315   2850085000
```

```
.........1.........2.........3.........4.........5.........6.........7.........8
PRØBLEM 7
INPUT CARD RECØRD
75925                    130095294                    0756515160983442561
37857                    364208841                    0574490097523336081
42252                    543012222                    0200000018500117001
21663                    543012223                    0203350018833118961
21074                    543012224                    0206683019167120911
37185                    543012225                    0210017019500122861
15296                    543012226                    0213350019835124811
42307                    543012227                    0216675020167126751
37418                    543012228                    0233333021833136501
63529                    543012229                    0243333022833142351
42630                    543012230                    0200017018500117011
63741                    543012231                    0200350018533117201
15375                    543012232                    0559417089503327261
63852                    543012233                    0246667023167144301
37963                    543012240                    0283333028000165751
21074                    543012241                    0283333028167165751
15185                    543012242                    0301667030167176481
75296                    543012246                    0316667031667185251
87307                    543012248                    0383333038333224251
15141                    543012249                    0433333043333253501
42902                    556320201                    0660761109683386541
21472                    557167782                    0614598120710359541
87524                    559109299                    0666908132430390141
63708                    578201141                    0564061062497329981
```

```
.........1.........2.........3.........4.........5.........6.........7.........8
PRØBLEM 7
INPUT TAPE RECØRD
     FØX WILLIAM          1300952941513030321967885l2
     PHILLIPS RØBERT      3642088411148980195047672l5
     JØHNSØN BEN          543012222040000003700023400
     WASHINGTØN GEØRGE    543012223040670003766723792
     MØNTGØMERY ALEX      543012224041336703833324182
     SMITH JØSEPH         543012225042003303900024572
     BRØWN WALLACE        543012226042670003967024962
     DUNIGAN HENRY        543012227043335004033525351
     JØNES WILLIAM        543012228046666704366727300
     DELANEY JERRY        543012229048666704566728470
     HALLECK FRANCES      543012230040003303700023402
     REID PATRICIA        543012231040070003706723441
     JACKSØN KENNETH      543012232111883317900765452
     ALEXANDER CHARLES    543012233049333304633328860
     HALL GEØRGE          543012240056666705600033150
     SIPLE CHARLES        543012241056666705633333150
     GØØDMAN HENRY        543012242060333306033335295
     CAMM FRED J          543012243063333306333337050
     DENTØN TERRENCE      543012248076666707666744850
     GØØDSALL PJILLIP     543012249086666708666750700
     SAWYER DAVID         556320201132152121936777309
     YØUNG SAMUEL         557167782122318824142071556
     HEPNER ELMER         559109299133381726486078028
     HØRNE ALBERT         578201141112812312499365995
```

```
.........1.........2.........3.........4.........5.........6.........7.........8
PRØBLEM 8
INPUT CARDS

12167207108257                                   0118980
12167207911234                                   0016806
12167207929031                                   0006300
12167209579992                                   0008774
12177203702965                                   1271692
12177200109002                                   0084217
12177200201179                                   0707112
12177200113605                                   0309272
12177200927654                                   0021790
12187209200390                                   0002700
12187205605006                                   0089732
12187218112125                                   0037198
12197210000298                                   0202060
12197202400106                                   0149473
12207200300256                                   0007953
12207200800652                                   0009518
12207209018569                                   0042115
12207213220106                                   0070642
12207200500321                                   0059010
```

```
.........1.........2.........3.........4.........5.........6.........7.........8
PRØBLEM 9
INPUT CARDS

525460      0115121                      001250004490      100012500
525198      0078345                      008255124490      080066041
544483      0072914                      000690704490      140009670
514152      0014983                      000110202513      060000661
513444      0712129                      010986002513      120131832
512136      0011110                      009850402513      080078803
```

```
PAGE  1
.........1.........2.........3.........4.........5.........6.........7.........8
PRØBLEM 10
INPUT CARDS

060675MDBALBALMDGN                                                      000100
060675CALA LA VLZJ                                                      000200
060675WAKNGSTLMWAG                                                      000300
060675ILCØKCHINFXL                                                      000400
060675ALSECNBIØBAY                                                      000500
060675MDPGSSSPPZCK                                                      000600
060675WAKNGSTLVAMN                                                      000717
060675CASCLSUNAQKD                                                      000889
060675ARABACALRBHE                                                      000929
060675CAØRALB SFCC                                                      000100
060675MDCARTØWDTUG                                                      000231
060675CAVENPH UBIE                                                      000321
060675DEBANØXAFTUM                                                      000443
060675CASD SD SJDG                                                      000542
060675MDBALPIKJHØP                                                      001001
060675DEBANRØNHWHL                                                      000655
060675FLSD AR QEIN                                                      000762
060675NECENFØNRXØJ                                                      000867
060675WNDUNREDKPYI                                                      000984
060675CALA LA VLZJ                                                      000200
060675WAKNGSTLMWAG                                                      000300
060675ILCØKCHINFXL                                                      000400
060675ALSECNBIØBAY                                                      000500
060675MDBALBALMDGN                                                      000100
060675MDPGSSSPPZCK                                                      000600
060675WAKNGSTLVAMN                                                      000717
060675CASCLSUNAQKD                                                      000889
060675ARABACALRBHE                                                      000929
060675CAØRALB SFCC                                                      000100
060675MDCARTØWDTUG                                                      000231
060675CAVENPH UBIE                                                      000321
060675DEBANØXAFTUM                                                      000443
06Q675CASD SD SJDG                                                      000542
060675MDBALPIKJHØP                                                      001001
060675DEBANRØNHWHL                                                      000655
060675FLSD AR QEIN                                                      000762
060675NECENFØNRXØJ                                                      000867
060675WNDUNREDKPYI                                                      000984
060675CALA LA VLZJ                                                      000200
060675WAKNGSTLMWAG                                                      000300
060675ILCØKCHINFXL                                                      000400
060675ALSECNBIØBAY                                                      000500
```

```
PAGE  2
.........1.........2.........3.........4.........5.........6.........7.........8
060675MDPGSSSPPZCK                                                      000600
060675WAKNGSTLVAMN                                                      000717
060675CASCLSUNAQKD                                                      000889
060675ARABACALRBHE                                                      000929
060675CAØRALB SFCC                                                      000100
060675MDCARTØWDTUG                                                      000231
060675CAVENPH UBIE                                                      000321
060675DEBANØXAFTUM                                                      000443
060675CASD SD SJDG                                                      000542
060675MDBALPIKJHØP                                                      001001
060675DEBANRØNHWHL                                                      000655
060675FLSD AR QEIN                                                      000762
060675NECENFØNRXØJ                                                      000867
060675WNDUNREDKPYI                                                      000984
060675CALA LA VLZJ                                                      000200
060675WAKNGSTLMWAG                                                      000300
```

```
060675ILCØKCHINFXL                                        000400
060675ALSECNBIØBAY                                        000500
060675MDPGSSSPPZCK                                        000600
060675WAKNGSTLVAMN                                        000717
060675CASCLSUNAQKD                                        000889
060675ARABACALRBHE                                        000929
060675CAØRALB SFCC                                        000100
060675MDCARTØWDTUG                                        000231
060675CAVENPH UBIE                                        000321
060675DEBANØXAFTUM                                        000443
060675CASD SD SJDG                                        000542
060675MDBALPIKJHØP                                        001001
060675DEBANRØNHWHL                                        000655
060675FLSD AR QEIN                                        000762
060675NECENFØNRXØJ                                        000867
060675WNDUNREDKPYI                                        000984
```

IBM American National Standard COBOL Reserved Words

No word in the following list should appear as a programmer-defined name. The keys that appear before some of the words and their meanings are:

(xa) before a word means that the word is an IBM extension to American National Standard COBOL.

(xac) before a word means that the word is an IBM extension to both American National Standard COBOL and CODASYL COBOL.

(ca) before a word means that the word is a CODASYL COBOL reserved word not incorporated in American National Standard COBOL or in IBM American National Standard COBOL.

(sp) before a word means that the word is an IBM function-name established in support of the SPECIAL-NAMES function.

(spn) before a word means that the word is used by an IBM American National Standard COBOL compiler, but not by this compiler.

(asn) before a word means that the word is defined by American National Standard COBOL, but is not used by this compiler.

	ACCEPT
	ACCESS
	ACTUAL
	ADD
(asn)	ADDRESS
	ADVANCING
	AFTER
	ALL
	ALPHABETIC
(ca)	ALPHANUMERIC
(ca)	ALPHANUMERIC-EDITED
	ALTER
	ALTERNATE
	AND
(xa)	APPLY
	ARE
	AREA
	AREAS
	ASCENDING
	ASSIGN
	AT
	AUTHOR
(xac)	BASIS
	BEFORE
	BEGINNING
	BLANK
	BLOCK
(ca)	BOTTOM
	BY

(xa)	CALL
(xa)	CANCEL
(xac)	CBL
(xa)	CD
	CF
	CH
(xac)	CHANGED
(xa)	CHARACTER
	CHARACTERS
(asn)	CLOCK-UNITS
	CLOSE
(asn)	COBOL
	CODE
	COLUMN
(spn)	COM-REG
	COMMA
(xa)	COMMUNICATION
	COMP
(xa)	COMP-1
(xa)	COMP-2
(xa)	COMP-3
(xa)	COMP-4
	COMPUTATIONAL
(xa)	COMPUTATIONAL-1
(xa)	COMPUTATIONAL-2
(xa)	COMPUTATIONAL-3
(xa)	COMPUTATIONAL-4
	COMPUTE
	CONFIGURATION
(sp)	CONSOLE
	CONTAINS
	CONTROL
	CONTROLS
	COPY
(xac)	CORE-INDEX
	CORR
	CORRESPONDING
(xa)	COUNT
(sp)	CSP
	CURRENCY
(xac)	CURRENT-DATE
(spn)	CYL-INDEX
(spn)	CYL-OVERFLOW
(sp)	C01
(sp)	C02
(sp)	C03
(sp)	C04
(sp)	C05
(sp)	C06
(sp)	C07
(sp)	C08
(sp)	C09
(sp)	C10
(sp)	C11
(sp)	C12
	DATA
(xa)	DATE
	DATE-COMPILED
	DATE-WRITTEN
(xa)	DAY
(ca)	DAY-OF-WEEK
	DE
(xac)	DEBUG
(ca)	DEBUG-CONTENTS
(ca)	DEBUG-ITEM
(ca)	DEBUG-LINE
(ca)	DEBUG-NAME
(ca)	DEBUG-SUB-1
(ca)	DEBUG-SUB-2
(ca)	DEBUG-SUB-3
(ca)	DEBUGGING
	DECIMAL-POINT
	DECLARATIVES
(xa)	DELETE
(xa)	DELIMITED
(xa)	DELIMITER
	DEPENDING
(xa)	DEPTH
	DESCENDING
(xa)	DESTINATION
	DETAIL
(ca)	DISABLE
(xac)	DISP
	DISPLAY
(xac)	DISPLAY-ST
(ca)	DISPLAY-n
	DIVIDE

DIVISION
DOWN
(ca) DUPLICATES
(xa) DYNAMIC

(xa) EGI
(xac) EJECT
ELSE
(xa) EMI
(ca) ENABLE
END
END-OF-PAGE
(xa) ENDING
ENTER
(xac) ENTRY
ENVIRONMENT
(xa) EOP
EQUAL
(ca) EQUALS
ERROR
(xa) ESI
EVERY
EXAMINE
(ca) EXCEEDS
(xa) EXCEPTION
(xac) EXHIBIT
EXIT
(xa) EXTEND
(spn) EXTENDED-SEARCH

FD
FILE
FILE-CONTROL
FILE-LIMIT
FILE-LIMITS
FILLER
FINAL
FIRST
FOOTING
FOR
FROM

GENERATE
GIVING
GO
(xac) GOBACK
GREATER
GROUP

HEADING
HIGH-VALUE
HIGH-VALUES
(ca) HOLD

I-O
I-O-CONTROL
(xac) ID
IDENTIFICATION
IF
IN
INDEX
(ca) INDEX-n
INDEXED
INDICATE
(ca) INITIAL
(ca) INITIALIZE
INITIATE
INPUT
INPUT-OUTPUT
(xac) INSERT
(ca) INSPECT
INSTALLATION
INTO
INVALID
IS

JUST
JUSTIFIED

KEY

LABEL
(xac) LABEL-RETURN
LAST
LEADING
(xac) LEAVE
LEFT
(xa) LENGTH

LESS
(ca) LIBRARY
LIMIT
LIMITS
(ca) LINAGE
(ca) LINAGE-COUNTER
LINE
LINE-COUNTER
LINES
(xa) LINKAGE
LOCK
LOW-VALUE
LOW-VALUES

(spn) MASTER-INDEX
MEMORY
(xa) MERGE
(xa) MESSAGE
MODE
MODULES
(xac) MORE-LABELS
MOVE
MULTIPLE
MULTIPLY

(xac) NAMED
NEGATIVE
NEXT
NO
(xac) NOMINAL
NOT
NOTE
(spn) NSTD-REELS
NUMBER
NUMERIC
(ca) NUMERIC-EDITED

OBJECT-COMPUTER
(ca) OBJECT-PROGRAM
OCCURS
OF
OFF
OMITTED
ON

OPEN
OPTIONAL
OR
(xa) ORGANIZATION
(xac) OTHERWISE
OUTPUT
(xa) OVERFLOW

PAGE
PAGE-COUNTER
(xac) PASSWORD
PERFORM
PF
PH
PIC
PICTURE
PLUS
(xa) POINTER
POSITION
(xac) POSITIONING
POSITIVE
(xac) PRINT-SWITCH
(ca) PRINTING
PROCEDURE
(ca) PROCEDURES
PROCEED
(ca) PROCESS
PROCESSING
(xa) PROGRAM
PROGRAM-ID

(xa) QUEUE
QUOTE
QUOTES

RANDOM
RD
READ
(xac) READY
(xa) RECEIVE
RECORD
(xac) RECORD-OVERFLOW
(xa) RECORDING
RECORDS

	REDEFINES
	REEL
(ca)	REFERENCES
(ca)	RELATIVE
	RELEASE
(xac)	RELOAD
	REMAINDER
	REMARKS
(ca)	REMOVAL
	RENAMES
(xac)	REORG-CRITERIA
	REPLACING
	REPORT
	REPORTING
	REPORTS
(xac)	REREAD
	RERUN
	RESERVE
	RESET
	RETURN
(xac)	RETURN-CODE
	REVERSED
	REWIND
(xa)	REWRITE
	RF
	RH
	RIGHT
	ROUNDED
	RUN
(ca)	SA
	SAME
	SD
	SEARCH
	SECTION
	SECURITY
	SEEK
(xa)	SEGMENT
	SEGMENT-LIMIT
	SELECT
(xa)	SEND
	SENTENCE
(xa)	SEPARATE
	SEQUENTIAL
(xac)	SERVICE
	SET
	SIGN
	SIZE
(xac)	SKIP1
(xac)	SKIP2
(xac)	SKIP3
	SORT
(xac)	SORT-CORE-SIZE
(xac)	SORT-FILE-SIZE
(ca)	SORT-MERGE
(xac)	SORT-MESSAGE
(xac)	SORT-MODE-SIZE
(spn)	SORT-OPTION
(xac)	SORT-RETURN
	SOURCE
	SOURCE-COMPUTER
	SPACE
	SPACES
	SPECIAL-NAMES
	STANDARD
(xa)	START
	STATUS
	STOP
(xa)	STRING
(xa)	SUB-QUEUE-1
(xa)	SUB-QUEUE-2
(xa)	SUB-QUEUE-3
	SUBTRACT
	SUM
(ca)	SUPERVISOR
(xa)	SUPPRESS
(ca)	SUSPEND
(xa)	SYMBOLIC
	SYNC
	SYNCHRONIZED
(sp)	SYSIN
(spn)	SYSIPT
(spn)	SYSLST
(sp)	SYSOUT
(spn)	SYSPCH
(sp)	SYSPUNCH
(sp)	S01

(sp)	S02
(ca)	TABLE
	TALLY
	TALLYING
	TAPE
(ca)	TERMINAL
	TERMINATE
(xa)	TEXT
	THAN
(xac)	THEN
	THROUGH
	THRU
(xa)	TIME
(xac)	TIME-OF-DAY
	TIMES
	TO
(ca)	TOP
(xac)	TOTALED
(xac)	TOTALING
(xac)	TRACE
(xac)	TRACK
(xac)	TRACK-AREA
(xac)	TRACK-LIMIT
(xac)	TRACKS
(xa)	TRAILING
(xac)	TRANSFORM
	TYPE
(ca)	UNEQUAL
	UNIT
(xa)	UNSTRING
	UNTIL
	UP
	UPON
(spn)	UPSI-0
(spn)	UPSI-1
(spn)	UPSI-2
(spn)	UPSI-3
(spn)	UPSI-4
(spn)	UPSI-5
(spn)	UPSI-6
(spn)	UPSI-7
	USAGE
	USE
	USING
	VALUE
	VALUES
	VARYING
	WHEN
(xac)	WHEN-COMPILED
	WITH
	WORDS
	WORKING-STORAGE
	WRITE
(xac)	WRITE-ONLY
(spn)	WRITE-VERIFY
	ZERO
	ZEROES
	ZEROS

COBOL Character Set

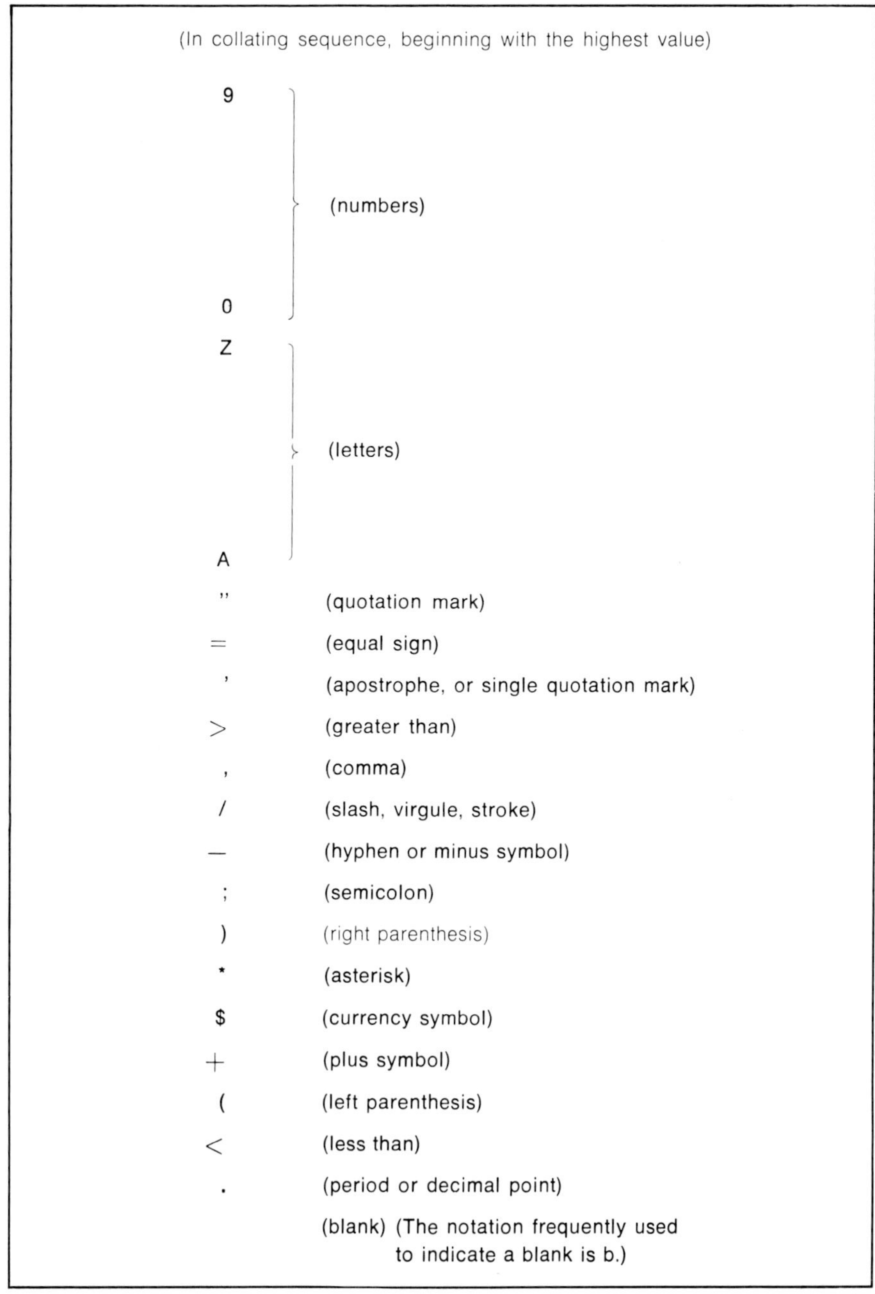

(In collating sequence, beginning with the highest value)

9	
	(numbers)
0	
Z	
	(letters)
A	
"	(quotation mark)
=	(equal sign)
'	(apostrophe, or single quotation mark)
>	(greater than)
,	(comma)
/	(slash, virgule, stroke)
—	(hyphen or minus symbol)
;	(semicolon)
)	(right parenthesis)
*	(asterisk)
$	(currency symbol)
+	(plus symbol)
(	(left parenthesis)
<	(less than)
.	(period or decimal point)
	(blank) (The notation frequently used to indicate a blank is b.)

Structure of System-Name in the ASSIGN Clause

IBM Disk Operating System (DOS)

SYSnnn-class-device-organization [-name]

nnn—a three-digit number between 000 and 221

class—a two-character field that represents device class.
Allowable combinations are:

DA mass storage
UT utility
UR unit record

device—a four- or five-digit number that represents device number.
Allowable numbers for each device class are:

DA 2311, 2321, 2314
UT 2400, 2311, 2314, 2321
UR 1442R, 1442P, 1403, 1404 (continuous forms only), 1443, 2501, 2520R, 2520P, 2540R, 2540P
(*R* indicates reader; *P* indicates punch.)

organization—a one-character field that specifies file organization.

Allowable characters are:

S for standard sequential files
A for direct files—actual track addressing
D for direct files—relative track addressing
U for direct files using REWRITE statement* —actual track addressing
W for direct files using REWRITE statement* —relative track addressing
I for indexed files
*When the file is opened INPUT or OUTPUT, U and W are equivalent to A and D.

name—a one- to seven-character field that specifies the external-name by which the file is known to the system.

IBM Operating System (OS)

class-device-organization-name

class—a two-character field that represents device class. Allowable combinations are:

DA mass storage
UT utility
UR unit record

device—a four- to six-character number that represents device number. Allowable numbers for each device class are:

DA 2301, 2302, 2303, 2311, 2314, 2321
UT 2301, 2302, 2311, 2314, 2321, 2400
UR 1403, 1404 (continuous forms only), 1442R, 1442P, 1443, 1445, 2501, 2520R, 2520P, 2540R, 2540P (*R* indicates reader; *P* indicates printer.)

organization—a one-character field that specifies file organization. Allowable characters are:

S for standard sequential organization files
D for direct organization files
W for direct organization files using REWRITE statement. When the file is opened as INPUT or OUTPUT, however, *W* is the equivalent of *D*.
R for relative organization files
I for indexed organization files

name—a one- to eight-character field that specifies the external-name by which the file is known to the system. It is the name that appears in the name field of the DD card for the file.

Symbols Allowed in the PICTURE Clause

Symbol	Meaning
A	Alphabetic character or space
B	Space insertion character
P	Decimal scaling position (not counted in size of data item)
S	Operational sign (not counted in size of data item)
V	Assumed decimal point (not counted in size of data item)
X	Alphanumeric character (any from the EBCDIC set)
Z	Zero suppression character
9	Numeric character
0	Zero insertion character
,	Comma insertion character
.	Decimal point or period editing control character
+	Plus sign insertion editing control character
–	Minus sign editing control character
CR	Credit editing control characters
DB	Debit editing control characters
*	Check protect insertion character
$	Currency sign insertion character

Summary of File-Processing Techniques and Applicable Statements and Clauses

This appendix summarizes the statements and clauses that may be specified for an IBM system 360/370 for each file-processing technique. In addition, each file-name must be specified in a SELECT clause in the Environment Division and must be defined by an FD entry in the File Section of the Data Division.

STANDARD SEQUENTIAL FILES—Required and Optional Entries

Device Type	Required Entries					Optional Entries						
	System-name	LABEL RECORDS	OPEN	CLOSE	Access Verbs	APPLY[3]	RESERVE	ACCESS	Other ENVIRONMENT DIVISION Clauses	BLOCK CONTAINS[4]	RECORDING MODE	USE
Reader	UR [-xxxx] -S-name	OMITTED	INPUT	[LOCK]	READ [INTO] AT END		{integer NO}	SEQUENTIAL	SAME [RECORD] AREA RERUN	[n TO] m	{F U V}	ERROR
Punch	UR [-xxxx] -S-name	OMITTED	OUTPUT	[LOCK]	WRITE [FROM] [{BEFORE AFTER} ADVANCING] [AFTER POSITIONING]	WRITE-ONLY (V-mode only)	{integer NO}	SEQUENTIAL	SAME [RECORD] AREA RERUN	[n TO] m	{F U V}	ERROR
Printer	UR [-xxxx] -S-name	OMITTED	OUTPUT	[LOCK]	WRITE [FROM] [{BEFORE AFTER} ADVANCING] [AFTER POSITIONING] — — — — — [END-OF-PAGE]	WRITE-ONLY (V-mode only)	{integer NO} — — — NO	SEQUENTIAL	SAME [RECORD] AREA RERUN	[n TO] m	{F U V}	ERROR REPORTING
Tape	UT-[-xxxx] -S-name	{STANDARD OMITTED data-name [TOTALING- TOTALED]}	INPUT [REVERSED NO REWIND] [LEAVE REREAD DISP]	[REEL] [LOCK NO REWIND POSITIONING DISP]	READ [INTO] AT END	WRITE-ONLY (V-mode only)	{integer NO}	SEQUENTIAL	SAME [RECORD] AREA RERUN MULTIPLE FILE TAPE	[n TO] m	{F U V S}	LABEL ERROR
			OUTPUT [NO REWIND] [LEAVE REREAD DISP]	[REEL] [LOCK NO REWIND POSITIONING DISP]	WRITE[1] [FROM] [{BEFORE AFTER} ADVANCING] [AFTER POSITIONING]							LABEL ERROR REPORTING
Mass Storage	UT DA [-xxxx] -S-name	{STANDARD OMITTED data-name [TOTALING- TOTALED]}	INPUT	[UNIT] [LOCK]	READ [INTO] AT END	RECORD-OVERFLOW (not for S-mode)	{integer NO}	SEQUENTIAL	SAME [RECORD] AREA RERUN	[n TO] m	{F U V S}	AFTER LABEL ERROR
			OUTPUT	[UNIT] [LOCK]	WRITE[1] [FROM] INVALID KEY WRITE[1] [FROM] [{BEFORE AFTER} ADVANCING] [AFTER POSITIONING]	WRITE-ONLY (V-mode only)						AFTER LABEL ERROR REPORTING
			I-O	[LOCK]	READ [INTO] AT END WRITE[2] [FROM] INVALID KEY REWRITE[2] [FROM] [INVALID KEY]							AFTER LABEL ERROR

[1]Create

[2]Update

[3]These APPLY clauses have meaning only for OUTPUT files; however, the compiler accepts them when the same file is opened for INPUT or I-O.

[4]Not for U mode

DIRECT FILES (mass storage devices only)—Required and Optional Entries

Required Entries							Optional Entries			
ACCESS	KEY	System-name	LABEL RECORDS	OPEN	CLOSE	Access Verbs	RECORDING MODE	APPLY[5]	Other ENVIRONMENT DIVISION Clauses	USE
[SEQUENTIAL]	[ACTUAL]	DA [-xxxx] -D-name	{STANDARD / data-name}	INPUT	[UNIT] [LOCK]	READ [INTO] AT END	{F / U / V / S}		SAME [RECORD] AREA RERUN	AFTER LABEL ERROR
[SEQUENTIAL]	ACTUAL	DA [-xxxx] -D-name	{STANDARD / data-name}	OUTPUT	[UNIT] [LOCK]	WRITE[1] [FROM] INVALID KEY	F	RECORD-OVERFLOW	SAME [RECORD] AREA TRACK-LIMIT RERUN	AFTER LABEL ERROR
							{U / V / S}			
RANDOM	ACTUAL	DA [-xxxx] -D-name	{STANDARD / data-name}	INPUT	[LOCK]	SEEK READ [INTO] INVALID KEY	{F / U / V / S}		SAME [RECORD] AREA RERUN ON RECORDS	AFTER LABEL ERROR
				OUTPUT	[LOCK]	SEEK WRITE[1] [FROM] INVALID KEY	F	RECORD-OVERFLOW		
							{U / V / S}		SAME [RECORD] AREA TRACK-LIMIT RERUN ON RECORDS	
				I-O	[LOCK]	SEEK READ [INTO] INVALID KEY WRITE[2] [FROM] INVALID KEY	{F / U / V / S}		SAME [RECORD] AREA RERUN ON RECORDS	
RANDOM	ACTUAL	DA [-xxxx] -W-name	{STANDARD / data-name}	INPUT	[LOCK]	SEEK READ [INTO] INVALID KEY	{F / U / V / S}		SAME [RECORD] AREA RERUN ON RECORDS	AFTER LABEL ERROR
				OUTPUT	[LOCK]	SEEK WRITE[1] [FROM] INVALID KEY	F	RECORD-OVERFLOW	SAME [RECORD] AREA TRACK-LIMIT RERUN ON RECORDS	
							{U / V / S}			
RANDOM	ACTUAL	DA [-xxxx] -W-name	{STANDARD / data-name}	I-O	[LOCK]	SEEK READ [INTO] INVALID KEY WRITE[3] [FROM] INVALID KEY REWRITE[4] (FROM) [INVALID KEY]	{F / U / V / S}		SAME [RECORD] AREA RERUN ON RECORDS	AFTER LABEL ERROR

[1]Create [2]Update and add [3]Add [4]Update [5]These APPLY clauses have meaning only for OUTPUT files; however, the compiler accepts them when the same file is opened for INPUT or I-O.

INDEXED FILES (mass storage devices only)—Required and Optional Entries

Required Entries							Optional Entries					
ACCESS	KEY	System-name	LABEL RECORDS	OPEN	CLOSE	Access Verbs	APPLY[4]	RESERVE	Other ENVIRONMENT DIVISION Clauses	RECORDING MODE	BLOCK CONTAINS	USE
[SEQUENTIAL]	RECORD	DA [-xxxx] -I-name	STANDARD	INPUT	[LOCK]	READ [INTO] AT END		{integer NO}	SAME [RECORD] AREA RERUN	F	m	ERROR
	RECORD NOMINAL					START [INVALID KEY]						
	RECORD [NOMINAL]					START USING KEY [INVALID KEY]						
	RECORD			OUTPUT	[LOCK]	WRITE[1] [FROM] INVALID KEY						
	RECORD			I-O	[LOCK]	READ [INTO] AT END REWRITE[2] [FROM] [INVALID KEY]						
	RECORD NOMINAL					START [INVALID KEY]						
	RECORD [NOMINAL]					START USING KEY [INVALID KEY]						
RANDOM	RECORD NOMINAL	DA [-xxxx] -I-name	STANDARD	INPUT	[LOCK]	READ [INTO] INVALID KEY	REORG-CRITERIA CORE-INDEX	NO	SAME [RECORD] AREA RERUN ON RECORDS	F	m	ERROR
				I-O	[LOCK]	READ [INTO] INVALID KEY WRITE[3] [FROM] INVALID KEY REWRITE[2] [FROM] [INVALID KEY]			SAME [RECORD] AREA TRACK-AREA RERUN ON RECORDS			

[1]Create [2]Update [3]Add [4]These APPLY clauses have meaning only for OUTPUT files; however, the compiler accepts them when the same file is opened for INPUT or I-O.

RELATIVE FILES (mass storage devices only)—Required and Optional Entries

Required Entries							Optional Entries		
ACCESS	KEY	System-name	LABEL RECORDS	OPEN	CLOSE	Access Verbs	Other ENVIRONMENT DIVISION Clauses	RECORDING MODE	USE
[SEQUENTIAL]	[NOMINAL]	DA [-xxxx] -R-name	{STANDARD data-name}	INPUT	[UNIT] [LOCK]	READ [INTO] AT END	SAME [RECORD]AREA RERUN	F	ERROR AFTER LABELS
				OUTPUT	[UNIT] [LOCK]	WRITE[1] [FROM] INVALID KEY	APPLY RECORD-OVERFLOW		
RANDOM	NOMINAL	DA [-xxxx] -R-name	{STANDARD data-name}	INPUT	[LOCK]	READ [INTO] INVALID KEY	SAME [RECORD]AREA RERUN ON RECORDS	F	ERROR AFTER LABELS
				I-O	[LOCK]	READ [INTO] INVALID KEY REWRITE[2] [FROM] [INVALID KEY]	APPLY RECORD-OVERFLOW		

[1]Create [2]Update

Basic Formats

IBM FULL ANS COBOL | **ANSI 1974 COBOL**

IDENTIFICATION DIVISION FORMATS

IBM FULL ANS COBOL

```
{IDENTIFICATION DIVISION.}
{ID DIVISION.            }
 PROGRAM-ID. program-name.
 AUTHOR. [comment-entry] ...
 INSTALLATION. [comment-entry] ...
 DATE-WRITTEN. [comment-entry] ...
 DATE-COMPILED. [comment-entry] ...
 SECURITY. [comment-entry] ...
 REMARKS. [comment-entry] ...
```

ANSI 1974 COBOL

```
 IDENTIFICATION DIVISION.
 PROGRAM-ID. program-name.
[AUTHOR. [comment-entry] ...]
[INSTALLATION. [comment-entry] ...]
[DATE-WRITTEN. [comment-entry] ...]
[DATE-COMPILED. [comment-entry] ...]
[SECURITY. [comment-entry] ...]
```

IBM FULL ANS COBOL | **ANSI 1974 COBOL**

ENVIRONMENT DIVISION FORMATS

IBM FULL ANS COBOL

```
ENVIRONMENT DIVISION.
CONFIGURATION SECTION.
SOURCE-COMPUTER. computer-name.
                                                    {WORDS     }
OBJECT-COMPUTER. computer-name [MEMORY SIZE integer {CHARACTERS}]
                                                    {MODULES   }
      [SEGMENT-LIMIT IS priority-number].
SPECIAL-NAMES. [function-name IS mnemonic-name] ...
      [CURRENCY SIGN IS literal]
      [DECIMAL-POINT IS COMMA].
```

ANSI 1974 COBOL

```
ENVIRONMENT DIVISION.
CONFIGURATION SECTION.
SOURCE-COMPUTER. computer-name [WITH DEBUGGING MODE].
OBJECT-COMPUTER. computer-name

                               {WORDS     }
      [, MEMORY SIZE integer   {CHARACTERS}]
                               {MODULES   }
      [, PROGRAM COLLATING SEQUENCE IS alphabet-name]
      [, SEGMENT-LIMIT IS segment-number].

[SPECIAL-NAMES. [, implementor-name

   {IS mnemonic-name [, ON STATUS IS condition-name-1 [, OFF STATUS IS condition-name-2]]}
   {IS mnemonic-name [, OFF STATUS IS condition-name-2 [, ON STATUS IS condition-name-1]]}
   {ON STATUS IS condition-name-1 [, OFF STATUS IS condition-name-2]                     }
   {OFF STATUS IS condition-name-2 [, ON STATUS IS condition-name-1]                     }] ...
```

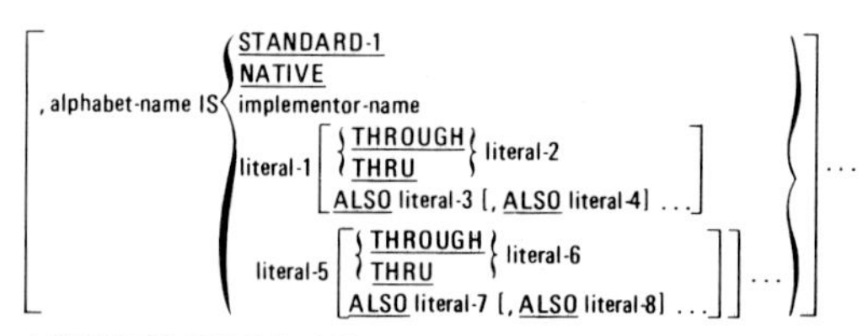

```
                       {STANDARD-1
                       {NATIVE
   [, alphabet-name IS {implementor-name
                       {           [{THROUGH} literal-2                  ]
                       {literal-1  [{THRU   }                            ]
                       {           [ALSO literal-3 [, ALSO literal-4] ...]
                       {
                       {           [{THROUGH} literal-6                  ]
                       {literal-5  [{THRU   }                            ] ...
                       {           [ALSO literal-7 [, ALSO literal-8] ...]   } ] ...
```

```
      [, CURRENCY SIGN IS literal-9]
      [, DECIMAL-POINT IS COMMA].]
```

ENVIRONMENT DIVISION FORMATS

IBM FULL ANS COBOL

```
INPUT-OUTPUT SECTION.
FILE-CONTROL.
     {SELECT [OPTIONAL] file name
     ASSIGN TO [integer-1] system-name-1 [system-name-2] . . .
          [FOR MULTIPLE {REEL | UNIT}]
     RESERVE {NO | integer-1} ALTERNATE [AREA | AREAS]
     {FILE-LIMIT IS | FILE-LIMITS ARE} {data-name-1 | literal-1} THRU {data-name-2 | literal-2}
          [{data-name-3 | literal-3} THRU {data-name-4 | literal-4}] . . .
     ACCESS MODE IS {SEQUENTIAL | RANDOM}
     PROCESSING MODE IS SEQUENTIAL
     ACTUAL KEY IS data-name
     NOMINAL KEY IS data-name
     RECORD KEY IS data-name
```

ANSI 1974 COBOL

```
INPUT-OUTPUT SECTION.
FILE-CONTROL.
FORMAT 1:
SELECT [OPTIONAL] file-name
     ASSIGN TO implementor-name-1 [, implementor-name-2] . . .
     [; RESERVE integer-1 [AREA | AREAS]]
     [; ORGANIZATION IS SEQUENTIAL]
     [; ACCESS MODE IS SEQUENTIAL]
     [; FILE STATUS IS data-name-1] .

FORMAT 2:
SELECT file-name
     ASSIGN TO implementor-name-1 [, implementor-name-2] . . .
     [; RESERVE integer-1 [AREA | AREAS]]
      ; ORGANIZATION IS RELATIVE
     [; ACCESS MODE IS { SEQUENTIAL [, RELATIVE KEY IS data-name-1]
                       | {RANDOM | DYNAMIC} , RELATIVE KEY IS data-name-1 }]
[; FILE STATUS IS data-name-2] .
```

ENVIRONMENT DIVISION FORMATS

IBM FULL ANS COBOL

```
TRACK-AREA IS {data-name | integer} CHARACTERS
TRACK-LIMIT IS integer [TRACK | TRACKS] .} . . .
```

ANSI 1974 COBOL

```
FORMAT 3:
SELECT file-name
     ASSIGN TO implementor-name-1 [, implementor-name-2] . . .
     [; RESERVE integer-1 [AREA | AREAS]]
      ; ORGANIZATION IS INDEXED
     [; ACCESS MODE IS {SEQUENTIAL | RANDOM | DYNAMIC}]
      ; RECORD KEY IS data-name-1
     [; ALTERNATE RECORD KEY IS data-name-2 [WITH DUPLICATES]] . . .
     [; FILE STATUS IS data-name-3] .

FORMAT 4:
SELECT file-name ASSIGN TO implementor-name-1 [, implementor-name-2] . . .
```

ENVIRONMENT DIVISION FORMATS

IBM FULL ANS COBOL

```
I-O-CONTROL
                                      integer RECORDS
     RERUN ON system-name EVERY                 REEL      OF file-name
                                      [END OF]  UNIT
     SAME [RECORD | SORT] AREA FOR file-name-1 {file-name-2} . . .
     MULTIPLE FILE TAPE CONTAINS file-name-1 [POSITION integer-1]
          [file-name-2 [POSITION integer-2]] . . .
     APPLY WRITE-ONLY ON file-name-1 [file-name-2] . . .
     APPLY CORE-INDEX ON file-name-1 [file-name-2] . . .
     APPLY RECORD-OVERFLOW ON file-name-1 [file-name-2] . . .
     APPLY REORG-CRITERIA TO data-name ON file-name
```

NOTE: Format 2 of the RERUN Clause (for Sort Files) is included with Formats for the SORT feature.

ANSI 1974 COBOL

```
I-O-CONTROL.
     [; RERUN [ON {file-name-1 | implementor-name}]
          EVERY { {[END OF] {REEL | UNIT} | integer-1 RECORDS} OF file-name-2
                | integer-2 CLOCK-UNITS
                | condition-name } ] . . .
     [; SAME [RECORD | SORT | SORT-MERGE] AREA FOR file-name-3 {, file-name-4} . . . ] .
     [; MULTIPLE FILE TAPE CONTAINS file-name-5 [POSITION integer-3]
          [, file-name-6 [POSITION integer-4]] . . . ] . . . .]]
```

IBM FULL ANS COBOL | **ANSI 1974 COBOL**

DATA DIVISION FORMATS

IBM FULL ANS COBOL

```
DATA DIVISION.
FILE SECTION.
FD file-name
    BLOCK CONTAINS [integer-1 TO] integer-2 {CHARACTERS / RECORDS}
    RECORD CONTAINS [integer-1 TO] integer-2 CHARACTERS
    RECORDING MODE IS mode
    LABEL {RECORD IS / RECORDS ARE} {OMITTED
                                     STANDARD
                                     data-name-1 [data-name-2] . . . [TOTALING AREA IS
                                       data-name-3 TOTALED AREA IS data-name-4]}
    VALUE OF data-name-1 IS {data-name-2 / literal-1} [data-name-3 IS {data-name-4 / literal-2}] . . .
    DATA {RECORD IS / RECORDS ARE} data-name-1 [data-name-2] . . .
```

NOTE: Format for the REPORT Clause is included with Formats for the REPORT WRITER feature.

ANSI 1974 COBOL

```
DATA DIVISION.
[FILE SECTION.
[FD file-name
    [; BLOCK CONTAINS [integer-1 TO] integer-2 {RECORDS / CHARACTERS}]
    [; RECORD CONTAINS [integer-3 TO] integer-4 CHARACTERS]
    ; LABEL {RECORD IS / RECORDS ARE} {STANDARD / OMITTED}
    [; VALUE OF implementor-name-1 IS {data-name-1 / literal-1}
        [, implementor-name-2 IS {data-name-2 / literal-2}] . . .]
    [; DATA {RECORD IS / RECORDS ARE} data-name-3 [, data-name-4] . . .]
    [; LINAGE IS {data-name-5 / integer-5} LINES [, WITH FOOTING AT {data-name-6 / integer-6}]
        [, LINES AT TOP {data-name-7 / integer-7}] [, LINES AT BOTTOM {data-name-8 / integer-8}]]
    [; CODE-SET IS alphabet-name]
```

IBM FULL ANS COBOL | **ANSI 1974 COBOL**

DATA DIVISION FORMATS

IBM FULL ANS COBOL

```
01-49 {data-name-1 / FILLER}
    REDEFINES data-name-2
    BLANK WHEN ZERO
    {JUSTIFIED / JUST} RIGHT
    {PICTURE / PIC} IS character string
    [SIGN IS] {LEADING / TRAILING} [SEPARATE CHARACTER]
    {SYNCHRONIZED / SYNC} [LEFT / RIGHT]
    [USAGE IS] {INDEX
                DISPLAY
                {COMPUTATIONAL / COMP}
                {COMPUTATIONAL-1 / COMP-1}
                {COMPUTATIONAL-2 / COMP-2}
                {COMPUTATIONAL-3 / COMP-3}
                DISPLAY-ST}
88    condition-name {VALUE IS / VALUES ARE} literal-1 [THRU literal-2]
          [literal-3 [THRU literal-4]] . . . .
66    data-name-1 RENAMES data-name-2 [THRU data-name-3].
```

NOTE: Formats for the OCCURS Clause are included with Formats for the TABLE HANDLING feature.

ANSI 1974 COBOL

```
FORMAT 1:
level-number {data-name-1 / FILLER}
    [; REDEFINES data-name-2]
    [; {PICTURE / PIC} IS character-string]
    [; [USAGE IS] {COMPUTATIONAL / COMP / DISPLAY / INDEX}]
    [; [SIGN IS] {LEADING / TRAILING} [SEPARATE CHARACTER]]
    [; {SYNCHRONIZED / SYNC} [LEFT / RIGHT]]
    [; {JUSTIFIED / JUST} RIGHT]
    [; BLANK WHEN ZERO]
    [; VALUE IS literal].

FORMAT 2
66 data-name-1; RENAMES data-name-2 [{THROUGH / THRU} data-name-3] .

FORMAT 3:
88 condition-name; {VALUE IS / VALUES ARE} literal-1 [{THROUGH / THRU} literal-2]
    [, literal-3 [{THROUGH / THRU} literal-4]] . . . .
```

IBM FULL ANS COBOL | **ANSI 1974 COBOL**

DATA DIVISION FORMATS

```
WORKING-STORAGE SECTION.
77      data-name-1
01-49 {data-name-1}
      {FILLER     }
        REDEFINES data-name-2
        BLANK WHEN ZERO
        {JUSTIFIED} RIGHT
        {JUST     }
        {PICTURE} IS character string
        {PIC    }
        [SIGN IS] {LEADING } [SEPARATE CHARACTER]
                  {TRAILING}
        {SYNCHRONIZED} [LEFT ]
        {SYNC        } [RIGHT]
                   INDEX
                   DISPLAY
                   {COMPUTATIONAL}
                   {COMP         }
                   {COMPUTATIONAL-1}
        [USAGE IS] {COMP-1         }
                   {COMPUTATIONAL-2}
                   {COMP-2         }
                   {COMPUTATIONAL-3}
                   {COMP-3         }
                   DISPLAY-ST
        VALUE IS literal.
88      condition-name {VALUE IS  } literal-1 [THRU literal-2]
                       {VALUES ARE}
             [literal-3 [THRU literal-4]] . . .
66      data-name-1 RENAMES data-name-2 [THRU data-name-3].
NOTE: Formats for the OCCURS Clause are included with Formats for the TABLE
      HANDLING feature.
```

```
[WORKING-STORAGE SECTION.
[77-level-description-entry]
[record-description-entry  ] . . . ]
```

IBM FULL ANS COBOL | **ANSI 1974 COBOL**

DATA DIVISION FORMATS

```
LINKAGE SECTION
77      data-name-1
01-49 {data-name-1}
      {FILLER     }
        REDEFINES data-name-2
        BLANK WHEN ZERO
        {JUSTIFIED} RIGHT
        {JUST     }
        {PICTURE} IS character string
        {PIC    }
        [SIGN IS {LEADING } [SEPARATE CHARACTER]
                 {TRAILING}
        {SYNCHRONIZED} [LEFT ]
        {SYNC        } [RIGHT]
                   INDEX
                   DISPLAY
                   {COMPUTATIONAL}
                   {COMP         }
                   {COMPUTATIONAL-1}
        [USAGE IS] {COMP-1         }
                   {COMPUTATIONAL-2}
                   {COMP-2         }
                   {COMPUTATIONAL-3}
                   {COMP-3         }
                   DISPLAY-ST
88      condition-name {VALUE IS  } literal-1 [THRU literal-2]
                       {VALUES ARE}
             [literal-3 [THRU literal-4]] . . .
66      data-name-1 RENAMES data-name-2 [THRU data-name-3].
NOTE: Formats for the OCCURS Clause are included with Formats for the TABLE
      HANDLING feature.
```

```
[LINKAGE SECTION.
[77-level-description-entry]
[record-description-entry  ] . . . ]
```

IBM FULL ANS COBOL

ANSI 1974 COBOL

PROCEDURE DIVISION FORMATS

IBM FULL ANS COBOL

```
{ PROCEDURE DIVISION.                                                 }
{ PROCEDURE DIVISION USING identifier-1 [identifier-2] . . . .        }
 ACCEPT Statement

FORMAT 1
                              { SYSIN          }
  ACCEPT identifier [FROM     { CONSOLE        } ]
                              { mnemonic-name  }

FORMAT 2
                              { DATE }
  ACCEPT identifier FROM      { DAY  }
                              { TIME }

ADD Statement

FORMAT 1
       { identifier-1 }  [ identifier-2 ]
  ADD  { literal-1    }  [ literal-2    ] . . . TO identifier-m [ROUNDED]
       [identifier-n [ROUNDED]] . . . [ON SIZE ERROR imperative-statement]

FORMAT 2
       { identifier-1 }  { identifier-2 }  [ identifier-3 ]
  ADD  { literal-1    }  { literal-2    }  [ literal-3    ] . . . GIVING
       identifier-m [ROUNDED] [ON SIZE ERROR imperative-statement]

FORMAT 3
       { CORRESPONDING }
  ADD  { CORR          }  identifier-1 TO identifier-2 [ROUNDED]
       [ON SIZE ERROR imperative-statement]

ALTER Statement
  ALTER procedure-name-1 TO [PROCEED TO] procedure-name-2
       [procedure-name-3 TO [PROCEED TO] procedure-name-4] . . .

CALL Statement

FORMAT 1
  CALL literal-1 [USING identifier-1 [identifier-2] . . .]

FORMAT 2
  CALL identifier-1 [USING identifier-2 [identifier-3] . . .]
          { literal-1    }  [ literal-2    ]
  CANCEL  { identifier-1 }  [ identifier-2 ]
```

ANSI 1974 COBOL

```
FORMAT 1:
PROCEDURE DIVISION [USING data-name-1 [, data-name-2] . . .].
ACCEPT identifier [FROM mnemonic-name]
                          { DATE }
ACCEPT identifier FROM    { DAY  }
                          { TIME }
ACCEPT cd-name MESSAGE COUNT
     { identifier-1 }  { , identifier-2 }
ADD  { literal-1    }  { , literal-2    } . . . TO identifier-m [ROUNDED]
   [, identifier-n [ROUNDED]] . . . [; ON SIZE ERROR imperative-statement]
     { identifier-1 }    { identifier-2 }  [ , identifier-3 ]
ADD  { literal-1    }  , { literal-2    }  [ , literal-3    ] . . .
   GIVING identifier-m [ROUNDED] [, identifier-n [ROUNDED]] . . .
   [; ON SIZE ERROR imperative-statement]
     { CORRESPONDING }
ADD  { CORR          }  identifier-1 TO identifier-2 [ROUNDED]
   [; ON SIZE ERROR imperative-statement]
ALTER procedure-name-1 TO [PROCEED TO] procedure-name-2
   [, procedure-name-3 TO [PROCEED TO] procedure-name-4] . . .
      { identifier-1 }
CALL  { literal-1    }  [USING data-name-1 [, data-name-2] . . .]
   [; ON OVERFLOW imperative-statement]
        { identifier-1 }  [ , identifier-2 ]
CANCEL  { literal-1    }  [ , literal-2    ] . . .
```

IBM FULL ANS COBOL | **ANSI 1974 COBOL**

PROCEDURE DIVISION FORMATS

IBM FULL ANS COBOL

CLOSE Statement

FORMAT 1

```
CLOSE file-name-1 [REEL|UNIT] [WITH {NO REWIND|LOCK}]
    [file-name-2 [REEL|UNIT] [WITH {NO REWIND|LOCK}]] ...
```

FORMAT 2

```
CLOSE file-name-1 [WITH {NO REWIND|LOCK|DISP}]
    [file-name-2 [WITH {NO REWIND|LOCK|DISP}]] ...
```

FORMAT 3

```
CLOSE file-name-1 {REEL|UNIT} [WITH {NO REWIND|LOCK|POSITIONING}]
    [file-name-2 {REEL|UNIT} [WITH {NO REWIND|LOCK|POSITIONING}]] ...
```

COMPUTE Statement

```
COMPUTE identifier-1 [ROUNDED] = {arithmetic-expression|identifier-2|literal-1}
    [ON SIZE ERROR imperative-statement]
```

DECLARATIVE Section

```
PROCEDURE DIVISION.
DECLARATIVES.
{section-name SECTION. USE sentence.
{paragraph-name. {sentence}...}...}...
END DECLARATIVES.
```

ANSI 1974 COBOL

```
CLOSE file-name-1 [ {REEL|UNIT} [WITH NO REWIND|FOR REMOVAL]
                    WITH {NO REWIND|LOCK} ]
    [, file-name-2 [ {REEL|UNIT} [WITH NO REWIND|FOR REMOVAL]
                    WITH {NO REWIND|LOCK} ]] ...
```

```
CLOSE file-name-1 [WITH LOCK] [, file-name-2 [WITH LOCK]] .
```

```
COMPUTE identifier-1 [ROUNDED] [, identifier-2 [ROUNDED]] ...
    = arithmetic-expression [; ON SIZE ERROR imperative-statement]
```

```
[DECLARATIVES.
{section-name SECTION [segment-number] . declarative-sentence
[paragraph-name.[sentence] ...] ...} ...
END DECLARATIVES.]
{section-name SECTION [segment-number] .
[paragraph-name. [sentence] ...] ...} ...
```

IBM FULL ANS COBOL | **ANSI 1974 COBOL**

PROCEDURE DIVISION FORMATS

IBM FULL ANS COBOL

DISPLAY Statement

DISPLAY { identifier-1 | literal-1 } [identifier-2 | literal-2] . . . [UPON { CONSOLE | SYSPUNCH | SYSOUT | mnemonic-name }]

DIVIDE Statement

FORMAT 1

DIVIDE { identifier-1 | literal-1 } INTO identifier-2 [ROUNDED]
[ON SIZE ERROR imperative-statement]

FORMAT 2

DIVIDE { identifier-1 | literal-1 } { INTO | BY } { identifier-2 | literal-2 } GIVING identifier-3
[ROUNDED] [REMAINDER identifier-4] [ON SIZE ERROR imperative-statement]

ENTER Statement

ENTER language-name [routine-name]

ENTRY Statement

ENTRY literal-1 [USING identifier-1 [identifier-2] . . .]

ENTER Statement

ENTER language-name [routine-name]

EXAMINE Statement

FORMAT 1

EXAMINE identifier TALLYING { UNTIL FIRST | ALL | LEADING } literal-1
[REPLACING BY literal-2]

FORMAT 2

EXAMINE identifier REPLACING { ALL | LEADING | FIRST | UNTIL FIRST } literal-1 BY literal-2

EXIT Statement

paragraph-name. EXIT [PROGRAM].

GOBACK Statement

GOBACK.

GO TO Statement

FORMAT 1

GO TO procedure-name-1

FORMAT 2

GO TO procedure-name-1 [procedure-name-2] . . . DEPENDING ON identifier

FORMAT 3

GO TO.

ANSI 1974 COBOL

DELETE file-name RECORD [; INVALID KEY imperative-statement]

DISABLE { INPUT [TERMINAL] | OUTPUT } cd-name WITH KEY { identifier-1 | literal-1 }

DISPLAY { identifier-1 | literal-1 } [, identifier-2 | , literal-2] . . . [UPON mnemonic-name]

DIVIDE { identifier-1 | literal-1 } INTO identifier-2 [ROUNDED]
[, identifier-3 [ROUNDED]] . . . [; ON SIZE ERROR imperative-statement]

DIVIDE { identifier-1 | literal-1 } INTO { identifier-2 | literal-2 } GIVING identifier-3 [ROUNDED]
[, identifier-4 [ROUNDED]] . . . [; ON SIZE ERROR imperative-statement]

DIVIDE { identifier-1 | literal-1 } BY { identifier-2 | literal-2 } GIVING identifier-3 [ROUNDED]
[, identifier-4 [ROUNDED]] . . . [; ON SIZE ERROR imperative-statement]

DIVIDE { identifier-1 | literal-1 } INTO { identifier-2 | literal-2 } GIVING identifier-3 [ROUNDED]
REMAINDER identifier-4 [; ON SIZE ERROR imperative-statement]

DIVIDE { identifier-1 | literal-1 } BY { identifier-2 | literal-2 } GIVING identifier-3 [ROUNDED]
REMAINDER identifier-4 [; ON SIZE ERROR imperative-statement]

ENABLE { INPUT [TERMINAL] | OUTPUT } cd-name WITH KEY { identifier-1 | literal-1 }

ENTER language-name [routine-name].

EXIT [PROGRAM].

GO TO [procedure-name-1]

GO TO procedure-name-1 [, procedure-name-2] . . . , procedure-name-n
DEPENDING ON identifier

IBM FULL ANS COBOL **ANSI 1974 COBOL**

PROCEDURE DIVISION FORMATS

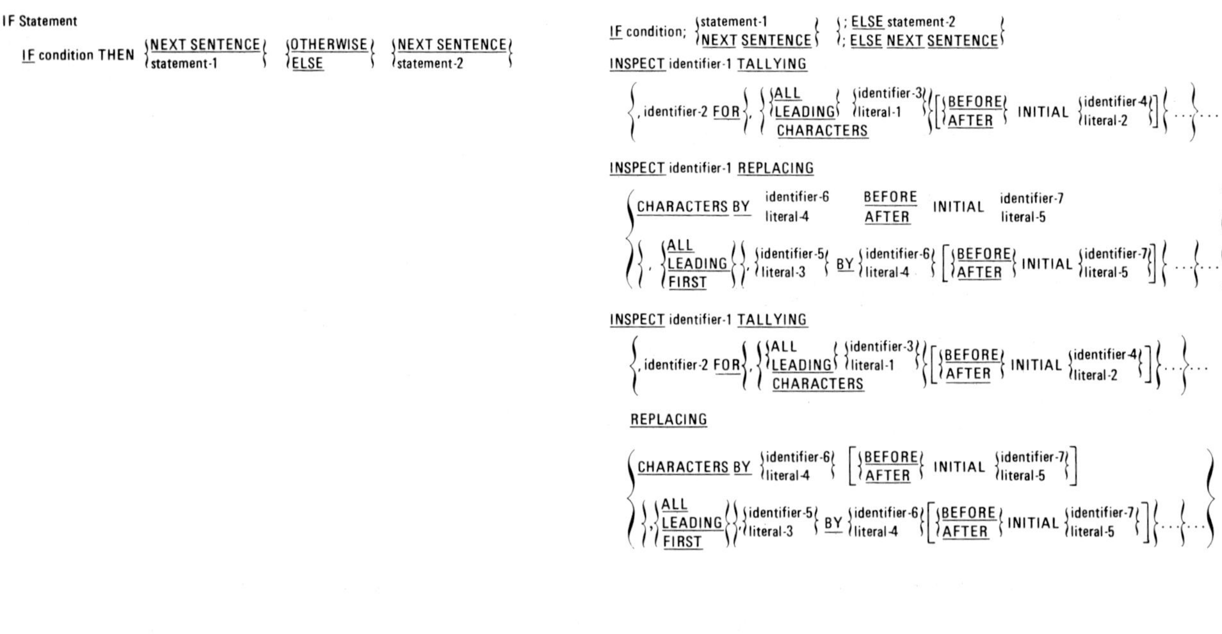
IF Statement

IF condition THEN {NEXT SENTENCE | statement-1} {OTHERWISE | ELSE} {NEXT SENTENCE | statement-2}

IF condition; {statement-1 | NEXT SENTENCE} {; ELSE statement-2 | ; ELSE NEXT SENTENCE}

INSPECT identifier-1 TALLYING

{, identifier-2 FOR {, {ALL | LEADING} {identifier-3 | literal-1} | CHARACTERS} [{BEFORE | AFTER} INITIAL {identifier-4 | literal-2}]} . . . } . . .

INSPECT identifier-1 REPLACING

{CHARACTERS BY {identifier-6 | literal-4} [{BEFORE | AFTER} INITIAL {identifier-7 | literal-5}]
| {ALL | LEADING | FIRST} {, {identifier-5 | literal-3} BY {identifier-6 | literal-4} [{BEFORE | AFTER} INITIAL {identifier-7 | literal-5}]} . . . } . . .

INSPECT identifier-1 TALLYING

{, identifier-2 FOR {, {ALL | LEADING} {identifier-3 | literal-1} | CHARACTERS} [{BEFORE | AFTER} INITIAL {identifier-4 | literal-2}]} . . . } . . .

REPLACING

{CHARACTERS BY {identifier-6 | literal-4} [{BEFORE | AFTER} INITIAL {identifier-7 | literal-5}]
| {ALL | LEADING | FIRST} {, {identifier-5 | literal-3} BY {identifier-6 | literal-4} [{BEFORE | AFTER} INITIAL {identifier-7 | literal-5}]} . . . } . . .

IBM FULL ANS COBOL **ANSI 1974 COBOL**

PROCEDURE DIVISION FORMATS

MOVE Statement

FORMAT 1

MOVE {identifier-1 | literal-1} TO identifier-2 [identifier-3] . . .

FORMAT 2

MOVE {CORRESPONDING | CORR} identifier-1 TO identifier-2

MULTIPLY Statement

FORMAT 1

MULTIPLY {identifier-1 | literal-1} BY identifier-2 [ROUNDED]
[ON SIZE ERROR imperative-statement]

FORMAT 2

MULTIPLY {identifier-1 | literal-1} BY {identifier-2 | literal-2} GIVING identifier-3
[ROUNDED] [ON SIZE ERROR imperative-statement]

NOTE Statement

NOTE character string

OPEN Statement

FORMAT 1

OPEN [INPUT {file-name [REVERSED | WITH NO REWIND] } . . .]
[OUTPUT {file-name [WITH NO REWIND] } . . .]
[I-O {file-name } . . .]

FORMAT 2

OPEN [INPUT {file-name [REVERSED | WITH NO REWIND] [LEAVE | REREAD | DISP] } . . .]
[OUTPUT {file-name [WITH NO REWIND] [LEAVE | REREAD | DISP] } . . .]
[I-O {file-name } . . .]

MERGE file-name-1 ON {ASCENDING | DESCENDING} KEY data-name-1 [, data-name-2] . . .
[ON {ASCENDING | DESCENDING} KEY data-name-3 [, data-name-4] . . .] . . .
[COLLATING SEQUENCE IS alphabet-name]
USING file-name-2, file-name-3 [, file-name-4] . . .
{OUTPUT PROCEDURE IS section-name-1 [{THROUGH | THRU} section-name-2] | GIVING file-name-5}

MOVE {identifier-1 | literal} TO identifier-2 [, identifier-3] . . .

MOVE {CORRESPONDING | CORR} identifier-1 TO identifier-2

MULTIPLY {identifier-1 | literal-1} BY identifier-2 [ROUNDED]
[, identifier-3 [ROUNDED]] . . . [; ON SIZE ERROR imperative-statement]

MULTIPLY {identifier-1 | literal-1} BY {identifier-2 | literal-2} GIVING identifier-3 [ROUNDED]
[, identifier-4 [ROUNDED]] . . . [; ON SIZE ERROR imperative-statement]

OPEN {INPUT file-name-1 [REVERSED | WITH NO REWIND] [, file-name-2 [REVERSED | WITH NO REWIND]] . . .
| OUTPUT file-name-3 [WITH NO REWIND] [, file-name-4 [WITH NO REWIND]] . . .
| I-O file-name-5 [, file-name-6] . . .
| EXTEND file-name-7 [, file-name-8] . . . } . . .

OPEN {INPUT file-name-1 [, file-name-2] . . .
| OUTPUT file-name-3 [, file-name-4] . . .
| I-O file-name-5 [, file-name-6] . . . } . . .

IBM FULL ANS COBOL | **ANSI 1974 COBOL**

PROCEDURE DIVISION FORMATS

IBM FULL ANS COBOL

PERFORM Statement

FORMAT 1

```
PERFORM procedure-name-1 [THRU procedure-name-2]
```

FORMAT 2

```
PERFORM procedure-name-1 [THRU procedure-name-2] {identifier-1} TIMES
                                                 {integer-1   }
```

FORMAT 3

```
PERFORM procedure-name-1 [THRU procedure-name-2] UNTIL condition-1
```

FORMAT 4

```
PERFORM procedure-name-1 [THRU procedure-name-2]

VARYING {index-name-1} FROM {index-name-2} BY {literal-3   } UNTIL condition-1
        {identifier-1}      {literal-2   }    {identifier-3}
                            {identifier-2}

[AFTER {index-name-4} FROM {index-name-5} BY {literal-6   } UNTIL condition-2
       {identifier-4}      {literal-5   }    {identifier-6}
                           {identifier-5}

[AFTER {index-name-7} FROM {index-name-8} BY {literal-9   } UNTIL condition-3]]
       {identifier-7}      {literal-8   }    {identifier-9}
                           {identifier-8}
```

ANSI 1974 COBOL

```
PERFORM procedure-name-1 [{THROUGH} procedure-name-2]
                          {THRU   }

PERFORM procedure-name-1 [{THROUGH} procedure-name-2] {identifier-1} TIMES
                          {THRU   }                   {integer-1   }

PERFORM procedure-name-1 [{THROUGH} procedure-name-2] UNTIL condition-1
                          {THRU   }

PERFORM procedure-name-1 [{THROUGH} procedure-name-2]
                          {THRU   }

  VARYING {identifier-2} FROM {indentifier-3}
          {index-name-1}      {index-name-2 }
                              {literal-1    }

     BY {identifier-4} UNTIL condition-1
        {literal-3   }

  [AFTER {identifier-5} FROM {identifier-6}
         {index-name-3}      {index-name-4}
                             {literal-3   }

     BY {identifier-7} UNTIL condition-2
        {literal-4   }

   AFTER {identifier-8} FROM {identifier-9}
         {index-name-5}      {index-name-6}
                             {literal-5   }

     BY {identifier-10} UNTIL condition-3]]
        {literal-6    }
```

IBM FULL ANS COBOL | **ANSI 1974 COBOL**

PROCEDURE DIVISION FORMATS

IBM FULL ANS COBOL

READ Statement

```
READ file name RECORD [INTO identifier]
     {AT END     } imperative-statement
     {INVALID KEY}
```

REWRITE Statement

```
REWRITE record-name [FROM identifier] [INVALID KEY imperative-statement]
```

SEEK Statement

```
SEEK file-name RECORD
```

START Statement

FORMAT 1

```
START file-name [INVALID KEY imperative-statement]
```

FORMAT 2

```
START file-name
    USING KEY data-name {EQUAL TO} identifier
                        {=       }
    [INVALID KEY imperative-statement]
```

ANSI 1974 COBOL

```
READ file-name RECORD [INTO identifier] [; AT END imperative-statement]
READ file-name [NEXT] RECORD [INTO identifier]
   [; AT END imperative-statement]
READ file-name RECORD [INTO identifier] [; INVALID KEY imperative-statement]
READ file-name RECORD [INTO identifier]
   [; KEY IS data-name]
   [; INVALID KEY imperative-statement]
RECEIVE cd-name {MESSAGE} INTO identifier-1 [; NO DATA imperative-statement]
                {SEGMENT}
REWRITE record-name [FROM identifier]
REWRITE record-name [FROM identifier] [; INVALID KEY imperative-statement]
SEND cd-name FROM identifier-1
                                       {WITH identifier-2}
SEND cd-name [FROM identifier-1]       {WITH ESI         }
                                       {WITH EMI         }
                                       {WITH EGI         }

   [{BEFORE} ADVANCING {{identifier-3} [LINE ]}]
   [{AFTER }           {{integer     } [LINES]}]
                       {                      }
                       {{mnemonic-name}       }
                       {{PAGE         }       }

                      [     {IS EQUAL TO      }          ]
                      [     {IS =             }          ]
START file-name       [ KEY {IS GREATER THAN  } data-name]
                      [     {IS >             }          ]
                      [     {IS NOT LESS THAN }          ]
                      [     {IS NOT <         }          ]
   [; INVALID KEY imperative-statement]
```

IBM FULL ANS COBOL — ANSI 1974 COBOL

PROCEDURE DIVISION FORMATS

IBM FULL ANS COBOL

STOP Statement

```
STOP {RUN | literal}
```

SUBTRACT Statement

FORMAT 1

```
SUBTRACT {identifier-1 | literal-1} [identifier-2 | literal-2] . . . FROM identifier-m [ROUNDED]
    [identifier-n [ROUNDED]] . . . [ON SIZE ERROR imperative-statement]
```

FORMAT 2

```
SUBTRACT {identifier-1 | literal-1} [identifier-2 | literal-2] . . . FROM {identifier-m | literal-m} GIVING identifier-n
    [ROUNDED] [ON SIZE ERROR imperative-statement]
```

FORMAT 3

```
SUBTRACT {CORRESPONDING | CORR} identifier-1 FROM identifier-2 [ROUNDED]
    [ON SIZE ERROR imperative-statement]
```

TRANSFORM Statement

```
TRANSFORM identifier-3 CHARACTERS FROM {figurative-constant-1 | nonnumeric-literal-1 | identifier-1}

    TO {figurative-constant-2 | nonnumeric-literal-2 | identifier-2}
```

ANSI 1974 COBOL

```
STOP {RUN | literal}

STRING {identifier-1 | literal-1} [, identifier-2 | , literal-2] . . . DELIMITED BY {identifier-3 | literal-3 | SIZE}

    [, {identifier-4 | literal-4} [, identifier-5 | , literal-5] . . . DELIMITED BY {identifier-6 | literal-6 | SIZE}]

    INTO identifier-7 [WITH POINTER identifier-8]
    [; ON OVERFLOW imperative-statement]

SUBTRACT {identifier-1 | literal-1} [, identifier-2 | , literal-2] . . . FROM identifier-m [ROUNDED]
    [, identifier-n [ROUNDED]] . . . [; ON SIZE ERROR imperative-statement]

SUBTRACT {identifier-1 | literal-1} [, identifier-2 | , literal-2] . . . FROM {identifier-m | literal-m}
    GIVING identifier-n [ROUNDED] [, identifier-o [ROUNDED]] . . .
    [; ON SIZE ERROR imperative-statement]

SUBTRACT {CORRESPONDING | CORR} identifier-1 FROM identifier-2 [ROUNDED]
    [; ON SIZE ERROR imperative-statement]

SUPPRESS PRINTING
```

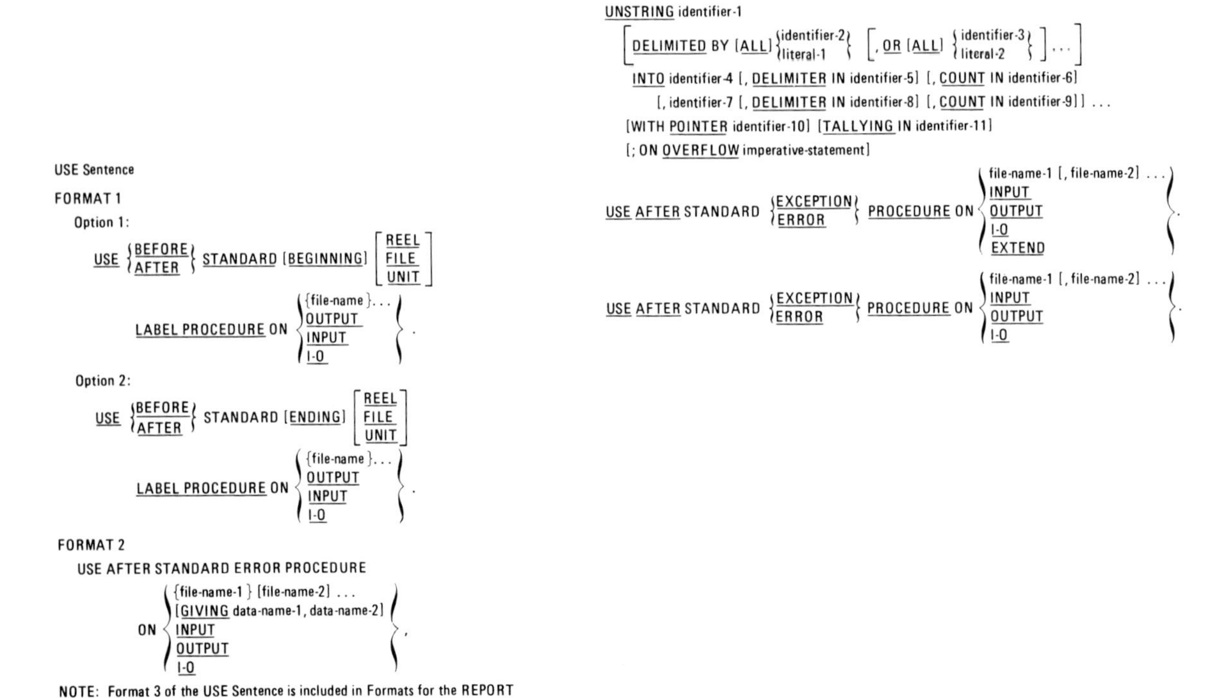

IBM FULL ANS COBOL — ANSI 1974 COBOL

PROCEDURE DIVISION FORMATS

IBM FULL ANS COBOL

USE Sentence

FORMAT 1

Option 1:

```
USE {BEFORE | AFTER} STANDARD [BEGINNING] [REEL | FILE | UNIT]

    LABEL PROCEDURE ON {{file-name} . . . | OUTPUT | INPUT | I-O} .
```

Option 2:

```
USE {BEFORE | AFTER} STANDARD [ENDING] [REEL | FILE | UNIT]

    LABEL PROCEDURE ON {{file-name} . . . | OUTPUT | INPUT | I-O} .
```

FORMAT 2

```
USE AFTER STANDARD ERROR PROCEDURE

    ON { {file-name-1} [file-name-2] . . .
         [GIVING data-name-1, data-name-2]
       | INPUT
       | OUTPUT
       | I-O } .
```

NOTE: Format 3 of the USE Sentence is included in Formats for the REPORT WRITER feature.

ANSI 1974 COBOL

```
UNSTRING identifier-1

    [DELIMITED BY [ALL] {identifier-2 | literal-1} [, OR [ALL] {identifier-3 | literal-2}] . . .]

    INTO identifier-4 [, DELIMITER IN identifier-5] [, COUNT IN identifier-6]
        [, identifier-7 [, DELIMITER IN identifier-8] [, COUNT IN identifier-9]] . . .
    [WITH POINTER identifier-10] [TALLYING IN identifier-11]
    [; ON OVERFLOW imperative-statement]

USE AFTER STANDARD {EXCEPTION | ERROR} PROCEDURE ON {file-name-1 [, file-name-2] . . . | INPUT | OUTPUT | I-O | EXTEND} .

USE AFTER STANDARD {EXCEPTION | ERROR} PROCEDURE ON {file-name-1 [, file-name-2] . . . | INPUT | OUTPUT | I-O} .
```

IBM FULL ANS COBOL

ANSI 1974 COBOL

PROCEDURE DIVISION FORMATS

IBM FULL ANS COBOL

WRITE Statement

FORMAT 1

```
WRITE record-name [FROM identifier-1] [ {BEFORE} ADVANCING]
                                        {AFTER }

    {identifier-2 LINES}                 {END-OF-PAGE}
  [ {integer LINES     } ] [AT           {EOP        } imperative-statement]
    {mnemonic-name     }
```

FORMAT 2

```
WRITE record-name [FROM identifier-1] AFTER POSITIONING {identifier-2} LINES
                                                        {integer     }

    [AT {END-OF-PAGE} imperative-statement]
        {EOP        }
```

FORMAT 3

```
WRITE record-name [FROM identifier-1] INVALID KEY imperative-statement
```

ANSI 1974 COBOL

```
WRITE record-name [FROM identifier-1]

  [                       {identifier-2} [LINE ] ]
  [ {BEFORE} ADVANCING    {integer     } [LINES] ]
  [ {AFTER }              {mnemonic-name}        ]
  [                       {PAGE         }        ]

  [; AT {END-OF-PAGE} imperative-statement]
  [     {EOP        }                     ]

WRITE record-name [FROM identifier] [; INVALID KEY imperative-statement]
```

IBM FULL ANS COBOL

ANSI 1974 COBOL

SORT FORMATS

IBM FULL ANS COBOL

Environment Division Sort Formats

FILE-CONTROL PARAGRAPH—SELECT SENTENCE

SELECT Sentence (for GIVING option only)

```
  SELECT file-name
    ASSIGN TO [integer-1] system-name-1 [system-name-2] . . .
          OR system-name-3 [FOR MULTIPLE {REEL} ]
                                         {UNIT}
              [RESERVE {integer-2} ALTERNATE [AREA ] ].
                       {NO       }           [AREAS]
```

SELECT Sentence (for Sort Work Files)

```
  SELECT sort-file-name
    ASSIGN TO [integer] system-name-1 [system-name-2] . . .
```

I-O-CONTROL PARAGRAPH

RERUN Clause

```
  RERUN ON system-name
```

SAME RECORD/SORT AREA Clause

```
  SAME {RECORD} AREA FOR file-name-1 {file-name-2}. . .
       {SORT  }
```

Data Division Sort Formats

SORT-FILE DESCRIPTION

```
  SD sort-file-name
    RECORDING MODE IS mode
    DATA {RECORD IS  } data-name-1 [data-name-2] . . .
         {RECORDS ARE}
    RECORD CONTAINS [integer-1 TO] integer-2 CHARACTERS.
```

Procedure Division Sort Formats

RELEASE Statement

```
  RELEASE sort-record-name [FROM identifier]
```

RETURN Statement

```
  RETURN sort-file-name RECORD [INTO identifier]
    AT END imperative-statement
```

SORT Statement

```
  SORT file-name-1 ON {DESCENDING} KEY {data-name-1}. . .
                      {ASCENDING }

  [ON {DESCENDING} KEY {data-name-2}. . .] . . .
      {ASCENDING }

  {INPUT PROCEDURE IS section-name-1 [THRU section-name-2]}
  {USING file-name-2                                      }
  {OUTPUT PROCEDURE IS section-name-3 [THRU section-name-4]}
  {GIVING file-name-3                                      }
```

ANSI 1974 COBOL

```
[SD file-name
   [; RECORD CONTAINS [integer-1 TO] integer-2 CHARACTERS]
   [; DATA {RECORD IS  } data-name-1 [, data-name-2] . . .]
   [       {RECORDS ARE}                                  ]
   {record-description-entry}. . .] . . .]

RELEASE record-name [FROM identifier]

RETURN file-name RECORD [INTO identifier] ; AT END imperative-statement

SORT file-name-1 ON {ASCENDING } KEY data-name-1 [, data-name-2] . . .
                    {DESCENDING}

         [ON {ASCENDING } KEY data-name-3 [, data-name-4] . . .
         [   {DESCENDING}

  [COLLATING SEQUENCE IS alphabet-name]

  {INPUT PROCEDURE IS section-name-1 [ {THROUGH} section-name-2 ]}
  {                                  [ {THRU   }                ]}
  {USING file-name-2 [, file-name-3] . . .                       }

  {OUTPUT PROCEDURE IS section-name-3 [ {THROUGH} section-name-4 ]}
  {                                   [ {THRU   }                ]}
  {GIVING file-name-4                                             }
```

REPORT WRITER FORMATS

IBM FULL ANS COBOL

Data Division Report Writer Formats

NOTE: Formats that appear as Basic Formats within the general description of the Data Division are illustrated there.

FILE SECTION—REPORT Clause

```
{REPORT IS   } report-name-1 [report-name-2] . . .
{REPORTS ARE }
```

REPORT SECTION

```
REPORT SECTION.
RD report-name
   WITH CODE mnemonic-name
   {CONTROL IS  }  {FINAL                                          }
   {CONTROLS ARE}  {identifier-1 [identifier-2] . . .              }
                   {FINAL identifier-1 [identifier-2] . . .        }
   PAGE [LIMIT IS  ] integer-1 {LINE }
        [LIMITS ARE]           {LINES}
     [HEADING integer-2]
     [FIRST DETAIL integer-3]
     [LAST DETAIL integer-4]
     [FOOTING integer-5]
```

ANSI 1974 COBOL

```
[; {REPORT IS  } report-name-1 [, report-name-2] . . .] .
[  {REPORTS ARE}                                      ]
[record-description-entry] . . .] . . .
[REPORT SECTION.
[RD report-name
   [; CODE literal-1]
   [; {CONTROL IS  } {data-name-1 [, data-name-2] . . .                 } ]
   [  {CONTROLS ARE} {FINAL [, data-name-1 [, data-name-2] . . .]       } ]
   [; PAGE {LIMIT IS  } integer-1 [LINE ] [, HEADING integer-2]
   [       {LIMITS ARE}           [LINES]
        [, FIRST DETAIL integer-3] [, LAST DETAIL integer-4]
        [, FOOTING integer-5] ] .
{report-group-description-entry } . . .] . . .]
```

REPORT WRITER FORMATS

IBM FULL ANS COBOL

REPORT GROUP DESCRIPTION ENTRY

FORMAT 1

```
01    [data-name-1]
                     {integer-1        }
   LINE NUMBER IS    {PLUS integer-2   }
                     {NEXT PAGE        }
                     {integer-1        }
   NEXT GROUP IS     {PLUS integer-2   }
                     {NEXT PAGE        }
          ( {REPORT HEADING}                                 )
          ( {RH            }                                 )
          ( {PAGE HEADING  }                                 )
          ( {PH            }                                 )
          ( {CONTROL HEADING}  {identifier-n}                )
          ( {CH             }  {FINAL       }                )
TYPE IS   ( {DETAIL}                                         )
          ( {DE    }                                         )
          ( {CONTROL FOOTING}  {identifier-n}                )
          ( {CF             }  {FINAL       }                )
          ( {PAGE FOOTING}                                   )
          ( {PF          }                                   )
          ( {REPORT FOOTING}                                 )
          ( {RF            }                                 )
```

USAGE Clause.

FORMAT 2

```
nn    [data-name-1]
   LINE Clause—See Format 1
   USAGE Clause.
```

ANSI 1974 COBOL

FORMAT 1:

```
01    [data-name-1]
   [; LINE NUMBER IS {integer-1 [ON NEXT PAGE]} ]
   [                 {PLUS integer-2          } ]
   [                   {integer-3     } ]
   [; NEXT GROUP IS    {PLUS integer-4} ]
   [                   {NEXT PAGE     } ]
      ( {REPORT HEADING}                          )
      ( {RH            }                          )
      ( {PAGE HEADING}                            )
      ( {PH          }                            )
      ( {CONTROL HEADING}  {data-name-2}          )
      ( {CH             }  {FINAL      }          )
      ( {DETAIL}                                  )
      ( {DE    }                                  )
      ( {CONTROL FOOTING}  {data-name-3}          )
      ( {CF             }  {FINAL      }          )
      ( {PAGE FOOTING}                            )
      ( {PF          }                            )
      ( {REPORT FOOTING}                          )
      ( {RF            }                          )
   [; [USAGE IS] DISPLAY] .
```

FORMAT 2:

```
level-number [data-name-1]
   [; LINE NUMBER IS {integer-1 [ON NEXT PAGE]} ]
   [                 {PLUS integer-2          } ]
   [; [USAGE IS] DISPLAY] .
```

IBM FULL ANS COBOL — **ANSI 1974 COBOL**

REPORT WRITER FORMATS

IBM FULL ANS COBOL

FORMAT 3

```
nn    [data-name-1]
      COLUMN NUMBER IS integer-1
      GROUP INDICATE
      JUSTIFIED Clause
      LINE Clause—See Format 1
      PICTURE Clause
      RESET ON {identifier-1}
               {FINAL       }
      BLANK WHEN ZERO Clause
      SOURCE IS {TALLY       }
                {identifier-2}
      SUM {TALLY       } {TALLY       } . . . [UPON data-name]
          {identifier-3} {identifier-4}
      VALUE IS literal-1
      USAGE Clause.
```

FORMAT 4

```
01    data-name-1
      BLANK WHEN ZERO Clause
      COLUMN Clause—See Format 2
      GROUP Clause—See Format 2
      JUSTIFIED Clause
      LINE Clause—See Format 1
      NEXT GROUP Clause—See Format 1
      PICTURE Clause
      RESET Clause—See Format 2
     {SOURCE Clause}
     {SUM Clause   }  See Format 2
     {VALUE Clause }
      TYPE Clause—See Format 1
      USAGE Clause.
```

Procedure Division Report Writer Formats

GENERATE Statement

```
   GENERATE identifier
```

INITIATE Statement

```
   INITIATE report-name-1 [report-name-2] . . .
```

TERMINATE Statement

```
   TERMINATE report-name-1 [report-name-2] . . .
```

USE Sentence

```
   USE BEFORE REPORTING data-name.
```

ANSI 1974 COBOL

FORMAT 3:

```
level-number [data-name-1]
   [; BLANK WHEN ZERO]
   [; GROUP INDICATE]
   [; {JUSTIFIED} RIGHT]
      {JUST     }
   [; LINE NUMBER IS {integer-1 [ON NEXT PAGE]}]
                     {PLUS integer-2          }
   [; COLUMN NUMBER IS integer-3]
   ; {PICTURE} IS character-string
     {PIC    }
 { ; SOURCE IS identifier-1                                      }
 { ; VALUE IS literal                                            }
 { {; SUM identifier-2 [, identifier-3] . . .                    }
 {    [UPON data-name-2 [, data-name-3] . . .] } . . .           }
 {    [RESET ON {data-name-4}]                                   }
 {              {FINAL      }                                    }
   [; [USAGE IS] DISPLAY].

GENERATE {data-name  }
         {report-name}

INITIATE report-name-1 [, report-name-2] . . .

TERMINATE report-name-1 [, report-name-2]

USE BEFORE REPORTING identifier.
```

IBM FULL ANS COBOL

TABLE HANDLING FORMATS

Data Division Table Handling Formats

OCCURS Clause

FORMAT 1

```
OCCURS integer-2 TIMES
    [{ASCENDING | DESCENDING} KEY IS data-name-2 [data-name-3...] ...
    [INDEXED BY index-name-1 [index-name-2] ...]
```

FORMAT 2

```
OCCURS integer-1 TO integer-2 TIMES [DEPENDING ON data-name-1]
    [{ASCENDING | DESCENDING} KEY IS data-name-2 [data-name-3] ...] ...
    [INDEXED BY index-name-1 [index-name-2] ...]
```

FORMAT 3

```
OCCURS integer-2 TIMES [DEPENDING ON data-name-1]
    [{ASCENDING | DESCENDING} KEY IS data-name-2 [data-name-3] ...] ...
    [INDEXED BY index-name-1 [index-name-2] ...]
```

USAGE Clause

```
[USAGE IS] INDEX
```

SEARCH Statement

FORMAT 1

```
SEARCH identifier-1 [VARYING {index-name-1 | identifier-2}]
    [AT END imperative-statement-1]
    WHEN condition-1 {imperative-statement-2 | NEXT SENTENCE}
    [WHEN condition-2 {imperative-statement-3 | NEXT SENTENCE}] ...
```

FORMAT 2

```
SEARCH ALL identifier-1 [AT END imperative-statement-1]
    WHEN condition-1 {imperative-statement-2 | NEXT SENTENCE}
```

SET Statement

FORMAT 1

```
SET {index-name-1 [index-name-2] ... | identifier-1 [identifier-2] ...} TO {index-name-3 | identifier-3 | literal-1}
```

FORMAT 2

```
SET index-name-4 [index-name-5] ... {UP BY | DOWN BY} {identifier-4 | literal-2}
```

ANSI 1974 COBOL

```
[; OCCURS {integer-1 TO integer-2 TIMES DEPENDING ON data-name-3 | integer-2 TIMES}
    [{ASCENDING | DESCENDING} KEY IS data-name-4 [, data-name-5] ...] ...
    [INDEXED BY index-name-1 [, index-name-2] ...]]

SEARCH identifier-1 [VARYING {identifier-2 | index-name-1}] [; AT END imperative-statement-1]
    ; WHEN condition-1 {imperative-statement-2 | NEXT SENTENCE}
    [; WHEN condition-2 {imperative-statement-3 | NEXT SENTENCE}] ...

SEARCH ALL identifier-1 [; AT END imperative-statement-1]
    ; WHEN {data-name-1 {IS EQUAL TO | IS =} {identifier-3 | literal-1 | arithmetic-expression-1}
            | condition-name-1}
        [AND {data-name-2 {IS EQUAL TO | IS =} {identifier-4 | literal-2 | arithmetic-expression-2}
            | condition-name-2}] ...
        {imperative-statement-2 | NEXT SENTENCE}

SET {identifier-1 [, identifier-2] ... | index-name-1 [, index-name-2] ...} TO {identifier-3 | index-name-3 | integer-1}

SET index-name-4 [, index-name-5] ... {UP BY | DOWN BY} {identifier-4 | integer-2}
```

IBM FULL ANS COBOL | **ANSI 1974 COBOL**

SEGMENTATION, LIBRARY, DEBUGGING FORMATS

SEGMENTATION—BASIC FORMATS

Environment Division Segmentation Formats

OBJECT-COMPUTER PARAGRAPH

SEGMENT-LIMIT Clause

SEGMENT-LIMIT IS priority-number

Procedure Division Segmentation Formats

Priority Numbers

section-name SECTION [priority-number].

SOURCE PROGRAM LIBRARY FACILITY

COPY Statement

COPY library-name [SUPPRESS]

[REPLACING word-1 BY {word-2 | literal-1 | identifier-1} word-3 BY {word-4 | literal-2 | identifier-2}] ...]

COPY text-name [{OF | IN} library-name]

[REPLACING {{==pseudo-text-1== | identifier-1 | literal-1 | word-1} BY {==pseudo-text-2== | identifier-2 | literal-2 | word}} ...]

Extended Source Program Library Facility

BASIS Card

BASIS library-name

INSERT Card

INSERT sequence-number-field

DELETE Card

DELETE sequence-number-field

DEBUGGING LANGUAGE—BASIC FORMATS

Procedure Division Debugging Formats

EXHIBIT Statement

EXHIBIT {NAMED | CHANGED NAMED | CHANGED} {identifier-1 | nonnumeric-literal-1} [identifier-2 | nonnumeric-literal-2]

ON (Count-Conditional) Statement

ON integer-1 [AND EVERY integer-2] [UNTIL integer-3]

{imperative-statement ... | NEXT SENTENCE} {ELSE | OTHERWISE} {statement ... | NEXT SENTENCE}

TRACE Statement

{READY | RESET} TRACE

Compile-Time Debugging Packet

DEBUG Card

DEBUG location

USE FOR DEBUGGING ON {cd-name-1 | [ALL REFERENCES OF] identifier-1 | file-name-1 | procedure-name-1 | ALL PROCEDURES}

[, cd-name-2 | [ALL REFERENCES OF] identifier-2 | file-name-2 | procedure-name-2 | ALL PROCEDURES] ....

GENERAL FORMAT FOR CONDITIONS

RELATION CONDITION:

```
{ identifier-1              }  { IS [NOT] GREATER THAN }  { identifier-2              }
{ literal-1                 }  { IS [NOT] LESS THAN    }  { literal-2                 }
{ arthimetic-expression-1   }  { IS [NOT] EQUAL TO     }  { arithmetic-expression-2   }
{ index-name-1              }  { IS [NOT] >            }  { index-name-2              }
                               { IS [NOT] <            }
                               { IS [NOT] =            }
```

CLASS CONDITION:

```
identifier IS [NOT] { NUMERIC    }
                    { ALPHABETIC }
```

SIGN CONDITION:

```
arithmetic-expression is [NOT] { POSITIVE }
                               { NEGATIVE }
                               { ZERO     }
```

CONDITION-NAME CONDITION:

condition-name

SWITCH-STATUS CONDITION:

condition-name

NEGATED SIMPLE CONDITION:

NOT simple-condition

COMBINED CONDITION:

```
condition { { AND } condition } ...
          { { OR  }           }
```

ABBREVIATED COMBINED RELATION CONDITION:

```
relation-condition { { AND } [NOT] [relational-operator] object } ...
                   { { OR  }                                    }
```

MISCELLANEOUS FORMATS

QUALIFICATION:

```
{ data-name-1    } [ { OF } data-name-2 ] ...
{ condition-name } [ { IN }             ]

paragraph-name [ { OF } section-name ]
               [ { IN }              ]

text-name [ { OF } library-name ]
          [ { IN }              ]
```

SUBSCRIPTING:

```
{ data-name      } (subscript-1 [, subscript-2 [, subscript-3] ] )
{ condition-name }
```

INDEXING:

```
{ data-name      } ( { index-name-1 [ {±} literal-2] }
{ condition-name }   { literal-1                     }

   [ , { index-name-2 [ {±} literal-4] }  [ , { index-name-3 [ {±} literal-6] } ] ] )
   [   { literal-3                     }  [   { literal-5                     } ] ]
```

IDENTIFIER: FORMAT 1

```
data-name-1 [ { OF } data-name-2 ] ... [(subscript-1 [, subscript-2
            [ { IN }             ]

   [, subscript-3] ] )]
```

IDENTIFIER: FORMAT 2

```
data-name-1 [ { OF } data-name-2 ] ... [ ( { index-name-1 [ {±} literal-2] }
            [ { IN }             ]     [   { literal-1                     }

   [ , { index-name-2 [ {±} literal-4 }  [ , { index-name-3 [ {±} literal-6] } ] ] ) ]
   [   { literal-3                    }  [   { literal-5                     } ] ]
```

Glossary

ACCESS: The manner in which files are referenced by the computer. Access can be sequential (records are referred to one after another in the order in which they appear on the file), or it can be random (the individual records can be referred to in a nonsequential manner).

Actual Decimal Point: The physical representation, using either of the decimal point characters (. or ,), of the decimal point position in a data item. When specified, it will appear in a printed report, and it requires an actual space in storage.

ACTUAL KEY: A key which can be directly used by the system to locate a logical record on a mass storage device. An ACTUAL KEY must be a data item of 5 to 259 bytes in length.

Alphabetic Character: A character which is one of the 26 characters of the alphabet, or a space. In COBOL, the term does *not* include any other characters.

Alphanumeric Character: Any character in the computer's character set.

Alphanumeric Edited Character: A character within an alphanumeric character string which contains at least one B or 0.

Arithmetic Expression: A statement containing any combination of data-names, numeric literals, and figurative constants, joined together by one or more arithmetic operators in such a way that the statement as a whole can be reduced to a single numeric value.

Arithmetic Operator: A symbol (single character or 2-character set) or COBOL verb which directs the system to perform an arithmetic operation. The following list shows arithmetic operators:

Meaning	*Symbol*
Addition	+
Subtraction	–
Multiplication	*
Division	/
Exponentiation	**

Assumed Decimal Point: A decimal point position which does not involve the existence of an actual character in a data item. It does not occupy an actual space in storage, but is used by the compiler to align a value properly for calculation.

BLOCK: In COBOL, a group of characters or records which is treated as an entity when moved into or out of the computer. The term is synonymous with the term Physical Record.

Buffer: A portion of main storage into which data is read or from which it is written.

Byte: A sequence of eight adjacent binary bits. When properly aligned, two bytes form a halfword, four bytes a fullword, and eight bytes a doubleword.

Channel: A device that directs the flow of information between the computer main storage and the input/output devices.

Character: One of a set of indivisible symbols that can be arranged in sequences to express information. These symbols include the letters A through Z, the decimal digits 0 through 9, punctuation symbols, and any other symbols which will be accepted by the data-processing system.

Character Set: All the valid COBOL characters. The complete set of 51 characters.

Character String: A connected sequence of characters. All COBOL characters are valid.

Checkpoint: A reference point in a program at which information about the contents of core storage can be recorded so that, if necessary, the program can be restarted at an intermediate point.

Class Condition: A statement that the content of an item is wholly alphabetic or wholly numeric. It may be true or false.

Clause: A set of consecutive COBOL words whose purpose is to specify an attribute of an entry. There are three types of clauses: data, environment, and file.

COBOL Character: Any of the 51 valid characters (see "Character") in the COBOL character set.

Collating Sequence: The arrangement of all valid characters in the order of their relative precedence. The collating sequence of a computer is part of the computer design—each acceptable character has a predetermined place in the sequence. A collating sequence is used primarily in comparison operations.

COLUMN Clause: A COBOL clause used to identify a specific position within a report line.

Comment: An annotation in the Identification Division or Procedure Division of a COBOL source program. A comment is ignored by the compiler. As an IBM extension, comments may be included at any point in a COBOL source program.

Compile Time: The time during which a COBOL source program is translated by the COBOL compiler into a machine language object program.

Compiler: A program which translates a program written in a higher level language into a machine language object program.

Compiler Directing Statement: A COBOL statement which causes the compiler to take a specific action at compile time, rather than the object program to take a particular action at execution time.

Compound Condition: A statement that tests two or more relational expressions. It may be true or false.

Condition:

- One of a set of specified values a data item can assume.
- A simple conditional expression: relation condition, class condition, condition-name condition, sign condition, NOT condition.

Conditional Statement: A syntactically correct statement, made up of data-names, and/or figurative literals, and/or constants, and/or logical operators, so constructed that it tests a truth value. The subsequent action of the object program is dependent on this truth value.

Conditional Variable: A data item that can assume more than one value; the value(s) it assumes has a condition-name assigned to it.

Condition Name: The name assigned to a specific value, set of values, or range of values, that a data item may assume.

Condition-name Condition: A statement that the value of a conditional variable is one of a set (or range) of values of a data item identified by a condition-name. The statement may be true or false.

CONFIGURATION SECTION: A section of the Environment Division of the COBOL program. It describes the overall specifications of computers.

Connective: A word or a punctuation character that does one of the following:

- Associates a data-name or paragraph-name with its qualifier
- Links two or more operands in a series
- Forms a conditional expression

CONSOLE: A COBOL mnemonic-name associated with the console typewriter.

Contiguous Items: Consecutive elementary or group items in the Data Division that have a definite relationship with each other.

Control Break: A recognition of a change in the contents of a control data item that governs a hierarchy.

Control Bytes: Bytes associated with a physical record that serve to identify the record and indicate its length, blocking factor, etc.

Control Data Item: A data item that is tested each time a report line is to be printed. If the value of the data item has changed, a control break occurs and special actions are performed before the line is printed.

CONTROL FOOTING: A report group that occurs at the end of the control group of which it is a member.

Control Group: An integral set of related data that is specifically associated with a control data item.

CONTROL HEADING: A report group that occurs at the beginning of the control group of which it is a member.

Control Hierarchy: A designated order of specific control data items. The highest level is the final control; the lowest level is the minor control.

Core Storage: Storage within the central processing unit of the computer, so called because this storage exists in the form of magnetic cores.

Data Description Entry: An entry in the Data Division that is used to describe the characteristics of a data item. It consists of a level number, followed by an optional data-name, followed by data clauses that fully describe the format the data will take. An elementary data description entry (or item) cannot logically be subdivided further. A group data description entry (or item) is made up of a number of related groups and/or elementary items.

DATA DIVISION: One of the four main component parts of a COBOL program. The Data Division describes the files to be used in the program and the records contained within the files. It also describes any internal Working-Storage records that will be needed (see "Data Division," chapter 7, for full details).

Data Item: A unit of recorded information that can be identified by a symbolic name or by a combination of names and subscripts. Elementary data items cannot logically be subdivided. Group data items are made up of logically related group and/or elementary items, and can be a logical group within a record or can itself be a complete record.

Data-name: A name assigned by the programmer to a data item in a COBOL program. It must contain at least one alphabetic character.

DECLARATIVES: A set of one or more compiler- directing sections written at the beginning of the Procedure Division of a COBOL program. The first section is preceded by the header DECLARATIVES. The last sec-

tion is followed by the header END DECLARATIVES. There are three options:
1. Input/output label handling
2. Input/output error-checking procedures
3. Report Writing procedures
Each has its standard format.

Delimiter: A character or sequence of contiguous characters that identifies the end of a string of characters and that separates the string of characters from the following string of characters. A delimiter is not part of the string of characters that it delimits.

Device-number: The reference number assigned to any external device.

Digit: Any of the numerals from 0 through 9. In COBOL, the term is not used in reference to any other symbol.

DIVISION: One of the four major portions of a COBOL program:
- IDENTIFICATION DIVISION, which names the program.
- ENVIRONMENT DIVISION, which indicates the machine equipment and equipment features to be used in the program.
- DATA DIVISION, which defines the nature and characteristics of data to be processed.
- PROCEDURE DIVISION, which consists of statements directing the processing of data in a specified manner at execution time.

Division Header: The COBOL words that indicate the beginning of a particular division of a COBOL program. The four division headers are:
- IDENTIFICATION DIVISION.
- ENVIRONMENT DIVISION.
- DATA DIVISION.
- PROCEDURE DIVISION.

Division-name: The name of one of the four divisions of a COBOL program.

EBCDIC Character: Any one of the symbols included in the eight-bit EBCDIC (Extended Binary-Coded-Decimal Interchange Code) set. All 51 COBOL characters are included.

Editing Character: A single character or a fixed two-character combination used to create proper formats for output reports.

Elementary Item: A data item that cannot logically be subdivided.

Entry: Any consecutive set of descriptive clauses terminated by a period, written in the Identification, Environment, or Procedure Divisions of a COBOL program.

Entry-name: A programmer-specified name that establishes an entry point into a COBOL subprogram.

ENVIRONMENT DIVISION: One of the four main component parts of a COBOL program. The Environment Division describes the computers upon which the source program is compiled and those on which the object program is executed, and provides a linkage between the logical concept of files and their records, and the physical aspects of the devices on which files are stored (see ''Environment Division,'' chapter 6, for full details).

Execution Time: The time at which an object program actually performs the instructions coded in the Procedure Division, using the actual data provided.

Exponent: A number indicating how many times another number (the base) is to be repeated as a factor. Positive exponents denote multiplication, negative exponents denote division, fractional exponents denote a root of a quantity. In COBOL, exponentiation is indicated with the symbol ** followed by the exponent.

F-mode Records: Records of a fixed length. Blocks may contain more than one record.

Figurative Constant: A reserved word that represents a numeric value, a character, or a string of repeated values or characters. The word can be written in a COBOL program to represent the values or characters without being defined in the Data Division.

FILE-CONTROL: The name and header of an Environment Division paragraph in which the data files for a given source program are named and assigned to specific input/output devices.

File Description: An entry in the File Section of the Data Division that provides information about the identification and physical structure of a file.

File-name: A name assigned to a set of input data or output data. A file-name must include at least one alphabetic character.

FILE SECTION: A section of the Data Division that contains descriptions of all externally stored data (or files) used in a program. Such information is given in one or more file description entries.

Floating-Point Literal: A numeric literal whose value is expressed in floating-point notation—that is, as a decimal number followed by an exponent which indicates the actual placement of the decimal point.

Function-name: A name, specified by the computer manufacturer, that identifies system logical units, printer and card punch control characters, and report codes. When a function-name is associated with a mnemonic-name in the Environment Division, the mnemonic-name may then be substituted in any format in which such substitution is valid.

Group Item: A data item made up of a series of logically related elementary items. It can be part of a record or a complete record.

Header Label: A record that identifies the beginning of a physical file or a volume.

High-Order: The leftmost position in a string of characters.

IDENTIFICATION DIVISION: One of the four main component parts of a COBOL program. The Identification Division identifies the source program and the object program and, in addition, may include such documentation as the author's name, the installation where written, date written, etc. (see "Identification Division," chapter 6, for full details).

Identifier: A data-name, unique in itself, or made unique by the syntactically correct combination of qualifiers, subscripts, and/or indexes.

Imperative-Statement: A statement consisting of an imperative verb and its operands, which specifies that an action be taken, unconditionally. An imperative-statement may consist of a series of imperative-statements.

Index: A computer storage position or register, the contents of which identify a particular element in a table.

Index Data Item: A data item in which the contents of an index can be stored without conversion to subscript form.

Index-name: A name, given by the programmer, for an index of a specific table. An index-name must contain at least one alphabetic character. It is one word (4 bytes) in length.

Indexed Data-name: A data-name identifier which is subscripted with one or more index-names.

INPUT-OUTPUT SECTION: In the Environment Division, the section that names the files and external media needed by an object program. It also provides information required for the transmission and handling of data during the execution of an object program.

INPUT PROCEDURE: A set of statements that is executed each time a record is released to the sort file. Input procedures are optional; whether they are used or not depends upon the logic of the program.

Integer: A numeric data item or literal that does not include any character positions to the right of the decimal point, actual or assumed. Where the term *integer* appears in formats, *integer* must not be a numeric data item.

INVALID KEY Condition: A condition that may arise at execution time in which the value of a specific key associated with a mass storage file does not result in a correct reference to the file (see the READ, REWRITE, START, and WRITE statements for the specific error conditions involved).

I-O-CONTROL: The name, and the header, for an Environment Division paragraph in which object program requirements for specific input/output techniques are specified. These techniques include rerun checkpoints, sharing of same areas by several data files, and multiple file storage on a single tape device.

KEY: One or more data items, the contents of which identify the type or the location of a record, or the ordering of data.

Key Word: A reserved word whose employment is essential to the meaning and structure of a COBOL statement. In this text, key words are indicated in the formats of statements by underscoring. Key words are included in the reserved word list.

Level Indicator: Two alphabetic characters that identify a specific type of file, or the highest position in a hierarchy. The level indicators are: FD, RD, SD.

Level Number: A numeric character or 2-character set that identifies the properties of a data description entry. Level numbers 01 through 49 define group items, the highest level being identified as 01, and the subordinate data items within the hierarchy being identified with level numbers 02 through 49. Level numbers 66, 77, and 88 identify special properties of a data description entry in the Data Division.

Library-name: The name of a member of a data set containing COBOL entries, used with the COPY and BASIS statements.

LINKAGE SECTION: A section of the Data Division that describes data made available from another program.

Literal: A character string whose value is implicit in the characters themselves. The numeric literal 7 expresses the value 7, and the nonnumeric literal "CHARACTERS" expresses the value CHARACTERS.

Logical Operator: A COBOL word that defines the logical connections between relational operators. The three logical operators and their meanings are:

OR (logical inclusive—either or both)
AND (logical connective—both)
NOT (logical negation)

(See "Procedure Division," chapter 8, for a more detailed explanation.)

Logical Record: The most inclusive data item, identified by a level-01 entry. It consists of one or more related data items.

Low-Order: The rightmost position in a string of characters.

Main Program: The highest level COBOL program involved in a step. (Programs written in other languages that follow COBOL linkage conventions are considered COBOL programs in this sense.)

Mantissa: The decimal part of a logarithm. Therefore, the part of a floating-point number that is expressed as a decimal fraction.

Mass Storage: A storage medium—disk, drum, or data cell—in which data can be collected and maintained in a sequential, direct, indexed, or relative organization.

Mass Storage File: A collection of records assigned to a mass storage device.

Mass Storage File Segment: A part of a mass storage file whose beginning and end are defined by the FILE-LIMIT clause in the Environment Division.

Mnemonic-name: A programmer-supplied word associated with a specific function-name in the Environment Division. It then may be written in place of the function-name in any format wherein such a substitution is valid.

MODE: The manner in which records of a file are accessed or processed.

Name: A word composed of not more than 30 characters, which defines a COBOL operand.

Noncontiguous Item: A data item in the Working-Storage Section of the Data Division which bears no relationship to other data items.

Nonnumeric Literal: A character string bounded by quotation marks, which means literally itself. For example, "CHARACTER" is the literal for, and means, CHARACTER. The string of characters may include any characters in the computer's set, with the exception of the quotation mark. Characters that are not COBOL characters may be included.

Numeric Character: A character that belongs to one of the set of digits 0 through 9.

Numeric Edited Character: A numeric character which is in such a form that it may be used in a printed output. It may consist of external decimal digits 0 through 9, the decimal point, commas, the dollar sign, etc., as the programmer wishes (see "Data Division" for a fuller explanation).

Numeric Item: An item whose description restricts its contents to a value represented by characters chosen from the digits 0 through 9; if signed, the item may also contain a + or − , or other representation of an operational sign.

Numeric Literal: A numeric character or string of characters whose value is implicit in the characters themselves. Thus, 777 is the literal as well as the value of the number 777.

OBJECT-COMPUTER: The name of an Environment Division paragraph in which the computer upon which the object program will be run is described.

Object Program: The set of machine language instructions that is the output from the compilation of a COBOL source program. The actual processing of data is done by the object program.

Object Time: The time during which an object program is executed.

Operand: The "object" of a verb or an operator. That is, the data or equipment governed or directed by a verb or operator.

Operational Sign: An algebraic sign associated with a numeric data item, which indicates whether the item is positive or negative.

Optional Word: A reserved word included in a specific format only to improve the readability of a COBOL statement. If the programmer wishes, optional words may be omitted.

OUTPUT PROCEDURE: A set of programmer-defined statements that is executed each time a sorted record is returned from the sort file. Output procedures are optional; whether they are used or not depends upon the logic of the program.

Overflow condition: In string manipulation, a condition that occurs when the sending area(s) contains untransferred characters after the receiving area(s) has been filled.

Overlay: The technique of repeatedly using the same areas of internal storage during different stages in processing a problem.

PAGE: A physical separation of continuous data in a report. The separation is based on internal requirements and/or the physical characteristics of the reporting medium.

PAGE FOOTINGS: A report group at the end of a report page which is printed before a page control break is executed.

PAGE HEADING: A report group printed at the beginning of a report page, after a page control break is executed.

Paragraph: A set of one or more COBOL sentences, making up a logical processing entity, and preceded by a paragraph-name or a paragraph header.

Paragraph Header: A word followed by a period that identifies and precedes all paragraphs in the Identification Division and Environment Division.

Paragraph-name: A programmer-defined word that identifies and precedes a paragraph.

Parameter: A variable that is given a specific value for a specific purpose or process. In COBOL, parameters are often used to pass data values between calling and called programs.

Physical Record: A physical unit of data, synonymous with a block. It can be composed of a portion of one logical record, of one complete logical record, or of a group of logical records.

Print Group: An integral set of related data within a report.

Priority-number: A number, ranging in value from 0 to 99, which classifies source program sections in the Procedure Division (see "Segmentation," Appendix C, for more information).

Procedure: One or more logically connected paragraphs or sections within the Procedure Division, which direct the computer to perform some action or series of related actions.

PROCEDURE DIVISION: One of the four main component parts of a COBOL program. The Procedure Division contains instructions for solving a problem. The Procedure Division may contain imperative-statements, conditional statements, paragraphs, procedures, and sections (see "Procedure Division," chapter 8, for full details).

Procedure-name: A word that precedes and identifies a procedure, used by the programmer to transfer control from one point of the program to another.

Process: Any operation or combination of operations on data.

Program-name: A word in the Identification Division that identifies a COBOL source program.

Punctuation Character: A comma, semicolon, period, quotation mark, left or right parenthesis, or a space.

Qualifier: A group data-name that is used to reference a nonunique data-name at a lower level in the same hierarchy, or a section-name that is used to reference a nonunique paragraph. In this way, the data-name or the paragraph-name can be made unique.

Random Access: An access mode in which specific logical records are obtained from or placed into a mass storage file in a nonsequential manner.

RECORD: A set of one or more related data items grouped for handling either internally or by the input/output systems (see "Logical Record").

Record Description: The total set of data description entries associated with a particular logical record.

Record-name: A data-name that identifies a logical record.

REEL: A module of external storage associated with a tape device.

Relation Character: A character that expresses a relationship between two operands. The following are COBOL relation characters:

Character	*Meaning*
>	Greater than
<	Less than
=	Equal to

Relation Condition: A statement that the value of an arithmetic expression or data item has a specific relationship to another arithmetic expression or data item. The statement may be true or false.

Relational Operator: A reserved word, or a group of reserved words, or a group of reserved words and relation characters. A relational operational plus programmer-defined operands make up a relational expression. A complete listing is given in "Procedure Division" (chapter 8).

REPORT: A presentation of a set of processed data described in a Report File.

Report Description Entry: An entry in the Report Section of the Data Division that names and describes the format of a report to be produced.

Report File: A collection of records, produced by the Report Writer, that can be used to print a report in the desired format.

REPORT FOOTING: A report group that occurs, and is printed, only at the end of a report.

Report Group: A set of related data that makes up a logical entity in a report.

REPORT HEADING: A report group that occurs, and is printed, only at the beginning of a report.

Report Line: One row of printed characters in a report.

Report-name: A data-name that identifies a report.

REPORT SECTION: A section of the Data Division that contains one or more Report Description entries.

Reserved Word: A word used in a COBOL source program for syntactical purposes. It must not appear in a program as a user-defined operand.

Routine: A set of statements in a program that causes the computer to perform an operation or series of related operations.

Run Unit: A set of one or more object programs which function, at object time, as a unit to provide problem solutions. This compiler considers a run unit to be the highest level calling program plus all called subprograms.

S-Mode Records: Records that span physical blocks. Records may be fixed or variable in length; blocks may contain one or more segments. Each segment contains a segment-descriptor field and a control field indicating whether it is the first and/or last or an intermediate segment of the record. Each block contains a block- descriptor field.

SECTION: A logically related sequence of one or more paragraphs. A section must always be named.

Section Header: A combination of words that precedes and identifies each section in the Environment, Data, and Procedure Divisions.

Section-name: A word specified by the programmer that precedes and identifies a section in the Procedure Division.

Sentence: A sequence of one or more statements, the last ending with a period followed by a space.

Separator: An optional word or character that improves readability.

Sequential Access: An access mode in which logical records are obtained from or placed into a file in such a way that each successive access to the file refers to the next subsequent logical record in the file. The order of the records is established by the programmer when creating the file.

Sequential Processing: The processing of logical records in the order in which records are accessed.

Sign Condition: A statement that the algebraic value of a data item is less than, equal to, or greater than zero. It may be true or false.

Simple Condition: An expression that can have two values, and causes the object program to select between alternate paths of control, depending on the value found. The expression can be true or false.

Slack Bytes: Bytes inserted between data items or records to ensure correct alignment of some numeric items. Slack bytes contain no meaningful data. In some cases, they are inserted by the compiler; in others, it is the responsibility of the programmer to insert them. The SYNCHRONIZED clause instructs the compiler to insert slack bytes when they are needed for proper alignment. Slack bytes between records are inserted by the programmer.

Sort File: A collection of records that is sorted by a SORT statement. The sort file is created and used only while the sort function is operative.

Sort-File-Description Entry: An entry in the File Section of the Data Division that names and describes a collection of records that is used in a SORT statement.

Sort-file-name: A data-name that identifies a Sort File.

Sort-key: The field within a record on which a file is sorted.

Sort-work-file: A collection of records involved in the sorting operation as this collection exists on intermediate device(s).

SOURCE-COMPUTER: The name of an Environment Division paragraph. In it, the computer upon which the source program will be compiled is described.

Source Program: A problem-solving program written in COBOL.

Special Character: A character that is neither numeric nor alphabetic. Special characters in COBOL include the space ( ), the period (.), as well as the following: + − * / = $, ; ") (

SPECIAL-NAMES: The name of an Environment Division paragraph, and the paragraph itself, in which names supplied by the computer manufacturer are related to mnemonic-names specified by the programmer. In addition, this paragraph can be used to exchange the functions of the comma and the period, or to specify a substitution character for the currency sign, in the PICTURE string.

Special Register: Compiler-generated storage areas primarily used to store information produced with the use of specific COBOL features. The special registers are: TALLY, LINE-COUNTER, PAGE-COUNTER, CURRENT-DATE, TIME-OF-DAY, LABEL-RETURN, RETURN-CODE, SORT-RETURN, SORT-FILE-SIZE, SORT-CORE-SIZE, and SORT-MODE-SIZE.

Standard Data Format: The concept of actual physical or logical record size in storage. The length in the Standard Data Format is expressed in the number of bytes a record occupies and not necessarily the number of characters, since some characters take up one full byte of storage and others take up less.

Statement: A syntactically valid combination of words and symbols written in the Procedure Division. A statement combines COBOL reserved words and programmer-defined operands.

Subject of entry: A data-name or reserved word that appears immediately after a level indicator or level number in a Data Division entry. It serves to reference the entry.

Subprogram: A COBOL program that is invoked by another COBOL program. (Programs written in other languages that follow COBOL linkage conventions are COBOL programs in this sense.)

Subscript: An integer or a variable whose value references a particular element in a table.

SYSIN: The system logical input device.

SYSOUT: The system logical output device.

SYSPUNCH: The system logical punch device.

System-name: A name that identifies any particular external device used with the computer, and characteristics of files contained within it.

Table: A collection and arrangement of data in a fixed form for ready reference. Such a collection follows some logical order, expressing particular values (functions) corresponding to other values (arguments) by which they are referenced.

Table Element: A data item that belongs to the set of repeated items comprising a table. An argument together with its corresponding function(s) makes up a table element.

Test Condition: A statement that, taken as a whole, may be either true or false, depending on the circumstances existing at the time the expression is evaluated.

Trailer Label: A record that identifies the ending of a physical file or a volume.

U-mode Records: Records of unspecified length. They may be fixed or variable in length; there is only one record per block.

Unary Operator: An arithmetic operator (+ or −) that can precede a single variable, a literal, or a left parenthesis in an arithmetic expression. The plus sign multiplies the value by +1; the minus sign multiplies the value by −1.

UNIT: A module of external storage. Its dimensions are determined by the computer manufacturer.

V-mode Records: Records of variable length. Blocks may contain more than one record. Each record contains a record length field, and each block contains a block length field.

Variable: A data item whose value may be changed during execution of the object program.

Verb: A COBOL reserved word that expresses an action to be taken by a COBOL compiler or an object program.

Volume: A module of external storage. For tape devices it is a reel; for mass storage devices it is a unit.

Volume Switch Procedures: Standard procedures executed automatically when the end of a unit or reel has been reached before end-of-file has been reached.

WORD:

1. In COBOL: A string of not more than 30 characters, chosen from the following: the letters A through Z, the digits 0 through 9, and the hyphen (-). The hyphen may not appear as either the first or last character.
2. In the IBM System/360 or System/370: A fullword is four bytes of storage; a doubleword is eight bytes of storage; a halfword is two bytes of storage.

Word Boundary: Any particular storage position at which data must be aligned for certain processing operations in the IBM System/360 or System/370. The halfword boundary must be divisible by 2, the fullword boundary must be divisible by 4, the doubleword boundary must be divisible by 8.

WORKING-STORAGE SECTION: A section-name (and the section itself) in the Data Division. The section describes records and noncontiguous data items that are not part of external files, but are developed and processed internally. It also defines data items whose values are assigned in the source program.

Index

The Formats and Reserved Words in this appendix have been printed in a specially reduced size with pages numbered in sequence to make up a pocket-sized reference booklet for use when coding IBM Full American National Standard COBOL programs. Although most readers may prefer to retain this reference material within the manual, the booklet can be prepared as follows:

- cut along trim lines.
- place sheets so that page numbers at lower right-hand corner are in ascending order in odd-number progression (i.e., 1, 3, 5, etc.); lower left-hand page numbers will then be in descending order in even-number progression (i.e., 20, 18, 16, etc.).
- fold trimmed sheets after collation.
- staple along fold if desired.
- punch for six-hole binder.

FOLD

TRIM HERE

WHEN
(xac) WHEN-COMPILED
WITH
WORDS
WORKING-STORAGE
WRITE
(xac) WRITE-ONLY
(spn) WRITE-VERIFY

ZERO
ZEROES
ZEROS

IBM

International Business Machines Corporation
Data Processing Division
1133 Westchester Avenue
White Plains, New York 10604
(U.S.A. only)

IBM World Trade Corporation
821 United Nations Plaza
New York, New York 10017
(International)

Printed in U.S.A. Extracted from GC28-6396-4
Not orderable separately.

IBM Reference Data

Operating System

IBM Full American National Standard COBOL

Appendix C: IBM Full American National Standard COBOL Format Summary and Reserved Words

The general format of a COBOL program is illustrated in these format summaries. Included within the general format is the specific format for each valid COBOL statement. All clauses are shown as though they were required by the COBOL source program, although within a given context many are optional. Several formats are included under special headings, which are different from, or additions to, the general format. Under these special headings are included formats peculiar to the following COBOL features: Sort. Report Writer, Table Handling, Segmentation, Source Program Library Facility, Debugging Language, Format Control of the Source Program Listing, Sterling Currency, Teleprocessing, and String Manipulation. Each of these features is explained within a special chapter of this publication — *IBM OS Full American National Standard COBOL*, Order Nos. GC28-6396-3, and -4.

Note: OS/VS COBOL formats are included and identified as OS/VS COBOL only.

TRIM HERE

FOLD

FOLD

TRIM HERE

IDENTIFICATION DIVISION — BASIC FORMATS

{IDENTIFICATION DIVISION.}
{ID DIVISION.}

PROGRAM-ID. *program-name.*

AUTHOR. [*comment-entry*] . . .

INSTALLATION. [*comment-entry*] . . .

DATE-WRITTEN. [*comment-entry*] . . .

DATE-COMPILED. [*comment-entry*] . . .

SECURITY. [*comment-entry*] . . .

REMARKS. [*comment-entry*] . . .

ENVIRONMENT DIVISION — BASIC FORMATS

ENVIRONMENT DIVISION.

CONFIGURATION SECTION.

SOURCE-COMPUTER. *computer-name.*

OBJECT-COMPUTER. *computer-name* [MEMORY SIZE *integer* {WORDS / CHARACTERS / MODULES}]

[SEGMENT-LIMIT IS *priority-number*].

SPECIAL-NAMES. [*function-name* IS *mnemonic-name*] . . .

[CURRENCY SIGN IS *literal*]

[DECIMAL-POINT IS COMMA].

INPUT-OUTPUT SECTION.

FILE-CONTROL.

{SELECT [OPTIONAL] *file name*

ASSIGN TO [*integer-1*] *system-name-1* [*system-name-2*] . . .

[FOR MULTIPLE {REEL / UNIT}]

RESERVE {NO / *integer-1*} ALTERNATE [AREA / AREAS]

{FILE-LIMIT IS / FILE-LIMITS ARE} {*data-name-1* / *literal-1*} THRU {*data-name-2* / *literal-2*}

[{*data-name-3* / *literal-3*} THRU {*data-name-4* / *literal-4*}] . . .

ACCESS MODE IS {SEQUENTIAL / RANDOM}

PROCESSING MODE IS SEQUENTIAL

ACTUAL KEY IS *data-name*

NOMINAL KEY IS *data-name*

RECORD KEY IS *data-name*

TRACK-AREA IS {*data-name* / *integer*} CHARACTERS

TRACK-LIMIT IS *integer* [TRACK / TRACKS] . } . . .

I-O-CONTROL.

RERUN ON *system-name* EVERY [*integer* RECORDS / [END OF] {REEL / UNIT}] OF *file-name*

SAME [RECORD / SORT] AREA FOR *file-name-1* {*file-name-2*} . . .

MULTIPLE FILE TAPE CONTAINS *file-name-1* [POSITION *integer-1*]

[*file-name-2* [POSITION *integer-2*]] . . .

APPLY WRITE-ONLY ON *file-name-1* [*file-name-2*] . . .

APPLY CORE-INDEX ON *file-name-1* [*file-name-2*] . . .

APPLY RECORD-OVERFLOW ON *file-name-1* [*file-name-2*] . . .

APPLY REORG-CRITERIA TO *data-name* ON *file-name* . . .

Note: Format 2 of the RERUN Clause (for Sort Files) is included with Formats for the SORT feature.

(ca) PRINTING
PROCEDURE
(ca) PROCEDURES
PROCEED
(ca) PROCESS
PROCESSING
(xa) PROGRAM
PROGRAM-ID

(xa) QUEUE
QUOTE
QUOTES

RANDOM
RD
READ
(xac) READY
(xa) RECEIVE
RECORD
(xac) RECORD-OVERFLOW
(xa) RECORDING
RECORDS
REDEFINES
REEL
(ca) REFERENCES
(ca) RELATIVE
RELEASE
(xac) RELOAD
REMAINDER
REMARKS
(ca) REMOVAL
RENAMES
(xac) REORG-CRITERIA
REPLACING
REPORT
REPORTING
REPORTS
(xac) REREAD
RERUN
RESERVE
RESET
RETURN
(xac) RETURN-CODE
REVERSED
REWIND
(xa) REWRITE
RF
RH
RIGHT
ROUNDED
RUN

(ca) SA
SAME
SD
SEARCH
SECTION
SECURITY
SEEK
(xa) SEGMENT
SEGMENT-LIMIT
SELECT
(xa) SEND
SENTENCE
(xa) SEPARATE
SEQUENTIAL
(xac) SERVICE
SET
SIGN
SIZE
(xac) SKIP1
(xac) SKIP2
(xac) SKIP3
SORT
(xac) SORT-CORE-SIZE
(xac) SORT-FILE-SIZE
(ca) SORT-MERGE
(xac) SORT-MESSAGE
(xac) SORT-MODE-SIZE
(spn) SORT-OPTION
(xac) SORT-RETURN

SOURCE
SOURCE-COMPUTER
SPACE
SPACES
SPECIAL-NAMES
STANDARD
(xa) START
STATUS
STOP
(xa) STRING
(xa) SUB-QUEUE-1
(xa) SUB-QUEUE-2
(xa) SUB-QUEUE-3
SUBTRACT
SUM
(ca) SUPERVISOR
(xa) SUPPRESS
(ca) SUSPEND
(xa) SYMBOLIC
SYNC
SYNCHRONIZED
(sp) SYSIN
(spn) SYSIPT
(spn) SYSLST
(sp) SYSOUT
(spn) SYSPCH
(sp) SYSPUNCH
(sp) S01
(sp) S02

(ca) TABLE
TALLY
TALLYING
TAPE
(ca) TERMINAL
TERMINATE
(xa) TEXT
THAN
(xac) THEN
THROUGH
THRU
(xa) TIME
(xac) TIME-OF-DAY
TIMES
TO
(ca) TOP
(xac) TOTALED
(xac) TOTALING
(xac) TRACE
(xac) TRACK
(xac) TRACK-AREA
(xac) TRACK-LIMIT
(xac) TRACKS
(xa) TRAILING
(xac) TRANSFORM
TYPE

(ca) UNEQUAL
UNIT
(xa) UNSTRING
UNTIL
UP
UPON
(spn) UPSI-0
(spn) UPSI-1
(spn) UPSI-2
(spn) UPSI-3
(spn) UPSI-4
(spn) UPSI-5
(spn) UPSI-6
(spn) UPSI-7
USAGE
USE
USING

VALUE
VALUES
VARYING

TRIM HERE

TRIM HERE

FOLD

FOLD

TRIM HERE

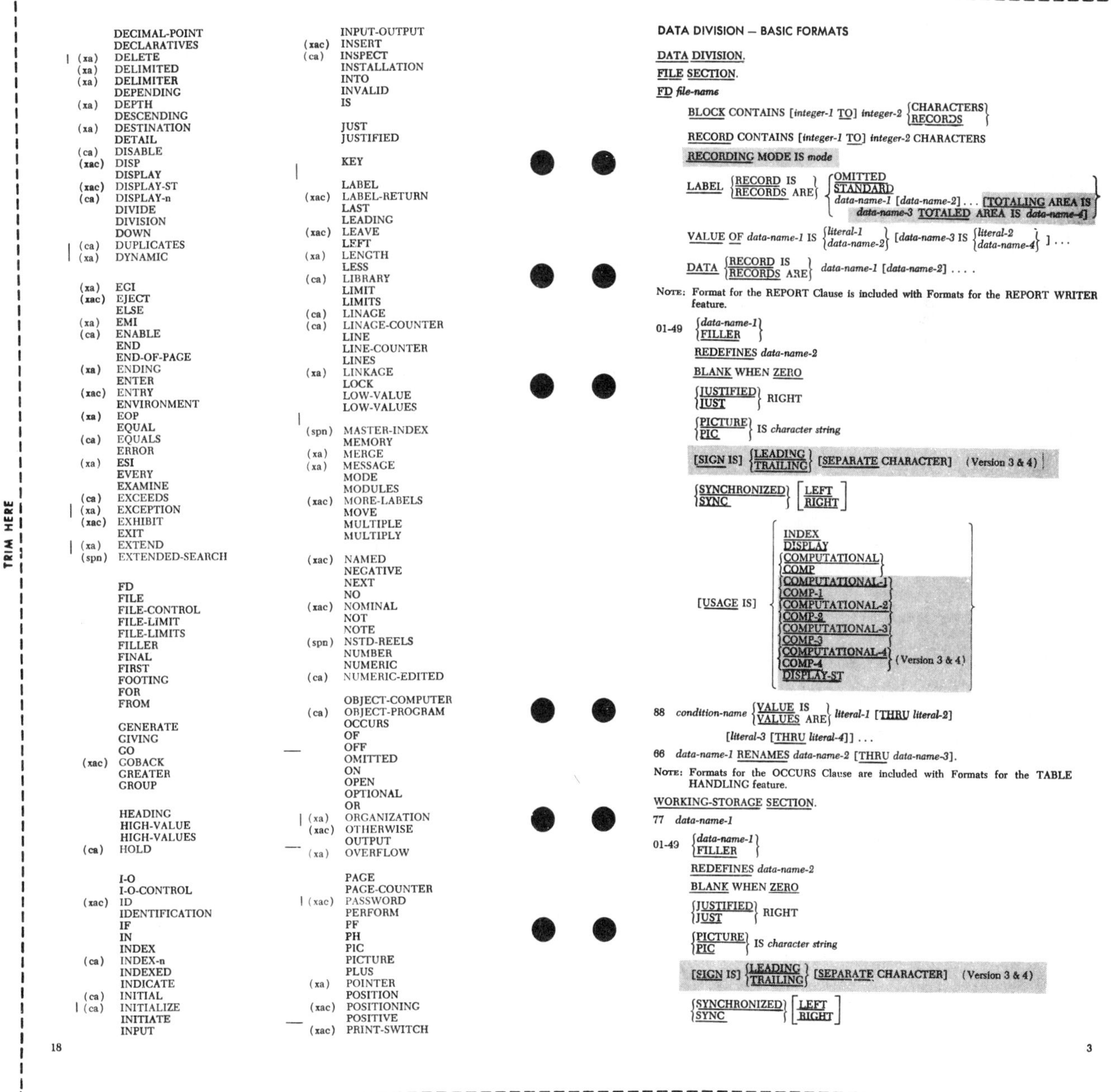

	DECIMAL-POINT
	DECLARATIVES
| (xa)	DELETE
(xa)	DELIMITED
(xa)	DELIMITER
	DEPENDING
(xa)	DEPTH
	DESCENDING
(xa)	DESTINATION
	DETAIL
(ca)	DISABLE
(xac)	DISP
	DISPLAY
(xac)	DISPLAY-ST
(ca)	DISPLAY-n
	DIVIDE
	DIVISION
	DOWN
| (ca)	DUPLICATES
| (xa)	DYNAMIC
(xa)	EGI
(xac)	EJECT
	ELSE
(xa)	EMI
(ca)	ENABLE
	END
	END-OF-PAGE
(xa)	ENDING
	ENTER
(xac)	ENTRY
	ENVIRONMENT
(xa)	EOP
	EQUAL
(ca)	EQUALS
	ERROR
(xa)	ESI
	EVERY
	EXAMINE
(ca)	EXCEEDS
| (xa)	EXCEPTION
(xac)	EXHIBIT
	EXIT
| (xa)	EXTEND
(spn)	EXTENDED-SEARCH
	FD
	FILE
	FILE-CONTROL
	FILE-LIMIT
	FILE-LIMITS
	FILLER
	FINAL
	FIRST
	FOOTING
	FOR
	FROM
	GENERATE
	GIVING
	GO
(xac)	GOBACK
	GREATER
	GROUP
	HEADING
	HIGH-VALUE
	HIGH-VALUES
(ca)	HOLD
	I-O
	I-O-CONTROL
(xac)	ID
	IDENTIFICATION
	IF
	IN
	INDEX
(ca)	INDEX-n
	INDEXED
	INDICATE
(ca)	INITIAL
| (ca)	INITIALIZE
	INITIATE
	INPUT
	INPUT-OUTPUT
(xac)	INSERT
(ca)	INSPECT
	INSTALLATION
	INTO
	INVALID
	IS
	JUST
	JUSTIFIED
|	KEY
	LABEL
(xac)	LABEL-RETURN
	LAST
	LEADING
(xac)	LEAVE
	LEFT
(xa)	LENGTH
	LESS
(ca)	LIBRARY
	LIMIT
	LIMITS
(ca)	LINAGE
(ca)	LINAGE-COUNTER
	LINE
	LINE-COUNTER
	LINES
(xa)	LINKAGE
	LOCK
	LOW-VALUE
	LOW-VALUES
| (spn)	MASTER-INDEX
	MEMORY
(xa)	MERGE
(xa)	MESSAGE
	MODE
	MODULES
(xac)	MORE-LABELS
	MOVE
	MULTIPLE
	MULTIPLY
(xac)	NAMED
	NEGATIVE
	NEXT
	NO
(xac)	NOMINAL
	NOT
	NOTE
(spn)	NSTD-REELS
	NUMBER
	NUMERIC
(ca)	NUMERIC-EDITED
	OBJECT-COMPUTER
(ca)	OBJECT-PROGRAM
	OCCURS
	OF
—	OFF
	OMITTED
	ON
	OPEN
	OPTIONAL
	OR
| (xa)	ORGANIZATION
(xac)	OTHERWISE
	OUTPUT
— (xa)	OVERFLOW
	PAGE
	PAGE-COUNTER
| (xac)	PASSWORD
	PERFORM
	PF
	PH
	PIC
	PICTURE
	PLUS
(xa)	POINTER
	POSITION
(xac)	POSITIONING
—	POSITIVE
(xac)	PRINT-SWITCH

DATA DIVISION — BASIC FORMATS

DATA DIVISION.

FILE SECTION.

FD *file-name*

BLOCK CONTAINS [*integer-1* TO] *integer-2* {CHARACTERS | RECORDS}

RECORD CONTAINS [*integer-1* TO] *integer-2* CHARACTERS

RECORDING MODE IS *mode*

LABEL {RECORD IS | RECORDS ARE} {OMITTED | STANDARD | *data-name-1* [*data-name-2*] . . . [TOTALING AREA IS *data-name-3* TOTALED AREA IS *data-name-4*]}

VALUE OF *data-name-1* IS {*literal-1* | *data-name-2*} [*data-name-3* IS {*literal-2* | *data-name-4*}] . . .

DATA {RECORD IS | RECORDS ARE} *data-name-1* [*data-name-2*] . . . .

NOTE: Format for the REPORT Clause is included with Formats for the REPORT WRITER feature.

01-49 {*data-name-1* | FILLER}

REDEFINES *data-name-2*

BLANK WHEN ZERO

{JUSTIFIED | JUST} RIGHT

{PICTURE | PIC} IS *character string*

[SIGN IS] {LEADING | TRAILING} [SEPARATE CHARACTER] (Version 3 & 4)

{SYNCHRONIZED | SYNC} [LEFT | RIGHT]

[USAGE IS] {INDEX | DISPLAY | {COMPUTATIONAL | COMP} | {COMPUTATIONAL-1 | COMP-1} | {COMPUTATIONAL-2 | COMP-2} | {COMPUTATIONAL-3 | COMP-3} | {COMPUTATIONAL-4 | COMP-4} (Version 3 & 4) | DISPLAY-ST}

88 *condition-name* {VALUE IS | VALUES ARE} *literal-1* [THRU *literal-2*]

[*literal-3* [THRU *literal-4*]] . . .

66 *data-name-1* RENAMES *data-name-2* [THRU *data-name-3*].

NOTE: Formats for the OCCURS Clause are included with Formats for the TABLE HANDLING feature.

WORKING-STORAGE SECTION.

77 *data-name-1*

01-49 {*data-name-1* | FILLER}

REDEFINES *data-name-2*

BLANK WHEN ZERO

{JUSTIFIED | JUST} RIGHT

{PICTURE | PIC} IS *character string*

[SIGN IS] {LEADING | TRAILING} [SEPARATE CHARACTER] (Version 3 & 4)

{SYNCHRONIZED | SYNC} [LEFT | RIGHT]

TRIM HERE

TRIM HERE

FOLD

[USAGE IS] { INDEX | DISPLAY | {COMPUTATIONAL | COMP} | {COMPUTATIONAL-1 | COMP-1} | {COMPUTATIONAL-2 | COMP-2} | {COMPUTATIONAL-3 | COMP-3} | {COMPUTATIONAL-4 | COMP-4} (Version 3 & 4) | DISPLAY-ST }

VALUE IS *literal*

88 *condition-name* {VALUE IS | VALUES ARE} *literal-1* [THRU *literal-2*]

[*literal-3* [THRU *literal-4*]] . . .

66 *data-name-1* RENAMES *data-name-2* [THRU *data-name-3*].

NOTE: Formats for the OCCURS Clause are included with Formats for the TABLE HANDLING feature.

LINKAGE SECTION.

77 *data-name-1*

01-49 {*data-name-1* | FILLER}

REDEFINES *data-name-2*

BLANK WHEN ZERO

{JUSTIFIED | JUST} RIGHT

{PICTURE | PIC} IS *character string*

[SIGN IS] {LEADING | TRAILING} [SEPARATE CHARACTER] (Version 3 & 4)

{SYNCHRONIZED | SYNC} [LEFT | RIGHT]

[USAGE IS] { INDEX | DISPLAY | {COMPUTATIONAL | COMP} | {COMPUTATIONAL-1 | COMP-1} | {COMPUTATIONAL-2 | COMP-2} | {COMPUTATIONAL-3 | COMP-3} | {COMPUTATIONAL-4 | COMP-4} (Version 3 & 4) | DISPLAY-ST }

88 *condition-name* {VALUE IS | VALUES ARE} *literal-1* [THRU *literal-2*]

[*literal-3* [THRU *literal-4*]] . . .

66 *data-name-1* RENAMES *data-name-2* [THRU *data-name-3*].

NOTE: Formats for the OCCURS Clause are included with Formats for the TABLE HANDLING feature.

PROCEDURE DIVISION — BASIC FORMATS

{PROCEDURE DIVISION. | PROCEDURE DIVISION USING *identifier-1* [*identifier-2*] . . .}

ACCEPT Statement

FORMAT 1

ACCEPT *identifier* [FROM {SYSIN | CONSOLE | *mnemonic-name*}]

FORMAT 2 (Version 4)

ACCEPT *identifier* FROM {DATE | DAY | TIME}

IBM AMERICAN NATIONAL STANDARD COBOL RESERVED WORDS

No word in the following list should appear as a programmer-defined name. The keys that appear before some of the words, and their meanings, are:

(xa) before a word means that the word is an IBM extension to American National Standard COBOL.

(xac) before a word means that the word is an IBM extension to both American National Standard COBOL and CODASYL COBOL.

(ca) before a word means that the word is a CODASYL COBOL reserved word not incorporated in American National Standard COBOL or in IBM American National Standard COBOL.

(sp) before a word means that the word is an IBM function-name established in support of the SPECIAL-NAMES function.

(spn) before a word means that the word is used by an IBM American National Standard COBOL compiler, but not this compiler.

(asn) before a word means that the word is defined by American National Standard COBOL, but is not used by this compiler.

ACCEPT
ACCESS
ACTUAL
ADD
(asn) ADDRESS
ADVANCING
AFTER
ALL
ALPHABETIC
(ca) ALPHANUMERIC
(ca) ALPHANUMERIC-EDITED
ALTER
ALTERNATE
AND
(xa) APPLY
ARE
AREA
AREAS
ASCENDING
ASSIGN
AT
AUTHOR

(xac) BASIS
BEFORE
BEGINNING
BLANK
BLOCK
(ca) BOTTOM
BY

(xa) CALL
(xa) CANCEL
(xac) CBL
(xa) CD
CF
CH
(xac) CHANGED
(xa) CHARACTER
CHARACTERS
(asn) CLOCK-UNITS
CLOSE
(asn) COBOL
CODE
COLUMN
(spn) COM-REG
COMMA
(xa) COMMUNICATION
COMP
(xa) COMP-1
(xa) COMP-2
(xa) COMP-3
(xa) COMP-4
COMPUTATIONAL
(xa) COMPUTATIONAL-1
(xa) COMPUTATIONAL-2
(xa) COMPUTATIONAL-3
(xa) COMPUTATIONAL-4
COMPUTE
CONFIGURATION
(sp) CONSOLE
CONTAINS
CONTROL
CONTROLS
COPY
(xac) CORE-INDEX
CORR
CORRESPONDING
(xa) COUNT
(sp) CSP
CURRENCY
(xac) CURRENT-DATE
(spn) CYL-INDEX
(spn) CYL-OVERFLOW
(sp) C01
(sp) C02
(sp) C03
(sp) C04
(sp) C05
(sp) C06
(sp) C07
(sp) C08
(sp) C09
(sp) C10
(sp) C11
(sp) C12

DATA
(xa) DATE
DATE-COMPILED
DATE-WRITTEN
(xa) DAY
(ca) DAY-OF-WEEK
DE
(xac) DEBUG
(ca) DEBUG-CONTENTS
(ca) DEBUG-ITEM
(ca) DEBUG-LINE
(ca) DEBUG-NAME
(ca) DEBUG-SUB-1
(ca) DEBUG-SUB-2
(ca) DEBUG-SUB-3
(ca) DEBUGGING

FOLD

TRIM HERE

TRIM HERE

REWRITE Statement

```
REWRITE record-name [FROM identifier]
    [INVALID KEY imperative-statement]
```

START Statement

```
                           { EQUAL TO       }
                           { =              }
START file-name [KEY IS    { GREATER THAN   } ]
                           { >              }
                           { NOT LESS THAN  }
                           { NOT <          }
    [INVALID KEY imperative-statement]
```

USE Sentence

```
USE AFTER STANDARD {EXCEPTION} PROCEDURE
                   {ERROR    }
       { file-name-1 [file-name-2] ... }
       { INPUT                         }
    ON { OUTPUT                        } .
       { I-O                           }
       { EXTEND                        }
```

WRITE Statement

```
WRITE record-name [FROM identifier]
    [INVALID KEY imperative-statement]
```

MERGE FACILITY FORMATS (OS/VS COBOL Only)

Environment Division — Input-Output Section

FILE-CONTROL Entry

```
FILE-CONTROL.
    {SELECT file-name
    ASSIGN TO system-name-1 [system-name-2] ... .} ...
```

I-O-CONTROL Entry

```
I-O-CONTROL.
         { SORT       }
    SAME { SORT-MERGE } AREA FOR file-name-1 [file-name-2] ... .
         { RECORD     }
```

Data Division — Merge File Description Entry

```
SD merge-file-name
    [RECORD CONTAINS [integer-1 TO] integer-2 CHARACTERS]
    [DATA {RECORD IS   } data-name-1 [data-name-2] ... ] .
          {RECORDS ARE }
```

Procedure Division — Merge Statement

```
MERGE file-name-1
    ON {ASCENDING } KEY data-name-1 [data-name-2] ...
       {DESCENDING}
    [ON {ASCENDING } KEY data-name-3 [data-name-4] ... ] ...
        {DESCENDING}
    USING file-name-2 file-name-3 [file-name-4] ...
    {GIVING file-name-5                                            }
    {OUTPUT PROCEDURE IS section-name-1 [THRU section-name-2]      }
```


ADD Statement

FORMAT 1

```
ADD {identifier-1} [identifier-2] ... TO identifier-m [ROUNDED]
    {literal-1   } [literal-2   ]
        [identifier-n [ROUNDED]] ... [ON SIZE ERROR imperative-statement]
```

FORMAT 2

```
ADD {identifier-1} {identifier-2} [identifier-3] ... GIVING
    {literal-1   } {literal-2   } [literal-3   ]
        identifier-m [ROUNDED] [ON SIZE ERROR imperative-statement]
```

FORMAT 3

```
ADD {CORRESPONDING} identifier-1 TO identifier-2 [ROUNDED]
    {CORR         }
        [ON SIZE ERROR imperative-statement]
```

ALTER Statement

```
ALTER procedure-name-1 TO [PROCEED TO] procedure-name-2
        [procedure-name-3 TO [PROCEED TO] procedure-name-4] ...
```

CALL Statement

FORMAT 1

```
CALL literal-1 [USING identifier-1 [identifier-2] ... ]
```

FORMAT 2 (Version 4)

```
CALL identifier-1 [USING identifier-2 [identifier-3] ... ]
```

CANCEL Statement (Version 4)

```
CANCEL {literal-1   } [literal-2   ] ...
       {identifier-1} [identifier-2]
```

CLOSE Statement

FORMAT 1

```
CLOSE file-name-1 [REEL] [WITH {NO REWIND} ]
                  [UNIT]       {LOCK     }
    [file-name-2 [REEL] [WITH {NO REWIND} ]] ...
                 [UNIT]       {LOCK     }
```

FORMAT 2

```
                        { NO REWIND }
CLOSE file-name-1 [WITH { LOCK      } ]
                        { DISP      }
                         { NO REWIND }
    [file-name-2 [WITH   { LOCK      } ]] ...
                         { DISP      }
```

FORMAT 3

```
                  {REEL}       { NO REWIND   }
CLOSE file-name-1 {UNIT} [WITH { LOCK        } ]
                               { POSITIONING }
                  {REEL}       { NO REWIND   }
    [file-name-2  {UNIT} [WITH { LOCK        } ]] ...
                               { POSITIONING }
```

COMPUTE Statement

```
                                         { identifier-2          }
COMPUTE identifier-1 [ROUNDED] =         { literal-1             }
                                         { arithmetic-expression }
        [ON SIZE ERROR imperative-statement]
```

DECLARATIVE Section

```
PROCEDURE DIVISION.
DECLARATIVES.
{section-name SECTION. USE sentence.
{paragraph-name. {sentence} ...} ... } ...
END DECLARATIVES.
```

DISPLAY Statement

```
                                                    { CONSOLE       }
DISPLAY {literal-1   } [literal-2   ] ... [UPON     { SYSPUNCH      } ]
        {identifier-1} [identifier-2]               { SYSOUT        }
                                                    { mnemonic-name }
```


TRIM HERE

FOLD

DIVIDE Statement

FORMAT 1

DIVIDE {identifier-1 / literal-1} INTO identifier-2 [ROUNDED]

[ON SIZE ERROR imperative-statement]

FORMAT 2

DIVIDE {identifier-1 / literal-1} {INTO / BY} {identifier-2 / literal-2} GIVING identifier-3

[ROUNDED] [REMAINDER identifier-4] [ON SIZE ERROR imperative-statement]

ENTER Statement

ENTER language-name [routine-name].

ENTRY Statement

ENTRY literal-1 [USING identifier-1 [identifier-2] . . .]

EXAMINE Statement

FORMAT 1

EXAMINE identifier TALLYING {UNTIL FIRST / ALL / LEADING} literal-1

[REPLACING BY literal-2]

ᶠORMAT 2

EXAMINE identifier REPLACING {ALL / LEADING / FIRST / UNTIL FIRST} literal-1 BY literal-2

XIT Statement

ıragraph-name. EXIT [PROGRAM].

OBACK Statement

GOBACK.

) TO Statement

RMAT 1

GO TO procedure-name-1

ıMAT 2

GO TO procedure-name-1 [procedure-name-2] . . . DEPENDING ON identifier

MAT 3

GO TO.

tatement

F condition THEN {statement-1 / NEXT SENTENCE} {ELSE / OTHERWISE} {statement-2 / NEXT SENTENCE}

'E Statement

AT 1

OVE {identifier-1 / literal-1} TO identifier-2 [identifier-3] . . .

AT 2

)VE {CORRESPONDING / CORR} identifier-1 TO identifier-2

IPLY Statement

T 1

LTIPLY {identifier-1 / literal-1} BY identifier-2 [ROUNDED]

[ON SIZE ERROR imperative-statement]

2

TIPLY {identifier-1 / literal-1} BY {identifier-2 / literal-2} GIVING identifier-3

[ROUNDED] [ON SIZE ERROR imperative-statement]

VSAM FORMATS (OS/VS COBOL Only)

Environment Division — File-Control Entry

FORMAT 1 — Sequential VSAM Files

FILE-CONTROL.

{SELECT [OPTIONAL] file-name

ASSIGN TO system-name-1 [system-name-2] . . .

[RESERVE integer [AREA / AREAS]]

[ORGANIZATION IS SEQUENTIAL]

[ACCESS MODE IS SEQUENTIAL]

[PASSWORD IS data-name-1]

[FILE STATUS IS data-name-2] . } . . .

FORMAT 2 — Indexed VSAM Files

FILE-CONTROL.

{SELECT file-name

ASSIGN TO system-name-1 [system-name-2] . . .

[RESERVE integer [AREA / AREAS]]

ORGANIZATION IS INDEXED

[ACCESS MODE IS {SEQUENTIAL / RANDOM / DYNAMIC}]

RECORD KEY IS data-name-3

[PASSWORD IS data-name-1]

[FILE STATUS IS data-name-2] . } . . .

Environment Division — I-O-Control Entry

I-O-CONTROL.

[RERUN ON system-name EVERY integer RECORDS

OF file-name-1] . . .

[SAME [RECORD] AREA

FOR file-name-2 [file-name-3] . . .] . . .

Data Division

LABEL RECORDS Clause

LABEL {RECORD IS / RECORDS ARE} {STANDARD / OMITTED}

NOTE: Other Data Division clauses have the same syntax for VSAM files that they have for other files.

Procedure Division

CLOSE Statement

CLOSE file-name-1 [WITH LOCK]

[file-name-2 [WITH LOCK]] . . .

DELETE Statement

DELETE file-name RECORD

[INVALID KEY imperative-statement]

OPEN Statement

OPEN {INPUT file-name-1 [file-name-2] . . . / OUTPUT file-name-1 [file-name-2] . . . / I-O file-name-1 [file-name-2] . . . / EXTEND file-name-1 [file-name-2] . . .} . . .

READ Statement

FORMAT 1

READ file-name [NEXT] RECORD [INTO identifier]

[AT END imperative-statement]

FORMAT 2

READ file-name RECORD [INTO identifier]

[INVALID KEY imperative-statement]

TRIM HERE

FOLD

FOLD

TRIM HERE

STRING MANIPULATION — BASIC FORMATS

STRING Statement

STRING {identifier-1 / literal-1} [identifier-2 / literal-2] . . . DELIMITED BY {identifier-3 / literal-3 / SIZE}

[{identifier-4 / literal-4} [identifier-5 / literal-5] . . . DELIMITED BY {identifier-6 / literal-6 / SIZE}] . . .

INTO identifier-7 [WITH POINTER identifier-8]

[ON OVERFLOW imperative-statement]

UNSTRING Statement

UNSTRING identifier-1

[DELIMITED BY [ALL] {identifier-2 / literal-2} [OR [ALL] {identifier-3 / literal-3}] . . .]

INTO identifier-4 [DELIMITER IN identifier-5]

[COUNT IN identifier-6]

[identifier-7 [DELIMITER IN identifier-8]

[COUNT IN identifier-9]] . . .

[WITH POINTER identifier-10]

[TALLYING IN identifier-11]

[ON OVERFLOW imperative-statement]

NOTE Statement

NOTE character string

OPEN Statement

FORMAT 1

OPEN [INPUT {file-name [REVERSED / WITH NO REWIND]} . . .]

[OUTPUT {file-name [WITH NO REWIND]} . . .]

[I-O {file-name} . . .]

FORMAT 2

OPEN [INPUT {file-name [REVERSED / WITH NO REWIND] [LEAVE / REREAD / DISP]} . . .]

[OUTPUT {file-name [WITH NO REWIND] [LEAVE / REREAD / DISP]} . . .]

[I-O {file-name} . . .]

PERFORM Statement

FORMAT 1

PERFORM procedure-name-1 [THRU procedure-name-2]

FORMAT 2

PERFORM procedure-name-1 [THRU procedure-name-2] {identifier-1 / integer-1} TIMES

FORMAT 3

PERFORM procedure-name-1 [THRU procedure-name-2] UNTIL condition-1

FORMAT 4

PERFORM procedure-name-1 [THRU procedure-name-2]

VARYING {index-name-1 / identifier-1} FROM {index-name-2 / literal-2 / identifier-2} BY {literal-3 / identifier-3} UNTIL condition-1

[AFTER {index-name-4 / identifier-4} FROM {index-name-5 / literal-5 / identifier-5} BY {literal-6 / identifier-6} UNTIL condition-2

[AFTER {index-name-7 / identifier-7} FROM {index-name-8 / literal-8 / identifier-8} BY {literal-9 / identifier-9} UNTIL condition-3]]

READ Statement

READ file-name RECORD [INTO identifier]

{AT END / INVALID KEY} imperative-statement

REWRITE Statement

REWRITE record-name [FROM identifier] [INVALID KEY imperative-statement]

SEEK Statement

SEEK file-name RECORD

START Statement

FORMAT 1

START file-name [INVALID KEY imperative-statement]

FORMAT 2 (Version 3 & 4)

START file-name

USING KEY data-name {EQUAL TO / =} identifier

[INVALID KEY imperative-statement]

STOP Statement

STOP {RUN / literal}

TRIM HERE

FOLD

SUBTRACT Statement

FORMAT 1

SUBTRACT {identifier-1 | literal-1} [identifier-2 | literal-2] . . . FROM identifier-m [ROUNDED]

[identifier-n [ROUNDED]] . . . [ON SIZE ERROR imperative-statement]

FORMAT 2

SUBTRACT {identifier-1 | literal-1} [identifier-2 | literal-2] . . . FROM {identifier-m | literal-m} GIVING identifier-n

[ROUNDED] [ON SIZE ERROR imperative-statement]

FORMAT 3

SUBTRACT {CORRESPONDING | CORR} identifier-1 FROM identifier-2 [ROUNDED]

[ON SIZE ERROR imperative-statement]

TRANSFORM Statement

TRANSFORM identifier-3 CHARACTERS FROM {figurative-constant-1 | nonnumeric-literal-1 | identifier-1}

TO {figurative-constant-2 | nonnumeric-literal-2 | identifier-2}

USE Sentence

FORMAT 1

Option 1:

USE {BEFORE | AFTER} STANDARD [BEGINNING] [REEL | FILE | UNIT]

LABEL PROCEDURE ON {{file-name} . . . | OUTPUT | INPUT | I-O}.

Option 2:

USE {BEFORE | AFTER} STANDARD [ENDING] [REEL | FILE | UNIT]

LABEL PROCEDURE ON {{file-name} . . . | OUTPUT | INPUT | I-O}.

FORMAT 2

USE AFTER STANDARD ERROR PROCEDURE

ON {{file-name-1} [file-name-2] . . . | INPUT | OUTPUT | I-O}

[GIVING data-name-1 [data-name-2]].

NOTE: Format 3 of the USE Sentence is included in Formats for the REPORT WRITER feature.

WRITE Statement

FORMAT 1

WRITE record-name [FROM identifier-1] [{BEFORE | AFTER} ADVANCING

{identifier-2 LINES | integer LINES | mnemonic-name}] [AT {END-OF-PAGE | EOP} imperative-statement]

FORMAT 2

WRITE record-name [FROM identifier-1] AFTER POSITIONING {identifier-2 | integer} LINES

[AT {END-OF-PAGE | EOP} imperative-statement]

FORMAT 3

WRITE record-name [FROM identifier-1] INVALID KEY imperative-statement

FORMAT CONTROL — BASIC FORMATS

EJECT Statement

1	Area B
	EJECT

SKIP1, SKIP2, SKIP3 Statements

1	Area B
	{SKIP1 | SKIP2 | SKIP3}

STERLING CURRENCY — BASIC FORMATS

Data Division Sterling Formats

Nonreport PICTURE Clause

{PICTURE | PIC} IS 9 [(n)] D [8] 8D {6[6] | 7[7]} [[V] 9 [(n)]] [USAGE IS] DISPLAY-ST

Report PICTURE Clause

{PICTURE | PIC} IS

[pound-report-string] [pound-separator-string] delimiter shilling-report-string [shilling-separator-string] delimiter pence-report-string [pence-separator-string] [sign-string] [USAGE IS] DISPLAY-ST

PROGRAM PRODUCT INFORMATION — VERSION 4

TELEPROCESSING — BASIC FORMATS

Data Division Teleprocessing Formats

CD Entry

FORMAT 1

CD cd-name FOR INPUT

[[[SYMBOLIC QUEUE IS data-name-1]
[SYMBOLIC SUB-QUEUE-1 IS data-name-2]
[SYMBOLIC SUB-QUEUE-2 IS data-name-3]
[SYMBOLIC SUB-QUEUE-3 IS data-name-4]
[MESSAGE DATE IS data-name-5]
[MESSAGE TIME IS data-name-6]
[SYMBOLIC SOURCE IS data-name-7]
[TEXT LENGTH IS data-name-8]
[END KEY IS data-name-9]
[STATUS KEY IS data-name-10]
[QUEUE DEPTH IS data-name-11]]
[data-name-1 data-name-2 . . . data-name-11]].

FORMAT 2

CD cd-name FOR OUTPUT
[DESTINATION COUNT IS data-name-1]
[TEXT LENGTH IS data-name-2]
[STATUS KEY IS data-name-3]
[ERROR KEY IS data-name-4]
[SYMBOLIC DESTINATION IS data-name-5].

Procedure Division Teleprocessing Formats

Message Condition

[NOT] MESSAGE FOR cd-name

RECEIVE Statement

RECEIVE cd-name {MESSAGE | SEGMENT} INTO identifier-1

[NO DATA imperative-statement]

SEND Statement

FORMAT 1

SEND cd-name FROM identifier-1

FORMAT 2

SEND cd-name [FROM identifier-1] {WITH identifier-2 | WITH ESI | WITH EMI | WITH EGI}

FOLD

TRIM HERE

SEGMENTATION — BASIC FORMATS

Environment Division Segmentation Formats

OBJECT-COMPUTER PARAGRAPH

SEGMENT-LIMIT Clause

SEGMENT-LIMIT IS *priority-number*

Procedure Division Segmentation Formats

Priority Numbers

section-name SECTION [*priority-number*].

SOURCE PROGRAM LIBRARY FACILITY

COPY Statement

COPY *library-name* [SUPPRESS]

[REPLACING *word-1* BY { *word-2* / *literal-1* / *identifier-1* } [*word-3* BY { *word-4* / *literal-2* / *identifier-2* }] . . .] .

Extended Source Program Library Facility

BASIS Card

BASIS *library-name*

INSERT Card

INSERT *sequence-number-field*

DELETE Card

DELETE *sequence-number-field*

DEBUGGING LANGUAGE — BASIC FORMATS

Procedure Division Debugging Formats

EXHIBIT Statement

EXHIBIT { NAMED / CHANGED NAMED / CHANGED } { *identifier-1* / *nonnumeric-literal-1* } [*identifier-2* / *nonnumeric-literal-2*] . . .

ON (Count-Conditional) Statement

FORMAT 1

ON *integer-1* [AND EVERY *integer-2*] [UNTIL *integer-3*]

{ *imperative-statement* / NEXT SENTENCE } { ELSE / OTHERWISE } { *statement* . . . / NEXT SENTENCE }

FORMAT 2 (Version 3 & 4)

ON { *integer-1* / *identifier-1* } [AND EVERY { *integer-2* / *identifier-2* }] [UNTIL { *integer-3* / *identifier-3* }]

{ *imperative-statement* / NEXT SENTENCE } { ELSE / OTHERWISE } { *statement* . . . / NEXT SENTENCE }

TRACE Statement

{ READY / RESET } TRACE

Compile-Time Debugging Packet

DEBUG Card

DEBUG *location*

SORT — BASIC FORMATS

Environment Division Sort Formats

FILE-CONTROL PARAGRAPH — SELECT SENTENCE

SELECT Sentence (for GIVING option only)

SELECT *file-name*

ASSIGN TO [*integer-1*] *system-name-1* [*system-name-2*] . . .

OR *system-name-3* [FOR MULTIPLE { REEL / UNIT }]

[RESERVE { *integer-2* / NO } ALTERNATE [AREA / AREAS]].

SELECT Sentence (for Sort Work Files)

SELECT *sort-file-name*

ASSIGN TO [*integer*] *system-name-1* [*system-name-2*] . . .

I-O-CONTROL PARAGRAPH

RERUN Clause

RERUN ON *system-name*

SAME RECORD/SORT AREA Clause

SAME { RECORD / SORT } AREA FOR *file-name-1* {*file-name-2*} . . .

Data Division Sort Formats

SORT-FILE DESCRIPTION

SD *sort-file-name*

RECORDING MODE IS *mode*

DATA { RECORD IS / RECORDS ARE } *data-name-1* [*data-name-2*] . . .

RECORD CONTAINS [*integer-1* TO] *integer-2* CHARACTERS

[LABEL { RECORD IS / RECORDS ARE } { STANDARD / OMITTED }] . (Version 4)

Procedure Division Sort Formats

RELEASE Statement

RELEASE *sort-record-name* [FROM *identifier*]

RETURN Statement

RETURN *sort-file-name* RECORD [INTO *identifier*]

AT END *imperative-statement*

SORT Statement

SORT *file-name-1* ON { DESCENDING / ASCENDING } KEY {*data-name-1*} . . .

[ON { DESCENDING / ASCENDING } KEY {*data-name-2*} . . .] . . .

{ INPUT PROCEDURE IS *section-name-1* [THRU *section-name-2*] / USING *file-name-2* }

{ OUTPUT PROCEDURE IS *section-name-3* [THRU *section-name-4*] / GIVING *file-name-3* }

REPORT WRITER — BASIC FORMATS

Data Division Report Writer Formats

NOTE: Formats that appear as Basic Formats within the general description of the Data Division are illustrated there.

FILE SECTION — REPORT Clause

{ REPORT IS / REPORTS ARE } *report-name-1* [*report-name-2*] . . .

TRIM HERE

FOLD

TRIM HERE

FOLD

TRIM HERE

REPORT SECTION

```
REPORT SECTION.
RD report-name
   WITH CODE mnemonic-name
   {CONTROL IS   } { FINAL                                          }
   {CONTROLS ARE } { identifier-1 [identifier-2] ...               }
                   { FINAL identifier-1 [identifier-2] ...         }
   PAGE [LIMIT IS  ] integer-1 {LINE }
        [LIMITS ARE]           {LINES}
      [HEADING integer-2]
      [FIRST DETAIL integer-3]
      [LAST DETAIL integer-4]
      [FOOTING integer-5].
```

REPORT GROUP DESCRIPTION ENTRY

Format 1

```
01  [data-name-1]
   LINE NUMBER IS { integer-1        }
                  { PLUS integer-2   }
                  { NEXT PAGE        }
   NEXT GROUP IS  { integer-1        }
                  { PLUS integer-2   }
                  { NEXT PAGE        }
           { {REPORT HEADING}                              }
           { {RH            }                              }
           { {PAGE HEADING}                                }
           { {PH          }                                }
           { {CONTROL HEADING}  {identifier-n}             }
           { {CH             }  {FINAL       }             }
   TYPE IS { {DETAIL}                                      }
           { {DE    }                                      }
           { {CONTROL FOOTING}  {identifier-n}             }
           { {CF             }  {FINAL       }             }
           { {PAGE FOOTING}                                }
           { {PF          }                                }
           { {REPORT FOOTING}                              }
           { {RF            }                              }
  USAGE Clause.
```

Format 2

```
nn  [data-name-1]
   LINE Clause – See Format 1
   USAGE Clause.
```

Format 3

```
nn  [data-name-1]
   COLUMN NUMBER IS integer-1
   GROUP INDICATE
   JUSTIFIED Clause
   LINE Clause – See Format 1
   PICTURE Clause
   RESET ON {identifier-1}
            {FINAL       }
   BLANK WHEN ZERO Clause
   SOURCE IS {TALLY       }
             {identifier-2}
   SUM {TALLY       } [TALLY       ] ... [UPON data-name]
       {identifier-3} [identifier-4]
   VALUE IS literal-1
   USAGE Clause.
```

Format 4

```
01  data-name-1
   BLANK WHEN ZERO Clause
   COLUMN Clause – See Format 3
   GROUP Clause – See Format 3
   JUSTIFIED Clause
   LINE Clause – See Format 1
   NEXT GROUP Clause – See Format 1
   PICTURE Clause
   RESET Clause – See Format 3
   { SOURCE Clause }
   { SUM Clause    }  See Format 3
   { VALUE Clause  }
   TYPE Clause – See Format 1
   USAGE Clause.
```


Procedure Division Report Writer Formats

GENERATE Statement

```
GENERATE identifier
```

INITIATE Statement

```
INITIATE report-name-1 [report-name-2] ...
```

TERMINATE Statement

```
TERMINATE report-name-1 [report-name-2] ...
```

USE Sentence

```
USE BEFORE REPORTING data-name.
```

TABLE HANDLING – BASIC FORMATS

Data Division Table Handling Formats

OCCURS Clause

Format 1

```
OCCURS integer-2 TIMES
      [ {ASCENDING } KEY IS data-name-2 [data-name-3 ...] ...
        {DESCENDING}
      [INDEXED BY index-name-1 [index-name-2] ...]
```

Format 2

```
OCCURS integer-1 TO integer-2 TIMES [DEPENDING ON data-name-1]
      [ {ASCENDING } KEY IS data-name-2 [data-name-3] ...] ...
        {DESCENDING}
      [INDEXED BY index-name-1 [index-name-2] ...]
```

Format 3

```
OCCURS integer-2 TIMES [DEPENDING ON data-name-1]
      [ {ASCENDING } KEY IS data-name-2 [data-name-3] ...] ...
        {DESCENDING}
      [INDEXED BY index-name-1 [index-name-2] ...]
```

USAGE Clause

```
[USAGE IS] INDEX
```

Procedure Division Table Handling Formats

SEARCH Statement

Format 1

```
SEARCH identifier-1 [VARYING {index-name-1} ]
                             {identifier-2}
      [AT END imperative-statement-1]
      WHEN condition-1 {imperative-statement-2}
                       {NEXT SENTENCE         }
      [WHEN condition-2 {imperative-statement-3} ] ...
                        {NEXT SENTENCE         }
```

Format 2

```
SEARCH ALL identifier-1 [AT END imperative-statement-1]
      WHEN condition-1 {imperative-statement-2}
                       {NEXT SENTENCE         }
```

SET Statement

Format 1

```
SET {index-name-1 [index-name-2] ...} TO {index-name-3}
    {identifier-1 [identifier-2 ] ...}    {identifier-3}
                                          {literal-1   }
```

Format 2

```
SET index-name-4 [index-name-5] ... {UP BY  } {identifier-4}
                                    {DOWN BY} {literal-2   }
```


TRIM HERE

TRIM HERE

FOLD